A Pipeline Runs Through It

KEITH FISHER

A Pipeline Runs Through It

The Story of Oil from Ancient Times to the First World War

ALLEN LANE
an imprint of
PENGUIN BOOKS

ALLEN LANE

UK | USA | Canada | Ireland | Australia
India | New Zealand | South Africa

Allen Lane is part of the Penguin Random House group of companies
whose addresses can be found at global.penguinrandomhouse.com

First published 2022
001

Copyright © Keith Fisher, 2022

The moral right of the author has been asserted

Set in 10.5/14pt Sabon LT Std
Typeset by Jouve (UK), Milton Keynes
Printed and bound in Great Britain by Clays Ltd, Elcograf S.p.A.

The authorized representative in the EEA is Penguin Random House Ireland,
Morrison Chambers, 32 Nassau Street, Dublin D02 YH68

A CIP catalogue record for this book is available from the British Library

ISBN: 978-0-241-55822-5

Contents

List of Illustrations	ix
Acknowledgements	xiii
1 Earth Oil	1
The Middle East and Central Asia	1
Africa	17
Asia	19
Europe	31
Latin America and the Caribbean	40
North America	50
2 Seneca Oil	74
Drilling in Ohio Country	74
Imperial Collision	79
Imperial Overstretch	86
Imperial Handover	90
'Trade of a Rising Empire'	92
'A Subdued People'	96
Seneca Chief	102
The $300,000 Question	105
3 Oildom	108
Barrels of Oil	108
Wartime Demand and Destruction	115
The Cleveland Nexus	122
From Pithole by Pipeline and Tank Car	124
Fire, Flood, Boom and Bust	128

4 Corporate Control   139

  Battle for the Oil Trade   139

  Pipeline to Monopoly   147

  Benson's Folly   151

  'Corporations Derive their Powers from the People'   157

  The Standard Oil Trust   169

  Standard Oil, Producers and Independents   175

5 A Global Industry   181

  'So Little Exploited by Man'   181

  Caucasian Battleground   195

  Reclaiming the Russian Market   202

  Across the Caucasus to the World   211

  International Competition   218

  Colonial Oil   222

  East and West of Suez   231

  'The Battle for All Christendom'   234

  The Markets of the World   242

  'Petroleum Craze at Baku'   249

  The Royal Dutch and the Shell   258

6 Oil for Power   266

  Oil Engines   266

  'The Teeming Millions of the Middle Kingdom'   275

  An Industry in Flux   281

  'Teeming Liquid Wealth'   289

  'The Virgin Oil Fields of Persia'   298

  Liquid Fuel and Sea Power   308

  Petroleum Spirit   324

  'The Destruction of the Communistic System'   333

7 A Volatile Mix   338

   The 'Asphalt War'   338

   Revolution in Russia – A 'Dress Rehearsal'   343

   Revolution in Persia, and the Plain of Oil   356

   British Supremacy over Baghdad   367

   Revolution in Mexico – 'Police the
   Surrounding Premises'   380

   America 'at the Threshold of a Revolution?'   388

   'A Paper Victory for the People'   398

   Economic Warfare   408

   'A Volley of Arrows'   422

8 Oil for War   430

   Oil 'for a Great War'   430

   Internal Combustion   447

   Spheres of Interest   455

   'Will Germany Control the Oil Supply of our Navy?'   469

   Britain's State Oil Company   480

   'Surrounded by Material Far More
   Inflammable than the Oil'   489

   The Turkish Petroleum Company   504

   The Sanctity of British Supremacy'   517

   'We Have Lost Control and the
   Landslide has Begun'   527

   Epilogue   539

   Notes   547

   Bibliography   639

   Index   715

List of Illustrations

1: Section from 'A map of the country between Albemarle Sound, and Lake Erie', engraved by S.J. Neele, in Thomas Jefferson's 1787 London edition of his *Notes on the State of Virginia*. General Collection, Beinecke Rare Book and Manuscript Library, Yale University. 1977 Folio 177. (https://collections.library.yale. edu/catalog/10563623).

2: F. Bartoli, Gayëtwahgeh (also known as Cornplanter, 1732/40– 1846), 1796. Oil on canvas, 30 x 25 in. Gift of Thomas Jefferson Bryan, New-York Historical Society, 1867.314. Photography ©New-York Historical Society.

3: Edwin Drake in front of the Seneca Oil Co.'s original well, photographed by John Mather in 1866. The Drake Well Museum, Pennsylvania Historical and Museum Commission. DW 676.

4: 'Benninghoff Farm 1865–1866'. The Drake Well Museum, Pennsylvania Historical and Museum Commission. DW 497.

5: 'Burning of the Imperial Refinery', one of two stereograph photographs by Frank Robbins, Oil City, PA, 1875. Library of Congress, Control No. 2008678995.

6: 'Columbia Oil Co. loading barrels of oil onto barges, Oil Creek, 1863–4'. The Drake Well Museum, Pennsylvania Historical and Museum Commission. DW 54.

7: The oil tank steamer *Zoroaster*. Photograph. National Archives of the Republic of Azerbaijan.

8: 'The Standard Oil Octopus', *The Daily Graphic* (New York) (4 February 1879). Fultonhistory.com / Old Fulton NY Postcards.

9: Map from Charles Marvin, 'Six Hundred Miles of Petroleum Pipe', *Pall Mall Gazette* (8 October 1886). Newspaper image ©

The British Library Board. The British Newspaper Archive (www.britishnewspaperarchive.co.uk).

10: 'Borysław: Łebaki ropne' (Borysław oil peasants), photograph, Ignacy Łukasiewicz Museum of the Oil and Gas Industry, Bóbrka, Poland.

11: 'Balakhani-Sabunchi oilfield', from photograph album of Nobel Bros. director Karl Wilhelm Hagelin. Teknik-och industrihistoriska arkivet / Tekniska museet (ARK-K175), TEKA0102298.

12: Still from the Lumière Brothers' *Puits de pétrole à Bakou. Vue de près* (*Oil Wells at Baku. Close View*), filmed by Alexandre Michon c. 1897 and first shown in August 1899. © Institut Lumière.

13: 'From Gloom to "Royal Daylight"', Anglo-American Oil Co., Ltd. advertisement, *Petroleum World* (2 January 1905).

14: 'Oil Wells at Summerland, Santa Barbara County, California'. Photograph by George H. Eldridge, 1902; in Ralph Arnold, 'Petroleum Resources and Industries of the Pacific Coast', 86, plate VIII, in *Nature and Science on the Pacific coast: A Guide-Book for Scientific Travelers in the West* (San Francisco, CA: Paul Elder & Co., 1915). For dating of photograph, see Ralph Arnold, *Geology and Oil Resources of the Summerland District, Santa Barbara County, California*, U.S. Geological Survey (Washington, DC: GPO, 1907), 7.

15: Car advertisements by the Electric Vehicle Co. and the Overman Automobile Co., *Automobile Topics* (21 June 1902), 410.

16: 'Armoured War-Automobiles Forging Their Way through the Ranks of the Enemy', 'The Motor Age: In 1946', *The Sketch* (28 March 1906), 339.

17: 'Piccadilly Traffic', May 1910. Photograph, Hulton Archive. Topical Press Agency/Stringer/Getty Images.

18: 'Cossack Patrol in Balakhany', in 'Blazing Baku', *The Sphere*, Supplement (30 September 1905), i.

19: 'Remains of Machinery Destroyed by Fire', in 'Blazing Baku', *The Sphere*, Supplement (30 September 1905), ii.

20: 'A group of Bakhtiari oil guards, 1908'. Photograph. Arash Khazeni (in his *Tribes and Empire on the Margins of Nineteenth-Century Iran* (Seattle: University of Washington Press, 2009), p.128) /

School of Oriental and African Studies archive, Papers of Lieutenant-Colonel David Lockhart Robertson Lorimer, PP MS 66.

21: The Anglo-Persian Oil Co.'s discovery well at Masjid i Suleiman, southwest Persia, where oil was struck in May 1908. Photograph. BP Archive. ARC179496. © BP plc.

22: Horse-drawn transport of sections of the Anglo-Persian Oil Co.'s pipeline in southwest Persia, circa 1911. Photograph. BP Archive. ARC64874_064. © BP plc

23: 'Laying the pipeline, 1910'. Photograph. BP Archive. ARC178512. © BP plc

24: 'Union Oil Co./Lakeview Oil Co. gusher, 1910'. Kern County Museum, Bakersfield, CA.

25: Map, 'Position of Oil Wells in Relation to International Boundaries', *The Times* (22 June 1914). The Times / News Licensing.

26: Map from 'The Petroleum Deposits of Mesopotamia', *Petroleum Review* (23 May 1914), 577.

27: '"Shell" at the Front!' advertisement, *Daily Mail* (19 November 1914).

Acknowledgements

For a project of this scope and length, it has been an unusually solitary affair, pursued in large part as a personal inquiry into some of the defining features of the modern human condition. Nevertheless, of course, many people were key to its genesis. For their early inspiration and encouragement I am deeply thankful to Helen Edwards, George Monbiot and Trevor Williams.

I especially thank Greg Muttitt for his generously offered, rich and insightful feedback that was, in retrospect, pivotal to the eventual completion of this book. Along the way, I received the greatly appreciated assistance of a number of people: Geoffrey Miller, Gary Staff, Stephen C. Cote, Rasoul Sorkhabi, Arash Khazeni and Susan Beates. Thank you also to Enrique Zapata-Bravo, Roos Geraedts and Esmé Carter for their careful translations, and to Michael Barrie for his advice on the rendition of Seneca Indian words.

The vast resources of the Bodleian Libraries in Oxford and their helpful staff were indispensable to my research, and I am particularly grateful to Craig Finlay, Jo Gardner and their great team at the Social Science Library.

I am for ever indebted to Anthony Werner, whose endorsement of my near-completed manuscript became instrumental in its eventual publication; and for their fulsome praise of my manuscript at that stage, I am hugely thankful to Michael Klare and Douglas Newton. I am immensely grateful to Simon Winder at Penguin for his enthusiasm for my endeavour and for his attentive and pertinent editorial input; and to the copy-editor, Richard Mason, for his most thorough fine-tuning of the text.

Among those whose enduring faith helped to keep the flame

burning, I feel particularly grateful to Enrique Zapata-Bravo, Flor Duran, Sophie Phipps, Bella Aris, Christina Sage, Jackie Nevin, Zita Hadju, Julian and Lucas Martin, Zoe Bicât, Debbie and Millie Hastings, Hannah Fisher and Janet Morgan; and the mysterious influences of my beloved cat, Hugo, and my fellow cat-loving neighbour, Wendy Davies, undoubtedly run through the book. The presence and guidance of Martin and Gail Aylward over this time has been of the most profound support, while dear Sara Jansson – whose life was cut so short – and her wonderful son, Joe, became my constant reminder of what is really most important in life. And I thank, from the bottom of my heart, Constantia Elia, whose revitalizing passion and inspiration helped me over the finishing line.

I dedicate the book to my parents and grandparents, without whom none of this would have been possible.

I

Earth Oil

Mineral oils naturally appear at the earth's surface in various forms, often with associated gases, in many parts of the world. As *Scientific American* informed its readers in 1856, 'There are a number of different kinds of it', from naphtha, the most liquid, volatile and flammable form of oil, to the thicker 'fluid bitumen [that] is of a dark color, and oozes from certain rocks and crevices in the earth, and becomes solid by exposure to the atmosphere', to the most dense 'asphalt, which is sometimes too hard to be scratched with the finger nail'.[1] There is archaeological evidence of bitumen having been used from as early as the Middle Palaeolithic era as an adhesive for tool-making,[2] and for millennia people have made use of petroleum in a wide variety of ways: as a waterproofing agent; as mortar for stone- or brick-work and as a sealant; for medicinal and ceremonial purposes; for art and craft-work; for embalming and mummification; as a lubricant; as a material for surfacing pavements and roads; and to burn for its heat and light.[3]

THE MIDDLE EAST AND CENTRAL ASIA

In the Middle East, evidence has been found of the use of bitumen for tool-making during the Neolithic and Palaeolithic eras – up to about 70,000 years ago – in modern-day Turkey, Syria, Iran and Kuwait. From about 3500 BC the beginning of urbanization and the rise of the Sumerian civilization in Mesopotamia (modern-day Iraq) led to an increase in demand for bitumen, sourced initially from Hit, then from Mosul, and later from southwestern Iran, as well as some from Kuwait. These sources supplied a trading region covering the length of the Tigris and Euphrates rivers and the northern Persian Gulf, while bitumen from the Dead Sea supplied a distinct trading network extending across modern-day Israel and Egypt.[4] According to the *Gilgamesh* epic, as translated from cuneiform stone tablets dated between 2100 and 600 BC, Noah's Ark, a hugely scaled-up version of the circular boats used on the Tigris and Euphrates, required about 270 tons of bitumen to waterproof its hull. Model boats made from bitumen were found in graves at the Sumerian city of Ur dated about 2500 BC, and the sixth-century BC Book of Genesis tells us that bitumen was used as mortar for building the Tower of Babel in Babylonia, in modern-day Iraq.[5]

In ancient texts, it is often unclear whether the author is writing about pitch derived from tree sap or from a petroleum source, a distinction expressed in about 5 AD by Ovid in his *Metamorphoses*: Byblis, deranged by desire, dissolves in her own tears, 'like resin-drops dotting the bark of a new-cut pine, or bitumen stickily oozing out of the oil-rich earth'.[6] However, it is evident that Herodotus, in 450 BC, was referring to petroleum when he told of how bitumen from Hit was used as mortar in the building of Babylon, south of present-day Baghdad:

> [W]ere you to leave Babylon and travel for eight days, you would come to another city, Is [Hit], which shares its name with the river, an insignificant tributary of the Euphrates, next to which it stands. The waters of the River Is are full of lumps of bitumen, which are brought to its surface – and this was the source of the bitumen used in the building of Babylon's wall.[7]

He also wrote of a well in the vicinity of Susa (Shush), near the Karun River, in present-day southwest Iran,

> which yields three distinct products. Asphalt, salt and oil: all of these are extracted from it. As for the means of extraction, the various substances are drawn up using a swipe to which half an animal skin has been attached to serve as a bucket. In goes the swipe and up comes a liquid, which is then poured into a container. From there, it is strained into a further container, and diverted along three different channels. The asphalt and the salt solidify immediately, but the oil . . . this *rhadinake*, as the Persians call it, is black, and gives off a revolting stench.[8]

Vitruvius, in about 30 BC, mentioned petroleum from the region of Babylon, as well as from modern-day Lebanon, eastern Turkey, north Africa and Ethiopia.[9] In about 20 AD Strabo, followed by Tacitus roughly eighty years later, wrote of how asphalt floating on the Dead Sea was gathered:[10]

> The asphalt is blown to the surface at irregular intervals from the midst of the deep, and with it rise bubbles, as though the water were boiling . . . [People] reach the asphalt on rafts and chop it and carry off as much as they each can.[11]

Writing of the oil found in Mesopotamia and Persia, Strabo contrasts the dense asphalt, used as mortar and for waterproofing boats, with 'black naphtha, liquid asphalt, which is burnt in lamps instead of [olive] oil'.[12] He elaborates:

> The liquid kind, which they call naphtha, is of a singular nature; for if the naphtha is brought near fire it catches the fire; and if you smear a body with it and bring it near to the fire, the body bursts into flames; and it is impossible to quench these flames with water (for they burn more violently), unless a great amount is used, though they can be smothered and quenched with mud, vinegar, alum, and bird-lime.[13]

In about 78 AD Pliny the Elder mentions the occurrences and use of petroleum in the Middle East: near the bank of the Euphrates in modern-day southeast Turkey; in southeast Iran; on the coast of Lebanon; in the Dead Sea; and in Iraq where, at Babylon, he wrote, oil skimmed from the surface of brine was used for burning in lamps.[14] In about 385 AD Ammianus Marcellinus wrote of how in Assyria, centred on the upper Tigris and Euphrates rivers,

> bitumen is found near the lake called Sosingites . . . Here naphtha also is produced, a glutinous substance which looks like pitch. This too is similar to bitumen, and even a little bird, if it lights upon it, is prevented from flying, sinks, and disappears utterly. And when fluid of this kind catches fire, the mind of man will find no means of putting it out, except dust. In these regions there is also to be seen a cleft in the earth, from which rises a deadly exhalation, which with its foul odour destroys every living creature that comes near it. If this pestilential stuff, rising

from a kind of deep well, should spread out widely from its opening before rising on high, it would by its fetid odour have made the surrounding country a desert.[15]

Plutarch, in about 75 AD, wrote of Alexander the Great, telling us that,

As he traversed all Babylonia, which at once submitted to him, he was most of all amazed at the chasm from which fire continually streamed forth as from a spring, and at the stream of naphtha, so abundant as to form a lake, not far from the chasm. This naphtha is in other ways like asphaltum, but is so sensitive to fire that, before the flame touches it, it is kindled by the very radiance about the flame and often sets fire also to the intervening air. To show its nature and power, the Barbarians sprinkled the street leading to Alexander's quarters with small quantities of the liquid; then, standing at the farther end of the street, they applied their torches to the moistened spots; for it was now getting dark. The first spots at once caught fire, and without an appreciable interval of time, but with the speed of thought, the flame darted to the other end, and the street was one continuous fire.

He goes on to embellish an earlier account, by Strabo, of how an attendant suggested to Alexander that he try an experiment with this fluid on a young singer, Stephanus:

'Wilt thou, O King, that we make a trial of the liquid upon Stephanus? For if it should lay hold him and not be extinguished, I would certainly say that its power was invincible and terrible.' The youth also, strangely enough, offered himself for the experiment, and as soon as he touched the liquid and began to anoint himself with it, his body broke out into so great a flame and was so wholly possessed by fire that Alexander fell into extreme perplexity and fear; and had it not been by chance that many were standing by holding vessels of water for the bath, the youth would have been consumed before aid reached him. Even as it was, they had great difficulty in putting out the fire, for it covered the boy's whole body, and after they had done so, he was in a sorry plight.[16]

Indeed, these deadly properties of petroleum came to be used in warfare when added to the traditional incendiary mixes of pine-tree resin, sulphur and quicklime. Thucydides described the deployment of burning pitch in the siege of Plataea in 429 BC during the Peloponnesian War, when the Greeks hurled a burning mixture of pitch and sulphur at the city's fortifications, a weapon that became known as 'Greek fire'.[17] But since the word 'pitch' conventionally referred to pine resin, it is unlikely that this and similar instances of the deployment of Greek fire – as related by Herodotus and Xenophon, for example – involved the use of petroleum. However, Pliny, writing in about 78 AD, made this particular innovation explicit in his account of a method by which, a few years earlier, soldiers of Mithridates VI of Pontus had defended their fortifications against a Roman attack. Near the banks of the upper Euphrates River there was

> a pool which discharges an inflammable mud, called Maltha. It adheres to every solid body which it touches, and moreover, when touched it follows you, if you attempt to escape from it. By means of it the people defended their walls against Lucullus, and the soldiers were burned in their armour.[18]

Cassius Dio, in about 220 AD, wrote that after many victories won by Lucullus' army, during a later attack on a Pontic city, '. . . the barbarians did him serious injury by means of their archery as well as by the naphtha which they poured over his [siege] engines; this chemical is full of bitumen and is so fiery that it is sure to burn up whatever it touches, and it cannot easily be extinguished by any liquid.'[19] Dio subsequently described how, in 198 AD, a Roman siege of the Mesopotamian city of Hatra was repelled by the Parthians, who

> inflicted the greatest damage on their assailants when these approached the wall, and much more still after they had broken down a small portion of it; for they hurled down upon them, among other things, the bituminous naphtha, of which I wrote above, and consumed the engines and all the soldiers on whom it fell.[20]

A little before 550 AD, Procopius, in his *History of the Wars*, recounted how the Romans' battering-rams were about to break through the walls of a Persian fort in the Caucasus, 'But the Persians hit on the following plan.' From the fort wall they filled pots with sulphur and bitumen and the substance which the Persians call 'naphtha' and the Greeks 'Medea's oil', and they now set fire to these and commenced to throw them upon the sheds of the rams, and they came within a little of burning them all ... for the fire kindled instantly whatever it touched, unless it was immediately thrown off.[21]

Petroleum was added to the mixtures used for flaming arrows and other incendiary projectiles that were deployed for setting fire to siege engines and wooden defences at a distance. During the mid-seventh century the Romans began to use incendiary weapons also at sea to repel Arab ships attempting to blockade their capital, Constantinople. Soon they were equipping their Byzantine galleys with quite sophisticated flame-throwers – sometimes also used in land warfare – that employed a piston mechanism to project flaming petroleum-based fluids. Emperor Constantine VII, in his manual on statecraft that he wrote for his son in about 950, cautioned his heir that 'the infidel and dishonourable tribes of the north' would make excessive demands to which he should not accede. 'Similar care and thought,' wrote Constantine,

> you must take in the matter of the liquid fire which is discharged through tubes, so that if any shall ever venture this demand too ... you may rebut and dismiss them in words like these: 'This too was revealed and by God through an angel to the great holy Constantine, the first Christian emperor, and concerning this too he received great charges from the same angel ... that it should be manufactured among the Christians only and in the city ruled by them, and nowhere else at all, nor should it be sent nor taught to any other nation whatsoever.'[22]

If anyone were found to have revealed this information, his son should 'dismiss him to a death most hateful and cruel'. Constantine listed several sources of petroleum around the Black Sea: firstly, near modern-day Taman on the Taman Peninsula, on the eastern side of the Kerch Strait between the Azov and Black seas; secondly, in the territory between Maikop and Sochi; and thirdly, in eastern Turkey. Due to the geology of the Taman Peninsula region, the oil found there at the surface was especially light and fluid, making it particularly suitable for use in flame-throwers. However, by the end of the twelfth century the

Byzantine navy was no longer being equipped with flame-throwe
perhaps because the Romans had by then lost easy access to these
sources of petroleum.[23]

Fragments of pitch found in the excavation of an early medieval
ship burial, located in a seventh-century cemetery at Sutton Hoo in
Suffolk, England, were initially thought to have been 'Stockholm tar'
rather than native pitch or bitumen. On later analysis the fragments
turned out, indeed, to be bitumen, but most likely from the Dead Sea
region, evidencing the extent of overseas trade routes in Anglo-Saxon
times, particularly for rare and valuable items that would be symbols
of high status.[24]

When, in about 760, the Persian Abbasids displaced the Syrian
Umayyad Caliphate, the Persians gained control over the oil springs
at Baku, a trading port on the western shore of the Caspian Sea in
the southern Caucasus, in present-day Azerbaijan. The Persians
levied a tax on these springs, and the new dynasty sourced bitumen
also from Ferghana (in present-day Uzbekistan).[25] Indeed, the Kur-
ramiyya revolt of 816–37 in Azerbaijan may largely have been a
conflict over whether the region's mineral resources would be controlled by local rulers or by a foreign empire.[26]

Meanwhile, during a civil war for control of the Caliphate of Baghdad, in 813 the invading forces of Al-Ma'mun captured a position
near al-Anbar Gate. One of Caliph Al-Amin's commanders launched
an ultimately unsuccessful counter-attack on a nearby district and
according to the Persian scholar Al-Tabari, 'It was at this time that he
commanded that al-Harbiyyah should be bombarded with naphtha
and fire and by *manjanigs* and *arradahs* [types of trebuchet]; people
coming and going there were killed as a result of them.'[27] In 855 the
Caliph of Baghdad, al-Mu'tamid, granted the revenues from the Baku
oil wells to the people of Darband (Derbent) just to the north, although
from about 900 the local governor began keeping these revenues
for himself.[28] In 943 the Baghdad-born historian and geographer
Abul'hasan al-Mas'ūdī wrote in his *Meadows of Gold and Mines of
Gems* of Baku's naphtha sources, 'from which fire issues perpetually,
throwing up a high flame'.[29] The following year, competing groups
fought for control over the oil wells, and in about 950 the widely
travelled Abū-Dulaf wrote that two wells were each being leased for

1,000 dirhams per day; two hundred years later the geographer Yāqūt claimed that a single well could generate this revenue daily.[30]

In about 950 the Persian geographer Abu Is'haq Farsi Istakhri wrote of a bitumen mine, the 'Kubbat-al-Mumiya', in a cave near Darabjird (Darab) in Fars province, which was kept under guard and reserved for royal use.[31] Thirty years later the geographer Ibn Hawqal wrote that near the border between the provinces of Khuzestan and Fars there was 'a mountain, from which fire issues at all times . . . and the general opinion is, that there is here a fountain of Naphta, or of pitch, which has taken fire'; to the west, in northern Mesopotamia, there were 'springs or fountains that yield . . . bitumen', and to the north of Persia, in the Ferghana Valley, there were 'springs of naphta, and of bitumen'.[32]

Muslim armies now adopted Greek fire, although not the Byzantine flame-thrower apparatus itself. The armies of the Abbasid Caliphate and the Seljuk Empire had contingents of naphtha-throwers who wore fire-protective clothing while hurling incendiary material at the enemy or to clear a pathway through vegetation.[33] During the Crusades, from 1097 to 1270, the Muslim soldiers frequently used petroleum incendiary weapons, mainly defensively against the Christians' wooden siege engines.[34] For example, the thirteenth-century Arab historian Ibn al-Athir related how, in a battle during the 1189–91 Siege of Acre, Saladin's army

> first threw pots with naphtha and other things, not kindled, against one of the [siege] towers . . . [then] at the right moment threw onto it a well-burning pot. At once fire broke out over the whole tower and it was destroyed. The fire was so quick that the Christians had no time to climb down and they and their weapons were all burnt up. The other two towers were similarly destroyed.[35]

A French Crusader, Jean de Joinville, described how, during the 1249 siege of Mansura in Egypt, their siege engines were attacked one night when some Saracens advanced with a trebuchet

> and filled the sling of the engine with Greek fire . . . [T]heir first shot landed just in front of us . . . [but] [o]ur firemen were ready to extinguish the fire . . . The way the Greek fire came was that in front it was

as large as a cask of vinegar, and the tail of fire that came out of it was as big as a great spear. It made such a noise as it came that it seemed like the thunder of heaven; it looked like a dragon flying through the air. It cast so much light that the camp could be seen as clearly as if it were day, due to the great mass of fire and the brilliance of its light ... [Later] they brought up their trebuchet in broad daylight ... and launched the Greek fire onto our siege engines ... As a result our two siege engines were burned.[36]

The deployment of petroleum as a weapon of war was accompanied by the first attempts to distil crude oil. Some petroleum springs yielded oil that was naturally light – volatile and easily flammable – such as was used for illuminating the great mosques of Mecca and Medina until 860 AD, after which candles began to be used instead.[37] However, most springs yielded heavier oil that, in order to be made more incendiary, had to be either combined with other flammable materials, such as pine sap, saltpetre, quicklime and sulphur, or distilled to separate out the lighter fractions.[38] The process of distillation – collecting the condensed vapours of a heated liquid – was first described in about 100 AD, but its use was confined to the production of essential oils and perfumes. It was not until 900 that the early Persian scientist Al-Râzî, in his *Book of the Secret of Secrets*, described the distillation of petroleum, which was '... an operation like the manufacture of water of roses. It consists in putting the thing into the alembic and lighting a fire under it, so that its "water" ascends in the alembic and flows off to the recipient and collects there.'[39]

Instructions for the distillation of petroleum are included in Al-Hasan al-Rammah Najim al-din al-Ahdab's 1285 *Treatise on Horsemanship and Stratagems for War*, which gives an extensive account of pyrotechnics, just as the technique of distillation was beginning to be taken up in Europe.[40] By the early fourteenth century the itinerant Syrian scholar Al-Dimashqi was the first to write of the regular distillation of petroleum in the Middle East, by Coptic Christians:

Many types of naft are water white by nature and so volatile that they cannot be stored in open vessels. Others are obtained from a kind of pitch (or bitumen) in a turbid and dark condition, but by further treatment they can be made clear and white by distilling them like rose-water.[41]

In 1168 a Crusader army under Amalric I of Jerusalem invaded Fustat, old Cairo, and according to several Arab historians such as Al-Maqrīzī the city was ordered by the vizier of Egypt, Shawar, to be burned down to deny it to the Franks as a strategic base; and it was said that an inferno lasting fifty-four days was ignited with 20,000 jars of naphtha and 10,000 torches. However, it seems that these accounts of the fire may have been greatly embellished and its causes misreported.[42]

The early thirteenth-century Persian writer Ibn al-Balkhi reported the use of high-quality lamp oil in the coastal town of Bandar Deylam near the head of the Persian Gulf, the petroleum most likely being sourced from the surrounding region.[43] There was also a naphtha spring that yielded a significant revenue at this time at Khanikin (Khanaqin), on the trade route between Baghdad and Tehran, so the inhabitants of these latter cities may have sourced oil from there.[44]

Oil continued to be drawn from wells around Baku and traded in significant quantities, as Marco Polo described towards the end of the thirteenth century in his account of his travels:

> On the confines towards Georgiana there is a fountain from which oil springs in great abundance, insomuch that a hundred shiploads might be taken from it at one time. This oil is not good to use with food, but 'tis good to burn, and is also used to anoint camels that have the mange. People come from vast distances to fetch it, for in all the countries round about they have no other oil.[45]

An Italian merchant, Cesar Frederick, reporting on his travels through Mesopotamia in 1563, wrote,

> I embarked on one of those small vessels which ply upon the Tigris between Babylon and Basora ... They use no pumps, being so well daubed with pitch as effectively to exclude the water. This pitch they have from a great plain near the city of *Heit* [Hit] on the Euphrates ... This plain full of pitch is marvellous to behold, and a thing almost incredible, as from a hole in the earth the pitch is continually thrown into the air with a constant great smoke ... The Moors and Arabs of the neighbourhood allege that this hole is the mouth of Hell.[46]

In 1569 Geoffrey Duckett, on a trading mission for the Muscovy Company, travelled from England via Russia to Persia; he informed his employer that in the province of Azerbaijan there was

> a towne called *Backo*, neere unto which towne is a strange thing to behold – for there issueth out of the ground a marveilous quantitie of Oyle, which Oyle they fetch from the uttermost bounds of all Persia; it serveth all the countrey to burne in their houses.

> This oyle is blacke, and is called *Nefte*; they use to cary it throughout all the countrey upon kine and asses, of which you shall oftentimes meete with foure or five hundred in a companie. There is also by the said towne of *Backo* another kind of oyle, which is white and very precious, and is supposed to be the same that here is called *Petroleum*. There is also, not farre from *Shamaky*, a thing like unto tarre, and issueth out of the ground, whereof wee made the proofe that in our ships it serveth well in the stead of tarre.[47]

The 'white' oil mentioned by Duckett – presumably a pale, almost colourless and very fluid form of naphtha – was also described by Adam Olearius, secretary to a delegation sent to Persia by the Duke of Holstein, who witnessed the drawing of oil from Baku's wells in 1638:

> We saw, as we pass'd by, within the space of five hundred paces, about thirty sources of *Nefte*, which is a kind of Medicinal Oil. There are, among the rest, three great ones, into which they go down by sticks, plac'd there to serve instead of a Ladder, fifteen or sixteen foot into the ground. A man, standing above at the pits mouth, might hear the Oil coming out in great bubbles, sending up a strong smell, though that of the white *Nefte* be incomparably more pleasant than that of the black: for there are two sorts of it.

He wrote of the extensive and heavy taxation that fed the Shah's extravagant wealth, adding, 'He farms out also the Fishing of the Rivers, the Baths, and Stoves, the places of publick Prostitution, and the Springs of *Nefte*.'[48]

In 1684 Engelbert Kämpfer, secretary to the Swedish delegation at the Persian court, wrote an account of the continuing thriving oil trade that he witnessed near Baku:

The work of drawing off the liquid is carried out manually by means of leather bottles lowered from above either by hand or by a quite simple winch. Only one particularly rich cavity has a rather large closed shed built over it. It was deeper and wider than the others, and the naphtha flowing forth audibly in a fierce torrent, was drawn off with the aid of a contrivance built over it and set in motion alternately by two horses moving in a circle ... The naphtha was transported in sheepskin containers to the towns of Schamachi and Baku, thence by canal to the whole of Media, and from there on by sea to Hyrcania, Usbek, Circassia, and Dagestan. It is all used as a lamp-oil and as a fuel for torches being capable of illumination which surpasses all expectations![49]

In the same year Leonty Kislyansky, appointed by Russia's Siberia Office as the chief administrator of Irkutsk, reported the existence of hot petroleum seeps nearby, and he envisioned extracting the oil on a significant scale, hoping, 'God willing, there will be no trouble from the unfriendly Mongols and Chinese, and ... I will dig and earn a living in all solicitude.' However, nothing came of his plans.[50]

In 1694 Pierre Pomet, a prominent French pharmacist, found that bitumen from the Dead Sea was still being used as in ancient times:

Judean bitumen or Asphaltum is a bitumen found floating on the surface of the lake which now covers the former cities of Sodom and Gomorrha ... [T]he inhabitants of this place, who are Arabs, derive great benefit from it, as they use it to caulk their boats, in the same way as the Northerners and ourselves use pitch.[51]

John Fryer, a physician working for the East India Company, published in 1698 an account of his travels, including through Persia's Fars province, in which he wrote of the continuing royal use of the bitumen source near Darab:

On the right hand of the King's Highway, between *Siras* [Shiraz] and *Gerom* [Jahrum], at *Derab*, on the side of a Mountain, issues the *Pissasphaltum* of *Dioscorides*, or Natural Mummy, into a large Stone Tank or Storehouse, sealed with the King's Seal, and that of the [mayors], and all the Noblemen of that City, and kept with a constant Watch, till at a stated Time of the Year they all repair thither, to open it for the King's Use, to prevent its being stole: Which notwithstanding, though it be Death if discovered, yet many Shepherds following their Flocks on these Mountains, by chance light on great Portions of the same Balsam, and offer it to Passengers to Sale, and sometimes play the Cheat in adulterating it.[52]

During the early eighteenth century Tsar Peter the Great endeavoured to facilitate the mining of minerals across the Russian Empire. He was well aware of the high cost of importing petroleum from Baku and was covetous of the significant revenue that its oil trade earned for the Persian government; and during a period of political upheaval in Persia, the Tsar sent his army into the southern Caucasus, taking over Baku in 1723.[53] However, by 1743, when an agent for the Muscovy Company, Jonas Hanway, went to assess the trading potential of Persia and the Caspian region, Baku had returned to Persian sovereignty under a treaty agreement with Russia. In his account of his travels Hanway gave a description of a Zoroastrian fire temple near Baku and of the area's oil resources:

[T]here is a little temple, in which the Indians now worship: near the altar about 3 feet high is a large hollow cane, from the end of which issues a blue flame, in colour and gentleness not unlike a lamp that burns with spirits, but seemingly more pure. These Indians affirm, that this flame has continued since the flood, and they believe it will last to the end of the world; that if it was resisted or suppressed in that place, it would rise in some other ...

If a cane or tube, even of paper, be set about 2 inches in the ground, confined and closed with earth below, and the top of it touched with a live coal, and blown upon, immediately a flame issues without hurting either the cane or paper, provided the edges be covered with clay; and this method they use for light in their houses, which have only earth for the floor: three or four of these lighted canes will boil water in a pot ...

Near this place brimstone is dug, and naptha springs are found.

The chief place for the black or dark-grey naptha is the small island Wetoy [Sviatoi, Holy Island], now uninhabited, except for at such times as they take naptha from thence. The Persians load it in bulk in their wretched vessels; so that sometimes the sea is covered with it for leagues together. When the weather is thick and hazy, the springs boil up the higher; and the naptha often takes fire on the surface of the earth, and runs in a flame into the sea, in great quantities, to a distance almost incredible. In clear weather the springs do not boil up above two or three feet: in boiling over, this oily substance makes so strong a consistency as by degrees almost to close the mouth of the spring; sometimes it is quite closed, and forms hillocks that look as black as pitch; but the spring, which is resisted in one place, breaks out in another. Some of the springs, which have not been long open, form a mouth of eight or ten feet diameter.

The people carry the naptha by troughs into pits or reservoirs, drawing it off from one to another, leaving in the first reservoir the water, or the heavier part with which it is mixed when it issues from the spring. It is unpleasant to the smell, and used mostly among the poorer sort of the Persians, and other neighbouring people, as we use oil in lamps, or to boil their victuals; but it communicates a disagreeable taste. They find it burns best with a small mixture of ashes: as they find it in great abundance, every family is well supplied. They keep it, at a small distance from their houses, in earthen vessels under ground, to prevent any accident by fire, of which it is extremely susceptible.

There is also a white naptha on the peninsula of Apcheron, of a much thinner consistency; but this is found only in small quantities. The Russians drink it both as a cordial and medicine.[54]

In 1811 the East India Company's Resident at Baghdad, Claudius Rich, visited the ruins of Babylon where he saw evidence of bitumen having been used as a mortar in stonework. In his account published several years later he wrote,

> There are two places in the pashalick of Bagdad where bitumen is found: the first is near Kerkouk . . . ; the next is at Heet, the Is of Herodotus, whence the Babylonians drew their supplies . . . The principal bitumen-pit [at Hit] has two sources, and is divided by a wall in the centre, on one side of which the bitumen bubbles up, and on the other oil of naphtha.[55]

John Kinneir, in 1813, published an extensive report for the East India Company on the commercial and strategic importance of Persia and Mesopotamia. He wrote of the 'enormous' quantities of oil collected from oil wells of Baku, which 'are, in a certain degree, inexhaustible,

as they are no sooner emptied than they again begin to fill'.[56] Of the
resources of Persia and Mesopotamia, he wrote,

> of all its natural productions, the *naft*, or *naphta*, is the most extraor-
> dinary, as well as the most useful. Of this mineral there are two kinds,
> the black and the white. The former, which is the bitumen so famous
> in the Babylonian history, and so often described by travellers, is, when
> taken from the pit, a thick liquid resembling pitch. To me it appeared
> to be similar, although of a finer quality, to the specimens which I had
> seen of the pitch taken from the lake in the Island of *Trinidad*. It is,
> undoubtedly, a most excellent substitute for pitch. The bottoms of most
> of the vessels which navigate the *Euphrates* and *Tigris* are covered with
> it; and it is also used in the lamps, instead of oil, by the natives. There
> are several fountains of this bitumen in *Irak Arabi*, and the lower *Kurd-
> istan*. The most productive are those in the vicinity of *Kerkook* . . .
> *Mendali*, and *Hit* on the banks of the *Euphrates*.[57]

Near Kirkuk, Kinneir wrote, there '. . . are a number of *naptha* pits,
which yield an inexhaustible supply of that useful commodity . . . The
naptha is here in a liquid state, and perfectly black: it is conveyed
from the bottom to the top in leathern buckets, then put in earthen
jars, and sent all over the neighbouring country.'[58]

In 1836 the surgeon to the British Residency at Baghdad, John Ross,
travelled up the Tigris, and the following year Lieutenant H. Blosse
Lynch of the Indian Navy surveyed the region. A map by Lynch,
accompanying their detailed reports published in the *Journal of the
Royal Geographical Society*, indicated 'Bitumen pits' near the Tigris
about forty miles south of Mosul and 'Naptha Springs' about five
miles northwest of Kirkuk. At Tikrit, Ross encountered 'people going
with skins (for rafts) to Járah, for the purpose of floating bitumen to
Baghdád'; a few miles northeast of Baghdad itself he came across a
place where 'the remains of an ancient bridge' had been dug up to be
used for building a house in the city: 'the bridge was built of bricks,
with cuneiform inscriptions, exactly similar to those of Babylon, and
cemented with bitumen'.[59] As Ross was journeying up the Tigris,
Major Sir Henry Rawlinson – soon to be appointed Britain's political
agent at Kandahar – was travelling through Persia from Zohab down
to Khuzestan province near the head of the Persian Gulf. In his report,

published in the same issue of the geographical journal, he noted, 'naphtha pits . . . are passed on the road from Shuster to Rám-Hormuz', while on the plain of Kír Ab, twenty-five miles north of Dezful, 'The liquid bitumen is collected at the present day in the same way as is related by Herodotus: the ground is impregnated with this noxious matter, and the waters are most unwholesome.'[60]

In 1840 Austen Layard, the soon-to-be famous archaeologist, was exploring southwest Persia when he became involved in a dispute and had to flee from Shuster with a Bakhtiari chief to a remote ravine in the hills where the chief happened to own 'some naphtha, or bitumen, springs'.[61] Layard explored Khuzestan province again in 1846, and his account of this journey includes descriptions of several oil springs in the region. There were 'bitumen-pits' south of the ruins of Masjid i Suleiman (Masjed Soleyman); to the east of Ram Hormuz was a range of hills 'containing the celebrated white naphtha springs and the bitumen pits of Meï Dáwud'. Some valleys on the eastern side of the Zagros Mountains, he wrote, '. . . contain but few springs of fresh water, and abound with pools of naphtha or petroleum, bitumen, and sulphureous or brackish water, and frequently, as at the naphtha-springs near Ram Hormuz, have a burnt and volcanic appearance'.[62] From Mesopotamia, Layard reported that boats plying the Tigris and Euphrates were waterproofed with bitumen,[63] and at Mosul in 1853 he entertained some Yazidi leaders '. . . in an arched hall, open to the courtyard, which was lighted up at night with *mashaals*, or bundles of flaming rags saturated with bitumen, and raised in iron baskets on high poles, casting a flood of rich red light'.[64]

On the evening of 28 January 1850 at Nimrud, on the Nineveh Plain twenty miles south of Mosul, Layard's workers had finalized the preparations for removing a pair of huge statues of the Sumerian deity Lamasu, a human-headed, winged lion, now bound for the British Museum in London. On a foundation composed partly of an inch-thick layer of bitumen, the statues had stood guard at the entrance to the throne room of the northwest palace of Ashurnasirpal for over 2,500 years. Layard wrote,

> they were ready to be dragged to the river-bank . . . It seemed almost sacrilege to tear them from their old haunts to make them a mere wonder-stock to the busy crowd of a new world . . . I had sent a party

of Jebours to the bitumen springs, outside the walls to the east of the inclosure. The Arabs having lighted a small fire with brushwood awaited our coming to throw the burning sticks upon the pitchy pools. A thick heavy smoke, such as rose from the jar on the sea shore when the fisherman had broken the seal of Solomon, rolled upwards in curling volumes, hiding the light of the moon, and spreading wide over the sky. Tongues of flame and jets of gas, driven from the burning pit, shot through the murky canopy. As the fire brightened, a thousand fantastic forms of light played amidst the smoke. To break the cindered crust, and to bring fresh slime to the surface, the Arabs threw large stones into the springs; a new volume of fire then burst forth, throwing a deep red glare upon the figures and upon the landscape. The Jebours danced round the burning pools, like demons in some midnight orgie, shouting their war-cry, and brandishing their glittering arms. In an hour the bitumen was exhausted for the time, the dense smoke gradually died away, and the pale light of the moon again shone over the black slime pits.[65]

AFRICA

One of the few early indications of the existence of petroleum in Africa comes in the form of a report, in a mid-nineteenth-century official Portuguese publication, stating that in 1767 the then Governor of Portuguese-colonized Angola, Francisco Inocêncio de Sousa Coutinho, had forty-nine barrels of petroleum shipped from Luanda to Lisbon. The oil was sourced from seeps at Libongo, near the mouth of the Dande River about thirty miles north of Luanda, and was used to caulk ships – an activity with a bleak significance: since the discovery of large gold deposits in Brazil in the late seventeenth century, followed by the growth of commodity plantation agriculture both there and, subsequently, in the Caribbean and the United States, Angola had become the largest single source of African slave labourers shipped to the Americas; the total number of captives shipped across the Atlantic from Angola would eventually reach almost 4.5 million.[66]

In 1783 a Brazilian-born naturalist, Joaquim José da Silva, was posted from Lisbon to Angola to be the colony's Government Secretary. Silva's other task was to conduct investigations into the region's natural resources, and early in the following year he explored the area around the mouth of the Dande River. In his report to the Minister for Overseas Colonies on 17 March 1784, he wrote,

> I asked the Government for permission to investigate a source of bitumen rumoured to be in the vicinity of the Dande River, and which was in fact located slightly southeast of it, some three leagues from the sea, or from the mouth of the Lifune River to the North of this same Dande. From the enclosed samples, one can clearly see that the aforementioned source contains genuine Petroleum running forth in droplets from the fissures ...; I have seen that they are so rich in Petroleum that with a little more effort, the harvesting and transportation process would be very easy and without great expense ... There are currently precious few barrels filled with this Petroleum, overseen by the Dembo Indoy who controls these lands, so much depends on force to gather this Petroleum for the use of the city's Royal Arsenal.[67]

In his first shipment of specimens that he sent back to Lisbon three days later – on a vessel also used to transport slaves to Brazil – Silva included 'A little bottle of Dande oil'. He was eventually assigned to the Portuguese garrison at Ambaca, on the main slave route inland from Luanda along which were forced around 10,000 slaves per year.[68]

In the early 1850s Dr David Livingstone – the Scottish physician, missionary and fierce critic of the barbarities of British imperialism – was beginning his famous explorations into central Africa, by which time European countries had formally abolished the slave trade, although planters in the United States found ways to skirt the ban. Dr Livingstone, referring to a map of the beginning of his route inland from Luanda, wrote,

> To the N. of 2 and 3 [just east of Bengo], near the river Dande, petroleum is reported, and so it is said to occur southwards of 5 [Ambaca], from under the dark red sandstone which forms the crust of the country. The spot reported is on the banks of the Coanza [Kwanza], and near Cambambe ... The government of the country may be described as a military one, and closely resembles that which Sir Harry Smith [Governor of the Cape Colony] endeavoured in vain to introduce among the Caffres [in the Eighth Xhosa War].[69]

ASIA

In China, there are accounts of the *shih yu*, or 'rock oil', from natural petroleum seepages dating from at least the fifth century. One described how

> south of Yen-shou among mountain rocks there oozes a liquid looking like uncoagulated fat. When burnt it generates an intense brightness, but it cannot be consumed as food (or used for frying). The local people call it 'mineral lacquer'.[70]

According to the Japanese chronicle *Nihongi*, during the seventh-century reign of Tenji, 'the province of Koshi presented to the Emperor burning earth and burning water'. The latter is presumed to have been petroleum from present-day Echigo province, where the oil was collected from excavated ditches.[71]

The ninth-century Chinese scholar Duan Chengshi wrote in his *Miscellanies of Youyang* that petroleum was being used in Shensi (Shaanxi) province: 'In the Weishui River of Gaonu County there is a greasy matter like paint flowing on the water. Local people get it to grease their wagons and burn it for lighting.'[72] By the tenth century, and perhaps earlier, the Chinese began to make use of the incendiary properties of petroleum in warfare and – quite likely inspired by the contemporary Roman Byzantine technology via Arab sea-traders to China – they soon developed their own flame-throwing weaponry.[73] The eleventh-century scientist Shen Kua wrote in his *Dream Pool Essays* that the petroleum of Shensi and Kansu 'can be easily burnt, but its smoke, which is very thick, makes the curtains all black'. He found that he could make high-quality ink from its soot and he anticipated that, as the forests were depleted, this ink would eventually replace that hitherto made with the soot of burned pine resin:

> The black colour was as bright as lacquer and could not be matched by pinewood resin ink. So I made a lot of it and called it Yen-Chhuan Shih I (Yen River Stone Juice). I think that this invention of mine will be widely adopted. The petroleum is abundant, and more will be formed in the earth while supplies of pinewood may be exhausted.[74]

Perhaps the first Chinese account of the intentional digging and working of oil wells is in the early fourteenth-century records of the Yuan dynasty:

> South of Yanchang district, at Yinghe, there is a stone oil well which has been cut open. Its oil can burn and also cures itches and ringworms of the six domestic animals. Annually, 110 *jin* [pounds] are handed in. Moreover, in Yongping village, 80 *li* northwest of Yanchuan district, there is another well, which annually procures 400 *jin*.[75]

There are reports, from at least the third century, of 'fire wells' where natural gas appeared at the surface. By the late tenth century, brine (from which salt was derived by evaporation) was being drawn from the ground by the use of quite advanced techniques for drilling deep boreholes, and these would sometimes hit pockets of natural gas, a 'poisonous mist' that could overwhelm and kill labourers. According to one account, 'If fire is thrown into the well, there will be [a sound like] roaring thunder gushing out and the mist will burst upwards, splashing out mud and dissipating stones. This is [indeed] extremely dreadful.'[76]

From the sixteenth century it became increasingly common in the salt-production industry – particularly in Szechuan (Sichuan) province – for the gas from these wells to be distributed through bamboo pipes and burned for heating the salt pans, especially in places where the forests and coal deposits were becoming depleted. In the 1750s it was written that from one well there was 'hot ether rising. People use ordinary fire to ignite it, whereupon the flames arise. The light [of the flames] exhibits a blue-green colour.'[77] As wells were bored to greater depths, more prolific 'fire wells' were struck – often also yielding liquid petroleum – so that by the nineteenth century a single well might supply gas to many hundreds of salt pans.[78]

In 1553 João de Barros, treasurer for the Portuguese Casa de India trading company, published the second volume of his history of the feats of the Portuguese in the Far East. He wrote of how in 1511, near an island in the Straits of Malacca, Sumatran sailors fended off an attack by a Portuguese ship

> by burning *olio da terra* [earth oil], found in great quantities near Pedir [in Aceh], where it flows forth from a fountain. The Moors call this oil

napta and doctors consider it remarkable and an excellent remedy for some illnesses. We obtained some and found it very useful for treating low temperatures and nervous tension.[79]

In 1596 the Dutch explorer Jan Huygen van Linschoten listed the many valuable commodities on Sumatra, adding, 'and they saye there is a fountaine which runneth pure and simple Balsame', in other words a naphtha spring.[80] Manuel Godinho de Erédia, an explorer and cartographer for the King of Spain, reported in 1600,

> Perlat [Perlak] is the place where they discovered the unceasing springs of Earth Oil; its situation lies on the Eastern coast of Sumatra in four degrees of North latitude, within the territorial limits of Achem [Aceh]. The soil in this area of Perlat is so 'oliferous' and full of oil that when it is raked or dug with mattocks, this Earth Oil called 'Minsat Tanna' wells up from underground in such quantities that several clay vessels or jars are filled daily, so that the whole of the Eastern coast to Jamboaer Point is supplied with oil for burning in the lamps at night.[81]

A French naval captain, Augustin de Beaulieu, wrote around 1620, in a memoir only published nearly fifty years later, that in Deli (North Sumatra), there was '... a well of oil, which they consider inextinguishable, should this oil take flame and burn on the sea. The King of Achem used it to set afire two Portuguese galleons against which he waged war some eight or ten years ago in the Straits of Malacca.'[82]

In 1636 the Dutch East India Company ordered its representatives at the court of the Sultan of Aceh to obtain a few pots of this 'oil from the earth' for the company's directors, who 'esteemed it highly and used it with great benefit for their stiff limbs'.[83] This Sumatran oil now began to be marketed in Europe alongside other supposedly medicinal petroleum oils. One of the company's physicians, Jacobus Bontius, wrote,

> we have, brought from Sumatra, an excellent kind of naphtha, called, by the Indians, minjac tannah (oil of earth), which, like the naphtha known in Europe by the name of ol. petræ, springs out of the earth, or drills into lakes and rivers from the contiguous rocks. This oil is held in so much esteem by the barbarians, that the king of Achen, the most powerful prince in that island, has prohibited the exportation of it under capital punishment; so that when any foreign vessel takes shelter

on that coast in stormy weather, it is common for the inhabitants to bring of it secretly to the ship under night. This oil, when rubbed upon the parts affected by the Barbiers [a kind of paralysis], affords wonderful relief.[84]

In 1755 Captain George Baker was exploring Burma (Myanmar) on behalf of the East India Company. When he reached Yenangyaung on the Irrawaddy River in the Kingdom of Ava, he made this brief mention of what seems to have been a well-established oil trade: 'At this Place there are about 200 Families, who are chiefly employed in getting *Earth-Oil*, out of Pitts, some five Miles in the Country.'[85] It was forty years later, in 1795, before another agent of the East India Company, Major Michael Symes, returned with a fuller account:

> After passing various sands and villages, we got to Yaynangheoum, or Earth-oil (petroleum) Creek, about two hours past noon . . . We were informed that the celebrated wells of petroleum, which supply the whole empire, and many parts of India, with that useful product, were five miles to the east of this place . . . The mouth of the creek was crowded with large boats, waiting to receive a lading of oil, and immense pyramids of earthen jars were raised within and around the village, disposed in the same manner as shot and shells are piled in an arsenal. This place is inhabited only by potters, who carry on an extensive manufactory, and find full employment. The smell of the oil was extremely offensive; we saw several thousand jars filled with it ranged along the bank; some of these were continually breaking, and the contents, mingling with the sand, formed a very filthy consistence. Mr. Wood had the curiosity to walk to the wells, but though I felt the same desire, I thought it prudent to postpone visiting them until my return, when I was likely to have more leisure, and to be less the subject of observation.[86]

He subsequently wrote,

> Doctor Buchanan partook of an early dinner with me; and when the sun had descended so low as to be no longer inconvenient, we mounted our horses to visit the celebrated wells that produce the oil, an article of universal use throughout the Birman empire . . . The evening being far advanced, we met but few carts; those we did observe were drawn each by a pair of oxen, and of a length disproportionate to the breadth

to allow space for the earthen pots that contained the oil. It was a matter of surprise to us, how they convey such brittle ware, with any degree of safety, over so rugged a road: each pot was packed in a separate basket, and laid on straw, notwithstanding which precaution, the ground all the way was strewn with the fragments of the vessels, and wet with oil; for no care can prevent the fracture of some in every journey ... [I]t was nearly dark when we reached the [pits], and the labourers had retired from work. There seemed to be a great many pits within a small compass: walking to the nearest, we found the aperture about four feet square, and the sides, as far as we could see down, were lined with timber; the oil is drawn up in an iron pot, fastened to a rope passed over a wooden cylinder, which revolves on an axis supported by two upright posts. Then the pot is filled, two men take the rope by the end, and run down a declivity, which is cut in the ground, to a distance equivalent to the depth of the well; thus when they reach the end of their track, the pot is raised to its proper elevation, the contents, water and oil together, are then discharged into a cistern, and the water is afterwards drawn off through a hole at the bottom. Our guide ... procured a well rope, by means of which we were enabled to measure the depth, and ascertained it to be thirty-seven fathom, but of the quantity of oil at the bottom we could not judge: the owner of the rope, who followed our guide, affirmed, that when a pit yielded as much as came up to the waist of a man, it was deemed tolerably productive; if it reached to his neck, it was abundant; but that which rose no higher than the knee, was accounted indifferent. When a well is exhausted, they restore the spring by cutting deeper into the rock, which is extremely hard in those places where the oil is produced. Government farm out the ground that supplies this useful commodity; and it is again let to adventurers who dig wells at their own hazard, by which they sometimes gain, and often lose, as the labour and expense of digging are considerable. The oil is sold on the spot for a mere trifle; I think two or three hundred pots for a tackal, or half-a-crown. The principal charge is incurred by the transportation and purchase of vessels. We had but half gratified our curiosity when it grew dark, and our guide urged us not to remain any longer, as the road was said to be infested by tigers, that prowled at night among the rocky uninhabited ways, through which we had to pass. We followed his advice, and returned with greater risk, as I thought, of breaking our necks from the badness of the road, than of being devoured by wild beasts.[87]

Two years later, the site was revisited by Captain Hiram Cox:

When a well grows dry, they deepen it. They say none are abandoned for barrenness. Even the death of a miner, from mephitic air, does not deter others from persisting in deepening them when dry. Two days before my arrival, a man was suffocated in one of the wells, yet they afterwards renewed their attempts, without further accident . . . The oil is drawn pure from the wells, in the liquid state as used, without variation, but in the cold season it congeals in the open air, and always loses something of its fluidity; the temperature of the wells preserving it in a liquid state fit to be drawn . . . The oil is of a dingey green and odorous; it is used for lamps, and boiled with a little dammer (a resin of the country), for paying the timbers of houses, and the bottoms of boats, &c. which it preserves from decay and vermin; its medicinal properties known to the natives is as a lotion in cutaneous eruptions, and as an embrocation in bruises and rheumatic affections . . .

The property of these wells is in the owners of the soil, natives of the country, and descends to the heirs general as a kind of entailed heriditament, with which it is said the government never interferes, and which no distress will induce them to alienate. One family perhaps will possess four or five wells, I heard of none who had more, the generality have less, they are sunk by, and wrought for the proprietors; the cost of sinking a new well is 2,000 tecals flowered silver of the country, or 2,500 sicca rupees; and the annual average net profit 1,000 tecals, or 1,250 sicca rupees . . . Each well is worked by four men, and their wages is regulated by the average produce of six days labour, of which they have one-sixth . . . and lastly the king's duty is a tenth of the production . . .

From the wells, the oil is carried, in small jars, by cooleys, or on carts, to the river; where it is delivered to the merchant exporter at two tecals per hundred viss, the value being enhanced three-eighths by the expense and risk of portage . . . There were between seventy and eighty boats average burthen sixty tons each, loading oil at several wharfs, and others constantly coming and going while I was there. A number of boats and men also find constant employment in providing the pots, &c. for the oil . . .

To conclude, this oil is a genuine petroleum, possessing all the properties of coal tar, being in fact the self-same thing, the only difference is, that nature elaborates in the bowels of the earth, that for the *Burmhas*, for which the *European* nations are obliged to the ingenuity of Lord Dundonald [who had commercialized a process for distilling tar products from coal].[88]

Meanwhile, in 1795 the scholar Tachibana Nankei included in the 'seven wonders' of Japan's Echigo (Niigata) province its oil and gas, used locally for heating and lighting, the oil being drawn from manually dug wells.[89]

During the First Anglo-Burmese War of 1824–6, the Burmese used petroleum-laden fire-rafts in defence against the Royal Navy's ships. During a battle near Rangoon (Yangon), according to a report of 1824,

the river above Kemmendine suddenly became illuminated by an immense mass of moving fire, and it was a grand sight, as raft after raft came floating down towards us, blazing high and brilliantly ... These fire-rafts are made of large beams of timber and bamboos tied together loosely, so that if the mass came athwart a ship's bows, it could swing round and encircle her. On this platform is placed every kind of combustible; fire-wood, and petroleum or earth oil (which abounds in Burmah, and ignites almost as quickly as gunpowder and slightly explodes on being fired) being the principal materials.[90]

Following the defeat of the Burmese forces – from which the British government gained large stretches of coastal territory and major concessions for its de facto subsidiary, the East India Company – Britain's Resident at Singapore, John Crawfurd, was sent as an envoy to the royal Burmese court at Ava. He described in some detail the flourishing trade in 'earth oil' that he encountered during his travels through northern Burma:

The wells occupy altogether a space of about sixteen square miles ... The surface gave no indication that we could detect of the existence of the petroleum. On the spot which we reached, there were eight or ten wells, and we examined one of the best. The shaft was of a square form, and its dimensions about four feet to a side. It was formed by sinking a frame of wood, composed of beams of the *Mimosa catechu*, which affords a durable timber. Our conductor, the son of the Myosugi of the village, informed us that the wells were commonly from one hundred and forty to one hundred and sixty cubits deep, and that their greatest depth in any case was two hundred ... A pot of the oil was taken up, and a good thermometer being immediately plunged into it, indicated a temperature of ninety degrees ... We looked into one or two of the wells, and could discern the bottom. The liquid seemed as if boiling; but whether from the emission of gaseous fluids, or simply from the escape of the oil itself from the ground, we had no means of determining ... The proprietors store the oil in their houses at the village, and there vend it to the exporters. The price, according to the demand, varies from four ticals of flowered silver, to six ticals per 1,000 viss; which is from five-pence to seven-pence halfpenny per cwt. The carriage of so bulky a commodity, and the brokage to which the pots are so liable, enhance the price, in the most distant parts to which the article is transported, to fifty ticals per 1,000 viss. Sesamum [sesame] oil will cost at the same place, not less than three hundred ticals for an equal weight; but it lasts longer, gives a better light, and is more agreeable than the petroleum, which in burning emits an immense quantity of black smoke, which soils every object near it. The cheapness, however, of this article is so great, that it must be considered as conducing much to the convenience and comfort of the Burmans.

Petroleum is used by the Burmans for the purpose of burning in lamps; and smearing timber, to protect it against insects, especially the white ant, which will not approach it. It is said that about two-thirds of it is used for burning; and that its consumption is universal, until its price reach that of sesamum oil, the only one which is used in the country for burning. Its consumption, therefore, is universal wherever there is water-carriage to convey it; that is, in all the country watered by the Irawadi, its tributary streams, and its branches. It includes Bassien, but excludes Martaban, Tavoy and Mergui, Aracan, Tongo, and all the northern and southern tributary states. The quantity exported to foreign parts is a mere trifle, not worth noticing ...

The celebrated Petroleum wells afford, as I ascertained at Ava, a revenue to the King or his officers. The wells are private property, and belong hereditarily to about thirty-two individuals. A duty of five parts in one hundred is levied upon the petroleum as it comes from the wells, and the amount realized upon it is said to be twenty-five thousand ticals per annum. No less than twenty thousand of this goes to contractors, collectors, or public officers; and the share of the state, or five thousand, was assigned during our visit as a pension of one of the Queens . . .

From the more accurate information which I obtained at Ava, it appears that the produce of these may be estimated at the highest, in round numbers, at about twenty-two millions of viss, each of $3^{65}/_{100}$ pounds, avoirdupois. This estimate is formed from the report of the Myo-Thugyi, who rents the tax on the wells, which is five in a hundred. His annual collection is 25,000 ticals; and he estimated, or conjectured, that he lost by smuggling about 8,000, making the total 33,000. The value of the whole produce, therefore, is 660,000 ticals.[91]

In 1828 a French Catholic missionary, Father Laurent Marie Joseph Imbert – who, ten years later, was beheaded in Korea during a period of anti-Christian violence – gave an account of the salt and gas wells at Ziliujing and Wutongqiao in China's Szechuan province:

The air that issues from these wells is highly inflammable. If when the tube full of water is near the top you were to present a torch at the opening, a great flame twenty or thirty feet in height would be kindled, which would burn the shed with the rapidity and explosion of gunpowder. This does happen sometimes through the imprudence of workmen . . . There are some wells from which fire only, and no salt, is obtained; they are called *Ho-tsing*, fire wells . . .

When a salt well has been dug to the depth of a thousand feet, a bituminous oil is found in it, that burns in water. Sometimes as many as four or five jars of a hundred pounds each are collected in a day. This oil is very fetid, but it is made use of to light the sheds in which are the wells and caldrons of salt. The Mandarins, by order of the Prince, sometimes buy thousands of jars of it . . .

In one valley there are four wells which yield fire in terrific quantities, and no water . . . These wells at first yielded salt water: and the water having dried up, about twelve years ago another well was dug to a depth of three thousand feet and more, in the expectation of finding water in abundance. This hope was vain, but suddenly there issued from it an enormous column of air, filled with blackish particles. I saw it with my own eyes. It does not resemble smoke, but the vapor of a fiery furnace, and it escapes with a frightful roaring sound, that is heard far off. It blows and respires continually, but never inspires . . .

The opening of the well is surrounded by a wall of freestone, six or seven feet high, to guard against the well-being set on fire by accident or malice, a misfortune which did really happen last August. The well is in the middle of an immense court, with large sheds in the centre, where the caldrons are placed for the boiling of the salt; and on that occasion, as soon as the fire touched the surface of the well, there arose a terrific explosion, and a shock as of an earthquake, and at the same moment the whole surface of the court appeared in flames . . . Four men, with great self-devotion, went and rolled an enormous stone over the surface of the well, but it was thrown up again immediately into the air. Three of the men were killed; the fourth escaped; but neither water nor mud would extinguish the fire. At length, after fifteen days' labor, a sufficient quantity of water was collected on a neighboring mountain, to form a large lake or reservoir, and this was let loose all at once upon the fire, by which means it was extinguished; but at a cost of thirty thousand francs, a large sum for China.

At the depth of a foot below the ground, four enormous bamboo tubes are fixed in the four sides of the well, and these conduct the inflammable air beneath the caldrons. More than three hundred are boiled by the fire from a single well, each of them being furnished with a bamboo tube, or fire conductor. On the top of the bamboo tube is one of clay, six inches long, with a hole in the centre six inches in diameter; this clay hinders the fire from burning the bamboo. Other tubes, carried outside, light the large sheds and the streets. There is such a supply of fire, that it cannot all be used; and the excess is carried by a tube, outside the inclosure of the salt-works, into three chimneys, out of the tops of which the flame leaps to a height of two feet.[92]

Meanwhile, an account of an exploration of the Indian province of Assam, undertaken by Lieutenant R. Wilcox in the late 1820s, was published in 1832, in which he described coal beds in the Burhi Dihing River at Supkong: '[T]he jungles are full of an odour of petroleum . . . In the middle, where bubbles of air are seen constantly rising to the surface, the mud is nearly white, and this is a more liquid state – on the edges green petroleum is seen floating, but it is not put to any use.'[93] In 1838 an agent to the regional governor reported that from a coal bed near the Disang River, '. . . several small springs of petroleum flow into the pools in the watercourse, and four or five seers (10 lbs.) of this oil were collected by my servants from their surfaces in a few minutes.'[94] Over the following decades, coal prospectors in northeast India reported petroleum also at Makum, and at Namchik in Arunachal Pradesh.[95]

In the Java War of 1826–30 the Dutch had defeated the popular resistance movement, incurring the deaths of around 200,000 Javanese.[96] Subsequently, in the early 1850s chemists and mining engineers surveying for minerals in the Dutch East Indies encountered natural oil seeps, as P.J. Maier reported from Karawang, West Java, in 1850: 'Usually the oil is spread as a thin layer over the water surface . . . There is one exception, a source located at the little Tjipanuan river, which produces a sizeable amount of oil.'

Of those on the island of Madura off the northeast coast of Java, Cornelis de Groot van Embden reported in 1852, 'The oil occurring on Madura is of petroleum quality (impure naphtha) . . . In the dry season the total daily yield from 3 seeps is 0.3 Dutch pounds . . . The Madurese sell the oil for 15 'duiten' (= c. 2 pennies) per bottle.' F.C.H. Liebert, searching for coal and mercury in the residency of Semarang, reported in 1853 that near the

> small dessah of Kedongbotok . . . is a small brook, into which a fairly large amount of oil oozes. The local population collects this oil and burns it in their lamps . . . At [Kampong Lantong] I found oil in large amounts; it was collected and used as fuel . . . Oil is found in many other locations in the Department . . . When it reaches the surface, the oil has a brownish color. When exposed to the air it becomes darker, more viscous and finally a mountain-tar mass . . .
>
> It is well-known that layers containing oil and bitumen develop hydrocarbon gas . . . This is the origin of the so-called 'eternal fires'. The 'sacred fire' or 'Mer-api' near Gubuk in Demak is nothing but a flow of hydrocarbon gas which has been lighted . . . It can easily be extinguished . . .

It is regrettable that so little use is made of the petroleum which is to be found in great quantities in the Department of Semarang. If efforts were made to collect this product in a suitable manner, a good deal of profit could undoubtedly be got out of it; for apart from its other uses, an oil can be prepared from it by distillation which, when burned in an Argan lamp, gives a seven-candle power light.[97]

In 1852 provocative actions by the Royal Navy triggered the Second Anglo-Burmese War in which Britain seized the southern province of Pegu, thereby incorporating Burma's entire coastal region into British Lower Burma. Captain Henry Yule was then sent on a diplomatic mission north to Ava, from where he reported that the oil trade continued much as before:

A rude windlass mounted on the trunk of a tree, laid across two forked stems, is all the machinery used. An earthen pot is let down and filled, and then a man or woman walks down the slope of the hill with the rope . . .

The petroleum from these pits is very generally used as a lamp-oil all over Burma. It is also used largely on the woodwork and planking of houses, as a preservative from insects, and for several minor purposes, as a liniment, and even as a medicine taken internally . . . There is now a considerable export of the article from Rangoon to England, and one of the Rangoon houses had a European agent residing on the spot. The demand in England is, I believe, for use to some extent as a lubricating oil, but it is also employed by Price's Company at Lambeth in the manufacture of patent candles, and has been found to yield several valuable products. It has sold in the London market at from 40*l.* to 45*l.* a ton . . .

The work of excavation becomes dangerous as the oily stratum is approached, and frequently the diggers become senseless from the exhalations. This also happens occasionally in wells that have been long worked. 'If a man is brought up to the surface with his tongue hanging out,' said one of our informants, 'it is a hopeless case. If his tongue is not hanging out, he can be brought round by hand-rubbing and kneading his body all over.' . . .

The ordinary price of the article used to be one tikal the hundred viss, or about sixteen shillings a ton. Lately, in consequence of the demand from Rangoon, it has risen to about thirty-five shillings a ton. As to the amount of revenue derived by the king from the petroleum we found it difficult to get definite information. One intelligent proprietor, who was *myo-ok* of the town, stated, that out of 27,000 viss, which formed the whole monthly yield of his wells, 9,000 went in payment to the work-people, 1,000 to the king, and 1,000 to the lord of the district.[98]

A British chemist had recently begun working with Price's Patent Candle Co. to develop a process for refining paraffin wax from Burmese crude, reporting that he had procured 'several tons of Rangoon tar, which was carefully collected at the source, and transmitted to Europe in well-secured vessels'.[99] Price's contracted for 2,000 barrels per month to be shipped to its Battersea factory – a five-month, 12,000-mile journey via the Cape of Good Hope – but the operation was a financial failure due the high shipping costs, the low yield of paraffin wax that could be extracted from the crude, and the reluctance of customers to adopt the new types of solvents, lubricants and lamp oil derived from this mineral source.[100]

Exploration for minerals in the Dutch East Indies continued to reveal more petroleum seeps. Mining engineer R. Everwijn, looking for coal in the Department of Pelambang in Sumatra in 1858, travelled down the Musi River and up the Manju-Assin River. Reaching Bali-Bukit, he found '. . . a small valley where a few oil seeps or tar seeps occur . . . The soil around the basins appears to be soaked with oil to a depth of about one foot, over an area of about 500m² . . . The local population concentrates the oil with long bamboo-poles, and then collects it.' Travelling along the Lematang River he looked for coal near Lahat and Muara Enim; near the Lalang tributary he found several oil seeps, and 'according to the local people the most prolific one yields some 3 liters of oil per day'.[101]

EUROPE

Evidence of one of the earliest uses of bitumen has been found in a cave in the Southern Carpathian mountain range in Romania near Râşnov in Braşov county where, about 30,000 years ago during the

Upper Palaeolithic era, Neanderthals appear to have used bitumen as an adhesive in tool-making.[102] Evidence for this use of bitumen has also been found in Italy from the Neolithic era.[103]

In about 450 BC, Herodotus witnessed oil springs on the Ionian island of Zakynthos (Zante):

> With my own eyes . . . I once saw pitch being brought to the surface of a lake in Zakynthos. There are quite a number of pools there, the largest of which is 70 feet across in every direction, and 2 fathoms deep. Into this same pool, someone will lower a pole with myrtle tied to the tip, and then, when it is brought back up, the myrtle will be coated in pitch. This, although it reeks like bitumen, is otherwise of much better quality than Pierian pitch. It is poured into a pit that has been excavated beside the lake, and then, when a substantial amount has been collected, they scoop it out of the pit and into jars.[104]

Vitruvius mentioned the oil of Zakynthos, as well as that of Apollonia in modern-day Albania, northeast of the port of Vlorë.[105] The latter was subsequently described by Strabo, by Dioscorides in about 60 AD, and by Pliny the Elder; indeed, there is evidence that as early as the Bronze Age asphalt was being exported across the Adriatic Sea to Italy.[106] Dioscorides wrote that 'there is something called pitch-asphalt, produced in Apollonia, near Epidamnos; it is carried down from the Ceraunian Mountains by the rush of the river and it is cast on the shores in condensed lumps.'[107] He described oil found on Sicily, followed by Pliny, who also mentioned oil from Zakynthos and from mainland Italy. According to Dioscorides, 'Up and down the land of the Agrigentes of Sicily, a liquid floats on wells which people use in their lamps instead of [olive] oil and which they call incorrectly Sicilian oil, for it is a type of liquid asphalt.'[108]

The Romans sourced asphalt from several locations along the Via Tiburtina road and the Pescara River, in the modern-day province of Pescara in Italy, from where it was transported to the coast for caulking boats and for export.[109] Passing through Italy in about 1170, Rabbi Benjamin of Tudela wrote that at Pozzuoli, 'A spring issues forth from beneath the ground containing the oil which is called petroleum. People collect it from the surface of the water and use it medicinally.'[110] In 1198 the mayor of Rouen registered with the Norman exchequer the purchase, among other items, of an ounce of petroleum;[111] by the fourteenth century the Florentine banker Francesco Pegolotti was listing asphalt and bitumen as articles of Mediterranean trade;[112] and in the fifteenth century oil from Tocco da Casauria, in Pescara province, was being gathered for its medical properties by German and Hungarian travellers.[113] It was from this period that Oil of St Catherine, from the province of Modena in northern Italy, became famed for its many medicinal powers, taken either externally or internally. One mid-sixteenth-century pamphlet extolled 'The virtue of the noble oil which is most useful and beneficial for several infirmities of the body . . . It is called balsam by several people on account of its great and outstanding virtue.'[114] It could apparently be used to heal wounds, burns, sores, bruises, bites, eye pain, broken bones, bladder complaints, constipation and gallstones, and to induce labour. The owners of the mountain from which this oil exuded, wrote Pierre Pomet in 1694,

> had channels or copper pipes prepared and placed directly so as to catch the oil coming out of the rock, and by means of these pipes the oil drops into . . . copper boilers from which it is collected . . . The white naphtha which we normally call white petroleum oil . . . [is] highly volatile and liable to catch fire, a fact that should serve as a warning to those who sell it and make them distrust it as though it were gunpowder.[115]

By the seventeenth century, bitumen and naphtha were being gathered from a number of locations in Italy. For example, from 1691 in the province of Piacenza to the northeast of Modena, Count Morando Morandi owned a profitable business selling the high-quality, light naphtha extracted from oil wells dug at Rallio di Montechiaro. After the owner of neighbouring land claimed that his mining interests were being impaired by these oil wells, a legal feud between them ran until 1715 when the count, and all his male heirs, were given exclusive rights to prospect for and extract oil in the whole of Piacenza. Petroleum continued to be gathered in Piacenza, Parma, Modena and Pescara into the nineteenth century.[116] In about 1800 naphtha from a spring at Amiano in Parma was adopted for lighting the streets of Genoa, in a specially designed oil lamp. 'By these means,' it was reported, 'the same quantity of light is obtained as with olive oil, and at a fourth of the expense, as the petroleum costs only two Genoese sous per pound, (which is less than a penny English).'[117]

In 1815 the royal physician and travel writer Sir Henry Holland visited the region around the Ionian Sea, at that time part of the Ottoman Empire. He reported that pitch from the mines of Selenitza (Selenicë) was one of the main exports being shipped from Vlorë, and that pitch from oil pools on Zakynthos was being used in significant quantities, mainly for caulking boats.[118]

A region straddling the modern-day German-Austrian border was, from the fourteenth century, the source of two medicinal petroleum oils promoted for their healing powers: Oil of St Quirinus from Tegernsee in Bavaria, and Blood of Thyrsus from Seefeld in Tyrol; however, neither gained the fame or widespread use of Modena Oil.[119] In the mid-sixteenth century the German mineralogist Georgius Agricola described how, to the east of Hanover, oil was collected from natural seeps, boiled, and put to use:

> Liquid bitumen sometimes floats in large quantities on the surface of wells, brooks and rivers and is collected with buckets or pots. Small quantities are collected by means of feathers, linen towels and the like. The bitumen easily adheres to these objects . . .
>
> [I]n Lower Saxony the peasants not only use the oil in lamps but also make marriage torches by dipping dry stems of the great mullein in it, besides lubricating their carts with it. They use it also to paint wooden posts to protect them from the effects of rain.[120]

By the eighteenth century the oil was being drawn from more actively maintained pits, reported the physician Johann Taube, who wrote of the oil's healing properties. By the early nineteenth century some shafts were over 100 feet deep, and in the mid-nineteenth century a dozen or more oil wells of this depth were drilled.[121]

In the south of France, at Gabian to the west of Montpellier, from the seventeenth century oil was collected for medicinal uses. Pomet wrote,

> This oil was formerly so abundant and copious that it was not highly valued, and a fairly large amount could be collected every day, but nowadays it is collected only on Mondays, and the place where it accumulates is surrounded by walls and guarded. I was assured at Gabian

that the Bishop of Béziers obtained a large income from it, though not so large as it used to be.[122]

In the Alsace region of northeast France oil seeps were reported from at least the fifteenth century, and by the seventeenth century the oil was being distilled into fractions and used for lubrication, as a wood preservative, for illumination, and as a treatment for a range of ailments.[123] From 1712 Eyrini d'Eyrinis, a doctor and teacher, began experimenting with asphalt found a few hundred miles to the south, in the Neuchâtel region of Switzerland, and he publicized its potential as being 'more suitable than any other material for joining together and tarring all manner of structures'.[124] While these deposits were commercialized initially on a small scale, mainly for distilled oil that was sold as a medicine, from 1745 the Alsace oil sands around Pechelbronn began to be mined and processed on a significant industrial scale. In the early nineteenth century bitumen deposits further south began to be exploited, near Seyssel in the Rhône valley, while oil shale deposits at Autun in Saône-et-Loire began to be distilled into products including lamp oil. As asphalt became used more widely as a mastic and for surfacing pavements and roads, the deposits of Neuchâtel similarly began to be commercially exploited.[125] Following the Franco-Prussian War of 1870–1, *Scientific American* would run a report arguing that 'The value of Alsace to Germany and the consequent extent of the loss to France, commercially considered, are alike enhanced by the probable development of a large petroleum industry in that celebrated province.'[126]

A late sixteenth-century encyclopedia of the geography of Great Britain described how near Wigan, Lancashire, there was a well – that became known as 'Camden's cooker' – from which

> breaks out a sulphurous vapour, which makes the water bubble up as if it boyl'd. When a Candle is put to it, it presently takes fire, and burns like brandy. The flames, in a calm season, will continue sometimes a whole day; by the heat whereof they can boyl eggs, meat &c. tho' the water itself be cold.[127]

In 1694 Martin Eele took out a patent, in the names of himself and his associates, for a method they had devised for manufacturing pitch at Broseley, just south of Ironbridge in Shropshire. It stated 'that he, after much paines and expences, hath certainely found out "A Way to Extract and Make great Quantityes of Pitch, Tarr, and Oyle out of a sort of Stone," of which there is sufficient plenty within our Dominions of England and Wales.'[128] By distilling local tar sands, they produced an oily substance that could be turned into pitch for waterproofing ships, which, following improved distillation techniques, was claimed to be a serviceable substitute for Swedish pine tar. However, a more lucrative product of Shropshire tar sands was 'Betton's British Oil', sold as a cure-all.[129] An account of 1764 described how at Broseley there was also 'a Well exhaling a sulphurous Vapour, which when contracted to one Vent by Means of an Iron Cover with a circular Hole, and set on Fire by a Candle, burns like the Spirit of Wine or Brandy, with a Heat that will even boil a large Piece of Beef in two Hours'.

About ten miles away, at Pitchford, there was 'a Well where a Liquid Bitumen floats, which the Inhabitants skim off, and use instead of Pitch; some think it good against the Falling Sickness and Wounds'.[130] In 1786, at Coalport near Broseley, a canal tunnel dug into the side of the Severn Gorge struck natural bitumen seeps from which hundreds of gallons of bitumen were gathered per day.[131] The following year, a visiting Italian aristocrat described the scene inside the tunnel:

> At length we arrived at the rock whence emerges the pitchy torrent in such copiousness that five or six barrels are filled with it every day. The workmen who gather it, as well as they who dig out the solid pitch, are, of a truth, like the imps described by Dante in his Inferno as gathering with a hook the souls of the damned into a lake of pitch – so horribly disfigured and begrimed are they.[132]

According to a wealthy Welsh naturalist writing in 1778, the coal miners of Flintshire were familiar with 'petroleum, or rock oil, [which] is esteemed serviceable in rheumatic cases, if rubbed in the parts affected. The miners call it *Ymenyn tylwyth têg*, or *fairies butter*.'[133]

In 1781 Lord Archibald Cochrane, 9th Earl of Dundonald, took out a patent for a commercial process for distilling coal tar from bituminous coal, having experimented with that mined on his Culross estate in Fife, Scotland. He formed the British Tar Co. in 1782 and by the late 1780s he had ten tar works in Scotland and England producing mainly tars and varnishes for preserving wood – particularly the hulls and other parts of wooden ships – as well as various other by-products. The business was profitable for some years but ran into trouble in the early nineteenth century, partly because the Royal Navy declined to regularize the use of his tar products, committing instead to copper sheathing to protect its ships' hulls. Lord Cochrane and his son, the future admiral Thomas Cochrane, also took out patents for distilling lamp oil from coal and for new designs of oil lamps.[134]

In 1847 a coal seam being dug at Riddings, Derbyshire, began to flood with brine which was discharged into the nearby canal. After locals found that the surface of the canal could catch fire, it was realized that there was petroleum in the mine water, and a Scottish chemist, James Young, came to investigate. He discovered that this crude oil could be refined into lamp oil, paraffin wax for making candles, and lubricating oil, which he produced and sold commercially for a while until his interest turned to the distillation of oil from coal.[135]

In the mid-nineteenth century, shale-oil works were established in Dorset in England, in Flintshire in Wales, and in West Lothian and the surrounding regions of Scotland, but only the Scottish shale-oil industry achieved lasting commercial viability.[136]

Back in 1530 a Polish aristocrat, the head of the Biecz district of southeastern Poland, attempted to find gold at the foot of Mount Chełm, just west of Ropa (Polish for 'oil'); but when the excavations became flooded with oil, fellow attendants of the royal court joked, 'He looked for gold in Ropa only to be doused in pitch.'[137] Four years later Stefan Falimierz, explaining that 'petroleum is an oil coming from rock', was the first of several Polish physicians of the sixteenth century to write of the healing properties of the surface oil found across this region: the northern fringes of the Carpathian Mountains, extending across the present-day Polish-Ukrainian border. At many places, from Gorlice and Krosno, to Borysław and Drohobycz, pits were dug for the oil which was used as an axle lubricant, as a roof sealant, for treating leather, and for lighting. From the early seventeenth century, Polish naturalists began to write about the geology of the region and to discuss distillation techniques, and by 1772, when the Habsburg monarchy annexed the region – now named Galicia – this small-scale oil industry was well established and well known.[138]

During the early nineteenth century, as the Habsburg government began to formalize its control over the petroleum deposits of Galicia, some attempts were made to distil the oil commercially for lighting or heating. However, it was only after Abraham Gesner in Canada, and James Young in Britain, managed to distil effective lamp oils from coal and asphalt in the late 1840s that, in 1853, a Polish pharmacist in Lviv (in present-day western Ukraine), Jan Zeh, and his apprentice Ignacy Łukasiewicz, succeeded in distilling the first commercially viable lamp oil from crude Galician petroleum, for which a local tinsmith designed lamps that could burn the oil effectively, with a reasonably clean, bright flame. This development sparked an oil rush: hundreds more oil wells were now dug and many small refineries were established across the region. It particularly drew in the local Jewish population who had struggled for decades under discriminatory employment restrictions imposed by the Habsburg authorities. By the late 1850s the Lviv General Hospital and some of the larger stations of the Emperor Ferdinand Northern Railway had turned to kerosene lamps for illumination.[139]

Whereas in western Galicia some wealthy landowners invested in somewhat larger-scale crude extraction, the very small-scale extraction methods practised in eastern Galicia made it difficult for refiners there to manufacture lamp oil in large quantities. One refiner described how he urged a peasant, whose oil seep yielded only three barrels of crude per week, to increase his production by digging deeper and using a bellows to dispel the toxic gases that would otherwise prevent him from working in his pit. However, in order to meet the increasing demand for lamp oil these refiners sourced additional crude from Russia and Romania.[140]

In Romania a mid-fifteenth-century report of petroleum seeps in the Bacău region of Moldavia was soon followed by accounts of its extraction both there and from south of Râşnov in the Prahova region of Wallachia. From the late eighteenth century, Romanian crude oil was exported via the Danube to Turkey and Austria; and a rudimentary oil refinery operating in Lucăceşti, Bacău, from 1840 was the first of several set up to produce lamp oil. In 1857 Bucharest became the first city to adopt kerosene on a large scale for street lighting, supplied from a refinery in Râfov, near Ploesti, Prahova. In one authoritative compilation of oil production statistics, the first registered national oil production was that of Romania: 275 tons in 1857 and 495 tons in 1858.[141]

In southwestern Russia two sources of petroleum in the vicinity of the Black Sea were mentioned by the Byzantine Emperor Constantine VII in about 950: on the Taman Peninsula, and in the region between Maikop and Sochi.[142] In northwestern Russia, according to the fifteenth-century *Dvina Chronicle*, people living near the Ukhta River, in the Timan-Pechora basin in the present-day Komi Republic, used oil that could be found floating on the river's surface; this was also mentioned by the Dutch diplomat Nicholas Witsen in his 1692 book on the region.[143] In 1718 Tsar Peter the Great looked into the potential for sourcing oil from the Terek-Sunzha region around Grozny in the north Caucasus – oil that was used at the time to lubricate cart axles – as it was costly to import from Baku which was then under Persian rule. His Mining Board also studied the prospects of obtaining oil from near the Ukhta River.[144] Eventually, in association with the Mining Board, a rudimentary oil works was set up on the Ukhta River in 1746 by an entrepreneur, Feydor Pryadunov, from which some of the refined oil was sent to Hamburg, where it was assessed to be comparable to Oil of Modena as a remedy for a range of ailments. The enterprise continued operating on a small scale under various owners for several decades, while the Mining Board investigated oil that was also found in the Volga-Urals region.[145] Studies of the north Caucasus, Volga and Ukhta oil regions continued throughout the reign of Catherine the Great and into the early nineteenth century. In 1792 Catherine granted her Cossack forces extensive land rights on the Taman Peninsula and around the Kuban River, in return for maintaining her imperial control over the region; they began to exploit the local oil resources commercially, and supplied oil products to the Black Sea fleet.[146]

On the Crimean Peninsula – progressively annexed by Russia from the Ottoman Empire in the late eighteenth century – a Russian geological survey of 1823 recorded several surface oil seeps. A French national, who became a lieutenant in Russia's Engineering Corps, saw the oil seeps on the Crimean Kerch Peninsula while he was involved in building Black Sea coastal fortifications, and in 1838 he was inspired by the sight of the asphalt streets of Paris to try to find a domestic Russian source of asphalt. Later that year, the regional governor provided him with the means to set up a distillation plant at Kerch that soon produced asphalt for surfacing the streets of that city and of Odessa.[147]

By the early 1840s Baku had been under Russian sovereignty for thirty years, and here mining engineer Nikolai Voskoboinikov, of the Bakinskii Corps of Mining Engineers, began promoting the use of drilling technology – long employed for water and brine wells – as a more effective method than digging oil wells. In 1845 the viceroy of the Caucasus, Mikhail Vorontsov, provided 1,000 roubles for the endeavour and in 1848, at Baku, the Mining Corps sunk what was perhaps the first drilled oil well, using a manually powered cable tool, or percussion drilling, rig.[148]

LATIN AMERICA AND THE CARIBBEAN

From about 1200 BC, people of the Olmec civilization – centred on what are now the Mexican states of Veracruz and Tabasco – painted bitumen, or *chapopote*, onto pottery artefacts and stored it in pottery vessels, and there is evidence that it was used as a sealant in stone-built troughs.[149] According to reports by the missionary Bernardino de Sahagún and by other conquistadors of Aztec- and colonial-era Mexico, native people used bitumen in many ways, for example as an adhesive, for ceremonial purposes, to chew and to scent tobacco.[150] Employing the native Nahuatl language, Sahagún wrote,

> Bitumen . . . comes from the ocean, from the sea . . . The waves cast it forth . . . When it comes forth [it is] like a mat, wide, thick. Those of the seashore, those of the coast lands gather it there . . . they pick it up from the sand . . . The pleasing scent of the tobacco with bitumen spreads over the whole land. As its second use, it is used by women; they chew the bitumen. And what they chew [is] named chicle.[151]

Bitumen was strongly associated with women in Aztec culture, and the archetypal feminine goddess Tlazolteotl was represented with a ring of *chapopote* around her mouth: She was called the 'filth eater' for eating people's sins to absolve them just before death.[152]

In 1526 the Spanish chronicler Gonzalo Fernández de Oviedo y Valdés, who took part in the colonization of the Caribbean and directed the Crown's gold-smelting operations, wrote that, on the island of Fernandina (Cuba),

> not far from the sea there flows from a mountain a liquid or bitumen-like pitch. It is sufficient in quantity and of good enough quality to tar ships. Since great quantities of this substance constantly flow into the sea, numerous rafts or patches of it float on the water from one place to another as moved by the wind or the currents along that coast.[153]

According to the explorer Alexander von Humboldt, in 1508 Sebastian de Ocampo, the first sailor to circumnavigate Cuba, had used its bitumen to careen his ships, naming the bay Puerto de Carenas, now known as Havana.[154] Oviedo continued, 'This pitch is found ... also in New Spain, recently discovered there in the province of Pánuco' (in modern-day Mexico).[155] In 1535 Oviedo published the first part of his much longer *General and Natural History of the Indies*, in which he wrote of a further five reported sources of petroleum in the Americas. There were two sources on the tip of the Santa Elena peninsula (in modern-day Ecuador), one of which 'looks like perfect turpentine'; there was a 'lake of bitumen in the province of Veneçuela', along with 'another form of bitumen' on the island of Cubagua, just off the Venezuelan coast; and, fifthly, there was 'another well of bitumen in the new kingdom of Granada, in the land of the wild Indians called Panches' (in present day Colombia, west of Bogota).[156] If the natives here were 'wild', it was generally because although, initially, they had typically greeted the newcomers with gifts, the Europeans soon plundered their gold, massacred and tortured them, or enslaved them to work exploiting the Cubagua pearl oyster beds ultimately to extinction.[157]

Oviedo wrote that he had been told by the Governor of Cuba and by a number of sailors that the bitumen found on the coast at Puerto del Príncipe, 'when mixed with a lot of tallow or oil, can be used to fix or caulk the ships ... A bit of such pitch was given to me by Diego Velazquez, and I took it to Spain in 1523, to show it there.'[158] The bitumen on the island of Cubagua was 'in the form of a natural liquor, which some call petroleum, and others call *stercus demonis* [devil's excrement]; the Indians give it other names.'[159]

In the second volume of his treatise, published in 1557, Oviedo provided two more detailed reports of petroleum in Venezuela and

Colombia. Since the formal establishment of Venezuela in 1528 – initially on lease to an enterprise of German colonists – plundering expeditions had been sent southwards around Lake Maracaibo. Oviedo wrote,

> All these people who live around the lagoon are poor people, warlike on the water and skilful with their arrows. There are in that province some wells or springs of bitumen, as a pitch or melted pitch, which the Indians call *mene*. In particular there are wells that originate on a hill, high up in the savanna, and many of them form ponds of more than a quarter of a league in perimeter. And from Maracaybo to these springs is twenty-five leagues.
>
> With the strength of the sun this bitumen or liquor seems to boil, emerging and running for a distance along the ground. It is quite soft and sticky during the day, but at night it cools with the nocturnal freshness and the absence of sunlight. In the morning you can walk on it without sinking or even sticking your feet, but when the sun is up, it is very sticky; and whoever passes by, on foot or on horseback, gets bogged down like someone trying to pass through slime or mud, and it can only be traversed with great difficulty. During the first incursion that Governor Ambrosio made inland, they took this route during the daytime, and they found a deer stuck in some wells and springs of this bitumen like a bird caught in a trap, and they took it as the animal could not escape. It is a very viscous substance, such that when it is found in the form described, after two or three hours exposed to the sun, it is like a pitch to caulk ships. Thus, from this incident with the deer they realized that they could kill many others, as they are innumerable in that territory – they would encircle them and force them to go through those deposits, where they are trapped and then they can be taken very easily; it is a very pleasurable hunting experience.[160]

In 1540 Jerónimo Lebrón led one of a number of expeditions up the Magdalena River to plunder gold and other valuables and to conquer and Christianize native people. The following year, two captains who had been among a 300-strong expeditionary force – which included native allies and West African slaves – told Oviedo what they had seen, over 200 miles inland from the mouth of the Magdalena, near the native village of La Tora (present-day Barrancabermeja, Colombia):

Among other peculiarities they testified that one day ahead of the village of La Tora, where the brigs disembark, there is a small source of bitumen that boils and runs out on the ground, and in the foothills of the mountains it exists in great quantity in the form of a thick liquor. And the Indians bring it to their homes and smear themselves with this bitumen, because they find it good to remove fatigue and strengthen the legs: and it is this black liquor, with a smell of pitch or worse, which Christians use to caulk the brigs.[161]

In his text of 1535 Oviedo concluded,

Thus, at present we know of seven sources or springs of bitumen in our Indies; each very different from each other. All our Spaniards, or most of them, have taken advantage of it to caulk ships, although according to what it is possible to know from the Indians, such liquors are appropriated for many sufferings, and they are medicinal . . . [A]nd I have no doubt that others will be found, because the Continent is another half of the world.[162]

A quantity of petroleum was sent to Spain the following year at the request of Queen Juana of Castile, who wrote to her officials at the port of Nueva Cádiz on Cubagua: 'Some people have brought to these Kingdoms a petroleum oil from a spring on said island and because here it has appeared to be of value, I command that all ships sailing from that island send me as much as possible.'[163]

In 1555 a Spanish official wrote, in his *History of the Discovery and Conquest of Peru*, that 'at a cape called Santa Elena by the Spaniards, there are some outflows of pitch or tar used by the natives to mend their boats'.[164] Father José de Acosta wrote in 1590 that here there was 'a spring or fountaine of pitch, which at *Peru*, they call *Coppey* . . . The Mariners use [it] to pitch their ropes and tackling.'[165] Five years later the Florentine merchant and explorer Francesco Carletti reported from here that 'A huge quantity of this bitumen is taken out, thus making a profit for the owners of the land that generates it'.[166] Meanwhile, a map of Tabasco province, Mexico, produced in 1579 by Melchor de Alfaro Santa Cruz noted 'springs of a kind of water which coagulates in the sun and forms a dark resin which can be used as pitch. It also occurs in other parts of this province.'[167]

In 1595 Sir Walter Raleigh – one of Queen Elizabeth I's raiders of

Spanish gold-carrying ships – had his ships treated with pitch on the coast of Trinidad. He wrote,

> At ... Tierra de Brea, or Piche, there is that abundance of stone pitch, that all the ships of the world may therewith be laden from thence, and we made trial of it in trimming our ships to be most excellent good, and melteth not with the sun as the pitch of Norway, and therefore for ships trading the south parts very profitable.[168]

Raleigh revisited the location in 1617, recording,

> This Terra de Bri is a piece of land of some 2 leagues long and a league broad, all of stone pitch or bitumen, which riseth out of the ground in little springs or fountains, and so running a little way, it hardeneth in the air and covereth all the plain ... Here rode at anchor, and trimmed our boats ...[169]

From the late 1500s Father Alvaro Alonso Barba had lived in Peru, and from 1624 he was based in Potosí (in present-day Bolivia), the economic capital of the huge Spanish silver-mining boom. The priest doubled as a metallurgist, and in 1637 he presented to the local authorities a report on the mining and chemical processing of the region's minerals. His treatise, published by the Spanish Crown in 1640, included a chapter on *Betunes*, or mineral 'pitches', which included 'asphalt', 'bitumen' and 'petroleum'. He reported that as the Spanish colonists had been

> occupied in the search for the wealth of Silver and Gold, but little attention has been paid to this and many other Curiosities. Nevertheless, an abundance of pitch-bearing materials exists in the Cordillera of Chiriguanes, on the frontier of Lemina. There is but little communication with this Place, because it is among the war-like Indians.[170]

In 1676 an English explorer, William Dampier, was sailing with pirates raiding Spanish possessions along the Tabasco coast; after a skirmish with a Spanish ship, he wrote, they

> searched the Bays for Munjack to carry with us for the Ship's use, as we had done before for the use both of Ships and Canoes.
>
> Munjack is a sort of Pitch or Bitumen, which we find in Lumps, from three or four Pounds to thirty Pounds in a Lump; washed up by the Sea, and left dry on all the Sandy-Bays on all this Coast: It is in Substance

like Pitch, but blacker; it melts by the Heat of the Sun, and runs abroad as Pitch would do if exposed, as this is, on the Bays: The smell of it is not so pleasant as Pitch, neither does it stick so firmly as Pitch, but it is apt to peel off from the Seams of Ships Bottoms; however we find it very useful here where we want Pitch; and because it is commonly mixed with Sand by lying on the Bays, we melt it and refine it very well before we use it; and commonly temper it with Oyl or Tallow to correct it; for though it melts by the Heat of the Sun, yet it is of a harsher Nature than Pitch. I did never find the like in any other Part of the World, neither can I tell from whence it comes.[171]

Eight years later, while on another raiding expedition, Dampier corroborated Spanish accounts of the bitumen seeps on the coast of Elena Peninsula, in modern-day Ecuador:

[C]lose by the Sea, about 5 paces from high-water mark, there is a sort of bitumenous matter boils out of a little hole in the Earth; it is like thin Tar: the Spaniards call it Algatrane. By much boiling it becomes hard like Pitch. It is frequently used by the Spaniards instead of Pitch; and the Indians that inhabit here save it in Jars. It boils up most at high Water; and then the Indians are ready to receive it . . . Their chief subsistence is Maiz, most of which they get from ships that come hither for Algatrane.[172]

In the eighteenth and early nineteenth centuries, during the later period of Spanish rule, royal monopolies were sometimes granted to the colonial owners of land in northern Peru, between Tumbes and Talara, for the exploitation of natural bitumen seeps, most notably at Amotape. The Spanish conquest had devastated the native Inca population with European diseases and through warfare, so the colonists used slaves from West Africa to work the bitumen pits. Following independence from Spain, the government of Simón Bolívar took ownership of these properties, but in 1826, under the pressure of war debt repayments, it sold them off to private citizens. However, since under Spanish law the state could sell land to individuals but retained the rights to subsoil minerals, it was now unclear who owned the petroleum resources of the Mancora and La Brea y Pariñas haciendas.[173] This basic tenet of Latin American petroleum law had been established in 1783 when King Carlos III issued a Mining Ordinance stating that under Spanish law the sovereign was the ultimate owner of all subsoil minerals, including what it termed 'bitumens and juices of the earth'.[174]

In the early seventeenth century one of the first English colonists of North America, John Smith, found on the island of Barbados 'a spring neer the middest of the Ile of Bitume, which is a liquid mixture like Tarre, that by the great raines falls from the tops of the mountaines, it floats upon the water in such abundance, that drying up, it remaines like great rocks of pitch, and as good as pitch for any use'.[175] From the mid-eighteenth century, Barbados oil became a popular treatment for a wide variety of ailments. According to one writer, this

> green Tar ... is of so inflammable a Nature, that it serves to burn in Lamps. As to its medicinal Qualities, it is chiefly made use of with great Success in paralytic and nervous Disorders, as well as in curing cutaneous Eruptions ... There is likewise another Species of Bitumen, of a solid Substance, here called *Munjack*. Where the liquid Kinds are thrown out of the Earth, the Surface of the Ground is one Continued Quagmire, bearing very little, if any Grass; and where the more solid is dug out, if the Veins are upon, or very near the Surface, scare any Vegetable grows upon it. If by Accident any of these Veins take Fire, they continue to burn a long time, tho' in a dull slow manner.[176]

Nearly a century later, the British medical journal *The Lancet* carried an endorsement of the healing properties of 'Barbadoes tar'.[177]

In 1789 a Scottish doctor and botanist, Alexander Anderson, published his account of how 'A most remarkable production of nature in the island of Trinidad, is a bituminous lake, or rather plain, known by the name of Tar Lake; by the French called La Bray, from the resemblance to, and answering the intention of, ship pitch.'[178] A geologist described Trinidad's Pitch Lake in 1811:

> This vast collection of bitumen might in all probability afford an inexhaustible supply of an essential article of naval stores, and being situated in the margin of the sea could be wrought and shipped with little inconvenience or expense ... I have frequently seen it used to pay the bottoms of small vessels, for which it is particularly well adapted, as it preserves them from the numerous tribe of worms so abundant in tropical countries. There seems indeed no reason why it should not when duly prepared and

attenuated be applicable to all the purposes of the petroleum of Zante, a well-known article of commerce in the Adriatic, or that of the district in Burmah, where 400,000 hogsheads are said to be collected annually.[179]

Sir Alexander Cochrane – brother of Lord Cochrane, 9th Earl of Dundonald, the Scottish coal tar manufacturer – while stationed in the Leeward Islands as a Royal Navy admiral, had tried to persuade the Navy to use this Trinidadian pitch for caulking its ships. 'But,' the geologist continued, 'whether it has arisen from certain perverse occurrences or from the prejudice of the mechanical superintendents of the Colonial Dock Yards, or really, as some have pretended, from an absolute unfitness of the substance in question, the views of the gallant admiral have I believe been invariably thwarted.'[180]

In 1828 Captain George Lyon, commissioner for two British mining companies operating in Mexico, published an account of his travels in the country in which he wrote of his journey up the Pánuco River from Tampico, in the southeast of the state of Tamaulipas:

> Passing for some time the banks of San Pedro, we came to the Estero de Chila, another extensive rancho ... On this estate, at about three or four miles from the river, is a large lake, from whence I understand that the petroleum which is brought in great quantities to Tampico is collected. It is here called Chapopote, and is said to bubble from the bottom of the lake and float in great quantities on the surface. That which I saw at different times was hard and of good appearance, and was used as a varnish, or for the covering of the bottom of canoes; the general price was four reals (half a dollar) for a quintal (100 pounds).[181]

An 1842–3 survey of Mexico's Isthmus of Tehuantepec, conducted by an Italian engineer for a proposed canal and rail link between the Gulf of Mexico and the Pacific, mentioned reports of petroleum at Sayultepec in the state of Oaxaca and at Moloacán in the state of Veracruz.[182] The prevalence of natural oil seeps south of Moloacán was corroborated in 1844 when Dr Federico Hechler alerted the government to a possibly imminent volcanic eruption in the vicinity, from a mountain surrounded by natural salt caverns belching sulphurous fumes. 'In my opinion the shell of the mountain is salt, and the interior consists of liquid chapopote that burns like spirits,' he wrote.[183] Forwarding this report to the central government, the secretary of the local government of Veracruz wrote, 'The whole terrain surrounding these natural excavations is chapopote, hardened in many places; in others, in an oily liquid state; but in such quantities that, in the hyperbole of the conveyor of this report, the world will end before this substance is exhausted.'[184] The government sent two eminent founding members of the Mexican Society of Geography and Statistics to investigate; they wrote,

> The tar or asphalt (which is already widely used in Tlacotalpam and other points of that coast to caulk the canoes, instead of tar) is already in great use in Europe for a variety of purposes; and without doubt, it being found in the abundance described and being so easy to extract means that it can be obtained here at a low price, and can be used for roofing, pavements, varnishes and in other no less profitable ways.[185]

The natural oil seeps near Moloacán were encountered during a subsequent survey of 1851–2, commissioned by the Tehuantepec Railroad Co. of New Orleans for a projected trans-isthmian rail connection. The surveying team began by way of the mouth of the Coatzacoalcos River, where in 1522 one of Hernan Cortés' officers had taken over the major Nahuas Aztec town of Coatzacoalcos to serve as a bridgehead in the Spanish conquest of Chiapas.[186] A short distance to the south, about ten miles east of Minatitlán not far from the Uxpanapa River, the surveyors found 'an extensive spring of petroleum, which covers an area of several acres. Of the value of this spontaneous product it is unnecessary to speak. No difficulty exists in the way of its transportation to the river; and the supply is said to be inexhaustible.'[187] Residents of Macuspana, Tabasco, collaborated in 1857 to invest in a metal tank to store oil from a local spring, which they were able to sell quite widely as an illuminant.[188]

Meanwhile, it had been reported from Cuba in 1837 that 'petroleum springs' at Guanabacoa 'have been known for centuries', and that 'Round a great portion of the bay of Havana, asphalt is still collected at low water, under the name of *Chapapote*, and is employed in the manner of tar, for paying vessels'.[189]

In 1850 Lord Thomas Cochrane, 10th Earl of Dundonald and Admiral of the North American and West Indian Fleet of the Royal Navy, was patrolling the British colonies in the region. Cochrane had pioneered the steam-powered warship after being made commander of the Chilean Navy in 1818 during its war of independence from Spain, and his family had long been involved in commercializing various products of coal tar.[190] In early 1850 he wrote in his diary that his patrol ship

> returned through the Dragon's Mouth, shaping our course for the great natural curiosity of Trinidad, the Pitch Lake, which I hoped might be rendered useful for fuel in our steam-ships – so important in the event of war – as fuel is only obtained at present from Europe. The United States and Nova Scotia are never resorted to; hence, could this pitch be rendered applicable as fuel, our vessels would be supplied when an enemy would be almost deprived of the use of steam in these seas.[191]

The following year Cochrane acquired a twenty-year lease on land at La Brea that included twenty-six acres of Pitch Lake. He consulted with the pioneering Canadian 'coal-oil' chemist Abraham Gesner and appointed a German immigrant to Trinidad, Conrad Stollmeyer, to commercialize his asphalt concession.[192] Trinidad's sugar refineries and the region's steamships burned imported British coal which, Cochrane and Stollmeyer hoped, the island's much cheaper asphalt might replace, or at least augment. This was already being practised in a crude manner nearby, according to an American scientist travelling the region in search of mining opportunities: 'It has been employed to advantage as fuel by the American steamers plying on the Orinoco. It is thrown in the furnaces among the wood, fusing too readily to be used alone.' He continued,

> With ten per cent of rosin oil it forms an excellent pitch for vessels.
>
> The Earl ... has instituted various experiments with the view of substituting the bitumen for India rubber and Gutta percha in the manufacture of water proof [sic] fabrics, covering of telegraph wires, &c. Judging from the specimens ... which were shown me by his agent at Port au Spain (Mr. C.F. Stollmeyer), these efforts bid fair to be quite successful.
>
> It seems only necessary that the required amount of intelligent enterprise should be directed to the subject in order to render this wonderful reservoir of bitumen a source of great individual profit and of essential service to mankind.[193]

In 1853 Stollmeyer wrote several articles for the weekly *Trinidadian*, of which he was the editor, promoting a new economic miracle based around Pitch Lake, and praising Lord Cochrane for 'the establishment of this new branch of industry [that] will become the most important event in the history of our island'. He envisaged the 'thousands and thousands who will be employed in the different manufactures to which our mineral treasure is susceptible', and he suggested a scheme for settling freed slaves from the United States in Trinidad, giving them 'a home ... where in a short time they may be able by their industry, and assisted by the productiveness of our soil, to acquire a position for themselves'. Ultimately, Stollmeyer asked his readers, 'May we not then cherish the hope that from this accumulation of wealth a radical change will take place throughout the civilized world towards the amelioration of suffering humanity ... ?'[194]

In March 1856 Lord Cochrane's son Arthur, a Royal Navy captain, issued a prospectus in London for the Trinidad Bituminous Fuel Co. on the basis of trials, in collaboration with an engineer sent by the Admiralty, with the steamer *Lady of the Lake* in which, it was claimed, a boiler could effectively burn equal proportions of coal and blocks of pitch mixed with wood shavings.[195] However, nothing seems to have come of this enterprise, nor of the Merrimac Oil Co. from the United States that, according to some reports, may have sunk a well near Pitch Lake in 1857 and struck oil. However, a few small companies such as the informally organized Trinidad Petroleum Co. began exporting small amounts of asphalt to the US and France.[196] Stollmeyer, writing to Lord Cochrane, entertained hopes that a Boston coal-oil manufacturer might purchase Trinidadian pitch as a raw material and that 'the influence of Americans and their ways would soon change La Brea into a second Newcastle'.[197]

NORTH AMERICA

There is archaeological evidence of bitumen being put to use from about 8,000 years ago, during the Early Holocene period, on the Californian coastal region of what would become the United States.[198] The first European reports of petroleum came from the Hernando de

Soto expedition in 1543 when, forced ashore by the weather onto the coast of Texas near Sabine Pass, they found oil floating on the sea and used it for caulking their boats.[199] Contemporaneous archaeological finds indicate that the Chumash Indians of southern California were processing bitumen before putting it to a number of uses, including waterproofing water containers – and it has been speculated that this may have had a negative impact on their health.[200]

Over a century later, several expeditions of French Jesuit missionaries to North America came across petroleum springs just south of the Great Lakes. In 1657, Fathers Paul Le Jeune and Jean de Quen were travelling through country inhabited by the Erie tribe, which they called the 'Cat Nation':

> As one approaches nearer to the country of the Cats, one finds heavy and thick water, which ignites like brandy, and boils up in bubbles of flame when fire is applied to it. It is, moreover, so oily, that all our Savages use it to anoint and grease their heads and their bodies . . .
>
> In the Lakes . . . they spear [fish] with a trident by the light of a bituminous fire, which they maintain in the bows of their canoes.[201]

In 1670 a team of French explorers, including two Jesuit priests, were again traversing this region in order to find strategic routes to facilitate French imperial ambitions. While they were in the vicinity of present-day Bristol, New York state, they encountered a 'bitumen spring' and one of the priests, René de Bréhant de Galinée, recorded,

> I went with M. de la Salle under the guidance of two Indians . . . to see an extraordinary spring . . . The water is very clear, but has a bad odor . . . He put a torch in it, and immediately the water took fire as brandy does, and it does not go out until rain comes. This flame is, amongst the Indians, a sign of abundance, or of scarcity when it has the opposite qualities.[202]

Further north, in Canada, the Hudson's Bay Company was engaged in fur trading, and on 27 June 1715 James Knight, governor of the company's York Factory – situated on the southern coast of Hudson Bay – made notes of that day's meeting with a group of Chipewyan Indians: 'Before they went I had some Discourse abt the Great River it runs into the Sea on the Back of this Country & they tell us there is a Certain Gum or pitch that runs down the river in Such abundance that they cannot land but at certain places.'[203] Four years later, a Cree Indian brought to the new governor, Henry Kelsey, a sample 'of that Gum or pitch that flows out of the Banks of the River', which was almost certainly the Athabasca River in present-day Alberta.[204]

During the mid-eighteenth century the French extended their presence northwards up the Mississippi and Ohio rivers and southwards from Lake Erie, while the British simultaneously expanded westwards from the East Coast. Consequently, the Ohio Valley and the region south of the Great Lakes was becoming a region of geopolitical collision. In 1720–2 a French Jesuit priest, Pierre-François-Xavier de Charlevoix, was sent to explore North America on behalf of the French Crown, and while he was in the area around the Allegheny River he spent some time with Captain Louis Thomas de Joncaire, who was, also in the name of France, operating a trading post in Seneca (Onödowá'ga:') Indian country. At a place called Ganos (present-day Cuba, New York state), Charlevoix wrote that this 'Officer worthy of Credit . . . assured me that he had seen a Fountain, the Water of which is like Oil, and has the Taste of Iron. He said also, that a little further there is another Fountain exactly like it, and that the Savages make Use of its Water to appease all Manner of Pains.'[205]

In 1755 the explorer and cartographer Lewis Evans published his ground-breaking *A General Map of the Middle British Colonies, in America*, which was reproduced in many editions – with only minor modifications – for sixty years as a standard reference to the region. The map indicated 'Petroleum' in two locations over a hundred miles apart: the first, about fifty miles south of the eastern end of Lake Erie near the Allegheny River, about ten miles upriver from Fort Venango (close to modern-day Franklin, Pennsylvania), near where French Creek flows into the Allegheny; the second location, to the southeast of the first, near the Ohio River, just upriver from modern-day Wheeling, West Virginia. The map was reproduced to accompany, for example, the 1771 English edition of *Travels into North America* by the Swedish-Finnish explorer Peter (Pehr) Kalm.[206]

David Zeisberger, a Moravian missionary, gave the first substantial account of the oil springs of the Allegheny River region in his journal of a trip there in 1767–8:

Of oil wells I have seen three kinds, (1) wells that have an outlet; (2) wells that have no outlet, but are stagnant pools; (3) wells in creeks and even in the Ohio at two different points, one hundred and fifty miles from one another.

In the wells which have an outlet, oil and water together exude from the earth and where these flow grass and soil become oily. When there is no means of discharge one sees nothing more than oil welling up, which, if none has been taken off for some time, floats an inch or more in depth on the water. In the creeks it is possible to see some of the places where the oil flows, others not, for often the current carries the oil with it at such rate that you see the oil over all the surface of the water ... Such places are usually revealed by the strong odor. Even though the water of a river keeps carrying away the oil, one may smell it at a distance of a quarter of a mile from the well. The soil near an oil well is poor, either a cold, clayey ground, or if it is near a creek, a poor quality of sand at the top. Neither good grass or wood will grow nearby, hardly anything more than a few stunted oak trees.

If the Indians wish to collect oil, which they prefer to do where the well has no discharge, as it is there most easily secured, they first throw away the old oil floating on top, as it has a stronger odor than that which wells up fresh from the earth ... [T]hey scoop it into kettles, and, as it is impossible to avoid getting some water, boil it and preserve it for use. They use it as a medicine in all sorts of cases for external application, thus for tooth-ache, head-ache, swelling, rheumatism, strained joints. Some also take it internally and it appears to have hurt no one in this way ... It is brownish in color and may be used in lamps, for it burns well.[207]

A fellow missionary reported that 'One of the most favorite medicines used by the Indians is the *Fossil oil* (petrolium) exuding from the earth ... Many [oil springs] have been found both in the country of the Delawares and the Iroquois ... The Indians sometimes sell it to the white people at four guineas a quart.'[208]

In 1768 the Spanish Crown sent several expeditions of troops and Franciscan missionaries to Alta California (modern-day California) in order to consolidate Spain's long-standing claims to the region against the possible encroachments of other European powers. In the initial expedition, under Gaspar de Portolá, Father Juan Crespí was a member of the first group of Europeans to reach San Francisco Bay. Crespí recorded in his compendious diary a number of natural petroleum seeps along the Californian coast, inhabited by indigenous peoples such as the Chumash and Tongva. In August 1769 he wrote, from a location near present-day Los Angeles City Hall,

> Our Captain and the scouts ... came upon volcanoes of pitch coming out of the ground like springs of water. It boils up molten and the water runs off one way and the pitch another. They reported having come across a great many of these springs and seeing very large swamplands of it, enough they said to have caulked many ships with ... We christened them *los volcanes de brea de la Porciúncula*, the pitch volcanoes of the Porciuncula.[209]

Crespí thought these 'pitch volcanoes' to be the cause of the earthquakes they frequently experienced.[210] Two weeks later, just before passing by a village at the location of present-day Carpinteria, he recorded, 'We saw, ... at a small ravine about a dozen paces from the sea, springs of pitch that had become solidified, half smoking.'[211] By early September they had reached present-day Pismo Beach – 'pismu' being a native word for bitumen – near which 'one comes upon a middling-sized, very swelling knoll some two hundred yards long that consists entirely of tar springs issuing molten out of the ground'.[212] They saw bitumen further along the coast at present-day El Capitán Beach, 'with the rocks in some spots as black and tarry as though large amounts of the tar had been poured over them'.[213]

A commander on the expedition, Lieutenant Pedro Fages, diverted from fighting Seri Indian resistance in Sonora to join the Portolá explorers, would soon become governor of Alta California. He recorded how the local Chumash Indians used liquid bitumen from seeps further inland to waterproof their plank boats, or *tomol*, and seal their water containers:

> [A]t a distance of two leagues from [the San Luis Obispo] mission there are as many as eight springs of a bitumen of thick black resin which they call *chapopote*; it is used chiefly by these natives for calking their small water-craft, and to pitch the vases and pitchers which the women make for holding water.[214]

Meanwhile, on the Atlantic coast a secessionist movement was afoot. George Washington and other men of wealth were challenging Britain's policy of limiting land claims beyond the Appalachian mountain range, designed to avoid provoking further conflict with native Americans. Washington was hoping to secure more land beyond the Blue Ridge Mountains in southern Ohio Country, and in 1774 – in one of a series of provocative acts that would lead to war with the local Shawnee and Cherokee Indians – his surveying team was working on a tract near Elk River, seven miles above where it meets the Kanawha River, near present-day Charleston, West Virginia. A member of the surveying party, Thomas Hanson, made note there of a gas spring:

> [T]he Burning Springs ... is one of the wonders of the world. Put a blaze of pine within 3 or 4 inches of the water and immediately the water will be in flames and continues so until it is put out by the force of the wind. The springs are small and boil continually like a pot on the fire; the water is black and has a taste of nitre.[215]

At the end of the American Revolution, in 1783 Benjamin Lincoln, one of Washington's most valued generals and a member of the American Academy of Arts and Sciences, gave an account of the oil and gas seeps of the war-torn regions of Pennsylvania and Virginia:

> In the northern parts of *Pennsylvania*, there is a creek, called *Oil-Creek*, which empties itself into the *Alleghana-river*, issuing from a spring, on the top of which floates an oil, similar to what is called *Barbadoes* tar, and from which may be collected, by one man, several gallons in a day. The troops, in marching that way, halted at the spring, collected the oil, and bathed their joints with it. This gave them great relief, and freed them immediately from the rheumatic complaints with which many of them were affected. The troops drank freely of the waters: they operated as a gentle purge.
>
> There is another spring in the western parts of Virginia, as extraordinary in its kind as the one just mentioned, called the *Burning-Spring*. It was known a long time to the hunters ... Some of them arrived late one night, and, after making a fire, they took a brand to light them to the spring. On their coming to it, some fire dropped from the brand, and, in an instant the water was in a flame, and so continued, over which they could roast

their meat as soon as by the greatest fire. It was left in this situation, and continued burning for three months without intermission. The fire was extinguished by excluding the air from it, or smothering it. The water taken from it into a vessel will not burn. This shews, that the fire is occasioned by nothing more than a vapour that ascends from the waters . . .

These accounts I have from the best authority. General *Washington*, from whom I had my information, as well as from others, owns the land around the *Burning-Spring*, which he bought for the sake of it.[216]

Some years after the Revolution, in 1787, Thomas Jefferson described this spring, in his *Notes on the State of Virginia*, as

a hole in the earth of the capacity of 30 or 40 gallons, from which issues constantly a gaseous stream so strong as to give to the sand about its orifice the motion which it has in a boiling spring. On presenting a lighted candle or torch within 18 inches of the hole, it flames up in a column of 18 inches diameter, and four or five feet height, which sometimes burns out within 20 minutes, and at other times has been known to continue three days, and then has been left still burning. The flame is unsteady, of the density of that of burning spirits, and smells like burning pit coal. Water sometimes collects in the bason, which is remarkably cold, and is kept in ebullition by the gas escaping through it. If the gas be fired in that state, the water soon becomes so warm that the hand cannot bear it, and evaporates wholly in a short time. This gaseous fluid is probably inflammable air, the hydrogene of the new chemistry, which we know will kindle on mixing with the oxygenous portion of the atmospheric air, and the application of flame . . . The circumjacent lands are the property of General Washington and of General Lewis. There is a similar one on Sandy River, the flame of which is a column of about 12 inches diameter, and 3 feet high. General Clarke, who informs me of it, kindled the vapor, staid about an hour, and left it burning.[217]

When making provision for his land in his will of 9 July 1799, Washington wrote that this tract 'was taken up by General Andrew Lewis and myself, for and on account of a bituminous spring which it contains, of so inflammable a nature as to burn as freely as spirits, and is nearly as difficult to extinguish'.[218]

In a map included in his London edition of *Notes on the State of*

Virginia, Jefferson also followed the tradition of the popular Evans map of 1755 by indicating 'Petroleum' just upriver from Venango, near the Allegheny River in northern Ohio Country (see Fig. 1).[219]

Meanwhile, from 1778 fur trader Peter Pond began exploring the Athabasca Lake region – straddling the present-day Canadian provinces of Alberta and Saskatchewan – a borderland between Cree and Chipewyan Indians, and in 1785 he presented a map of the area to the US Congress. To note number 20 on the map, immediately to the west of the 'Arabosca River', the accompanying text stated, 'Along the banks of this river one finds Springs of Bitumen that flows on the ground.'[220] Five years later, the Hudson's Bay Company sent an expedition to the area where a team member, Peter Fidler, recorded, 'Found great quantities of Bitumen a Kind of liquid tar oosing out of the Banks on both sides of the river, in many places which has a very sulphurous smell & quite black like real Tar, & in my opinion would be a very good substitute for that useful Mineral . . .'[221]

Between 1788 and 1793 Alexander Mackenzie explored Canada for the North West Company, another fur-trading enterprise. To the south of the 'Lake of the Hills' (Athabasca Lake) there was a fork where the 'Pelican River' (Clearwater River) met 'Elk River' (Athabasca River). Mackenzie recorded,

> At about twenty-four miles from the Fork, are some bitumenous fountains, into which a pole of twenty feet long may be inserted without the least resistance. The bitumen is in a fluid state, and when mixed with gum, or the resinous substance collected from the spruce fir, serves to gum the canoes. In its heated state it emits a smell like that of sea-coal. The banks of the river, which are there very elevated, discover veins of the same bitumenous quality.[222]

In 1793 the first Lieutenant Governor of Upper Canada (Ontario), Lieutenant-Colonel John Graves Simcoe of the British Army, led an expedition from Niagara to Detroit via the southwestern tip of Ontario just north of Lake Erie to reconnoitre strategic areas along the Thames River for settlement by British immigrants and Loyalists who had fought on Britain's side since the American Revolution. As they followed the river's route southwest of London, they passed some Chippewa/Ojibwe Indian hunting parties and encampments between Muncey and Moraviantown, and a member of Simcoe's team wrote that here 'the Indians discovered a spring of an oily nature, which upon examination proved to be a kind of petroleum'.[223]

In the same year, Captain George Vancouver was commanding a Royal Navy expedition sent to assert British fur-trading interests at Nootka Sound off the coast of Canada against Spanish claims. Sailing through the Santa Barbara Channel off the Californian coast he recorded the presence of oil, emanating from natural seabed oil seeps:

> The surface of the sea, which was perfectly smooth and tranquil, was covered with a thick slimy substance, which when separated, or disturbed by any little agitation, became very luminous, whilst the light breeze that came principally from the shore, brought with it a very strong smell of burning tar, or of some such resinous substance. The next morning the sea had the appearance of dissolved tar floating upon its surface, which covered the ocean in all directions within the limits of our view.[224]

In Virginia in 1804 a newspaper reported on the imminent construction of a road from 'Gennessee [Gennesee] country' in New York state to Pittsburgh, and one consequence, it argued, would be easier access to the oil spring located in the former territory of the Seneca Indians – who had, until recently, occupied the region from central New York state to northwestern Pennsylvania:

> Not far from the source of the principal fork of the Olean, is the Oil Spring, which produces the Seneca Oil; so esteemed by the Indians, that, in the sale of their territory, they reserved a square mile to include the spring. Difficulty of access to this spring, has hitherto opposed a scientific examination of all its qualities. This obstacle will, in a great degree, be removed: the road from Gennessee and Angelica passing near the spring. Hence forward, the curious may indulge their laudable disposition for research, and perhaps by the result, confer a benefit on society.[225]

The following year, a Harvard librarian and author published an account of his travels west of the Allegheny Mountains in which he reported on the oil found near Pittsburgh:

The *Seneca Indian Oil* in so much repute here is *Petroleum*; a liquid bitumen, which oozes through fissures of the rocks and coal in the mountains, and is found floating on the surface of the waters of several springs in this part of the country, whence it is skimmed off, and kept for use. From a strong vapour which arises from it when first collected, it appears to combine with it sulphureous particles. It is very inflammable. In these parts it is used as a medicine; and, probably, in external applications with considerable success. For chilblains and rheumatism it is considered as an infallible specific. I suppose it to be the bitumen which Pliny describes under the name of Naptha.[226]

An Irish writer travelling along the Ohio River in 1806 found that

Not far from Pittsburg is a well which has its surface covered with a bituminous matter resembling oil; and which the neighbouring inhabitants collect, and use in ointments and other medicinal preparations. The vapour rising from this well is inflammable . . . The medical men of Pittsburg profess to have analyzed this oil; and to have discovered in it a variety of virtues, if applied according to their advice.

Further downriver near Georgetown, he wrote, 'a few yards from the shore, a spring rises from the bottom of the river, which produces an oil nearly similar to Seneca oil'. The town's postmaster told him that 'the well was much frequented by the Indians previously to their retreat to the back countries, and that the neighbouring whites used the oil as a friction when suffering with rheumatism, and as an unction when afflicted with sores'.[227]

In 1809 another traveller passing the same spot wrote that 'the virtues of the Seneca oil are similar to those of the *British oil*', referring to the medicinal Betton's British Oil of the day, sourced from Shropshire, England: 'Large quantities [are] collected on Oil Creek, a branch of the Allegheny River, and sold at from one dollar and a half to two dollars per gallon ... [T]wenty or thirty gallons of pure oil can be obtained in two or three days by one man.'[228] Indeed, Tobias Hirte of Philadelphia had been advertising and selling Seneca Oil as a cure-all, an early American 'snake oil', since at least 1792.[229] It was from around this time that some Seneca Indians and settlers near Oil Creek began to take the oil on horseback seventy-five miles down to Pittsburgh, presumably then a cheaper round trip for an individual than by river.[230]

In the early nineteenth century, as Europeans settled west of the Allegheny Mountains in increasing numbers – many sending lumber, and later pig iron, downriver to Pittsburgh – they began to exploit the salt deposits of the headwaters of the Ohio River, eventually introducing well-boring techniques to drill brine wells. In 1808, near the mouth of Campbells Creek, a tributary of the Kanawha River in western Virginia, the brothers David and Joseph Ruffner drilled the first successful brine well in the region, setting off a salt-drilling boom. The drilling was at first powered by hand, but soon equipment was adopted that harnessed horse power for the work. Inconveniently, however, the drillers frequently struck gas or oil; a subsequent account described how,

> In many wells, salt water and inflammable gas rise in company with a steady uniform flow. In others, the gas rises at intervals of ten or twelve hours, or perhaps as many days, in vast quantity and with overwhelming force, throwing the water from the well to the height of fifty, or a hundred feet in the air, and again retiring within the bowels of the earth to acquire fresh power for a new effort. This phenomenon is called 'blowing,' and is very troublesome and vexatious to the manufacturer. The explosion is sometimes so powerful as to cause the copper tube which lines the upper part of the well to collapse, and to entirely misplace and derange the fixtures about it. By constant use this difficulty is sometimes overcome, by the exhaustion of the gas, and in others the well has been abandoned as hopeless of amendment. A well on the Muskingum, ten miles above McConnelsville, at six hundred feet in depth, afforded such an immense quantity of gas, and in such a constant stream, that while they were digging, it several times took fire, from the friction of the iron on the poles against the sides of the well, or from scintillations from the auger; driving the workmen away, and communicating the flame to the shed which covered the works. It spread itself along the surface of the earth, and ignited other combustible bodies at the distance of several rods. It became so troublesome and difficult to extinguish, when once ignited, being in this respect a little like the 'Greek fire'.[231]

Of the many salt wells in the region that struck oil, one drilled in 1814 near Duck Creek, north of Marietta, Ohio,

> furnishes the greatest quantity of any in this region . . . The oil from this well is discharged periodically, at intervals of from two to four days, and from three to six hours duration at each period. Great quantities of gas accompany the discharges of oil, which for the first few years, amounted to from thirty to sixty gallons at each eruption. The discharges at this time are less frequent, and diminished in amount, affording only about a barrel per week, which is worth at the well from fifty to seventy-five cents a gallon. A few years ago, when the oil was most abundant, a large quantity had been collected in a cistern holding thirty or forty barrels. At night, someone engaged about the works approached the well-head with a lighted candle. The gas instantly became ignited, and communicated the flame to the contents of the cistern, which giving way, suffered the oil to be discharged down a short declivity into the creek, whose waters pass with a rapid current close to the well. The oil still continued to burn most furiously; and spreading itself along the surface of the stream for half a mile in extent, shot its flames to the tops of the highest trees, exhibiting the novel, and perhaps never before witnessed spectacle of a river actually on fire.[232]

DeWitt Clinton, future governor of New York state, delivered a lecture that year in which he mentioned how petroleum from Amiano was used for street lighting in Italy; and, he speculated, 'It might be of considerable consequence to discover whether the petroleum of our springs might not be used for like beneficial purposes.'[233] Newspapers reported in 1818 that a well sunk in Medina County, Ohio, twenty miles from Cleveland, was producing 'from 5 to 10 gallons of Seneca oil per day',[234] while a salt well drilled near the Big South Fork of the Cumberland River in Kentucky, a few miles south of Rock Creek, encountered oil, which overflowed into the river. The drillers decided to convey barrels of the oil downriver in order to sell it as a medicinal product, but after they twice lost their cargo in the rapids at Devil's Jump, they took an overland route instead. Oil from the well continued to pollute the river for several years and some local women, who sold feathers from geese living on the river for use as down, took the drillers to court for damages.[235]

Comments made by Dr S.P. Hildreth in 1819 – though only

published in 1826 – extended Clinton's speculation to this frequent by-product of the salt wells of Ohio. One such well in Washington County near Little Muskingum Creek, twelve miles from Dr Hildreth's home town of Marietta,

> discharges such vast quantities of petroleum, or, as it is vulgarly called 'Seneka oil' ... is subject to such tremendous explosions of gas, as to force out all the water, and afford nothing but gas for several days, that they make but little or no salt. Nevertheless, the petroleum affords considerable profit, and is beginning to be in demand for lamps, in workshops and manufactories. It affords a clear brisk light, when burnt in this way, and will be a valuable article for lighting the street lamps in the future cities of Ohio.[236]

In 1820 Reverend Timothy Alden, the recent founder of Allegheny College at Meadville, reported that

> On the flats of Oil Creek, twenty-eight miles southeasterly from Meadville, many long pits have been dug several feet deep, from the bottom of which the Seneca oil, or petrolium, oozes and floats on the surface of the water, with which they are partially filled ... By extending the operation, this oil, called by the Senecas, au nūs´, might be collected so as to become a profitable article of commerce. Fifteen barrels were once taken, in one season from a single pit. It was formerly sold at two dollars a gallon. The common price is now one dollar fifty cents ... It burns well in lamps, and might be advantageously used in lighting streets. If, by some process, it could be rendered inodorous, it would become an important article for domestick illumination.[237]

Meanwhile, in 1819 a Royal Navy expedition, led by Captain John Franklin, was again hoping to discover and map a Northwest Passage through Canada. They revisited the region south of Athabasca Lake, near modern-day Fort McMurray, and on the map included in his account of the expedition Franklin labelled the area as containing 'Petroleum & Slaggy mineral pitch'. The surgeon accompanying Franklin's expedition saw

> a peaty bog whose crevices are filled with petroleum. This mineral exists in great abundance in this district. We never observed it flowing

from the limestone, but always above it, and generally agglutinating the beds of sand into a kind of pitchy sandstone. Sometimes fragments of this stone contain so much petroleum as to float down the stream.[238]

Similarly, the presence of oil springs in southwestern Ontario, just north of Lake Erie, was confirmed by a travel writer in 1821. Riding through Long Woods in the southwest of Middlesex County – where the British and their Native American allies had fought US forces in a significant battle in 1814 – he came to the River Thames:

> At one of the houses where I stopped to feed my horse, they showed me a specimen of mineral oil, that is found in considerable quantities upon the surface of the Thames. It flows from an aperture in the bank of the river, and three or four pints can be skimmed off the water daily. It very much resembles petroleum, being of thick consistence and black colour, and having a strong penetrating odour. The people employ it medicinally; and I was told, that its external application proved highly beneficial in cases of cramp, rheumatism, and other complaints of a similar kind. In this house there was a woman afflicted with acute rheumatism. She had tried the mineral oil without receiving any benefit from it, and consequently had been induced to put herself into the hands of one of the doctors of the settlement.[239]

The early nineteenth century saw the beginning of the widespread use of town gas. In 1792 William Murdoch had lit his office in Redruth, in Cornwall, England, with gas that he distilled from coal, conveyed seventy feet through tinned iron and copper tubes. Murdoch then worked for Boulton & Watt, one of several engineering companies that, over the next few decades, installed larger, commercial gasworks piping gas – distilled from coal or from whale and vegetable oils – around single factories and large buildings; and in 1813 the Gas Light & Coke Co. was the first enterprise to distribute coal gas on a more extensive scale in London via under-street iron pipes, similar to the water-supply network. In 1825 – by which time companies in Baltimore, Boston and New York were piping town gas to the public – William Hart was inspired to drill for natural gas that was bubbling from a water well at Fredonia in western New York state, just south of Lake Erie. Fredonia, on a commonly travelled route westward from New York, soon became quite famous for supplying its inhabitants with natural gas – initially through hollowed-out wooden logs, later upgraded to lead piping.[240] A scientific journal described how 'A gazometer was ... constructed, with a small house for its protection, and pipes being laid, the gas is conveyed through the whole village. One hundred lights are fed from it more or less, at an expense of one dollar and a half yearly for each. The flame is large, but not so strong, or brilliant as that from gas in our cities.'[241] A few years later, Hart was also supplying natural gas to a lighthouse at the Lake Erie port of Barcelona, conveyed through a wooden log pipeline from a source on Tupper Creek, almost a mile away.[242]

A letter published in the *Pittsburgh Gazette* in 1828 commended the city authorities for their decision to light the city with coal gas; but, the author advised,

> I will tell you what is the cheapest, best and most economical light you can use; it is what is called the West Seneca oil, which is petroleum. This substance, were there a ready market for it, might be supplied at your very doors to an almost unlimited extent ... The price of it is very low, because a few barrels glut the demands of the apothecaries; but if the city would take a large quantity ... I think it could be supplied at 25 cents per gallon ... Let anyone who doubts that it is a perfectly good oil for lamps send to the apothecary's for half a pint, and burn it one night in a lamp of any kind, precisely as fish or spermaceti oils are burned.[243]

On the Californian West Coast at this time, a Yokuts Indian who had traded in bitumen from Coalinga in the San Joaquin Valley now saw that Europeans were taking the asphalt, for use particularly as a roofing material. The region's indigenous population had long put the substance to practical uses such as for water-proofing boats and baskets, and as an adhesive; it was also used decoratively and for religious and ceremonial purposes, featuring centrally in a winter solstice ceremony, and it was smeared on the face as a symbol of mourning.[244] In 1828 a Kentucky trapper, James Pattie, travelling through the present-day Los Angeles area wrote,

My next advance was to a small town, inhabited by Spaniards, called the town of The Angels. The houses have flat roofs, covered with bituminous pitch, brought from a place within four miles of the town, where this article boils up from the earth. As the liquid rises, hollow bubbles like a shell of a large size, are formed. When they burst, the noise is heard distinctly in the town. The material is obtained by breaking off portions, that have become hard, with an axe, or something of the kind. The large pieces thus separated, are laid on the roof, previously covered with earth, through which the pitch cannot penetrate, when it is rendered liquid again by the heat of the sun.[245]

The following year, in Pattie's home state a brine well being drilled north of Burkesville, near the confluence of Little Renox Creek and the Cumberland River, struck flowing oil that shot twelve feet into the air. It was reported that

The discharges were by floods, at intervals of from two to five minutes, at each flow vomiting forth many barrels of pure oil ... These floods continued for three or four weeks, when they subsided to a constant stream, affording many thousand gallons per day. This well is between a quarter and a half mile from the Cumberland river, on a small hill down which it runs into the river. It was traced as far down the Cumberland as Gallatin, in Sumner county, nearly 500 miles – for many miles it covered the whole surface of the river, and its marks are now found on the rocks on each bank. About two miles below the point at which it touched the river, it was fired by a boy – and the effect is said to have been grand beyond description. An old gentleman who witnessed it says he has seen several cities on fire, but that he never beheld any thing like the flames which rose from the bosom of the Cumberland to touch the very clouds (his own words).[246]

The author of one of many books of the era on the pros and cons of emigrating from Britain to Canada wrote of how, in late 1826, he passed by an oil spring on the Thames River near Moraviantown, Ontario:

A singular spring of oil issues out of the banks of the river near here, on the land belonging to the Indians. It is of the consistence and colour of tar, with a peculiar smell. It is generally supposed to be coal tar

(*petrolium*) arising from a bed of coal said to run across the country . . .
The oil is gathered from the surface of the water (by Indians and
others), by blankets extended and lightly dropped on the surface, when
they absorb the oil. It is sold from 2s.3d. to 4s.6d. per quart, and sent
to all parts of the province, and even the States, as a cure for rheumatism, sprains, &c., and is sometimes taken internally, in small quantities,
for strengthening the tone of the stomach, and other complaints.[247]

In 1833 the editor of the scientific journal that had published the
report on the Fredonia gas well, Professor Benjamin Silliman Sr, published an article of his own in which he described 'a Fountain of
Petroleum, called the Oil Spring' near Cuba, New York state. From
the surface of the spring,

> They collect the petroleum by skimming it, like cream from a milk
> pan . . . It is used, by the people in the vicinity, for sprains and rheuma-
> tism, and for sores on their horses, it being, in both cases, rubbed upon
> the part. It is not monopolized by anyone, but is carried away freely, by
> all who care to collect it, and for this purpose the spring is frequently
> visited . . . The history of this spring is not distinctly known: the Indians
> were well acquainted with it, and a square mile around it is still reserved
> for the Senecas.

The medicinal Seneca Oil – sold widely in the East Coast states by
1830, displacing Betton's British Oil – was sourced, Silliman continued,
from around the Allegheny River in Venango County, Pennsylvania. He
added, 'I have frequently distilled it in a glass retort, and the naptha
which collects in the receiver is of a light straw colour, and much lighter,
more odorous and inflammable than the petroleum.'[248]

By the 1830s there was a booming salt industry in western Virginia
centred on the Kanawha River Valley, where slaves were put to work
mining coal and operating the coal-fuelled salt furnaces.[249] John Hale,
a prominent salt manufacturer, recalled some decades later that, of
the brine wells,

> Nearly all . . . have contained more or less petroleum oil, and some of
> the deepest wells a considerable flow. Many persons now think, trust-
> ing to their recollections, that some of the wells afforded as much as
> twenty-five to fifty barrels per day. This was allowed to flow over from

the top of the salt cisterns, on the river, where, from its specific gravity, it spread over a large surface, and by its beautiful iridescent hues, and not very savory odor, could be traced for many miles down the stream. It was from this that the river received the familiar nickname of 'Old Greasy'.[250]

In the early 1840s natural gas began to be used for the first time in the US as fuel in a manufacturing process for the evaporation of water from brine in salt production, when the brine wells were now being drilled to depths of around a thousand feet. This did not come without its risks, however, as evidenced by one such well sunk in 1843 by the Dickinson & Shrewsbury company. As Hale recalled,

So great was the pressure of this gas, and the force with which it was vented through this bore-hole, that the auger, consisting of a heavy iron sinker, weighing some 500 pounds, was shot up out of the well like an arrow out of a cross-bow. With it came a column of salt water, which stood probably 150 feet high. The roaring of this gas and water, as they issued, could be heard under favorable conditions for several miles.

Passing travellers stopped to witness this spectacle:

On one occasion a Professor from Harvard College . . . being a man of an investigating and experimenting turn of mind, he went as near to the well as he could get . . . and lighted a match to see if the gas would burn. Instantly the whole atmosphere was ablaze, the Professor's hair and eyebrows singed, and his clothes afire. The well-frame and engine-house also took fire, and were much damaged. The Professor, who had jumped into the river to save himself from the fire, crawled out, and back to the stage[coach], as best he could, and went to Charleston, where he took to bed, and sent for a doctor to dress his burns.[251]

A report in the *Lexington Gazette* that year described how in the previous year, 1842, William Tompkins had bored a brine well to a similar depth:

[H]e struck a crevice in the rock, and forth gushed a powerful stream of mingled gas and salt water . . . [T]he water gushed out so forcibly, that instead of applying the pump, he only lengthened his tube above the well . . . [He then] inserted the end of the spout from which the water and gas flowed, into a large hogshead, making a hole in the bottom to let out the water into the cistern. Thus the light gas was caught in the upper part of the hogshead, and thence conducted by pipes to the furnace, where it mingled with the blaze of the coal fire. It so increased the heat as to make very little coal necessary.[252]

At this time, the sons of a wealthy New Yorker, William Rathbone, began to sell oil – as 'nature's wonder cure' – from a salt well southeast of Parkersburg, western Virginia, near Burning Springs Run, a tributary of the Little Kanawha River. From about 1844 oil from another salt well in the region, near the Hughes River, was sold by George Lemon. A land title dispute between Lemon and Bushrod Creel was settled in favour of the latter, who, from 1847 to 1860, sold the oil – at between 33 and 40 cents per gallon – to the trading company Bosworth, Wells & Co. of Marietta, Ohio, which sold it on to customers in Pittsburgh and East Coast cities.[253]

However, this salt well oil would begin to be commercialized as more than just a cure-all. The industrialization of the era and the quest for ever-increasing productivity had raised the demand for lubricants for steam-powered metal machinery and for lamp oil for cheap, artificial illumination: together, these would enable both machines and labourers to work longer hours. Bio-oils from plants and animals, along with gaslight, were being produced on an industrial scale. The finest quality lubricant and lamp oil of the day was spermaceti, oil from a cavity in the head of sperm whales, but as the whales were hunted almost to extinction supplies of the oil became limited and increasingly expensive. In the United States, camphene replaced it as a cheaper yet equally effective illuminant, although it was also dangerously explosive – a mixture of alcohol and North Carolina turpentine produced from pine resin by slave labour, alongside the enslaved workers who produced most of the south's coal and gas and who laboured in the Kanawha Valley salt industry.[254]

From 1845 a Pittsburgh cotton factory owner, whose machinery was reliant on expensive spermaceti spindle lubricant, found the local light petroleum to be a good, cheap alternative, and he contracted with a salt well owner in Tarentum, Allegheny County, Pennsylvania, for two barrels per week.[255] At the same time the region's petroleum also began to be used in a crude manner for lamp light – though bad-smelling and smokey – in local Allegheny Valley homes and sawmills and in workplaces further afield. Residents of Tarentum recalled that this had been inspired in part by a conflagration on the canal as a result of all the oil running off from the salt wells:

> [T]he quantity of grease flowing into the canal was quite large. The boatmen complained about it . . . They said it greased their tow-lines, making their ropes hard to handle, besides soiling the sides and decks of their canal boats . . . One day the boys in the neighbourhood threw a burning brand into the canal, and there was an astonishing result. The whole canal, for a distance of half a mile or more seemed to burst into flames . . . We could do nothing but just let it burn itself out, protecting adjacent property. But, sir, it demonstrated one valuable fact to us . . . that this greasy stuff that was bothering the salt works would undoubtedly burn, and that it would make both light and heat.[256]

In about 1850 Samuel Kier, a canal boat operator and salt merchant living in Tarentum, on the Allegheny River, had begun selling the petroleum that polluted his salt wells as his own version of the medicinal Seneca Oil. 'Kier's Rock Oil' soon came to be marketed across the United States, from Pennsylvania to Michigan and Louisiana, as a cure for virtually any ailment. One of Kier's advertising flyers pictured a salt well derrick with an explanation that the oil came from a salt well near the Allegheny River and that the oil was skimmed from the surface of the salt water in storage vats.[257] Others began to imagine the commercial possibilities of this petroleum, as evidenced in an Ohio newspaper report of 1851 on an unusually large volume of petroleum that was flowing from a salt well drilled recently nearby: 'The uses to which Petroleum may be applied are various, and from the immense quantity discovered here, we have no doubt but that it might be made a source of great profit.'[258] Indeed, Kier had far more petroleum than he could sell as small bottles of medicine and he sent a sample for analysis to a prominent Philadelphia chemist, who suggested distilling it and selling it either as a solvent or as lamp oil. Accordingly, Kier, sometime in 1851–2, set up a simple refining apparatus in Pittsburgh and began selling his 'carbon oil' illuminant at the relatively low price of $1.50 a gallon. The lamps currently available could burn this oil fairly well, but they were not ideal for burning it with the brightness, cleanliness and efficiency to which people were accustomed. However, the emergence of a new 'coal-oil' industry would prompt design innovations that led to lamps well-suited for burning mineral oil.[259]

Over the preceding few decades, British and French scientists had discovered that bitumen, coal and residual coal tar from gasworks could be distilled to yield good-quality lamp oil, and the process had begun to be commercialized in both countries. It was now taken up in the United States, at the vanguard of which was the Canadian Abraham Gesner, one of the inventors of the process. In 1854 an agent of Gesner's Kerosene Oil Co. brought to the US a type of lamp already in use in Vienna for burning Galician petroleum-based illuminating oil, and this, along with other new US lamp designs, opened the way for the commercialization of coal oil. The bituminous deposits owned by the Breckenridge Cannel Coal Co. of Kentucky were found to be a particularly good source, according to an analysis conducted by the Yale chemist Professor Benjamin Silliman, Jr – whose father, in 1832, had published an account of the distillation of petroleum from a spring in Cuba, New York state, into serviceable lamp oil.[260] By 1856 the retorts and stills of the Breckenridge Coal and Oil Co. at Cloverport on the Ohio River were approaching the capacity for the manufacture of 19,500 barrels per year of lamp oil and other products, including lubricating oil, benzole, naphtha and paraffin wax. According to a report in the *New York Times* late that year,

> we have watched with great interest the progress of these experiments as of truly national importance, as giving a new impulse to the development of the mineral resources of the country, and as establishing the fact that we have within ourselves an inexhaustible supply of light, to be obtained with far less the expense, and labor, and danger, than the animal oil now in use. The whale, upon which we now depend for oil, is rapidly being driven by the energy of our fishermen into inaccessible seas, and with before many years, at the present rate of destruction, entirely disappear. By the discovery of the presence of a true illuminating and lubricating fluid residing in certain descriptions of coal, we have

become independent of such a misfortune, and the whole whaling fleet
might be laid up to rot, and we should still have light.

The merits of this lamp oil were its 'cheapness, brilliancy and entire
absence of danger of explosion, which makes the use of camphene so
hazardous'. One caveat, however, was that

> Since the experiments of the Breckinridge [*sic*] Company were made
> with such a successful result the whole of the country has been explored
> for oil-bearing coals, but thus far the experiments have resulted in dis-
> appointment. No coal has been yet found which could be made to yield
> much more than one-half the results of the Breckinridge [*sic*].[261]

Meanwhile, an initial survey report for the Pacific Railroad in 1855
had remarked on the bitumen seeps of California:

> It is an interesting fact, which I believe is not generally known, that
> there are numerous places in the Coast Mountains, south of San Fran-
> cisco, where *bitumen* exudes from the ground, and spreads in great
> quantity over the surface. These places are known as *Tar Springs*, and
> are most numerous in the vicinity of Los Angeles. It is also common
> to meet with large quantities of this material floating on the Pacific,
> west of Los Angeles, and northward towards Point Conception. I
> have seen it, when, passing this point, floating about in large black
> sheets and masses. They are probably the product of submarine
> springs; or they may be floated down by small streams from the
> interior.[262]

Other surveys of the 1850s found

> a large number of holes, where the Indians obtained their earthgrease,
> as they call it by name . . . At [site 176] the Indians have a burning sta-
> tion, where they prepare their earthgrease for medical purposes. They
> rub it over their persons to keep away the insects . . . Run a rod into the
> ground, and upon pulling it out, gas will escape which will burn for
> some time . . .
>
> We call [site 81] Lightening Camp, from the fine fireworks we had
> during the evening. The men stationed themselves at the different gas
> jettings, and touching them off, they would discharge themselves like
> fireworks . . .

Quite a lake of bitumen [at site 133]. The natives use it for the roofs of their houses it is an excellent substitute for cement or mortar, and I cannot see what reason it cannot be sent to San Francisco, to roof the houses and pave the streets. It would be of vast importance if the inhabitants of San Francisco could be induced to use it ...

The Indians use this yellow greasy material [found at site 618] for pains in the limbs, and we can all of us recommend it for the same purpose, from practical experience of its efficacy ... [At site 614] the out-burst of gas and oily matter is enormous, and will well repay a visit by the curious, as it comes out of the ground beneath the spur of the rocks. Here we made our camp, so as to avail ourselves of the material for camp purposes and light. We called this 'Camp Illumination'.[263]

Andrés Pico – the commander of the defeated Mexican forces in California during the Mexican-American War (1846–8), who had signed the 'Capitulation of Cahuenga' in 1847 – is credited as being the first Californian to refine the local petroleum, in the early 1850s, for use on his San Fernando Mission ranch. By 1855 the streets of San Francisco were being surfaced with asphalt, and from 1857 illuminating oil began to be refined from bitumen at Carpinteria on the Santa Barbara coast, while heavy oil was drawn from hand-dug wells in Los Angeles.[264]

Petroleum deposits were first mentioned in Canada's official geological surveys from 1850 when Thomas Hunt – who had studied under Benjamin Silliman – recorded the occurrence of 'asphaltum, or mineral pitch' in Enniskillen Township, Lambton County, southwestern Ontario. Since the amount of funding that Parliament would grant to the Geological Survey of Canada was at least in part dependent on the commercial potential of its findings, Hunt continued,

> The consumption of this material in England and on the Continent for the construction of pavements, for paying the bottoms of vessels, and for the manufacture of illuminating gas, to which it is eminently adapted, is such that the existence of deposits of it in this country is a matter of considerable importance.[265]

As the Geological Survey proceeded with further investigations at Enniskillen, a local foundry foreman, Charles N. Tripp, in early 1854 bought a plot of land around Bear Creek containing asphalt beds and oil springs. He sent samples to a consultant chemist in New York whose report prompted Tripp that December to incorporate the International Mining and Manufacturing Co. to market asphalt. His business folded in 1857, however, due to the limited transportation infrastructure, refining difficulties and lack of capital. Between 1856 and 1858 a number of businessmen, including James M. Williams, bought up hundreds of acres of asphalt- and oil-rich land in the township, including those of Tripp. Williams' workers struck oil in August 1858 while digging a water well, and he was soon successfully selling good-quality lamp oil at between 70 cents and $1.50 a gallon, facilitated crucially by the opening of the London-Sarnia branch of the Great Western Railway later that year. His business expanded and in 1859 he set up his J.M. Williams and Co. refinery in Hamilton.[266]

Thus, Canada had now joined the United States, Austria-Hungary and Romania in demonstrating that there existed a demand for mineral – coal- or petroleum-based – lamp oil.

2

Seneca Oil

DRILLING IN OHIO COUNTRY

The petroleum of the western reaches of New York, Pennsylvania and Virginia, either found escaping in small quantities from natural oil springs or accidentally struck when wells were sunk, was relatively 'light': it contained plenty of the more fluid, volatile oil suitable for lamp light. Furthermore, in contrast to many other parts of the world with natural petroleum seeps, by the mid-1850s western New York and Pennsylvania had a fairly extensive transportation infrastructure in the form of both the Erie Canal system and an expanding network of railways: the Sunbury & Erie Railroad (later the Philadelphia & Erie), the Lake Shore Railroad, the Meadville Railroad (later the Atlantic & Great Western), the New York Central Railroad, the Erie Railroad, the Pennsylvania Railroad, and the advancing Baltimore & Ohio Railroad.[1] Thus, it is not surprising that it was here that petroleum was first produced and commercialized on a large industrial scale. As Samuel Kier and a few other small entrepreneurs sold their new, cheap, petroleum-based lamp oil in relatively small quantities, a number of wealthier East Coast businessmen began eyeing up the oil of the Upper Allegheny River region as a potential investment opportunity.

By the mid-nineteenth century many logging and lumber companies were operating along the tributaries of the Allegheny River, rafting timber down to Pittsburgh, further down the Ohio River to Cincinnati, and even down the Mississippi as far as New Orleans. One of these firms was Brewer, Watson and Co., which operated its water-powered sawmills in and around Titusville on Oil Creek in Crawford County, Pennsylvania. Petroleum oil gathered from nearby springs was used by workers in the local sawmills for lighting and for lubricating the machinery, though the oil was still mostly valued for its supposed medicinal properties.[2] In 1851 the son of one of the firm's partners, Dr Francis B. Brewer, a physician practising on the East Coast, moved to Titusville and in 1853 took a sample of the oil with him on a visit back to relatives in New Hampshire. Two scientists at Dartmouth College, where Brewer had received his medical training, became curious, and the oil sample subsequently came to the attention of another Dartmouth alumnus, George H. Bissell, a New York commercial lawyer. Perhaps inspired by the newly emerging coal-oil industry or by Kier's advertisements featuring a well-drilling rig, they wondered if the oil might have commercial value as a raw material for lubricants or lamp oil. In November 1854 Bissell, along with his law practice partner J.G. Eveleth, persuaded several investors, including the president of the New Haven City Savings Bank, James M. Townsend, to buy some oil-rich land from Brewer, Watson and Co. This was a 100-acre tract known as Hibbard Farm in Cornplanter Township, Venango County, fifteen miles downriver from Titusville near the confluence of Oil Creek and the Allegheny River. In order to assess the petroleum's commercial potential, and then to attract investors, they commissioned a thorough analysis by Professor Benjamin Silliman, Jr, who was also consulting for the Breckenridge Coal and Oil Co., a new coal-oil company. They soon received an encouraging preliminary report from Silliman suggesting that their crude was rich in valuable solvents, and on 30 December 1854 Brewer, Eveleth and Bissell incorporated the Pennsylvania Rock Oil Company in New York state.[3]

By the following February the three entrepreneurs were becoming concerned at the length of time Silliman was taking over his full analysis and at the increasing cost of his experimentation. He appears to have exploited the occurrence of an explosion in his laboratory to demand more money for new apparatus and to have used the investors' funds also to pursue his own research into gas lighting. He charged an extortionate sum for his report, knowing that the Pennsylvania Rock Oil Co. would be unable to attract any significant investment without his eventual endorsement. But this they finally received in May, in Silliman's *Report on the Rock Oil, or Petroleum, from Venango County, Pennsylvania, with Special Reference to Its Use*

for Illumination and Other Purposes. His experiments had convinced
him that the crude petroleum taken from Oil Creek could be refined
relatively easily, by fractional distillation, to yield marketable solvents
and lubricants, and that it had 'a much higher value as an illuminator
than I had dared to hope . . . In conclusion, gentlemen, it appears to me
that there is much ground for encouragement in the belief that your
Company have in their possession a raw material from which, by
simple and not expensive process they may manufacture very valuable
products.'[4]

Within a few days the three entrepreneurs had distributed a print
run of Professor Silliman's report, copies of which, Brewer was
informed by one of his associates, were now 'in the hands of the
monied men in the city'. Eveleth wrote to Brewer, 'The report will
make a stir. Stock will then sell. Things look well.'[5] Investors were
likely to be wary following a recent spate of New York joint-stock
company frauds, yet the business trio regarded Silliman's report to be
so promising that they seem to have decided to keep their next moves
secret to avoid attracting competitors.[6] Nevertheless, Silliman's findings were reported in some detail within a few days in the *Boston
Evening Transcript* and several other newspapers, which informed
readers,

> It is expected that the arrangements for distilling the rock oil will be
> completed by next fall, and the article sold at a moderate price. With
> sperm oil at two dollars and a half a gallon [and] candles at a propor-
> tionate price . . . a new illuminator, which shall combine the advantages
> of cheapness, safety and brilliancy of light, cannot fail to come into
> extensive use. The tract of land – about one hundred acres – now
> owned by the company, will yield an average supply of about five hun-
> dred gallons per day; but arrangements will probably be made for
> increasing the supply by more extended operations.[7]

Since the incorporation laws of Connecticut granted greater secrecy
and fewer restrictions on land ownership, the Pennsylvania Rock Oil
Co. was reincorporated there in mid-1855, while some remaining
wariness among interested investors was addressed by sending two of
the company's associates to conduct a more thorough two-week survey of their 'Oil Lands'. They reported back that they had

made some 25 or 30 excavations at different points scattered over some 15 or 20 acres of land ... Not more than 2 or 3 excavations were sunk more than 4 feet. In every instance we found undoubted marks of oil. In some more & in others less. After digging down into the hard pan, the oil mixed with water seemed to rise up as though there was a pressure beneath. The deeper the excavation, the more abundant the oil appeared. So far as we could judge, the excavations would each daily yield from 2 qts up to 4 gallons & they might be increased to almost any number. From our investigations, it is evident that the oil can be found any where you are disposed to dig over a large tract of land in Oil Creek Valley ... There cannot be any doubt ... but that there will be a yield of oil sufficient to justify our company in going to work with all the force we may be capable of bringing into the field.[8]

In 1856 the company arranged for a Wall Street real estate firm to dig for oil in commercial quantities, on a fifteen-year lease of Hibbard Farm and a 12-cent royalty per gallon of crude produced; but the firm was hit by the financial Panic of 1857 and was forced to surrender the lease on a legal technicality. This was just as the market for petroleum-based lamp oil was beginning to grow, on the back of a sales campaign by a New York City businessman, A.C. Ferris, and the appearance in Pittsburgh of new firms, alongside Kier's, that began to refine 'carbon oil' from the Allegheny River Valley crude.[9]

James Townsend now took it upon himself, much to the other stock-holders' initial consternation, to engage the services of the unlikely figure of Edwin L. Drake, a retired railway conductor, again on a fifteen-year lease of the farmland but at a royalty rate of one-eighth of production. Drake was, nevertheless, subsequently contracted formally, at the previous 12-cent royalty per gallon plus a $1,000 a year salary, by a third company that the entrepreneurs now incorporated, again in Connecticut, on 23 March 1858: the Seneca Oil Company, capitalized at $300,000 divided into 12,000 shares.[10]

In May 1858 Drake travelled with his family on the New York Central Railroad, bound for the sleepy hamlet of Titusville deep in the wooded hills of northwestern Pennsylvania near the Allegheny River. They passed through Syracuse where Drake, on his first visit to Titus-ville on behalf of the company a few months earlier, had stopped to see the city's salt wells and processing plants. The family continued west to Erie, following the Erie Canal system through the gap between the Great Lakes to the north and, to the south, one end of the long Appalachian and Allegheny mountain ranges. From the southern shore of Lake Erie a horse-drawn stagecoach took them along the rough road south, via Waterford at the head of French Creek, to their destination on the banks of Oil Creek. Drake was now in Titusville and his mission, on behalf of the Seneca Oil Co., was to try to find petroleum in large quantities.[11]

The peace and quiet of this small settlement was, for now, punctuated only by the tapping of a woodpecker, the distant chopping and sawing of lumberjacks and, in places, the whine of waterwheel- or steam-powered sawmills. Exactly a century earlier, by stark contrast, these woodlands had been a war zone, echoing with musket fire and screams as two imperial powers, Britain and France, fought decisive battles for control of North America – while the Indian tribes caught in between struggled in vain to defend their lands. Edwin Drake had stepped onto centre stage of world history.

How had it come about that this region of North America, until quite recently inhabited by myriad indigenous tribes, was now considered to be the almost exclusive sovereign territory of the United States of America, home to a burgeoning population of Europeans – who, furthermore, spoke mainly English rather than French or Spanish?

We shall see how, due to a particular confluence of factors – geographical, historical, political, economic and geological – the banks of Oil Creek in northwestern Pennsylvania would, in 1859, become the setting for the first appearance of large-scale, industrialized oil production. Yet a key precondition for this, the near complete eradication of the Native American population of the region just over fifty years earlier, typically goes unmentioned in accounts of the world's first oil boom.[12] This omission is symptomatic of a common pattern: standard historical narratives typically efface many of the darker sides of the oil industry and downplay its close association with imperialist violence from its earliest days. The people who inhabit areas that become regions of oil and gas production, or who live along the associated transit routes, tend to be overlooked, their interests and their values largely ignored. Thus, in light of the current environmental devastation and potential catastrophic breakdown of our global ecosystem – due in large part to our burning of fossil fuels, including petroleum, and our intensive use of petrochemicals – it is of profound significance that the emergence of the modern oil industry was premised on the near total destruction of the Native American nations of the region, of people who lived essentially in harmony with nature.

This grim early North American backstory also exemplifies how geopolitical conflict on a global scale – in this case between Britain and France, and then between Britain and its American colonies – impacts on local populations. At the same time, it conveys clearly the crucial strategic and economic importance of transit routes; and in some striking instances we see how several of the lines of communication that were pivotal to these eighteenth- and early nineteenth-century conflicts went on to feature centrally during the later nineteenth century in the emergence and growth of the oil industry in and around the 'Oil Region' of Pennsylvania and New York states. Subsequently, we see how transit routes continue to be key to the evolution and expansion of the industry globally.

Looking further forwards into the twentieth century, it would be the dependency of Britain and its allies on the United States for their oil supplies during both world wars that was to be one of the primary reasons why the US would take over Britain's position as the world's superpower.[13] Thus, it will become apparent that the origins of this shift can be traced back to the American War of Independence when Britain lost territory that, from the mid-nineteenth to mid-twentieth centuries, would become by far the world's largest source of petroleum.

IMPERIAL COLLISION

A century before Edwin Drake was making his way to Titusville, in the winter of 1757 the region was in the tenuous grip of French troops and their uneasy local allies, Shawnee and Delaware Indians. This was known as Ohio Country, territory south of Lake Erie, between the Allegheny Mountains to the east and the Mississippi River to the southwest. Three years earlier, competing British and French claims to sovereignty over the region had led to military confrontation, sparking off a global conflagration between the major European powers. The French and Indian War in North America had escalated into the global Seven Years War, or Great War for Empire (1756–63).[14]

Ohio Country, in 1754, had become a geopolitical flashpoint. It was not just that British and French fur traders were competing here for business with the Indians, or that the Indians' hunting grounds of the Ohio River valley contained huge agricultural potential. It was also that the Allegheny and Ohio rivers flowing through the region offered the French the opportunity to connect their territorial possessions in the south and west – from the Gulf of Mexico at New Orleans all the way up the Mississippi River – to their northeastern colony of New France (Canada) with its waterways extending from the Great Lakes to the St Lawrence River and the North Atlantic. Securing this huge transcontinental arc of communications, it was thought, would bolster the now faltering French trade in North America while at the same time drawing the surrounding Indians more into the French political orbit and away from the British. By controlling the Ohio Country transit link, France could facilitate the movement of its troops and supplies, and could raise a barrier to the further westward expansion of Britain's colonies – whose populations were already twenty times greater than those of the French. Conversely, British control of the Ohio River valley would dash this French dream.[15] Given the immense strategic importance of the region, the stakes were high – perhaps as high as the eventual control over the whole of North America.

Attempts were made by the British and French governments to resolve the tension by a negotiated settlement, but this was made difficult by the militarists on both sides – and by the activities of the Ohio Company of Virginia.[16] This speculative venture had been set up in 1747 by wealthy Virginian planters who understood the potential profits to be made by whoever could win legal title to land in Ohio Country. Its members included Robert Dinwiddie, lieutenant governor of Virginia; John Hanbury, a prominent London merchant; and several members of the Washington and Lee families. Dinwiddie's close friendship with the Earl of Halifax, president of Britain's Board of Trade and a hawkish imperialist, along with Hanbury's powerful political connections and his effective lobbying in London, gave the Ohio Company immense behind-the-scenes influence. Accordingly, in 1749 the British government awarded this group of speculators a grant of half a million acres of Ohio Country.[17]

The Virginians attempted to clear away any obstacles to their Ohio land grant by inducing some Indian chiefs in 1752 to sign the Treaty of Logstown. Like most treaties signed between the colonists and the Indians, the treaty was a sham. Firstly, in the oral culture of the Indians it was the spoken word, the speech, that was held sacrosanct; they found it hard to understand that words written on paper should be given more weight than verbal discussions, even when that text seriously diverged from the white man's apparently reassuring, complimentary and earnest speeches. Secondly, the interpreters and intermediaries were often corrupt, quite willing for some personal gain to deceive the Indians into believing that the treaty was a good deal that fully respected their people's rights, concerns and interests. Thirdly, the Indian signatories to these treaties hardly ever genuinely spoke for all of the tribal groups living on the land in question. If an impressive array of gifts, including alcohol, and what was really a derisory sum of money persuaded some chiefs to sign a piece of paper, then that was the job done as far as the colonist was concerned, who could walk off claiming legal ownership of vast tracts of land. The treaty conference held at Logstown – an Indian settlement on the Ohio River – was fatally flawed in all three of these ways, its legitimacy in no way revived by the fact that the Virginians met with the chiefs in an Indian town. In many other instances, the Indians were negotiating under duress in a fort, in the shadow of guns and cannon.[18]

Around the Logstown conference fire, the Virginian commissioners effused, 'Brethren, be assur'd that the King, our Father, by purchasing your Lands, had never any Intention of *takeing them from you*, but that we might live together as one People, and *keep them from the French*, who wou'd be bad Neighbours.'[19]

What the written treaty actually contained was a reaffirmation of the 1744 Treaty of Lancaster which, the colonial government insisted, allowed European colonization deep inland, west of the Allegheny Mountains. The Indians, by contrast, were of the understanding that the new treaty would limit settlement to the east of the Alleghenies.[20] Since the interpreter at Logstown was secretly in the pay of the Ohio Company, it was not in his interests to ensure that the Indian delegates were made fully aware of the implications of what they were being asked to sign up to. Furthermore the signatories, belonging to the Iroquois (Hodenosaunee) league of Six Nations, were very far from being representatives of all the Ohio Country's inhabitants. The Iroquois, after suffering crushing defeats north of the Great Lakes to the French, had become the ambivalent allies of the British. The British now, ironically, capitalized on the imperialist tendencies of the Iroquois Grand Council, which claimed that the Shawnee and Delaware Indians of Ohio Country were its subjects. The Virginians were only too happy to accept Iroquois signatures and to translate this political fiction into a legally binding land cession, despite the fact that no Delaware even signed the Logstown Treaty: the Delawares had already been blatantly defrauded of their lands in Pennsylvania's Walking Purchase of 1737 and then harried off to Ohio Country by the British-allied Iroquois.[21]

Thus, by 1752 the Ohio Indians were distinctly suspicious of the British, particularly when the Ohio Company built a trading post on the Monongahela River, a tributary of the Ohio, and when a road that the company had opened up to improve access to the site attracted more traders and settlers into the area. The local Shawnee and Delaware were already beginning to suspect Virginian Governor Dinwiddie of what he was then admitting in private: 'I have the success and prosperity of the Ohio Company much at my heart.'[22]

The Indians were no happier when the French began, in early 1753, to proceed with their plans to construct a line of four forts from Lake Erie to the Ohio River, asserting French control over the strategic line of communications between the Great Lakes and the Mississippi. The construction of Fort Presque Isle (future Erie) and Fort Le Boeuf (future Waterford) secured the short overland 'carrying place' between Lake Erie and the head of French Creek (see Fig. 1).[23] And just as French troops became bogged down along this portage as it churned to mud, so would Edwin Drake's stagecoach, a century later, struggle through this same winter quagmire on its way to Titusville.

As the French finished building their third fort, Fort Machault (future Franklin), where French Creek meets the Allegheny River, Governor Dinwiddie assigned a new officer of his Virginia militia to deliver an ultimatum to the French to withdraw. The young Major George

Washington – whose family was prominent in the formation of the Ohio Company – returned from Fort Le Boeuf, however, with a message that the French were declaring sovereignty over the area. In response, Dinwiddie ordered his militia, in early 1754, to help the Ohio Company to build a fort of its own on the Ohio Forks, where the Monongahela and Allegheny rivers join to form the Ohio River (and future site of Pittsburgh). This was just where the French intended that their fourth and final fort should complete their Erie-Mississippi link. To support the Ohio Company's now urgent operations, Hanbury sent Dinwiddie £10,000 from London, while he successfully lobbied the British government to take the developments seriously and to provide greater assistance. This prompted a pamphleteer later to denounce Hanbury for conspiring, for private financial gain, to embroil his country in a war.[24]

Before the Virginians could complete their fort, a superior force of French troops arrived at the Ohio Forks, sending the Ohio Company and militia running and enabling the French to build their own fourth and final Fort Duquesne. The destruction of the Ohio Company's Monongahela trading post by the French and the defeat of Washington's militia at the Battle of the Great Meadows (3 July 1754) provoked the British into sending reinforcements under the command of General Edward Braddock to drive the French back. Thus began the so-called French and Indian War that would escalate globally into the Seven Years War. Braddock's arrogant faith in the superior nature of his troops, however, and his disdainful rejection of Shawnee and Delaware offers of assistance – declaring that he would be fighting for British, not Indian, sovereignty over Ohio Country – ensured his inglorious defeat. The alienated Shawnee and Delaware Indians now turned to the French as the most likely to prevail and to respect their land rights. They were even joined by Mingo (Ohio Seneca) Indians, thereby fracturing the Iroquois League between these neutral or pro-French western tribes and the pro-British eastern tribes. French and Indian war parties terrorized western Pennsylvanian and Virginian settlers into fleeing in great numbers back eastward across the Allegheny and Blue Ridge mountains. Atrocities were committed by both sides, encouraged on the British side by Pennsylvania Governor Robert H. Morris' declaration of a general bounty for enemy Indian prisoners and scalps. In 1756, for example, land surveyor Colonel John Armstrong led 300 irregulars in destroying the Delaware village of Kittanning on the Allegheny River, burning down homes with many of their inhabitants still inside. The British and other European settlers became engaged in a brutal racial war in which violence perpetrated by Indians was met with righteous horror and outrage, whereas British troops, who routinely committed atrocities, were lauded as heroes.[25]

It was at this time, in 1755, that Lewis Evans published by far the most accurate and detailed map of the region to date – including its indications of 'Petroleum' near the Allegheny and Ohio rivers – that became the standard reference to Ohio Country for several decades.[26] In a pamphlet written to accompany the map, Evans wrote,

> it is to be hoped, that . . . his Majesty of Great-Britain is no longer to be kept unacquainted with the Consequence of the Country between the British Settlements and Missisippi [sic]; which must one Day determine, whether the Southern Colonies shall remain the Property of the British Crown; or the Inhabitants, to prevent the entire Defection of their Slaves, which the French will encourage, as the Spaniards now do at St. Augustine, be obliged to fall under the Dominion of France. Let not the Public think this a remote Contingence: If the French settle back of us, the English must either submit to them, or have their Throats cut, and lose all their Slaves.[27]

In February 1756 George Washington sent 350 militia, troops and Cherokee, led by Major Andrew Lewis, to attack Shawnee villages near where the Big Sandy River meets the Ohio River. As the ill-fated and nearly starving Sandy Creek Expedition retreated, they stopped at Burning Spring near Tug Fork, where they used the burning gas to cook strips of buffalo hide.[28]

French territorial gains peaked in 1757, followed by a long period of retreat. British command of the seas and the overwhelming population and economic power of the British American colonies helped Britain to muster the largest fighting force yet seen in the colonies' history. Nevertheless, even this would not be enough to take on the French and Indian alliance across vast expanses of difficult terrain where slow-moving wagon trains found the going hard enough even without the deadly harassment of Indians.[29] Brigadier General John

Forbes, however, charged with leading an expedition against Fort Duquesne, stood out among the British commanders for the unusual respect he had for the strategic importance of the Indians. While preparing for his attack by opening up another road to the Ohio Forks, Forbes worked assiduously to bring about a peace agreement with the Ohio Indians, although it would be a challenge to overcome their understandable scepticism.[30] A delegation of Delaware chiefs asked,

> It is plain that you white people are the cause of this war; why do not you and the *French* fight in the old country, and on the sea? Why do you come to fight on our land? This makes everybody believe, you want to take the land from us by force, and settle it.[31]

Months of complex, delicate negotiations involving Quaker peace activists, the Iroquois League and trading and land-speculating interests eventually led to the pivotal 1758 Treaty of Easton. Under its terms, Ohio Country was returned to the Iroquois, while the Ohio Indians themselves would from now on be recognized by the British as negotiating parties independent of the Iroquois Grand Council. Crucially, the British promised the local Shawnee and Delaware that, after the war, permanent white settlement in Ohio Country would be prohibited. With this turnaround, the European balance of power in North America shifted decisively in favour of the British.[32] Ironically, the Quakers, having helped to bring peace to Pennsylvania's backwoods, thereby enabled the British and their newly won-over Indian allies to take decisive military action against the French at Fort Duquesne on the Ohio Forks. Colonel Henry Bouquet wrote, 'After God the success of this Expedition is intirely due to the General, who by bringing about the Treaty of Easton, has struck a blow which has knocked the French in the head.'[33] With their supply routes being cut both to the northeast (via the Great Lakes) and now to the southwest (via the Ohio and Mississippi), time was up for the French in the region.

Had the Ohio Indians been able to see into the post-Treaty of Easton future a century on, they would have been shocked to discover how feeble the treaty's protections would turn out to be. They would have seen Edwin Drake, in the vicinity of a permanent white settlement, Franklin (near the site of former Fort Machault), on Oil Creek, drilling a deep hole into the rocky ground; and they would have seen how great quantities of lumber were being taken from the local wood-lands and rafted down French Creek and the Allegheny River to the growing metropolis of Pittsburgh on the Ohio Forks – with hardly an Indian in sight.[34]

IMPERIAL OVERSTRETCH

The promise given at Easton by the Anglo-Americans to prevent white settlement in Ohio Country was immediately broken. The British maintained a military presence far larger than the French had ever stationed there and they behaved in a much more imperious manner towards their recent Indian allies. Large numbers of settlers established farms around Ohio Country's many heavily garrisoned forts, and aggressive settlement was even encouraged by General Jeffery Amherst, who, typically of the colonial army, favoured military solutions in Indian affairs.[35] In the trans-Appalachian west, mutual hatred rose to such intensity that battle was waged by each side with shocking brutality.[36]

In February 1763, after seven years of war, the European powers signed the Treaty of Paris that, by granting to Britain sovereignty over Ohio Country, ignored British promises made to the Ohio Indians in the Treaty of Easton. Three months later there arose the largest ever pan-Indian rebellion against illegal white encroachment. The western tribes, no longer able to turn to the French for protection and to play one European power against the other, now found strength in unity against a single common enemy. This unprecedented degree of inter-tribal cohesion across the Great Lakes and Ohio regions – including the Seneca, Mingo, Shawnee, Delaware, Wyandot, Miami, Ottawa and other tribes – was enhanced by a spiritual resistance movement in which the harassed, displaced and humiliated Indians found new meaning in their peril. The Delaware prophet Neolin and his disciples spread the word that their paradise had been lost because of the white man, a message now readily accepted. Indians were exhorted to stop trading with the colonists and to break their growing dependency on European goods that led Indians to lose their traditional skills. They should abandon the white man's ways and, above all, abstain from his poison, alcohol.[37]

The uprising began when an Ottawa chief, Pontiac, laid siege to Fort Detroit in the first of a wave of defensive assaults across Ohio Country and beyond. Senecas expelled the British-Americans from around the Allegheny River, routing the garrisons of Forts Venango (formerly Machault), Le Boeuf and Presque Isle, thereby cutting off the heavily defended Fort Pitt (formerly Fort Duquesne, and future Pittsburgh; see Fig. 1).[38] The uprising was seen by the colonials as a French-orchestrated conspiracy headed by the evil propagandist Pontiac, as if the Indians themselves had no legitimate grievance when faced with continual deception and illegal settlement. The governor of New York feigned innocence, writing to Britain's Board of Trade,

> I am surprised to find it repeatedly asserted in the English Newspapers, that the present insurrection has been occasioned by the Indians having been cheated of their lands by the English in America. I can assure your Lordships that there is not the least ground for this assertion and that, as to this Province it has happened without any provocation on our part so far as I have heard at least, and I believe to be true.[39]

Yet even General Philip Sheridan – the Plains Indians eradicator, infamous for his quip that 'The only good Indian is a dead Indian' – would later concede, 'We took away their country and their means of support, broke up their mode of living, their habits of life, introduced disease and decay among them, and it was for this and against this they made war. Could anyone expect less?'[40]

But General Amherst was in no mood for fair dealings with the Indians, 'their Extirpation being the only Security for our future safety'. He regarded them as 'the Vilest Race of Beings that ever Infested the Earth & whose Riddance from it must be Esteemed a Meritorious Act for the Good of Mankind', and he ordered his troops to 'Destroy their Huts and Plantations, putting to Death everyone of that Nation that may fall into your Hands.'[41] Amherst relished the power he felt 'to punish the delinquents with entire destruction, which I am firmly resolved on whenever any of them give me cause';[42] and he suggested to Colonel Bouquet at Fort Pitt that they could infect their enemies with smallpox. Bouquet expressed enthusiasm for the idea, and Amherst later added, 'You will do well to try to inoculate the Indians by means of blankets, as well as to try every other method, that can serve to extirpate this execrable race.'[43] In fact, a trader at the fort recorded in his diary that he had already given some Delawares 'two Blankets and a Handkerchief out of the Small Pox Hospital. I hope it will have the desired effect.'[44] Bouquet further suggested to Amherst, 'As it is a pity to expose good men against them I wish we would make use of the Spanish Method to hunt them with English Dogs . . . who would I think effectually extirpate or remove that Vermin.'[45]

The globally fought Seven Years War had pushed Britain heavily into debt, forcing it to adopt the more defensive policy of consolidating its enlarged imperial presence in North America. Now, the eruption of the Pontiac resistance against the colonial violation of the Treaty of Easton hastened Britain's announcement of the Royal Proclamation of 1763, which belatedly reinstated the ban on settlement and speculative land claims beyond the Appalachian and Allegheny Mountains.[46] The Proclamation was intended to forestall costly frontier conflict and to rein in Britain's eastern seaboard colonies that, by westward expansion inland, were beginning to stretch the limits of colonial administrative control. In practice, however, illegal western settlement continued, along with military strikes across the Proclamation Line. Only the following summer General Amherst ordered punitive attacks on Delaware, Shawnee and Mingo villages, extracting from the Indians peace treaties that gave the British control of the chain of forts located along the strategic communications routes around Lake Erie and down the Allegheny River.[47]

White settlers benefited most from the Royal Proclamation: the formal ban on their westward movement was only lightly enforced and, if they were prepared to risk having to expel Indians themselves, they could occupy land for free. This appalled America's now established landed gentry, including George Washington, who had been making their fortunes by using their political connections to acquire first legal title to huge tracts of land for next to nothing. The land companies, organized by this new economic elite, continued nevertheless to speculate in future western territories while their lobbyists, such as Benjamin Franklin, peddled influence in London for land grants beyond the Proclamation Line.[48] In 1768 Virginia, Pennsylvania and their respective land companies, with help from Franklin, managed to engineer a treaty-signing at Fort Stanwix whereby the Iroquois again presumed to sell rights to land in Ohio Country that was actually inhabited by the Delaware, Shawnee and Mingo Indians. The consequent clear danger of Indian resistance more widespread even than that of 1763, as formerly antagonistic tribes set aside their differences to confront a common threat, forced the imperially overstretched British authorities to backtrack by voiding the speculators' land surveys and refusing large land grant requests from the likes of George Washington, Thomas Jefferson and Patrick Henry.[49] It was this wealthy, well-connected, land-speculating class that, in frustration, became only too happy to see its colonies' newspapers and pamphlets fill with rhetoric against the British who, only recently, had fought to expel the French from Ohio Country and then from most of North America.[50] Thomas Paine, after making the acquaintance of Franklin in London, emigrated to Philadelphia where he became these land speculators' famously effective political agitator.[51]

The final straw for this increasingly revolutionary-minded aristocracy came in June 1774 when the Quebec Act received royal assent. Under this Act, all American speculation in land in Ohio Country was to be stopped by giving jurisdiction over the region to Quebec, which was entrusted to preserve the Ohio Indians' hunting grounds for the fur trade. Richard Henry Lee, whose family included founding members of the Ohio Company and who would submit the resolution for American independence, told the First Continental Congress that the Quebec Act was 'the worst grievance' that Britain had committed against America.[52] Almost as soon as the Act had been passed, Pennsylvania and Virginia, whose governing elites were making competing claims over Ohio Country, each sent state militia into the area. Agents of the Ohio Company provoked the Indians into conflict, giving Virginian governor Lord Dunmore (of the Wabash Land Company) and Colonel Andrew Lewis (of the Loyal Land Company) the pretext to wage a campaign of disproportionate revenge; and after the Battle of Point Pleasant (10 October 1774) and further terrorism of Indian villages, the illegal 1768 Fort Stanwix treaty was forcibly reimposed on the Shawnee and Mingo.[53]

IMPERIAL HANDOVER

In Ohio Country the American Revolution resembled a re-enactment of the French and Indian War, the Canadian British taking over the role played previously by the French. The broad military objective of controlling the Mississippi-Great Lakes arc of communications was again hijacked by the big land-speculating interests of the political elite who claimed only to be concerned for the safety of frontier settlers. Again, the Indians were crucial to the balance of military power between the British Loyalists and the American Patriots.[54]

Since British strategy, at least for now, was to control and limit the American colonies' territorial expansion, most tribes, including those of the Iroquois Six Nations, backed the Loyalists. The Patriot cause, ironically, nearly defeated itself precisely because the thirteen states and their respective land companies were for years unable to resolve their mutually conflicting land claims. This held up ratification of the Articles of Confederation, and hence proper coordination of the Americans' revolutionary war effort, until total financial collapse and imminent defeat were staring Congress in the face. It took the intervention of the minister of France to the United States, the Chevalier de la Luzerne, finally to push Congress in 1781 into ratifying the Union as a precondition for further French financial and military support.[55]

Throughout the War of Independence the Indians deep in Ohio Country, supplied and assisted by the British, held their ground against several Patriot military expeditions: 'The federal government seemed quite unable to deal with the red menace,' as one later account would put it.[56] The largest of these expeditions was led by Lieutenant Colonel George Rogers Clark in 1778. Overtly, it was intended to strike at British posts north of the Ohio River and to secure an American supply line to the region from New Orleans via the Mississippi. More covertly, it was aimed at reinforcing Virginia's territorial claims to the area.[57] During the campaign, the governor of Virginia, Thomas Jefferson, suggested to Clark what his policy toward the Shawnee might be, whether 'the end proposed should be their extermination, or their removal beyond the lakes or Illinois river. The same world will scarcely do for them and us.'[58]

The revolutionary war hit Iroquoia and eastern Ohio Country even harder than the previous conflicts, leaving it a wasteland. Here, the success of Loyalist Indian attacks, particularly from Fort Niagara into the Mohawk Valley, provoked General Washington into ordering the more concerted Sullivan, Clinton and Brodhead campaigns of 1779, aimed at throwing the Indians out of the rich agricultural region of central and western New York state and, if possible, breaking the British-held Great Lakes-Allegheny supply route. Washington instructed Major General John Sullivan to terrorize the Iroquois there into submission, be they warrior or civilian, earning Washington, among the Iroquois, the *nom de guerre* of 'Town Destroyer':[59]

> The immediate objects are the total destruction and devastation of their settlements and the capture of as many prisoners of every age and sex as possible ... [P]arties should be detached to lay waste all the settlements around, with instructions to do it in the most effectual manner, that the country may not be merely *overrun*, but *destroyed* ... But you will not by any means, listen to any overture of peace before the total ruin of their settlements is effected ... Our future security will be in their inability to injure us; [the distance to wch. they are driven] and in the terror with which the severity of the chastizement they will receive will inspire them. Peace without this would be fallacious and temporary ... When we have effectually chastized them we may then listen to peace and endeavour to draw further advantages from their fears.[60]

From Fort Pitt on the Ohio Forks of western Pennsylvania, Colonel Daniel Brodhead led 600 troops of the Continental Army on a tour of plunder and destruction against the Seneca and Delaware villages of the Allegheny Valley. Through Buckaloons and Conewango, across the New York state boundary up to Bucktooth (future Salamanca) and back south across French Creek via Venango, the Indians' homes and crops were burnt to ashes. Only a very few Seneca ever returned to live here, the future centre of America's first oil rush.[61] A report in the *New York Gazette* described how Brodhead's troops advanced on a region of Indian towns

> with extensive cornfields on both sides of the river, and deserted by the inhabitants on our approach. Eight towns we set in flames ... The corn,

amounting in the whole to near six hundred acres, was our next object, which in three days we cut down and piled into heaps, without the least interruption from the enemy.

Upon our return, we several times crossed a creek about ten miles above Venango, remarkable for an oily liquid which oozes from the sides and bottom of the channel and the adjacent springs ... After burning the old towns of Conauwago and Mahusquachinkocken, we arrived at Pittsburgh ... with the scalps we had taken, and three thousand dollars' worth of plunder.[62]

In the 1783 Treaty of Paris, formally ending the War of Independence, the British and Americans omitted any mention of their wartime Indian allies. The United States was declared to have gained sovereignty over all territory east of the Mississippi except for Florida which, along with territory west of the Mississippi, went to Spain. The US, now an independent imperial force on the world stage, had won Ohio Country and Iroquoia – on paper at least – at the negotiating table at Versailles. John Dickinson, the chief drafter of the Articles of Confederation, now recommended that unless the Indians give up all of this land peacefully, 'we will instantly turn upon them our armies that have conquered the king of Great Britain ... and extirpate them from the Land where they were born and now live'.[63] George Washington predicted that even without further military action, 'the gradual extension of our Settlements will as certainly cause the Savage as the Wolf to retire; both being beasts of prey tho' they differ in shape'.[64]

'TRADE OF A RISING EMPIRE'

Now that the newly independent American ruling class had opened the western frontier to its land speculations, it inherited the very problem that had ultimately defeated its British predecessors: how to keep the westward-expanding empire under its control. Settlers in the Midwest beyond the Appalachians were exporting agricultural surplus southwards down the Mississippi to Spanish New Orleans; those in the Northwest Territory (later known as the Old Northwest) and around the Great Lakes were trading through British Canada. Might westerners turn away, economically and then politically, from their East Coast masters?[65] The solution seemed to be to tie east and west together into economic interdependency through trade. George Washington was already in 1770 hoping to persuade his fellow 'monied Gentry' to invest in a road and river development, via the Cumberland Gap and Potomac River to his western landholdings, to be 'the Channel of conveyance of the extensive and valuable Trade of a rising Empire'.[66] In 1780, during the recently concluded war, Thomas Jefferson had ordered Colonel Clark to take Fort Detroit on Lake Erie, thus denying Canada that region's trade and supply routes:

> If that Post be reduced ... we shall divert through our own Country a branch of commerce which the European States have thought worthy of the most important struggles and sacrifices, and in the event of peace on terms which have been contemplated by some powers we shall form to the American union a barrier against the dangerous extension of the British Province of Canada and add to the Empire of liberty an extensive and fertile Country thereby converting dangerous Enemies into valuable friends.[67]

In 1784 Washington was arguing that it was

> of great political importance ... to prevent the trade of the Western territory from settling in the hands, either of the Spaniards or British. If either of these happen, there is a line of seperation [sic] at once drawn between the Eastern and Western Country. The consequences of which may be fatal ... [I]t is by the cement of interest only, we can be held together. If then the trade of that Country should flow through the Mississippi or St. Lawrence. If the Inhabitants thereof should form commercial connexions, which lead, we know, to intercourse of other kinds, they would in a few years be as unconnected with us, indeed more so, than we are with South America; and would soon be alienated from us.
>
> It may be asked how we are to prevent this? Happily for us the way is plain, and our *immediate* interests, as well as more remote political advantages, points to it; ... Extend the inland navigation of the Eastern waters, communicate them as near as possible (by excellent Roads) with those which run to the westward. Open these to the Ohio, and such others as extend from the Ohio towards Lake Erie; and we shall

not only draw the produce of the western Settlers, but the Fur and pel-
try trade of the lakes also, to our Ports (being the nearest, and easiest of
transportation) to the amazing encrease of our Exports, while we bind
those people to us by a chain that can never be broken.

. . . [N]ot only the produce of the Ohio and its waters . . . but those
of the lakes also . . . may be brought to the Sea Ports in the United
States by routs shorter, easier and less expensive than they can be carried to Montreal or New Orleans . . .[68]

In the days when water transport was by far the easiest and most
economical way to move goods over long distances, the immense
economic – and often geopolitical – importance of large canal projects was well appreciated. Jefferson, in 1788, sent to Washington
notes he had made, while visiting the south of France, on the construction of the Languedoc Canal, the most ambitious stretch of the
Canal du Midi. Linking the Mediterranean and the Atlantic, this was
designed to facilitate regional and export trade, and to open up an
alternative waterway to the long journey around Spain via the Straits
of Gibraltar.[69]

However, it would not be Washington's Virginia but New York state,
with its geographic advantages and greater political unity, that would
win this economic prize. Throughout the decades of fur trading and
warfare around the Great Lakes and the upper Ohio, Americans had
followed ancient Indian trails that made the most of rivers and lakes in
a way that reduced overland haulage to a minimum. The 'carrying
places' between waterways had been hotly contested because of their
strategic importance, with Forts Le Boeuf, Presque Isle, Niagara,
Schlosser and Stanwix built to protect them.[70] However, in 1768, following the Fort Stanwix Treaty grab for fertile lands in Ohio Country,
the governor of New York state, Sir Henry Moore, alerted the Assembly to the limited access to the region. Comparing the navigation
works required to those of the Canal du Midi, he informed the house,

It has been observed by all who are concerned in the Indian trade, that
the great inconvenience and delay, together with the expense attending
the transport of goods at the carrying places, have considerably diminished the profits of the trader, and called for the aid of the Legislature,
which, if not timely exerted in their behalf, the commerce with the

interior parts of the country may be diverted into such channels as to deprive this colony every advantage which could arise from it ... [I] recommend ... the improvement of our inland navigation, as a matter of the greatest consequence to the province.[71]

Now, post-independence, ambitious plans were afoot to open up these portages and construct a vast system of inland water transport. It was predicted that by linking the Hudson and Mohawk rivers with the sprawling inland waterways of the Great Lakes region, New York City could profit from exporting the agricultural surplus of the huge expanse of the trans-Appalachian and -Allegheny Midwest as well as that from central and western New York State. In 1788 Elkanah Watson, an ambitious land speculator who made it his business to fraternize with the rich and powerful, and who would become a leading promoter for an Erie Canal system, saw the commercial possibilities:

> In contemplating the situation of Fort Stanwix, at the head of batteaux navigation on the Mohawk river, within one mile of Wood Creek, running west, I am led to think this situation will, in time, become the emporium of commerce between Albany and the vast western world above ... By [navigation improvements], the state of New York have it within their power, by a grand stroke of policy, to divert the future trade of Lake Ontario and the great lakes above, from Alexandria and Quebec, to Albany and New-York.[72]

The remaining New York Iroquois stood no chance against the powerful private land companies, whose members held high office in state and federal assemblies that, in turn, were desperate for revenue from land sales due to post-war bankruptcy and to mutiny from their unpaid war veterans. Native Americans living on the projected canal routes or at the locations of intended ports – at Buffalo, in the Erie Triangle around Presque Isle (future Erie) and in Oneida territory – were therefore of particular annoyance to the land companies, whose real estate speculations stood to be much more lucrative with the development of transport infrastructure that would facilitate settlement and trade.[73]

'A SUBDUED PEOPLE'

You are a cunning People without Sincerity, and not to be trusted, for after making Professions of your Regard, and saying every thing favorable to us, you ... tell us that our Country is within the lines of the States. This surprizes us, for we had thought that our Lands were our own ...

(Red Jacket (Sagayewatha), New York Seneca Indian chief, 21 September 1796, Fort Niagara)[74]

Defeated and weakened by war, the Iroquois tribes – even those, like the Oneida, who had fought with the Patriots – were easily pushed aside. At the second Treaty of Fort Stanwix (renamed Schuyler) in 1784, Virginian commissioner Arthur Lee told the Iroquois they were 'a subdued people', later saying of them, 'They are Animals that must be subdued and kept in awe or they will be mischievous, and fear alone will effect this submission.'[75] The post-independence land cession treaties continued the long history of fraudulence as private, state and federal interests competed to secure land titles by finding the most dubious ways to go through the legal formalities of obtaining signatures – from various unrepresentative Indians, and by whatever combination of threat, deception, alcohol and bribery it took. While the Indian commissioners reassured tribes that their land was safe, surveyors were dividing it up for townships, soldiers' land bounties, large speculative land deals and projected canal developments.[76] Even commissioner Timothy Pickering admitted to Washington a few years later, 'Indians have so often been deceived by white people that *White Man* is, among many of them, but another name for *Liar*.'[77] Meanwhile, in 1786 one future president, James Madison, wrote to another, James Monroe, expressing regret at not having bought up more Oneida territory, for 'my private opinion is that the vacant land in that part of America opens the surest field of speculation in any of the U.S.'[78]

Almost inevitably, the accumulation of broken promises, deceptions and territorial incursions by the Americans provoked conflict. As Monroe read his letter from Madison, warfare was breaking out across the Northwest Territory, and an escalation in attacks by white settlers on

Native Americans and attacks by the latter on white traders was making life around Pittsburgh perilous.[79] When, in late 1786, Mohawk leader Joseph Brant (Thayendanegea) emerged to rally a new Indian confederation – the United Indian Nations, or the Northwest Indian Confederacy – against the tide of illegal settlement and disturbance of hunting grounds, the Americans blamed British incitement. In fact, their post-conquest mistreatment of the Indians was in itself enough to alienate even the more compliant moderates like Seneca Chief Cornplanter (Kayéthwahkeh; see Fig.2), foremost among the chiefs to have been strung along at Fort Stanwix.[80] The East Coast elite now moved, in mid-1787, to formalize an effectively pre-existing public-private partnership between Congress and the land companies to project their sovereignty across the Appalachians by facing down this native alliance.[81] They strengthened their presence on the Allegheny by constructing Fort Franklin in place of the old British Fort Venango (see Fig.1), where citizens of nearby Meadville took refuge during a period of military campaigns whose objective was to subdue the Western Confederacy of Indians in the northwest of Ohio Country. In early 1791 Chief Cornplanter travelled with other Seneca chiefs to Philadelphia, new seat of the US government, and protested to George Washington that the Americans at Fort Stanwix had acted 'as if our want of strength had destroyed our rights', and that the treaty had been negotiated 'while you were too angry at us, and, therefore, unreasonable and unjust'.[82]

Meanwhile, as the Seneca of New York became confined to a few small reservations in the west of the state, the Pennsylvania legislature attempted to placate Cornplanter by deeding him personally some consolation tracts of land. These included 600 acres around his wife's home village of Jenuchshadego, or Burnt House – his new residence along with several hundred of his tribe – on the Allegheny River five miles upriver from Kinzua Creek, just south of the boundary with New York state; and 300 acres – with an oil spring – near the mouth of Oil Creek, about a hundred miles downriver from Burnt House and a few miles up from Fort Franklin.[83]

In April 1791 Secretary of State Jefferson wrote to Monroe, 'I hope we shall drub the Indians well this summer & then change our plan from war to bribery.'[84] Such hopes were initially disappointed by resounding defeats in Ohio Country suffered by General Josiah

Harmar and then by General Arthur St Clair.[85] Nevertheless, the demand for provisions from federal troops was a welcome boost for settler merchants, one of whom hoped that reports were true that 'Indians intend[ed] to bring war into our Country this spring with all their force ... I wish they may ... [F]or my part I don't care [if] the war lasts this twenty years or my lifetime [because] while the war lasts we have a ready sale for our beef, flour, bacon and old horses.'[86]

Cornplanter and several other Seneca chiefs continued to be so accommodating towards the white Americans that they remained neutral during this assault on the Ohio Indians and even briefly sided against them. But as it became more apparent just how much land they had been misled into ceding, and as the Seneca people, especially the young men, became increasingly angry at their chiefs' excessive compliance, they began to take a more resistant and assertive stance. In the summer of 1794 a Pennsylvanian surveyor, John Adlum, arrived to stake out land in the vicinity of Burnt House: about a million acres around present-day Meadville, for the Holland Land Company and Judge James Wilson, chief legal architect of the US Constitution. Although Adlum was given a formally polite reception by Cornplanter and other Seneca leaders, they impressed on him that any further encroachments would provoke them into taking up arms.[87] 'I knew there was not one man present who would shrink from death,' Adlum recorded, but he warned them that to take on the now better trained and supplied US Army under General 'Mad' Anthony Wayne would be to invite their 'certain destruction'.[88]

Indeed, General Wayne was at that very time leading an assault into northwestern Ohio Country – although, the general complained, local 'Merchants ... charge a most exhorbitant advanced price for every Article.'[89] This inflicted the drubbing that Jefferson was hoping for when the Indian coalition, let down by the refusal of any British support, was defeated at the Battle of Fallen Timbers (20 August 1794). General Wayne reported that his troops were 'laying waste the Villages & Corn fields for about Fifty miles on each side of the Miamis [river]', and that the natives' 'future prospects must naturally be gloomy & unpleasant'.[90] The victors' peace terms, imposed upon the Ohio and Northwestern Indians at the Treaty of Greenville (3 August 1795), brought the dreams of land speculators a step closer; and canal promoter Elkanah Watson could now anticipate the realization of his vision of 'those fertile regions, bounded west by the Mississippi, north by the great lakes, east by the Allegheny mountains, and south by the placid Ohio, overspread with millions of freemen . . . commanding the boldest inland navigation on this globe . . .'[91]

After the Ohio Indians' defeat at Fallen Timbers, the New York and Pennsylvania Iroquois were left almost entirely defenceless against the powerful land companies. The Holland Land Co., the Pennsylvania Population Co. and the Pulteney Associates, among others – backed by large sums of speculative capital from Europe and from many of the northeast's leading political figures such as Aaron Burr, James Wilson and Thomas Mifflin – now swallowed up most of the defeated and vulnerable tribes' remaining territory, under the Canandaigua and Big Tree treaties of 1794 and 1797 respectively. The Seneca's tribal land was now reduced to ten small reservations. Two of these were just northeast of Cornplanter's Burnt House tract at the headwaters of the Allegheny River across the New York state border: Allegany reservation and the Oil Spring reservation near the town of Cuba. Meanwhile, in 1795 the Pennsylvania legislature began to formally establish towns in the northwest of the state such as Warren and Franklin, downriver of Burnt House, and towns including Waterford and Presque Isle (later Erie) in the Erie Triangle that now connected Pennsylvania to the Great Lakes.[92] The following year the most prominent of the New York Seneca chiefs, Red Jacket, expressed his distrust of the land companies to the US commander of Fort Niagara during a conference meeting. He singled out the particularly voracious and unscrupulous Robert Morris of the Holland Land Co., well known as one of the main financiers of the Patriot side in the American Revolution: 'We are much disturbed in our dreams about *the great Eater with a big Belly* endeavoring to devour our Lands – We are afraid of him, believe him to be a Conjurer, and that he will be too cunning and hard for us, therefore request Congress will not license nor suffer him to purchase our Lands.'[93]

The year after signing the Big Tree treaty, Cornplanter lamented (as rendered in biblical language by a local Quaker missionary),

[Europeans] have taken away our Lands, so that we have but very little left of all that our forefathers possessed, and be it moreover known unto you, that in days which are over and gone, a certain Man Named Robert whose Sirname is Morris purchased our Countrey, and we fear he will not deal honestly with us, but if he deal honestly with us, we have yet sufficient inheritance remaining therein ... [O]ur Forefathers once lived in ease and plenty, but the White People have not dealt honestly with us, they have taken away our Lands, and drove us into the Wilderness, and now we have but little Land left ... You know Brethren there are some bad people among us, and you know we have been deprived of the lot of our inheritance and that makes us bad and our minds uneasy ... [I]f the White inhabitants of the land had dealt honestly with us, we should have been a very rich people, and had every thing in plenty, and then our minds would be easy & we could think upon the great Spirit.[94]

There were still very few white settlers in the Upper Allegheny region, greatly outnumbered by the remaining Seneca Indians, including the 400 or so residing at Burnt House. At the heart of the village stood a large carved wooden figure of the Iroquois deity Táöyawá'göh, or Sky Holder, the focal point of religious festivals and ceremonies for this matrilineal society in which women had high status; and it was from this village that a half-brother of Cornplanter, Handsome Lake (Kanyotaiyo'), would soon emerge as the famous Iroquois revivalist prophet.[95]

Cornplanter installed a water-powered sawmill near the village to sell lumber to the Holland Land Co. and to the US Army at Fort Franklin where, in 1797, the storekeeper William Wilson recorded '3 Kegs Senica Oil 50 Dllrs' in his account book.[96] Ironically, given Cornplanter's pioneering of mechanically powered lumbering in the area, the slowly increasing numbers of white settlers would clear the forest to open up farmland, often by the slash-and-burn method, and to engage in larger-scale lumbering, thereby progressively destroying the Seneca's traditional hunting and gathering ecosystem.[97]

Although many, such as David A. Ogden of the Ogden Land Co., advocated the outright removal of the remaining indigenous population, the confinement of the Iroquois tribes to ever smaller reservations effectively cleared the way for the Western Inland Navigation Co. – the precursor to the state-financed Erie Canal project – to open up New York state's inland waterways.[98] The construction of transport infrastructure westward – greatly scaling up the Native Americans' traditional trails and river channels – encouraged yet further land acquisitions in the Northwest and Midwest. In 1803, while Governor William Henry Harrison of Indiana Territory was provoking native anger, yet again, by his unscrupulous land appropriations, President Jefferson was persuading himself, Congress, and the world at large, that the United States had the Indians' best interests at heart: 'I trust and believe we are acting for their greatest good.'[99] A few weeks later Jefferson wrote more candidly to Harrison in a letter that he instructed should be kept secret: the President proposed setting up trading posts to run natives into debt and force them to sell off their land in repayment, with the aim of acquiring all territory east of the Mississippi.[100] Years of such cynical treatment of the Native Americans by the US authorities was complemented by the fact, as Harrison reported, that 'a great many of the Inhabitants of the Fronteers consider the murdering of the Indians in the highest degree meritorious'.[101]

There emerged again, all too predictably, a broad-based Indian resistance, headed this time by the commanding figure of Shawnee Chief Tecumseh and animated by another wave of native religious revival, led by 'The Prophet', that rejected the whites and their ways.[102] Although this succeeded briefly in halting further fraudulent land purchases and settlement, it gave Harrison the excuse to launch a military offensive against the Northwestern Indians. Meanwhile, Harrison's fellow western 'War Hawks' blamed the British for being behind the native resistance, thereby tipping the political balance in Congress in their favour and leading President Madison to support the extension of Harrison's local campaign into a fully fledged expansionist war into both British Canada and Spanish Florida.

The War of 1812 (1812–15) was, like most wars, fought for control over economic resources – in this case for agricultural land – and trade routes: around the Gulf of Mexico for southerners and around the Great Lakes and the Atlantic for northerners.[103] The Americans took advantage of the control they already held over the Allegheny River, French Creek, the Waterford portage and Erie by using this strategic transit route to move construction materials, arms and supplies to Presque Isle Bay on Lake Erie. This enabled naval Lieutenant Oliver Perry to construct a fleet strong enough to wrest control of

Lake Erie from the British in September 1813 – a decisive turning point in the war.

Strategic military and economic interests being almost identical, the northerners' prosecution of their disastrous attempt to invade Canada followed precisely the contours of the land speculation and trading interests of New York's power elite. The stratagems employed by 'War Hawk' Congressman Peter B. Porter, for example – as an original member of the New York State Board of Canal Commissioners from 1810 and as chairman of the House Select Committee on Foreign Relations from 1811 to 1812 – flowed directly from his land and trading concerns at Black Rock on Lake Ontario and his consequent preference for New York's proposed canal to terminate there rather than at Buffalo on Lake Erie. The latter route, however, was favoured by the overwhelming majority of New York's big business interests, such as the Holland Land Co., who wanted to see trade from central and western New York State – and now from the fully conquered Northwest and Midwest – funnelled through New York City.[104]

Whatever the Native Americans may have wanted, wrote land speculator David Ogden, they should no longer 'be considered as independent nations. Our view of their interests, and not their own, ought to govern them.'[105] President Monroe, in his First Annual Address of 1817, celebrated the rapid settlement of formerly disputed frontier lands. The Native American, he exclaimed:

> ... yields to the more dense and compact form and greater force of civilized population; and of right it ought to yield, for the earth was given to mankind to support the greatest number of which it is capable, and no tribe or people have a right to withhold from the wants of others more than is necessary for their own support and comfort.[106]

SENECA CHIEF

New York's landowning elite would see the value of their holdings rise as internal improvements – epitomized by the Erie Canal system – encouraged more immigration, increased agricultural and other economic activity, and boosted regional and foreign trade. In 1818 the New York

Society for the Promotion of Internal Improvements argued that an Erie Canal system would, as Washington and Jefferson had urged, bind the nation together economically and politically, thereby completing the foundations for a great new empire:

> [Internal improvements] cement together a widespread community, not only by the strong ties of interest, but also by every social tie ... Our mountains must be politically annihilated. Our sectional barriers must be swept away by a moral arm, whose power is resistless ... Nothing but this, can perpetuate that union which is to guarantee our future national greatness.[107]

The proposed waterway development would drastically reduce the 'monstrous expense of land transportation' between Buffalo, at the eastern end of Lake Erie, and Albany on the Hudson River. This would connect with New York on the Atlantic coast, 'the greatest chain of internal seas upon the face of the globe, diversified by interior Lakes and tributary systems', harnessing the trade of a huge region from New York to the western frontier: '[It] will enable us to grasp at the whole trade of the Lakes.'[108] America, the canal proposers argued, should take this opportunity to strike an economic blow against its deadly geopolitical rival. As a cheaper, more convenient alternative to the Canadian route, a Buffalo-New York export outlet would bring about:

> ... the diversion of a growing and important trade from Great Britain to ourselves ... [Otherwise,] a branch of the most profitable trade, flows to a British market, and enriches our enemy, the arrogant usurper, that would domineer over the whole world ... It has frequently been avowed in the British Ministerial Gazettes and Journals, since the late war, that the North American Colonies should be fostered and protected, as a check upon the alarming commercial greatness of the United States.[109]

In any future war with Britain – precipitated precisely, perhaps, by this fierce economic competition – the envisioned canal development would enhance America's security. Firstly, the transport improvements, by attracting more immigrants to the frontier regions, would drive back resistant Native Americans:

[W]e have nothing to expect from these savage tribes but continual depredation, while hanging on our borders, with feelings of jealousy and revenge, and ready to rush on murder and devastation, at the beck of England or Spain. Physical force on the frontiers, by means of settlements ... must correct these evils, and remove these dangers from us.[110]

Secondly, whereas the lack of transport infrastructure to Lake Erie had presented a major strategic obstacle during the recent war against the British in Canada, now the construction of an Erie Canal system would greatly facilitate

the transportation of every thing of a naval and military nature, connected with the defence of our extensive frontiers ... [s]hould the day ever come, when the interest and policy of the United States and Great Britain should clash again; or should that great crisis ever arrive, when the empire of the maritime world shall be decided by the two nations, and the Lakes again become the splendid theatre of naval warfare.[111]

Ultimately, construction of this east-west transport route, by harnessing the economic potential of such a large landmass, would create the conditions by which America could grow to become a world power:

The import and export trade of New-York, through this great channel, will hereafter astonish the nation and the world ... [A] system of internal communication, that will rival those of England and Holland ... [will turn New York into] the emporium of a country at the west, of itself sufficient to form an empire ... [New York will become] the London of America – a monument of magnificence, worthy to attract and command the wondering admiration of the world ... New York is to become a commercial emporium, second to none on the globe. Nature has willed it; let not man trifle with her mighty designs.[112]

In anticipation of the future, the canal's promoters remarked on the region's mineral resources: 'The country from the Hudson to the Lakes is possessed of the richest mineral treasures.'[113] Supporters of the canal project could look forward to being able to exploit and market salt, gypsum and perhaps coal on a grand scale.[114] And, they speculated, 'what other discoveries may yet be made, time will soon reveal'.[115]

The following year, the Empire State, with the help of European investment and loans, duly took over the funding and construction of the Erie Canal, completing the main route in 1825. For the opening ceremony, New York's dignitaries travelled from Lake Erie to the Atlantic on a flotilla of boats – led, perversely, by one named *Seneca Chief*.[116] Meanwhile, the actual Seneca Chief, Cornplanter, was marooned at Burnt House, a small tract of land on the Allegheny River, where he resided until his eventual death in 1836.[117]

THE $300,000 QUESTION

Now great quantities of agricultural surplus, timber and commodities from the west could be cheaply transported, traded and exported. Increasing trade stimulated further transport development in roads, canals and then railways that, with state and federal support, attracted large sums of European investment through merchant banks such as N.M. Rothschild & Sons, Baring Brothers & Co. and George Peabody & Co.[118] Under President Andrew Jackson's Removal Act of 1830, most of the Native Americans still in the northwest, vulnerable and politically powerless, were expelled by the usual brutal and duplicitous methods.[119] In his second annual message to Congress that year, Jackson hoped that the Indians would

cast off their savage habits and become an interesting, civilized, and Christian community ... Humanity has often wept over the fate of the aborigines of this country, and Philanthropy has been long busily employed in devising means to avert it, but its progress has never for a moment been arrested, and one by one have many powerful tribes disappeared from the earth ... [T]rue philanthropy reconciles the mind to these vicissitudes as it does to the extinction of one generation to make room for another. Philanthropy could not wish to see this continent restored to the condition in which it was found by our forefathers. What good man would prefer a country covered with forests and ranged by a few thousand savages to our extensive Republic, studded with cities, towns, and prosperous farms, embellished with all the improvements which art can devise or industry execute, occupied by more than 12,000,000 happy people, and filled with all the blessings of liberty, civilization, and religion?[120]

By the early 1850s what was now the northeast of the US – from the eastern seaboard to Chicago – was covered by an extensive transportation matrix focused primarily on New York as the gateway for foreign exports and imports. By commanding the international trade from intense economic activity across a vast expanse of territory, New York gained the commercial lead nationally, according it an increasing international prominence.

At the geographical centre of this economically booming region flowed the Allegheny River, and in May 1858 Edwin Drake arrived in Titusville to attempt to extract oil in large, commercial quantities on land acquired by the Seneca Oil Co. near the mouth of one of the Allegheny's tributaries, Oil Creek. As a few small-scale entrepreneurs were demonstrating, there was a demand for the new 'carbon oil' illuminant – distilled from petroleum that often emerged from the region's water and brine wells – and for cheap lubricants for the industrial age. The Allegheny River provided a ready transport outlet from Oil Creek; the area was within reach of the Erie Canal system; and rail branch lines were penetrating deeper into the Allegheny Valley. The only real uncertainty was whether petroleum would be found only in small, discrete pockets or – as the investors in the Seneca Oil Co. hoped – it could be found in large underground reservoirs from which it might be extracted in large quantities from a single well, like water or brine. The company's $300,000 capitalization was riding on the outcome.

How might this new petroleum-based 'carbon oil' fare in competition with the equally new coal-based 'coal oil' that was already competing successfully against whale oil? By 1858 there were about fifty coal-oil manufacturing firms in the United States processing the particularly oil-rich cannel coal from the Ohio Valley: in western Pennsylvania, and in Kentucky and western Virginia where slave workers cut production costs. Cannel coal was even being imported from Nova Scotia and Britain for the new industry, as *Scientific American* paid homage to this new mineral illuminant:

> The most degraded savage stands infinitely above the most intelligent of the brute species by the use of two discoveries, viz., fire and artificial

light. The Esquimaux, in his dreary clime, cheers his ice-tent, during his long wintry night of six months, with light from the blubber of the whale; the Indian, in the dark tangled forest, lights up his wigwam with the blazing pine knot, or fat of the deer; and the civilized white man illumines his houses and cities with a subtle gas made from coal obtained from the bosom of the earth, or with some of the numerous hydro-carbon fluids. Human life cannot be enjoyed without artificial light; if man were deprived of this agent, he would become a brute. In proportion as the means of obtaining artificial light are improved, and rendered accessible to the multitude, so, in proportion, is the mass benefited and elevated in a social capacity . . . [Coal oil] is fast coming into popular favor . . . [and] [v]ast beds of the rich coal from which this oil can be obtained exist in Pennsylvania, Ohio and Kentucky, affording sources of supply for thousands of years to come.[121]

It seemed likely, however, that lamp oil and other products could be distilled far more easily and cheaply on an industrial scale from petroleum – if it could be reliably sourced in large enough quantities. But this would not be achieved with manually dug wells of the kind sunk by Drake and his hired team, from which they were able to separate out about 10 gallons of oil per day from the water. He struggled to find an experienced salt-well borer in the neighbourhood willing to engage in what most locals saw as a crazy attempt to try drilling intentionally for oil – although whether Drake and the Seneca Oil Co. knew it or not, several other entrepreneurs were now making similar attempts to drill for oil in western Virginia southeast of Parkersburg, where the region's brine-well drillers had often also encountered oil. In the spring of 1859, after nearly a year of failure, the company gave up on Drake, but he persisted nevertheless. He borrowed money himself to secure the services of William A. 'Uncle Billy' Smith, who had worked with Samuel Kier drilling salt wells in Tarentum, Pennsylvania, and in August 1859, after a month of initial groundwork, they fired up their steam-powered drilling rig (see Fig. 3). On 27 August they were about to finish work when, at a depth of sixty-nine feet, the drill abruptly sank by six inches. When Smith returned to the well the following day he saw that it was now full of oil – which he was able to draw up in larger quantities than anyone in the area had seen before.[122]

3

Oildom

BARRELS OF OIL

Having spent the past winter at Oil Creek, (the Eldorado of Oildom) I
may claim [that] California, Washoe, or Pike's Peak, never witnessed a
greater or a more motley crowd. Oil . . . oil barrels, oil crude, oil refined,
price per gal., oil pumps, high water, low water, etc., are the topics of
conversation in place of the Secession question, which, in other lati-
tudes, is the all absorbing question under discussion.

('Letter from Cleveland', *Rutland Weekly Herald*, 4 April 1861)

Jonathan Watson, of the lumber firm from which the Seneca Oil Co.
had leased land, sped off on horseback as soon as he heard that Edwin
Drake had struck oil. Racing along Oil Creek, Watson proceeded to
snap up cheap land leases before the locals learned how valuable their
holdings might now be. The Seneca Oil Co., it seems, went further by
conspiring with the editor of the local *Venango Spectator* to suppress
news of Drake's oil strike, giving the company more time to sign bargain leases. Nevertheless, the story was eventually reported just over
two weeks later, on 13 September 1859, in a short article on the fifth
page of the *New-York Daily Tribune*:

Last week, at the depth of 71 feet, [Drake] struck a fissure in the rock
through which he was boring when, to the surprise and joy of every
one, he found he had tapped a vein of water and oil, yielding 400 gallons of pure oil to every 24 hours (one day).

The pump now in use throws only five gallons per minute of water
and oil into a large vat, when the oil rises to the top and the water runs
out from the bottom. In a few days they will have a pump of three times

the capacity . . . and then from ten to twelve hundred gallons of oil will be the daily yield.

The springs along the stream, I understand, have been mostly taken up or secured by Brewer & Watson, the parties who formerly owned the one now in operation.

The excitement attendant on the discovery of this vast source of oil was fully equal to what I ever saw in California, when a large lump of gold was accidentally turned out.[1]

This report was reprinted in several regional newspapers over the following weeks, and a horde of speculators and labourers descended on Titusville, which was now being transformed into the first oil boom town as a forest of oil derricks began to rise up on the rapidly deforested tracts of land along Oil Creek.[2]

The initial production of about 400 gallons of crude oil per day by Drake's team, using a manually operated pump, was a huge increase on the rates of production attained by the few others who were already sourcing petroleum from brine wells and oil pits. Predictably, Drake had no immediate outlets for production on this scale and they worked frantically to find containers and build vats to store the oil. Despite a temporary interruption of output when a fire destroyed Drake's entire stored production and operating equipment, in September he began shipping his oil to the two established refiners of the new 'carbon oil' illuminant in Pittsburgh, Samuel Kier and MacKeown & Finley. The following month he contracted for the sale of 1,000 gallons per week and by the following March he was selling around thirty-five barrels per week at the going rate of about $12 per barrel (each containing the standard 42 US, or 35 British, gallons). While Drake and other oil producers toured the cities of the northeast seeking buyers, total output from the now numerous wells along Oil Creek and the Allegheny River rose to hundreds of barrels per day.[3] Even the earliest shallow wells could flow out of control. In April 1860 it was reported of one, 'The ground in the vicinity was completely saturated with oil that had ran so fast as to prevent the operators from saving it.'[4]

Then, from late 1860 to early 1861, just as the Civil War was breaking out, several big-flowing wells were struck.[5] In April 1861, at Buchanan Farm near Oil Creek in Cornplanter Township, Henry

Rouse struck flowing oil at a depth of 300 feet as his drillers punctured a pocket of high-pressure oil and gas, releasing a gush of oil 60 feet into the air. A crowd came to witness the sight, but cries of joy and awe were suddenly cut short by a huge explosion that sent the bystanders and all the derricks and other buildings in the vicinity up in a huge shower of flames, killing nineteen people. According to one newspaper report,

> Scores were thrown flat, and for a distance of twenty feet, and numbers horribly burned, rushing blazing from the hell of misfortune, shrieking and screaming in their anguish. Just within the circle of the flame could be seen four bodies boiling in the seething oil, and one man who had been digging at a ditch to convey the oil to a lower part of the ground was killed as he dug, and could be seen, as he fell over the handle of the spade, roasting in the fierce element. Mr. H.R. Rouse ... a gentleman largely interested in wells in this locality, and whose income from them amounted to $1,000 a day, was standing near the pit and was blown twenty feet by the explosion. He got up and ran about ten or fifteen feet further, and was dragged out by two men, and conveyed to a shanty some distance from the well. When he arrived not a vestige of clothing was left upon him except his stockings and boots. His hair was burned off as well as his finger nails his ears and his eyelids, while the balls of his eyes were crisped to nothingness. In this condition he lived nine hours ... The fire was so intense that no one could stand within 150 feet without scorching their skin or garments ... It seems that the earth is really on fire, and its elements about to melt with fervent heat.[6]

The well continued to gush at a rate of up to 3,000 barrels a day, and it took three days to extinguish the fire. More flowing wells came roaring in over the summer – the Fountain well, Empire well, Phillips No. 2 and Elephant No. 2 – completely overwhelming the existing storage and transport capacity.[7]

In September 1860 the editor of the *Titusville Gazette* reported on his experience of an oil strike:

> We have no language at our command by which to convey to the minds of our readers any adequate idea of the agitated state at the time we saw it. We are told it continues the same. The bore of the pipe which the oil

and gas escape is five inches in diameter. The gas from below, from a depth of 145 feet was forcing up immense quantities of oil in a fearful manner and attended with a noise that was terrifying at times, in pulse like throes, the gas threw up a stream of oil the full size of the pipe to the height of three feet above its top, and continued a boiling, tossing and surging commotion for a minute throwing out, seemingly, a half barrel of oil at a single toss of the gas ... [W]hen the gas subsided for a few seconds, the oil rushed back down the pipe with a hollow, gurgling sound, so much resembling the struggle and suffocating breathings of a dying man, as to make one feel as though the earth were a huge giant seized with the pains of death and in its spasmodic efforts to retain a hold on life was throwing all nature into convulsions. During the upheavings of the gas it seemed as if the very bowels of the earth were being all torn out and her sides must soon collapse. At times the unearthly sounds, and the seemingly struggling efforts of nature, imperceptibly drew one almost to sympathize with earth as though it were animate.[8]

All available barrels were full in no time, and improvised pits and dams extending over many acres were constructed in an attempt to contain the rising lakes of oil. Oil Creek turned into a river of oil and the surrounding fields became a muddy quagmire resembling a battlefield.[9] According to a report of the scene in October 1861, 'So much oil is produced it is impossible to care for it, and thousands of barrels are running into the creek; the surface of the river is covered with oil for miles below Franklin ... Fears are entertained that the supply will soon be exhausted if something is not done to prevent the waste.'[10]

Why were producers pumping out more oil than they could realistically deal with? On the one hand, under the 'rule of capture' oil production was essentially unregulated. The rule is traceable back to Roman law and was seen as a natural extension of French civil law and English case law on water wells: an oil producer became the owner of any oil they could extract from their point of access on the surface. But this meant that there was nothing to stop neighbours from tapping a reservoir situated mainly under one's own land and extracting all of the oil for themselves. Consequently, every producer hurried to pump out as much oil as they could as quickly as possible, usually from multiple wells, fearful that surrounding competitors would extract most of it before them, perhaps even emptying the reserve entirely. Land-owners who leased their land for drilling contributed to the production frenzy by stipulating a minimum output in order to guarantee for themselves an income from royalties, ranging from one-eighth to over a half of production.[11]

On the other hand, as the production of crude oil increased so did the strain on the available storage facilities and on the existing transit routes from the oilfields to refineries. In these early days, oil had to be moved from wellheads in barrels and hauled away by teams of horses. From the wells around Oil Creek and the Allegheny River, much of the oil was initially taken to the nearest navigable waterway for shipment, in barrels or in bulk, 60 miles downriver to Pittsburgh. However, due to the seasonal limitations of the Allegheny River system, and as crude production exceeded the transit capacity in that direction, barrels of oil were also hauled by road either 40 miles to Erie or 20 miles to one of the stations on the Sunbury & Erie (soon renamed Philadelphia & Erie) Railroad. Transhipment by rail then took the crude on to refineries – either newly built, or adapted coal-oil refineries – in the Atlantic seaboard cities, mainly New York, Philadelphia, Boston and Baltimore. It was the initial muddy, cross-country haulage by teamsters, from well-heads to the nearest railway, waterway or town, that proved to be the greatest obstacle to conveying the crude to refineries, and the teamsters could exploit their fortuitous stranglehold on the trade by charging the oil buyers extortionate carriage rates.[12] As production increased, it was recounted, these wagon trains became as long as

> six thousand two-horse teams and wagons. No such transport-service was ever before seen outside of an army on a march . . . Travellers in the oil-regions seldom lost sight of these endless trains of wagons bearing their greasy freight to the nearest railroad or shipping-point. Five to seven barrels – a barrel of oil weighed three-hundred-and-sixty pounds – taxed the strength of the stoutest teams. The mud was practically bottomless. Horses sank to their breasts and wagons far above their axles. Oil dripping from innumerable barrels mixed with the dirt to keep the mass a perpetual paste, which destroyed the capillary glands and the hair of the animals. Many horses and mules had not a hair below their eyes. A long caravan of these hairless beasts gave a spectral aspect to the

landscape ... Many a horse fell into the batter and was left to smother ...
Teamsters would pull down fences and drive through fields whenever
possible, until the valley of Oil Creek was an unfathomable quagmire.[13]

The transportation bottleneck around the oilfields and the high cost
of teamster haulage therefore dissuaded many lamp-oil manufacturers from switching from coal to crude oil as their raw material, or
from building new oil refineries. In any case, many doubted whether
such a flow of crude oil could last, given the unprecedented rates of
production and the high proportion of dry wells. Although some
entrepreneurs were quick to turn to refining petroleum, in 1860 the
manufacture of coal-oil illuminant continued to increase.[14] Nevertheless, a coal-oil manufacturer visiting Titusville voiced in alarm, 'If this
business succeeds, mine is ruined.'[15] As the Pennsylvania Geological
Survey would later explain, 'our oil ... contains a greater proportion
of the lighter products than any surface oil found elsewhere. Nature
has distilled it for us from the shales, and we take directly from the
well an equivalent to the first product of the coal oil still, but containing also a much greater proportion of those members which possess
illuminating properties.'[16]

Indeed, by early 1862, *Scientific American* was declaring the fleeting coal-oil era to be over:

> If El Dorado was a myth of the olden time, *Oil* Dorado is a shining
> reality of the present ... It will readily be appreciated how the *coal-oil*
> business has been extinguished by the petroleum oil wells, as about
> fifty gallons of crude oil was obtained from a tun of good cannel coal,
> costing from two dollars per tun at the mines to twelve and sixteen
> dollars in New York and other Eastern cities, whereas one well now
> delivers daily 20,000 gallons, equal to the product of 400 tuns of coal,
> and all this without the expense for coal or first distillation. The many
> coal works which were fitted up at great expense in various places,
> have been converted into petroleum refineries – the only way to save
> them from extinction.[17]

The booming oil patch at the mouth of Oil Creek encompassing the
original Drake well on Hibbard Farm and the Cornplanter tract –
once owned by the famous Seneca chief who had lived the latter part

of his life further up the Allegheny River – was renamed Oil City. Conversely, the name 'Cornplanter' was given to a well on Clapp Farm, part of which was bought up by the Cornplanter Oil Co.[18] A reporter for the London *Morning Post* described the proliferation of oil wells in the town's vicinity:

> Derricks peered up behind the houses of Oil City like dismounted steeples, and oil was pumping in the backyards. Every foot of land on the creek is considered 'good boreable territory', and one reason alleged by the inhabitants for not improving the town is the fact that some day the houses will be torn down and the streets bored in search for oil . . . [A] dead man was buried in a convenient lot. Unfortunately that lot was sold in a few days as oil territory, and the body was removed to another place. The second place of burial was also sold as the site for an oil well, and the body was at length shipped to Rochester, New York, to prevent its being bored through in the search for oil . . . From Titusville, the present head of the actual oil-producing portion of [Oil Creek], to Oil City, is about 20 miles, and along the whole of this distance the ground is punched full of holes, and on most of it the derricks stand as thick as trees in a forest. Steep bluffs, ranging from 300 to 600 feet high, bound the narrow valley on either side . . . Derricks throng the low marshy bottom land, derricks congregate on the sloping banks, derricks even climb the precipitous face of the cliffs, establishing a foothold wherever a ledge of rock projects or a recess exists . . . The air reeked with the scent of petroleum and gas, the mud under foot was greasy and slippery, the standing pools had the appearance of pure oil, and even the water of the Creek was hidden beneath a mask of gorgeous hues formed by the waste oil floating on its surface. Not a bright green thing was to be seen in the valley; everything was black, sooty and oily.[19]

A report in the *Venango Spectator* told of how in Oil City, 'everything that addresses itself to the eye or nose is full of grease. The plank pavements are saturated with oil.'[20] Catastrophic fires were a constant danger, the *New York Times* impressing on its readers that 'Persons who have not spent some time in the oil regions can scarcely form an idea of the frightful effects of such conflagrations in localities where producing wells are so numerous.'[21]

The crude market was flooded as oil wells continued to flow, often uncontrollably. One visitor wrote, 'The owners were bewildered. It was truly "too much of a good thing". The real value of petroleum had not yet been discovered, and the market for it was limited.'[22] The escalating production sent the price of oil at the wellheads plummeting to 10 cents per barrel by the end of 1861. There were, as yet, only relatively few oil refineries, and oil buyers were stymied by an acute shortage of barrels and by the limited transport infrastructure, while the teamsters continued to charge $3–$4 freight per barrel. Some landowners and producers responded by setting up the Oil Creek Association, a collective attempt to control production according to the level of demand and to set a minimum oil price. However, the oil price rose of its own accord when production declined as the financially weaker producers went out of business, and when demand for crude and its refined products began to increase.[23]

WARTIME DEMAND AND DESTRUCTION

The abundance of cheap crude catalysed the construction of more refineries. Despite the initial difficulties of getting machinery and materials into the heart of the Oil Region, many refineries were built close to the source of crude supply to minimize the financial impact of the transportation bottleneck from the wellheads. Several, such as the Abbott and Humboldt refineries, were built in Titusville. Twenty miles to the north, Samuel Downer – a Boston coal-oil manufacturer who had been quick to take up refining crude oil – built a strategically placed refinery on the Philadelphia & Erie Railroad at Corry in Pennsylvania, to where the Atlantic & Great Western Railroad had just extended its line 60 miles east from Salamanca in New York state.[24]

The directors of the Atlantic & Great Western – and their British financial backers Peto & Betts and the Bank of England – had seen a lucrative opportunity in this new Pennsylvanian industry. In January 1861 the railway's president sent samples of oil to his chief contractor in London, James McHenry, and to Britain's Board of Trade. McHenry wrote in June, 'all depends on the cost of oil placed in New York or London . . . If the oil comes cheap the trade will be fabulous.'[25] In a report to its shareholders in March, the railway's chief engineer, Thomas Kennard, added that 'the new discovery of oil must not be overlooked, if we could only believe what we see, and only a tenth part of what we hear, this traffic alone would pay the whole of [the] cost of our road in three years . . . [I]t is very possible this traffic may realize a fabulous amount.'[26]

Indeed, the prospects for both national and international trade looked more commercially promising by the day. The Oil Region received an economic boost from the Civil War by supplying lamp oil to the Union army, which could no longer source either the camphene derived from North Carolina turpentine or the spermaceti from the now largely sunk or commandeered whaling fleet. At the same time, the Union government sought to boost the production and export of this new commodity to fund its war against the Confederate South, which now retained all the revenue from cotton exports. American consuls across Europe promoted their country's oil, A.W. Crawford – himself a native of the Oil Region – reporting from Antwerp in October 1861 that 'some of the numerous oil springs of the Allegheny and Kanawha Valleys might profitably let their lights shine in this direction.'[27]

A smaller oil boom in Canada, north of Lake Erie in Enniskillen township, Lambton County, Ontario, followed a similar pattern to that around Titusville. Oil strikes at Oil Springs and Petrolia triggered a frenzy of land speculation; large oil spills polluted the creeks and fires were frequent; and the difficulty of transporting crude ten miles along the muddy track to the village of Wyoming on the route of the Great Western Railway curtailed the industry's expansion until a consortium of investors constructed a plank road, completed in early 1863. The head of the petroleum section of the Geological Survey of Canada wrote that one well was

> flowing over into the adjacent creek . . . Meanwhile there is no market for the oil, and many thousands of barrels are stowed up in tanks and pits awaiting purchasers . . . The results of the last ten days in this region have surpassed the dreams of the most sanguine as to the supply of oil, and judging from present appearances, the wells of Enniskillen will rival those of Burmah and Persia, which have for centuries supplied the East with petroleum.[28]

A group led by oil producer John Henry Fairbank built a railway spur from Petrolia to Wyoming that would open in late 1866; but Enniskillen's crude output proved less prolific and less reliable than that of the Allegheny Valley region, and the type of crude left an unpleasant odour in the refined lamp oil, limiting its marketability.[29]

In November 1861 the brig *Elizabeth Watts* arrived in Britain with the first large shipment of oil from America – 3,000 barrels of crude – and in December a Liverpool broker's report in the London *Times* eulogized the new commodity:

> If the rocks and wells of Pennsylvania, Canada, and other districts continue their exudation at the present rate of supply, the value of the trade in this oil may even approach that of American cotton ... The oil gas distilled from the raw petroleum is immensely superior, and much more brilliant than our own coal gas. For years we have sent coals to America for her gasworks, and it will be a singular freak of events if she and Canada should now supply us with a better expedient.

Extolling the many useful products, in addition to lamp oil, that could be refined from crude oil – such as lubricants, candle wax, solvents, dyes and paints – the broker concluded, 'American hostilities and the ice in the St. Lawrence (although we still have St. John's, New Brunswick) may stop supplies to some extent, but that the future will vindicate the best expectation requires little prescience to affirm.'[30]

Britain's coal-oil manufacturers, in response, denounced liquid petroleum as unsafe to handle and they lobbied for an import duty on this new competitor to match the equivalent tariff the United States had placed on imported kerosene. Peto & Betts, by contrast, envisioned financing a petroleum refining industry in Britain in order to boost demand for the crude that their railway was transporting out of the Oil Region, and in 1863 they had three transatlantic, sail-driven oil tankers constructed on the Tyne: the *Ramsey*, the *Atlantic* and the *Great Western*. Their ambitions were set back, however, by a sudden boom in the refining of Scottish shale oil that caused a serious drop in demand for imported US crude.[31]

Whereas the oilfields of western Pennsylvania were spared the fighting, the Civil War temporarily killed off similar emergent oil industries to the south, around Burning Springs in western Virginia and in Kentucky.[32] In April 1863 General Robert E. Lee sent Confederate cavalry forces from Virginia to sabotage a section of the Baltimore & Ohio Railroad and to steal supplies, during which one detachment destroyed the oil installations at Burning Springs – the first time in history that industrialized oil production was impacted directly by military conflict. Brigadier General William E. 'Grumble' Jones reported back to Lee,

> we moved on to Oiltown [Burning Springs], where we arrived on May 9. The wells are owned mainly by Southern men, now driven from their homes, and their property is appropriated either by the Federal Government or Northern men. This oil is used extensively as a lubricator of machinery and for illumination. All the oil, the tanks, barrels, engines for pumping, engine-houses, and wagons – in a word, everything used for raising, holding, or sending it off was burned. The smoke is very dense and jet black. The boats, filled with oil in bulk, burst with a report almost equaling artillery, and spread the burning fluid over the river. Before night huge columns of ebon smoke marked the meanderings of the stream as far as the eye could reach. By dark the oil from the tanks on the burning creek had reached the river, and the whole stream became a sheet of fire. A burning river, carrying destruction to our merciless enemy, was a scene of magnificence that might well carry joy to every patriotic heart. Men of experience estimated the oil destroyed at 150,000 barrels. It will be many months before a large supply can be had from this source, as it can only be boated down the Little Kanawha when the waters are high.[33]

The Atlantic & Great Western Railroad, the main transport route out of Pennsylvania's Oil Region, retained the backing of its British financiers, allies of the Union.[34] The relatively good northern transport routes remained open, and rail links were extended further into the Region. While the Atlantic & Great Western – with foreign investment fast approaching $50 million – constructed a branch line from Corry 40 miles down to Meadville and then a further 25 miles on to Franklin, the locally financed Oil Creek Railroad linked Titusville with Corry, also 25 miles apart.[35] The once quiet Titusville, though safe from Confederate troops, was suddenly invaded by the modern industrialized world, as described by a visiting reporter:

The habits of the people of Titusville and vicinity had become so settled, and they had so long lived an isolated life, that even after the great
discovery in the oil region they could not conceive that a railroad could
be built to tap that wonderful district, and to open it up to the commerce of the world; and they were accordingly startled from their
propriety when, on the 1st of October, 1862, a locomotive dashed into
their midst, dispelled their prejudices, awakened their energies, and
taught them that they were the citizens of a progressive world.[36]

Oil exports – aided by a tax exemption and a favourable exchange
rate – helped to maintain the Union's balance of trade during the Civil
War, and the increasingly mass-produced and affordable American
kerosene began to displace candles, rapeseed oil, olive oil, linseed oil,
coal oil, lard oil and whale oil as the world's staple artificial illuminant.[37]
In the summer of 1860 the editors of *Living Age* magazine predicted 'a
good time coming for whales', and the following year a cartoon in *Van-
ity Fair* showed sperm whales throwing a ball to celebrate the birth of
the petroleum industry – although, in fact, the arrival of steam-powered,
and later diesel, whaling ships led to a huge increase in whaling.[38] Several years later, according to the London *Morning Post*:

An old whaling captain living at Oil City – and, by the way, the number
of whale catchers and former dealers in whale oil now engaged in the
petroleum business is somewhat remarkable – explains the deposit of
oil by the hypothesis that a large shoal of whales were stranded in
Western Pennsylvania at the time of the subsidence of the flood.[39]

In the autumn of 1862 the *Venango Spectator* described the economics of supply and demand: 'The great depression in the market prices
of crude Petroleum during the past year, while it has almost ruined all
the operators of limited means ... has also been the means of introducing the product to all parts of the world and made it as much of a
necessity as any single article of human want.'[40] By 1863 foreign
demand for American crude equalled domestic demand, and exports
of refined oil overtook domestic consumption. That year the US consul in St Petersburg reported on the popularity in Russia of American
lamps and oil: 'The people are becoming accustomed to it and they
will not do without it in the future. It is, therefore, safe to calculate

upon a large annual increase of the demand from the United States for several years to come.'[41]

From Hamburg the US consul wrote in 1865,

> The importation and consumption of petroleum are rapidly increasing. Consumed in lamps of American manufacture, or invention at least, this our last and noblest gift is largely contributing to the enlightenment of the German public. In a country where tallow, wax, common oil, and other substances, used for dispelling the darkness of the long winter evenings and the gloom of the short winter days of this climate, are so very dear, it is hard to imagine how the common people got along before petroleum was discovered.[42]

From France, which was largely importing crude from the United States to be refined domestically, the US consul at Le Havre wrote, 'petroleum oil is now the principal article of importation from the United States, and, from the steadily increasing demand . . . it bids fair in a few years to vie in importance with the great southern staple, cotton.'[43] Indeed, on the profits to be made from refining, the US consul in Venice argued,

> I can see no reason why petroleum, which is coming into general use in southern Europe, might not be shipped direct to these ports instead of, as at present, passing through the refineries of Great Britain, paying to the English refiner a profit, and to the British shipper a toll for re-shipment, when it might be refined at the place of production at a great advantage over the foreign refiner, and then shipped direct by a much shorter route direct to the Italian consumer.[44]

The US consul at Beirut reported that lamp oil was being 'received in considerable quantities by British steamers via Liverpool . . . Beirut and Damascus use this oil in preference to the olive oil of Syria, because of its superior quality and cheapness; and its use is becoming more general throughout the country.'[45]

In January 1865 the *Oil City Register* offered this eulogy to the booming oil industry of the Allegheny Valley:

> It adds, moreover, to the national wealth, gives employment to railroad, mechanical, and shipping interests, and takes the place of gold in the

purchase of European goods and the settlement of foreign claims. As an illuminator and lubricator it is but little more than half the price of sperm and whale oils, and its use is daily being extended in every department of mechanics. It is used largely in the manufacture of soaps and all the finer toilet articles, while some of the most beautiful shades of color ever known are being, by chemical process, extracted from it. And besides all this, the residium is now being brought into use as a steam generator, to navigate the ocean, with most flattering promise of success.[46]

As Thomas Gale, a resident of Oil Creek, wrote excitedly in one of the first books to be published on the Pennsylvanian oil rush,

One gallon of it will raise more steam than many times its weight and volume of coal or coke ... And if the ingenious gentleman in Pittsburgh, or any other can get up a first rate apparatus for generating steam from this oil, with the saving of fuel and tonnage claimed, his patent will be worth more than the richest well in Oildom.[47]

By the end of the Civil War in the Spring of 1865, oil ranked sixth among US exports, amounting to nearly 750,000 barrels, 80 per cent of which was refined.[48] Congressman and future president James A. Garfield caught the oil-speculation fever, investing in land in Big Sandy Valley, Kentucky, thought to be rich in oil. In February he replied to his colleague who had proposed the investment to him,

I have conversed on the general question of oil with a number of members who are in the business, for you must know the fever has assailed Congress in no mild form ... [T]he territory is as yet unoccupied and you and I have knowledge of it, which will in the first place be valuable in selecting locations and making negotiations and in the second place, should we conclude to form a company, the fact that we had been in military occupancy of the country would give us a power over capitalists that we could hardly expect others to have ... Oil, not cotton is King now, in the world of commerce, and it is a beautiful thought that oil is found only in the free states and in the mountains of slave states where freedom loves to dwell.[49]

As the Titusville *Morning Herald* saw it,

The revenue derived from petroleum during the years '64 and '65 was greater than that yielded by both iron and coal during the same period. Constituting a medium of exchange with foreign countries, in the opinion of importers, petroleum alone enabled this country to successfully carry on and terminate our civil war.[50]

And Sir Samuel Morton Peto similarly effused,

At a moment of civil war, when the balance of trade was against the nation, when gold was necessarily going out, and there was a heavy drain upon the natural resources of the country, petroleum sprung up, from lands previously valueless, in quantities sufficient to make a sensible diversion in the national commerce ... It is difficult to find a parallel to such a blessing bestowed upon a nation in the hour of her direst necessity.[51]

THE CLEVELAND NEXUS

The Lake Erie port of Cleveland, Ohio, had by the early 1850s become a central transportation link between east and west. The completion of the Ohio Canal in 1833 had connected Cleveland with the Ohio River, and in 1851 the Cleveland, Columbus & Cincinnati Railroad came into operation. When civil war cut off the North from the Mississippi trade route to the South, trade from the Midwest was diverted along the eastward river, canal and rail routes. Cleveland thus became one of the North's main strategic transportation hubs, crucial to the Union's economic and military strength.[52]

One of many businesses to profit from Cleveland's commercial boom was the trading company Clark & Rockefeller. John D. Rockefeller, from his early experience as a bookkeeper, had become adept at calculating meticulously the relative costs of different lake, canal and rail routes for shipments to and from Cleveland, aided by the revolutionary new communication technology, the magnetic telegraph; 'my eyes were opened to the business of transportation,' he later recalled.[53] In early 1859 Rockefeller combined his savings with money borrowed from his father, an unscrupulous travelling snake-oil salesman, to set up a trading business of his own in partnership with Maurice Clark, which they expanded with the help of loans from Rockefeller's banking friends at the local Baptist church. At the outbreak of war, Rockefeller joined many other well-off men of fighting age in paying a substitute to serve in his place in the Union army – from which a heavy demand for provisions further boosted the Clark & Rockefeller business.[54]

Then, in 1863 there arose for Clevelanders the prospect that their city could become a major oil-refining and distribution centre. The imminent arrival of the Atlantic & Great Western Railroad trunk line would bring them into direct rail connection with the Oil Region. Proximity to the oilfields had made Pittsburgh an early refining centre, but its eastward and westward transport routes for the distribution of refined products were limited compared with those of Cleveland. Competition between Cleveland's transport links, in particular the eastward routes – the Erie Canal system and the two rail systems of the Atlantic & Great Western-Erie network and the Lake Shore-New York Central network – looked likely to keep the city's distribution freight rates low. Pittsburgh, by contrast, was connected to the east only by the Pennsylvania Railroad, which had a freight-rate policy that was disadvantageous to the city. In addition, Cleveland's transport links to other actual or potential sources of supply, in Canada or in other parts of the Midwest, may have assuaged recurrent concerns that the oil wells of western Pennsylvania and New York state might soon run dry.[55]

Consequently, Clark & Rockefeller, among other local entrepreneurs, decided to enter the oil-refining business. Along with an experienced refiner, Samuel Andrews, the partners in 1863 organized as Andrews, Clark & Co. and built a refinery, the Excelsior Works, at a carefully selected location adjoining the tracks of the Atlantic & Great Western as well as a river connection to Lake Erie. Reorganizing in February 1865 as Rockefeller & Andrews, from 21 February 1865 they advertised themselves frequently in the Cleveland *Daily Leader*, highlighting their main product, 'Carbon Oil', as the new lamp oil was still generally known. A total of twenty refineries were operating in Cleveland by mid-1863, and in 1865 the city sold nearly 130,000 barrels of refined oil domestically and almost 25,000 barrels abroad. By the end of 1866 Cleveland had fifty refineries, now exporting two-thirds of their output and ranking the city second only to Pittsburgh as a refining centre.[56]

Rockefeller borrowed heavily for investment in the early years and opened a second refinery, the Standard Works, at the end of 1865. The following year he sent his brother William to New York to run their exporting firm Rockefeller & Co. and to negotiate further loans from larger Wall Street banks, which would have been attracted by Rockefeller's ruthless monopolizing aspirations as an almost guaranteed path to high profits. Indeed, this was probably a major reason for Rockefeller's meteoric rise to dominance of the Cleveland oil-refining industry. Certainly, he stood out for his single-minded drive, his astute choice of partners, his attention to detail in planning and operations, and his company's innovation such as refining greater proportions of the crude oil into marketable products and making wooden barrels in-house so as to be immune to periodic barrel shortages. Yet many other refiners were evolving along similar lines. Rockefeller's banking connections gave him an extra edge that enabled him to expand ahead of competitors and to ride out recessions, while taking over those companies that sank.[57]

FROM PITHOLE BY PIPELINE AND TANK CAR

Surging crude production from the flowing wells of 1861 had overwhelmed the existing capacity for barrel-making, transportation, refining and marketing. The consequent rock-bottom oil price that sent many smaller producers out of business then led to a slump in output. As domestic and foreign consumption of kerosene, lubricants and other refined products of petroleum rapidly developed, and as low crude prices made refining an attractive enterprise, demand for crude increased and, in response, crude production began to rise again.[58] Crucially, however, the increasing refinery output across northeast America through the 1860s would have been impossible without a corresponding increase in the capacity for crude oil to be transported out of the Oil Region.

The transit of many thousands of barrels of oil – by teamsters from wellheads, then by rail flatcar from trackside loading points – made for a continual bottleneck, or 'choke point', in the passage of oil from well to lamp. The teamsters could charge oil buyers monopoly haulage rates, barrels were leaky, often in short supply and expensive, and there was a limit to how quickly railway depots could load and unload such a large number of unwieldy containers. In response, this bottle-neck was gradually opened up by the gradual phasing out of the use of wooden oil barrels – that, ultimately, would be retained only in abstraction as a standard unit measure of oil volume. Oil would soon be stored and moved in bulk, in large tanks and through pipelines, eventually rendering the 42-US-gallon barrel obsolete.

Pipelines were not a new invention. The Chinese had used bamboo pipelines in the process of salt production, to convey both brine to evaporating pans and gas to heat them.[59] In North America, iron pipe-lines were already in common use for water and gas systems in urban centres, while in more remote rural areas wooden pipelines, crafted from hollowed-out logs, conveyed water on farms and brine in salt-works. Now, in 1861, one of the first proposals to the Pennsylvania legislature for an oil pipeline charter was for the construction of a 6-mile, 4-inch-diameter wooden pipeline from Tarr farm to Oil City. It was rejected, however, as legislators feared losing the votes of the approximately 4,000 teamsters whose livelihoods would be threat-ened. Several charters for the construction of iron pipelines were subsequently passed, but some teamsters became so fiercely protective of their lucrative business that they sabotaged the first pipelines that were laid. They also held up the extension of the Oil Creek Railroad beyond Titusville, deeper into the Oil Region, as an ominous incur-sion into their transit territory.[60]

Poor-quality piping, leaky joints and inadequate pumping machin-ery plagued the early pipelines, which hardly needed interference from teamsters to put them out of action. Technical improvements, however – such as the introduction of the rotary pump – meant that by late 1864 oil pipelines seemed genuinely feasible. The ambitious oil buyer Samuel Van Syckel later recounted, 'I saw, when I reached Titus-ville, that the most money was to be made in shipping oil.'[61] Then, in January 1865, as the railways of the Oil Region were expanding their transport capacity out of the area, another ground-breaking oilfield was opened up at Pithole Creek, a tributary of the Allegheny River running roughly parallel to Oil Creek about 4 miles to the east. Now on an even greater scale than before, a hectic rush for riches and

frantically competitive drilling under the rule of capture led to a stag-
gering level of production. The immense output from Pithole, and its
remote location, now prompted oil traders to invest in the budding
alternative to the limited-capacity, high-cost teamsters.[62] As the British railway developer Sir Morton Peto saw for himself during a visit
to Pennsylvania's oilfields in late 1865, 'The great difficulty which has
hitherto attended the oil trade has been transport. It was easy enough
to produce oil, but far more difficult to convey it to the consumer.'[63]
From Pithole, there germinated the modern oil pipeline.

A host of entrepreneurs followed the example of Van Syckel, who
organized a $100,000 pipeline venture – combining capital from inves-
tors in New York, Cleveland and the Oil Region with a $30,000 loan
from the First National Bank of Titusville – to lay a 5-mile, 2-inch-
diameter pipeline from Pithole to Miller Farm on the Oil Creek
Railroad. Van Syckel's first pipeline was opened in October 1865, soon
followed by a second. Powered by several steam pumps along their
length and coordinated using a telegraph line, they operated at a total
capacity of 2,000 barrels a day. To protect the pipelines from teamsters'
attempts to plough them up, they were buried two feet underground
for most of the way and guards armed with carbines were deployed to
patrol them.[64] One of Van Syckel's employees later recalled, 'All the
officials of the company, including the writer, were threatened by the
teamsters with transportation to a warmer climate.'[65] The conflict came
to a head the following year when a venture run by Henry Harley built
a pipeline from Bennehoff/Benninghoff Run 2.5 miles to Shaffer on the
Oil Creek Railroad. Teamsters impeded construction, ruptured the
pipeline, and set fire to the oil storage tanks and trackside loading gear.
But after Harley's hired army of Pinkerton detectives successfully infil-
trated and broke up the saboteur groups, it was clear that power and
influence were shifting to the pipeline companies.[66]

Demand for this cheaper, high-volume mode of transit for crude oil
to a local shipping point or refinery was such that those with capital
to invest in pipeline construction often made very high returns. The
pipeline companies were able to set their shipping prices at just below
those of the teamsters, so that producers themselves were financially
little better off.[67] Following the money, numerous pipelines were rapidly extended from the Oil Region's river and rail shipping points to

the oilfields. For a while, teamsters were still employed to carry barrels of oil from individual wellheads to 'dump tanks' at the terminus
of 'gathering' pipelines, but soon 'accommodation' pipelines were
extended from the dump tanks directly to wellheads, thereby eventually eliminating the teamsters entirely. Van Syckel's oil buyer at the
prolific Pithole field, Alfred W. Smiley, is credited with laying the first
accommodation lines in 1866, and with introducing the 'run' ticket.
This duplicate record, for the producer and the pipeline company,
certified the volume of oil disgorged into the pipeline from the producer's wellhead storage tank, and it would become the basis for the
negotiable oil certificates traded on the first oil exchanges.[68]

In turn, at the shippers' end of the pipeline the barrel was superseded
by the construction of immense storage tanks and by the development
of the rail tank car – up to which time the Atlantic & Great Western
and the Oil Creek railroads had trundled away northwards over ten
million barrels of oil. The first innovation beyond the barrel-stacked
flatcar or boxcar appeared in mid-1865 when a firm of Oil Region
crude buyers dispensed with barrels by mounting two forty-five-barrel
capacity wooden tanks on an Atlantic & Great Western-owned flatcar.
The following year, the rail companies began rolling out their own,
purpose-built wooden tank cars for the bulk carriage of oil. With the
simultaneous emergence of pipelines, larger storage tanks and rail tank
cars, the *Titusville Morning Herald* reported that

> in the past year or two there has been a complete revolution in the
> manner of transporting oil, and that instead of piles of barrels and
> many teams employed, one sees miles of pipe lines and great numbers
> of tank cars. The only difficulty experienced with tank cars for the ship-
> ment of oil is the tendency, if the tank be full, of explosions or straining
> of the tank and consequent leakage and loss of oil.[69]

Scientific American informed its readers,

> Through these pipes the oil flows almost literally out of the wells aboard
> of the railroad car. The propelling agent is, of course, the steam pump.
> Thus oil, 600 feet below the surface of the earth, at Oil Creek, reaches
> Jersey City, a distance of over 400 miles, without having been touched
> by the hand of man.[70]

FIRE, FLOOD, BOOM AND BUST

In October 1865 the entire front page of the *New York Times* was devoted to 'Petroleum', exhorting the nation to move on from the Civil War by taking on new challenges:

> Now that we have come out triumphant from our great struggle, and are called upon to concentrate all our energies for the development of our material resources ... in surveying the wide fields of our yet undeveloped wealth, men's eyes rest naturally for a time upon that wonderful region in Western Pennsylvania which has contributed so largely to our revenue during the past four years, and which promises to yield so much more largely in the future.[71]

Many Civil War veterans, seduced into trying their luck in Pennsylvanian oil, must have wondered whether they had simply been redeployed from one war zone to another. Most shrewd, calculating businessmen such as Rockefeller again stayed at a safe distance from the front line, keeping close company with the bankers and manoeuvring for more predictable and larger profits. In the Oil Region itself, the invading multitude transformed the once quiet backwoods into a toxic, perilous wasteland where chaos ruled, and where one's fate, particularly that of the smaller producer, was a wild lottery. According to one report, the three main hotels in Oil City were

> all crowded to overflowing, and many persons are crowded together in one room. The accommodations, says an old Californian, are worse than he ever saw in his country. But it cannot be helped. – People will come, people will buy, people will make money, people will lose it, people will get crazy – and the consequences are, discomfort, ill-temper, bad beds, worse meals, and worser whiskey. The town hardly looks as if it was built to stay.[72]

Drilling for oil, being the least capital-intensive end of the industry and the easiest to enter, enticed a throng of speculators with dreams of making a quick fortune into descending on the Oil Region. An oil strike would trigger a headlong rush of drillers into its vicinity who, under the rule of capture, would devastate the natural resources more thoroughly than a plague of locusts. The desperately competitive and uncontrolled rate of extraction would produce a sudden glut of oil on the market; and as this pushed the oil price down, producers responded simply by extracting more frenetically than ever to make up for declining profits. This frenzied, haphazard drilling irreparably damaged the structures of the subterranean oil reservoirs causing, far sooner than otherwise, an abrupt decline in output. Many producers, having sold thousands of barrels of oil at rock-bottom prices, would be left with dry wells and nothing to sell while they watched oil prices shoot back up again. In late 1861 the Oil Creek Association was formed in an attempt to organize self-imposed production restraints; the *Pittsburgh Gazette* reported the

> strenuous efforts . . . being made by the leading oil dealers on Oil Creek, for the formation of an association by means of which the price and supply of crude could be so regulated as to ensure remunerating rates to the well owners, and at the same time prevent the immense waste from flowing wells. The great bulk of the profits now resulting from the oil trade, go into the coffers of the refiners, who manage to keep up the price of the refined article, while crude oil is now selling on Oil Creek *at less than one half cent per gallon!* The flowing wells which spout forth five, six, seven and eight hundred barrels daily, without the expense of engines, fuel, etc., have had the effect of closing up entirely pumping wells, which, under more favourable circumstances, would pay a handsome profit on the investment. At the present ruinous rates, however, it does not pay to operate pumping wells. The owners of the flowing wells have been striving to undersell one another, until the merely nominal fraction above mentioned has been reached; and, if nothing is done to prevent it, the price will go down still further.[73]

However, the disunity fostered between the many producers by the rule of capture doomed repeated communal attempts to steady production levels, and crude output lurched between boom and bust.[74]

Woodlands were laid waste for the erection of a forest of drilling rigs (see Fig. 4), and the earth was churned to an oily swamp as approximately a third of the crude extracted was lost to spillage and leakage. The combined effect was to greatly increase the risk of floods and mudslides. Oil City, at the mouth of Oil Creek on the Allegheny

River, became especially prone to flooding. In early 1865 heavy rains swept away many of Oil Creek's wooden storage tanks and put the streets of Oil City under four feet of water. Oil barges that had lost their moorings combined with debris to block the town's bridge and the torrent rerouted itself down Main Street taking the wooden buildings with it. In addition to the loss of oil on land, about a further third of the crude was lost during river transit: Oil Creek was constantly heavily polluted due to the number of boats laden with oil – loose in the hull, or in barrels – that collided and broke up on their way downriver, spilling thousands of gallons of crude into the water (see Fig. 6). In December 1862 *Scientific American* reported, 'petroleum to the value of $100,000 was lost . . . [T]wenty boats broke loose, and these swept a large number of others from their moorings, and fifty-six were wrecked. About 10,000 barrels were lost and all the cargoes that were in bulk.'[75]

Six months later, the Allegheny River from Oil City to Franklin became a river of flames after a boat filled with loose oil was engulfed by fire that spread to nearby boats which then broke free. Generally, so much oil floated on Oil Creek that it was even possible to make money by constructing improvised dams downriver, gathering the oil from the surface and selling it in barrels, while 'oil dippers', many of them children, would move along the riverbank with small containers, skimming off the oil.[76]

Immense fires were so common that, *Scientific American* dared its readers, 'The *blasé* gentleman who looks down the crater of Vesuvius and finds nothing in it, may still hope here to experience a new sensation. If he is not moved by what may on any day be observed, let him wait a little while for a flood or a conflagration!' (see Fig. 5)[77] As described in the *New York Times,*

> The ground is . . . forested with derricks, shanties, tanks, &c., all saturated with petroleum, and only awaiting the falling of a spark or the scratch of a match, to blaze up with fury . . . [A]t the moment when oil is struck, when there is apt to be a rush of gas and petroleum to the surface, a single spark from the engine or a tobacco-pipe comes into contact with this column, what can the consequence be but a conflagration? In a single second the blaze may leap upward fifty feet, seizing

upon derrick, engine-house, and every other inflammable object in the neighbourhood.[78]

Many of the Oil Region's worst conflagrations occurred at Pithole, such as the immense fire of August 1865 when the overflowing Grant Well ignited. After suffering a fire a week between December 1865 and January 1866, the townspeople of Pithole finally adopted some firefighting measures. But a short-lived fire brigade and the ubiquitous 'No Smoking' signs failed to prevent fires from sweeping through Pithole and its oilfields over the summer of 1866.[79] Sir Morton Peto, in his report on the Oil Region, endorsed this testimony: 'I have spent many hours in great powder magazines, yet, on the whole, I would rather pass a month in them than a day by the great wells of Pithole, which are simply as dangerous as powder stores without one of their precautions.'[80]

As the English visitor walked the gauntlet through Pithole, the dangers of gunpowder itself, and later of nitroglycerine, also arrived in the Oil Region. Drillers began to set off underground explosions in 'torpedoes', as a method of opening up or increasing an oil flow, and their introduction resulted in heavy loss of life. According to one account of the torpedoing of an oil well in the Bradford district some years later,

> Those who stood ... and saw the [oil well] torpedoed, gazed upon the grandest scene ever witnessed in oildom. When the shot took effect, and the barren rock, as if smitten by the rod of Moses, poured forth its torrent of oil, it was such a magnificent and awful spectacle that only a painter's brush or poet's pencil could do it justice. Men familiar with the wonderful sights of the oil country were struck dumb with astonishment as they gazed upon this mighty display of nature's forces ... [W]ith a mighty roar the gas burst forth. The noise was deafening. It was like the loosing of a thunderbolt ... For over an hour [a] grand column of oil, rushing swifter than any torrent, and straight as a mountain pine, united derrick floor and top. In a few moments the ground around the derrick was covered inches deep with petroleum ... [T]he dams overflowed and were swept away before they could be completed ... People living along Thorn creek packed up their household goods and fled to the hillsides ... It was literally a flood of oil.[81]

The *Titusville Morning Herald* reported one explosion of a wagon loaded with 400lb of nitroglycerine

> ... on the public highway leading from Titusville. The driver, Mr. Charles C. Clark, who is one of Roberts & Co.'s most careful agents, was blown to atoms, as also was the horse and wagon. A large circular excavation, about four or five feet deep, and twelve feet in diameter, was made in the middle of the road; the fences and trees for a considerable distance on either side were shattered into fragments ... The butt end of the driver's whip was driven a distance of nearly a quarter of a mile, passed through a window in Mr. Arnamine's house, and knocked his wife senseless. Several persons a remote distance from the scene of the disaster were so stunned by the shock that they are confined in bed ... The shock was like that of an earthquake, extending for miles and being felt throughout the whole surrounding country ... Clark's face was found almost entire, without the skull. One eye was blown out, and the other was open, glaring and transfixed in death ... All the remaining portions of the body were scattered so widely and torn into such diminutive fragments that it was difficult to tell which belonged to the man and which to the horse ... The woods on the upper side of the road were set on fire in several places, but it was subsequently extinguished. The appearance of what was once the horse and the wagon beggars all description ... [The local inhabitants were] all forced to the conclusion that the transportation of nitro-glycerine over the public highway, and through populous villages was a villainous crime of the deepest dye, and worse than murder, as it momentarily renders it quite probable that hundreds of souls may instantly be launched into eternity without the slightest warning.[82]

Fires, floods, leakage and evaporation caused such a loss of oil from the increasing number and size of wooden storage tanks that oil producers and traders began to construct more fire- and leak-proof iron tanks. In addition, huge iron tanks would enable traders to store oil purchased at a low price during a production boom for future sale at a higher price if – they hoped – production declined. Following the general increase in crude production through the 1860s, iron storage tanks with capacities of ten, twenty and thirty thousand barrels became common. For rail transit, horizontal, boiler-type iron tanks were introduced, although the cheaper wooden tank cars were only phased out after years of blazing wreckages in 1871 when they were banned by New York state following a conflagration on the Hudson River Railroad that caused many fatalities.[83]

Fires followed the oil to the refineries, which in Cleveland were eventually expelled to beyond the city limits, while the city's wealthier citizens moved to cleaner locations upwind of the oil installations. Rockefeller later recalled, 'I was always ready, night and day, for a fire alarm from the direction of our works. Then proceeded a dark cloud of smoke from the area, and then we dashed madly to the scene of the action ... When the fire was burning I would have my pencil out, making plans for the rebuilding of our works.'[84] Another prominent Cleveland refinery owner, Mark Hanna, who had known Rockefeller since childhood, woke up one morning in 1867 to find that his refinery had burned down.

The refineries also became a major source of pollution. One by-product of refining lamp oil was great quantities of highly volatile gasoline which, at that time, was a largely useless waste product. Many refiners allowed it to flow off into the Cuyahoga River, the surface of which itself became highly flammable. Rockefeller recalled, 'We used to burn it for fuel in distilling the oil, and thousands and hundreds of thousands of barrels of it floated down the creeks and rivers, and the ground was saturated with it, in the constant effort to get rid of it.'[85]

Refiners dumped a cocktail of other poisonous chemical agents into the watercourses, and by the late 1860s Cleveland's water supply was as noticeably contaminated as the air. Following a public outcry, the *Cleveland Leader* warned its readers, however, of the economic impact of imposing restrictions on the refining industry, arguing 'it is certainly better that an oily taste should occasionally get into the drinking water than that any course should be taken which would seriously impede the oil interest of the city.'[86] The city authorities were slow to impose any effective environmental regulation, and subsequent engineering works on the water system – built at public expense – meant that the symptoms of the water pollution were addressed far more than its causes.[87]

The storage of petroleum further afield, whether in its crude or refined state, led to numerous, sometimes devastating, fires that often resulted in loss of life and serious injury. For example, in the early hours of 8 February 1865 a fire broke out in a Philadelphia warehouse storing several thousand barrels of kerosene in a densely populated area. Soon after the fire was discovered, the warehouse and a coal yard located opposite 'were a mass of ruins; immense masses of thick black smoke rolled up . . . The burning oil escaping from the barrels streamed down Washington avenue and around into Ninth street, pursuing its course southward, and constituted a perfect river of fire.'[88] Many local residents died in the inferno, which destroyed about a hundred buildings: 'The scene was frightful. Men, women and children were burned alive in escaping from burning dwellings.'[89] It was later reported,

> The 'petroleum' fire in Philadelphia proves to have been more disastrous than at first supposed. Not one of the eleven members of the Wright family have since been heard of, and under the ruins of the house it is supposed lie their calcinated bones. Seven of Captain Ware's family were also burned, and the only one who escaped cannot survive his terrible injuries. There is also missing an entire family named Scott. There are thus far nineteen persons known to be lost, and probably many more perished.[90]

An editorial in the *Philadelphia Inquirer* complained that

> Demands have been made by many citizens that municipal ordinances should be passed to prevent the storage of petroleum in any inhabited neighbourhood . . . We waited for the Legislature, and up to this time no action has been taken upon the subject at Harrisburg. We hope that there will be no more delay upon the matter.[91]

At Antwerp, one of Europe's main conduits for American petroleum imports, in August 1866 a fire broke out in a quayside warehouse holding several thousand barrels of oil,

> [to which] may be attributed the dreadful consequences that have befallen the city . . . [The fire] then reached a magazine or depot where there was stored some 10,000 barrels of petroleum. The local firemen, police and military strove every exertion to stay its progress . . . but the explosions and vehement fury of the flames compelled them to retreat . . . The fire then increased in magnitude tenfold, and the explosions that followed shook the whole city, and brought down many houses, while many people are reported to have been killed.[92]

At the end of 1866 *Scientific American* declared,

> Shortly after the destructive fire at the Erie Railway Docks, in Jersey City, caused by the ignition of stored petroleum, we drew attention to the fact that in its crude state that substance is highly inflammable and explosive, and suggested what we then considered a remedy. Subsequent events and investigations have assured us that our opinion was well founded ... We hope the attention of our legislators will speedily be directed to this subject, which must be conceded to be one of the first importance.[93]

At the same time, the popular excitement over the commercial potential of this new industry made it ripe for speculative investment, particularly following big oil strikes on Cherry Run, a tributary of Oil Creek, in 1864 and along Pithole Creek in 1866. Surrounding land prices suddenly skyrocketed as producers and speculators clamoured for plots, leases and wells, many grasping at the tiniest stake in anything. The speculative fever, which spread to become a national 'oil stock company epidemic', was a gold mine for fraudsters who tricked the rash and naive out of large sums of money.[94] As the *Titusville Morning Herald* reported in May 1866,

> In very many cases we know of parties who have sunk from fifty thousand to one hundred thousand dollars, without the slightest hope of ever seeing one cent of the cash again ... A very heavy proportion of the loss has fallen upon parties with straitened financial means, and ruin has in thousands of cases followed the attempt to make a fortune in a month by dabbling in bogus oil stocks ... The inside history of many 'operations' on the 'ground floor', etc., would be sad yet astonishing. Men of high standing in church and State would be proved to have been liars, hypocrites, reckless gamesters, and utterly devoid of all humanity or honesty. Other people, supposed to be shrewd business men, would be shown to have been the silliest dupes imaginable.[95]

A lucky few did walk away with that dazzling fortune, sustaining the dream in countless others.[96] As a correspondent for the London *Morning Post* wrote, 'The richest members of the "oil aristocracy" were not three years since as poor as church mice ... Strange freaks of fortune occur every few days.'[97] In his commercial report on the Oil Region,

Sir Peto shared the story – or parable, perhaps – of 'the Widow M'Clintock':

> Her wretched, half-scratched land was valuable enough for oil. She got a very large sum for it, which, with the characteristic business habits of a poor old country widow, she would receive in nothing but green-backs, which were accordingly paid to her in a bundle as big as a bolster. She hoarded them, still living amid the noise of derricks and the gas of wells in the shingle-hut in which she had for years been accustomed to dwell, till one night her petroleum lamps exploded, lighting her and her wooden house at once, and from the sudden fire neither she nor her greenbacks were saved; nothing, in fact, but some $80,000 which she had been persuaded to invest in United States' securities. All Petrolia is full of anecdotes like these.[98]

The term 'wild catting' would duly make its first appearance in the context of the oil industry in mid-July 1870 when the *Titusville Morning Herald* reported,

> There is perhaps more 'wild catting', or independent operating in untried and apparently valueless territory being conducted at present than at any time previous since the discovery of oil. In the vicinity of the Allegheny river between Tidioute and Tionesta several 'wild cat' or test wells have been finished or are being drilled.
>
> These generally prove worthless, though as in the case of the 'Venture' well, an occasional paying and large producing well is discovered, which opens a new and often extensive field for operations.[99]

Alongside speculative investment in oil operations came speculation in the commodity itself based on fluctuations in the oil price. One of the simplest ways of profiting from oil trading would be to purchase oil from a producer who lacked the storage capacity to contain their own output and who therefore could be pressured into selling it at lower than market price. A trader with the necessary storage facilities could then sell this crude on at a profit. Speculation entered when oil traders endeavoured to predict future changes in the oil price, to reduce the risk of having to sell on at a lower price, but ideally to be in a position to sell on at a higher price. In order to make accurate predictions about future movements in the oil price, traders needed to gather the best information they could about the many factors affecting it – information that should be as comprehensive and up to date as possible. A trader who bought oil before learning of a huge oil strike or a drop in the demand for crude would lose money; conversely, a trader who bought in the knowledge – before it became widespread – of a significant decline in production or an increase in the demand for crude would make money on their subsequent sale of oil.[100]

Trading depended on the analysis of accurate information, so important developments in the oilfields were often kept a close secret, and insider dealing, the spreading of misleading rumours and deceptions were common; according to one news item, well 646 on Cherry Grove was 'boarded up as tightly as ever and under constant guard ... The report has been spread that there is oil in the hole and that it has been plugged to allow the owners to acquire more territory in the vicinity of the well.' A few days later it was reported, 'The mystery on 646 still guarded and would-be intruders are warned away with the contents of shot guns whistling about their ears.'[101] Further down the line, pipeline companies and traders made attempts to 'corner' the market by using their oil storage facilities to buy up and withhold most of the available crude in order to artificially reduce the supply, enabling them to sell their oil at an inflated price.[102]

In the 1860s trading was conducted where buyers and sellers directly met up – on the streets of shipping points such as Titusville, Oil City or Franklin, or on a special Oil Creek Railroad carriage – where deals would be struck for 'spot' oil (immediate payment and delivery), 'regular' oil (payment and delivery within ten days), or 'future' oil (payment and delivery on a specified future date). Where traders had access to a telegraph service, such as in a hotel or post office, they could maintain close contact with their clients and keep up to date with market signals. By 1870 trading was conducted in more formal oil exchanges and was dominated by large brokerage firms with international scope.[103] The equivalent of many days' total production was now traded each day, and it was here that the biggest fortunes were won and lost. American crude and lamp oil were being delivered to a large and growing global oil market and, according to the *Titusville Morning Herald*, 'the oil buyer ... watches eagerly for the daily two o'clock Antwerp quotations'.[104] By 1875 the Geological Survey of

Pennsylvania was giving an upbeat assessment of the international dominance of the US oil industry, albeit with a hint of caution:

> Our refined Petroleum has penetrated to the most distant parts of the world, it brightens the long winter nights of Sweden and Norway, and even Iceland, it is sent to Australia and New Zealand, to China and Japan, to Russia, Germany, Austria, France and Great Britain, in the face of the fact that in every one of the countries named, surface oil or bituminous shales, exist in quantities that would seem fatal to competition . . . Whether the drill in other countries as with us, would find a light oil . . . is a question that some day may interest us.[105]

4

Corporate Control

BATTLE FOR THE OIL TRADE

The railway networks of the nineteenth century were unprecedented in their scale and scope. Their impact on the world was revolutionary, and they ushered in the modern age of big business. Not only were the railway companies the first of the large industrial corporations, but they were also the engine that powered the rise of large-scale enterprise across the economy generally. Individual businesses could now acquire raw materials and distribute goods in such quantities as to increase enormously the potential scale of their economic activity, making it possible for them to grow into corporate giants.[1] Particularly following the Civil War, America's economy was increasingly dominated by the interests of high finance, organized largely around the railways. During the war, the railways had been crucial for the rapid movement of large numbers of troops, arms and provisions, and close cooperation between government and the railway companies, financed by a newly centralized banking system, had given the more industrialized North a decisive military advantage over the agrarian South. The rail and financial corporations – backed by large sums of foreign investment, mainly British, Dutch and German – emerged from the war as its biggest victors, with corporate bosses such as Thomas A. Scott of the Pennsylvania Railroad holding sway over state and federal governments.[2] Many Americans would feel disturbed by these developments for decades to come, as expressed in this impassioned statement:

> Yes, we can all congratulate ourselves that this cruel war is drawing to a close. It has cost a vast amount of treasure and blood. The best blood

of the flower of American youth has been freely offered upon our country's altar that the nation might live. It has been a trying hour for the republic, but I see in the near future a crisis arising which unnerves me and causes me to tremble for the safety of my country. As a result of the war, corporations have been enthroned, and an era of corruption in high places will follow, and the money power of the country will endeavor to prolong its reign by working upon the prejudices of the people until all wealth is aggregated in a few hands and the republic is destroyed. I feel at this time more anxiety for the safety of my country than ever before, even in the midst of the war. God grant that my suspicions may prove groundless![3]

Whether these really are the words of Abraham Lincoln, or are a politically inspired forgery, they articulate a common response to the rising power of private business corporations in the US.[4] Indeed, President Andrew Johnson declared two years after the end of the Civil War,

an aristocracy, based on over two thousand five hundred millions of national securities, has arisen in the Northern States; to assume that political control which the consolidation of great financial with political interests formerly gave to the slave oligarchy of the lately rebel States. The aristocracy based on negro property disappears at the Southern end of the line, but only to reappear in an oligarchy of bonds and national securities in the States which suppressed the rebellion . . . The war of finance is the next war we have to fight.[5]

The shifting economic landscape was described simply by the *Commercial and Financial Chronicle* soon after war's end: 'There is an increasing tendency in our capital to move in larger masses than formerly. Small business firms compete at more disadvantage with richer houses, and are gradually being absorbed into them.'[6]

The railway companies of the Gilded Age took advantage of the unregulated economic and political environment to become private industrial enterprises of extraordinary size and power. Since only those railways with the strongest political clout, the largest sources of finance and the most extensive infrastructure could survive the ferocious competition, they tended to consolidate progressively into a few massive corporations.

Along the way, when it seemed to be in their mutual interest, they organized privately among themselves to limit the effects of free competition, ending their periodic rate-cutting wars by coming to collective agreements, apportioning traffic between themselves and setting higher freight rates. Most controversially, they would secretly strike special rate reduction deals with the biggest enterprises in return for guaranteed regular, large shipments of freight. This put smaller businesses at a serious disadvantage: only the larger firms, with access to the capital needed to survive recessions and to expand and industrialize their operations to take the greatest advantage of economies of scale, had the power to negotiate reduced freight rates from the railways. These anti-competitive practices therefore systematically discriminated against small businesses in favour of larger, high-volume shippers at locations – mainly the big cities – served by several competing railways. Popular debate over the impact of the railways intensified during the 1870s. Were the rail corporations purely private enterprises, free to strike whatever deals they liked? Or, particularly since they enjoyed the privileges of vast land grants, rights of eminent domain and other state subsidies, were the railways more like public services with a duty to act in the general interest as 'common carriers' of freight and to treat all customers equally?[7]

It was in this context of an expanding, ever-changing railway industry, continually lurching between fierce competition and cartel-like combination, that oil producers and refiners struggled to conduct business in their own highly turbulent industry. By 1865 national and international demand for kerosene was being met by surging production of crude from the Oil Region, now transported to refining centres in increasing bulk by gathering, or feeder, pipelines and by railway tank cars. The strong demand for refined products coupled with the plentiful supply of cheap crude made refining an attractive business to enter. The additional refining capacity, however, soon began to outstrip product demand. During the late 1860s and 1870s, an over-supply of kerosene on the American market sent prices crashing, causing recessions, and refiners found themselves competing simply for survival. Crucially, the costs of transporting their raw crude and their refined products could now make the difference between profit and loss; and since around three-quarters of American kerosene went for export, the cost of rail shipment to the eastern seaboard was particularly significant.[8]

Meanwhile, the transport matrix shifted in a continual flux as the railways competed by laying strategic extensions to their networks. The structure of the rail industry in the United States encouraged fierce competition for freight between different railway companies. They struggled to meet their high fixed running and maintenance costs and to keep up regular payments on their large proportion of bonded debt, making them desperately dependent on a constant stream of income from a regular, high-volume turnover of freight. In order to capture valuable markets ahead of their competitors, the rail companies borrowed more for network expansion, either for the construction of new track, rolling stock and freight-handling facilities, or for the takeover of existing lines, as described by a rail industry journal in 1869: 'Where a link is strategically significant and traffically essential to a company's operating plan and income prospect, it must be built not at option and pleasure, but in advance of seizure of the location by a covetous party in opposition . . . Thus companies are often undertaking more works than at first apparent.'[9]

In 1864 the near monopoly over the shipment of crude out of the Oil Region, enjoyed so far by the Atlantic & Great Western and Erie Railroads, was broken by the Pennsylvania Railroad after it financed the completion of the Philadelphia & Erie Railroad to Warren on the northeast edge of the Region – about ten miles from the old Seneca village of Burnt House where Chief Cornplanter had lived out his final years. This new rail extension opened a nearly two-decade-long battle between the railways around the Oil Region as they competed for the traffic in crude oil and refined products to and from refining centres. The following year, while the Atlantic & Great Western defended its dominance within the Oil Region itself by extending its line from Franklin to Oil City, the Pennsylvania Railroad began to secure its own sources of crude by moving to take over the local, independent Oil Creek Railroad.[10]

It was not long after Van Syckel's pioneering pipelines that the railways recognized the crucial importance to their oil freight business of the systems of feeder pipelines that were now conveying oil to rail depots across the Oil Region. After all, the routing of pipelines and the piping rates charged now effectively determined the railways' access to supplies of oil freight. In order to secure the necessary pipeline access, the railways therefore moved into the pipeline business, through their 'fast-freight' or 'transfer' companies. These companies, common across the rail industry, were set up to manage the effective transport of goods across several rail and canal networks. This helped to attract business as the shipper, who would otherwise have had to negotiate with many parties across a complex transport route, instead only had to purchase one ticket to send freight the full distance. Now, with pipelines added to the oil freight matrix, the Pennsylvania Railroad in 1866 set up its Empire Transportation Co., which in turn bought up the Titusville Pipe Co., and by July of that year the Empire was freighting crude oil from the end of its feeder pipelines in the Oil Region direct to New York. Likewise, the Erie-Atlantic & Great Western Railroad combination managed their oil freight through the Allegheny Transportation Co., which had its origins in the early, pioneering pipeline system of William Abbott and Henry Harley, and was destined to become the first of the great pipeline companies.[11]

The oil refiners of Cleveland were blessed with the lowest freight rates across the industry as a result of the particularly intense competition between the many rail and canal routes passing through this hub of the east-west transport corridor. The extension of the Lake Shore Railroad, which gave Cleveland yet another rail link with the Oil Region and eastern markets, presented the city's refiners with an opportunity to negotiate even lower freight rates. Three of Cleveland's biggest refiners, including Rockefeller, now coordinated their bargaining power to extract preferentially lower rates in return for their combined high volume of oil freight, first from the Lake Shore Railroad. This refiners' cartel then gained further leverage against the railways by going into the pipeline business themselves, buying a large stake in the Allegheny Transportation Co. In 1868 the company proposed financing the construction of pipelines to depots on the Lake Shore Railroad, now extending beyond Franklin to Oil City. In response, the Atlantic & Great Western granted the Cleveland refiners further special rate reductions, in return for which the railroad retained their oil freight and received a promise from the refiners that they would not finance feeder pipelines to the Lake Shore Railroad.[12] By exploiting such transportation advantages, in 1869 Cleveland overtook Pittsburgh to become America's dominant refining centre, the *Cleveland Daily Leader* announcing in February 1870 that 'petroleum is now more than ever before a source of national wealth and prosperity'.[13]

Over-capacity across the refining industry, however, kept kerosene prices down, making profit margins precipitously narrow. Moreover, the Cleveland cartel risked losing its crucial freight-rate advantages. The increasingly powerful Pennsylvania Railroad had now gained control over much of the industry's oil transit by absorbing the Oil Creek & Allegheny River Railroad and by taking over the Allegheny Transportation Co. (reorganized as the Pennsylvania Transportation Co.), giving it a near monopoly over the Oil Region's pipelines. Independent oil producers and refiners, meanwhile, sought to break free of the monopoly and cartel system that was engulfing them by planning pipelines of their own. The imminent arrival of the Baltimore & Ohio Railroad in Pittsburgh was also sure to change the transportation calculus. Finally, early indications of big new oilfields to the south of Venango County, in Butler, Armstrong and Clarion counties, close to the Pennsylvania Railroad's rail and pipeline infrastructure, further unnerved the Cleveland refiners – especially in view of an anticipated decline in output from the old Oil Creek fields which, a geologist warned, were being 'excessively and wastefully depleted ... [F]rom this time forward we have no reasonable ground for expecting ... that these oil fields will continue to supply the world with cheap light as they have in the past and that therefore it is wise to pause and consider how best to husband our remaining resources and make the most of them.'[14]

Rockefeller and his refinery cartel partners decided, therefore, to build on their current advantages by going on the offensive. They set out to reduce local competition by targeting rivals for takeover or inclusion in their cartel. This would enable them to increase profit margins: firstly by restraining refinery production levels; secondly by reducing local competition; and thirdly by strengthening further their bargaining power with the railways for preferential freight rates. As Rockefeller would later recall, their solution to the 'ruinous competition', caused by 'over-development of the refining industry', was to replace unpredictable, unprofitable chaos with controlled, profitable efficiency, through further combination: 'We had to do it in self-defence. The oil business was in confusion and daily growing worse ... It was the battle of the new idea of cooperation against competition.' However, 'to buy in the many refineries that were a source of overproduction and confusion we needed a great deal of money.'[15] The money was found, in January 1870, by incorporating the allied partnerships and selling shares in the new joint-stock company: the Standard Oil Company of Ohio. This tighter legal structure also gave Rockefeller, as president of the company, stronger administrative control over the cartel's extensive operations, covering many refineries in Cleveland, New York and elsewhere, accounting for 10 per cent of national refinery output, a giant leap for the new conglomerate.

While Rockefeller's Standard Oil combination manoeuvred to take control of Cleveland's entire refining industry, the region's railways planned to fend off potential threats to their oil trade from the possible construction of independently owned pipelines and from the Baltimore & Ohio Railroad, a branch line of which arrived in Pittsburgh in 1871. Thomas Scott, president of the Pennsylvania Railroad, wielded such power over the Pennsylvania legislature that he succeeded in blocking legislation that would have granted broader rights for the construction of competing pipelines in the state. Then, in an attempt to avoid the disruption of yet more competitive rate-cutting between the railways, Scott initiated an ambitious new oil-transit cartel scheme, composed of the established oil-trafficking railways and the big refining concerns in Pittsburgh and Cleveland, including Standard Oil. Through the South Improvement Co., under a charter obtained by Scott from the Pennsylvania legislature in late 1871, the railways and refiners within the cartel would keep the oil trade to themselves by secretly agreeing mutually preferential terms. Rockefeller signed up to the scheme with enthusiasm and used it to threaten rival refiners in Cleveland and beyond with the option of either selling up to the Standard Oil combination or face the inflated freight rates and uncertain access to pipelines and tank cars endured by non-cartel refiners, and therefore likely bankruptcy: 'If you don't sell your property,' Rockefeller told one refiner, 'it will be valueless, because we have advantages with the railroads.'[16] By such tactics, in what became known as the 'Cleveland Massacre', Standard Oil took over nearly the entire refining capacity of the city, now bringing under its control around a quarter of national output.

The South Improvement cartel arrangement of 1872 was not new in principle, but it became particularly notorious for the new scale of its ambition and because the scheme was accidentally revealed by a railway clerk. The discovery of this smoking gun triggered a wave of protest across the Oil Region against the kind of restraint of free trade and open competition that the independent producers and refiners had long suspected the transit companies and dominant refiners of orchestrating. For a few brief months the Region's many oil producers, organizing under the umbrella of the Petroleum Producers' Union, overcame their mutual competition enough to regulate their output for a successful boycott of crude to the cartel refineries. The railways and their pipeline affiliates capitulated and announced equal access and fair rates for all, while the Pennsylvania legislature revoked the South Improvement charter and granted limited rights for independent pipeline construction.[17]

But it was a fleeting and hollow victory for the independents and producers. The mass of producers were only very rarely able to set aside their mutual competition – enshrined in the rule of capture – to maintain a united front against the small number of dominant transit and refining concerns. As Rockefeller recounted, 'So many wells were flowing that the price of oil kept falling, yet they went right on drilling.'[18] The big players in transportation and refining continued to wield decisive economic power and political influence: free from formal, effective regulation, they simply conducted business as usual and continued along the path of consolidation. They regularly entered into their familiar self-protective cartel, or pooling, arrangements which, although inherently unstable and temporary, put independent outsiders at a profound disadvantage. The *Titusville Courier* raged, on behalf of the long-suffering independents, 'The people will not part with their sovereign rights, nor allow themselves to be ruled by King Pool.'[19] But under the prevalent economic, legal and political system, King Pool was destined to reign. When the ever-shifting economic forces upset a cartel arrangement between the big pipeline, railway and refining interests, a new outbreak of competition would resolve itself through another round of outright takeover and consolidation. Thus, ever fewer, more powerful corporations came to control the oil-transit and refining industries, which in turn dominated the Oil Region's producers.

PIPELINE TO MONOPOLY

Through the 1870s the producers and stalwart independent refiners continued to resist the advancing corporate cartel system in the state and federal legislatures, in the courts, in the oilfields and on the transit routes. But the disunity among the hordes of producers and the independents' relative lack of capital and political influence kept both on the back foot, while Standard Oil increased its command over the industry by extending its control over pipelines.

The Pennsylvania Railroad was acutely aware that the gathering pipelines feeding oil to its rail depots were the Achilles heel of its control over the transit of crude oil from the Oil Region. Accordingly, in 1872–3 its fast-freight affiliate, the Empire Transportation Co., extended its ownership of gathering lines, through its Union Pipe Line Co., across the new prolific oilfields further down the Allegheny River in Butler, Armstrong and Clarion counties. As production there surged, the *Titusville Morning Herald* asked 'Is Petroleum a Necessity?', answering, 'The production of petroleum has now become of such commercial and social importance to the world that if it were suddenly to cease no other known substance could supply its place, and such an event could not be looked upon in any other light than of a widespread calamity.'[20] The producers operating in the Lower Oil Region now found themselves encircled by the Empire's expensive transit monopoly. A reporter for the *Pittsburgh Daily Dispatch* wrote, 'The whole producing region is brought in contribution to the pipe line companies, whose lines extend in every direction to the extreme limits of the territory ... It will be easily seen that these corporations have in themselves the controlling power of the oil regions.'[21]

In the context of prevailing over-production and excess refining capacity that kept both crude oil and kerosene prices low, transportation remained one of the industry's biggest costs. This spurred Butler County producer David Hostetter, along with colleagues from the Petroleum Producers' Union, into implementing an idea widely mooted: the construction of a long-distance trunk pipeline to take crude directly out of the Oil Region, completely bypassing the cartel's discriminatory feeder pipelines and railways.[22]

In the summer of 1874, Hostetter's Columbia Conduit Co. began laying its pipeline to Pittsburgh, 36 miles to the south. The pipeline was supported by Lower Oil Region producers and by Pittsburgh refiners, but it was opposed by refiners in the Oil Region and in Philadelphia who would see their crude supply going instead to Pittsburgh. Then, Pittsburgh refiners had second thoughts about their support when they learned of a deal between Columbia Conduit and the Baltimore & Ohio Railroad that would take the crude all the way to refineries in Baltimore.

The Pennsylvania Railroad fought tooth and nail to protect its pipeline and rail monopoly. It exercised its influence over the Pennsylvania legislature to exclude Allegheny County – Pittsburgh's county – from the state's free pipe law; and when Columbia Conduit tried to lay its pipeline under any Pennsylvania Railroad track, the railroad sought court injunctions arguing that the pipeline company did not have 'eminent domain' rights of way. One morning in November 1874, violence between pipeline and railway employees was only narrowly averted following Columbia Conduit's overnight laying of 100 feet of pipeline under a rail culvert. In December rail workers tore up a section of the Conduit's pipeline, causing an extensive oil spill; the railway then constructed small forts and deployed guard patrols along strategic lengths of its track.

The new threat of independent trunk pipelines even brought together momentarily the Pennsylvania Railroad with its arch-rival Standard Oil against this common enemy – particularly when producers in Clarion County, organizing as the Atlantic Pipe Line Co., built an independent trunk pipeline that took two-thirds of Clarion's output to the Baltimore & Ohio Railroad and then considered the possibility of extending the pipeline all the way to the Atlantic seaboard.[23]

Empire Transportation's aggressive advance into feeder pipelines and the independents' bold counter-moves with trunk pipelines in the Lower Oil Region threatened the northwards flow of crude to Standard Oil's refining base in Cleveland and New York. Standard's response was to protect the supply of oil to its allied northern railways – the Erie and the New York Central – by entering fully into the feeder pipeline business itself, rapidly consolidating pipeline systems such as the United Pipe Line Co. into a newly formed fast-freight company of its own, the American Transfer Co. In an outflanking move, Standard also began to subvert the associated threat from a growing independent refining industry to the southeast by buying up refineries in Pittsburgh and Philadelphia and along the Baltimore & Ohio Railroad.[24]

Now the Pennsylvania Railroad, in addition to feeling the impact of competition from the Baltimore & Ohio Railroad, began to feel surrounded by Standard's unfriendly refinery cartel. Joseph Potts, president of the Pennsylvania's Empire Transportation subsidiary, recalled,

> We reached the conclusion that there were three great divisions in the petroleum business – the production, the carriage of it, and the preparation of it for market. If any one party controlled absolutely any one of those three divisions, they practically would have a very fair show of controlling the others. We were particularly solicitous about the transportation, and we were a little afraid that the refiners might combine in a single institution ... We therefore suggested to the Pennsylvania Road that we should do what we did not wish to do ... become interested in one or more refineries.[25]

While Standard Oil's United Pipe Line had the capital to take over the Columbia Conduit Co. – an oil lifeline for the Baltimore & Ohio Railroad and its associated independent refiners – the harried, overstretched and mutually competing railways were close to financial breaking point. They finally broke in 1877 after cutting their workers' wages, triggering a wave of mass strikes that brought services to a halt. At this opportunity, Standard Oil pounced, running heavily into debt itself to buy up the entire oil-transit system – 520 miles of feeder pipelines, storage tanks and loading facilities – of the now floundering Pennsylvania Railroad.[26] Rockefeller raced from bank to bank demanding loans, seeing that this acquisition could win him almost complete control over the US oil industry: 'I must have all you've got! I need it all! It's all right! Give me what you have!'[27] Later that year, an independent oil broker told the *New York Tribune*, 'I can ship no oil from the Oil Region now without paying a royalty to the Standard Oil Company [which] now controls all the pipe lines.'[28]

By gaining control over nearly all of the Oil Region's approximately 3,500 miles of gathering pipelines, Standard Oil had won the power to dictate terms across the oil industry. Since there now existed no alternative to Standard's United Pipelines for the transit of crude from most producing areas, the vast majority of refiners were absorbed into the Standard Oil conglomerate or were finished in the oil trade, and even the previously all-important railways were now almost entirely subservient to Standard for their oil traffic.

Despite the unprecedented scale and scope of Standard Oil's business empire, however, the corporation was never able to achieve complete control over the entire industry. There were always a few independent-minded refiners who looked for their financial reward outside the Standard Oil conglomerate, and who were prepared to take the risk. The fluidity and complexity of the industry kept Standard on the move, always working to fill gaps in its defences. When new oilfields were opened up, Standard's United Pipe Line Co. struggled to extend its network of pipelines ahead of all competition; while Standard dominated the kerosene market, niche markets for an ever-widening range of specialist refined products forced it into conceding that around a fifth of total refining activity would remain independent; and the undoubted success of Standard's intensive lobbying and outright bribery of legislators was never quite enough to defeat all political opposition.[29]

In 1876–7 the new Bradford oilfield, in Warren and McKean counties at the northeastern edge of the Oil Region, began smashing previous production records. While Standard's United Pipe Line summoned its resources to construct gathering lines and storage tanks, again tens of thousands of barrels of oil ran to waste or went up in flames. As Bradford district's streams became 'literally rivers of oil', there resounded across 'Oildom' the popular ditty, 'Oil, oil everywhere, and no untainted water to drink.'[30] Again, the rule of capture fostered unrestrained production that far exceeded the local storage and transit capacity out of the area, a check on demand for crude that sent prices, already down at $2.50 per barrel, down further to under $1 a barrel. The thousands of producers, swarming through makeshift boom-towns, were unable to organize a self-defensive bargaining position towards the highly unified Standard Oil refining and transit combine; and producers were faced with no alternative but to send their crude down United Pipe Line's gathering system at the behest of Joseph Seep, the Bradford agent for Standard's crude-oil purchasing affiliate, J.A. Bostwick & Co.

Under these non-competitive conditions – technically, a monopsony, where there is a single dominant buyer – the oil exchanges became irrelevant. In the Bradford field, Seep could now ignore the almost superfluous oil traders by going directly to producers at the wellhead and declaring a distress sale price for 'immediate shipment', take it or leave it. As virtually the sole buyer, Seep was able to get away with offering as low as 25 cents less per barrel than the day's quoted price on the Region's oil exchanges.[31]

Although Seep, eight months later, reverted to purchasing Bradford crude on the exchanges – following a combination of rising popular outcry against Standard Oil's market power and the easing of the storage and transit bottleneck – he would subsequently apply his Bradford experiment again, this time right across the US oil industry. In 1884 Standard centralized all of its crude purchasing in the Joseph Seep Agency, which, from 1886, began to reassert its dominance by again going directly to the producers with a set price, midway between the day's high and low on the exchanges.[32]

BENSON'S FOLLY

In 1877, now that Standard Oil had gained control over the established system of oil transit around the Oil Region all the way to the Atlantic seaboard, the producers and independent refiners faced the greatest test of their business resourcefulness: however, their most notable innovation, the construction of long-distance trunk pipelines, would immediately be adopted by Standard to extend even further its monopoly control over the US oil industry.

The only alternative to selling out to Standard was to think even bigger than the Columbia Conduit scheme, to bypass the Standard-dominated railways entirely by constructing independent trunk pipelines all the way to the East Coast refiners that were secure in independent hands. The first such attempt, by the operators of the Conduit pipeline for a trunk pipeline from the Butler oilfields to Baltimore, was soon squashed by Standard Oil. A Maryland refiner, J.N. Camden, who had secretly joined the Standard fold, blocked the proposed route by bribing his state legislature into denying any applications for pipeline charters;

Camden informed Standard's freight negotiator, Henry Flagler, 'The price is nominally $40,000.'[33] Another attempt by a group of Bradford independents, organized as the Equitable Petroleum Co. under the leadership of Lewis Emery Jr, endured continual harassment from Standard but achieved more success. With the support of the Petroleum Producers' Union they opened up a transit escape route that began with a seven-mile trunk pipeline from the Bradford field to Coryville, McKean County; from there the Buffalo, New York & Philadelphia Railroad, outside the Standard orbit, took their oil on to the Erie Canal for the final water-borne leg to New York. However, although this worked out at half the cost of the Standard-controlled all-rail route to the seaboard, this route was closed during the Erie Canal's winter freeze.[34]

The prolific Bradford field also became the new focus of the persistent Columbia Conduit veterans, led by Byron D. Benson. Incorporating their enterprise would have required persuading the Pennsylvania assembly to grant them valuable corporate charter rights for pipeline construction, a move that the powerful Standard Oil and rail lobbies would quite easily have blocked.[35] At the time, Flagler was making available to his lobbyist in the state assembly tens of thousands of dollars for bribing state legislators and was complaining that the rail companies should also be contributing to the fund: 'We have spent a large sum of money to squelch Seaboard Pipe Line Charters.'[36] However, the pipeline promoters were able to take advantage of recent changes to the state's commercial law to form, in 1878, a limited liability partnership, the Tidewater Pipe Line Co. Ltd. Although this reduced their powers to obtain rights of way – they would have to purchase or lease land rights on an entirely private basis – any financial liabilities would be limited to no more than the amount invested. Furthermore, since shares in the partnership were technically non-transferable and gave no automatic voting rights on the board, Tidewater was less vulnerable to a Standard takeover.[37]

Tidewater then teamed up with Equitable Petroleum, which had managed to lay a 2,500 'barrel-per-day system of gathering lines to its Bradford wells ahead of Standard Oil's momentarily overstretched United Pipe Line, and which was looking for additional independent transit capacity out of the area to supplement its insufficient Buffalo route. With the involvement of some prominent Wall Street bankers,

Tidewater raised over $500,000 to construct a trunk pipeline 110 miles from Coryville to Williamsport on the Reading Railroad, for rail shipment on to Philadelphia and New York.

Standard Oil and the established oil-trafficking railways did all they could to sabotage Tidewater's ambitious trunk-pipeline construction effort. They bought or leased long strips of land, 'dead-lines', intended to block the route, but were outwitted by Tidewater's strict secrecy and fake decoy surveys. They conspired to deny Tidewater permission to run pipelines under railway tracks, but when one railway tore up a section of pipeline laid under a culvert, the Tidewater management, drawing on its similar experiences with the Conduit, was ready with an effective court injunction against the rail company. Pressure was put on rail tank-car manufacturers not to deal with Tidewater, which needed tank cars for the Reading Railroad, and Standard suddenly took an interest in environmental issues, investing in newspapers and planting news stories warning farmers of the dangers of pipeline leaks. *Railway World*, whose owner had close connections with the Pennsylvania Railroad, ran an invective against pipelines that concluded, 'in short, a pipe line, whether upon the road or farm, is evil, nothing but evil, and every wise man will keep it off his property at all hazards.'[38]

Indeed, as oil pipelines proliferated across the country their environmental risks became increasingly apparent. The issue would surface prominently in 1883 when independent oil refiners led by Senator Lewis Emery Jr introduced a bill in the Pennsylvania legislature designed to make it easier to gain rights for laying long-distance pipelines. By granting them railroad-like powers of eminent domain, the independents presented this as an anti-monopoly bill, arguing that Standard Oil's greater financial resources gave it an unfair advantage when acquiring the land and access rights for its own pipelines or to block access to others.[39] In response, the bill's opponents – which included Standard – bolstered their case by highlighting environmental concerns, receiving particular attention from one Pennsylvania newspaper after a leak, ironically, from a Standard pipeline at Leavittsburg, Ohio. This incident

gave an emphatic affirmation of one of the most serious objections to the pipe line system of transporting oil through a thickly-settled country. One of the pipelines of the Standard monopoly burst at that point, discharging one hundred barrels of oil per hour. It ran into a stream and some miscreant set fire to it, the entire vicinity being filled with smoke and flame ... [I]t is sufficient proof of the liability to damage to surrounding property, which the advocates of pipe lines have hitherto ridiculed as an idle foreboding.[40]

The newspaper later wrote that the bill's advocates

tried to make us believe pipes would neither leak nor burst, differing in these particulars from all pipes ever made. But we have the pipes in this and Chester counties, and the truth, as well as the oil, is gradually leaking out ... [N]ow the West Chester papers give the details of a burst oil pipe on the farm of Mr. Maris Woodward, of East Bradford township. Another occurred on the large farm of Sarah Young ... Through a leak at one of the joints about fifteen barrels of oil quickly made their appearance on the surface, covering and ruining the land for agricultural purposes, and then, finding its way to a creek on the premises, ran down, so fouling the water that the cattle could not drink it.[41]

Senator Thomas V. Cooper argued that 'the bill ... proposes to confer. no great public good. The only one who would profit by it would be those engaged in the oil business. Owners of land under which it would pass would be punished by it by having their property damaged, as I learn that land soaked with oil becomes non-productive for years.' Senator Emery replied, 'Why, oil is the greatest fertilizer in the world', and at a hearing of the Industrial Commission he subsequently accused opponents of the bill of distributing handbills designed to stoke opposition among farmers: 'There was pushed under my arm a paper which read: "These people are endeavoring to pass a law that will destroy the springs on your farms; it will blow up your houses; it will create havoc in your fields, when a pipe bursts, by killing the grass. The most dangerous of all laws." '[42]

The Tidewater scheme became known as 'Benson's Folly', due not only to the level of corporate and political resistance but also to the engineering challenges that appeared to many to be insurmountable. The hydraulic pressures needed to convey crude oil through a 6-inch-diameter pipeline across the Allegheny Mountains – initially ascending

69 feet over 22 miles, then 480 feet over 8 miles – would require significant advances in pumping and piping technologies. Yet, against all the odds, in June 1879 the completed Tidewater pipeline began sending 6,000 barrels of crude per day to its Williamsport terminus. Railway tycoon William H. Vanderbilt now predicted, 'we won't any of us have the oil business long; they [Standard Oil] will build their own pipe lines to the sea board.'[43] The consequence of the profitability of trunk pipelines, despite charging lower freight rates than the railways, was clear to the *Railroad Gazette*: 'The railroad oil traffic seems finally doomed. Not that there will not always be a vast traffic in the distribution of refined oil ... but that the enormous traffic in carrying crude oil to the refining centres at Cleveland and Pittsburgh and to the sea-ports where oil is exported will henceforth go mostly by pipe-lines.'[44]

John D. Archbold, a Standard Oil executive on his way to becoming the company's vice-president, had once been an independent Titusville refiner, but now it was his job to oppose independent pipelines: 'The tide water line was a vastly disturbing element in the trade this year,' he wrote.[45] Standard's pipeline boss, Daniel O'Day, felt inclined towards violence: 'I feel extremely satisfied that the Tidewater Pipe Line can be stopped and torn up if it is thought best to do it. I also think that the sooner the Tidewater knows this the better, as it might have a healthy effect upon them.'[46]

True to Vanderbilt's prediction, the Standard Oil alliance realized that simply attempting to block the construction of independent trunk pipelines was a strategy no longer guaranteed to safeguard its transit monopoly. The railways continued to obstruct the laying of all trunk pipelines; and they countered the Tidewater pipeline by reducing their crude freight rates, although this had the side effect of benefiting Standard Oil, the biggest oil shipper by far. Standard, by contrast, launched into a massive programme of long-distance pipeline construction. By the end of 1879 it was laying a 100-mile, 5-inch-diameter pipeline from Butler County to Cleveland, completed the following March at a cost of about $6,000 per mile. June 1880 saw the completion of a 100-mile seaboard-bound pipeline in direct competition with Tidewater, and August the completion of a 3-inch-diameter, 63-mile pipeline from the Bradford field to Buffalo. In November 1881 Standard formed the National Transit Co. as the parent company to consolidate its now sprawling pipeline empire, and two months later it completed a 6-inch-diameter pipeline from Bradford nearly 400 miles to its Bayonne and Weehawken refineries in New Jersey, near New York. After the capacity of this pipeline was doubled with a parallel pipeline in January 1883, Standard had the combined capacity to deliver 32,000 barrels of crude per day to New York. To these pipelines was added another seaboard pipeline from McKean County to Philadelphia, from which a branch was later extended to Baltimore. By 1888 over 70 per cent of crude shipments in the United States would be by pipeline, almost entirely by Standard's National Transit Co.[47]

Standard Oil, through the control it wielded over crude transit, was now able to dictate which refineries would survive and how large they could grow. The only way to operate independently of the Standard alliance was by adopting Tidewater's strategy of 'vertical integration' of operations, from production, through transit, to refining, designed to prevent Standard from holding all hostage by controlling either of the latter two stages. Tidewater had to establish its own refineries, the Ocean Oil Co. and the Chester Oil Co., after several independents that its pipeline was intended to serve, such as the new Solar Oil Co. at Williamsport, were taken over by Standard. Even Tidewater had to resort to entering into compromising business arrangements with the Standard behemoth, in return for being allowed to retain circumscribed independent operations. In 1883 Tidewater narrowly held on to its independent status when a Venango county court judge ruled that a concerted takeover attempt by Standard had been 'farcical, fraudulent and void' – to the evident relief of the *Philadelphia Press*:

> Without any reference to the legal merits of the controversy, it is enough that the decision will keep the Tidewater Pipe Line free from the control of that overshadowing corporation of the State of Ohio which has swallowed up or strangled every other competitor in the business of transporting, storing and refining oil but this.[48]

'CORPORATIONS DERIVE THEIR POWERS FROM THE PEOPLE'

The oil-storage and shipment crisis of 1877 that had prompted the ambitious construction of independent trunk pipelines also elicited a political response from the Petroleum Producers' Union which protested that Standard Oil had abused its monopoly control over oil storage and transit in order to unfairly reduce the price of crude. The producers claimed that United Pipe Line was declaring its tanks full when they were not and that, in league with the railways, the pipeline company was continuing to run a transit cartel that charged unfair and discriminatory freight rates. Since the state legislature had issued charters of incorporation to the pipeline and railway operators, charters that came with special privileges such as public land grants and rights of eminent domain, was it not clear that these corporations had a duty to serve the public interest and to assist rather than to hinder the efforts of oil producers to make a fair profit from their labour? The Union adopted a resolution declaring that 'the system of freight discrimination by common carriers is absolutely wrong in principle and tends to the fostering of dangerous monopolies, and that it is the duty of Government, by legislative and executive action, to protect the people from their growing and dangerous power'.[49] In an appeal to Pennsylvania Governor John Hartranft in 1878, the petroleum producers employed language that was common in nineteenth-century America:

> By the theory of the law, corporations derive their powers from the people of the commonwealth in General Assembly convened; they have no powers not delegated to them by the people; they take nothing by implication; they are public servants, invested for the public benefit with extraordinary privileges, and their charters may be taken from them when they cease to properly perform the duties of their creation. The railroad and pipeline companies are common carriers of freight for all persons, are bound to receive it when offered at convenient and usual places, and to transport it for all, for reasonable compensation, without unreasonable compensation in favor of any. These are but simple statements of well established legal principles . . .

Yet the people who granted these special privileges are now upon the defensive, their rights denied by these corporations, and they are challenged to enter the courts to establish them, while in the meantime they are inoperative, to the irreparable injury of their business ... These corporations have made themselves the interested tools of a monopoly that has become the buyer, the carrier, the manufacturer, and the seller of this product of immense value ... That monopolies are dangerous to free institutions is a political maxim so old as to have lost its force by irrelevant repetition, but if anything were needed to awaken the public sense to its truth, the immediate effect of this giant combination is before us ... [W]e ask that immediate steps be taken to enforce, by legislative enactment, the wise provisions of our State constitution ... and by such legal processes as are necessary, compel obedience to law and the performance by chartered companies of their public duties.[50]

National public opinion was now beginning to recognize the oil industry as another sector of the economy – alongside others such as sugar-refining, tobacco-processing and meat-packing – where large corporations were emulating the existing industrial behemoths of the railway, steel and coal industries. On 4 February 1879 New York's *Daily Graphic* ran a full front-page cartoon entitled 'The Standard Oil Octopus'(see Fig. 8), and an article on the following page called for federal and state legislation 'to confine these monopolies':

In the striking cartoon on our first page our artist has well figured the great Standard Oil monopoly as a devil fish, whose many tentacles are wound around the whole oil business and are used as conduits to draw to the central 'purse' all the profits which should go to enrich producers and consumers.

Against the oil octopus competition is out of the question. Its suckers have been attached to the wells, to the pipe lines, to the refineries, and nothing but the death of the monster will make him release his hold ... This great advantage permits the managers of the monopoly not only to crush out the producers by purchasing crude oil at starvation rates, but it enables them also to charge what figure they please to the retailer and to the public. In this way that grasping monopoly was enabled to make $10,000,000 in profits in four months.

It is idle to think of trusting these people. Might as well trust a rattle-snake. Monopoly is always monopoly. It is always a law unto itself. Its managers are believers in the 'higher law', such higher law being their own greedy instincts.[51]

Two years later the *Atlantic Monthly* magazine published the first comprehensive analysis of the rise of Standard Oil in 'Story of a Great Monopoly' by Henry Demarest Lloyd, which made a great impression on the reading public. Lloyd argued that the company had achieved its market dominance

... by conspiracy with the railroads ... The Standard killed its rivals, in brief, by getting the great trunk lines to refuse to give them transportation. Commodore Vanderbilt is reported to have said that there was but one man – Rockefeller – who could dictate to him ... So closely had the Standard octopus gripped itself about Mr. Vanderbilt that even at the outside rates its competitors could not get transportation from him.[52]

Standard Oil had already exerted a corrupting influence on inquiries into the rail companies' setting of freight rates, Lloyd claimed: 'The Standard has done everything with the Pennsylvania legislature, except refine it.' Indeed, exactly a year after Lloyd wrote this article, Standard executive J.N. Camden wrote to Rockefeller, 'I have arranged to kill the two bills in Md. [Maryland] legislature at comparatively small expense.'[53] In 1888, after the former Republican governor of Ohio, Charles Foster, wrote unabashedly to Rockefeller requesting a campaign contribution of $1,200, Rockefeller duly obliged, though with a firm quid pro quo: '[W]e have not received fair treatment from the Republican Party, but we expect better things in the future.'[54] Simultaneously, Rockefeller was bribing Detroit politicians in order to secure the city's municipal natural gas franchise.[55]

The rise of powerful business corporations across America during the nineteenth century presented a particular conundrum to a society founded – in the popular imagination, at least – on the ordinary individual's freedom from oppressive economic or political power.[56] As the possibilities of large-scale marketing were opened up by a growing transportation network, powerful cartels with increasing market dominance were beginning to appear, with profits concentrating in a small number of big businesses that came to control transport, processing, manufacturing, packaging and marketing. It appeared that profitable economic activity was gravitating towards urban centres and becoming dominated by a small number of business oligarchs from the railways and allied industries and, ultimately, from Wall Street.[57] This expanding network of railroad and big business interests came to be popularly portrayed as an octopus, a growing monster with tentacles reaching out to, and controlling, every important sphere of economic and political life. Yet it was widely believed that the American Revolution had been fought for independence, in large part, from Britain's mercantile trading companies that had monopolized and controlled much of the colonies' economic activity – through privileges contained in those companies' corporate charters, granted at first by the monarch and later by Parliament. That home-grown business corporations were now controlling huge swathes of economic activity and beginning to wield great political power struck many American citizens as little short of a domestic counter-revolution. In some confusion, they found themselves returning to the question of the 'corporation', an old battleground in the perennial contest over political and economic sovereignty.[58]

Under English common law, charters of incorporation were grants by the sovereign power of official rights of self-government. Early recipients of corporate privileges included various ecclesiastical, educational and charitable groups, trade guilds, and municipalities such as towns and boroughs. In principle, corporate status and privileges were granted in the interest of effective, delegated administration: a corporate 'body' took on an officially recognized legal existence that could continue beyond the lifetimes of its individual members; it could own collective property; and it could enter into contracts or appear in legal proceedings in its own name. The intention was that the groups of people who were granted these powers and freedoms of self-government should exercise their autonomy in a way that was consistent with the public good. In 1437 the English Parliament explicitly legislated that activities of incorporated bodies should not cause 'common damage to the people', and a similar enactment of 1504 was, according to Francis Bacon, intended 'to restrain the by-laws or ordinances of corporations, which many times were against the prerogative of the King, the common law of the realm, and the liberty of the subject, being fraternities in evil'.[59]

In 1659 the first English treatise on corporations stated that their powers should not be 'repugnant to the Lawes of the Nation, against the publick and common good of the people'; and the legal authority Edward Coke argued that a corporate power 'ought to be made in furtherance of the public good and the better execution of the laws, and not in utter prejudice of subjects or for private gain'.[60]

The powers of self-government awarded under corporate charters, alongside the similar 'letters patent', were the formal, legal expression of the current distribution of power in society, in an ongoing contest for political and economic sovereignty.[61] For example, the early incorporation of towns and boroughs, where considerable economic activity had begun to be conducted outside the traditional manor system of the feudal nobility, were officially recognized transfers of sovereignty; so was the incorporation in these commercial centres of guilds of crafts people and merchants that, often controversially, came with monopoly trading rights.[62] Since charter privileges granted by the monarch were generally acts of patronage, rewards for allegiance, they were the legal manifestation of contests for power and sovereignty.[63]

From the late sixteenth century, beginning with Queen Elizabeth I, English monarchs had awarded charters of incorporation, including monopoly privileges, for the formation of increasingly ambitious overseas trading enterprises. Although they were driven by private gain, they were seen also as operating in the state's imperial and commercial interests. Furthermore, in order to attract the funds required for such capital-intensive, high-risk and long-term speculative endeavours, these enterprises – most notably the East India Company – began to combine the familiar corporate form of organization with a public, joint-stock method of financing. The more lucrative these trading concerns became, the more those invested in them became protective of their privileges from the dictates of the sovereign, the monarch. At the same time, however, those excluded from economic activity by exclusive corporate monopoly rights often became hostile towards the monarch; and these mutually opposing antagonisms toward the Crown became a significant factor behind Parliament's more assertive claim to sovereignty that led to the outbreak of the English Civil War in 1642.[64]

Under their charters, England's overseas trading and colonial corporations were delegated sovereign powers, thereby authorizing their directors and agents to act as almost autonomous branches of the British state when operating far from home shores for many months or years at a time.[65] Thus, the administrators of English transatlantic enterprises of the early seventeenth century onwards were authorized, under various forms of royal charter, to colonize specified regions of North America. The colonial administrators, in turn, issued their own corporate charters: for the official creation of municipalities such as boroughs and towns; for non-profit organizations such as hospitals, churches, colleges and charities; and for profit-making, joint-stock enterprises for public works such as the construction of roads, canals and bridges that might not otherwise be built without various corporate privileges. However, the exclusive rights granted to transatlantic trading corporations, and the associated restrictions imposed on Americans' economic freedom, led American revolutionaries in the 1760s to begin to argue for complete independence from both the English Parliament and the Crown. Yet the migration of the English common-law tradition regarding corporations across the Atlantic informed the terms of debate among opponents of revolution who argued, by contrast, that since the colonies were founded on the basis of royal charters, the king was sovereign over them and he should reclaim, on their behalf, the powers that England's Parliament – in which they had no representation – had usurped and exerted over them. Nevertheless, following full independence, when the legislatures of the thirteen newly independent states took over full sovereign powers, there remained an acute awareness that the privileges associated with incorporation had long been abused at the people's expense. Any similar infractions, therefore, could now be seen as jeopardizing the republic's newly won egalitarian distribution of power.

Thus, it was a basic principle of American republicanism that commercial corporations could become problematic, given their historic tendency to deviate from serving the public interest – as Americans had felt towards England's transatlantic trading companies.[66] Consequently, there was profound resistance from those of a more egalitarian persuasion when a political faction in Congress, spearheaded by Alexander Hamilton, passed a bill in 1791 to federally incorporate a national bank, the Bank of the United States, the ultimate aim of which was to bind the Union more closely and turn it into a financial and industrial power to rival Britain and other European nations.[67] At the state level, when in the following year New Jersey granted a corporate charter to an organization of manufacturing businesses, a critic asked,

> can it be doubted, whether it violates the spirit of all just laws? Whether it subverts the principles of that equality, of which freemen ought to be so jealous? Whether it establishes a class of citizens with distinct interests from their fellow citizens? Will it not, by fostering an inequality of fortune, prove the destruction of the equality of rights, and tend strongly to an aristocracy?[68]

In 1801 some wealthy New York merchants petitioned for a charter to form a city-wide bread company. This led independent bakers, whom the prospective company hoped to employ, to fear that by becoming dependent on a large business with the power to control labour and reduce wages, 'the independent spirit, so distinguished at present in our mechanics, and so useful in republics, will be entirely annihilated'.[69]

However, between 1807 and 1812, escalating disputes with Britain cut off America's imports of British manufactured goods, and even the more agrarian traditionalist Thomas Jefferson now accepted Hamilton's arguments for promoting industrialization in order to maintain the national economic independence needed to wage war successfully. State legislatures now began, as a matter of public policy, to facilitate the issuance of many more corporate charters for transportation projects and manufacturing enterprises.[70] Although this shifted the notion of the 'corporation' towards the private sphere, these charters usually came with strict limitations on the amount of capitalization, the kinds of economic activity that could be engaged in and the duration of the charter. As the Supreme Court of Virginia spelled out in 1809, corporate charters were only to be granted for the public benefit:

> With respect to acts of incorporation, they ought never to be passed, but in consideration of services rendered to the public . . . [O]ur bill of rights interdicts 'all exclusive and separate emoluments or privileges from the

community, but in consideration of public services'. It may be often *convenient* for a set of associated individuals, to have the privilege of a corporation bestowed upon them; but if their object is merely *private* or selfish; if it is detrimental to, or not promotive of, the public good, they have no adequate claim upon the legislature for the privilege.[71]

In 1816, four years after the 1791, twenty-year charter of the First Bank of the United States had expired, Congress incorporated the Second Bank of the United States to further stimulate a financial and investment environment conducive to larger industrial enterprises, and increasing numbers of private manufacturing businesses took advantage of the opportunity to incorporate. However, legislatures were initially cautious and did not, at first, routinely grant limited liability, which was associated more with public works.[72] Nevertheless, many still saw this as a fundamental threat to the young republic's economic and political balance of power. If their union of states had fought the War of Independence to free themselves of the British monarchist plutocracy, they now began to feel vulnerable to a growing domestic republican plutocracy. Some months after the chartering of the Second Bank, Jefferson wrote of his views of the present state of England, of 'the ruin of its people' by its 'hereditary aristocracy': 'I hope we shall take warning from the example and crush in its birth the aristocracy of our monied corporations which dare already to challenge our government to a trial of strength and bid defiance to the laws of our country.'[73]

The threat from private business corporations seemed more acute after the US Supreme Court ruled, in the *Dartmouth College* case of 1819, that a charter of incorporation was legally a contract between a private group and the state legislature and was, therefore, protected by the contract clause of the US Constitution.[74] This meant that once a state had created a corporation, the state's sovereign powers over its creation were then strictly limited. The ruling provoked protests against the federal court's attack on state – and by implication the people's – sovereignty. Thomas Earle complained in a pamphlet, 'It is aristocracy and despotism, to have a body of officers, whose decisions are, for a longtime, beyond the control of the people', while Massachusetts senator David Henshaw declared, 'Sure I am that, if the American people acquiesce in the principles laid down in this case, the

Supreme Court will have effected what the whole power of the British Empire, after eight years of bloody conflict, failed to achieve against our fathers.'[75]

These protesters would have been disheartened to read Judge Thomas Cooley's assessment of the significance of this case fifty years later: 'It is under the protection of the decision of the Dartmouth College case that the most enormous and threatening powers in our country have been created; some of the great and wealthy corporations actually having greater influence in the country at large and upon the legislation of the country than the States to which they owe their corporate existence.'[76]

The new possibilities for mass trading created by canals and the ever expanding railways were exploited particularly by those businesses that – like the large railroad corporations themselves – strove for aggressive expansion, a process that was facilitated by the large-scale sale of stocks and bonds conducted by the growing and concentrating banking sector.[77] Faced with these economic developments, what was to become of the traditional republican values of political egalitarianism, economic individualism, autonomy and self-reliance? This question defined political debate during the Jacksonian era, which Louis Brandeis, a future Supreme Court Justice, described as a time of

> Fear of encroachment upon the liberties and opportunities of the individual. Fear of the subjection of labor to capital. Fear that the absorption of capital by corporations, and their perpetual life, might bring evils similar to those which attended mortmain. There was a sense of an insidious menace inherent in large aggregations of capital.[78]

Independent carriage-makers, for instance, opposed an application for incorporation on the grounds that 'incorporated bodies tend to crush all feable enterprise and compel us to worke out our dayes in the Service of others.'[79] Should private business corporations even be permitted at all, when the common-law partnership or association had sufficed for centuries? In 1835 a group of tradesmen announced, 'We entirely disapprove of the incorporation of Companies, for carrying on manual mechanical business, inasmuch as we believe their tendency is to eventuate in and produce monopolies, thereby crippling the energies of individual enterprise.'[80]

Two years later, the proceedings of Pennsylvania's Constitutional

Convention recorded the opinion of law professor James M. Porter that

> the subject of corporations [he said] had not yet, in this country, received that consideration to which it was entitled. Our courts had followed the rule laid down in England, which he did not consider a proper precedent for this country to pattern after. There they … conferred special favors upon corporations on the ground that it was so much power taken from the sovereign and conferred upon the people. Now in this country the sovereignty is vested in the people themselves, and whatever power is granted to corporations, is so much abstracted from the people themselves.[81]

Pennsylvania's governor, Francis Shunk, concurred. Corporations were politically 'behind the times' and belonged 'to an age past' when charters of incorporation had brought a new degree of freedom to many inhabitants of medieval English towns and boroughs:

> The time was, in other countries, where all the rights of the people were usurped by despotic governments, when a grant by the King to a portion of his subjects, of corporate privileges, to carry on trade, or for municipal purposes, was a partial enfranchisement, and made the means of resuming some of their civil rights. Then and there, corporations had merits, and were cherished by the friends of liberty. But, in this age and country, under our free system, where the people are sovereign, to grant special privileges, is an invertion of the order of things. It is not to restore, but to take away from the people, their common rights, and give them to a few. It is to go back to the dark ages for instruction in the science of government, and having found an example, to wrest it from its original purpose, and to make it the instrument of restoring the inequality and despotism, which its introduction tended to correct.[82]

William Gouge, a prominent egalitarian economist, charged that 'Against corporations of every kind, the objection may be brought that whatever power is given to them is so much taken from either the government or the people. As the object of charters is to give to members of companies power which they would not possess in their individual capacity, the very existence of monied corporations is incompatible with equality of rights.'[83]

To these prevalent concerns were added widespread accusations that corporate charters for private businesses were routinely being awarded on the strength of the applicant's political connections. Gone were the trade monopolies and other arbitrary regulations that came with the British Empire's mercantile control and that were designed for the benefit of London's banking houses. Yet by the mid-nineteenth century there were signs of newly concentrating economic and political power as the granting of corporate charters once more became the currency of political patronage; indeed, similar developments were generating unease back in Britain itself. However, in a curious political twist, instead of binding incorporation to the public interest and endeavouring to combat political corruption, states made private incorporation easier. In what was presented as a progressive, democratizing shift, incorporation would no longer be a privilege reserved to a few but would be made more freely available: From the mid-1830s, general, or free, incorporation laws were introduced that, through incremental liberalization, allowed a widening range of businesses to incorporate simply by fulfilling a few registration requirements. In the name of freedom and equality, the chartering process began to lose its traditional constitutional moorings in legislative scrutiny and the public benefit. The right to form a corporation had, for centuries, been a special privilege granted by the prevailing sovereign power explicitly to serve the public interest. Now, the legal and financial advantages of the corporate form were coming to be seen as a natural right, almost automatically available to any private business venture regardless of its public utility.[84] For a legislature to give up this central element of sovereignty was to enter uncharted legal territory, particularly when even those corporations that had been formed by special charter were running beyond the control of the democratic legislature. As future president General James Garfield warned in 1873,

> In most States each legislature has narrowed and abridged the powers of its successors, and enlarged the powers of the corporations; and these, by the strong grip of the law, and in the name of property and vested rights, hold fast all they have received. By these means, not only the corporations, but the vast railroad and telegraph systems, have

virtually passed from the control of the State. It is painfully evident, from the experience of the last few years, that the efforts of the States to regulate their railroads have amounted to but little more than feeble annoyance. In many cases the corporations have treated such efforts as impertinent intermeddling, and have brushed away legislative restrictions as easily as Gulliver broke the cords with which the Liliputians [*sic*] attempted to bind him. In these contests the corporations have become conscious of their strength, and have entered upon the work of controlling the States. Already they have captured several of the oldest and strongest of them; and these discrowned sovereigns now follow in chains the triumphal chariot of their conquerors.[85]

In the legislatures and in the courts, the newly emerging big business corporations – the most prominent of which, at this stage, were the railways – pushed relentlessly against receding checks on their power, and they would readily flout the law where its enforcement came only from the few private litigants who could afford to battle against teams of highly paid corporate lawyers.[86] Without active state or federal enforcement of their public interest duties, the railway companies were routinely able to aid the larger businesses with secretly reduced freight rates in return for their regular, high-volume shipments. As it became clear how much political and economic power these big business cartels were acquiring, the ordinary small business people or farmers began to feel correspondingly disempowered and vulnerable. The protests made during the late 1870s by the Petroleum Producers' Union were typical of the times: big business corporations were widely seen as taking control over economic activity in a manner that was anathema to popular values of revolutionary republicanism.

In the mid-1880s, when regulations eventually began to appear on the statute books that, it was thought, would check the power and influence of big business, the defence of the public interest already often sounded like a belated, lost cause. Senator Lewis Emery Jr, the crusading independent oil producer, clamoured in the Pennsylvania assembly against the corporate interests behind the prevailing oil-transit cartel: 'What right has that company to steal our business? It is an octopus that has driven me three times to bankruptcy . . . Do we own ourselves? If we do let us give ourselves some rights.'[87]

THE STANDARD OIL TRUST

In 1882 Standard Oil took corporate consolidation a stage further. At that time, one of the remaining obstacles to corporate expansion was the legal prohibition against the 'holding company' – a corporation holding the stock of another corporation – which was a major reason why corporate takeovers or mergers were implemented as relatively informal alliances. By 1882 Standard's sprawling network of allies had become so extensive, and the associated legal and administrative complexities so acute, that its chief lawyer, Samuel C.T. Dodd, devised a way to sidestep the stock-ownership prohibition by setting up a trust – whose trustees were Standard alliance's top management – to centralize ownership of the vast array of stock.[88] The Standard Oil trust thereby consolidated its control over a vast business empire in a way that would have been illegal for the Standard Oil corporation itself. When other large business combinations followed Standard's example, it seemed – in what became known as the problem of 'trusts' – that corporations were escaping from their legal confines and threatening to take over the nation. As Henry Demarest Lloyd wrote in the *Atlantic Monthly*, 'The time has come to face the fact that the forces of capital and industry have outgrown the forces of our government.'[89]

It now appeared to many that private business corporations needed reining in; and since corporate domination of whole sectors of the economy had been made possible in large part by freight-rate conspiracies between the railways and big business, there erupted a 'fever for railroad regulation', as the *Philadelphia Press* put it.[90]

However, egalitarian republicanism was in retreat as the American economy increasingly became one of corporations rather than of individuals. From the 1880s corporate lawyers began to persuade the most senior judges – many of whom were themselves former corporate lawyers – that the convention of calling a corporation a 'person' for the purpose of legal proceedings implied that, under the US Constitution, a corporation was entitled to the legal rights of a real person. Judges began to overlook that it was precisely the constitutional function of the corporation to provide a legal entity that was distinct from any real person. Constitutionally, for hundreds of years corporate charters had been granted to groups of people as a privilege for specific purposes, and the charter would define and customize the corporation's functions, responsibilities and rights in ways that were, by definition, different from those of any real person. Nevertheless, business corporations would increasingly come to be regarded as fundamentally equal to real, flesh-and-blood people in the eyes of the law, despite such obvious differences as that a corporation cannot be executed or imprisoned, was most likely to be vastly more powerful than any individual, and could effectively be immortal.[91] As Judge Charles Walton of the Maine Supreme Court encapsulated the issue, 'Men are mortal, and their combinations short lived, but corporations are immortal and their combinations and acquisitions may go on forever; they may add field to field, wealth to wealth, and power to power, till they become too strong for the government itself; all experience shows that such accumulations of wealth and power are dangerous to the public welfare.'[92]

In response, defenders of business corporations – who included most academic economists of the day – argued that their growth was beneficial, necessary and, moreover, simply inevitable.[93] Certainly, easier nationwide transportation had made the widespread rise of big business possible, and the consequent capital-intensive industrialization and nationwide competition had clearly led to business combination, which Rockefeller, for one, argued was indeed necessary:

> To perfect the pipe-line system of transportation required in the neighbourhood of fifty millions of capital. This could not be obtained or maintained without industrial combination. The entire oil business is dependent upon this pipe-line system. Without it every well would shut down and every foreign market would be closed to us.[94]

For Rockefeller, business consolidation was justified by the economic order, efficiency and prosperity that he believed it brought. He eulogized big business as a benevolent, quasi-communist form of capitalism that, in 'the battle of the new idea of cooperation against competition' had defeated 'dreadful competition'.[95] Rockefeller's outlook reflected the vision of a popular utopian socialist movement of the day that took its inspiration from one of the best-selling novels of the Gilded Age, *Looking Backward: 2000–1887*, in which its author, Edward

Bellamy, foresees a time when all economic activity is rationally and efficiently subsumed under one 'Great Trust'.[96]

Yet there was widespread scepticism about whether the growth of such large-scale, private business corporations was either inevitable, necessary or beneficial. If business corporations were contingent legal constructs, how could their rise be inevitable and, by implication, beyond the reach of democratic control? Perhaps big businesses prevailed and expanded more because of their privileged access to finance and negotiating power in the market than by being more efficient and selling better products. And even if large business corporations were beneficial in some respects, what might be the social and political costs?[97]

Although regulation supposedly designed to contain corporate power was introduced in the late nineteenth century, it came too late, was largely ineffective, and was only rather sporadically enforced. Federal regulation of the railways, under the Interstate Commerce Act of 1877, began only after the giant corporations had already achieved dominance by exploiting deregulation, and the Act now gave the railways a legal mechanism for ending their recurrent rate-cutting wars which had reduced profitability across the industry. Limited popular understanding of how crucial the control over pipelines was to Standard Oil's market power meant that these escaped the common carrier provisions of the Interstate Commerce Act until 1914, too late to have any real effect.[98]

On the floor of Congress in March 1890, Senator John Sherman gave vent to a rising tide of public opinion:

> If the concentrated powers of [a] combination are intrusted to a single man, it is a kingly prerogative, inconsistent with our form of government, and should be subject to the strong resistance of the State and national authorities. If anything is wrong, this is wrong. If we will not endure a king as a political power we should not endure a king over the production, transportation, and sale of any of the necessaries of life. If we would not submit to an emperor we should not submit to an autocrat of trade, with power to prevent competition and to fix the price of any commodity . . .
>
> [The Senate] . . . has always been ready to preserve, not only popular rights in their broad sense, but the rights of individuals as against associated and corporate wealth and power . . .

[N]ow the people of the United States as well as of other countries are feeling the power and grasp of these combinations, and are demanding of every Legislature and of Congress a remedy for this evil, only grown into huge proportions in recent times. They had monopolies and mortmains of old, but never before such giants as in our day . . . Society is now disturbed by forces never felt before.

The popular mind is agitated with problems that may disturb social order, and among them all none is more threatening than the inequality of condition, of wealth, and opportunity that has grown within a single generation out of the concentration of capital into vast combinations to control production and trade and to break down competition. These combinations already defy or control powerful transportation corporations and reach State authorities. They reach out their Briarian arms to every part of our country.[99]

The Sherman Antitrust Act was trumpeted as a bulwark against the rising economic and political power of large corporations, and thereby protective of the 'industrial liberty' of the ordinary citizen and of small businesses.[100] However, in reality it was more a smokescreen designed to neutralize public protest; as Sherman himself – who had received campaign contributions from Rockefeller – hinted to the Senate: 'You must heed their appeal or be ready for the socialist, the communist and the nihilist.'[101] Most senators believed that the Act's wording was too vague to add anything to existing common law, thus rendering it virtually legally inert; and it was seen as coming too late to prevent a corporate takeover that many felt had already occurred. According to Senator James George, the legislation was not only inadequate to its supposed task, but it could also potentially be used to attack labour unions and workers' cooperatives as being unlawful. He lamented:

It is a sad thought . . . that the present system of production and of exchange is having that tendency which is sure at some not very distant day to crush out all small men, all small capitalists, all small enterprises. This is being done now. We find everywhere over our land the wrecks of small independent enterprises thrown in our pathway . . . Is production, is trade, to be taken away from the great mass of the people and concentrated in the hands of a few men who, I am obliged to add, by the policies pursued by our Government, have been enabled to aggregate to themselves large, enormous fortunes?[102]

Rockefeller nevertheless argued publicly that Sherman's bill was 'of a very radical and destructive character'.[103] During the 1890s, where the trust device itself was ruled illegal this was usually achieved in state courts under traditional corporation law, on the grounds that if a corporation's directors relinquished control over its operations to an outside board of trustees they were acting beyond the powers conferred by their corporate charter. In 1892 the Standard Oil trust was dissolved in just this way by the state supreme court of Ohio.[104] Judge Thaddeus A. Minshall wrote, in the unanimous decision, that even if the rise of a large corporation apparently brings some benefits such as cheaper goods,

> A society in which a few men are the employers and a great body are merely employees or servants is not the most desirable in a republic; and it should be as much the policy of the laws to multiply the numbers engaged in independent pursuits or in the profits of production, as to cheapen the price to the consumer.[105]

This dissolution of the Standard trust had almost no practical effect, however, as the same clique of businessmen continued to centrally manage the combination, at first informally and soon on a new legal footing. In 1896, New Jersey – enticed by the promise of incorporation fees that, for a small state, amounted to a sizeable income – greatly liberalized its general incorporation law. Crucially, the state removed the prohibition against a corporation holding the stock of another corporation, and Standard Oil, along with hordes of other businesses, newly incorporated in the state. By making the holding company device generally available, New Jersey transformed the corporation into a recursive building block that could legally merge and proliferate indefinitely, clearing the way for the building of modern business empires.[106] For liberating private corporate power to such a degree, effectively tearing up the other states' corporate restrictions in a 'race to the bottom', New Jersey became known as the 'Mother of the Trusts' and was branded the 'Traitor State'.[107] (In everyday language the word 'trust' came to refer to any market-dominating big business, even after the trust itself was superseded as the typical legal device for business combination.)

When action was taken under the Sherman Act against a sugar monopoly, the corporate defence lawyers were able to run rings around plaintiffs and even federal prosecutors. Similarly, as one senior Standard Oil executive advised Rockefeller in 1888, 'I think this anti-Trust fever is a craze, which we should meet in a very dignified way & parry every question with answers which while perfectly truthful are evasive of bottom facts! I would avoid the preparation of any statistics.'[108] Senator Sherman himself had admitted, 'A citizen would appear in such a suit at every disadvantage, and even the United States is scarcely the equal of a powerful corporation in a suit where a single officer with insufficient pay is required to compete with the ablest lawyers encouraged by compensation far beyond the limits allowed to the highest government officer.'[109] The sugar company's victory, in 1895, was taken as a green light for corporate consolidation by a holding company, marking the start of a decade-long wave of mergers.[110] The associate editor of the American Law Register wrote, 'if this decision stands, and it is true that the national government is powerless to protect the people against such combinations as this . . . then this government is a failure, and the sooner the social and political revolution which many far-sighted men can see already darkening the horizon overtakes us, the better.'[111]

When the Supreme Court later decided that holding companies were not immune to anti-monopoly proceedings, it adopted a definition of 'monopoly' so narrow that it failed to cover even some of the very largest corporations.[112] As the influence of big business threaded seamlessly through the body politic, private corporations knew that they could defend their interests, against those of the public, by relying on the American reverence for private property – ironically, originally intended to protect the individual from overbearing power. Notionally free individuals were becoming the subservient dependants of a corporate economic system, presided over by a few vast banking and investment houses that commanded a large chunk of the people's accumulated assets.[113] A kind of unelected, corporate shadow government had taken root, effectively with its own private powers to tax and spend. In the words of Justice John Marshall Harlan,

All who recall the condition of the country in 1890 will remember that there was everywhere, among the people generally, a deep feeling of unrest. The nation had been rid of human slavery ... but the conviction was universal that the country was in real danger from another kind of slavery sought to be fastened on the American people, namely, the slavery that would result from aggregations of capital in the hands of a few individuals and corporations controlling, for their own profit and advantage exclusively, the entire business of the country.[114]

STANDARD OIL, PRODUCERS AND INDEPENDENTS

Through the 1880s and into the 1890s, Standard Oil continued to dominate the US oil industry by maximizing its control over oil transit – the key to its control over producers and refiners.[115] Standard was quick to pre-empt the emergence of competitors who, following the vertically integrated example of Tidewater, might successfully forge an unbroken chain from production, through transit, to refining and marketing to achieve full operational independence. In 1883 Standard Oil trustee W.G. Warden, concerned about competition from independent gathering pipelines, wrote to Rockefeller, 'We must prevent all such enterprises from making any headway and make it as discouraging as possible for any such to start – No difference how small such an enterprise may start, we must prevent it from taking root as far as we can do so by legitimate & fair means.'[116] Thus, to prevent the Baltimore & Ohio Railroad from establishing an integrated oil business, in 1885 Standard made a rare, early purchase of a production site in West Virginia. When, in the same year, commercial quantities of oil were discovered near Lima, Ohio, Standard's National Transit was quick to organize a subsidiary, the Buckeye Pipe Line Co., to pre-empt the emergence of competition by rapidly constructing pipelines and storage facilities across the new producing areas and buying up independently laid pipelines.[117]

The heavy and sulphurous, or 'sour', Ohio crude could initially only be marketed as a new type of fuel for steam boilers. As Thomas

Gale, an Oil Region resident, had predicted as early as 1860, 'rock oil' was destined to become a major new source of fuel:

> One gallon of it will raise more steam than many times its weight and volume of coal or coke . . . And if the ingenious gentleman in Pittsburgh, or any other can get up a first rate apparatus for generating steam from this oil, with the saving of fuel and tonnage claimed, his patent will be worth more than the richest well in Oildom.[118]

Promoted for its advantages over coke, coal and wood for use in locomotives, steamships and various manufacturing processes, fuel oil was taken up in nearby Chicago, to which Buckeye laid an 8-inch-diameter pipeline in 1888, and then in Cleveland, Detroit and beyond: 'Oil, we are convinced, is the fuel of our business for the future,' wrote one factory owner.[119] It was not long before this type of crude was also being refined into kerosene, after Standard patented a process, devised by Herman Frasch, for removing the sulphur, for which independent refiners developed their own methods.

In 1887, following the continuing high overall levels of production from the Appalachian fields – including New York, Pennsylvania, West Virginia, Kentucky and southeastern Ohio – and the rapid rise of central Ohio production, the price of crude dropped to as low as 54 cents a barrel, the lowest since 1861. Pennsylvanian producers attempted to address the situation initially through their state legislature by introducing a bill aimed at trimming the powers of National Transit. With little or no competition in most oilfields, the Standard Oil pipeline subsidiary was accused of profiting unreasonably, at the expense of the producers, from excessive crude transit and storage fees. National Transit should be prevented from abusing its monopoly position, they argued, by being regulated as a common carrier of crude.[120] However, the Billingsley Bill was defeated – with the help of senators bribed by Standard Oil – whereupon the more determined producers resolved to establish that ever-elusive unity needed for the more effective promotion of their interests, and they reaffirmed 'the fact that the Legislature of the State has the power to regulate, control, and if necessary for the public good, destroy this monster corporation'.[121] They promptly formed the Producers' Protective Association,

To include in one organization of all producers of petroleum, and those who are engaged in industries incidental thereto, and known to be friendly to the producers' interests, in order that they may, by united action and all honorable means, protect and defend their industry against the aggressions of monopolistic transporters, refiners, buyers, and sellers of their products, in order that the producers may reap the just reward of their capital and labor, and to this end encourage and assist as far as possible the refining and marketing of their product and sale direct to the consumer by the producer.[122]

They aimed to organize large, integrated oil companies of their own and, somewhat ironically, they now worked in cooperation with their arch enemy Standard Oil to address the over-production of crude. If the producers had coordinated production restraint alone, Standard would simply have released onto the market its own crude – of which it now had a large inventory in storage – for an increased profit at the producers' expense. But Standard had its own motives for cooperating to lift crude prices. It had decided to move rapidly into the production phase itself: firstly as an insurance against a predicted decline in traditional Appalachian output; secondly to ensure that its capital-intensive pipelines were kept working to capacity; and thirdly as a way of forestalling the emergence of independent integrated oil companies, as a higher crude price would stimulate the necessary investment in Standard's new crude production. On 1 November 1887 the *Titusville Morning Herald* headlined with 'A New Era in Oil, The Great Shutdown Movement Commences Today', and although recurrent over-production would remain an intrinsic feature of the oil industry, this one-year agreement proved almost uniquely successful at restraining production and raising crude prices. At the year's end, drilling activity and crude output increased and prices fell again.[123]

By 1891 Standard Oil was producing 56 per cent of the heavy, sour Ohio-Indiana crude and 9 per cent of the light, sweet Appalachian crude. Standard concentrated its refining operations in fewer, larger refineries, strategically located: at Cleveland; at Whiting, Indiana, for the western and southern markets; and along the east coast, in New Jersey, New York and Philadelphia for the eastern and export markets. The Buckeye Pipe Line system, covering the new Ohio-Indiana field, was linked to Whiting via the Connecting Pipe Line and the Indiana Pipe Line, all of which were then connected via the Cygnet Pipe Line to Standard's original pipeline system across the Oil Region. In 1890 National Transit entrenched its coverage of the entire Appalachian field when its Eureka Pipe Line brought production from Kentucky and northern Tennessee into the system, and its Southern Pipe Line connected West Virginian production with the East Coast.[124]

It was only where Standard was unable to extend its pipelines, and thus its control over new producing areas, quickly enough that producers and independent refiners stood a chance of setting up independently of the oil giant. For example, several independent oil companies emerged during a flurry of production from the new McDonald oilfields on the Pennsylvania-West Virginia border. The Producers' Protective Association raised $500,000 to form the Producers' Oil Co. Ltd. that, a year later in 1892, was operating a 15-mile pipeline from McDonald to the Coraopolis rail depot northwest of Pittsburgh. When rate discriminations hampered their rail transport of crude to the Columbia Oil Co., an independent refiner and exporter at Bayonne on the East Coast, they laid a pipeline to independent refiners in the Oil Region at Titusville and Oil City, jointly organizing as the Producers' and Refiners' Oil Co. Ltd.[125] In order then to overcome the obstacle of conveying independently refined kerosene to the East Coast for export, Lewis Emery Jr resurrected the idea, first implemented in 1865, of constructing a long-distance kerosene pipeline. After a year of weaving around Standard Oil's obstructive land purchases, wrestling with the Standard-allied railways in court over rights of way, and running the gauntlet of repeated armed conflict with rail workers, the United States Pipe Line Co. eventually, in mid-1893, completed a 200-mile parallel kerosene and crude pipeline from the Oil Region to Wilkes-Barre, Pennsylvania, from where the Central Railroad of New Jersey transported the oil to the seaboard.[126] Ida Tarbell, a popular critic of Standard Oil and supporter of the independents, recognized that 'a new advance had been made in the oil industry – the most substantial and revolutionary since the day the Tidewater demonstrated that crude oil could be pumped over the mountains'.[127]

Simultaneously, William L. Mellon, of the wealthy Pittsburgh banking family, organized the Crescent Pipe Line Co., which, after being blocked at several stages, managed in late 1892 to complete a 5-inch-diameter pipeline 270 miles from the McDonald oilfields to its independent refinery and docks at Marcus Hook, Philadelphia. However, even this enterprise, like so many others, was sold to Standard after, in 1895, Pennsylvania repealed an 1883 law limiting pipeline consolidation.[128] Governor Robert E. Pattison, who had originally ratified the law, vetoed a first attempt to repeal it, arguing, 'The inevitable effect would be to drive competing lines into consolidation or to put the shippers of this important product at the mercy of the great monopolies which might be able to secure and hold the controlling interest in the stock or bonds of competing lines.'[129]

The successes of Mellon's and Emery's vertically integrated oil enterprises were, however, notable exceptions to the rule of Standard Oil's huge dominance over oil transit, refining and marketing. Since in most producing areas Standard was the sole purchaser of crude oil, local oil trading and the oil exchanges became virtually irrelevant, to the extent that from January 1895 Standard's crude oil purchaser, Joseph Seep, took to dictating to producers a daily 'posted price'. (This was the precursor to the emergence, in the early 1930s, of a government-supported system of crude production quotas and prices, regulated by the Texas Railroad Commission, that would evolve eventually into a global system of controlled oil production run by a cartel of dominant American and European companies – a system that prevailed until the 1970s when the world's largest crude-exporting countries came together to organize a producers' counter-cartel, the Organization of the Petroleum Exporting Countries, or OPEC.)[130] Standard set a price for crude that narrowed the profit margins of independent refiners to perilously low levels, prompting those who did not now sell out to Standard to consolidate themselves into a sizeable oil company, the Pure Oil Co., vertically integrated from upstream crude production all the way through to downstream domestic and foreign marketing.[131] In 1896 United States Pipe Line allied with the Pure Oil Co. to extend its independent pipeline, following yet more armed skirmishes and court battles.[132]

From 1880 to the turn of the century Standard Oil controlled approximately 85 per cent of refinery capacity and sales in the United States while it steadily extended its control over the Canadian industry, eventually taking over the large, integrated, Imperial Oil Co. in 1898. US refinery throughput nearly tripled over the period, from roughly 18 to 52 million barrels of crude per year, with lubricants and fuel oil each rising to about 10 per cent of product output. The US was exporting increasing volumes of kerosene – around 70 per cent to Europe and 20 per cent to the Far East – along with appreciable quantities of crude.[133] Standard's foreign representative, William Herbert Libby, wrote to the US Minister to Turkey that petroleum had 'forced its way into more nooks and corners of civilized and uncivilized countries than any other product in history emanating from a single source'.[134]

While oil production remained concentrated in a region centred around Lake Erie and Pittsburgh, Standard Oil reigned supreme. It was only when commercially significant volumes of crude began to be produced further afield, in locations where Standard was unable to extend its control, that the oil giant's dominance came seriously to be challenged. Significant competition of this kind first appeared in 1883, when a booming new Russian oil industry began to compete against the US industry for a share of the world's markets.

5

A Global Industry

The oil industry of the northeastern United States, plus an adjacent region of Canada just to the north, had by the 1870s demonstrated the potentially great profitability of producing crude oil on a large scale, and particularly of refining crude into kerosene and other petroleum products. Around the world, oil lamps of continually improving design were burning with a cleaner, brighter flame – almost all of which were filled with Standard Oil's cheap, mass-produced, mass-marketed illuminant. Now, however, Standard became the victim of the very success of its aggressive global marketing. The burgeoning international demand for kerosene, which the conglomerate had worked hard to stimulate and from which it had profited so hugely, now presented a major opportunity for competitors to enter the market. Within the US itself, Standard's control over the pipelines serving the oilfields of the Northeast and Midwest, coupled with its huge financial clout, stifled virtually all serious competition. But beyond its national sphere of influence Standard's hegemony was much more vulnerable, and the further one looked from the refining and export centres of the US East Coast, the lower the all-important transport costs would be for anyone thinking of exploiting local petroleum reserves on an industrial scale.

From the 1850s, settlers in California had extracted oil from hand-dug wells and tunnels to use for illumination and asphalting streets.[1] In the early 1840s this had been Mexican territory, but Americans feared that Britain had designs on the large northern Mexican province of Alta California which was mortgaged to British creditors; so, from 1846 to 1848 the United States waged a major, expansionist war against Mexico, based on a relatively minor pretext and justified by the notion that it was America's 'Manifest Destiny' to be the vanguard of human evolution and progress. The US seized almost half of Mexican territory, encompassing an expanded Texas plus New Mexico, Arizona, Utah, Nevada and California.[2] However, within this area there lived Native American tribes such as the Comanches, Apaches, Pueblo and Navajo of Texas, New Mexico and Arizona, while in California there resided a multitude of small tribes such as the Modoc, Pomo, Miwok, Chumash and Tongva. Between 1846 and 1873 – during which time, in 1848, gold was discovered in great quantities – the natives of California were subjected to one of the clearest examples of genocide in the bloody history of the westward expansion of America.[3] As usual, the largely one-sided violence was justified as the inevitable eradication of an inferior race by technologically and morally superior white people. Governor Peter Burnett, in his 1851 annual message to the state legislature, declared, 'That a war of extermination will continue to be waged between the races, until the Indian race becomes extinct, must be expected.'[4]

In 1864 Professor Benjamin Silliman Jr, of Yale University, arrived in California as a highly paid consultant working initially for gold, silver and mercury mining companies, and then for an oil prospector who had acquired rights to 100,000 acres east of San Buenaventura (Ventura). Silliman quickly concluded that the surface oil found there was thick and bituminous due to being baked in the sun and that lighter oil would be found by drilling.[5] Another interested party in Californian minerals was Thomas A. Scott, director of the Pennsylvania Railroad, and in return for a consultancy fee of $1,500 Silliman sent Scott effusively optimistic reports on the commercial potential of oil on his land there:

> Suffice it to say, that having made the first researches on the products of Oil Creek, long before any wells were bored there, I am of the opinion that the promise of a remarkable development at Buenaventura is far better than it was in the Pennsylvania or Ohio regions – since so famous ... Every mine of metals is a magazine of limited supply ... Not so with petroleum. It flows on year after year, and still the source of supply seems unimpaired.[6]

Back on the East Coast, Silliman gave presentations on the great commercial potential of Californian oil, while acting as a consultant to oil companies intending to extract oil on further tracts of land that Scott had acquired. But the head of the official California Geological Survey was extremely sceptical, writing, 'If Silliman's reports are correct, I am an idiot and should be hung when I get back to California.'[7] Indeed, by the end of 1865 none of the oil companies drilling in California had managed to strike flowing oil, and the heavy crude was not easy to refine into marketable products. By early 1867 the investments in these companies had evaporated to nothing, leading to denouncements of Silliman from fellow scientists and to lawsuits against him and the oil companies for whom he consulted; and in 1870 Yale University relieved Silliman of his teaching post. In any case, whatever the character of Californian oil, the rudimentary transport infrastructure would have hampered the movement of large volumes of crude to a refinery and of products to customers. The transcontinental rail connection was not completed until 1869, and even then the high cost of rail freight meant that equipment for drilling and refining had to be shipped from the East Coast all the way round Cape Horn.[8]

One of the earliest uses of modern mechanical drilling methods outside North America was undertaken in Peru, at Zorritos in the vast Mancora hacienda situated in the northern tip of the country, southeast of Tumbez. In 1863 the commercial agent of the estate's owner, Diego de Lama, called upon the Peruvian government to send one of its civil engineers, A.E. Prentice, to carry out exploratory drilling for commercial quantities of oil of a kind suitable for refining into lamp oil.[9] The results were encouraging enough that George H. Bissell – a founding member of the pioneering Seneca Oil Co. – was moved to set up the *Compañia Peruana de Petroleo*, which by 1866 had a share capital of $5 million. In a report of that year the company argued that, due to the proximity to the coast of its proven oil wells, the low transportation costs as contrasted with Pennsylvanian oil would enable it not only to capture the markets of Peru, Chile and Ecuador but also to compete successfully with American oil on the West Coast of the US, in Australasia and in Europe. Over the next few years several other entrepreneurs set up oil companies at Zorritos, while the American railway magnate Henry Meiggs exploited oil further down the coast at

Negritos, near Talara, and built a refinery at Callao. However, Meiggs' coastal installations were destroyed by the Chilean Navy's *Amazonas* warship following the outbreak of the War of the Pacific in February 1879.[10]

Further north, in Mexico City, attempts were made between 1860 and 1862 to sink an oil well next to the Church of Our Lady of Guadalupe, which had long used the naphtha from a natural seep for ritual and medicinal purposes; and in 1863 a priest, Manuel Gil y Sáenz, began to collect oil from seeps near Tepetitlán in Veracruz and sell it on a small scale as an illuminant.[11] But in January 1865 John McLeod Murphy returned to the province, with greater commercial aspirations, on a five-month mission to investigate the potential of its petroleum resources. Over a decade earlier he had been a member of a surveying team, tracing a route for a possible railway across the Isthmus of Tehuantepec, that had encountered large oil seeps east of Minatitlán in the southern tip of the state of Veracruz. Now, Mexico was under a four-year occupation by France which, with the support of conservative Mexican monarchists, had installed Archduke Ferdinand Maximilian of the Austrian branch of the Habsburg monarchy as emperor of the country. A significant number of petroleum deposits were registered at this time in its eastern regions, from Tamaulipas to Tabasco, and the Minister of Public Works helped Murphy and his associate, George S. Drew, to register theirs,

> to place [us] in possession of the lands that I might indicate; and, in view of the political difficulties prevailing in the country, the usual time of ninety days, accorded for the commencement of the work, was extended to one year. This was done by a special order of the Emperor, who regarded these discoveries as calculated to exercise an important influence upon the productive industry of the country.[12]

A mile from San Cristóbal, on the Coachapa River, Murphy and his team found 'extensive salt and sulphur springs, in which there are large quantities of petroleum and other bituminous constituents constantly brought to the surface', and the owner of a saltworks there found that these were 'a great source of trouble, by compelling the workmen to skim the wells'. The 'most extensive deposit on the Isthmus' was at the nearby 'Laguna of Alquitran', which was like the pitch lake of Trinidad ... The odours evolved from the lakes are precisely similar to those emitted from the oil wells and gum beds of Michigan and Canada. If too closely or constantly inhaled, they produce nausea, giddiness, and a debilitating sense of faintness. We found a large hawk stuck fast in the lake, and upwards of a dozen large turtles, besides snails, small birds, snakes, &c., which had perished in attempting to cross. Mr. Montalva informed me that the cattle grazing in the prairie often fall into this place and are lost. In this respect, as in many others, the Isthmian deposits closely resemble the great pools of petroleum near Santa Barbara, in California ...

At rare intervals (about once a year) the Lake of Alquitran is spontaneously ignited and the whole surface is covered with a sheet of flame, which is accompanied by volumes of dense smoke, impregnating the atmosphere with powerful bituminous odors ... The heat rising from the flames was very great, and the sky was darkened by clouds of black smoke that rose up above the lake, recalling the descriptions given of the Caspian 'Field of Fire'.[13]

Murphy reported that bitumen was 'used extensively for caulking vessels. Mixed with linseed oil, and boiled over a slow fire, it forms excellent pitch.' Asphalt was found on long stretches of the coastline: 'By the laws of Mexico, the entire sea-coast for a league inland belongs to the Government. This chapapote is, therefore, public property; and, like the beach sand with which it is mingled, it may be collected *ad libitum*.'[14] Murphy documented many petroleum seeps in southern Veracruz and Tabasco, but chemical analysis of the crude showed that it was very heavy, containing only a small proportion of oil suitable for lamp light; furthermore, the transportation infrastructure was inadequate, and the country was politically unstable.[15] Nevertheless, a New York oilman, Wedworth Clarke, in April 1865 lobbied the Mexican envoy to the United States, Matías Romero, for oil concessions from Veracruz to Campeche, arguing that it would 'greatly augment the commercial importance of Mexico and contribute to the aggrandizement of that country'.[16] Clarke was unsuccessful, but Romero wrote back to his Foreign Minister,

Oil will soon replace coal and wood, and will become the only source of fuel. This new source of immense wealth discovered in this country has made those most entrepreneurial of speculators, based upon their understanding of the Republic's geological structures, think that there must be richer veins in Mexico than those of Pennsylvania.[17]

With the ending of the American Civil War that very month, the US government felt strong enough to enforce its policy – the Monroe Doctrine, formally announced in 1823 – of denying European political influence in Latin America. Secretary of State William H. Seward now adopted the position towards France that its intervention in Mexico, he wrote,

> now distinctly wears the character of a European intervention to overthrow that domestic republican government, and to erect in its stead a European, imperial, military despotism by military force. The United States, in view of the character of their own political institutions, their proximity and intimate relations towards Mexico, and their just influence in the political affairs of the American continent, cannot consent to the accomplishment of that purpose by the means described. The United States have therefore addressed themselves, as they think, seasonably to the government of France, and have asked that its military forces, engaged in that objectionable political invasion, may desist from further intervention and be withdrawn from Mexico.[18]

The French troops were withdrawn and the royalist forces in Mexico were soon defeated; in June 1867 Emperor Maximilian was executed by republican forces and Murphy's oil concessions died with him.[19] Two months later, a *New York Times* editorial publicized Murphy's explorations of the Isthmus:

> Among the sources of wealth which must soon be made to contribute, in one way or other, to the trade of the world, there is one which exists in our own country, but is found in Mexico on a scale worthy of the tropical grandeur of the region ... [S]ixteen years ago, there were discovered large deposits of petroleum, then regarded only as a natural curiosity. Within the last two years, extensive and accurate explorations of these oil-bearing regions have been made, and large purchases effected in the districts nearest to navigable waters, by Col. John McLeod Murphy ... These deposits, together with the other immense natural resources of Mexico, will pour their wealth into the commerce

of the world as soon as the United States shall determine in what way most wisely to aid in the redemption of that magnificent and miserable country.[20]

The American consul at Minatitlán, Rollin Hoyt, applied to the Minister for Public Works for a concession there, and in an official report of 1868 Hoyt wrote,

Petroleum is sufficiently abundant in this district to supply the world. Indications of its locality exist everywhere, and in many places it comes to the surface and forms small lakes and springs to such an extent that it can be dipped up in large quantities. In fact, the whole of this side of the isthmus is a vast lake of petroleum, in my estimation; and from the explorations I have made I believe it can be found almost anywhere ... Localities where the oil is found may be [registered with] the proper authorities, in conformity with the mineral laws, which, strange to say, amid all the revolutions and commotions of this distressed country, have remained immutable ... [B]ut the want of confidence in the government deters capitalists in the United States from investing in this profitable business.[21]

The following year two businessmen formed the *Compañía Explotadora de Petróleo del Golfo Mexicano* in an attempt to use American oil technology to exploit oil seeps near Papantla, midway between Veracruz and Tampico, that were already being worked by indigenous people, probably the local Totonac. However, neither this nor several other early initiatives in Mexico led to commercial production.[22] Instead, refined lamp oil was imported into Mexico by Standard Oil's marketing affiliate for the southern United States, Waters-Pierce & Co., which in the 1880s even set up refineries in Veracruz and Mexico City that were supplied with US crude.[23]

Meanwhile, in 1864 the Trinidad Petroleum Co. was registered in London with the intention of refining Trinidadian asphalt to produce oil, either in Trinidad or in Britain, and the director reported that they had acquired rights to large tracts of Pitch Lake – apparently including those of Lord Cochrane, Earl of Dundonald – in order 'to protect themselves and command market'.[24] However, the enterprise faltered from the beginning. In early 1865 the US-incorporated West Indies

Petroleum Co. sent a former Civil War captain on the Union side, Walter Darwent, to Trinidad to drill for oil near Pitch Lake. He struck good-quality oil the following year, and the operation was re-formed as the Paria Petroleum Co., introducing local investment. Lord Cochrane's agent, Conrad Stollmeyer, was extremely sceptical that drilling would be profitable and argued that oil could be more cheaply distilled from the already accessible asphalt, writing to the *Trinidad Chronicle* in 1866 that 'it would be safer to speculate upon the great prize in the Frankfurt lottery than upon the finding of oil-wells in Trinidad'.[25] The Trinidad Petroleum Co. also drilled for, and struck, oil in small quantities, but went into liquidation soon afterwards. Darwent remained optimistic about the prospects of Paria Petroleum and in 1868 he provided the Royal Navy with asphalt briquettes for trials of this alternative fuel in the boilers of HMS *Gannet*, based at Port of Spain. The Governor of Trinidad, Sir Arthur Gordon, reported in a despatch to the Duke of Buckingham, Britain's Secretary of State for the Colonies, that the trial was successful when asphalt and coal were burned in the proportion of 35 per cent to 65 per cent: '[A]n asphalt fuel may safely and usefully be employed, whilst, as it can be delivered here at certainly half the price of coal, there is no doubt that its employment, even in these proportions, would insure [*sic*] a considerable saving of expense ... Contrary to expectation, the smoke produced by the fuel proved to be lighter in colour and less dense in volume than that of ordinary coal.'[26] However, Darwent died of a fever soon afterwards and Paria Petroleum followed Trinidad Petroleum into liquidation. The following year the governor was visited by his friend Charles Kingsley, novelist and private tutor to the Prince of Wales, and he came across Trinidad Petroleum's abandoned well:

> Suddenly a loathsome smell defiled the air ... [A]cross the path crept, festering in the sun, a black runnel of petroleum and water; and twenty yards to our left stood, under a fast-crumbling trunk, what was a year or two ago a little engine-house. Now roof, beams, machinery, were all tumbled and tangled in hideous and somewhat dangerous ruin, over a shaft, in the midst of which a rusty pump-cylinder gurgled, and clicked, and bubbled, and spued, with black oil and nasty gas; a foul ulcer in Dame Nature's side, which happily was healing fast beneath the tropic

rain and sun. The creepers were climbing over it, the earth crumbling into it, and in a few years more the whole would be ingulfed in forest, and the oil-spring, it is to be hoped, choked up with mud.[27]

Across the Atlantic in Europe, Austria-Hungary continued to source much of its lamp oil domestically, as it had since the early 1850s when Ignacy Łukasiewicz had pioneered the kerosene lamp and built a chain of refineries near the Galician oilfields on the northern edge of the Carpathian Mountains (in present-day southeast Poland and western Ukraine). In the mid-1850s some of the harsh restrictions that had long been imposed on the Jewish community began to be lifted, allowing Jews now to engage in the oil industry.[28] A railway engineer described the oilfields in 1865:

> Whoever comes to Borislav sees before him a second California, a scene of brisk activity ... one windlass beside another, one man next to another man; between them hustle the buyers and sellers of wax and oil. The hubbub and clamour is like a carnival. Here a trouble-maker is pummelled with fists, a naphtha thief chased. Women pick at a pile of rocks with their bare hands to glean whatever wax might be left. Two workers, followed by their Jewish overseer, carry the mining machinery and tools that consist of just a hemp rope and a windlass crank, in which there is not one particle of metal, in order to mount it in the next moment over a shaft to produce oil. The ventilation system, often only an old blacksmith's bellows or a winnowing fan, is clipped to a 2- to 3-inch lead pipe and operated to bring fresh air every now and then to the workers in the shaft ... The central character in all this hustle and bustle is the Polish Jew in his long caftan; for the most part, he is the owner or the overseer of the mine, as well as the exclusive buyer and seller of the exploited product. His poorer fellow believer stands as a day labourer at the windlass for the bellows, but only on rare occasions does he go to work in the mine.[29]

Apart from a few instances of mechanized drilling, the Galician oil industry remained largely pre-industrial until the mid-1880s when the construction of the Transversalbahn railway through many of the producing areas made rapid, high-volume transit possible, and after restrictions on foreign investment were lifted. It was then that modern drilling rigs began to proliferate alongside the traditional manually dug pits.[30]

From the adjacent Romania a parallel industry was exporting illuminating oil some distance; here also, a lack of transportation infrastructure doomed early attempts, by Romanian entrepreneurs and by British, French, German and Austrian investors, to industrialize oil production.[31]

From 1864 an American who had worked in the Romanian oilfields and some Italian entrepreneurs went into business gathering bitumen on the two branches of the Arollo River at Tocco da Casauria, in Italy's Pescara province. Some commercial success was achieved, and a geologist proposed that it might be possible to manufacture lamp oil from the raw material.[32] In 1868 an English geologist prefaced his book on Italy's petroleum resources by asserting that many indications of petroleum in the country

> ... especially in the provinces of Modena and Reggio ... appeared to me to be identical with those of America, and, in some spots, even much more favourable ... In considering the enormous revenue obtained in America from this precious material, and the almost fabulous fortunes made by private individuals thereby in an incredibly short space of time, I have been led to believe most sincerely that the like splendid results would be obtained in Italy, were it possible to induce either the Italians themselves or foreign capitalists to engage in an enterprise which offers such fair hopes of a successful issue.[33]

Meanwhile, in 1865 the Zante Petroleum Co., formed in London in 1862 to exploit petroleum on the island of Zakynthos in the Ionian Sea, placed an advert in *The Times*:

> The Company's land is saturated with petroleum, and there are two valuable springs of it, which there is good reason to believe have yielded this product for more than 2,000 years ... The Directors propose to bore for the reservoirs from which this natural flow proceeds, and they confidently expect that the yield will be equal to that of the best oil wells in America ... The Company possesses the very great advantage of having its oil springs within a few hundred yards of safe anchorage; thus will be saved the heavy cost of land transit.[34]

Further south, evidence that oil might be found in West Africa was conveyed in reports from Angola by a Portuguese colonial official and mining engineer, Joachim Monteiro, published in 1875. Along the coastal region of present-day Zaire province, from which most of the indigenous population had been driven inland by the Portuguese settlers, Monteiro had noted sandstone at Musserra that was 'strongly impregnated with bitumen, so strongly, indeed, that it oozes out in the hot season'. It could be seen at other coastal locations, 'and a few miles to the interior, a lake of this mineral pitch is said to exist, but of course the natives will not allow a white man to visit the locality to ascertain the fact'.

Indeed, it was taboo among the locals here even to mention the existence of bitumen, and of malachite deposits, 'for fear of annexation of the country by the white men'.[35] Further south, Monteiro travelled a few miles inland from the mouth of the Lifune River to the Libongo garrison:

> Libongo is celebrated for its mineral pitch, which was formerly much used at Loanda for tarring ships and boats. The inhabitants of the district used to pay their dues or taxes to government in this pitch. It is not collected at the present time, but I do not know the reason why ... Although it was very interesting to see a rock so impregnated with pitch as to melt out with the heat of the sun, I was disappointed, as from the reports of the natives I had been led to believe that it was a regular spring or lake.[36]

Meanwhile, in the Far East, Burma and parts of India had, since 1858, been supplied with lamp oil from a refinery in Rangoon in Lower Burma, annexed by the British in the Second Anglo-Burmese War of 1852–3. From 1871 the refinery had traded as the Edinburgh-registered Rangoon Oil Co. and exported its products as far as London and even New York, but the scale of the company's operations was limited by the fact that the locally sourced crude oil was extracted by hand from the long-established wells near Yenangyaung up the Irrawaddy River. Here, an attempt was made to convey the oil to riverboats by means of a bamboo duct: '[T]he oleiduct now in the course of construction ... [is] a decided improvement upon the old method of transport by chatties in carts. Wholly made out of bamboo, supported by wood stages (the inside lacquered), it runs with a gentle slope from the wells down to the river bank: a great loss of oil by evaporation is inevitable.'[37] The duct was so leaky, however, that it was almost immediately abandoned. Mechanical boring techniques were first tried in Burma during the mid-1870s on the islands of Arakan but failed to achieve commercial levels of production (although small-scale, manually operated oil wells are still worked on the islands to this day). Concurrently, it was thought that commercial production of petroleum might be possible in Assam, to the northwest.[38]

Further east, oil was being drawn from manually dug wells at Echigo (Niigata) in Japan, and an attempt was made to drill for oil on the island of Formosa (Taiwan). A British tea merchant, John Dodd, wrote a series of articles for the *Hong Kong Daily Press* during the Sino-French War of 1884–5, in which he reported,

> in the neighbourhood of Mount Sylvia there are extensive cisterns of petroleum. If the country were systematically explored by geologists or mining experts the hilly country might be found to be of much more value than it is really known to be. So long, however, as the island belongs to China, the Government will reserve to itself all the treasures of the earth that may be brought to light ... Petroleum was discovered twenty years ago by a foreigner in savage territory but separated by a river only from Chinese territory. They tacitly allowed the foreigner to own for several years the wells; a Chinese headman of the district who had leased his right to the foreigner was caught and beheaded, and it was made so hot for every one that the place had to be evacuated. So it will be with everything so long as the Chinese remain masters of Formosa.[39]

According to a later American consul in Formosa, it seems that this 'foreigner' was Dodd himself, who failed to mention that in 1877 the Chinese authorities had brought to the island two highly experienced oil drillers from Titusville, Pennsylvania, along with $30,000 worth of equipment and machinery.[40] One of the drillers later recalled that they worked 'in a ravine in the Head Hunters' territory', under the protection of 200 soldiers and with a similar number of 'coolies'; but 'then the Head Hunters were chopping off heads and we had to take care of them, and made a contract with the nearest tribe to watch the tribes farther back and not allow them to come down and interfere with us – we paid them \$50 a month and the next month they didn't pay them so they started chopping off Chinese heads so they had to make another contract.'[41] The drillers struck oil but, given the challenging working conditions, they returned to Pennsylvania the following year.

Meanwhile, in the Dutch East Indies (Indonesia) the colonial government was concerned about how dependent its navy and merchant shipping were on British coal. This prompted the colonial authorities to organize local geological surveys in search of what they called the 'political minerals', and in May 1863 Edouard Henri von Baumhauer, a professor of chemistry, wrote to the Dutch Minister of Colonies expressing his conviction that petroleum was

> destined to play an important role in economic development ... As more than one eye-witness has testified, this substance has also been found in our East Indian possessions, and it is probable that its present limited quantity could be considerably increased by adequate well-drilling, as was also the case in America ...
>
> [I]n view of the use of petroleum not only in the preparation of illuminating gas but also in the production of steam, it may well be the instrument by which industry and steam navigation may soar to unknown heights in these regions, so lavishly endowed by nature but so little exploited by man.[42]

In August that year the botanist and explorer Franz Wilhelm Junghuhn offered his prediction that it would soon be more economic for Java simply to import oil from the Caspian Sea region. Nevertheless, with security of supply in mind the Minister of Colonies advised the king, 'In view of the amazing petroleum yields from drilling operations in the United States, it is the duty of the Netherlands Government also to promote and encourage the production of petroleum, which occurs so plentifully, especially in Java.'[43] But it would be another eight years before a wealthy trader, Jan Reerink, brought a drilling team of Americans and Canadians to West Java in 1871 to drill for oil near natural seepages in the Tjibodas valley, south of the port of Cheribon. However, although they struck flowing oil and began to sell kerosene, the enterprise folded in 1876, partly due to its reliance on ox-drawn carts for transport over difficult terrain.[44]

Junghuhn's prediction proved accurate. Until at least the early

1890s there was only one region, the eastern Caucasus by the Caspian Sea, that would produce oil and export kerosene in quantities that threatened the North American industry's – effectively Standard Oil's – virtual global monopoly. For in Russia, from around Baku on the western shore of the Caspian, oil production surged from the mid-1870s to the turn of the century when it would momentarily overtake US production.

Since the thirteenth century there had been a thriving trade in oil extracted from around Baku in Azerbaijan – heartland of Persia's ruling Safavid dynasty, though its western reaches were contested by the Ottoman Empire. In the early eighteenth century Azerbaijan descended into chaos as internal feuding among the Safavids and local rebellions weakened the Persian Empire, and in 1722 Afghan rebels, having ejected their Persian rulers, went on to overthrow the Safavid monarchy itself. Peter the Great took advantage of Persia's weakness, pushing his army southwards to occupy northern Azerbaijan, including Baku, prized by the Russians as a valuable and strategic east-west trading port. The tsar was also interested in the revenue that he might accrue from the oil extracted on the nearby Apsheron Peninsula; he instructed Lieutenant General Matyushkin to take an inventory of the oil wells and the associated installations, and to send him 'a thousand poods of white oil, or as much as possible' – a pood being a measure of weight of just over 36lbs, about an eighth of a 42-US gallon barrel of oil.[45]

Just over a decade later, a revived Safavid monarchy recovered its losses in the southeastern Caucasus. By the early 1750s, when an agent for the Muscovy Company, Jonas Hanway, came to the region to assess the company's silk trade at Ghilan and the trade routes to India via northern Persia and the Caspian Sea, Baku was again solidly within the Persian Empire. Hanway described how the locals collected oil from nearby oil springs, separated it from the water that flowed up with it, and shipped it away: 'The Persians load it in bulk in their wretched vessels; so that sometimes the sea is covered with it for leagues together.'[46]

CAUCASIAN BATTLEGROUND

The Ottoman, Persian and Russian empires continued to battle for control of Transcaucasia, each side vying for the allegiance of the myriad tribes of the region. Through the late eighteenth century the Russians, allied with Georgian tribes and facilitated by the construction of the Georgian Military Road, made incremental advances into the central and western Caucasus against the Ottoman army and local Karbadian, Circassian and Chechen tribes. Similarly, in the eastern Caucasus neither the army of Persia's new Qajar dynasty nor the local khans could halt a cumulative Russian advance. The khanate of Baku fell in 1806, giving the Russian army a strategic corridor from the Caspian Sea to allied Georgia, previously accessible only across the high Caucasus range along the perilous Military Road to Tiflis (Tbilisi), which was easily cut off by Chechens and Ossetians.

Under the 1812 Treaty of Bucharest, the Ottomans ceded to Russia much of the Black Sea coast and hinterland, including Abkhazia, although they retained the fortified ports of Anapa and Poti as well as strongholds threatening the Russians' strategic route through the Suram Pass, while Circassia in the northwest still held out against the Russian army. The following year, under the Treaty of Gulistan, Persia relinquished to Russia much of its Caspian coastline from Derbent down to Lenkoran, including the khanate of Baku, and conceded exclusive naval shipping rights on the Caspian Sea. Dagestan and Chechnya in the northeast, however, continued to resist Russian expansion.[47]

Some tribes made tactical alliances with the Russians in the hope of receiving their protection, but the belligerence of General Paul Tsitsianov, the self-aggrandizing commander-in-chief for the Caucasus, earned him the hostility of the indigenous majority. Regularly describing Persians in words such as 'scum', the general wrote to the Russian Foreign Minister in 1805 explaining the need for a policy of military aggression against their new Muslim subjects, because 'among the Asians, nothing works like fear as the natural consequence of force'.[48]

Over the decades, the extraction of oil from around Baku continued – traded across the region for its traditional uses as a cheap source of light and heat, as a lubricant, and for its medicinal applications – with reportedly about fifty wells in operation in 1735 and about seventy in 1771.[49] In 1813, just as sovereignty over the oilfields was formally transferred, under the Treaty of Gulistan, from the Baku Khanate to the Russian Crown, John Kinneir, an officer in the private army of the East India Company, published a treatise on Persia based on information gathered by himself and other British agents tasked with ascertaining 'the nature and resources of those countries through which an invading European army might advance towards Hindostan'.[50] Kinneir wrote of Baku,

> The quantity of *naphtha* produced in the plain to the S.E. of the city is enormous . . . [I]t is drawn from wells; some of which have been found, by a computation of the inhabitants, to yield from a thousand to fifteen hundred pounds a day. These wells are, in a certain degree, inexhaustible, as they are no sooner emptied than they again begin to fill, and the *naphtha* continues gradually to increase, until it has attained its former level. It is used by the natives as a substitute for lamp-oil, and when ignited, emits a clear light, with much smoke and a disagreeable smell.[51]

The tax revenue from Baku's oilfields now went to the Russian government, which from 1820 began auctioning monopoly production rights to the highest bidder every four years.[52] Meanwhile, the indigenous population was faced with Russian military occupation under the command of General Aleksei Yermolov, who established a series of strongly fortified outposts, the most important of which was at Grozny in Chechnya where the local oil wells came under the control of the Caucasus Cossacks.[53] In response to a rebellion in Guria, Georgia, Yermolov wrote 'Their settlements were destroyed and burnt down. Their orchards and vineyards were hacked to the root, and for many years these traitors will not reach primitive conditions. Extreme poverty will be their punishment.'[54] He described his colonial occupation methods thus:

> I desire that the terror of my name should guard our frontiers more potently than chains of fortresses, that my word should be for the natives a law more inevitable than death . . .
>
> [C]ondescension in the eyes of Asiatics is a sign of weakness, and out of pure humanity I am inexorably severe. One execution saves

hundreds of Russians from destruction and thousands of Muslims from treason.[55]

Whereas the Christian populations of Georgia and Armenia tended to acquiesce more readily to co-religious Russian rule, the Muslim populations and fiercely independent mountain tribes were driven by Yermolov's terror into seeking refuge in a militant form of Naqshbandi Sufism, which sustained and galvanized them through decades of armed resistance.[56] Shaykh Muhammad al-Yaraghi preached,

> As long as we remain under the supremacy of the infidels we are covered with shame ... All your ablutions, prayers and pilgrimages to Mecca, your repentance and sacrifices, all your holy deeds are invalid as long as the Muscovites supervise your life. Moreover, even your marriages and children are illegal as long as the Muscovites rule over you. So how can you serve God if you are serving the Russians?[57]

Thus, in many parts of the Caucasus, particularly the mountains, Russian control remained tenuous, as a Russian correspondent reported: 'In Chechnya only that area belongs to us where our detachment is stationed. The moment the detachment moves on, this area immediately passes into the hands of the rebels.'[58]

In 1826 an eruption of anti-Russian rebellions, along with the return of the Persian army in a surprise invasion, threatened Russia's presence in the southern Caucasus. However, the Russian army, under the new Caucasian command of General I.F. Paskievich, soon turned the situation around. After defeating the Persians at Ganja (Elizavetpol), the Russians pushed further south into Persia, taking Erivan (Yerevan) and advancing as far as Tabriz. By 1828 Persia was forced into signing the Treaty of Turkmenchai that reasserted the terms of the Treaty of Gulistan and conceded to Russia two more khanates – Erivan and Nakhicheyan, thus splitting Azerbaijan definitively into two – as well as war reparations and a say in Persian political affairs. The ceasefire with Persia now enabled Russia to drive the Ottoman army from its remaining positions in the eastern and southern Caucasus and to go on to take the Turkish garrison town of Erzurum. At this point, Russia halted Paskievich's troops. Advancing on Constantinople would have imperilled Britain's control of the eastern Mediterranean and its trade and communications routes across the Ottoman Empire, almost certainly provoking a wider conflict between the European powers. Under the 1829 Treaty of Adrianople (Edirne), Turkey renounced any claims to Georgia, eastern Armenia and Circassia – relinquishing the important Black Sea ports of Anapa and Poti – and granted to Russia full trading rights throughout the Ottoman Empire, including free passage through the Bosphorus and Dardanelles Strait for its merchant shipping.[59]

When, in 1833, Egypt rose up to take on its Ottoman master by invading Syria, Russia came to Turkey's aid, ahead of the hesitant British, resulting in the Treaty of Hünkâr İskelesi creating a Russo-Turkish alliance that included an agreement by Turkey to close the Straits to foreign warships when requested by Russia. Britain now perceived its economic interests in the Ottoman Empire – crucial communications routes with India and increasingly valuable export markets – to be at risk, and Russia's expansion into the Caucasus and Central Asia seemed to threaten the British Empire's imperial jewel itself. It was in the Caucasus, therefore, that the 'Eastern Question' – the geopolitical consequences of the decline of the Ottoman Empire – merged into the 'Great Game' – the geopolitical struggle in Central Asia for control over channels of communication, trade routes and lines of approach to British India, extending from Persia and the Caucasus in the west, through Afghanistan, trans-Caspian Central Asia, Punjab and Kashmir, to Tibet in the east. British foreign policy now became premised on hostility towards Russia and on shoring up, and maintaining close relations with, the Ottoman Empire and, to a lesser degree, Persia. Propagandists in Britain stoked Russophobia, pointing to Catherine the Great's dream that, by controlling the Caspian region and Central Asia, Russia might divert east-west trade away from British routes, and by portraying Russia's large military base in the Caucasus, at Tiflis, as a launchpad for a future invasion of India.[60]

Throughout the 1830s Russia's Corps of Mining Engineers introduced various improvements to the way Baku's oil wells were dug and maintained, augmented the oil storage facilities, and set up several refineries, one of which, in the village of Balakhani, used natural gas for heating the stills. However, the refineries' lamp oil was of such poor quality that they fell into disuse; and the Corps' plans to replace buckets with pumps for drawing crude up from the wells, and to improve the transit and port facilities, did not come to fruition.[61]

While Russia may have gained sovereignty over the Caucasus on paper, native populations, however – left out of international treaty conferences – denied it real control on the ground for decades to come. In the western Caucasus Russian troops barely held the Black Sea shoreline against Circassians fighting to retain their independence. As an Englishman in the area at that time, Edmund Spencer, wrote in the late 1830s, 'along the whole line of coast from Kouban-Tartary to the port of Anakria [Poti] in Mingrelia, the Russian government does not possess a foot of land, with the exception of the forts, or rather mud entrenchments we visited, and these are constantly besieged by the indefatigable mountaineers.'[62]

In Tiflis, Prince Viktor Kochubey voiced an attitude typical of the Russian occupiers, telling an American traveller that 'These Circassians are just like your American Indians – as untamable and uncivilised – and that, owing to their natural energy of character, extermination only would keep them quiet, or that if they came under Russian rule, the only safe policy would be to employ their wild and warlike tastes, against others.'[63]

Meanwhile, in the central and eastern Caucasus – Chechnya and Dagestan – an independence movement led by Shaykh Shamil resisted Russian occupation ferociously.[64] The Dubinin brothers, who ran most of the Russian-Cossack oil operations in the region, petitioned the viceroy of the Caucasus about the various economic and security pressures they were working under, proclaiming 'We, the Dubinin family, have been producing oil for more than 20 years amid incessant danger from enemy attacks by mountain peoples, and have made continuous efforts to please the government.'[65] Indeed, soon after they sent this petition, raids by Shamil's fighters led to the demise of the Dubinin enterprise; and after the repeated failure of the Russian army's scorched-earth policy, in 1842 War Minister Prince Alexander Chernyshov suggested, 'The system of our activity, being based exclusively on the use of force of arms, has left political means completely untried. The English have been able to consolidate their power in India by political means. They thus preserved their forces and gained time in subduing that country. Should we not try this system as well?'[66]

In the same year, the oil industry of the southern Caucasus was flourishing, with 136 wells around Baku producing up to 27,500 barrels of oil per year and peaking at 36,000 during the next few years. When production began to fall, however, from 1847, the Minister of Finance, Fedor Vronchenko, instructed the Mining Corps of Engineers – which oversaw the area's oil and salt wells – to explore for oil using mechanical drilling techniques. After this proved unsuccessful, however, from 1850 the government began to lease the oilfields to private entrepreneurs.[67]

As the British became ever more distressed at Russia's naval presence on the Black Sea – which, noted Karl Marx who had moved to England in June 1849, threatened Britain's trading supremacy through the Dardanelles Strait – the Caucasian resistance movements took on a wider geopolitical significance. In that same year Major General Sir Henry Rawlinson, based in Baghdad as Britain's political agent to Ottoman Arabia, argued for the importance of Caucasian independence in order to contain Russia and maintain the balance of power in Europe: 'Moderate support of Shamil might still, perhaps, save the Danubian principalities, and as long as his banner floats from the summits of the Caucasus, so long is Persia safe from the hostile invasion of a Russian army.'[68]

In 1853 armed conflict broke out between Turkey and Russia, simultaneously on the Danube and in the Caucasus. A virulently anti-Russian press soon pushed Britain, along with France and Austria-Hungary, into war, with the chief aim of eliminating Russia's naval power on the Black Sea by destroying its coastal bases, foremost of which was at Sevastopol on the Crimean Peninsula. The British bombardment of Kerch in May 1855 happened to destroy the city's asphalt plant, which produced about 36 tons of the material in 1854 up to the time that it was put out of action.[69] During the Crimean War the British Prime Minister Lord Palmerston wrote, 'The best and most effectual security for the peace of Europe would be the severance from Russia of some of the frontier territories acquired by her in late times, Georgia, Circassia, the Crimea, Bessarabia, Poland and Finland . . . She could still remain an enormous Power, but far less advantageously posted for aggression on her neighbours.'[70]

When the Ottoman army attempted an advance from the Black Sea port of Batum towards Tiflis, Russian generals feared that, if it was accompanied by Muslim uprisings in Circassia, Chechnya, Dagestan and Azerbaijan, they would lose the southern Caucasus. However, the Ottoman forces – faced with the new, determined governor general of the Caucasus, Prince Aleksandr Baryatinsky, and with British and French efforts focused on Sevastopol – managed only to cut off Poti from the Russian fleet and to help a Circassian prince take the port of Novorossiysk. In the autumn of 1855 Palmerston argued for extending the war into Circassia and Georgia, as well as into the Baltic region, but Britain lacked the strength to pursue these further aims alone and failed also, in the Treaty of Paris the following year, to have Circassia and Mingrelia established as buffer states between the Russian and Ottoman empires. However, Britain achieved its primary war aim: Russia – after surrendering at Sevastopol and with its army overstretched in Poland, the Baltic provinces, Galicia and the Caucasus – was forced to accept the demilitarization of the Black Sea, thus protecting Britain's eastern Mediterranean route to India from a possible Russian naval threat.[71]

The end of conflict in the west allowed Russia to intensify its subjugation of the Caucasus, and in 1859 General Baryatinsky eventually captured the legendary resistance leader Shamil. Large numbers of natives fled the Caucasus, many ruthlessly expelled by the Russian army which gradually brought the Black Sea coast under its control, forcing the port of Tuapse into submission in 1864.[72] A British diplomat reported, 'The Russians in order to compel the natives of Netauchee and Shapsik to abandon the country and emigrate to Turkey have lately destroyed the whole of that part of Circassia, burning down the houses and crops of the people and thus obliging them to fly.'[73] Thus continued a pattern of terrorism, forced migrations and population exchanges across the shifting boundaries as, throughout the nineteenth century, Muslims withdrew behind retreating Persian and Ottoman borders and Armenian Christians were displaced to newly Russian territory, to be joined later in the century by an influx of Russian Orthodox Christians.[74]

In 1877 armed conflict between Russia and Turkey flared up again over the Danube and in the Caucasus where, for example, Ajars, Muslim Georgians north of Batum, rose up in revolt on the side of the

Turks. Russian troops advanced to within a few miles of Constantinople but, in a replay of the war of 1828, Russia withdrew its by then over-extended army to avoid threatening British interests and risking a wider war. In the subsequent Treaty of Berlin, the Trebizond (Trabzon)-Erzurum-Tabriz trade route was, as in 1828, kept out of Russia's hands, but Turkey ceded to Russia the important Black Sea port of Batum.[75] The rich resources of the Caucasus – its fertile agricultural land, petroleum deposits, and its strategic military and trading location – now lay firmly within the Russian Empire.

RECLAIMING THE RUSSIAN MARKET

From the 1830s to 1860s the extraction of oil continued around Baku from manually dug wells, although in 1848 Russia's Corps of Mining Engineers carried out some exploratory mechanical drilling using a manually powered rig. Production alternated between being conducted as a state- and privately run monopoly, with output tending to increase when worked privately, particularly from 1862 under Ivan Mirzoyev, a wealthy Armenian merchant, who additionally set up a smaller oil production and refining operation at Grozny. As crude production around Baku rose, so did the number of refineries. One of the largest was operated by the Transcaspian Trading Co., run by Russian business partners V.A. Kokorev and P.I. Gubonin, whose state-supported shipping company, the Caucasus & Mercury Co., dominated the one economically viable route for transporting Baku's kerosene to Russia's population centres: over the Caspian Sea to Astrakhan and then by barge up the Volga River.[76]

Despite the relative proximity of Baku, the growing demand for kerosene's light and heat from Russian population centres continued to be met by American imports. The methods of production, refining and transportation in the Caucasus remained pre-industrial, and wealthier Russians were prepared to pay a little more for the superior-quality American oil. When, in 1872, Russian troops working under British engineers completed a strategic rail link between Tiflis – Russia's military and administrative headquarters in the Caucasus – and Poti on the Black Sea, it actually became cheaper for the inhabitants of Tiflis to transport their kerosene 8,000 miles eastwards from America than 340 miles westwards from Baku. The domestic product was, however, beginning to make inroads into Standard Oil's dominant market share, which fell from 88 to 82 per cent between 1869 and 1872.[77]

In 1866 more advanced mechanical percussion drilling methods were brought from the United States for oil exploration in the north Caucasus about 200 miles west of Maikop near the Kudako River, a tributary of the Kuban, in an area inhabited by Circassians. Here, a team led by Colonel Ardalion Novosiltsev struck Russia's first flowing well at a depth of 124 feet. To mark the opening up of what became the Kuban oilfield following a second strike two months later, thousands of barrels of oil flowed into the Kudako River; one writer described 'the pungent smell throughout the valley, the river bends filled with combustible, black, thick liquid, giving it the appearance of one of the rivers of an ancient Tartarus'.[78] The American consul at Odessa reported in December of that year,

> The discovering of oil in such quantities found them all unprepared, and as it was thought not prudent to stop the flow, a great quantity of it ran to waste. The well continues to supply at the present time from 3,000 to 5,000 gallons per day. The coal and petroleum hid in the bowels of the earth to the north and east of the sea of Azoff is probably inexhaustible.[79]

In 1869, Colonel Novosiltsev used state-of-the-art British equipment to build a successful refinery on the Kerch Strait, which one petroleum engineer argued 'could serve as a model for refineries in the Baku area'.[80] Meanwhile, exploratory drilling began to be conducted in the Ukhta region of northwest Russia by the Pechora Company.[81]

Simultaneously with the opening up of the Kuban oilfield, Colonel Charles Stewart, an officer in the Indian Army, travelled through Baku where he visited the ancient Zoroastrian fire temple of Ateshgah at Surakhani, located near one of Baku's two refineries:

> At this refinery was situated what was known as 'The Temple of the Everlasting Fire', which was one of the sights of Baku. The petroleum refinery had been placed here for the purpose of utilizing the natural petroleum gas which rose from the fissures in the soil. For ages a so-called

everlasting fire had been kept burning and watched by Hindu priests from India. The spot where the gas rose from the ground had been enclosed by a wall, and a small temple built in the midst ... After the murder of the Abbot one of the surviving priests fled, but the third remained to tend the fire, which was merely a pipe in the ground connecting with the naturally rising gas.[82]

Throughout the 1860s voices were raised by the business community protesting that Baku's oil industry was being stifled by the state's four-year monopoly contract system for crude production. In 1872, following a report by Russian chemist Dmitri Mendeleev, the government decided to introduce long-term leasing and American-style competition. At the end of that year the oil-bearing properties were sold off by sealed bids to the highest bidder, in small plots to encourage competition between many small capitalists. However, the Armenian production monopolist Mirzoyev and the Kokorev and Gubonin partners – now with a 7.5 million rouble joint-stock company, the Baku Oil Co. – had the capital to dominate the auction, outbidding others for the most prized leases on the Balakhani oilfield, where Mirzoyev had just sunk Baku's first commercially successful mechanically drilled well.[83]

On the Apsheron Peninsula was now replayed the familiar script of the Pennsylvanian oil rushes, as producers competed to maximize output from their plots. Around Baku, as modern drilling machinery began to be introduced alongside the traditional manual digging and extraction methods, flowing oil was struck more easily at much shallower depths than in the American Oil Region. Crude production suddenly far exceeded the existing storage, transit, refining and product marketing capacity, and a huge oil glut was the almost immediate consequence. In this arid, treeless landscape, wooden storage vats were a costly investment, and barrels were soon more valuable than their contents, which added to the expense of slow, inadequate horse-drawn transit. A predictable environmental disaster ensued, particularly after big oil strikes such as one made in July 1873 by the Armenian-owned Khalafi Co., whose 'fountain' gushed oil uncontrollably for three months, creating a vast swamp of noxious crude – a decidedly mixed blessing, as related by a prominent English analyst of the Russian oil industry: '[T]he unhappy proprietors, who were quite unprepared for such an event, were inundated with demands for compensation from neighbours who had their ground buried in oil and sand, and were obliged to go into liquidation.'[84] Indeed, some decades later a British military report on the Transcaucasus would note that Baku's fresh water was 'impregnated with naphtha; consequently, nearly the whole of the drinking water in use is distilled from the sea.'[85]

When the price of crude dropped from 45 to 30 kopeks per pood (approximately 36lbs), Mirzoyev and the Baku Oil Co. sought, unsuccessfully, to introduce production restraints. The oil price fell to as low as half a kopek per pood the following year and was kept low by oil strikes on the new Sabunchi oilfield, where 'drilling rigs began to pop up with incredible speed, like mushrooms after a spring rain', wrote the Russian industrialist Viktor Ragozin.[86] Nevertheless, the means of extraction at the wellheads often remained quite rudimentary, as Arthur Arnold, an English newspaper editor and future MP, witnessed in 1875:

> From the first well we visited, a small steam engine with most primitive gear, was lifting about four hundred and fifty thousand pounds' weight of petroleum in a day. The oil is of greenish colour, and as it is drawn from the earth, is emptied into a square pit dug in the surface soil, from whence men take it in buckets and pour it into skins or barrels, the charge at the wells being at the rate of 1½d. per fifty pounds' weight of oil ... At all the wells, the oil is now raised in circular tubes about nine feet long and as many inches in diameter, with a valve at the lower end which opens on touching the ground and closes when the tube is lifted. This cylinder is lowered empty and raised again when filled with oil in less than two minutes. A man pulls the full tube towards a tub, into which its contents are poured, and through a hole in the tub the oil runs into the pit from which the skins and barrels are filled.[87]

Arnold saw also how the 'eternal flames' of the fire temple at Surakhani were now lit not for worship

> but – to what base uses may gods come! – in order to burn lime for Baku, and to purify the oil raised from the natural reservoir in which the gas is generated ... Before us stood the priest of a very venerable religion ... ready for half a rouble to perform the rites of his worn-out

worship, and there also was the object of his life-long devotion set to work as economic firing. Such a rude encounter of the old and the new, of ideality and utility, of the practical and the visionary, was surely never seen elsewhere. I suspect that, as a Yankee would say, the worship of '*le feu éternel*' at Baku is almost played out.[88]

The refining industry was stimulated by the rock-bottom price of crude but enjoyed only a brief honeymoon of high profitability. The Americans quickly responded to the new competition by flooding the Russian market, in 1873–4, with kerosene far more cheaply manufactured and transported, and Baku's lamp oil struggled to compete. Firstly, at Baku the transit of crude in barrels by horse-drawn carts from wellhead to refinery was not only expensive and slow but also precluded the economies of scale associated with bulk transit and processing. Secondly, Baku's primitive refineries – now confined to 'Black Town', a few miles outside Baku itself, after the townspeople protested at the horrendous pollution – manufactured rather poor-quality lamp oil. The local heavy crude would yield only around 30 per cent kerosene, in contrast to the 75–80 per cent from light Pennsylvanian crude, described by Mendeleev as 'impure kerosene';[89] and refiners were incentivized by the way an excise tax on kerosene was calculated to refine their lamp oil too quickly and rather carelessly. Finally, the cost of transporting kerosene from Baku's refineries to Russian consumers added significantly to its cost. As there was no rail link from Baku to the Tiflis-Poti railway, the oil had to be shipped northwards in barrels by sailing boat and barge, and even this route was frozen over for much of the winter, precisely the period of highest demand.[90]

Refiners and merchants sought ways to increase profitability. Initially, refiners had regarded the large volumes of oil residue left from refining kerosene as waste, allowing it to stream into the Caspian Sea or to accumulate in great pools that were then burned. But now it began to be marketed as a fuel for steam boilers as a cheaper alternative to wood or imported coal. By 1874 most steamships on the Caspian and Volga were oil-fired, and some of the fuel oil being shipped from Baku began to be carried in bulk, in large tanks or in the boat's hull itself. Over the same period, Baku's business community, supported by Mendeleev, successfully lobbied the government to follow more closely the example of American economic policy by abolishing the excise tax on oil products to stimulate domestic refining and by erecting a tariff barrier of 55 kopeks per pood on imports to protect the nascent domestic industry somewhat from foreign competition. As James D. Henry, the most prominent English analyst of the Russian industry, wrote, 'Without Customs protection the home industry, then in the first stage of development, would have been crushed out of existence by American competition.'[91] However, only the industrialization of oil transit and the associated upscaling of its refineries – the keys to the commercial success of the American industry – would enable Baku to compete seriously, and this would require large-scale capital investment.[92]

In March 1873 Robert Nobel – whose family had access both to capital and to the highest echelons of Russian officialdom (and whose younger brother Alfred later bequeathed his fortune to establish the Nobel Prize) – arrived in Baku and promptly bought a small refinery. He was not entirely new to the oil industry, having struggled ten years earlier in Finland, against superior American competition, to sell Russian oil lamps and kerosene. Now, injections of capital by the Nobels and by the Russian state would catalyse Baku's oil industry into taking on the Americans at their own game.[93]

The rise of Standard Oil was a story familiar to anyone engaged in the oil industry, but the moral drawn depended on one's business interests. It inspired Robert Nobel and his brother Ludwig to emulate Rockefeller and served as the model on which they based their ambitious business plans. On the other hand, the same story forewarned the majority of Baku's small oil businesses of the dangers of monopolization if any one firm were to gain too much control over oil transit or refining. Thus, in 1875 the Nobels failed to draw other firms into their plan for a jointly financed pipeline from the Balakhani and Sabunchi oilfields to the refineries of Black Town, and in 1877 the government likewise refused the brothers permission to privately construct a rail connection from the oilfields. Instead, it built a public railway, opened in early 1879, complete with a number of bulk oil-carrying tank cars and with feeder and gathering pipelines to the oil wells. Nevertheless, in 1877 the Nobels gained permission to lay a 6-mile, 5-inch-diameter pipeline to Black Town with a capacity of

80,000 poods of crude per day, along with iron storage tanks capable of holding over 100,000 poods. However, such was the intensity of opposition from the thousands of oil carriers, now faced with the demise of horse-drawn transit, that the Nobels posted Cossacks in watchtowers erected every few hundred yards along the pipeline to guard against sabotage. After a year of successful operation, the transit fee of 5 kopeks per pood had paid for the pipeline's £10,000 construction and security costs, spurring the construction of other pipelines: by the Baku Oil Co., by Mirzoyev, by the major crude producer G.M. Lianozov, and by the Caspian Petroleum and Trade Co.[94]

In 1878 the Nobels sought to raise more investment capital by incorporating as a joint-stock company, with the aims of moving upstream into crude production, expanding and upgrading their refining operations and, crucially, increasing their capacity to distribute kerosene, fuel oil and other petroleum products across Russia. In their successful application to the Ministry of Finance for permission to form the Nobel Brothers Petroleum Co., they explained,

> One of the most important reasons for the slow development of the Russian petroleum business in the Baku region ... is the lack of suitable means to transport crude oil from the wells to the refineries and refined products from Baku to the Russian market ... [W]e were forced to focus special attention on the construction of rational methods of handling our goods, which require many complex technical installations. To reach this goal, and apart from the improvement of our refinery, we have built a pipeline to move crude oil from the wells to the refinery and [then] our finished products from the refinery to the banks of the Caspian Sea; we have constructed special ships with reservoirs for the bulk transport of oil products; and on the banks of the Volga River we have installed special iron tanks for the storage of kerosene.
>
> On all of these installations we have already spent about a million roubles. Taking these installations forward to the point where they would constitute a single, organically united whole, capable of putting Russia's own kerosene on the Russian market in successful competition with the American product, requires investments which are twice as large as those which have already been made.[95]

The introduction of bulk transit at the production and refining stages was rendering obsolete the use of scarce and expensive wooden barrels. But until 1878 – with little prospect yet of any alternative outlet – barrels were still indispensable for the 500-mile sea journey from Baku to the mouth of the Volga and for transport 300 miles upriver by barge to the first major rail depot at Tsaritsyn (Stalingrad/Volgograd). With additional finance, however, the Nobels now worked to phase out barrels almost entirely by reviving the long-stalled development of the water-borne bulk shipment of oil.

Since 1866, when the failure of an oil-refining venture in Britain had brought an end to its transatlantic bulk shipments of American crude, only the *Charles*, an 800-ton wooden ship fitted out with fifty-nine iron tanks, had performed this task, and then only from 1869 to 1872. Too many tankers were being lost at sea due to fire and explosions or by capsizing, and in 1873–4 three tank steamers built for the purpose – the *Vaderland*, *Nederland* and *Switzerland* – were used instead to ship other cargoes. Now, however, Nobel Brothers, capitalized at three million roubles, invested in the development and construction of a new generation of oil tankers, an innovation necessary to extend the bulk movement of kerosene all the way to Russia's towns and cities – which Standard Oil had been able to do in America by pipeline and rail tank car alone.[96]

In January 1878 Nobel Brothers signed a contract with a Swedish shipbuilder for a much-improved design of tank steamer built from the new Bessemer steel, with twenty-one vertical tanks capable of holding a total of 250 tons of oil. The *Zoroaster* (see Fig. 7) successfully entered into service later that year, followed close in its wake by the *Buddha* and the *Nordenskjöld*. At the mouth of the Volga the oil was transferred in bulk to specially built tank barges that were tugged upriver to a tank farm at Tsaritsyn. From here, the Nobels' own tank cars sent oil over the Russian Empire's rail network to numerous new storage and distribution centres, the largest of which was near Orel in central Russia, making possible the uninterrupted supply of kerosene to consumers all year round; and by 1880 around 85 per cent of Nobel Brothers kerosene was sold wholesale in bulk rather than in barrels. Along with improved refining techniques by which they matched the Americans in quality, the Nobels' increasing economies of scale and decreasing freight costs now enabled them to beat the Americans on price and to win a growing share of Russia's domestic market, itself growing as kerosene became cheaper. As Nobel Brothers raised a further two million roubles investment, it ordered the first of a more advanced design of tanker, the *Moses*, that used the shell of the vessel itself for the tanks' outer walls, and within a year a further seven were on order. The company suffered bad publicity in 1881 when the *Nordenskjöld* exploded while loading at Baku, killing half the crew, followed by a similar incident with one of its Volga barges; the problem was overcome, however, by replacing the rigid loading pipes with flexible ones that would stay secure and leak-proof despite movements of the tanker.[97]

Nobel Brothers, having begun to integrate upstream production into its sprawling operations, made its first big oil strike at this time, contributing to the company's rapid rise to producing a quarter of Baku's total crude output. By 1883 Russian kerosene, over half of which was shipped by the Nobels, had the lion's share of the domestic market.[98]

In 1881 Colonel Stewart returned to Baku and revisited its fire temple; but, he wrote, 'I found the fire out and no priest. The engineer in charge of the neighbouring petroleum refinery accompanied me over the temple, of which he held the key. He relit the fire, and when leaving carefully extinguished it, as he said he wanted all the natural petroleum gas for heating the furnaces of his own works.'[99] Travellers such as John Osmaston came to visit this new industrial wonder of the world, where a few hugely prolific square miles of oilfields on the Caspian shore yielded a sizeable proportion of the volumes of crude extracted from the many thousands of square miles of American oilfields, and that sold at a fraction of the price. Nevertheless, just a short distance offshore from the filth and frenzy of Black Town's refineries – 'these stretch along the bay, and belch forth smoke like a concentrated Birmingham', as one prominent account at the time described it[100] – the English visitor found tranquillity of a kind:

> We left the quay just as the sun went down . . . Gas was bubbling up in several places near the boat, the water looking as if it were boiling . . . A strong smell of naphtha permeated the air. [On being ignited], the waves were wrapped for several yards in flame. It was quite dusk, so we

saw it beautifully. It was a most extraordinary sight, the sea as though it were on fire ... We rowed round it, and then away, but the flame could still be seen dancing up and down with the waves till we had gone nearly a mile distant. The wind blew stronger, and extinguished it, for it suddenly disappeared.[101]

ACROSS THE CAUCASUS
TO THE WORLD

As Nobel Brothers raised another four million roubles to fund yet further investment in refining capacity and technology, the Russian oil industry stood precariously on the threshold of a new era. Due to its very success in the early 1880s, this burgeoning industry had fast outgrown itself as ever cheaper, mass-marketed Russian kerosene saturated and then flooded the domestic market. Thus, along with victory over Standard Oil for the home market came a crisis of overproduction, plummeting crude and kerosene prices, and recession for Baku's oil industry, compounded by a global economic downturn.

It was widely agreed that a primary cause of the domestic glut was Baku's lack of capacity for exporting its surplus to the wider world market – which was growing 8 per cent a year and totalled six times Russia's kerosene output by 1883. As Charles Marvin, a prominent contemporary chronicler of the Russian oil industry put it,

> The first thought that strikes the observer as he surveys the lakes of oil before him is – Why Baku, having the richest supply of petroleum in the world, worked ages before the American oil was touched, should have nevertheless allowed the United States to take possession of the markets of both hemispheres, including for a time the very important one of Russia itself. The matter is susceptible of easy explanation. Until the Russians completed railway communication between the Black Sea and the Caspian, Baku was severed from the world ... [W]hen [the Volga] was frozen over in winter, Baku was practically cut off from the European system of communications. The only way to reach the place was to proceed to Vladikavkaz or Tiflis by rail, and post the rest of the distance through the Caucasus to Baku. This was not encouraging for capitalists,

especially if it be remembered that it was not until 1878 that the last traces of independence were crushed out of the Caucasus, and the region delivered from further fear of tribal insurrection.[102]

Only relatively insignificant quantities of Baku's kerosene were reaching central Europe beyond the Nobel-dominated northern route, either overland by rail or across the Baltic Sea by tanker from the port of Libau (Liepāja). But could the far more direct route to larger world markets – via the Black Sea, the Dardanelles Strait and the Mediterranean – be opened up by breaching the Caucasus Mountains to the west? This route would indeed be opened up, but only slowly. Progress would be hindered firstly by a lack of the kind of capital needed for major infrastructure development, itself a consequence of precisely the lack of export outlets; secondly by conflicting interests within the industry, between the smaller independents and the larger operators, and between producers and refiners; and thirdly by the conflicting priorities of different government departments.[103]

Following Russia's capture of the Black Sea port of Batum from Turkey in 1878, two crude-oil producers, A. Bunge and S. Palashkovskii, set about raising investment to extend the Tiflis-Poti railway into a complete trans-Caucasus rail link connecting Baku with the newly acquired Black Sea port. This would provide Baku's oil companies not only with a much-needed export route – as a government commission had recommended in 1874 – but also with an alternative to selling out to Nobel Brothers, who were seen to be rapidly monopolizing refining and transit, just as Standard Oil had done in the United States. The government gave its usual support for the construction of railways deemed of economic importance and backed the Tiflis-Poti railway extension additionally for its military value: it would 'enable Russia to throw her military resources with equal facility towards the Caspian or Black Sea', as Marvin put it.[104] Thus, when Baku's struggling producers and refiners failed to raise enough investment capital, they and the government were willing, in 1880, to grant favourable terms to the Paris Rothschilds – experienced railway investors who were by now trading in kerosene and financing a refinery at the Austro-Hungarian port of Fiume on the Adriatic coast (present-day Rijeka, Croatia) – for a $10 million loan to continue the Tiflis-Poti project.[105]

The American oil industrialist William Brough, in a report on his visit to Baku, warned his colleagues,

> When this [rail]road shall be completed, it will furnish an outlet for Baku oil to the markets of Europe, and will bring it into direct competition with American oil in those markets. The work of building this road is, if measured by the Russian standard, progressing rapidly ... and it is expected that the whole road will be completed by August, 1882.[106]

In May 1883 the first oil trains began moving along the completed track, as Marvin was writing his account of 'the Kerosine factor in the Central Asian problem':

> From the present year will probably date a fresh epoch in the petroleum industry – the Batoum period. Up to the summer of 1883 Caspian petroleum only found its way to Europe via the Volga and Western Russia, traversing more than 2,000 miles in steamers and tank-cars before reaching the holds of foreign vessels. The construction of the Batoum line reduced this distance to 560 miles at a stroke, and laid the industry open to the civilized world.[107]

The arrival of the Rothschilds triggered a major shift in Russia's oil industry: as well as opening up a channel for Russian oil to begin competing seriously on the worldwide market, they dashed the Nobels' hopes of becoming the Standard Oil of Russia. In 1882–3 the Nobels attempted to monopolize Baku's entire oil industry in a manner reminiscent of Rockefeller's 'Cleveland Massacre' a decade earlier: they planned to use their market dominance – particularly over transit – to take advantage of the recession by herding the struggling independents into one conglomerate in order to gain even greater economies of scale and to regulate production and refining so as to maintain oil prices at profitable levels. Due to the very conditions of recession that the Nobels intended to exploit, they – like Rockefeller – had to borrow heavily: from their younger brother Alfred, the Russian State Bank, Crédit Lyonnais, Deutsche Bank and the Rothschilds. But when the latter saw an opportunity here for themselves, the Nobels had to contend with a powerful rival – which the equally vulnerable Rockefeller did not – that would thwart their ambitions of becoming the 'Russian Rockefellers'. The Nobels now no longer dominated oil transit, and many smaller independents sold out to the Rothschilds rather than to the Nobels.[108]

Although many involved in the Baku oil industry were glad of the Rothschilds' investments, there was some disquiet. The new Minister for State Property, Mikhail Ostrovskii, wrote to a colleague, 'Rothschild has had a steadily growing influence on the oil industry in the Absheron Peninsula, influence which has given him almost monopoly control over its future, and he holds many local people working in the oil industry in his powerful hands.'[109] Independents began calling for state intervention to support their ailing enterprises and to protect them from unfair discrimination by the transit monopolies of Nobel Brothers and, now, of the Batum Oil Refining and Trading Co. through which Bunge and Palashkovskii operated the Baku-Batum railway. In response, Ostrovskii supported the industrialists' formation of the Baku Petroleum Association; he held the Baku-Batum railway to its duties as a 'common carrier' by introducing a system for rationing tank cars fairly; and he ordered the construction of commercial port facilities at Batum.[110]

However, it had been clear all along that this one railway would have nowhere near the capacity to export Baku's huge surplus output. On the absorption of Batum into Russia, entrepreneurs were as quick to propose linking the port to Baku by pipeline as by rail (see Fig. 9). Before the ink on the 1878 Treaty of San Stefano between Russia and the Ottoman Empire was dry, the American oil industrialist Herbert Tweedle and an official in the Russian Ministry of Finance, K.A. Bodisko, had teamed up to submit four different plans for a privately financed Baku-Batum pipeline in return for a thirty-year monopoly. The ministry, however, was persuaded by Baku's industrialists that a privately run pipeline 'would amount to giving one firm a monopoly on all the most important petroleum deposits'.[111] Secondly, opinion was deeply divided over whether to prioritize the export of crude or kerosene. Baku's producers, who clamoured for crude oil pipelines, were staunchly opposed by refiners – including the well-connected Nobels – who lobbied for a kerosene pipeline, lest Baku's crude be largely exported. This, they argued, would push up crude prices in Baku and would stimulate foreign refining; foreign demand for their kerosene would be reduced and Baku would see the greater profits from refining disappear abroad. 'A strong feeling prevails in Russia against allowing the oil to be exported except in manufactured form, so as to retain for Russia the profit of refining it,' wrote Charles Marvin.[112] The prominent oil industrialist Ragozin asked, 'Will petroleum production be Russian or will we yet again surrender ourselves to the hands of Europe?'[113] Thirdly, pipeline construction was impeded due to resistance from the Ministry of Finance against providing it with state support: the ministry feared losing revenue from the railway and was seeking generally to keep public spending in check.

Yet as prolific crude production continued, the case for augmenting export capacity became ever clearer to Baku's industrialists and by late 1883 at least six applicants were proposing terms for permission to construct a Transcaucasian pipeline. One applicant, for example, A.P. Khanykov, asked for a fifteen-year monopoly in return for financing a crude-oil pipeline privately and for paying, as a royalty to the government, the difference between the railway's freight rate and his lower piping rate. He pressed his case that the industry was suffering because, 'having overfilled the demand of the Russian internal market, it does not have an outlet abroad'.[114] Tweedle was again in the running and had by now already built a pipeline from French-operated wells in the north Caucasus to the nearby Black Sea port of Novorossiysk.[115] In January 1884 Minister Ostrovskii impressed on the Ministry of Finance the current inadequacy and expense of export transit: 'In this case we must follow the example of America, where iron rails laid down in oil-bearing regions were replaced by pipelines, which permitted cost reductions for kerosene.'[116] In late 1886, with still no pipeline agreed, and after yet further record-breaking oil strikes, two producers, G.M. Lianozov and M.I. Lazarev, pleaded with Ostrovskii,

> Tagiev, the owner of the Bibi-Eibat gusher, is at present able to supply twice as much as the whole Baku region demands and offers his oil like a gift. This can't go on: only ruin and liquidation of the business remain for us if there is no hope for the authorization of the Transcaucasian pipelines by the higher administration.[117]

Even British naval architects that year were sure of the need for Baku-Batum pipelines:

[I]f the industry is to be adequately developed it will be necessary to lay one or more pipes between the two towns, just as the Pennsylvanian oil wells are connected by the pipe system of the National Transit Company with New York, Philadelphia, Baltimore and other large towns, and to pump the supply into reservoirs at Batoum ready for shipment.[118]

But with the Nobels, along with Baku's other refiners, opposed to the construction of crude pipelines that would facilitate the emergence of competing refiners elsewhere, and with the Russian government fearful of losing revenue from the railway and of allowing the formation of a private, probably foreign-owned monopoly over export transit, pipeline applications were stalled and rejected. The government opted instead for increasing as far as possible the capacity of the railway. It authorized private firms to operate their own additional kerosene- and crude-carrying tank cars; it accepted a bid by the Nobels to lay a 40-mile, 4-inch-diameter kerosene pipeline over the steep Suram Pass bottleneck; and, using 400 tons of Alfred Nobel's dynamite, it constructed a tunnel through the Pass, eventually completed in late 1889.[119] On the Russian government's preference for a railway, William Libby, Standard Oil's main foreign business agent, observed, 'in a portion of that country growing daily in strategic and political importance, the Government would naturally reflect that while a tunnel might facilitate the meeting of some great military crisis, army maneuvering in a pipeline would be slightly complicated.'[120]

Nevertheless, having made little impression outside the empire before the railway opened in 1883, five years later Russia was supplying approximately 23 per cent of the world's growing demand for kerosene – where Standard Oil had once commanded an almost total global monopoly. As soon as the Baku-Batum railway was opened, the Rothschilds formed the Caspian and Black Sea Petroleum Co. (known as 'Bnito', from its Russian initials) to expand Bunge and Palashkovskii's storage, packing and shipping facilities at Batum; and in this expansion drive they were soon followed by the other major Baku refiners for their trade via Batum, the largest of whom, the Nobels, formed the Russian Steam Navigation and Trading Co.[121]

The Nobels immediately applied their experience of bulk shipping on the Caspian Sea to the problem of the shortage of barrels at the

Black Sea port when their Belgian agent that year commissioned a British shipyard to convert the steamer *Fergusons* into a kerosene tanker. The following year an Englishman, Alfred Suart, ordered a new tank steamer, *Bakuin*, from another British shipyard, and by 1890 he had sixteen tankers in service on the Black Sea. In 1885 the Nobels' regular Swedish shipbuilder launched the Black Sea tanker *Sviet* that, along with their *Petrolea*, made the first bulk deliveries of kerosene to London through the shipping agent Lane & Macandrew – who subsequently switched to dealing, instead, with the Rothschilds. By 1888 Britons were buying nearly 30 per cent of their lamp oil from Russia, up from 2 per cent in 1884. At the major eastern Mediterranean trading port of Beirut, Russian rapidly replaced American kerosene and Lane & Macandrew began shipping Russian oil to India via the Suez Canal.[122]

Not surprisingly, incidents of fires and explosions multiplied with the increasing trade in these highly flammable and volatile liquids – whether in barrels, in more secure metal cases or in bulk in metal tanks – and demands for legislation covering the storage and transport of oil increased accordingly. The captain on the *Bakuin*'s maiden voyage died along with nine others in an explosion on another of Alfred Suart's tankers, *Petriana*, at Birkenhead in 1886, and the *Bakuin* itself would be destroyed by fire in 1902 while docked at Callao, Peru. In 1887 the *Fergusons* exploded at Rouen, and the following year the French tanker *Ville de Calais* was wrecked in a huge explosion; the *Times* reported,

> The greatest excitement prevailed in Calais throughout the night. People who believed that an earthquake had occurred left their houses in terror and took to the streets ... The force of the explosion was so great that the hull of the vessel was torn into ribands ... Some pieces of machinery weighing many tons have been hurled three-quarters of a mile or more ... At present it is thought that four persons have been killed. Fragments of what are supposed to be three persons – the fire engineer, one of the ship's officers, and a seaman – have been recovered.[123]

INTERNATIONAL COMPETITION

For twenty years, North America had been the world's only major exporter of petroleum, crude and refined. Now Russia had arisen as a serious rival, and other countries also appeared to have the potential to become significant oil producers. Whereas Standard Oil had once enjoyed the freedom of an easy, near global monopoly, now several major international oil companies and a few smaller localized ones competed in the many markets of the world. All sought to reduce their burdensome transportation costs – a key competitive variable – by moving further into the bulk handling of oil.

The crucial role of the new Black Sea oil tankers in bringing huge volumes of new Russian oil onto the world market prompted a revival of transatlantic tankering. In 1884 America was exporting two-thirds of its 15 million barrels of total kerosene output, manufactured largely by East Coast refineries that were supplied with crude by a vast seaboard-focused pipeline system.[124] Thus, phasing out barrels and moving to the use of oil tankers would bring huge savings in export costs that were now more significant in an increasingly competitive international market – and barrels would eventually remain only as a standard unit of measurement. In 1879–80 three Norwegian sailing tankers, *Jan Mayn*, *Nordkyn* and *Lindernoes*, and a tank steamer *Stat*, had brought bulk shipments of American crude to Europe, but these had been withdrawn from service after the French sailing tanker *Fanny* was lost at sea in late 1880. Now, following recent advances in tanker design, in 1885 two converted sailing ships, the American kerosene tanker *Crusader* and the German crude tanker *Andromeda*, were put into service on the Atlantic. The next year a German oil trader brought American kerosene from New York to Bremen in the British-built tanker *Glückauf* – 'as in some of the Caspian steamers, the oil extends to the skin of the vessel', as a naval architect described this new, advanced tanker design[125] – and by 1888 the trader had five transatlantic tank steamers in operation. In 1887 one of Alfred Suart's tankers, *Chigwall*, was the first British-owned tanker to make the Atlantic crossing.[126] London's imports of American oil were sometimes interrupted when sailing ships were held up by headwinds in the English Channel: this created the conditions for one of the earliest tanker trades when a tanker of Russian oil, bound for a Continental port, was diverted to London to take advantage of the resulting momentarily higher prices there.[127]

As crude production and refinery throughput continued to increase during the 1880s in both North America and Russia, competition for the world's export markets intensified and the oil companies marketed their products more actively. They forged closer ties with distribution agents, became more involved in retail and marketing, and in most cases eventually integrated these extensive downstream operations as company subsidiaries. For example, in Britain – the second-largest European importer after Germany – during the spring of 1888 the Rothschilds, the Nobels and Standard Oil each consolidated the marketing and distribution of their products. The Rothschilds' Bnito joined Fred Lane, of Lane & Macandrew shipping brokers, to form the Kerosene Co. Ltd. for importing, and the Tank Storage and Carriage Co. for storage and distribution; and the Nobels appointed Bessler, Waechter & Co. as their exclusive agent in Britain. Standard Oil formed the Anglo-American Oil Co., capitalized at £500,000; and later, the vertically integrated Pure Oil Co. and Bear Creek Refining Co. – Standard's only really significant export rivals – would similarly expand their downstream operations in order to maintain competitiveness. These import and distribution companies invested in tank steamers, tank barges, rail tank cars and storage facilities, and began to eliminate barrels even at the retail end by delivering kerosene from horse-drawn tank wagons and tank carts. This spread of bulk handling of oil brought protests from those whose interests were tied to the barrel, from coopers to the smaller, independent kerosene retailers; and when the tanks of the *Glückauf* were loaded at New York harbour by pipe and pump, the Knights of Labor union demonstrated fiercely against the loss of dockyard manual work. By 1892 Standard's Anglo-American owned three transatlantic tank steamers, and its German affiliate, the Deutsch-Amerikanische Petroleum-Gesellschaft (DAPG) had five, including the *Glückauf*. Meanwhile, American diplomats, at the instruction of the State Department, promoted Standard Oil's business abroad and pressed foreign governments to grant easy access to their markets for American oil.[128] As the

Economist saw it, 'nothing . . . has happened to mar the even progress of the industry and consumption of an article which has grown to be as much a necessary of daily life as bread and coals, or water itself.'[129]

While Standard Oil parried the assaults from Russian oil on its worldwide sales, there appeared on the horizon other possible new centres of large-scale oil production and refining, also with the potential to undercut Standard's prices in their respective localities through lower transportation costs. Particularly in these early days of petroleum geology, it was extremely difficult to know where or when significant new oil production might turn up. To keep abreast of any developments and to seek opportunities to establish refining or production outposts of its own, Standard's agents – most notably William Libby – travelled the world gathering intelligence, complemented by reports from the many American consuls. As Standard trustee W.G. Warden put it to Rockefeller, 'We should have all the information that it is possible to get, which can be obtained in a proper way.'[130]

In Europe, even before the arrival of serious Russian competition, Standard Oil's concern to maintain its position in the important British market prompted the company's Manufacturing Committee in 1883 to report on the continuing success of the Scottish shale oil industry.[131] The emergence of new competition from Europe looked imminent when, in 1881, a particularly productive manually dug well in Austro-Hungarian Galicia generated a surge of renewed interest in the industry there. Similarly, a sizeable oil strike at Ölheim near Hanover seemed to indicate the potential for another significant European oil region, and that year John Simeon Bergheim, a British engineer, persuaded a Canadian driller, William Henry McGarvey, to leave the southwestern Ontario oil town of Petrolia to join him to try their luck in the German town. But their attention was soon drawn further east to Galicia, where recent changes in the law and improved transport infrastructure had prepared the ground for large-scale foreign investment and the industrial mechanization of crude production. Indeed, from 1884, when Bergheim and McGarvey introduced modern mechanical drilling methods to its oilfields, the Galician industry expanded rapidly to dominate the surrounding central and eastern European markets in the face of strong American and Russian competition.

Nevertheless, for some years much of Galicia's oil continued to be drawn up manually from pits and shafts (see Fig. 10) by labourers who suffered from the lack of enforcement of safety and employment legislation; the resulting tensions led to a number of violent labour strikes in which Christians and Jews polarized along religious and ethnic lines.[132] The American consul in Vienna reported that labourers were forced to pay most of their wages back to their employers in return for squalid board and lodging:

> These petroleum slaves, clad in miserable rags, may be seen in droves any day at Boryslaw, and the terrible fate of these men is really deplorable. They must work hard and drag along a miserable existence in spite of the not inconsiderable rates of wages, while unconscionable speculators pocket the profits of their labor ...
>
> Whether a laborer breaks his neck, or dies miserably, matters not; there are always unfortunates who step into their places, and this seems to be regarded as the most important consideration. The labourer, like the draft animal, is only considered of value here while he retains the capacity to work ...
>
> The laborer who falls sick is to be commiserated most. He is left to his fate in the most cruel manner, and, as it is claimed, secretly transported beyond the Boryslaw territory, so that he may not become a burden to the proprietors of the petroleum mines.[133]

Concurrently, to the southeast mechanical drilling was being introduced at the Romanian end of the Carpathian oil region where, for example, Prince Cantacuzino's workers struck a prolific well on his estates in the Prahova valley.[134]

Just as Russian oil began to reach the populous Asian markets – which consumed around a quarter of American exports[135] – there appeared indications that Standard Oil might soon meet further competition from sources of supply in British and Dutch colonies within the region itself. Here, in British India and the Dutch East Indies (Indonesia), oil and conflict were beginning to mix more thoroughly.

COLONIAL OIL

In 1876 David Cargill, a wealthy Glasgow businessman with prominent East Asian trading interests, had bought up the failing Rangoon Oil Co. for £15,000 in what looked like a foolhardy investment. Although demand for the new products of petroleum was growing around the world, and Burma lay right next to the large Indian market, the company's refinery at Dunneedaw near Rangoon had struggled ever since its establishment, in 1858, to secure a regular and affordable supply of crude from the manually worked oil wells at Twingon and Beme near Yenangyaung on the Irrawaddy River. Following the Second Anglo-Burmese War of 1852–3, Burma had lost to the British its trading port of Rangoon and other coastal areas – leaving what remained landlocked – as well as its principal rice-growing region, which sent Upper Burma into a period of economic and social decline. Faced with a severe revenue crisis, King Mindon had nationalized several of his country's industries, most significantly timber, gemstones and oil. For the Rangoon Oil Co. this meant being dependent on his royal monopoly agent, who demanded bribes and delivered only irregular supplies of crude now at higher prices.[136]

During the early 1880s Rangoon's British business community, with the support of British chambers of commerce, clamoured for the acquisition of Upper Burma, or at least for deposing the current King Thibaw and replacing him with a pliant ruler under a British protectorate. In October 1884, after a meeting in Rangoon Town Hall, they sent a petition to the government, forwarded by Sir Charles Bernard, Chief Commissioner for British Burma. Bernard's covering letter explained,

> It has been thought that such annexation would promote trade and also improve the prosperity of the people by securing the exploitation of the coal, teak, earth-oil and other valuable products ... and by opening a direct route for trade with the western provinces of China. It may be accepted that these consequences would certainly follow from annexation; it may be accepted that, so far as military obstacles are concerned, annexation could be easily effected ... Mandalay would surrender to British arms within a single week and with little loss of life.[137]

Bernard did not believe 'that the British have any right to annex, or have at present any sufficient pretext for the annexation of the kingdom'. But relentless lobbying from commercial interests, combined with strategic fears of possible French influence in Upper Burma – following France's conquest of Tonkin, in northern Vietnam, and its transformation into a French protectorate – ensured that a pretext would soon be found.[138] One of the pleas received by the Secretary of State for India, Lord Randolph Churchill, came from the owner of the Rangoon Oil Co., David Cargill, who called for annexation or a protectorate, 'being largely interested commercially with British Burmah, and having considerable interests at stake there', he wrote.[139] With Churchill's close friend, Lord Rothschild, offering £400,000 a year for thirty years for all of Burma's mineral resources, including oil, the Secretary of State came to the view, not surprisingly, that annexation would be 'a most remunerative investment'.[140] In late October 1885 the London Chamber of Commerce petitioned Churchill to annex Upper Burma to enable 'the carrying on of industrial operations, such as the cultivation of rice, the "exploitation" of the teak forests and the development of the other natural resources with which the country abounds'.[141]

Indeed, the British were at that moment issuing an ultimatum to King Thibaw, following Upper Burma's imposition of a hefty fine on the Bombay Burmah Trading Corporation for exporting teak without paying royalties. Britain's demands entailed such a loss of Burmese sovereignty that the Burmese ruler's predictable refusal to accede to them in full became the pretext for war, although several colonial officials warned of the dangers of popular resistance: Sir Charles Aitchison, Lieutenant Governor of the Punjab, argued that 'the annexation of Upper Burma is not a thing to be undertaken by Lord Randoph Churchill in the easy-going, jaunty way, in which he seems to have announced it'.[142] Nevertheless, an invasion was launched in November that year, backed by a campaign of propaganda in the British press that demonized and ridiculed King Thibaw and the Burmese generally. British and sepoy troops, carried by the Irrawaddy Flotilla Co. – which looked forward to burgeoning trade, not least from the transportation of crude oil downriver – took Upper Burma within two weeks: the British commander, Lieutenant General Harry Prendergast, described this Third Anglo-Burmese War as 'a military picnic'.[143]

Even before the British ultimatum had been rejected by King Thibaw, Cargill instructed his crude-oil purchasing agent, Finlay Fleming & Co., to apply to the Chief Commissioner for oil concessions at Yenangyaung: '[W]e trust Government will give us the power to sink wells ... and will grant us the concession of three or four square miles at present untouched.'[144] Permission was granted, and in July 1886 the Burmah Oil Co. Ltd., with Cargill as its managing director, was incorporated in Edinburgh to exploit the oilfields of Upper Burma.[145] As an editorial in *The Times of India* had noted a month earlier, British India had imported nearly half a million gallons of oil from America in the previous year: 'But doubtless Sir Charles Bernard is keeping watch upon the subject of the industrial development of Burmah, and when once the dacoits have been repressed, some steps may be taken towards investigating the mineral and other resources of the province.'[146]

A year later, the impassioned commentator on British foreign policy and petroleum affairs, Charles Marvin, published a pamphlet, *England as a Petroleum Power*, in which he wrote excitedly that 'by the acquisition of the famous Burmese Oil Fields, England had rendered herself the third Petroleum Power of the world.'[147] On Marvin's promotion of Burmese oil, *The Times of India* reported,

> If, for instance, it could be used for the Quetta Railway in a district where there is no fuel, the saving would be enormous. There are deposits of petroleum at Rawul Pindi, near Peshawar, at Khatan, near Quetta, in Assam, on some islands off the Burmese coast, and, above all, at Yenangyoung, in Upper Burmah ... [T]he Indian authorities rest their hopes mainly on Yenangyoung. If England really is a 'Petroleum Power', it is the annexation of Burmah that has made her so.[148]

Britain's 1885 invasion of Upper Burma – justified by the use of time-honoured propaganda techniques – was perhaps the first war in the modern industrial era fought, at least in part, for access to oil; and it would take British suppression of the Burmese resistance movement to enable the Burmah Oil Company to bring industrialized production to the Yenangyaung oilfields. Yet in the most prominent historical account of the oil industry these events are buried beneath an aside, when recounting the Royal Navy's initial adoption of fuel oil in 1904, that 'Burma was regarded as a secure source owing to its annexation into India in 1885.'[149]

The invasion became a less 'remunerative investment' than originally hoped for – at least for the ordinary British taxpayer. Until mid-1886 Burma was peaceful; but, as a British civil servant there later wrote,

> And then the trouble came. The people had by that time, even in the wild forest villages, begun to understand that we wanted to stay, that we did not intend going away unless forced to ... And as the people did not desire to be governed – certainly not by foreigners, at least – they began to organize resistance ... The whole country rose, from Bhamo to Minhla, from the Shan Plateau to the Chin Mountains. All Upper Burma was in a passion of insurrection, a very fury of rebellion against the usurping foreigners ... To overthrow King Thibaw was easy, to subdue the people a very different thing ... Of desperate encounters between our troops and the insurgents I could tell many a story.[150]

In October, Herbert Thirkell White, Secretary for Upper Burma, wrote to the Chief Commissioner, 'The people of this country have not, as was by some expected, welcomed us as deliverers from tyranny.'[151] The number of British and Indian troops and military police was increased to 40,000, and they proceeded to mete out treatment – including summary executions and the destruction of hundreds of villages – of a kind generally denounced as 'terrorist' when perpetrated by the Burmese resistance.[152] The editor of the *Bombay Gazette* wrote critically,

> It was in too many cases accepted as an axiom that all that was necessary was to shoot out of hand, whoever was found under circumstances of suspicion, and thereby establish a terror which would produce the immediate submission of the population ... The idea fades out that the population which has to be brought into subjection by terror, has any claim whatever to be regarded as possessing human rights. The one virtue is to inspire fear, and anything not calculated to produce that effect is regarded as ... evidence of weakness, which will interfere with the effect to be produced.[153]

In mid-1886 *The Times of India* reported that 'on account of the unsettled state of the country [oil] production has been much reduced'.[154] From late 1886 to early 1887, Burmah Oil Co.'s 'operations at the oilfields were attended with some danger',[155] the company's crude-oil agent informed the Chief Commissioner, and crude shipments downriver were interrupted for some weeks while the boats of the Irrawaddy Flotilla Co. were used for transporting troops. By 1889 British forces still numbered 18,000, distributed across 233 police and military posts. Under the protection of armed guards, work at the oilfields was able to continue and crude production increased considerably – although the insurgency would continue in some areas for several decades.[156] Charles Marvin concluded that prospects again looked good: 'Now the country is pacified, the railway runs to within 60 miles of the Oil fields, and no obstacles exist to their development by British enterprise and capital.' He suggested that a 300-mile pipeline be constructed to carry Yenangyaung crude down to Rangoon: after all, 'the pipe line from Olean, in the Bradford Oil field, to New York, is 300 miles long'.[157] The *Financial Times* complained,

> With ample supplies in Burmah, comparatively close at hand, it seems odd that India should be the largest consumer of Russian petroleum . . . She took 2,000,000 poods in 1887, and 400,000 poods in the first month of the present year . . . It should be a source of dissatisfaction, if not of irritation, to English merchants to see Russia reaping a commercial harvest in our Indian possessions by the supply of a commodity of which an abundance is to be had for the gathering, without recourse to the foreigner at all . . . We presume that no one will deny that it is to our interest, whether politically or commercially, to do all we can to develop Burmah now that we have become possessed of that country. We have railway communication to within 60 miles of the petroleum wells – why cannot we provide tank cars, tank ships, pipes and other appliances to take this trade into our own hands, and not allow Russia to step in and take the bread out of our mouths?[158]

The newspaper, along with Marvin, pointed out other petroleum deposits with commercial potential in British India, such as in Assam – taken from Burma during the First Anglo-Burmese War of 1826 – where, following the completion of the Upper Assam rail system, the Assam Railways and Trading Co. in 1887 began drilling at Borbhil, or 'Digboi' as it later became known.[159]

Alongside the rise of Russian oil exports and the commencement of industrial-scale oil extraction in British-held Burma, Standard Oil was also alert to emerging threats to its global dominance from developments in the Dutch colonial possessions of the Indonesian archipelago. Five years after Reerink had given up drilling on Java, the 1882 Annual Report of the Department of Mines noted 'a petroleum concession granted by the Pangeran of Langkat (northeast Sumatra) to a European'.[160] In 1880 Aeilko Jans Zijlker, who was managing plantations near the Lepan River for the East Sumatra Tobacco Co., noticed that the locals routinely used oil skimmed from natural pools as an illuminant for torches and for caulking boats. Realizing that this oil was rich in kerosene, Zijlker obtained land and drilling concessions from the regional sultan and gained financial and political support from his compatriots.[161]

In 1884 Zijlker managed to enlist the Groundwater Development section of the Department of Mines to begin drilling at the most accessible of the oil pools on his Telaga Said concession. After encouraging early results, they struck high-quality flowing oil in mid-1885. Work might then have been abandoned, however, due to a lack of capital and expertise. But Zijlker's small consortium, with close connections to the governing council, attracted the interest and support of the new Governor General, Otto van Rees. The Dutch government decided to continue to support the enterprise because of 'the general interests involved in the establishment of a petroleum industry in the Netherlands Indies' and 'the desirability of giving the mining personnel an opportunity to acquire knowledge concerning the occurrence and the working of petroleum wells'.[162] In 1886 a groundwater engineer, Adriaan Stoop, was sent by the government to the United States and Canada to learn more about deep-drilling techniques and about the oil industry generally. On his introduction by the Dutch consul general at New York to some leading figures in Standard Oil, the latter flatly refused to offer assistance, wishing in no way to encourage new competition in the region, and Stoop carried out his in-depth study without them. The Dutch government also commissioned a detailed report from mining engineer Reinder Fennema on how a Sumatran oil industry might be made economically viable.[163]

At Telaga Said, deeper drilling necessitated opening up a new road from the Lepan River for the transportation of heavier equipment. Work was extremely slow, however, partly because they were operating right on the edge of a conflict zone – and the accompaniment of twenty soldiers had drawn the attention of Acehnese independence fighters.[164]

During the mid-nineteenth century the Sultanate of Aceh, at the northwestern end of Sumatra, was a prosperous trading centre supplying half the world's pepper. Just beyond the fringe of effective Dutch control, its independence was backed, in theory, by the British; as late as 1867 a Foreign Office minute recorded, 'The independence of Acheh should be guaranteed.'[165] But the geopolitical landscape was changing. From the opening of the Suez Canal in 1869 and the turn from sail to steam, the quickest route for sea trade between Europe and China now passed through the Malacca Straits between the Malay Peninsula and Sumatra rather than the Sunda Straits further south between Java and Sumatra. This increased the importance of the Aceh coastline of northeastern Sumatra at a time when other powers, particularly France and the United States, were in the ascendant in the region and might have the will and the means to take possession of Aceh themselves.[166] In 1868 the British Colonial Office signalled its new geopolitical calculation that 'Acheen had much better be in the hands of Holland than of France or the United States',[167] and by 1871 Britain and Holland were ready to swap Aceh – regardless of native wishes – for the Dutch-held part of the Gold Coast and some trade concessions. The consequent Aceh and Ashanti wars became issues in the run-up to the British general election of 1874. A *Times* editorial charged that 'the concession to a foreign state of a right to annex territory in exchange for dominions ceded to us is hardly consonant with modern English notions of our proper relations to inferior races', and Conservative candidate Thomas Bowles proclaimed, 'Upon England rested the bloodshed which had taken place in Atchin.' Criticism of the deal by Benjamin Disraeli, leader of the Conservative opposition, was dismissed by Prime Minister William Gladstone, for whom Aceh was 'about as far off as the kingdom of Brobdingnag'.[168]

The 3,000-strong Dutch invasion force of 1873 was dramatically repelled by the Acehnese, demonstrating that European power was not invincible – though the invaders took note that the ruler of Perlak had prepared to defend his harbour by sending oil-laden fireships downstream to collide with the Dutch sloops. The indigenous guer-rilla resistance movement, unified under local Islamic leadership, fought for the defence of their homeland and of Islam in a conflict that continued for decades. The Aceh War, in addition to emptying the coffers of the Netherlands East Indies government, led to the deaths of up to 100,000 Acehnese and thousands of Dutch troops, and – in what seemed to be the intention behind the Dutch scorched-earth tactics – to the shattering of Acehnese society. Hundreds of villages, along with crops, were destroyed, and many tens of thousands of civilians were displaced.[169]

In the province of Langkat, straddling an undefined border region between Aceh and the province of Sumatra, popular resistance con-tinued even though the local sultan had been co-opted by the Dutch. Among the tobacco plantations at Telaga Said, Zijlker had managed to carry out further drilling, but work again ground to a halt when the supportive van Rees resigned as Governor General after a renewed wave of rebellion swept across the islands. However, Zijlker, armed with the favourable conclusions of the Fennema Report, in 1890 secured finance from some high-profile investors with the encourage-ment of formal backing from King William III. The venture, reorganized as the Royal Dutch Company for the Exploitation of Oil Wells in the Dutch East Indies, was four and a half times oversub-scribed on its stock-market flotation. Johannes Arnoldus de Gelder, one of the company's ten co-founders alongside Zijlker, wrote,

> We are indebted for this success to the names of the persons who have taken an interest in this enterprise and, further, to the high favor shown by His Majesty the King in granting us the rare privilege of conferring the royal title upon our Company – a favor which may largely be attributed to the national character of this Company. The Indies yearly import from America to the tune of ten million guilders, by which amount the national wealth of the Indies is burdened. It is therefore considered of the greatest importance to our colonies that this oil should now become a home product.[170]

A few months later Zijlker died of a tropical disease, whereupon the direction of Royal Dutch's operations, guided by the detailed recom-mendations set out in the Fennema Report, was taken over by Jean

Baptiste August Kessler. A road was cut from the drilling site 8 miles through the jungle to Pangkalan Brandan on Babalan Bay opening onto the Malacca Straits, along which was laid a pipeline. For days on end, workers, under torrential rain, were waist deep in swamps and flood waters and, wrote the company's first official historian, 'in his despair Kessler was sometimes tempted to thrash the Chinese out of their living quarters with a stick'.[171] Nevertheless, in February 1892, with 'a roar as of a mighty storm,' wrote Kessler, the pipeline began disgorging crude oil into the tanks of the new bay-side refinery.[172]

Meanwhile, as investors back in the Netherlands became increasingly frustrated at the continuing civil conflict, the Acehnese resistance launched a wave of attacks on colonial business interests in the rich plantation districts. At the end of May 1893 Royal Dutch's road between the drilling site and the refinery was cut off and the 150 troops sent in response to Kessler's calls for military protection were barely able to defend the oil installations, let alone allow work to continue; several lives were lost and some buildings set on fire.[173] In a statement from the Netherlands the company directors protested, 'Either Langkat must be entirely cleared of Achinese and be made permanently secure from their raids or the enterprise must be allowed to go to ruin!' Standard Oil, which had been involved in negotiations for a possible oil concession on Sumatra, decided it was time to pull out.[174] Standard was now beginning to face the competition of Dutch as well as British enterprises in the Far East. In Langkat, on Sumatra, Royal Dutch received the political and military backing to enable it to rapidly expand its operations, just beyond the reach of the continuing Acehnese resistance. Thus, the emergence of Royal Dutch from 1886 as a major new oil company was predicated on the invasion of Aceh in north Sumatra by the Dutch colonial army and the brutal subjugation of its people during what would be the forty-year-long Aceh War – a genocidal campaign that is dismissed in the most widely available account of the origins of the company as a 'native revolt' or 'rebellion' with no apparent cause.[175]

Meanwhile, in eastern Java, Stoop, having written up his comprehensive 'Report on the Petroleum Industry of North America' for the Dutch government, applied his knowledge to develop oil concessions around Surabaya. With the help of financial backing from wealthy family and friends, by 1892 Stoop's Dordtsche Petroleum Co. had pipelines sending crude from two oilfields to its refinery at Wonokromo, and the following year it began developing an oilfield in Rembang province. Here, the locals tapped the company's pipeline for its oil, which they had always regarded as a free, common resource like water, owned by no one and everyone.[176]

EAST AND WEST OF SUEZ

In 1884 Asian demand for lamp oil had reached 2.1 million barrels, supplied almost entirely from the United States to the tune of $9.2 million, accounting for a fifth of its total kerosene exports. Concerns for the safety of oil tankers if sent into high equatorial temperatures, and the absence of bulk storage facilities in the Far East, led to the oil being sealed in tin cans and packed into wooden cases for the journey of four and a half months by sail from the East Coast of the US round the Cape of Good Hope. By the late 1880s, however, this American monopoly was being challenged by the arrival of Russian case-oil, shipped from Batum in just one month – thanks to the Suez Canal.[177]

By linking the waters of the Mediterranean and the Red Sea, the Suez Canal made it possible for shipping to bypass the long route south around Africa's Cape of Good Hope and to cut the journey time between Europe and Asia by up to half. This dramatically shorter route became commercially viable only because of simultaneous developments in steamship technology: firstly, the paddle wheel was being superseded by the screw propeller which was more effective on the high seas; and secondly, the more efficient compound steam engine was greatly reducing the coal consumption of steamers, allowing them to cover greater distances before re-coaling and freeing up space for more cargo. Crucially, unlike the sailing ship, a steamer could readily navigate the often treacherous conditions in the Red Sea and Indian Ocean, and could move under its own steam through the Canal.[178]

Prior to the opening of the Suez Canal in 1869, one of the basic parameters of international commerce and military strategy had been that any large movement of goods or troops between Europe and the Far East had to go by the Cape route. The two main so-called 'overland routes' – from the eastern Mediterranean across Ottoman territory to either the Red Sea or the Persian Gulf – were, by contrast, used for more local trade, for the speedier movement of passengers and for more rapid communications by postal service which, more recently, had been augmented by the electric telegraph. Britain, at the height of its imperial power, jealously guarded these lifelines to India. Maintenance of the overland routes – the more difficult to shield from French or Russian encroachments – involved Britain in a string of complex diplomatic, commercial and military relationships with the transit territories and the coastal regions of the respective sea lanes: the Ottoman Empire, Persia, the Arabian Peninsula and northeast Africa. Most blatantly, in 1839 the East India Company had simply captured Aden, on the south-western tip of the Arabian Peninsula (in present-day Yemen), for a coaling station between Bombay and the northern end of the Red Sea at Suez, at that time accessed overland from Alexandria on the Mediterranean coast 200 miles by donkey and camel.[179]

From 1840, when the Peninsula & Oriental Steam Navigation Co., formed three years earlier, gained a Royal Charter to run a unified steamship service between Britain and India, the British eventually gained permission to build and operate a railway across the Suez Isthmus, completed in 1858. Egypt, a semi-autonomous province of the Ottoman Empire, became the target of increasingly intense international intrigue aimed at winning preferential access to the overland route between Alexandria and Suez. When Ferdinand de Lesseps, former French consul at Alexandria, began lobbying for permission to construct a canal, the British governments of Palmerston and Disraeli consistently opposed him, fearing that they would lose control of the route and would hand east-west trade to France, Austria and Italy, whose ports were closer to Alexandria.[180] However, opposition MP William Gladstone dissented, arguing that, if a canal were built,

> Who would have control of the Red Sea? Who has now got control of that Sea at its southern issue? Who has occupied Aden on one side and Perim on the other? – What is the power that would readily possess the canal if it were opened? Is it not a canal which would necessarily fall within the control of the first maritime power of Europe? It is England and no foreign country, that would obtain the command of it.[181]

Ten years later, as de Lesseps' canal neared completion – like the railway before it, using forced labour – the degree to which British shipping dominated the sea legs of the Suez route lent support to Gladstone's prediction. A historian wrote then of P&O,

> The importance of this enterprise of a private company to the interests of the mother-country, and her eastern dependencies, it would be difficult to overrate. It has given a character of solidity and compactness to the British empire in the Eastern world, which enables us to contemplate its expansion without any feeling of apprehension. It has linked the most distant countries of the east with the European world ... given full effect to the views of Alexander the Great when he founded Alexandria, and destined it to be the highway between Europe and Asia. It has covered the Red Sea with steamers, and converted it into an English lake.[182]

Indeed, from the Canal's opening British shipping supremacy flowed through it, the new all-sea route to the east triggering a boom for Britain's shipyards and shipping companies. As the *Economist* commented at the time, 'the Canal has been cut by French energy and Egyptian money for British advantage', and for many years over three-quarters of Suez traffic would be British.[183] An eminent American journalist wrote effusively of how steam-powered ships and locomotives were circling the globe:

> Coal, the stored-up sunlight of a million years, is the grand agent. Liberty lights the fire, and Christian civilization is the engine which is taking the whole world in its train. There are but three aggressive nations, – England, America, and Russia, – and together they are to give civilization to six hundred millions of the human race.[184]

The Egyptian government's finances were in dire straits, however, and by 1875 Khedive Ismail was unable to meet the high interest payments on foreign loans and was facing bankruptcy, although the economy was basically sound. He was forced into accepting dual French and British control over his national finances, and Disraeli took the opportunity to buy up the Khedive's 45 per cent stake in the Suez Canal Co. for £4 million, arranged with a loan from Lord Rothschild. Disraeli wrote to Queen Victoria, 'It is vital to Your Majesty's authority and power at this critical moment that the Canal should belong to England.'[185]

As the Suez route became pivotal to British imperial policy – both for its commercial and military value – it thereby also became Britain's Achilles heel. The route had been of clear military value even before the Canal. In 1855 two regiments of cavalry had been sent via Suez from India to the Crimea, and in 1857 several regiments of infantry had been rushed the other way in response to the Indian Mutiny.[186] So it followed that when Russia declared war on Turkey in April 1877, the British government issued a warning that its routes to India, including the Suez Canal, must not be threatened and it sent six warships to Port Said: 'An attempt to blockade or otherwise interfere with the Canal or its approaches would be regarded . . . as a menace to India, and as a grave injury to the commerce of the world.'[187] A year later, after the despatch of an Indian military contingent to Malta via the Canal, it was poignantly observed, 'Henceforth, England and India are one for purposes of offensive or defensive warfare, and are definitely leagued together against all possible antagonists, whether on this side or on the other side of the Suez Canal.'[188]

'THE BATTLE FOR ALL CHRISTENDOM'

During the period of Anglo-French Dual Control, Egyptians felt ever more acutely their subservience to the European powers. The population laboured under a heavy tax burden, two-thirds of which the dual-controllers devoted to servicing foreign debts at the expense of domestic creditors and of civil service and army wages; Europeans enjoyed legal extraterritoriality, in other words they were subject to their national law rather than Egyptian law, which in practice often meant simple impunity; the army was run by a corrupt and incompetent Turkish-Circassian officer elite that systematically discriminated against its Egyptian personnel; and in 1879 the Europeans pressured the Sultan into substituting the independent-minded Khedive Ismail with his son, Tewfik, as a more pliant puppet ruler. Britain had considered armed intervention, but in addition to the fact that its military was currently bogged down in Afghanistan and Zululand, outright invasion might provoke Egyptian nationalism. In the words of Foreign Secretary Lord Salisbury, 'The Musselman feeling is still so strong that I believe we shall be safer and more powerful as wire-pullers than as ostensible rulers.'[189] Under the influence of the dual-controllers, Tewfik removed his new constitutionalist leading minister, Sherif Pasha, in favour of a more authoritarian figure, and sold off the country's remaining 15 per cent share of the profits of the Suez Canal Co.[190]

Egyptians' deep resentment at increasing European control and exploitation of their country now began to well up as a broad-based demand for political change.[191] One vocal supporter of Sherif Pasha and of representative, constitutional government was Sayyid Jamal al-Din al-Afghani, a progressive cleric, denounced by religious conservatives as an atheist. Al-Afghani urged the need for national unity, based around Islam, as the only way to resist an expansionist Christian Europe. When, by mid-1879, he appeared to be 'obtaining an amount of influence over his hearers which threatened to become dangerous', in the judgement of British Consul General Frank Lascelles, he was arrested, charged with leading a secret group intent on violence aimed at 'corrupting religion and the world', and expelled from the country.[192]

Another supporter of constitutionalist, democratic reform was Colonel Ahmed Urabi, a hero of the recent Abyssinian campaigns and the only native Egyptian to have risen to such high rank in the face of persecution by the Turkish-Circassian officers. He was drawn to the ideas of political reform, democracy and justice expounded by al-Afghani and by American staff officers, Civil War veterans from the Union side.[193] The American consul general, Simon Wolf, wrote of Urabi a few years later,

> He knows all about the United States, its history and struggles, and told me repeatedly that his desire was to have a constitutional form of government. When I presented him with a copy of the biography of Garfield in French, he at once said he would have it translated into Arabic, so that the youth of Egypt could profit by it. At my suggestion he ordered the translation of the Constitution of the United States into Arabic.[194]

Urabi came to prominence in May 1880 when, with the backing of his regiment and the new National Party, he hand-delivered to the

Khedive a petition protesting about military laws that discriminated against native Egyptian soldiers. When, early the next year, the Khedive attempted to have him arrested for court martial and exile, his troops protected him and he became a national figurehead for political change. Initially, the almost universal opinion of observers was that this was, indeed, but one manifestation of a genuine and widespread movement for democratic reform.[195] Even the British controller, Auckland Colvin, officially backing the Khedive, recognized that 'what gains [the army] support among great numbers of the more respectable Egyptians is that there is a great deal of truth in their complaints'.[196] British Foreign Minister George Granville wrote to his consul general in Egypt, Edward Malet, 'The Government of England would run counter to its most cherished traditions of national history, were it to entertain a desire to diminish that liberty or to tamper with the institutions to which it has given birth.'[197]

However, the British government soon decided to do precisely this. Rather than risk losing the control it had gained over Egypt – with its increasing commercial and strategic value – in mid-1882 British officials engineered a false pretext to invade the country and crush its budding constitutionalist democratic movement. Although this episode is a rather faded memory in mainstream historical awareness – particularly the degree of British deception and subterfuge that opened the way for the invasion – it is the headwaters of the later Suez Crisis when, in 1956, British Prime Minister Anthony Eden would order another invasion of Egypt that, he would claim to US president Dwight Eisenhower, was necessary in order to maintain Anglo-American control over Middle Eastern oil supplies.[198] Indeed, Britain's 1882 invasion and occupation of Egypt prefigures in many respects numerous subsequent political and military interventions in the Middle East region – along with the facilitating propaganda campaigns – up to the present day: from the MI6- and CIA-orchestrated overthrow of Prime Minister Mohammed Mossadegh of Iran in 1953 to the 2001 invasion and occupation of Afghanistan, the 2003 invasion and occupation of Iraq, and the 2011 bombing of Libya.[199]

In 1882 the British government worried, despite overwhelming evidence to the contrary, that their bondholders might no longer be paid and that other business interests in Egypt might be harmed; Prime

Minister Gladstone himself held £51,500 in Egyptian stock worth £40,500, amounting to 37 per cent of his investment portfolio.[200] The government was concerned that a successful independence movement in Egypt might give encouragement to similar movements across the empire.[201] It felt also that the French, who had just invaded Tunis, might be gaining too much power in the region and – again, contrary to the evidence – that the Egyptians might close the Suez Canal, which was now overtaking the Cape route in shipping tonnage. After an emergency Cabinet meeting following the French seizure of Tunis, Foreign Office Under-Secretary Sir Charles Dilke recorded,

> Steps were taken ... to see whether now that France had knocked another bit out of the bottom of the Ottoman Empire by her attack on Tunis, we ought to try to get any compensation in Egypt for ourselves [and] to consider how we stood with reference to the dispatch of troops through Egypt in the event of (1) a rising in India, (2) an invasion of India by Russia.[202]

Accordingly, the British press – most of which was more or less directly controlled by interested parties such as the Rothschilds and the Corporation of Foreign Bondholders – launched a campaign of propaganda, fed by a steady supply of misinformation from British officials, aimed at demonizing Colonel Urabi and preparing public opinion for strong measures to be taken against him. In defiance of the fact that life in Egypt, despite the obvious political tremors, carried on much as ever and that, as Urabi had urged, free elections to the National Assembly had just been held, newspapers such as *The Times* successfully created the impression that Egypt was descending into chaos and that Urabi was becoming a military dictator. At the beginning of 1882 Britain's ambassador in Paris admitted that an official warning from Britain and France to the Egyptian nationalists needed to be worded strongly, 'otherwise we shall have the bondholders, French and British, on our backs again'.[203]

Privately, concerns were expressed whether a Liberal government, with an avowed ethical foreign policy, could get away with adopting such an aggressive stance. Edward Malet, for example, wrote to George Granville, 'I own to having a repugnance to a war engaged on behalf of bondholders and which would have for effect to repress the first attempt of a Mussulman country at Parliamentary Government. It seems unnatural for England to do this.'[204] But, the *Spectator* informed its readers, 'the more general opinion, especially among those experts who have come much in contact with prominent Asiatic statesmen', was that

> such a feeling as patriotism does not exist in any [of them] ... [H]is impelling motive is always either ambition, or pride, or fanaticism. [They] ridicule the notion that a man like Arabi Pasha can be governed by anything like a 'nationalist' feeling. He may be, they admit, a Mussalman fanatic, or a devotee of the Khalifate ... or even an 'Asiatic', that is, a man who loathes European ascendancy, but he cannot care enough for Egypt to make Egyptian interest, as he conceives it, the guiding star of his policy.[205]

In this vein Colvin wrote, 'The Egyptians being, in my opinion, incapable of conducting the administration of affairs, I think we are rapidly approaching a state of ... anarchy.'[206] Similarly, Gladstone confessed, 'I am much surprised at this rapid development of a national sentiment and party in Egypt. The very ideas of such a sentiment and the Egyptian people seemed quite incompatible. How it has come up I do not know.'[207]

In February Colonel Urabi became Minister of War after his predecessor took up the Presidency of the National Assembly, which now reclaimed management of the debt from the dual-controllers. At this point British and French warships arrived in Alexandria and a reactionary Egyptian Cabinet minister, with some Circassian officers, attempted to assassinate Urabi. When Malet persuaded the Khedive to refuse to sign the culprits' sentencing papers and, rather, to eject Urabi from the Cabinet and exile him, Egyptians rallied more fervently behind Urabi and against foreign intervention. With the support of the elected Assembly, Urabi now effectively became head of state, for which Gladstone – with the cynicism of a politician who had lambasted the previous Conservative government for its aggressive imperialism and had recently supported Italian and Bulgarian independence struggles – branded him a 'usurper and dictator, who betrayed the true aspirations of Egyptian liberty'.[208] As the American consul general in Cairo, Elbert Farman, explained a few years later, '[Arabi's] popularity with the populace made him unacceptable to those [the British] who dictated policy to the government – Arabi has no real personal ambition – just a passionate desire to be freed from an oppressive foreign domination.'[209] Britain, now clearly committed to overthrowing the Urabi government, just needed a pretext for military action. Malet wrote to Granville, 'I believe that some complication of an acute character must supervene before any satisfactory solution of the Egyptian question can be attained, and that it would be wiser to hasten it than retard it.'[210]

There now played out a suspicious sequence of events. While Urabi warned the British that the arrival of its Mediterranean Fleet in Alexandria harbour was creating fear and panic in the city, Britain was secretly arming its resident Maltese Greeks with rifles. Then, on 11 June 1882, after a Greek merchant stabbed to death an Egyptian donkey boy, there ensued a riot in which an Egyptian mob beat to death some fifty Europeans, mostly Maltese Greeks who, being well-armed, shot dead hundreds of Egyptians. Britain's press and government now went into propaganda overdrive, calling the riot a 'massacre of Christians by Moslem fanatics', ignoring the many more Egyptians killed and blaming Urabi for instigating it – quite falsely, as was later generally acknowledged.[211] Urabi's army quickly restored order, but with such damning 'evidence' now piling up against him, the *Economist* declared,

> the contingency of anarchy in Egypt, in which event Lord Granville has throughout stipulated for perfect freedom of action, has arrived. Those who were credulous enough to imagine that Arabi was the sincere exponent of a genuine and innocent popular movement must be rudely deceived by the facts disclosed in the papers recently presented to Parliament.[212]

The *Illustrated London News* pitched in:

> Suddenly arises a military adventurer with a peculiar audacity and cunning such as Oriental races can alone produce, who has been able, step by step, and in the face of a wondering world, to establish, without let or hindrance and out of the most contemptible materials, a military despotism which threatens to depose the Khedive, and which defies, with impunity, the Western Powers and their ironclad fleet.[213]

At the hastily convened Constantinople Conference on Egypt, Britain's bellicose denunciations of Urabi did not square with the reality as seen by other diplomats. Britain not only failed in its efforts to win international backing for an invasion but, at the beginning of July, other European powers, along with the Sultan, began moves to recognize the Urabi government. Yet the British position, encapsulated by *The Times*, remained clear: 'If Arabi is allowed to prevail, the country must go from bad to worse, and no European interest in it, not even the Canal itself, will be safe.'[214] Aware of the imminence of an invasion, Urabi wrote to Gladstone that he was open to negotiation but that Britain, if it made war,

> may rest assured that we are determined to fight, to die martyrs for our country – as has been enjoined on us by our Prophet – or else to conquer and so live independently and happy. Happiness in either case is promised to us, and when a people is imbued with this belief their courage knows no bounds.[215]

A week later, Gladstone ordered Admiral Frederick Seymour to prepare to bombard Alexandria, again under a false pretext: that his ironclads were threatened by Egyptian gun emplacements overlooking the harbour which, in any case, Seymour himself had intentionally anchored his squadron close to. Seymour launched the pre-emptive strike on 11 July, rejected an Egyptian surrender, and proceeded with a bombardment using over 3,000 shells, many fired indiscriminately into the city itself. On the morning of 13 July Seymour reported, 'There was an immense conflagration in Alexandria last night', but, contrary to most other accounts, British reports perversely blamed the fires, death and destruction on Egyptian mobs.[216]

A small number of opposition MPs within Gladstone's party denounced the bombardment as a crime against humanity, committed for the benefit of Britain's bondholders. In the House of Commons, Sir Wilfrid Lawson called it 'an act of international atrocity . . . a cowardly, a cruel, and a criminal act'. The government had sent its fleet into Egyptian waters

> To overrule the people of that country and to establish a Government which they were in favour of, but which there was no evidence at all to show that the people of Egypt themselves were in favour of. What should they have thought if the Germans had sent their Fleet into the

Thames and demanded the dismissal of his right hon. Friend the Minister for War? They would have had riots in England.[217]

Sir Charles Dilke, by contrast, remarked in a letter to the Viceroy of India, Lord Ripon, 'The bombardment of Alexandria, like all butchery, is popular' – probably correctly, given the virulent anti-Urabi propaganda in the British press – and General Garnet Wolseley was appointed to command a full British land invasion.[218] A member of the Cabinet, John Bright, resigned in disgust, recording, 'Painful to observe how much of the "jingo" or war spirit can be shown by certain members of a Liberal Cabinet.'[219] After meeting with Gladstone he wrote that the Prime Minister

> urged as if all that had been done in the Egyptian case was right and even persuaded himself that he is fully justified in the interest of Peace ... He seems to have the power of convincing himself that what seems to me to be glaringly wrong is evidently right ... He even spoke of our being able to justify our conduct in the great day of account.[220]

The *New York Times* predicted 'A Holy War':

> There has never been so promising an opportunity for a grand Mohammedan uprising against England as there is at the present moment. With Egypt and the Suez Canal in the power of a Mohammedan leader already at war with England, with the Mohammedans of India longing to repeat the sepoy rebellion, and with the Sultan urged on to join with Arabi in order to forestall the possible proclamation of a rival Caliph at Mecca, there is abundant reason to fear that [Turkey] will unite all the Mohammedan races in a desperate struggle to rid themselves of British rule and Christian influence.[221]

However, the British and Indian invasion force of 40,000 troops – assisted by Thomas Cook & Son's Nile cruise steamers – defeated the under-resourced Egyptian army easily within a matter of weeks, costing the lives of fifty-seven British troops and an estimated 2,000 to 10,000 Egyptian deaths. Malet wrote to the Foreign Secretary,

> I congratulate you and Her Majesty's Government on the complete success which has crowned your Egyptian policy. It has been a struggle between civilization and barbarism. Had Her Majesty's Government

held back and allowed Arabi to gain the upper hand in Egypt, the country would have been thrown back a hundred years and the lives of Christians in all Mussulman states would have been in danger. You have fought the battle for all Christendom and history will acknowledge it. May I also venture to say that it has given the liberal party a new lease of popularity and power.[222]

Against the wishes of Gladstone and Queen Victoria that Colonel Urabi be summarily hanged, his few British friends secured for him at least a fair trial; on finding virtually no evidence against him, he was instead exiled to the British colony of Ceylon (Sri Lanka). Britain was soon drawn into a war between Egypt and Sudan, and the British continued to occupy Egypt as a de facto protectorate – the 'Veiled Protectorate' – for the next forty years, despite numerous promises to withdraw. In acting so unilaterally over Egypt, with little regard to the diplomatic checks and balances of the Concert of Europe, Britain entered an era of imperious isolation from the other European powers, and its relationship with Ottoman Turkey would never recover from this violation of the Sultan's sovereignty.[223]

THE MARKETS OF THE WORLD

By taking the route via the Suez Canal, Russian case-oil became a major new feature in Far Eastern trade. In 1887 cargo ships took a million cases of Russian kerosene through the canal, surging to 7.5 million by 1891. Standard Oil thereby lost its almost total monopoly over the Oriental markets – mainly China, India and Japan – although burgeoning demand across the region enabled it nevertheless to increase its yearly sales to over 15 million cases, or around 3 million barrels.[224]

At this time, shipping operator Fred Lane, who was now running regular tanker shipments of Russian kerosene to Europe for the Rothschilds' Bnito, travelled to Batum with another of his clients, Marcus Samuel, to evaluate the Russian oil industry. Samuel headed a successful Far Eastern trading syndicate, M. Samuel & Co., with an established distribution network. He was considering adding kerosene to his trade, and he had been recommended to the French Rothschilds by several

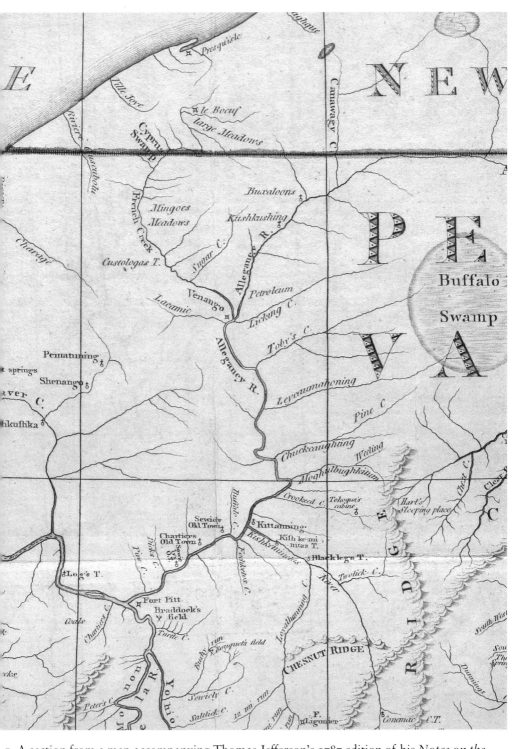

1. A section from a map accompanying Thomas Jefferson's 1787 edition of his *Notes on the State of Virginia*, indicating 'Petroleum' near Fort Venango on the Allegheny River, to the west of 'Buffalo Swamp' (in northwestern Pennsylvania). The tributary running north-south just above the word 'Petroleum' is Oil Creek. The forts, from north to south, are Fort Presque Isle (future Erie) near the shore of Lake Erie; Fort Le Boeuf (future Waterford); Fort Venango (formerly Machault, later Franklin); and Fort Pitt (formerly Duquesne, future Pittsburgh).

2. Seneca Indian Chief Cornplanter (Kayéthwahkeh). Twenty-three years after his death in 1836, his tribe's former territory would see the emergence of the world's first large-scale, industrialised oil production.

3. A photograph of Edwin Drake (on the right, foreground) in front of the Seneca Oil Co.'s original well near Titusville, Pennsylvania; taken in 1866, after the well's production had ceased in 1862.

4. Oil wells on Bennehoff Run, Venango County, Pennsylvania, in 1866. The photograph was taken shortly before a lightning strike that led to the destruction of these oil derricks. About 20 wells, including several large flowing wells, and about 13,000 barrels of oil went up in flames. 'The oil was a foot deep as it ran down the run to Oil Creek.' ('The Fire Fiend on Bennehoff Run!', *Titusville Morning Herald*, 9 July 1866).

5. Smoke billowing from two oil storage tanks after they caught fire at the Imperial Refinery near Oil City, 1875.

6. Workers of the Columbia Oil Co. loading barrels of oil onto barges at Oil Creek, Venango County, Pennsylvania, c.1863.

7. One of the first true oil tankers, the *Zoroaster*, launched by Nobel Bros. on the Caspian Sea in 1878.

THE STANDARD OIL OCTOPUS

8. In America, from the mid-1870s big business corporations began to be likened to octopuses with tentacles extending out in all directions to take control over an industry. This is perhaps the first such depiction of Standard Oil, in 1879.

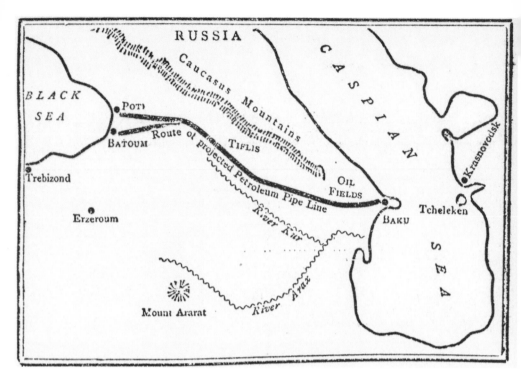

9. A map of the proposed Baku-Batum pipeline in the *Pall Mall Gazette*, 1886.

10. Peasants in the Drohobycz district of Borysław, in the Austro-Hungarian province of Galicia around the turn of the 20th century, scooping crude oil from shallow pits near more modern drilling rigs in the background.

11. A camel train on a road through the Balakhani-Sabunchi oilfield near Baku, with pipelines criss-crossing the terrain, circa 1900.

12. A still from the Lumière Brothers' circa 1898 film, *Puits de pétrole à Bakou. Vue de près* (*Oil Wells at Baku: Close View*), showing an oil well blaze at the Balakhani-Sabunchi oilfield. The person dressed in white, standing on the embankment in front of the derrick in the foreground, gives an indication of the scale of the conflagration.

13. A 1905 advert by Standard Oil's British subsidiary, the Anglo-American Oil Co., for its 'Royal Daylight' lamp oil.

other leading trading houses, including Jardine, Matheson & Co. As Lane and Samuel made their economic and geographic calculations, they perceived an opportunity to profit by capturing more of Standard Oil's Asian market share. If they could implement in the Far East the kind of bulk shipment and storage methods that had proved so successful in the European trade, the savings in transport and distribution costs would enable them to undercut American kerosene prices across much of the region. By complementing Rothschilds oil with Samuel's marketing network, they figured that the requisite investment in new shipping and port facilities for the bulk handling of oil would pay off handsomely.[225]

The biggest obstacle, requiring significant investment, was to build a more advanced tanker to meet the Suez Canal Co.'s strict safety criteria. Both Standard Oil and Bnito had already been refused applications for tanker access, so Samuel commissioned a leading tanker designer, Fortescue Flannery, to produce a blueprint for a new Suez-compliant tanker. In mid-1891 the design was rated as safe by Lloyd's insurers and accepted by the Suez Canal Co. as meeting its regulations. Samuel signed an exclusive nine-year contract with the Rothschilds to sell Bnito kerosene east of Suez, and he placed an order for one of the new tankers with William Gray & Co. of West Hartlepool in northeast England.[226] In early 1892, as the *Murex* neared completion, the *Economist* proclaimed,

> The new scheme is one of singular boldness and great magnitude ... Instead of sending out cargoes of oil in cases, costly to make, expensive to handle, easy to be damaged, and always prone to leak, the promoters intend to ship in tank-steamers via the Suez Canal, and discharge wherever the demand is greatest into reservoirs ... and if the sanguine anticipations of the promoters are realised, the Eastern case-oil trade must needs become obsolete.[227]

In the face of intense opposition and lobbying from those involved in the case-oil trade – the other oil companies, shipping companies and tin-plate manufacturers – the *Murex* left Hartlepool docks on 26 July, filled its tanks with 4,000 tons of kerosene at Batum and passed through the Suez Canal on 24 August. Arriving at Singapore's Freshwater Island, it unloaded part of its liquid cargo and then continued on to Bangkok with its remaining 1,500 tons. By the end of the following year the shipyards of Hartlepool, Sunderland and Newcastle had built for the Samuels a further ten tankers for the Suez route, all named after seashells, such as *Conch*, *Clam*, *Bullmouth*, *Volute*, *Turbo*, *Trocas*, *Spondilus* and *Cowrie*. By the end of 1895, tankers had passed through the Canal sixty-nine times, all but four owned or chartered by Samuel; for their return trip they were cleaned out and filled with bulk cargoes of tea, coffee, sugar, rice and tapioca.[228] In a presentation to the Society of Arts in London, Samuel described his Tank Syndicate's expanding Far Eastern port installations:

> The facilities for landing and handling oil necessarily differ very much at various ports. The surroundings, too, of the tanks are essentially different, many being placed amidst scenes of natural beauty, with deep water right up to the walls of the embankment, such as Nagasaki, Japan ... where it will be seen Nature admits of the steamer lying next to the wharf, whilst [at] ... the installation at Kobe, Japan ... a pier had to be constructed, it being impossible for a steamer to get alongside. [At the] Port of Madras a breakwater of almost a mile in length [had to be] constructed before water sufficiently deep to allow a steamer to get near of the size employed in this trade could be found, and even then ... a contrivance had to be constructed to connect the discharging pipe of the steamer with the breakwater. In spite, however, of these drawbacks, it is found that a steamer can easily discharge into the tanks, placed at one mile distant from the ships, at the rate of fully 200 tons an hour ...
>
> The [Singapore] Government, not understanding how free the business was from danger, would not allow the tanks to be erected on the island of Singapore itself, and compelled us to go to the picturesque spot shown, and, as a final example of the progress of liberal ideas, I show the installation at Bombay, where permission to land the oil was only given some two years ago, and after experience had shown the immunity from danger attending the transport of oil in bulk, and it will be seen that these tanks are placed almost in the middle of the shipping, railway sidings have been taken right up to them, whence oil is pumped into the tank waggons for conveyance all over India.[229]

Marcus Samuel's Tank Syndicate was thus instrumental in raising the Russian share of the Oriental kerosene market through the 1890s to the point of outstripping American sales. This was despite the limited transit capacity of the Baku-Batum railway and the associated high freight rates, which stalled the growth of Russia's overall global market share at one-third.[230]

During the early 1890s Russian production continued to surge, sending crude prices there plummeting, but refiners as well as producers struggled. The growth in domestic demand for kerosene ground to a halt due to a severe economic crisis in 1891–3 and to the imposition by the government of a heavy tax, as a revenue-generating measure, on the domestic sale of kerosene. Foreign demand was burgeoning but the high freight rates of the Baku-Batum bottleneck limited access to this market and many smaller producers and refiners faced collapse or takeover. Finance Minister Sergei Witte had opted to keep the Baku-Batum oil export transit infrastructure under state ownership in order, firstly, to prevent the emergence of a private monopoly along the lines of Standard Oil; secondly, to direct investment into the more strategically useful railway; and thirdly, to maintain a reliable stream of revenue from the freight charges, at little additional cost to the state. Thus, when an 1893 report by the Russian Ministry of Transportation yet again presented the case for a Baku-Batum kerosene export pipeline, Witte rejected its recommendations due to straitened state finances. Only a short stretch of pipeline was laid, two years later, after a section of the railway near Batum was washed away during flooding, necessitating the temporary re-routing of oil tank cars from Baku – via Petrovsk, Beslan and the Vladikavkaz Railway – to the port of Novorossiysk. There was even talk at heated meetings of the Petroleum Congress in Baku of building a pipeline across Persia, from the Caspian Sea to the Persian Gulf. The promoter of this plan, the crude-oil producer Palashkovskii, made the geopolitical case for creating an alternative to the Black Sea export route: 'In Turkey we have to deal with the whole of Europe but in Persia, where we can easily reach the ocean through the Gulf of Oman, we will have to deal only with England.'[231] The proposal was rejected, however, reported the British consul at Batum in 1895, due to 'the great risks and the many difficulties, both commercial and political, with which the execution of the works and the maintenance of the concern in a lawless country like Persia would, beyond doubt, be accompanied'.[232] It would be another decade before Baku's refiners finally saw the 560-mile trans-Caucasus kerosene pipeline completed.[233]

As well as being constricted by its export bottleneck while facing renewed competition from the United States, the Russian oil industry was exhausting itself with the kind of 'ruinous' intensity of competition characteristic of the American industry during the 1860s and 1870s. Standard Oil, by its adept manipulation of the economics of oil transit, had integrated most of its serious competitors into a more coordinated and cooperative corporate giant, thereby restraining the frenzied competition. By contrast, whereas Nobel Brothers held a near monopoly over Russia's domestic kerosene trade, a balance of power prevailed in the export-oriented side of the industry between the Nobels, the Rothschilds and several prominent independent operators, all with relatively equal access to the state-owned Baku-Batum railway. Without the degree of consolidation arrived at in the US industry, Russia's oil industrialists throughout the 1890s attempted to ameliorate their mutual competition through voluntary agreements. However, these arrangements quickly broke down, precisely due to the absence of the kind of corporate unity – the mutual self-interest and legal bonds that came with interlocking share ownership – achieved by the Standard Oil combination. An 1892 agreement signed between the Nobel Bros. and others, for example, soon collapsed, during which time many failing businesses were taken over by the Rothschilds' Bnito and a third big player, Alexander Mantashev, a prominent Armenian businessman. Over the next few years the Russian government encouraged the formation of a union of producers and refiners by offering its members preferential freight rates on the Baku-Batum railway, echoing one of the main methods by which Standard Oil had absorbed competitors. The combination of 85 per cent of the industry – encompassing the Nobel Bros., Bnito, Mantashev and others – set up a fourteen-member commission to agree production quotas, determine prices, and divide foreign markets among its members: Nobel Bros. would be confined to Europe, Mantashev to India, Syria and Egypt, and Bnito to its established contracts in the remaining areas. But the critical level of voluntary cooperation across the industry was barely achievable and the union collapsed after three years.[234]

The perceived threat from Russian oil to Standard Oil's eastern market share was such that in 1892 the American chargé d'affaires in Constantinople wrote to John D. Rockefeller, 'in one or two interior districts of Mesopotamia in the desert oil has been found ... [T]his gov't is inclined to be liberal and there may be a chance to pipe oil to the sea and wipe the Russians out of all this Eastern trade.'[235] Throughout the 1890s the major American and Russian oil industrialists attempted to agree to a sharing out of the world's markets in order to contain their recurrent price-cutting wars that were damaging to profits. In 1891 the Berlin *Vossische Zeitung* reported rumours about the 'formation of a ring by the two great controllers of the international petroleum business (the Standard Oil Company of New York and the Paris Rothschild house) for a division of the markets of the world and the fixing of a selling price for this article, so necessary for the use of the masses'.[236] A few months later the *New York Times* opined,

> It is interesting to learn that the European agent of the Standard Oil Company has prepared a report in order to disabuse the German Government of the notion that that corporation is aiming at a monopoly. The agent declares that the policy of the company is 'simply that of competitive commerce' and denies that it acts in collusion with the Russian producers, 'or any combination to exact an artificial price'. This may all be true, but a solid and substantial doubt is cast upon it by the further remark of the agent that such a combination would be in conflict with the history and policy of the company at home and abroad. This assertion is calculated to stagger those Americans who have paid any attention to the history and policy of the company at home, and to convince them that the agent either does not know much about that history and policy, or else that he does not care much what he says.[237]

Indeed, the following year one of Standard Oil's senior directors, John D. Archbold, was keeping Rockefeller informed of just such secret negotiations with Baron Alphonse de Rothschild: 'We reached a tentative agreement with them ... I need hardly report again that it seems desirable on all sides that this matter be kept exceedingly confidential.'[238]

In 1893 *Bradstreet's* reported discussions between Standard Oil, the Nobels and the Rothschilds on 'a scheme for parcelling out between them the whole of the refined oil markets of the world', whereby Germany and Britain would be reserved for American oil and Asia would take exclusively Russian oil.[239] Two years later the *New York Times* wrote,

> The reported division of the European oil markets by agreement between Nobel Brothers of Russia and the Standard Oil Company of the United States has engaged the attention of diplomatic and Consular representatives of this Government for some months, and several communications in regard to it have been received at the State Department. The latest of these ... is from Mr. Alexander, Minister to Greece. He says: '... American petroleum is still preferred, but I am informed that American companies will no longer sell it here because an agreement has been made by the Russian companies and themselves that each shall furnish petroleum only to certain countries, and Greece is one of the countries allotted to the Russian companies.'[240]

In the event, aside from a few local, tactical truces of this kind, talks aimed at a global market carve-up finally broke down in the spring of 1895. Neither side was able to bring its country's exports fully under control. Just as the Russians had failed to corral a critical mass of their industry into a voluntary union, so a significant volume of America's exports eluded Standard Oil's grasp. Its command over the domestic industry – through its dominance in crude storage, transit, refining and product marketing – was such that in January of that year Standard's crude-oil purchasing arm issued a circular stating that it would no longer deal in the oil certificates traded on the exchanges but would set its own daily 'posted price' at which it would pay producers in cash.[241] But Standard did not hold such total sway over the export trade, where several large independents – primarily the Bear Creek Refining Co. and the Pure Oil Co. – were able to undermine Standard's claim to be able to control the destination of American oil abroad.[242]

The quiet diplomacy had, in reality, been the strategic pursuit of vigorous competition at the global scale. As American exports to the Far East began to dip after years of ascent, Standard Oil became set on breaking up the winning Bnito-Tank Syndicate partnership. Could Marcus Samuel's Oriental shipping and distribution network be brought over to distributing American instead of Russian oil products?

Although Samuel's tanker oil trade had been profitable, his Tank Syndicate was tied to the terms of the nine-year exclusive contract with Bnito, while independent shippers, such as Alfred Suart, who was running sixteen Black Sea tankers, could benefit fully from the declining price of kerosene at Batum. Samuel, however, resisted Standard's tempting takeover offers: he did not want to see his family enterprise disappear into the Standard Oil behemoth. Concurrently, therefore, Standard sought to bring the Rothschilds and the Union of Russian Producers into an alliance that would leave the Tank Syndicate high and dry and force Samuel into accepting its terms. But Russia's Finance Minister, while keen to attract foreign investment, did not want to see his country's oil industry swallowed up by the giant Standard Oil either, and he vetoed any such deal.[243]

In parallel, Standard Oil sought to reverse its decline in the Far East by improving its sales and marketing techniques. However, this would always be an uphill struggle against its high transport costs from America's East Coast. Furthermore, raging domestic demand in the US, in contrast to that in Russia, threatened to reduce the volume and increase the price of Standard's surplus output available for export; occasionally, it was even having to buy oil at Batum to meet its Asian export commitments. It thus became clear to Standard Oil's executives that they would have to begin to source and refine crude more locally to their distant markets.[244] But due to its worldwide reputation for ruthless monopolizing, Standard would face years of suspicion from foreign enterprises and governments.

'PETROLEUM CRAZE AT BAKU'

In late 1889 George Curzon, Marquess of Kedleston – who, a few years later, would become Under-Secretary of State for India – set off from England to Persia as a correspondent for *The Times*, with the aim of impressing on his readers the increasing importance of Persia to Britain's geopolitical interests in the east, centred around the maintenance of India as the keystone of the British Empire. From Constantinople, Curzon was ferried to Batum on the *Lux*, a British-owned oil tank steamer:[245]

> Military necessities dictated to Russia the occupation of the only decent port on the eastern coast of the Black Sea; but petroleum ... has made Batum, and petroleum is its life blood ... Over 5,000 tank-cars run between Baku and Batum, the largest owners being Messrs. Nobel and Rothschild, the former of whom, with the enterprise for which they have long been notorious, have procured a concession for a pipe line over the difficult Suram mountain on the railway line nearer Tiflis.[246]

Curzon then traversed the Caucasus by rail:

> At every station, where are sidings, long lines of tank-cars stored with oil crawl by like an army of gigantic armoured caterpillars, and disappear down the stretch of rails just vacated ... As the line ascends, clinging closely to the bed of the stream almost to its source in the watershed that separates the Caspian and Black Sea drainage, the scenery becomes more imposing ... The locomotives ... are entirely propelled by residual naphtha, or *astatki*, as it is called, driven in the form of a fine spray into the furnace.[247]

Several days later, he arrived at the heartland of Russia's oil industry:

> Baku, with its chimneys and cisterns and refineries, with its acres of rails outside the station covered with tank-cars, its grimy naphtha-besprinkled streets, its sky-high telegraph poles and rattling tramcars, its shops for every article under the sun, its Persian ruins and its modern one-storeyed houses, its shabby conglomeration of peoples, its inky harbour, its canopy of smoke, and its all-pervading smells – Baku, larger, more pungent, and less inviting than ever, was reached on the evening of the day after I had left Tiflis. The population is now estimated at no less than 90,000, a growth which is almost wholly that of the last fifteen years, and is the exclusive creation of the petroleum industry.[248]

The bankruptcies of the early 1890s in the Russian oil industry, compounded by a cholera epidemic, caused production to fall and crude prices to rise again, while demand for oil products not only remained strong but escalated. As the Baku-Batum railway continued to operate at full capacity exporting kerosene, the domestic demand for fuel oil, in the form either of refinery residue or raw crude, exploded. Ever since the early 1870s Russians had pioneered the development of oil burners for steamships and railway steam engines to fuel Russia's rapidly industrializing economy. Thus, when sales of kerosene by Baku's oil companies hit a plateau they were easily able to stimulate a new market for fuel oil, promoted as a cheaper and more convenient alternative to coal.[249]

This new and growing domestic fuel-oil market was seen as an attractive, additional incentive to invest in the Russian oil industry which, with the encouragement of Minister of Finance Sergei Witte, attracted a flood of foreign investment from the mid-1890s. However, the Russian government's fears of Standard Oil, and its opposition to German proposals for a European oil monopoly, meant that this new investment would be disproportionately French and British.[250]

As the Rothschilds established their Mazut Co. in the 1890s, specializing in 'residual fuel oil', a wave of British investors followed in the wake of the pioneering Alfred Suart's successful tankering business. Suart went into crude production on the Mamakai oilfield in Chechnya, eight miles northwest of Grozny, the site of an old Cossack fortress, on territory belonging to the Cossack Army of the Don, the Russian government's regional imperial enforcers. Here, in late 1893, Suart's drillers, alongside those of the Russian Ivan Akhverdov, struck several huge gushers.[251] A prominent English petroleum engineer involved in the Russian oil industry, Arthur Beeby Thompson, described the consequences for Suart:

> The unexpected striking of a powerful gusher at shallow depth proved disastrous to Suart, for all efforts at control were futile, and the well flowed wild, drenching all the countryside with oil. As the well was located on the crest of a hill, the ejected oil flowed down its sides damaging pasturage, polluting streams, disfiguring dwellings and inundating low land. Claims for compensation were on such a scale that he was constrained to sell for a mere song property which later became known as the Akverdoff and proved to be the richest spot along the main Grosny anticline.[252]

The increasing size of oil strikes, and of lakes of stored oil, also raised the risk of fire, one of which at Baku in 1900 destroyed 200 derricks (see Figs. 11 and 12). Thompson advised,

Immediately a well starts spouting, guards should be placed all round to prevent any one with a lighted cigarette, or an evil-disposed person, from approaching; and if the well is near a public road, watchmen should be stationed to prevent smoking, to enforce a slow speed for vehicles, and to put out all carriage or cart lamps at night ... In the spring of 1901, an immense fountain was struck by Messrs. Mantasheff ... The derrick was blown to splinters at the first outbreak, and a vertical column of oil and sand rose to a height of several hundred feet, flinging stones in all directions and drenching the neighbourhood with oil ... Every fire within a radius of several hundred yards had to be extinguished ... Houses in the immediate vicinity were flooded out and rendered uninhabitable, and within a large circuit the drain-spouts from the roofs of buildings were pouring streams of oil into the roadways. The main roads were flooded with oil several feet deep in places, and the oil in some cases was more than a foot deep in the rooms of dwelling-houses, which the inhabitants had been forced to vacate speedily, and the roofs of which had in many cases collapsed from the weight of sand upon them ... An even more destructive fountain was one of the Baku Oil Co ... Oil spray was falling over Baku three miles away, and sheets of note-paper exposed to the air near the railway station at almost five miles' distance were spotted with oil particles. The ships in the harbour did not escape, and the gunboats of the Caspian fleet, which are painted white, were blackened and made unrecognisable by the drifting oil spray. Although some 15,000,000 poods of oil were secured, the high duty payable to the Government, and the enormous demands for compensation from persons who suffered damage – for practically the whole of the village of Baieloff had to be cleaned and the houses repainted – left but little surplus for profit ... One of the most terrible occurrences in an oil field is an outbreak of fire, which will, in the space of a few hours, if allowed to spread unchecked, devastate acre after acre, and wipe away hundreds of derricks, thousands of pounds' worth of machinery, and render hundreds of employees homeless ... No questions are needed to learn the whereabouts of the conflagration, for in the day-time a colossal cloud of black smoke will be observed rising, totally obscuring the light wherever it floats, and at night a lurid glare, and illuminated sky, thousands of sparks, and floating pieces of burning wood will disclose its situation.[253]

Those attempting to maintain a fountain wellhead, continued Thompson, had to

> work in relays in order that they may not be overcome by the gas. The work is of a most laborious and dangerous nature, for which it is usual to pay men double or treble their ordinary wages; if a man shows signs of giddiness or insensibility, he should be at once removed to a place where there is a free current of fresh air, where he will probably soon recover ... although it is often followed by a strong drowsiness, in fact, an almost irresistible desire to sleep. If a workman does not recover from the asphyxiation in a short time, artificial respiration should be practised and continued until signs of life appear.[254]

Although it was reported some years later that there were many work-related injuries in the industry, Thompson wrote,

> The regulations relating to labour are particularly strict, and it is doubtful whether a workman is better protected by legislation in any country in Europe. The severity of the labour laws often leads to imprisonment of the responsible engineer when workmen are killed or injured, whilst insurance against employers' liability is even more necessary than in England.[255]

At the same time, he added, 'The cost of labour on the Russian oil fields is very small compared with most European countries.'[256]

The Grozny oil strikes immediately took production there to the not insignificant 2 per cent of Baku's production and, following the construction of a pipeline to the Vladikavkaz railway and the completion of a rail connection between Beslan and Petrovsk, exports from the North Caucasus were soon being shipped from the port of Novorossiysk.[257] In 1896 Suart – having financially weathered the Grozny pollution disaster – set up the first British oil production enterprise in Baku, the European Petroleum Co., for whom Thompson became the chief engineer, and he was soon followed by the other major British oil traders who also sought independent sources of Russian crude. Henry Neville Gladstone, son of the former British Prime Minister and head of the prominent trading concern Ogilvy Gillanders and Co., formed the Russian Petroleum and Liquid Fuel Co. in 1897 and the Baku Russian Petroleum Co. the following year, which soon controlled more production than the Rothschilds. In a game of diplomatic quid pro quo the Russian government, which was keen to obtain loans from the City of London, had smoothed the way for Henry Gladstone in late 1897 to take over the large Baku producer Tagieff and Co. This made the Russian Petroleum and Liquid Fuel Co., with a director of the Bank of England, Evelyn Hubbard, as its vice-chairman, the largest of the British oil companies in Russia. By contrast, the Russian government had just recently blocked the takeover of Tagieff and Co. by the Rothschilds and had obstructed their formation of Mazut Co.; the obstruction disappeared, however, after the French banking house facilitated a loan to the Russian government. At the same time, Marcus Samuel, who was also keen to secure independent supplies of crude, took over Schibaieff and Co. via his ubiquitous agent Fred Lane, who also represented Rothschilds interests at Baku. Around the turn of the century about 30 per cent of the investment in Russian oil was foreign: nearly 70 per cent of this was British, responsible for 15 per cent of the production at Baku and 50 per cent of that from Grozny – this at a time when Russia's total oil production, in the period from 1899 to 1901, surpassed that of the United States.[258] The burgeoning trade in both Russian and American oil was hailed as a 'remarkable development' in a vivid newspaper report from Manchester in mid-1899 on the importation of oil by the Anglo-Caucasian Oil Co. and the Anglo-American Oil Co. (see Fig. 13):

> The four petroleum steamers that arrived in the canal last week discharged altogether more than four million gallons of oil into the storage tanks at Eccles and Mode Wheel. The first was the Circassian Prince, from Batoum, with 813,000 gallons of Russian petroleum, which she discharged at Eccles wharf. The Beacon Light, which was waiting to take her place alongside the wharf, discharged 3,864 tons of petroleum from the same port immediately afterwards. On Thursday the Elbruz arrived at Mode Wheel wharf and discharged 516,388 gallons of petroleum and 740,510 gallons of residuum from New York, and just before midnight on Friday the Phosphor was berthed at the same wharf, and at once began to discharge a cargo amounting to 750,000 gallons of petroleum from Philadelphia ... Already another tank steamer, the Broadmayne, is due at Eccles with a full cargo exceeding 1,000,000 gallons of refined

Russian petroleum from Batoum, and she is to be followed shortly by the
Baku Standard or Rocklight, from the same port.[259]

Thompson later described Baku as 'the nursery of British oilmen';[260]
and, indeed, such was the infusion of British capital into Russian oil
that in April 1898 the American consul at Baku wrote, 'Foreigners are
to be seen everywhere, so much so that Baku will soon be transformed
into an English town ... [N]o one will be surprised if in the near
future the whole naphtha trade gets into the hands of the English.'[261]
As the *Financial Times* put it the following year, 'Within the past few
years Russian enterprises have come very much to the front as openings for the investment of British capital, and the petroleum industry
of Southern Russia has proved very successful in attracting our surplus millions.'[262] However, the newspaper offered a note of caution
regarding what it called the 'Petroleum Craze at Baku'.[263] It queried
the impressions that these companies' promoters in London were giving to investors, and it asked whether these enterprises might be far
more risky than was thought. Not only was it forgotten, perhaps,
what a 'large element of speculation there is in the industry', in which
'a very bad year may easily follow a good one',

> But if there be any continental industry to which the maxim 'caveat
> emptor' should be severely applied, it is that of Russian petroleum ...
> A regular boom in land has meanwhile been started in the Caucasus,
> and the proprietors, rendered avaricious by the gains of their neigh-
> bours, are only too anxious to sell their holdings to the foreigner at
> exorbitant figures, while the latter kindly takes his chance of finding
> petroleum there or not.

Indeed, most of the new Anglo-Russian oil companies were excessively
speculative and badly managed, and were therefore extremely vulnerable to unanticipated economic or political challenges; many were,
ultimately, financial failures. As Thompson himself judged the situation
five years later, in 1904, the British 'have invested more than £6,500,000
in Russian oil, a great deal of which will never be recovered, and it is
doubtful whether more than a small proportion will ever pay a profitable return, on account of the high prices paid for partially exhausted
grounds'.[264] The *Financial Times*, at a similar moment, continued,

and it seems strange that investors here who are so chary about placing their savings on the Continent, in view of the risk of war, should enthusiastically subscribe towards industrial undertakings situated in a country with which we are possibly more likely to go to war than any other and with which we have actually been on the brink of war more than once in the past five years.[265]

Furthermore, Finance Minister Witte's openness to foreign investment – albeit somewhat restricted and controlled – was seen by broad sections of Russian society as a threat to their sovereignty and interests. The conservative landowning gentry feared for their traditional economic and political dominance. Just as the American consul was remarking on how the British were wading into Russian oil, the Tsar received a letter from his brother-in-law, Grand Duke Aleksandr Mikhailovich, warning that the British were attempting to 'seize control of the oil industry', and that granting their further recent requests for more oilfields would be 'equivalent to transferring the entire petroleum industry to the exclusive utilization of the English and equivalent to taking the entire petroleum industry out of Russian hands'.[266] A year later, the Minister of Foreign Affairs, Mikhail N. Muravyov, argued that along with foreign capital, 'in the form of joint-stock enterprises . . . will come ideals and strivings endemic to the capitalist order, [which will] penetrate the general population, and groups of foreigners – with the assistance of trusts and syndicates – will take control of the country's natural resources'.[267] In any case, in the context of their recent colonization, the Russian oil regions could not be said to be particularly stable; as Thompson put it,

> Police and military were everywhere, for the occupation of the Caucasus was only made effective by force . . . Armed gendarmes accompanied every train, for attacks by robbers or bandits were not uncommon in these lawless districts where the inhabitants did not accept Russian rule with complacency . . . Brigandage, robberies, murders and political assassinations were then very common . . . Professional assassins, it was known, could be hired to remove enemies or rivals. On more than one occasion I found the city in a turmoil over the assassination of a police officer or Governor, by a bomb being flung into the carriage he

was using. Attacks on high officials were invariably demonstrations of anti-Government feelings, and rarely due to personal enmity. Wholesale arrests usually followed, with harsh treatment alike of guilty and innocent suspects.[268]

More generally, the politically unrepresented and oppressed peasants reeled at the sudden, rapid pace of industrialization, which in western Europe had evolved more slowly over many generations, and reports of the time convey an acute sense of dislocation and alienation.[269] Witte nevertheless continued with his policy of shock industrialization and, in order to safeguard the 'English capital flow into the oil business', moved quickly to dispel any 'lack of trust in the stability of rules determining the status of foreign industrialists and businessmen in Russia'.[270] He elaborated,

> International competition does not wait. If we do not take energetic and decisive measures so that in the course of the next decades our industry will be able to satisfy the needs of Russia and of the Asiatic countries which are – or should be – under our influence, then the rapidly growing foreign industries will break through our tariff barriers and establish themselves in our fatherland and in the Asiatic countries ... and drive their roots into the depths of our economy. This may gradually clear the way also for triumphant political penetration by foreign powers.[271]

Meanwhile, shut out of the Caucasus, Standard Oil's agents and consultants turned elsewhere. They began to look to Romania as an approximate geographically equivalent source of oil to the Caucasus, on the opposite coast of the Black Sea. They briefly considered Alaska; and they scoured the Orient, from China's Szechwan (Sichuan) province to Russia's Sakhalin Island, where their efforts soon focused on gaining a share of production from the more proven reserves of the Dutch East Indies, now bustling with prospectors and speculators.[272] 'We are living here in a real petroleum fever,' wrote the Dutch groundwater engineer Adriaan Stoop from Java.[273]

THE ROYAL DUTCH AND THE SHELL

In 1895, three years after the Acehnese insurgency in northeast Sumatra had deterred Standard Oil from seeking concessions there, the company's foreign agent, William Libby, approached Royal Dutch for a share in its production in Langkat, a regency in northernmost Sumatra, adjacent to Aceh. But the flourishing new Dutch oil company had ambitions of its own – for the Far East and beyond – and did not wish to put its future in jeopardy by aiding its most powerful and feared competitor. As Royal Dutch's sales manager at the time, Henri Deterding, later recalled, 'The fields producing our Oil were so much nearer than theirs to the areas where we traded. Compared with Pennsylvania, Sumatra and Java was "just around the block" from Shanghai and Hong Kong.'[274]

Royal Dutch's yearly sales were about to surpass 3 million barrels and it began to build up a tanker fleet of its own and a network of storage tanks at ports across the region; its production, meanwhile, was on the way to reaching 10,000 barrels per day, more than it could currently process. Standard Oil now began to rapidly lose its Asian market share to the new, locally produced, Royal Dutch kerosene as well as to the still relatively nearby Russian kerosene. Whereas Marcus Samuel's Tank Syndicate could ship its oil in bulk by steam tanker, as Royal Dutch was also beginning to do, Standard continued to ship its kerosene from America's East Coast to Asia in tins by sailing boat via the Cape of Good Hope; although this journey took between four and eight months, compared to the Tank Syndicate's approximately forty days from Batum, it did, however, keep down Standard's shipping costs. Eager, nevertheless, for a local source of production, Standard approached Royal Dutch again with the lure of much-needed infrastructure investment; but Standard's terms suggested that it was aiming at a complete takeover and the offer was rejected. Furthermore, in the view of a prominent Dutch financier, M.J. Boissevain, who had acquired several oil concessions in Palembang, the establishment 'of an American combine, which is a powerful influence in American politics, in the immediate neighbourhood of Achin, which was hostile to the Government of the Netherlands

Indies, might easily lead to foreign political interference, on justifiable or unjustifiable grounds'.[275]

Meanwhile, oil exploration spread to southeast Sumatra. Natural oil and gas seeps were well-known in both the Jambi and Palembang residencies where the petroleum was used to caulk boats, but attention was focused on the latter region which was under the control of the Dutch authorities and where the Musi and Lalang rivers facilitated access inland.[276] Royal Dutch prospectors occupied an abandoned native hut as their base camp from where a local led them to petroleum seeps that seemed a favourable site for exploratory drilling. Chinese labourers were brought in to move drilling equipment through the swampy jungle; according to Deterding's secretary and later the official company historian, Frederik Gerretson, this was because 'The aboriginal Kubus are extremely timid hunting nomads who roam in small bands through the boundless swamps. Though friendly and gentle in disposition, they were completely unsuited for work in a European industry.'[277]

Profusely flowing oil was struck in early 1897 and the oilfield was linked, by road and then pipeline, to a refinery on the Lalang River. Simultaneously, several other prospecting syndicates carried out drilling in the area. One, exploring a concession near the confluence of the Lematang and Enim rivers, was welcomed by the local chief in the belief that the oil industry would be profitable for the region.[278] The syndicate was unsuccessful in its attempt to sell its concession to Standard Oil, which was only interested in thoroughly proven, producing fields; however when, between late 1896 and early 1897, the venture brought in several highly productive wells and formed the Moeara Enim Petroleum Co. to construct a pipeline and refinery, Standard became much more interested. Palembang was now beginning to look more commercially promising than Langkat.[279]

By February 1898 Standard Oil was on the point of buying up Moeara Enim when a rumour to this effect, circulating on the Amsterdam stock exchange, provoked an outcry in the press. How could the Netherlands allow such an important, fledgling industry to be exposed to the predations of a foreign giant? From a bridgehead at Palembang, Standard Oil would surely be able to deal a savage blow to Royal Dutch itself.[280] As Gerretson described the situation,

So many were already beholden to the Company for their rise, so influential had already become those interested in the young industry, that a threat to its existence could be felt in a wide circle as a threat to the general welfare . . . Warnings were raised against the dangers that might encompass the Indies as the result of the establishment of so powerful a company. Examples were recalled – and they happened to have been very numerous in recent years – of great Powers which had misused economic interests as a pretext for interfering with small nations. It was feared that difficulties might be deliberately stirred up for the purpose of summoning the help of foreign Governments.[281]

The supposedly hallowed principles of free trade and investment were trumped here by the national interest, it was argued, and the Dutch oil companies of the East Indies must join together to fend off foreign assaults. 'United we stand, divided we fall,' wrote the *Bataviaasch Nieuwsblad*.[282] As Deterding would later explain,

> You may call it a refined form of throat-cutting, a stranglehold, a dog-fight or by any more appropriate or opprobrious name you like. But competition is most certainly not the name to give to trade rivalry so misguided that a trader's only chance of survival – and a very remote chance – depends on just how low he can cut prices. Annihilation, not competition, is then the right word. You can't compete with a man, nor he with you, if all the while you are both bent on squeezing each other to death.[283]

This happened to mirror the stance of the government, which had been briefed on the matter by the head of Royal Dutch, Jean B.A. Kessler. The Minister of Colonies, Jacob T. Cremer, promptly held discussions with the directors of Moeara Enim. They should take note, he emphasized, of

> the danger to which the vigorously growing petroleum industry in the Netherlands Indies would be exposed if a powerful combine such as the Standard ensconced itself in Sumatra . . . [considering] the methods to which this trust did not hesitate to resort in order to monopolise production and the market . . . [T]he interests of the Netherlands and of the Indies would have been incomparably better served if the principal petroleum companies in the Indies had endeavoured to co-operate. When contact had to be sought, for purposes of sales abroad, with

powerful foreign combines such as the American and the Russian petroleum trusts, the Netherlands Indies industry would have been able to
negotiate upon an equal footing and it would not have been necessary
for it to become the obedient handmaiden of one of these trusts.[284]

Pressure was applied to the directors – they were reminded of the
Governor General's discretionary power to refuse exploitation rights
to the holders of a prospecting licence – and consequently they called
off the sale. Standard Oil complained but, unusually, it received little
support from its own government, which was currently embroiled in
a rapidly escalating conflict with Spain.[285] The political momentum
was now with the Dutch colonial authorities to actively pursue a policy of ensuring that East Indies oil interests remained securely in
Dutch hands and of inducing cooperation between its oil companies.
The government encouraged Moeara Enim to form an alliance with
Royal Dutch, while this national figurehead was empowered to adopt
a protectionist company by-law that created a privileged tier of controlling shares whose holders had to be Dutch and whose transfer
required the approval of the other preference shareholders.[286]

At the same time, the Tank Syndicate – sharing Standard Oil's desire
to reduce the transport costs involved in its Far Eastern sales – was
also looking to source oil from the Dutch colony. The Syndicate additionally wanted to diversify away from what it felt was becoming an
increasingly risky total dependence on Russian supplies, and it was
partly with the aim of integrating its Russian suppliers more closely
with its oil shipment and sales operations that in October 1897 Marcus Samuel incorporated the 'Shell' Transport and Trading Co.[287] As a
novice to oil production and refining, it did not provoke in the Dutch
the kind of hostility shown to Standard Oil, and in early 1897 the English trading concern began investing in exploration being undertaken
by Jacobus H. Menten in the Kutei region of Kalimantan, in Dutch
Borneo – the European colonists having crushed the native resistance
there between 1850 and 1863.[288] 'Coolies' – often subjected to gross
ill-treatment and many of whom died of fever – were set to work in the
dense jungle, and in April 1898 they struck a powerful, uncontrollable
gusher of heavy crude near Balikpapan Bay. 'For months the oil poured
into the sea, where it was visible ten miles from the coast,' recalled

Gerretson, and the by-products from refining the crude into lamp oil were disposed of by burning off the gasoline and discharging the heavy residue into the sea.[289] Concurrently, the newly formed Shell Co. was compelled, in line with Dutch government policy, to curtail any designs it might have on Royal Dutch, and it settled for a temporary marketing agreement aimed at fostering cooperation across the Dutch East Indies oil industry. Samuel wrote to Kessler, 'We are very desirous of causing you to avoid the error of putting up opposition installations all over the East, because we feel quite certain that in the long run terms must be arranged between us, or ruinous competition to both will take place.'[290] Indeed, without government protection Royal Dutch would at this stage have been quite vulnerable to a takeover. Its Langkat oilfield began issuing more brine than oil, its production collapsed and its share price plummeted. It needed new oil, and quickly.[291]

While Royal Dutch geologists surveyed the archipelago in search of new oilfields and ordered its drillers to go deeper, several speculators followed the scent of oil on Sumatra further north into Aceh. The Dutch had known of the existence of natural oil seeps around Perlak since their first military assaults in the 1870s when the Acehnese had deployed petroleum-laden fireships in an attempt to block their advance upriver.[292] In mid-1895 a mining engineer, sent to Perlak by a group of concession hunters, reported indications of a rich oilfield; on the political dimension he added,

> My personal impression was that the present Rajah of Perlak is a powerful chief over his people and he understands well his interest in doing all that he can to assist the peaceful development of a petroleum industry. As long as he lives and as long as he is satisfied with the revenue which the exploitation will bring him, I do not think there will be any danger of raids or larger attacks from the Achinese.[293]

In early 1896 Dutch forces were sent to Aceh to open up the area for unhampered exploitation; according to Gerretson, 'It was now realised that it would be necessary to make short shrift without counting the cost [and] by merciless chastisement ... to establish security.'[294] A special counter-insurgency force, composed of non-Acehnese Indonesians under the command of Dutch officers, adopted the tactics of guerrilla warfare to relentlessly target the largely Islamic leadership of the resistance, 'so that fear [would] prevent the Acehnese from joining with those gang-leaders', according to Christiaan Snouck Hurgronje, one of the main authors of the policy.[295] As Royal Dutch became desperate to discover new oil, in mid-1898 Major Joannes B. van Heutsz was tasked with finally forcing the Acehnese into total submission.[296] The Dutch offensive became increasingly brutal and – it eventually emerged – numerous atrocities were committed, particularly under the notorious Major Gotfried van Daalen, who even Snouck Hurgronje described as having 'a deeply rooted contempt towards everything related to the native people'.[297] Among those singled out for criticism was a first lieutenant in the Dutch infantry, Hendrikus Colijn, who would go on to become head of Royal Dutch Shell and Prime Minister of the Netherlands. Four years earlier, while taking part in the Dutch invasion of Lombok, an island valuable for its rice-exporting economy, Colijn had written home to his wife,

> I saw a woman, with a child about half a year old in her left arm and a long lance in her right hand, charging towards us. One of our bullets killed both mother and child. From then on we could grant no more mercy. I had to gather together 9 women and 3 children, who were begging for mercy, and they were all shot. It was unpleasant work, but there was no alternative. Our soldiers happily skewered them with their bayonets.[298]

The Aceh War would go on to span forty years, cost the lives of an estimated 75,000–100,000 Acehnese along with 12,500 colonial soldiers and 25,000 of their 'coolies'; a total of 125,000 Indonesians died as a result of the Dutch government's colonial war for control over the whole archipelago.[299]

Now, in 1898, as thousands of Acehnese were fleeing their homeland, Standard Oil's man on the spot, John F. Fertig, saw in the situation a possible opportunity to gain access, at last, to Dutch oil, writing to John Archbold that September, 'It would be a very good idea if the American fleet could come to Sumatra, and surely the war which the Dutch have been waging against the Achinese, should be sufficient excuse in the cause of humanity.'[300] However, the Spanish-American War – justified by this same humanitarian pretext – had only just ended, leaving the United States' armed forces with the task of occupying extensive, newly acquired Caribbean and Pacific territories.

In late 1899, while Van Heutsz was still conducting counter-insurgency operations in Perlak, he helped Royal Dutch to induce the local Rajah to allow exploratory drilling expeditions into the area; they were still accompanied by armed protection, however, as the Sultanate as a whole was still formally at war with the Dutch. In December a drilling team struck oil, the company's share price shot up, and the government moved quickly to exclude foreign interests from developing the concession.[301] Royal Dutch planned to construct a 128-kilometre pipeline back to its Langkat refinery at Pangkalan Brandan, but Van Heutsz had hoped that a refinery would be constructed in Perlak itself. To the Governor General he argued that this was

> a necessity for the sake of the native chiefs, who would otherwise be more likely to witness the impoverishment of their country, as a result of this development, than its prosperity. Such a state of things should be allowed to occur least of all in Achin where, in the future, security, order and tranquillity required to be ensured not only by military power, but also and principally by means of the influence of the chiefs. These native rulers must learn to take an interest in the progress of their country and all this can only be achieved when they see that welfare and prosperity go hand in hand with an increase in order, tranquillity and the security of persons and property. The example of Perlak, if the oil were simply to be pumped away and the country were left in the same condition as previously obtained, could not fail to make a highly unfavourable impression upon the Ulëebalangs of other territories, who at present entertain such high hopes with regard to the development of their own country by means of the petroleum industry. What impression would it make upon the Radjah of Perlak, if, for instance, he saw a flourishing settlement rise up at Lho Seumawé for the development of petroleum wells in Tjunda, while in his own land there was nothing but a dead pipe-line?[302]

The government decided in favour of Royal Dutch, however, and the pipeline finally became operational over the entire distance in January 1901.[303]

The Dutch now became more determined to extend their colonial rule into the neighbouring sultanate of Jambi. In May, following a number of applications for oil-exploration licences in the region from companies, including Royal Dutch and Moeara Enim, the Resident of Jambi, I.A. Van Rijn van Alkemade, proposed that it should be opened up for mineral extraction as soon as possible; and in December, Snouck Hurgronje wrote that mineral extraction there should urgently be pursued 'as one of the main peaceful means of firmly establishing our influence in that region'. However, exploration was stalled. After a Dutch newspaper claimed that oil companies were offering bribes to officials and army officers 'if a military victory was achieved and concessions granted', the government became concerned, as did Van Heutsz, that the oil companies could not be trusted, that it was not receiving a fair share of the oil companies' profits, and that it might be preferable to conduct oil operations in Jambi as a state enterprise. Then, after a confused licence-application round, in July 1904 Kessler claimed that some of the licence-winners were Standard Oil front companies, at which point the government halted the process.[304]

Until Royal Dutch's new oilfields came fully on-stream, and to combat Standard Oil's aggressive marketing in the region – it was accused of 'dumping' lamp oil at below cost to maintain and expand its market share – the Dutch company traded in Russian oil in a temporary alliance, formed in 1899 with Henry Gladstone and Alfred Suart, and named the Eastern Oil Association.[305] Meanwhile, Shell was manoeuvring in the opposite direction, reducing its dependence on Russian oil by drawing on new East Indies oil, and in late 1898 it signed a contract to buy the entire output of Moeara Enim, whose negotiations with Royal Dutch had broken down. Given how close Standard Oil had come to buying up Moeara Enim, the American consul promptly declared the latter's intimate relationship with a British oil company to be 'a clear case of discrimination against American interests'.[306]

At the same time, in Dutch Borneo, Shell struck more heavy crude suitable for fuel oil, for which there was a growing demand, stimulated further by the company's active marketing. In August 1898 it shipped its first cargo of Borneo crude to Singapore to be used unrefined by its steam tanker fleet; and the following February it despatched a tanker of the liquid fuel to the British Navy, which was now conducting trials into the possible use of fuel oil in its warships.[307]

6

Oil for Power

OIL ENGINES

Heavy refinery residue had been burnt as fuel on a small scale since the earliest days of industrialized oil production, initially by the oil industry itself: to heat refinery stills and to power steam engines – with specially adapted oil burners – that were used for mechanized drilling and for pumping oil from the ground and through pipelines. Then, during the early 1860s, the Downer refinery in Boston began trying refinery residue as a marine boiler fuel; and several years later the US Navy conducted early experiments in its use in ironclad warships and in the gunboat *Palos* in Boston Harbour. The results of the latter, the *New York Times* predicted in 1867, were 'destined to work a revolution in the material employed for generating steam in boilers ... [In] the fire-room ... were found two men doing the work of eight persons, and in the place of ashes and dust were cleanliness and ventilation. To the surprise of old firemen the floor was carpeted.'[1]

However, this initial excitement was extinguished by an official Navy report warning of the dangerous accumulation of highly explosive petroleum vapours, and in Britain similar experiments carried out by the Admiralty at Woolwich Dockyard were also soon abandoned. It was only in places where coal was expensive while oil was readily available and cheap that the impetus existed to develop oil-burning technology. Thus, it was pioneered from the early 1870s around Baku in Russia's south Caucasus, where wood was scarce and imported coal costly but where refinery residue suddenly became plentiful. As oil-burning systems improved, the Grazi-Tsaritsyn Railway and Caspian steam ships, including Russia's Caspian Fleet, turned to fuel oil, prompting Colonel C.E. Steward in 1886 to warn British naval strategists that a Russian oil-fired warship 'could keep at sea for a very long period, and do our shipping interests incalculable damage'.[2] Then, from the mid-1880s, as heavy crude from the newly opened Ohio-Indiana oilfields became cheaply available, Russian oil-burning technology began to be adopted in the United States, initially in the towns and cities of the Midwest.[3]

Several of America's growing population centres of this period were effectively cut off from the coal and oil regions of the industrialized Northeast. However, in California, a dry region where, as at Baku, neither was wood so abundant, the local oil deposits would make good fuel. After years of numerous small-scale, usually unsuccessful, ventures, in 1876 – as the Southern Pacific Railroad reached the region – the first really profitable well was sunk in southern California by the Pacific Coast Oil Co. at Pico Canyon in the San Fernando Mountains at the edge of Los Angeles County. The company laid a 6-mile, 2-inch-diameter pipeline in 1879 to its refinery at Newhall on the Southern Pacific Railroad; but the crude was so heavy and thick that fires had to be lit under the pipeline during winter to keep the oil flowing. In 1883 a 30-mile extension laid to a pipeline operated by an oil storage and transit company was ruptured by floods; and it was torn up by a local rancher, although he was eventually restrained by a court order. The pipeline soon paid for itself, however, as it gave Pacific Coast Oil the bargaining power over the Southern Pacific Railroad to obtain lower oil freight rates.[4]

Californian production surged in the late 1880s, with major oil strikes made by three oil companies that would merge in 1890 to form the Union Oil Co. of California (later Unocal). Following a large strike in 1888 in Adams Canyon, Ventura County, according to company records, 'the oil shot up to a height of nearly 100 feet and flowed at the rate of 800 to 900 barrels a day. Before it could be controlled, it sent a stream down the canyon for a distance of seven miles.'[5] A $40,000 coastal oil tanker with a capacity for 36,000 barrels of crude, the *W.L. Hardison*, named after one of the companies' directors, was, however, destroyed by fire in June 1889 only five months after its launch. The state's crude output that year reached nearly 700,000 barrels but then fell back for a few years, sparking fears that Russian

and Peruvian imports might capture the far western market, particularly as Congress was considering removing the 20 per cent tariff on
oil imports.[6] Following Russia's example, steam engines in Peru, until
recently powered by expensive imported British coal, were now being
built to be fired with domestically produced fuel oil, albeit by a British
firm – the London and Pacific Petroleum Co. – that would likely also
seek markets abroad.[7] W.H. Tilford, president of Standard Oil of
Iowa, telegraphed his chief executive for the western markets, warning that 'free petroleum might let Russian oil into California' and told
him to organize a lobbying campaign against the removal of the tariff.[8] George Loomis, head of Pacific Coast Oil, wrote to a fellow oil
executive, 'It is of course to the advantage of all of us to keep as heavy
a duty as possible on all foreign oils.'[9] Their objective was achieved by
a Senate amendment that set a 40 per cent tariff on imports from
countries that imposed an import duty on American oil. However,
Californian production took off again in the mid-1890s, centred in
and around Los Angeles city, following the first commercial oil strike
there by Edward L. Doheny, and presaged by another explosive oil
strike in Adams Canyon by the Union Oil Co. in 1892 that sent tens
of thousands of barrels of oil down the canyon, into the Santa Clara
River and out to sea.[10]

The ready availability of cheap fuel oil, which fell to 30 cents per
barrel in late 1895, facilitated enough industrial activity to trigger a
spiral of increasing demand and persuaded the western railways, led
by the Southern California section of the Santa Fe Railroad, to adopt
the liquid fuel. Pacific Coast Oil teamed up with its now far larger
rival, Union Oil, to construct the first ocean-going Pacific tank steamer,
the *George Loomis*, to run between San Francisco and Ventura, to
where it also laid a direct 44-mile, 3-inch-diameter pipeline along the
Santa Clara Valley; the pipeline was poorly constructed, however, and
it leaked oil badly over the countryside. By 1898 Union Oil was supplying the bulk of the 100,000 barrels of crude per year used by the
region's sugar refineries and the 100,000 per month used in San
Francisco. It was at this time that near the coastal settlement of Summerland, founded by a community of spiritualists a few miles southeast
of Santa Barbara, prospectors found oil ever nearer the beach and
then out beyond the shoreline. This led to the construction of derricks

built on wooden piers, in what was probably the first instance of off-shore oil drilling (see Fig. 14). As the *Boston Globe* reported from Summerland beach,

> Drilling wells in the sea is the latest and most remarkable, as well as the most picturesque, feature of the great crude oil industry, . . . and there the precious brown fluid is now pumped from beneath the Pacific ocean at the rate of about 100 carloads per month, from scores of holes, while many more wells are being bored . . . But in the improbable event of a violent storm . . . all traces of this novel oil field – wharves, derricks and machinery – would be swept away and the loss would be enormous . . . Often when new wells are opened the gas pressure is so strong that streams of oil spurt out upon the water until the surface is black, in strange contrast to the distant blue.[11]

Similarly in Texas, oil along with a great deal of associated gas had been encountered across the state for many years. Water wells that did not become polluted with oil often at least bubbled with gas, and some residents piped gas from their backyard into their homes for cooking, lighting and heating. A few commercially unsuccessful oil wells sunk in the Oil Springs area from the mid-1860s portended a more commercial level of production from forty wells sunk by a team of Pennsylvanian drillers. From 1890 their output was enough to justify laying the first significant oil pipeline in Texas, a 14.5 mile, 3-inch-diameter pipe to a 2,000-barrel storage tank next to the Houston, East & West Texas Railway, west of Nacogdoches. As in California, this crude oil was mostly heavy and sulphurous, difficult to transform into lamp oil but more suitable for fuel and lubricants, for which there existed demand from the region's agricultural economy and the associated processing industries and railways. In 1894, after a protracted period of low cotton prices, business leaders in Corsicana, Navarro County, southeast of Dallas, decided to try to diversify their local economy towards industry, to which end they drilled wells to augment their all-important water supply. In doing so, they encountered oil and persuaded a renowned wildcatter, John H. Galey, to investigate further. Initial minor oil strikes presaged a sudden burst of production from 1897 on a significant commercial scale. Without adequate pipeline and storage infrastructure, however, much oil ran to waste, but the mayor of Corsicana managed to attract investment from Pennsylvanian oilman Joseph S. Cullinan and two Standard Oil directors. Standard's local marketing affiliate, Waters-Pierce, handled their refinery's output while Cullinan campaigned to convert the local utilities, businesses and railways to the use of fuel oil and natural gas.[12]

Already in 1888, the *Financial Times* had predicted,

> Many departments of manufacturing industry will be revolutionised by the substitution of petroleum for coal, to an extent of which the ordinary observer cannot have the least conception. These things are only fully known to the scientists and capitalists who will shortly offer to the public the chance of participating in the profits of some of the more obvious applications of this oil to the wants of daily life.[13]

Indeed, sales of fuel oil and of refinery residuum in the United States rose from around 2 million barrels in 1889 to around 10 million in 1899, half of which was sold in the country's Eastern markets, a third in the Midwest and about a sixth on the Pacific Coast, almost entirely sourced from the Ohio-Indiana oilfields. These figures were still dwarfed by those for coal and, furthermore, a significant proportion of the growing demand for heavy oil came from the coal-gas companies for enriching their manufactured town gas with the volatile vapours from 'gas oil' to increase its luminance in gas lamps. Nevertheless, as had already occurred in Russia, many thousands of steam engines and furnaces across the US were converted from coal to fuel oil.

The new oil-burning technology was also proving itself on the high seas in ships built either with fuel-oil tanks alone or as an auxiliary to coal bunkers. By the time one of Alfred Suart's tankers, the *Baku Standard*, in 1894 became the first oil-fired steamer to cross the Atlantic, the Russian, Italian and German navies were beginning to adopt fuel oil. Between 1888 and 1894 the German Navy's *Siegfried* class of coastal cruisers and its *Kurfürst Friedrich Wilhelm* class of battleships – totalling ten warships – were built to be dual coal- and oil-fired, while the British Navy was carrying out a few tentative trials. In 1898 the US Congress appropriated $15,000 to advance its navy's experiments with fuel oil, and the British Admiralty's senior engineers attended a demonstration by Shell of its oil-fired tank steamer *Haliotis* on the Thames.[14] *The Times* reported,

When the furnace is in full blast there is very little, if any, smoke produced – a point which might commend the system for use in men-of-war, while the fact that oil can be stowed in places which are not practicable for coal bunkers, suggests that its use might ease matters a little in the closely-packed interiors of torpedo-boat destroyers ... [But] in connexion with its use for the Navy one fact of supreme importance must be remembered. None of the great petroleum fields of the world are on British territory or under British control, and it would obviously be sheer folly to make our ships depend on a fuel the supply of which could not be absolutely assured.[15]

The advantages of fuel oil for raising steam were indeed compelling. The flow of oil to the boiler could be controlled merely by the turn of a tap, which eliminated the need for a team of stokers to shovel coal from the ship's bunkers and also increased the engine's responsiveness for acceleration or deceleration. By the same token, refuelling a steamer with liquid fuel was very much quicker and easier. The higher calorific value of oil increased a ship's maximum speed and enabled it to remain at sea for longer before refuelling. Oil of the right specification was not only, like coal, safe to store ashore but, when burned efficiently, would also emit far less smoke. However, two important disadvantages were always quickly pointed out. For shipping generally, there did not yet exist a network of fuel-oil bunkers around the world as there did for coal. And, for most European navies – the British, in particular – any reliance on fuel oil would mean exchanging the advantage of secure domestic supplies of coal for much less secure foreign supplies of oil. Indeed, the British Navy had on order from Shell a consignment of fuel oil from Dutch Borneo, just at a time when Anglo-Dutch relations were strained over the Boer War in South Africa. There was some concern, also, that the loyalty of Shell's Marcus Samuel might be compromised by his close financial ties with the Japanese government.[16]

Soon after the Royal Navy received its first delivery of Shell's Borneo fuel oil, in June 1899 another potential military application for oil's motive power was being given its first public demonstration on London's Richmond Park. Probably inspired by Edward J. Pennington's 'fighting motor car', illustrated in an 1896 issue of the American monthly *The Horseless Age*, Frederick Simms rolled out his 'Motor Scout', built simply by fitting a Maxim machine gun and a protective metal shield to a De Dion-Bouton quadricycle, powered by a 1.5 horsepower internal combustion engine.[17]

The basic principle of the internal combustion engine can be traced back to a seventeenth-century idea of harnessing the movement of a cannonball when fired upwards inside a cannon by the explosion of gunpowder. The principle of the piston was subsequently advanced with the development of the steam engine, whereby the energy released in combustion – as heat – is converted into movement indirectly via steam's expansion and contraction in the piston cylinder. It was not until the late eighteenth century that inventors returned to the idea of harnessing the power of combustion – now as an explosive force – directly within the piston. Some early experimenters used liquid petroleum as fuel, vaporized on a hot element to transform it into an explosive gas. However, from the 1820s the prevalence in industrialized regions of piped town gas, generated from coal in gasworks, provided later developers of internal combustion engines with a ready source of gaseous fuel; hence, when these first came into commercial production as stationary engines in the 1860s and 1870s, they were commonly called 'gas engines'. There existed a demand for these new machines particularly from smaller manufacturers who wanted modest mechanical power, only intermittently but to be immediately available, from an engine that was easy to operate – all in contrast to the relatively cumbersome steam engine. One promoter of the popular Otto-Langen gas engine portrayed it as an economic and social leveller by making accessible to ordinary small businesses artificial power that had previously been almost entirely the preserve of large factory enterprises with their capital-intensive steam engines.[18]

Just as the steam engine had been extended in its use to provide locomotive power – on specially built tracks, on water, and sometimes on public roads – so too were the internal combustion engine and, contemporaneously, the electric motor. It was clearly going to be impractical to fuel a moving vehicle powered by an internal combustion engine from town gas pipes. However, technologies had been evolving that would eventually enable such an engine to be fuelled independently of the urban gasworks, in a development that was partly motivated by the expense of fuelling a gas engine with illuminating gas in the relatively large volumes required. While some inventors devised relatively small, stand-alone, units for producing gas more cheaply from anthracite, others turned to the development of techniques for vaporizing, or 'carburetting', liquid fuels into a gaseous, explosive state. Through the 1860s various methods had been invented for vaporizing volatile naphtha, or 'gas oil', either to produce neat illuminating gas for places not connected to a gaswork's pipeline network or to enrich town gas. Most notably, from 1870 the Pintsch gas-lighting system, which drew on a tank of compressed petroleum gas, was widely adopted for illuminating railway carriages. Thus, by the early 1870s engineers were familiar with a variety of methods for generating a steady stream of combustible vapour from liquid petroleum, and in 1873 the principle was applied successfully to the gas engine: simultaneously by George Brayton in Philadelphia and by Julius Hock in Vienna. The former used commonly available kerosene and even heavier oils, while the latter used the more volatile 'benzoline' as fuel.[19] As was reported from London's International Exhibition of 1874,

> Among the new machines shown ... is one called a petroleum motor, by Mr. Hock of Vienna, in which power is generated by inflaming within a cylinder a mixture of air and petroleum spray in the same way in which a gas-engine is worked by inflaming a mixture of air and gas within a cylinder.[20]

The following decade saw the gradual refinement of techniques for carburetting a range of petroleum liquids, from kerosene to the lighter petroleum fractions such as gasoline – already widely used as a solvent – and plant-based alcohol.

Since gas was available in urban areas from gasworks or from small, self-contained gas producers, there was little commercial incentive to develop liquid petroleum engines, and it was only more than a decade later, with the introduction of the Priestman engine in 1888, that stationary petroleum engines became more widely adopted. At about the same time, some experimental engineers felt driven to develop this engine as a new source of automotive power, alongside the steam engine and the electric motor. From the early 1880s several German inventors – Gottlieb Daimler, Wilhelm Maybach and Carl

Benz – embarked on further development of the internal combustion engine in order to make one that was both powerful and lightweight. By 1885 Daimler and Maybach were wheeling out their first motor bicycle, and Benz the following year took his motor tricycle onto the road. The high power-to-weight ratio now achievable by the internal combustion engine suited it also for propelling small boats and – it was already being speculated – perhaps for powered flight.[21]

In 1888 the *Financial Times* declared,

> Vast as is the petroleum industry to-day, it is probably but in its infancy. Regarded as fuel for steam generating and in metallurgical operations, as well as a source of gas for illuminating purposes, petroleum has undoubtedly a great future before it. Moreover its adaptability as a direct source of power in petroleum engines has been practically demonstrated. From these points of view, this may emphatically be termed a petroleum age, and the efforts on the one hand of those who seek to extend and cheapen supplies, and on the other hand of those who devote their ingenuity to devising fresh uses for the material will doubtless be richly rewarded.[22]

From 1890, French businesses took the lead in commercializing the new petrol vehicle, under licence from Daimler and Benz, beginning with Peugeot and Panhard & Levassor, followed by numerous other companies; and by 1900 the former steam-car manufacturer, De Dion-Bouton, had become the front-runner. In 1895 commercial production spread to the United States – where, however, steam- or electric-powered vehicles would remain more popular for some years (see Fig. 15) – and in Britain production was jump-started in 1896 with the repeal of the so-called 'Red Flag' Act. This legislation had effectively outlawed mechanically propelled vehicles by stipulating that they must be preceded on the road by someone waving a red warning flag. Road safety concerns expressed in London's *Engineering* journal were mocked by *The Horseless Age*:

> Its peace it disturbed by visions of slaughtered pedestrians and shattered vehicles, because of the latitude the [Local Government] Board has given as regards the speed at which a motor vehicle may be driven. In its imagination the new vehicle grows into a veritable juggernaut,

sweeping through the streets at top speed, regardless of the conditions of the highway and the rights of other users thereof.[23]

Two months later, *The Horseless Age* was quick to counter another 'decidedly novel objection to the petroleum road motor' presented before the Automobile Club of France:

> The learned gentleman is haunted by the fear that if this fuel comes into general use for vehicle propulsion, we shall sooner or later be brought face to face with an oil famine, because only 8,000,000 tons of oil are annually taken from the earth, while 400,000,000 tons of coal are annually mined ... [However,] the production of oil is now increasing as its uses multiply. Under the stimulus of a great demand from a vast and newly created industry we may expect that new fields will be discovered and improved methods of raising the liquid be introduced, so that in the present generation, at least, an oil famine seems impossible.[24]

Certainly, this was what the oil industry was banking on.

'THE TEEMING MILLIONS OF THE MIDDLE KINGDOM'

The massacre by the 7th US Cavalry Regiment of 300 Lakota Sioux Ghost Dancers, including women and children, at Wounded Knee in 1890 brought to an end military conflict with Native Americans and, along with that year's US census, signalled the disappearance of the western frontier. At the same time, with rapid industrialization the United States was becoming a major exporting nation.[25]

An economic panic and depression in 1893, along with ensuing labour unrest, led the business community and allied economists to argue that the underlying cause was essentially agricultural and industrial over-production, alongside an excess of capital, or profits, apparently with nowhere to be invested. The remedy they prescribed was neither to increase domestic consumption and welfare by reducing economic inequality, nor to reallocate resources and avoid over-production; the solution they proposed, rather, was to open up further outlets for the surplus production in foreign markets and to seek opportunities for the investment of surplus capital in foreign countries. In articles such as 'The Economic Basis of "Imperialism"' by the prominent journalist and economist Charles A. Conant, and in the works of the influential naval strategist Captain Alfred T. Mahan, it was explained to Americans that economic and political imperialism were absolutely necessary in order to preserve the nation's economic and political stability. Consequently, in order to open up the necessary foreign commercial outlets, a stronger, larger US navy and military would be required to secure the associated sea lanes and territorial access. Any popular unease that might be felt towards this project could be assuaged by dressing it up as the continuation of America's purported world-historic 'civilizing' mission, as its benevolent 'Manifest Destiny'.[26] In 1898, just before becoming a US senator, Albert J. Beveridge declared,

> American factories are making more than the American people can use; American soil is producing more than they can consume. Fate has written our policy for us; the trade of the world must and shall be ours. And we will get it as our mother has told us how. We will establish trading-posts throughout the world as distributing-points for American products. We will cover the ocean with our merchant marine. We will build a navy to the measure of our greatness. Great colonies governing themselves, flying our flag and trading with us, will grow about our posts of trade. Our institutions will follow our flag on the wings of our commerce. And American law, American order, American civilization, and the American flag will plant themselves on shores hitherto bloody and benighted, but by those agencies of God henceforth to be made beautiful and bright.[27]

America's business class was set on following the example of Britain by becoming an imperial power. Indeed, the United States had been following in its mother's footsteps since the early nineteenth century. Just as Britain in the eighteenth century had taken protective measures such as the Navigation Acts to shield its domestic economy and imperial commerce from open competition, so post-revolutionary America had erected tariff barriers behind which it could nurture its own 'infant industries' in order to gain greater economic independence from Britain. This protectionist policy of the US, spearheaded by

Alexander Hamilton, was an inspiration to the later hugely influential German economist Friedrich List, who agreed with the many Americans who saw that Britain had only ever promoted free trade in areas where it was already strong enough to beat the competition.[28] The US followed suit by maintaining high tariff barriers through the 1890s and into the twentieth century, to which Britain and other European powers replied by raising theirs.[29]

Thus, although Europe was America's largest export market in the 1890s, the US business community looked towards the relatively unexploited regions of Latin America and East Asia – China in particular – as holding out the greatest potential for future trade and investment. However, in contrast to its own Latin American backyard, the US found it difficult to make economic and political incursions into China. Britain, Russia, Germany, France and other European powers, along with Japan, were taking advantage of the weakness of China's tottering Qing, or Manchu, dynasty by carving out their own spheres of influence over the crumbling empire, which they accomplished by obtaining exclusive port treaties, trading rights and railway construction projects – either by military force or the threat of its use, or by the inducement of loans and then by the consequent leverage this gained over their debtor.

Yet American industrialists became convinced that their significant exports to China, particularly of cotton textiles and lamp oil, represented just the beginning of a vast – and vital – future export trade. The US Minister to China, Charles Denby, hoped that Christian missionaries might help to open up this market: 'Missionaries are the pioneers of trade and commerce. Civilization, learning, instruction breed new wants which commerce supplies.'[30] However, Standard Oil's monopoly over the sale of lamp oil in China had been broken and its market share was plummeting due to the appearance of competing kerosene from Russia and now, especially, from the Dutch East Indies. Standard's top executives declared of Royal Dutch in early 1897, 'In the whole history of the oil business, there has never been anything more phenomenal than the success and rapid growth of the R. D. Co.'[31] If Russia were to extend its Trans-Siberian Railway into Manchuria, as seemed likely, a further influx of Russian kerosene could be expected; and Standard's market might also be threatened by Germany's control of Kiaochow (Jiaozhou) Bay, including the port of Tsingtao, on the Shantung (Jiaodong) Peninsula.

Furthermore, the American-China Development Co. – backed by many of Wall Street's leading banking houses, including Rockefeller interests – failed to acquire railway concessions in China, whereas British, French and German syndicates won both railway concessions and oil exploration rights. It was in this climate that, in January 1898, Standard Oil's James McGee, an executive of long standing and responsible for exports, presided over the founding of the Committee on American Interests in China (soon renamed the American Asiatic Association) to lobby the State Department to take more assertive action to promote US business interests there. However, American demands for unrestricted trade and investment opportunities in China – for an 'Open Door' policy – were not taken seriously by the other powers. The US lacked the economic and political influence in East Asia, and its pleas for an 'Open Door' were easy to dismiss in view both of its own long-held policy, the Monroe Doctrine, of firmly excluding European influence in Latin America, and of its highly protectionist domestic economy. How, then, was the US to project its power across to China?[32]

Since the mid-nineteenth century, American geostrategists had envisioned the US gaining economic and naval dominance over the Caribbean and Pacific regions, linked by a trans-isthmian canal, and gradually taking over the imperial positions of the Spanish and the British, while resisting any similar incursions by Japan, France or Germany. With this grand project in mind, in the 1860s and 1870s the United States began to stake out coaling stations and naval bases in the Pacific by extending its control over the Midway Islands and Pearl Harbor in the Hawaiian archipelago and over the eastern islands of Samoa.[33] Meanwhile, the National Association of Manufacturers, among others, called for an isthmian canal, arguing that it would reduce export shipping costs from the East Coast to Pacific markets; furthermore, it would stimulate US-dominated regional trade and would dramatically raise the importance of North America for global trade. Crucially, it was pointed out, a canal across the Central American isthmus would also nearly double the efficacy of the US Navy, which was presently divided between the Atlantic and the Pacific; and, it was argued, a much stronger navy was precisely what would be required to open up and protect foreign export and investment outlets, particularly in China. The US would therefore also need to secure the necessary naval bases, coaling stations and cable relay points in order to guard the canal and to project its power across the Pacific. This, indeed, the US accomplished in mid-1898 by going to war against Spain, justified in part by the dubious attribution of Spanish responsibility for an explosion that sank the battleship USS *Maine* in Havana harbour. In the Caribbean, the US took possession of Puerto Rico and turned Cuba, prized for Guantanamo Bay as a naval base, into a virtual protectorate, while in the Pacific it took over Guam and the Philippines.[34] As the US Army unleashed an orgy of rape, massacre and torture in a protracted war against the indigenous Filipinos, the New York *Sun* editorialized,

> The question of the Philippine Islands is bound up inseparably with our commercial position in China, and cannot be considered as a thing apart. To protect our merchants in the far East we need what every other nation already has, a sufficient fleet and a permanent naval station ... Our commerce in China must be preserved, and the first necessity of its preservation is the retaining of the Philippines.[35]

The newspaper subsequently carried an article from the *Insurance Advocate*:

> The markets of India are open to us, and, unless we are fools and let slip this wonderful opportunity to possess a splendid coign of vantage close to the shores of eastern Asia, the teeming millions of the Middle Kingdom will soon buy largely of us. One-third of the human race within easy distance of us, coaling stations on the road, and Manila as the Hong Kong of Uncle Sam's alert and keen merchant traders! Think of it![36]

Woodrow Wilson would later recall, while campaigning for the presidency in 1912, 'when we reached the year 1890 there was no frontier discoverable in America ... [W]e needed a frontier so much that after the Spanish war we annexed a new frontier some seven thousand miles off in the Pacific.'[37]

From Canton, in early 1899, the American consul reported on the discovery of oil near Chungking (Chongqing) on the Yangtze River:

> The recent discovery of petroleum in the very heart of China, in the rich and populous province of Szechuan ... is very important, and in

view of the oil wells in the Japanese island of Formosa [Taiwan], and of other oil wells near the German possession of Kyaochau, in the province of Shantung, our trade in American kerosene is bound to be affected unless prompt efforts are made to control the output and, if possible, the trade in the Chinese product.[38]

Following the Spanish-American War, the US could begin to act more forcefully in China from its newly acquired Pacific outposts, although it still remained a minor player there relative to the other imperial powers. In 1900 the US Army diverted 5,000 troops from its bloody counter-insurgency in the Philippines to join a loose international coalition force in China to brutally suppress a revolt – the Boxer Uprising – against extensive foreign intervention and influence.[39] Meanwhile, the American consul in Canton advised,

> The points whence American oil is shipped to the Far East are thousands of miles away. Would it not be advisable to select some point on the Pacific coast for the export of oil to China? We could ship oil in tank steamers across the Pacific to the Chinese markets quicker and cheaper than is possible with sailing ships, as at present.[40]

Indeed, Standard Oil now embarked on a campaign to reverse its market decline in China by investing $20 million in new tankers and storage facilities needed for the bulk shipment and distribution of lamp oil there – as Shell and Royal Dutch had been doing for some years.[41] In 1901 Standard decided to build a large refinery at Point Richmond, across the bay from San Francisco, that would bring to realization the long-cherished dream of supplying Oriental markets from the West Coast of the US.[42] In 1905 Standard Oil encountered difficulty in obtaining permission to construct storage tanks at Chinkiang (Zhenjiang), but after the US Navy paid a visit the Chinese officials became more accommodating. The naval commander recorded this successful exercise of 'gunboat diplomacy':

> The unexpected appearance of our three vessels and the uncertainty as to their intentions were sufficient ... and the representative of the Standard Oil Company is now free to improve his property and by so doing to enlarge the trade of this American corporation ... the *argumentum baculicum* [*sic*] [appeal to force] is what [the Chinese] best

appreciate, anything more considerate they look upon as weakness and correspondingly despise.[43]

The American colonial expansion into Guam and the Philippines, in addition to Hawaii, would eventually become the geopolitical context in which Japan, in 1941, attacked the US naval base at Pearl Harbor, bringing America into the Second World War; and, in a macabre moment of historical repetition, in 2004 a US general would refer to his country's bloody occupation of the Philippines as a model for the establishment of American bases in Iraq.[44]

AN INDUSTRY IN FLUX

By the turn of the twentieth century the petroleum industry had become a significant component of the modern global economy. Crude oil was being extracted from the earth, refined and distributed on a hitherto unimaginable scale to supply the world with artificial light, with lubricants for machinery, with fuel for steam engines and furnaces, and with an expanding range of other petroleum products for an increasing number of uses such as surfacing streets, enriching town gas, for manufacturing paints, dyes and varnishes, for dry-cleaning, and for powering the new internal combustion engine.

For about twenty-five years from the early 1860s, the United States held a virtually uncontested global monopoly over the oil industry, until it encountered competition from other parts of the world where large-scale, industrialized production, refining and distribution were adopted: initially – and primarily – from Russia, and more recently – to a lesser degree – from Austria-Hungary, Romania, British Burma and the Dutch East Indies. Although US crude production had risen from 35 million barrels in 1889 to 57 million a decade later, over the same period Russia's had surged from 20 million barrels to overtake the US, in 1899, with a production of 62 million barrels. However, compared to the light crude extracted from most of America's oilfields, the heavier crude from around Baku contained far less of the valuable illuminating oil for which there was such a worldwide demand, and the Russian industry consequently focused on domestic distribution of the great volumes of refinery residue for use as fuel. Additionally, by contrast to the vast network of pipelines to the US Atlantic coast – totalling some 6,800 miles of interstate crude-oil trunk pipelines – the Trans-Caucasus Railway to the Black Sea seriously limited Russia's export capacity. By 1900, only 143 miles of the projected 560-mile Baku-Batum kerosene pipeline had been completed, to relieve the railway bottleneck over its section of steepest gradient. Therefore, Russia's lead in crude production translated into only an approximately one-third share of the global refined oil market. That even this was possible without an extensive inland system of pipelines was due to the astonishingly concentrated output of Russia's oilfields. In 1900 its 1,400 wells, concentrated around Baku and Grozny, yielded nearly 176,000 barrels of crude per day, the average well producing at around sixty times the rate of each of the approximately 83,000 wells in the US, spread across a huge swathe of the American Northeast and two small pockets in southern California and east Texas.[45]

Now, with the dawn of the twentieth century, came a period of great change and uncertainty for the oil industry as both the pattern of crude supply and the profile of product demand began to shift significantly. In the United States, output from the original Appalachian field peaked in 1900 – the states of Pennsylvania and New York had peaked about a decade earlier – and entered a period of gradual decline. Although production was increasing in Ohio, West Virginia and Indiana, so was total domestic and foreign demand increasing by at least as much. This presented a challenge to the Standard Oil behemoth, whose power was based on its near monopoly control over the transport and refining of oil across the American Northeast and Midwest. At the same time strong competition from Russian oil was being joined by increasing competition from eastern Europe and the Far East. These competitors clearly held the advantage of relative proximity in their respective local regions – the cost of transportation being a high proportion of the eventual oil price – but Standard Oil's biggest foreign market, Europe, was also well within their reach. Although demand for kerosene in America's hitherto most important domestic and western European markets was still increasing, it looked set to level off and even contract, as coal gas and electricity were adopted for lighting; Britain, in 1902, became the first country whose kerosene demand peaked and went into decline. But regarding fuel oil and gasoline, would these remain merely relatively niche oil products?[46]

In the late 1890s Standard Oil had begun to look beyond its home area of domestic operations – its network of pipelines and strategically located refineries across the sprawling Appalachian and Lima-Indiana fields – by making relatively small investments in a refinery at Neodesha in southeast Kansas and another at Corsicana, in east Texas: these were two new fields that had begun to produce kerosene-rich crude in paying quantities. Standard was less interested in the new production of low-grade crude, first from southern California and now from southeast Texas following a huge strike in January 1901 at Spindletop, just outside Beaumont, about 20 miles from the Gulf coast. The ground-breaking Lucas well, drilled by the famous roving oil explorers James M. Guffey and John H. Galey, burst forth with an initial flow of over 175,000 barrels per day and was certainly America's most prolific gusher yet. The American consul at Batum, however, felt compelled to denounce as 'ridiculous' newspaper claims that it was the world's largest well;[47] as the US government's official yearly report on the industry commented more soberly of the Lucas well, 'Only a few of the Russian wells at Baku, flowing from larger diameters, have produced more petroleum in the same length of time.'[48] Nevertheless, the Spindletop strike was huge and triggered an oil rush and production boom. By the end of the year the field had yielded more than 6 million barrels of crude that, due to the lack of ready storage and transportation facilities, sold for as little as 3 cents per barrel. Most of it was held in huge earthen-work reservoirs and a network of ditches, and much of it was lost through ground seepage and evaporation, the latter creating a serious fire hazard; from 1901 to 1904 an estimated 1.5 million barrels 'were wasted and consumed by fire', according to a US government report.[49]

Standard Oil was deterred, however, from making the kind of investments on the Gulf coast of Texas to match the volumes of new production, and – even more than in California – the field was left unusually open for the emergence of new oil companies outside the Standard fold. Firstly, the crude was very low-grade, with a high sulphur and low kerosene content. Secondly, Standard's current sales operations in Texas had been hampered by the enforcement of the state's strict antitrust laws against its marketing affiliate Waters-Pierce, and there was little sign that Texan hostility to large, particularly out-of-state, corporations would abate. The investments that Standard did make in Texas would be made by stealth, through 'blind tigers': these were front companies that presented and advertised themselves as independent enterprises but were actually controlled by Standard, a practice that it was also deploying in Missouri.[50] Thirdly, and most significantly, the barriers to entry into the industry in Texas were lower than elsewhere. The proximity of this new flush production to Port Arthur and Sabine on the coast of the Gulf of Mexico – and thereby to cheap water-borne access, by tanker, to the East Coast and world markets – freed producers and refiners from the dependency on long-distance rail and pipeline transit, control over which was key to Standard's hold over the industry in the north. Furthermore, lower levels of refinery investment were required because of the relative ease with which the crude could be turned into fuel oil; it could even simply be left, open to the air, for the most volatile fractions to evaporate of their own accord. Nevertheless, new companies that quickly came to prominence, such as the J.M. Guffey Petroleum Co. (later Gulf Oil) and the Texas Co. (later Texaco), still needed major investment for pipeline and storage facilities; J.M. Guffey Petroleum, which in the spring of 1901 built the first pipeline from Spindletop to Port Arthur, was backed by the Mellon banking family. By the end of 1902 three other pipelines had been laid over the distance, only one of which was under the control of Standard, through its Security Oil Co. affiliate.[51] By the end of 1904, $34 million had been invested in the Texas oil industry that included $2.7 million for 514 miles of pipelines; the state's production for the year exceeded 22 million barrels – second only to California's 30 million – 70 per cent of which was shipped to the East Coast and 10 per cent exported.[52]

Just six months after the first Spindletop strike, Shell stepped in with a bold and somewhat rash move: an attempt to take advantage of the sudden appearance of this chink in Standard Oil's armour by signing a long-term contract with Guffey Petroleum for large volumes of fuel oil.[53] 'For all practical purposes,' gushed the *Petroleum Review*, 'this places in [Shell's] hands the chief liquid fuel sources in the world.'[54] Sir Marcus Samuel, head of Shell, had become a passionate believer in the

superiority of fuel oil over coal for powering steamships, having witnessed it being used successfully in Russia, and having struck heavy
crude in Borneo which he had put to effective use in his own ships. By
the turn of the century fuel oil was being taken up by some British railways, despite the ready availability of high-quality Welsh steam coal,
due to its high calorific content, efficiency of use and lower emissions of
smoke. However, British shipping companies and the Navy were under-
standably reluctant to adopt a new fuel that could not be domestically
sourced and lacked an established system of global distribution and
bunkering. As the Admiralty's engineer-in-chief noted in October 1902,
'It is considered that oil obtained from British and colonial sources
should be preferred . . . because it will be presumably easier to maintain
supplies from within the British Empire in time of war.'[55] The following
year, Britain's most prominent petroleum expert, Boverton Redwood,
explained in evidence to an inquiry into coal supplies,

> Those who talk lightly of petroleum replacing coal cannot realise what is
> involved in such a change. Hitherto, exported petroleum products have
> been chiefly used by the pint as a source of light, or in small quantities for
> the lubrication of machinery; and to supply the markets of the world
> with a commodity which is to be burned by the ton, is another matter.[56]

But in this Shell saw an opportunity. By deploying the world's largest
tanker fleet it might take the lead in the distribution of fuel oil and, at
the same time, establish itself in the European oil markets from which
it had been excluded under the terms of its contract with the Rothschilds' Bnito for Russian kerosene.[57] As Shell began increasing its
European sales presence by opening up a fuel oil market there, Standard Oil became worried, its senior director John Archbold writing to
Rockefeller in October 1901, 'This company represents by all means
the most important distributing Agency for Refined Oil throughout
the World, outside of our own interests.'[58] For, indeed, Shell had also
just managed to break Standard's monopoly of the small but growing
gasoline market in Britain.[59] Having struck a rich field of very light
crude in Borneo, in late 1900 Shell had inaugurated bulk tanker deliveries of gasoline to Thames Haven, giving 'so great an impetus to the
motor-car industry', lauded the *Petroleum Review* some years later.[60]
At the end of 1901 Shell also began to deliver gasoline from Sumatra

produced by Royal Dutch, intimating further collaboration between the two companies and causing more concern for Standard.[61]

Standard Oil fought back: it attempted to buy up Shell completely, but Samuel rejected the offer; it bought up the Pacific Coast Oil Co. so as to compete for Oriental fuel oil sales; and it cut its kerosene and gasoline prices.[62] At the same time, Standard continued its quest for sources of supply and refinery locations nearer to the most contested foreign markets in order to curtail the high transportation costs incurred from the United States. Russia had been under obvious consideration since 1900, when one of Standard's senior executives wrote to Archbold, 'What is the best method for the Standard Oil Company to engage in the Russian oil business – having in view not only the direct profits to be derived therefrom, but the advantages which might be secured for the protection of [our] American interests and World's trade, with the least hazard to capital?'[63]

A consortium headed by the Standard Oil-allied National City Bank of New York, and including J.P. Morgan, worked to promote an American economic presence in Russia. However, Standard Oil's approaches were met with hostility from the Russian government, which was concerned that the company intended to take over its oil industry – and National City Bank's apparent threat to extend a loan to Russia's Far Eastern rival, Japan, hardened the government's attitude.[64] Standard was therefore limited to purchasing kerosene from Batum in order to make up its commitments to surrounding markets. It set up a refinery in Japan, which had domestic production enough to supply 20 per cent of its local demand; but it proved cheaper to import refined American products into Japan, and after seven years Standard sold this refinery to the Nippon Co.[65]

In Romania, the government in December 1900 withdrew its grant to Standard of a concession for the construction of a pipeline from the Câmpina oilfield to the port of Constanța, as a result of nationalist, anti-American opposition that favoured German investment. In 1903 the largest Romanian oil company, Steaua Română, was bought up by Deutsche Bank whose head, Arthur von Gwinner, was congratulated by the German Chancellor, Bernhard von Bülow, for gaining control over the provision of 'the domestic market of the entire German economy with petroleum'.[66] The German Navy was increasing its use of fuel oil, and a 1902 report by the British Navy noted the challenge that its German rival faced in reaching the Atlantic without risking passing through the Straits of Dover; this involved taking the much longer route around the north of Scotland from which, according to the report, 'The nearest German station where coaling would be possible is the Cameroons . . . It is probable that some such consideration as this has led the Germans to place a supply of oil fuel upon all their new ships.'[67]

Later that year Standard Oil did, however, manage to acquire rights to private, rather than state-owned, land in Romania at Prahova, and it formed a local subsidiary, Româno-Americană S.A., which built a refinery at Ploesti, the output of which was not subject to Romania's protective import tariff on petroleum products.[68] Similarly, Standard incorporated the Vacuum Oil Co. A.G, an Austro-Hungarian subsidiary, in response to the Dual Monarchy's high import tariffs, where its two refineries were able to take advantage of a production boom from 1902 at Borysław in the Galician oil region – a boom that was accompanied by the typical environmental and social costs.[69]

Standard Oil continued to be excluded from the Dutch East Indies, however, and it was now also refused entry into Burma by the country's British authorities. In 1900 the Burmah Oil Co. had grown – despite the prevalence of manual production methods unchanged since the late eighteenth century – to the point where it overtook Standard's share of the large Indian kerosene market. Burmah Oil benefited not only from being geographically local but also from a protective tariff on foreign oil imports. Therefore, if Standard were to establish itself in Burma it could avoid this tariff barrier in addition to securing a source of supply for the wider region. However, all of its applications from early 1902 to lease territory there – through a newly formed subsidiary the Colonial Oil Co. and through its well-known, British-registered Anglo-American Oil Co. – were rejected by the government of India, despite the assiduous representations of senior Standard executives. The British argued, as had the Dutch, Russians, Romanians and many Americans themselves, that large trusts such as Standard Oil were ruthless monopolists: Standard might gain control over the Burmese oil industry only to run it down, due to its relatively high exploration and production costs, in favour of developing other, more profitable oil regions. Boverton Redwood, adviser to both Burmah Oil and the British government, had already been urging the government of India to increase the size of Burmah Oil's concession so that it might 'successfully repel the dangerous attacks of powerful foreign competitors'.[70] Meanwhile, in Parliament an MP voiced the 'anxiety prevailing among those who have invested capital in the oilfields of Burma'.[71] Redwood argued that only a British company with a virtual monopoly would have both the incentive and the financial resources to build a strong oil industry in Burma. Around the world, most 'successful petroleum undertakings [were] in the hands of powerful corporations or firms having ample capital at their disposal'. The petroleum industry was 'essentially one which in the public interest must be carried out on a large scale, and this is especially so since the system of storage, transport and distribution in bulk has been so extensively substituted for the employment of barrels and cases'.[72]

The Viceroy of India, Lord Curzon, agreed with Redwood and refused to discuss Standard Oil's requests with the American consul, merely replying, 'It is not desirable for an American company or subsidiary to work the petroleum fields of Burma.'[73] Accordingly, access was denied to Standard, while the concessions granted to Burmah Oil were significantly increased; and the protective import tariff would be maintained, the India Office decided, until the British Burmese oil industry could 'stand alone'.[74]

Standard Oil's forays into Europe and the Far East for new, foreign bases of operations reflected its desire not only to cut transportation costs but also to relieve the pressure on its domestic capacity to meet the rising home demand for kerosene and now gasoline. As Redwood observed in April 1903, 'There is to some extent a scarcity of petroleum in America; there is a difficulty in getting from America sufficient petroleum spirit to satisfy the demands of those who are using it now in unexpected quantities for road vehicles.'[75] Standard even began to supply its Californian market with gasoline obtained from its great rival, the Asiatic Petroleum Co. This was a new marketing combine – one of the most significant of many market-sharing agreements of the era – organized in 1902 by Royal Dutch, Shell and the Rothschilds' Bnito. It was designed to curtail the 'ruinous competition' between the three companies, as the Rothschilds' representative Fred Lane put it;

to face the challenge arising in the Far East from Burmah Oil; and to unite against the ever-present threat from Standard.[76] The Asiatic Petroleum Co. vied for the Oriental markets by using bulk tankering and storage wherever possible and by upgrading its sales networks, particularly across China.[77] Standard would continue to draw on large quantities of Asiatic's gasoline for its US West Coast market for the rest of the decade, while it also purchased crude from Peru to supply its new Californian refinery.[78] However, by 1904, as the production of heavy Californian crude surged, there were signs of potentially major new sources of high-grade, light crude from Kansas, Oklahoma and Indian Territory, the kernel of what would become America's huge Mid-Continent field.[79]

'TEEMING LIQUID WEALTH'

At the end of 1903, *Petroleum World* declared that several years of oil strikes, from Beaumont to Borysław,

> increases the mass of incontrovertible evidence which proves the existence of teeming liquid wealth beneath the earth's crust. These oil fields surpassing in their output the smaller neighbouring territories show that the oil world is capable of almost unlimited expansion. If we carry this theory of the augmentation of oil fields of increasing size to its logical consequence we must come to the conclusion that Borysław, Beaumont and even Baku, the greatest of them all, will some day be eclipsed by larger producing territories, and be classed amongst the small pioneer oil fields of the world.[80]

Where would this theory of 'the augmentation of oil fields' and the vision of the industry's 'almost unlimited expansion' lead? At the beginning of the twentieth century, geologists, prospectors and wildcatters, financiers, entrepreneurs and speculators were presented with a world of possibilities, both in the established major producing regions and beyond.

In the Americas, prolific oilfields were being opened up in newly discovered oil regions in the United States. Exploratory teams were even sent to Alaska – one was shipwrecked but another struck oil – and there was evidence that Standard Oil might be behind the search:

'Knowing the present shortage of high gravity oil, the oil men think that this big corporation has discovered an oil field and is preparing to obtain a plentiful supply of crude.'[81] In Peru, the production of heavy crude from the Zorritos and Talara oilfields, by Italian, British and French companies, was by 1904 approaching half a million barrels, from which fuel oil was being used by railways and in smelting furnaces, both domestically and in Chile.[82] Canada had a similar level of production to Peru, mainly from Ontario and more recently from New Brunswick, where the industry was dominated by the Standard Oil subsidiary, Imperial Oil Co.[83]

In the British West Indies, asphalt had been exported in large quantities for several decades, mainly for surfacing roads and pavements, from the island of Trinidad, dug from Pitch Lake at La Brea, hailed as 'the Baku of the asphalt industry'; the output of the entire lake was under the control of the Trinidad Asphalt Co. of New Jersey, a subsidiary of the Barber Asphalt Paving Co.[84] In 1901 two entrepreneurs – an Englishman, Randolph Rust, and a Chinese resident, John Lee Lum – began exploring for oil near Pitch Lake and on the opposite side of the island on Lee Lum's land at Guayaguayare, hoping to supply the Royal Navy with fuel oil, and from 1902 they made a series of small oil strikes.[85] Meanwhile, across the narrow Bocas del Dragón (Dragon's Mouths), Venezuela was emerging as an additional source of asphalt for American street-paving companies, dug from a huge pitch lake, Largo La Brea, in the northeastern state of Sucre near Guanoco, and from another in the state of Zulia at Inciarte, to the west of Maracaibo.[86] To the southwest of Lake Maracaibo there were known to be 'numerous springs of petroleum'.[87] Indeed, to the south of Zulia in the state of Táchira, since 1878 a group of entrepreneurs had operated as the *Compañía Petrolea del Táchira* to extract and refine oil from a pit on the La Alquitrana hacienda near Rubio, an operation soon expanded with the use of modern drilling and distillation equipment brought from Pennsylvania in 1880.[88] Just over the Colombian border, in 1905 the government granted an oil concession to General Virgilio Barco to refine oil from a seepage near the village of Petrolea in Norte de Santander state.[89] On Barbados, a British Caribbean colony, the West India Petroleum Co. was extracting some oil and had laid pipelines and built a refinery.[90]

In 1896 Dr Manuel Cuéllar, a prominent surgeon in Sucre, and a particularly strident Bolivian nationalist among the capital's white and mestizo oligarchy, organized an expedition to the lowland plains of the Chaco Boreal after being warned by his brother that Paraguayan troops were entering the contested region. At Mandiuti – in the province of Azero in the southwestern tip of the state of Santa Cruz, about 140 miles southeast of the capital, Sucre – the guides suggested a remedy for the sores that developed on their heavily laden mules: the application of a dark, viscous liquid that oozed from the surrounding earth, some samples of which Dr Cuéllar collected to take back to Sucre.[91]

Although the doctor's expedition found no Paraguayan troops, it was however attacked by natives near the town of D'Orbigny. The expedition was in a region where, just a few years earlier in 1892, at the Battle of Kuruyuki, the local Guaraní (known to the Spanish as 'Chiriguano') had been brutally defeated by Bolivian troops and their native, Christian-convert, allies. The Guaraní had already lost large tracts of land to cattle barons in the Huacaya War of 1877, at which a young man, Chapiaguasu, witnessed the massacre of his fellow tribes-people, including perhaps of his own mother. Cattle ranchers continued to encroach on the Guaraní settlement of Ivo, near the Santa Rosa mission at Cuevo where Chapiaguasu had lived briefly as a child – about five miles from the Mandiuti oil seeps – and in 1891 he emerged as the leader of a spiritual revivalist movement, renaming himself Apiaguaiki Tupa. A Guaraní man, Juan Ayemoti Guasu, who had lived at the mission but left to join the movement, wrote to one of the Franciscan fathers to explain Apiaguaiki's motives:

> From all these sufferings, the rage came to him which he has for the *caraises* [whites and mestizos] because they have been bad to us, but not the fathers, who always give food in town and [hold] services and do not let the *caraises* finish us all off, which is what they want ... [H]e says nothing bad against Your Lordship, only about those who took their land from the *avas* [Guaraní] and kill for the joy of killing and steal our things. He says that he will ask the fathers to help him so that the *caraises* give back to the people what they took away, that if it is not in a good way, it will be in a bad way.[92]

The mission, suspecting an imminent uprising, requested reinforce-ments of Creole militia. An attack was, indeed, triggered by the rape and murder of a local Guaraní woman by a Bolivian official, whereupon the Guaraní laid siege to the mission, though this was successfully repelled. A few weeks later, in late January 1892, with further rein-forcements from Santa Cruz, 140 Creole militia and 50 Bolivian Army regulars – most armed with Winchester repeater rifles – along with 1,500 native allies drawn from other missions, launched an attack on about 5,000 Guaraní, dug in at Kuruyuki, who fought to the death armed only with bows and arrows and knives. Compared to four mis-sion Indians killed and thirty wounded, and just five Creole wounded, over 600 Guaraní were killed. Those who gave themselves up were massacred at the Santa Rosa mission by the settlers who had sought refuge there, and over the following days and weeks the militia and troops went on a spree of massacre and looting in the surrounding villages. They killed nearly 2,000 more Guaraní men and boys, and the surviving women and children were distributed to sell as slaves in the cities of Sucre, Santa Cruz and Tarija. Apiaguaiki and Ayemoti were later captured and taken for a military trial at Monteagudo where they were tortured for possible information and killed, their bodies then put on public display in the plaza.[93] In 1896, as Dr Cuél-lar's expedition traversed the area, the regional prefect wrote,

> At present, relative tranquility can be guaranteed to new settlers, owing
> to the military colonies and forts founded ... [These settlers] little by
> little are removing to a distance the settlements of the unconquered
> savage hordes, making for effective and evident national sovereignty in
> those faraway regions, so often disputed by the neighbouring republics
> of Paraguay and Argentina. This has been accomplished at no greater
> cost to the state than the small expense of rifle ammunition and the
> awarding of public land to the colonists.[94]

On arriving back in Sucre, Dr Cuéllar had his samples of the oily liquid analysed and it was found to be high-quality, light crude suit-able for lamp oil. He formed *Cuéllar y Cía*, along with other wealthy investors, and in 1899 they acquired two petroleum concessions from the government, with the rather impractical idea of conveying barrels of crude on the backs of mules from Mandiuti to the city. They subsequently re-formed the enterprise as the Incahuasi Petroleum Syndicate that, some years later, alongside other entrepreneurs, began drilling in the region around Mandiuti, northeast of Tarija, for petroleum in commercial quantities.[95]

In 1897 an attempt was made to drill for oil in Brazil, at Bofete about 100 miles northeast of São Paulo. Eugênio Ferreira de Camargo had acquired land there, in 1892, that included a coal concession, and he hired a Belgian scientist, Auguste Collon, to investigate its potential. Using a steam-powered drilling rig, operated by an American oil engineer, they encountered oil at a depth of almost 450 metres, but the well only produced about two barrels-worth.[96]

An oil well in Cuba, at Motembo in Santa Clara province, had been yielding small quantities of crude since being drilled by Manuel Cueto in 1881, and in 1903 the roaming British petroleum engineer, John Bergheim, came to the area to supervise the drilling efforts of the Cuban Petroleum Co. However, at a company meeting in London, it was reported, Bergheim drew on his years of experience to caution shareholders:

> [A]s far as oil was concerned, he did not believe there was any such thing as an oil expert, in regard, at any rate, to virgin properties. The drill was the only thing which could tell them where there was oil, and the pump was the only thing which could guide them as to quantity; those were the only two oil experts he knew of.[97]

In Mexico, President Porfirio Díaz, who was keen to encourage foreign investment, relaxed the country's mining laws to make its national resources more attractive for private commercial exploitation. As long ago as 1867 a *New York Times* editorial had brought attention to the abundant petroleum seeps in the northern parts of the Isthmus of Tehuantepec:

> These deposits, together with the other immense natural resources of Mexico, will pour their wealth into the commerce of the world as soon as the United States shall determine in what way most wisely to aid in the redemption of that magnificent and miserable country. The next grand engineering achievement of the world, after the Atlantic Cable and the Pacific Railroad, must be the excavation of a ship canal across

either Tehuantepec or Darien. The European Powers, standing aloof from any interference whatever, either hostile or friendly, with this continent, leave the way open for our country to construct and control this noble work. When this is done, Oriental commerce, reversing its secular tide, will flow from West to East across both our lines of transit, and in its course will regenerate the torpid life of Mexico.[98]

As the American consul general in Mexico had affirmed in 1883, 'A variety of bituminous substances, from the pure, hard asphaltum to the fluid naphtha and petroleum, are known to exist in immense quantities along the coast of the Gulf of Mexico, chiefly in the States of Tamaulipas, Vera Cruz and Tabasco.'[99]

In northern Veracruz some indigenous Téenek – who used *chapopote* for waterproofing, painting and ceremonial burning – innocently led geologists to some natural oil seeps, where local cattle ranchers of Hispanic descent had fought an endless battle against jungle regrowth, lost their cows in the oil pools, and lived with the threat of insurrection by the Téenek who had been dispossessed of their land. The ranchers were therefore only too happy to sell off quite cheaply huge swathes of potential oil territory to some wealthy foreign businessmen.[100] In 1900 Edward L. Doheny – who had made a fortune from California's Los Angeles oilfield – bought a total of 448,000 acres of hacienda in northern Veracruz, five miles inland from Tampico, for just over a dollar an acre, later recalling, 'We felt that we knew ... that we were in an oil region which would produce in unlimited quantities that for which the world had the greatest need – oil fuel.'[101] The following year, Weetman Pearson – a British engineering tycoon responsible for several major projects in Mexico such as the construction of the Tehuantepec Isthmus railway, and with close ties to the Díaz regime – was inspired by the Spindletop strike to buy up estates on the isthmus and in Tabasco that contained oil seepages.[102] Pearson cabled his agent in southern Veracruz to 'secure an option on oil land with all land for miles around', because 'oil deposits frequently extend over big areas, so the oil rights must extend over a large district to be really valuable'.[103] Small oil strikes made by Doheny's Mexican Petroleum Co. at El Ebano in northern Veracruz from mid-1901 led to a larger strike in April 1904; however, the crude was very heavy and was mainly sold as asphalt. In addition to the established production from Peru, this new Mexican production, and early exploration in Bolivia and Brazil, there were reports of potential oil reserves in several other South American countries from Colombia to Argentina.[104]

In Europe, alongside new production in Austria-Hungary and Romania, much smaller yet commercially viable levels of production continued in Germany at Weitze near Hanover, and in Alsace-Lorraine. Asphalt was produced in France, where a lamp-oil refining industry thrived behind a protective tariff on imported oil products; its raw materials were vegetable oils, shale oil and semi-refined oil from the United States that was classified as crude oil. In 1903 French alcohol producers, who also manufactured lamp oil, made an assault on oil-refining competitors by getting the protective tariff removed, which stimulated French investors into joining the Rothschilds to seek their own foreign sources of crude oil. In Britain, the Scottish shale-oil industry continued, and asphalt was being produced in Italy, both countries also producing small amounts of oil and natural gas.[105] On Europe's eastern fringe, new fields were being opened up in Russia's north Caucasus, mainly near Grozny, and oil production began to extend to the other side of the Caspian Sea as far east as Ferghana (in present-day Uzbekistan).[106] Yet further east, there were indications of oil on Sakhalin Island, *Petroleum World* reporting, 'the Viceroy of the Far East has decidedly declared himself against the retention of the island of Saghalin as a place of deportation, and now proposes to throw the island open to colonists ... [T]here ought to be a great future in store for the extensive oil sources known to exist in this part of the world.'[107]

In the Far East, while new oil was being struck in the Dutch East Indies, in British India increasing Burmese production was joined by new production from Digboi and Makum in Upper Assam by the Assam Oil Co., made economically feasible by the new railway of the Assam Railways and Trading Co.; and small levels of production had been achieved intermittently in Punjab and Baluchistan.[108] Japan, with more than 40 miles of modern pipelines, was producing well over a million barrels of crude per year, mainly from the Echigo oilfield, about 200 miles northwest of Tokyo in present-day Niigata prefecture. Oil had also been found on the island of Formosa (Taiwan), recently ceded by China to Japan following the First Sino-Japanese War of 1894–5.[109]

In the Philippines, the American Army had been brutally suppressing local insurgents since 1898 when the United States had annexed the archipelago in order to secure a naval base at Subic Bay as a strategic outpost close to China.[110] For most Filipinos, it was reported, the price of imported illuminating oil was beyond their means: 'the natives work on the plantations or about the homes of the richer classes for a small pittance'.[111] But 'the American capitalist has now come to the rescue,' wrote the *Petroleum Review*.[112] Refineries were built to take locally produced crude that was manually pumped through bamboo pipelines, from which locals were paid to prevent leaks. According to the trade journal *Paint, Oil and Drug Review*,

> As natives can be hired for 10 to 15 cents (American) per day, to attend to pipe joints, and as a native can readily attend to 4 miles of piping, it does not require much labor or expense to keep up the pipes. Besides acting as repairers to the joints, these natives are guards to the lines and are really patrols, for they are always on the move. They help to keep off the guerillas, who, if they can not rob, will burn and loot, and, who prior to American occupation, would think nothing of setting fire to an oil-pipe line for the sake of watching the affair burn. But the American garrisons are so freely distributed through all of the islands at present that the freebooters have no opportunity to commit depredations as of old, and ... there have been no recent interferences with the pipe lines.[113]

In the late 1890s Standard Oil sent an agent to assess the oil production potential of China's Sichuan province, where small quantities of oil were being extracted manually, and between 1898 and 1902 railway and mining concessions that included oil rights were acquired by British and French speculators.[114]

In New Zealand some oil was being produced in the district of Taranaki,[115] and there was some excitement in Portuguese Timor over the potential for oil production in the mountains east of Dilli. In 1891 this port was on the itinerary of a new cruise for tourists, described by a journalist for a Sydney newspaper:

> [O]ne of the principal attractions, apart from the lovely scenery, shooting and fishing, will be the wonderful petroleum springs, which are said

to have been flowing for the past 30 years. The oil is so pure that the natives use it for all lighting purposes without any further preparation. A company has been floated to work the springs, and already an 8-in. pipe proves barely adequate to conserve the natural overflow.[116]

A concession was granted at this time for commercial exploitation of the oil but, it was reported, 'the uncivilised character of the inhabitants has hitherto always been deemed a bar to their successful development'.[117] Indeed, a well had been sunk and oil encountered, but the concession was abandoned until 1901 when some Australians from Darwin applied to work it, while another group from Sydney drilled for oil near the southern coast of East Timor at Aliambata in Viqueque district.[118]

In West Africa, oil was struck in small quantities from drilling in the Apollonia district of the Gold Coast (present-day Ghana) at Takinta near the Tano River; a British Royal Commission heard that 'oil suitable for use as fuel was found at no great depth ... and it is contemplated to use the oil as fuel in connection with the gold-mining industry of that country'.[119] According to the British colony's former Director of Education, writing between the Fourth Ashanti War and the Ashanti Uprising,

> No proper survey either geological or geographical has ever been made of this country, but when it is, I confidently expect that it will be found worthy of European capital and enterprise. Bitumen has been obtained some distance inland from Newtown, and petroleum has recently been discovered in the neighbourhood of Half Assini, though the latter has been found to be more of a lubricant than an illuminant.[120]

In north Africa a few minor producing wells had been sunk in Algeria following French occupation and, it was declared, 'the existence of petroleum in commercial quantities has been proved.'[121] In Egypt, a few years after Britain's 1882 invasion and occupation some successful exploratory drilling had been carried out on the Red Sea coast, and more recent drilling at Gebel Zeit and Gemseh showed signs of commercial potential.[122] At around the same time natural oil seeps found in Mesopotamia – at the eastern reaches of the Ottoman Empire – and in Persia were beginning to receive outside attention.[123]

*

In May 1904, *Petroleum World* announced the dawning of a new age for the oil industry: 'The geography of the petroleum world is no longer confined to two great centres – Baku and the northern oil fields of America; the sun never sets on the territorial dominion of oil.'[124]

'THE VIRGIN OIL FIELDS OF PERSIA'

From ancient times, petroleum and bitumen had been found across a wide region west of the Zagros Mountains, from northern Mesopotamia to southwest Persia; the most well-known occurrences were those near Hit and Kirkuk in Mesopotamia and those close to Zohab and Ram Hormuz in Persia (see Fig. 26). Between 1849 and 1852 William Loftus – working for a joint British, Russian, Turkish and Persian boundary commission set up with the intention of defining the hazy Turkish-Persian frontier – conducted a broad geological survey of the region, and he witnessed the exploitation of oil from springs near an ancient temple, Masjid i Suleiman, in the Bakhtiari Mountains:

> The stench of sulphur in the ravine is almost unbearable. From these springs are collected annually about 2000 mauns (or 12,000lbs. English) of liquid naphtha and prepared bitumen; which, including the expense of collecting, of manufacturing, and of carriage, are sold in Shúster at the following rate: – Liquid naphtha, 1½ Keran per maun, or 2½d. per lb., – Bitumen, at 2 Kerans par maun, or 3¼d. per lb. There might be collected 7000 or 8000 mauns annually, if there were sufficient demand.[125]

Since the early seventeenth century the rising value of the East India Company's commercial operations had led Britain increasingly to see Persia as having strategic importance for the protection of India and the associated trade and communications routes. To this end, throughout the nineteenth century Britain had pursued a policy of maintaining the decaying Ottoman and Persian empires as buffers against perceived threats from Egypt, France and, primarily, Russia.[126] 'Were it not for our possessing India, we should trouble ourselves but little about Persia,' Lord Salisbury, while Prime Minister, would put it some years later.[127] The boundary commission on which Loftus had served was but one manifestation of the heightening geopolitical significance of the region, as the British increased their presence in the south while Russia extended its reach in the north into Ottoman, Persian and Central Asian territory.

The Anglo-Persian War of 1857, fought by the British to maintain Afghanistan as a separate buffer state against Russia on India's Northwest frontier, followed by the Indian Mutiny a few months later, highlighted for the British the slowness of communications over such vast distances. When Britain consequently decided to extend the European and Ottoman electric telegraph system all the way to Karachi, the British interest in Persia was increased further. Two telegraph cables were constructed from Baghdad, in case one or the other were sabotaged. One went straight down, via Basra, to Fao at the head of the Persian Gulf, and the other extended across Persia via Tehran. In 1863 the attempt to determine where the latter line left Ottoman and entered Persian territory triggered another in a long series of Ottoman-Persian border disputes; the British, always keen to avert any outright armed conflict between Turkey and Persia that would only benefit Russia, mediated a compromise solution that left the precise point undefined – somewhere between Khanaqin and Qasr-i Shirin, on the well-trodden Baghdad-Tehran route, and coincidentally near known natural oil springs.[128]

Neither Persia's rulers nor its general population were happy with the loss of sovereignty implied by the increasing presence and influence of two great European powers. However, competing threats, offers of assistance and personal financial inducements reduced the shahs of the period to playing off one power against the other, while ceding significant ground to both. Britain proclaimed its support for a strong, independent Persia, by which it meant, in reality, a Persia compliant with British rather than Russian interests. For example, when in 1870 Persia decided to create its own navy and approached France to acquire a warship, London pressurized the Shah into abandoning his plans, the government of India protesting that it was 'neither necessary nor expedient that the Shah should maintain in the Persian Gulf any number of armed craft'.[129] Britain, after all, claimed a monopoly of sea power in the Gulf.

The Shah put up some resistance by thwarting for as long as he could opening up the Karun River to a British steamship company, subsidized for the purpose by the Indian government. The Karun was of key strategic value as the only navigable waterway into the Persian interior. Furthermore, it flowed into the confluence of the Tigris and Euphrates, the Shatt al Arab delta at the head of the Persian Gulf, of disputed sovereignty between Turkey and Persia. In 1846 the British archaeologist Austen Layard had persuaded Ambassador Stratford Canning at Constantinople that Britain ought to occupy Mohammerah (Khorramshahr), in the heart of the delta, as 'the military and commercial Key' to the region.[130] British forces subsequently did so during the 1857 war as a base from which to advance up the Karun.[131] Thus, the Indian Intelligence Department reiterated that gaining access to the Karun up to Shuster 'would give England, as the country which would most largely employ the new route, paramount influence in Southern Persia, while the possibility of bringing troops within a few hundred miles of the most important Persian towns naturally would contribute largely to the re-establishment of British influence at Tehran'.[132]

Naser al-Din Shah was suspicious of Britain's evolving relations with the independent tribes of the southwest such as the Bakhtiari – which, indeed, came to be seen as a possible counter-force to the Russian-trained Cossack Brigade in Tehran – and was concerned that British commercial incursions into the region might presage an eventual takeover of the country, on the familiar pattern of the East India Company.[133] However, Persians' equal fear of Russian designs, combined with the eagerness of officials from the Shah downwards to be persuaded by bribes, created openings for British enterprise. In one extreme case, Baron Paul Julius von Reuter deployed tens of thousands of pounds to gain a concession, in 1870, for railway construction and commercial monopolies that even the arch-imperialist George Curzon – future Viceroy of India and Foreign Secretary – described as 'the most complete and extraordinary surrender of the entire industrial resources of a kingdom into foreign hands that has probably ever been dreamed of, much less accomplished, in history'.[134] The Shah returned from a trip to Europe to such domestic protest, as well as Russian opposition, that he was forced to revoke the concession. Widespread discontent simmered against an authoritarian and corrupt elite that failed to administer effective government while it enriched itself on foreign loans – to which the British and Russians attached humiliating conditions and whose burden fell on the general population – and on bribes from foreign concession hunters. Curzon wrote,

> Whilst applauding the policy of assisting Persia by foreign capital where she cannot assist herself, and in enterprises of unquestioned stability, I am of the opinion that she is more likely to lose than to gain from the indiscriminate gift of commercial concessions, and that her best advisers should check any premature zeal in this direction. The [concession-seeker] is not uncommonly an adventurer, and sometimes a rogue. By the failure of such bogus undertakings, good capital is frightened away from the country, and the natives themselves form an unfavourable impression of European conduct and honesty.[135]

The arrival of a new ambassador, Sir Henry Drummond Wolff, in Tehran in April 1888 inaugurated a more assertive British foreign policy, to push for a greater economic presence in Persia in order to counter Russian influence. The overarching purpose remained geopolitical, a position stated clearly at the time by Curzon, then Under-Secretary of State for India:

> [I]n the contemplation of the kingdoms and principalities of Central Asia, no question, to my mind, is comparable in importance with the part which they are likely to play or are capable of playing in the future destinies of the East. Turkestan, Afghanistan, Transcaspia, Persia – to many these names breathe only a sense of utter remoteness or a memory of strange vicissitudes and of moribund romance. To me, I confess, they are the pieces on a chessboard upon which is being played out a game for the dominion of the world. The future of Great Britain, according to this view, will be decided, not in Europe, not even upon the seas and oceans which are swept by her flag . . . but in the continent whence our emigrant stock first came, and to which as conquerors their descendants have returned. Without India the British Empire could not exist. The possession of India is the inalienable badge of sovereignty in the eastern hemisphere. Since India was known its masters have been lords of half the world . . . [Our] . . . countrymen . . . while no longer the arbiters of the West . . . remain the trustees for the East, and are the rulers of the second largest dark-skinned population in the world . . . It ought not to

be difficult to interest Englishmen in the Persian people. They have the same lineage as ourselves. Three thousand years ago their forefathers left the uplands of that mysterious Aryan home from which our ancestral stock had already gone forth ... They were the first of the Indo-European family to embrace a purely monotheistic faith ...[136]

Towards this grand geopolitical project, Prime Minister Salisbury, for example, argued for the construction of a nearly 500-mile railway from Quetta, through Baluchistan, to Sistan in eastern Persia to fill a perceived gap in the defence of India: 'It would not pay: but its military & political effect would be prodigious.'[137] Wolff managed at last to open up the Karun River and revived the Reuter concession. It had been reported by the British consul at Bushire that Russia was making its mark on the kerosene trade of the Gulf region, via the Suez Canal and India: 'the cheapness of the Russian oil enables it to compete against superior American qualities.'[138] Now, however, under the renewed Reuter concession, the Persian Bank Mining Rights Corporation was formed with exclusive rights for sixty years to exploit the country's mineral resources, including oil. In May 1890 Curzon – a director of the new company and soon to be appointed Under-Secretary of State for India – delivered a paper to the Royal Geographical Society on the commercial potential of the Karun River region, noting, 'naphtha springs are well known in the neighbourhood of Shushter and Ram Hormuz' and that their production could 'be many times multiplied were communication rendered simultaneously more easy and more secure'.[139] A few weeks later, ambassador Wolff conveyed to Salisbury a report claiming that Russia's oilfields were becoming exhausted whereas 'the virgin oil fields of Persia promise a good future as they may be made to engage the whole of Western markets in a short time if there be sufficient oil.'[140]

Concurrently, a friend of Wolff acquired a monopoly over Persia's entire tobacco trade. This, more than any other commercial concession, had a more visible, tangible impact on the population, particularly merchants, and it became the trigger for widespread protests against the government and its perceived servitude to foreign money and power. In a traditional society without mass education or political freedom, the mosque became the obvious forum for political ideas, and proponents of civil constitutional reform saw that the most effect-ive way to mobilize the masses was to deploy familiar religious language and justifications. Thus, an array of anti-establishment cler-ics became instrumental in a popular movement against the tobacco monopoly – most significantly, backing a boycott of tobacco by endorsing a *fatwa*, or religious edict, against its use – that was, at root, a protest against the entire authoritarian regime and its venal submis-sion to foreign interests.[141]

The first instincts of the Shah and of the British were to forcefully suppress demonstrations and to blame the uprising entirely on Rus-sian agitators or on religious fanaticism; but when it appeared that the authority of the Shah and the stability of the government were coming under threat, British officials found it increasingly difficult to deny the underlying political realities. In December 1891 Sir Frank Lascelles, replacing Wolff as ambassador in Tehran, wrote to Salis-bury, 'it would have been impossible for the Mollahs to have obtained this power had it not been for the general discontent which prevails throughout Persia which has led the people to hope that by following their advice some remedy may be found for the grievances from which they undoubtedly suffer.'[142] As a report produced by the Foreign Department of the government of India a few months earlier had made clear,

> The sole object of the Monarch, as of every Prince or Official, is to accumulate as much money as possible entirely regardless of the means adopted ... The population have fully appreciated the character of their rulers and understood the value of the pretext upon which the policy of granting concessions has been entered upon (to the effect that it is for the national benefit).[143]

Salisbury was informed, furthermore, of a widespread fear that Britain 'meant under cover of trade to seize Persia as they did India and Egypt'[144] and, he was warned by Lascelles, 'if the Mollahs succeed in asserting their power and introduce a fanatical and anti-European regime, we should give up all hope of seeing the regeneration of Persia by means of commercial enterprize [*sic*].'[145] With mounting concern that Russia might take advantage of the worsening civil disorder to inter-vene in the north, Salisbury began to counsel caution and to countenance the annulment of the tobacco concession: 'We do not wish to assume the invidious position of urging rigorous measures in order that foreigners may make money.'[146] He later wrote to Lascelles, 'We have to guard against the suspicion that we are not labouring for the development, but only for the exploitation, of Persia.'[147] The Persian Bank Mining Rights Corporation went into liquidation, having been defeated by not only the climate and the difficulties of transport and communications, but also the local hostility encountered by its personnel, including those exploring for oil on Kishm Island, at Ram Hormuz and at Simnan near Tehran.[148] The tobacco monopoly, the immediate trigger for the protests, was withdrawn; but the underlying grievances remained. Three years later, Britain's chargé d'affaires wrote, 'There is no law, no administration, and no army. The poor do not know where to turn for justice . . . The root of the evil is the Shah.'[149]

Russia profited from the anti-British feeling, and the Secretary of State for India, the Earl of Kimberley, came to the view that Persia was no longer an effective buffer for India: 'Russia will, without actually taking possession of the North Eastern Provinces, control them virtually, and . . . the Shah will become a vassal of the Tsar.'[150] Russian oil spread from northern to central Persia, the British consul in Isfahan reporting in 1896, 'Kerosene oil is imported entirely from Baku, via Resht and Kasvin. It sells here at 5¼ kran per 3lbs. weight, or 1¼ gallons at 2s.2½d. Leakage is generally made good by filling up the tins with the crude stuff from Shustar; hence the oil here is not very good.'[151] At the same time, American kerosene was being imported from the south, competing with both local and Russian lamp oil.[152]

That year, as Britain renewed its efforts to penetrate south and west Persia, the Shah was assassinated, prompting serious consideration of how to deal with a break-up of the country. Nevertheless, once the Reuter oil concession rights had lapsed a new round of bargaining began, initiated by General Antoine Kitabgi, holder of the lucrative post of Persia's head of customs, who had also been involved in arranging most of the more controversial concessions. As such, he had been denounced during the tobacco protests as one of the main thieves of the nation's wealth.[153] In October 1900, while peddling for concessionaires in Paris, Kitabgi met former ambassador Wolff, who put him in contact with William Knox D'Arcy, an Englishman who had recently become very wealthy from a successful gold-mining venture in Australia. A French geological study, published in 1892 and republished in 1895, had reinforced earlier claims of indications of a 600-mile belt of oil running southeastwards from Kirkuk in Ottoman Mesopotamia to the Karun River;[154] and in 1897 Captain F.R. Maunsell had provided a compelling overview of the occurrences of petroleum in the region for the *Geographical Journal*:

> The existence of such a large number of sites in a well-defined area, and the considerable quantity of bitumen and naphtha which now comes to the surface, would suggest a large supply of oil if properly tapped, and the establishment of a commercial industry of great importance. The petroleum-bearing belt commences near Mosul, and extends south-east in a broad band skirting the base of the Kurdish and Persian frontier hills as far as Shuster.[155]

These reports were now corroborated by a new survey commissioned by D'Arcy from Boverton Redwood, Britain's foremost petroleum consultant and adviser to the Burmah Oil Co. and to the Admiralty. In April 1901 a team comprising Kitabgi, Edouard Cotte – a fellow key figure in the Reuter and tobacco concessions – and D'Arcy's representative Alfred L. Marriott began negotiations in Persia.[156] Crucially, they had the support of the Foreign Secretary, Lord Henry Lansdowne, and his minister in Tehran, Sir Arthur Hardinge, who conveyed to Lansdowne Marriott's opinion that 'petroleum will be found in great quantities and that an industry may be developed which will compete with that of Bakou'.[157]

Given the deep unpopularity among Persians of their corrupt and extravagantly wealthy governing elite that had a record of enriching themselves by selling commercial concessions to foreigners, D'Arcy assiduously tried to avoid drawing public attention to his activities, writing, 'the less that is published about Persia the better'.[158] In Tehran's corridors of power, meanwhile, D'Arcy's team was able to soothe the fears of the new Shah and his ministers about the reaction from Russia, with whom they were negotiating for a loan, by disbursing £50,000 of discreet bribes, with another £40,000 to follow, and by increasing their share in the profits from 10 per cent to 16 per cent.[159] Accordingly, on 28 May Mozaffar al-Din Shah obliged by signing a concession that granted D'Arcy exclusive rights for sixty years for the exploration, production and export of petroleum from a 500,000 square mile concessionary area; this covered all of Persia except for, at D'Arcy's suggestion, the five northernmost provinces – so as 'to give no umbrage to Russia'.[160] D'Arcy, thanking Lansdowne, expected the concession to yield dual benefits: the 'enterprise will prove advantageous to British Commerce, and to the influence of this country in Persia'.[161]

Russia, by contrast, was furious, primarily over a pipeline monopoly apparently bestowed under Article 6 of the concession, which stated that the Persian government 'shall not grant to any other person the right of constructing a pipeline to the southern rivers or to the South Coast of Persia'.[162] Since the 1870s Baku's oil industrialists had dreamed of having better export channels, and the idea of constructing a trans-Persian kerosene pipeline had been in the air since the early 1880s.[163] In January 1901, with crude production surpassing that of the United States, prices on the Baku oil market plummeted to new lows. Thus, D'Arcy's Persian pipeline monopoly was awarded precisely when Baku was feeling its limited export capacity most keenly. In August, with the backing of Russia's Finance Minister Sergei Witte, a new trans-Persian pipeline proposal was submitted to the Tsar. As well as justifying the project on economic grounds, the proposal played on Russian geopolitical aspirations: 'The laying and operating of the kerosene pipeline would in any case create in the Persian Gulf real Russian commercial interests, which no Power would have the right to ignore [and] it would therefore serve to increase our influence in Persia and on the shores of the Indian Ocean.'[164] A six-month diplomatic tussle ensued, as Witte demanded from the Shah pipeline transit rights as a condition for a new loan. Hardinge complained,

> The leverage thus acquired by [Witte] is being used without scruple or mercy in the present instance for the purpose of extorting from Persia a preposterous Concession for pipe-lines to the Persian Gulf, which will probably never be carried into actual effect, but will, nevertheless, afford an excuse for covering Southern Persia with surveyors, engineers, and protecting detachments of Cossacks, and preparing a veiled military occupation.[165]

Russian lawyers argued that the terms of the D'Arcy concession did not apply to Russian refined oil in transit, while British lawyers argued that they applied to any oil pipeline. However, the Russian ambassador in Tehran, K.M. Argyropoulo, was concerned that it would be difficult, under the tense circumstances, to maintain the security of a pipeline against 'the attempt of any voluntary or bribed malefactor who, on a dark night and in a remote place, would put under it a small packet of dynamite or gunpowder'.[166] D'Arcy, whose investment would be imperilled by the easy access of Russian oil by pipeline to the Persian Gulf, finally, in February 1902, decided the argument in his favour by providing the Shah with a further £300,000 loan.[167] For the British this was a geopolitical victory for the maintenance of their supremacy in the Gulf, officially proclaimed in the Lansdowne Declaration the following year. Second Sea Lord Sir John Fisher – known within the Admiralty as the 'Oil Maniac' for his passionate insistence on the Royal Navy's rapid adoption of fuel oil – wrote to a friend while on holiday at Marienbad, 'Mr. D'Arcy is here – the Mount Morgan gold-mine millionaire. He has just bought the south half of Persia for OIL. He has sent for all the papers and maps for me to see. He thinks it's going to be a great thing, and, politically, it will capsize Russia, as this oil concession, for which he has already paid ¼ of a million, will practically make the country English.[168] Commercially, however, Lord Curzon, now Viceroy of India, who had been a director of the previous British oil concession enterprise, counted on this also ending in failure.[169]

D'Arcy's new syndicate began drilling at Chiah Surkh, 10 miles from Qasr-i Shirin – in the long-disputed Ottoman-Persian border region of Zohab – near Khanikin on the ancient Baghdad-Tehran trade route (see Figs. 25 and 26). For at least forty years, crude oil, manually baled from deep pits, had been sold to travellers on this road without any sign of depletion.[170] However, its distance from the coast and from any waterway or railway made equipping and supplying the site extremely difficult; shipping equipment 300 miles up the Tigris from Basra to Baghdad was the easy part. Moreover, local tribes who did not recognize the Shah's suzerainty but were at war with each other demanded agreements of their own with D'Arcy. A pipeline was surveyed down to Mohammerah, but only a trickle of oil, by commercial standards, emerged at Chiah Surkh, while D'Arcy's funds dried up.

The Paris Rothschilds and Standard Oil began to circle the floundering enterprise as a potential bargain, and Lloyd's Bank asked for the oil concession as security for D'Arcy's £177,000 overdraft.[171]

By the end of 1903, Curzon and Hardinge became worried lest the enterprise be bought out, perhaps by Russian oil companies who could then lay their pipeline to the Gulf after all; this 'would be most dangerous to British interests in Southern Persia', Hardinge wrote to Lansdowne.[172] With the Admiralty, in addition, embarking on the use of fuel oil in its steamships, British officials began to consider whether there might now be reasons enough to provide D'Arcy's syndicate with stronger government backing.

LIQUID FUEL AND SEA POWER

During the first decade of the twentieth century the surging output of fuel oil, primarily in Russia and the United States, was only really taken up – by industrial plants, railways and steamships – around the immediate region of production, beyond which it was unable to compete with coal in price and quantity. The long-established solid fuel was being mined in far greater quantities, mainly in Britain and the northeastern US, and tens of millions of tons of British coal were being exported annually around the globe. Despite the advantages of the new liquid fuel – associated with its ease of handling and its high energy content – the relative lack of transportation and storage capacity for fuel oil made it expensive to distribute. Thus, its supply was not yet dependable; and neither was its production, a reputation well and truly secured in Britain when, in mid-1902, output from the original Texas Spindletop oilfield began to fall precipitously. Shell, which had been relying on the field, via its long-term contract with the J.M. Guffey Petroleum Co., for a mass conversion of Britain to liquid fuel, ended up turning four empty transatlantic tank steamers into cattle transport ships. In Britain, demand for fuel oil existed only for specialist purposes where the advantages of the liquid fuel were deemed to be worth the extra cost, such as in the pottery industry where high temperatures and precise temperature control in kilns were required.[173]

However, around the turn of the century a range of strategic and technical considerations began to attract greater interest in fuel oil from the world's naval authorities. The adoption of steam power had led navies into a great dependency on coal. Britain held the advantage of having sizeable indigenous reserves of high-grade steam coal that could also be extracted comparatively economically by a large coal-mining industry, transported efficiently to ports and, by its vast merchant marine, on to its worldwide network of coaling stations, located mainly in its far-flung colonies. By contrast, two aspiring world powers, Germany and the United States, had coal but lacked the necessary merchant marine and coaling bases; a captain on the General Board of the US Navy described the steam engine as 'an insatiable monster' that, if not continuously fed, rendered the ship 'an inert mass'.[174] At the same time, Britain's advantages came at a great cost. The empire's tiny island centre was hugely dependent on sea-borne imports and exports to sustain its industrial economy and its high population density, and its navy laboured under the logistical and financial strain of maintaining and defending all its sea lanes and coaling stations.[175] And, at a time when new design features were making ships more expensive to build and rendering them quickly obsolete, Britain's naval budget was leading the government 'straight into financial ruin', the Chancellor of the Exchequer warned the Cabinet in October 1901 – a fact that, by 1903, even the Admiralty had to concede.[176] In the context, furthermore, of demands that the Indian Army be greatly increased in size to meet a possible Russian invasion of India, the First Lord of the Admiralty, the Earl of Selborne, wrote, 'It is a terrific task to remain the greatest naval power when naval powers are increasing in numbers and in naval strength, and at the same time, to be a military power strong enough to meet the great military power in Asia.[177] As the Secretary of State for the Colonies, Joseph Chamberlain, told an Imperial Conference, 'The weary Titan staggers under the too vast orb of its fate.'[178] Such was the sense of imperial overstretch that Britain felt forced to abandon its so-called 'splendid isolation': by entering into a naval agreement with Japan; by cultivating a rapprochement with its old rival France; and, eventually, by entering into a placatory agreement with its other long-term rival, Russia – all designed ultimately to counter the emergence of Germany as a commercial, military and naval power.[179]

The adoption of fuel oil by navies was urged as a much-needed economizing measure while at the same time delivering great technological advantages. Compared to coal, the greater thermal efficiency of fuel oil offered all navies the possibility of increasing the range of their ships before they had to refuel, thus reducing their reliance on coaling stations – which were either too costly to maintain or simply nonexistent. Fuel oil's greater energy content also enabled ships to reach higher speeds and, because of its fluid nature, to accelerate more rapidly, as extra fuel could be introduced more quickly than stokers could shovel coal. A ship could be replenished with liquid fuel far more speedily and easily, in principle even while at sea from a tanker. Far fewer crew were therefore required to work the steam engines; and that this labour was far easier and cleaner significantly increased the crew's morale. In addition, fuel oil could be stored in places inaccessible for coal storage, freeing up valuable space aboard ship. The engineer-in-chief of the US Navy went as far as to argue, 'in coming naval conflicts the question of victory may be quite as much dependent upon the battle of the boilers as the contest between the guns.'[180]

In 1898 the British Navy had begun a series of fuel-oil trials in the torpedo boat destroyer *Surly*, and in mid-1901 the Admiralty sent its chief inspector of machinery to obtain 'all possible information' on the use of fuel oil by the Italian Navy, which had introduced it from the late 1880s as an auxiliary to coal in order to raise steam more rapidly and to maintain high speed for longer.[181] The Admiralty's great oil enthusiast Sir John Fisher, currently Commander-in-Chief of the Mediterranean Fleet, had written to a journalist in early 1901, 'Oil fuel will absolutely revolutionize naval strategy. It's a case of "Wake up, England!" but don't say anything about it, or they will certainly say that I have been writing to you on the subject.'[182]

The British Navy subsequently launched into a period of intensive experimentation with fuel oil, which Selborne called 'the most important investigation now in progress'.[183] In the accelerating global naval arms race, primarily between Britain and Germany, any technological advantage could prove crucial, and from May 1902 the Admiralty's development programme became shrouded in the strictest secrecy, while as much intelligence as possible was gathered on other navies' use of fuel oil; indeed, by this time a number of ships in the German fleet, including battleships, were dual coal- and oil-fired;[184] and in March, Fisher had written to Selborne, 'The new Russian battleship *Rostislav* burns fuel oil *alone*, and her Captain assured Prince Louis personally that it was a success.'[185] Meanwhile, on the other side of the Atlantic, in July 1902 the US Navy's Bureau of Steam Engineering set up the US Naval Liquid Fuel Board to carry out an extensive investigation into the use of fuel oil, prompted by the new seriousness with which the British and others were developing the technology and by the sudden surge in the production of heavy crude in California and Texas.[186]

The Royal Navy now rapidly advanced its oil-firing technology – though not without the occasional setback – in experiments with the cruisers *Bonaventure* and *Bedford*, the destroyer *Spiteful*, and the battleships *Mars* and *Hannibal*, and by 1905 Britain's torpedo boats, destroyers and battleships were being fitted with auxiliary oil burners as standard.[187] Indeed, under the direction of Admiral Fisher, appointed First Sea Lord in 1904, HMS *Dreadnought* was built with auxiliary oil-firing, oil composing a third of its total fuel capacity. As the first of a new generation of fast, steam-turbine-powered battleships armed with larger guns, *Dreadnought* rendered all previous battleships virtually obsolete and stoked a global naval arms race.[188] From 1905, new battleships of the Austrian Navy were built to be dual coal- and oil-fired, and in 1906 the Romanian Navy began to adopt oil, both countries being keen to find uses for their burgeoning output of residual fuel oil.[189]

Oil companies saw the potential for long-term, lucrative fuel-oil contracts with these important and influential consumers; reciprocally, navies began to develop a need for reliable supplies of the fuel.[190] Shell's sales of fuel oil to the German Navy and the Hamburg-Amerika Line were providing a foundation for its entry into the European oil market, and it had been wooing the British Admiralty for several years.[191] It sent the Admiralty samples of its fuel oil from Dutch Borneo, gave demonstrations of its oil-fuelled tank steamers, and in 1902 Sir Marcus Samuel even offered the Admiralty a controlling stake in Shell, although the offer was turned down. Similarly, several American oil companies offered free fuel to their navy for its trials.[192] However, at this relatively early stage in the distribution of fuel oil, deliveries of even small experimental quantities could not necessarily be counted upon. Even in the United States – the source of most of the British Navy's early supplies of fuel oil – it was sometimes only through the personal intervention of Standard Oil executives and prominent figures in the Californian oil industry that the navy could secure deliveries of fuel oil to its Atlantic ports; the US Naval Liquid Fuel Board reported,

> As the facilities for the transportation of oil are practically controlled by a few corporate interests, the board soon discovered that in order to provide for an adequate and reliable supply of such fuel it would be essential to deal only with those who could guarantee the timely shipment of the oil requisite for experimental needs ... The Texas oil was procured from the Standard Oil Company. The board is under special obligation to the officials of that corporation for assistance in keeping the experimental plant supplied with fuel.[193]

A symbiotic relationship thus began to develop between navies and oil companies. In 1903 the West India Petroleum Co. asked for a loan from the British Admiralty to fund its prospecting on Barbados, in return for the navy's first rights to its oil.[194] Shell, likewise, hoped to form a close commercial relationship with the Royal Navy; however, not only were the company's oilfields and refineries under Dutch sovereignty but it was also presenting a considerable competitive threat to the Burmah Oil Co., the only significant British oil company whose oilfields were entirely under British sovereignty.[195]

Although, noted the Royal Navy's engineer-in-chief in October 1902, 'oils from Texas, Borneo and Russia' had been found to be suitable in case of war, a supply 'from British and Colonial sources' would be preferable.[196] In August 1903 the Admiralty established its Oil Fuel Committee, mandated to investigate actual and possible sources of supply in 'British territory or within British spheres of influence'.[197] The Committee based its deliberations on 'the general principle that it is inexpedient to depend in peace time upon resources which would probably fail in war time ... [when] oil fuel would probably be declared contraband'.[198] On this basis, Russia was ruled out, as was Romania, because its oil industry was dominated by German firms and most of its exports went to Germany – either by river up the Danube or by sea via the Dardanelles from the Black Sea port of Constanţa.[199] The *Petroleum World* commented,

> a general substitution of petroleum for steam coal ... is bound to have important effects, both political and industrial ... The eyes of naval men would turn from South Wales to Texas and Borneo, to South Russia and Roumania; and the speculative student of foreign politics may let his imagination play upon the future of Roumania, and wonder whether that kingdom would become stronger by the change, or, not being strong enough, whether she might become an object of perilous covetousness.[200]

Even during peacetime, Russia only exported a small fraction of its fuel-oil output and, likewise, domestic consumption in the United States seriously limited the volumes available for export, especially during its coal strike of 1902. Consequently, the Oil Fuel Committee concluded, Britain should develop 'those supplies upon which reliance must be placed in time of war'.[201]

Apart from a potential small domestic supply from the Scottish shale-oil industry, only Burma was both under British control and looked even close to becoming a source of fuel oil on the scale of the projected quantities needed, which would certainly be more than the approximately 25,000 tons of Texas fuel oil purchased during the navy's 1904 trials, up from 1,200 tons in 1902.[202] In mid-1903 the Admiralty had requested a sample of fuel oil from Burmah Oil, which replied that presently it could supply only small quantities; it would be able to turn its operations towards the production of larger quantities of fuel but would expect something in return for the additional refinery investment required, especially since production of kerosene was far more profitable than that of fuel oil. The Oil Fuel Committee recognized that, in this respect, the navy's fuel interests could actually conflict with an oil company's commercial interests.[203] Thus, the medium-term outlook for the security of supply of fuel oil to the Royal Navy was not good, and the Committee arrived at a correspondingly cautious conclusion that reflected the current pros and cons: '[T]he quantities of fuel oils available and in sight are much less than has hitherto been supposed, and there can be little doubt that, at the best, only sufficient will be obtainable for use as an auxiliary to coal. Oil must be regarded as a supplement, not a substitute.'[204]

By June 1904 the Committee's president, Ernest Pretyman, was writing confidentially, and somewhat urgently, to the Secretary of State for India, 'There is no doubt that a large quantity of fuel oil will be required in the near future and it is very important that the supply should come from within the Empire, should be available in war as well as peace and should not be under the control of any foreign Trust or syndicate.'[205] The Admiralty was easily persuaded by its main petroleum consultant, Boverton Redwood, that Burmah Oil – which also employed Redwood as an adviser – should receive special treatment as it was an entirely British concern. 'I am of the opinion,' he wrote to the Admiralty in February 1904, 'that it would be better to furnish the Burmah Oil Company with such inducements as may be necessary to bring about an adequate extension of its operations than to seek to encourage competitive work by others . . . '[206]

Pretyman agreed that 'foreign trusts' should be denied access to British oilfields and that Burmah Oil should not be left to be 'beaten out of the field' by the competition.[207] The Admiralty accordingly made its case to the relevant government departments. In May, it conveyed to the Secretary of State for India its need 'to obtain oil fields situated in British Territory, the right of pre-emption of suitable residual oils for Naval use, and to secure a supply of such residual oils by requiring that the refining of the crude oil should be carried out in British Territory'.[208] To the Colonial Office it wrote, in July, that the empire faced two stark options: 'One to allow the colonial oilfields to be developed and worked by any capitalists who are prepared to pay a fair royalty, leaving them free to deal with the output as they please. The other to reserve the concessions for exploitation by British money and in the interests of the Colony and of the Navy.'[209]

Over the summer of 1904 the Admiralty and the Colonial Office convened a Joint Committee, initially to decide on the terms for leasing Crown Lands in the West Indies, occasioned by early oil production on Trinidad and Barbados; but the Committee elaborated a colonial oil policy intended to be generally applicable across the empire.[210] The Joint Committee took the view that competition from the 'great syndicates' would be detrimental to the 'development of the oil industry in the West Indian colonies unless the support and the assistance of the Imperial Government is forthcoming'.[211] It did not suggest direct state involvement, 'unless urgent necessity should arise', but private oil operators should be governed by the following principles: the enterprise should be British; the company must actively work its concession; refining should take place locally; the refinery should be geared towards the output of fuel oil to Admiralty specification; and the government should have the right to buy the oil at an acceptable price, to demand the expansion of naval fuel oil production and, if necessary, to take complete control of the concession.[212]

As the Oil Fuel Committee was compelled to consider also sources merely within British spheres of influence, in July 1904 Pretyman wrote to the Foreign Office, having been advised by Redwood that Persian territory being explored by D'Arcy's First Exploitation Co. – for whom Redwood was also an adviser – held the prospect of being hugely prolific: it would be 'most desirable that this concession should remain in British hands and especially from the point of view of supplies for the navy of the future'.[213] In contrast to Burmah Oil, which was rapidly heading towards capturing nearly half of the large Indian kerosene market, the First Exploitation Co. had not even struck crude in commercial quantities. But when D'Arcy began to seek outside investment to enable it to continue its exploratory drilling – the French Rothschilds and Standard Oil were among the possible candidates – the British government promptly engineered a rescue of the company. The Admiralty, for its part, was anxious to keep this potential source of fuel oil in British hands. The Foreign Office and the Indian government, by contrast, were not nearly so optimistic about the concession's prospects. However, their original geostrategic reasons for supporting the venture in 1901 still applied: namely, to maximize the British presence in Persia as a check on the perceived Russian threat to Britain's supremacy in the Persian Gulf and the approaches to India. Securing the D'Arcy concession would also scotch plans by oil companies in Russia – where the Rothschilds figured large – for a much-needed trans-Persian export pipeline from the Caspian Sea to the Gulf that may have threatened Burmah Oil's protected Indian market.[214] Pretyman, as he later recalled,

> ascertained that Mr. D'Arcy was, at that moment, in the Riviera negotiating for a transfer of his concession to the French Rothschilds. I therefore

wrote to Mr. D'Arcy explaining the Admiralty's interest in petroleum development and asking him, before parting with his concessions to any foreign interests, to give the Admiralty an opportunity, of endeavouring to arrange for its acquisition by a British Syndicate.[215]

The wealthy Lord Strathcona of Mount Royal was brought on board by the First Lord of the Admiralty: 'Lord Selborne was very insistent indeed, so I consented to do it ... really from an imperial point of view,' recalled Lord Strathcona.[216] Pretyman's 'British Syndicate' turned out to be Burmah Oil, which was induced into taking over the Persian concession, in part due to its dependency on the British authorities for the protection it was receiving in Burma, and in part as an insurance against a continuation of its recent failures to strike new oil in Burma. It would formally reorganize its new Persian enterprise in May 1905, as the Concessions Syndicate Ltd.[217]

To the Under-Secretary of State at the India Office, the Admiralty in January 1905 conveyed its need for 'a trustworthy source of supply from British territory ... free from Foreign control and possibility of interference in time of war'.[218] Currently, only Burma appeared to hold any real, imminent prospect of becoming a major source of fuel oil from within the British Empire. The Committee of Imperial Defence concurred, stating in a communiqué, 'it seems essential that effective measures should be taken to insure that the Burmah field remains under British control.'[219]

The government of India, in turn, became persuaded of the 'supreme importance' of gaining a secure supply of naval fuel oil from Burmah Oil and thus of the need to offer 'such inducements to the Company as will justify their entering into the proposed agreement with the Admiralty and erecting the necessary works'.[220] To this end, in 1905 the Secretary of State for India adopted the policy of refusing 'oil concessions to foreigners & to companies which are not British in origin & control',[221] thereby giving Burmah Oil a de facto monopoly.[222] He wrote to the Viceroy, Lord Curzon, that the Indian government should at the same time be careful to 'free itself from any suspicion that the Burmah Oil Company enjoys its favour in a special manner, and can rely on peculiar privileges and support against other British competitors'.[223]

Already in 1902, applications for concessions in Burma by Standard Oil – by its British-registered subsidiary, Anglo-American Oil Co., and via the affiliated Colonial Oil Co. of New Jersey – had been rejected, based on the Indian government's view 'that their main object is to obtain a monopoly of the oil trade of Burma ... Having once obtained full control of the oil-fields they would then be in a position to ... abandon at their discretion the working of the fields and thus to put an end altogether to the Burma oil industry.'[224] Subsequent applications by Standard in 1905 to build a refinery and even just storage tanks were again rejected, from the fear that it might seek to crush Burmah Oil by flooding India with cheap imported kerosene. In June, John D. Rockefeller complained to Lord Curzon, 'We are unable to believ[e] that the embargo placed upon our exploiting petroleum territories in Burma will be extended to denying us there the ordinary privileges of storage, manufacture and commerce, and which privileges are not denied us in any other country of the world.'[225]

Between 1903 and 1905 the Shell-Royal Dutch-Rothschild combine – the Asiatic Petroleum Co. – similarly tried to check the rise of Burmah Oil in India, but concession applications from Asiatic and Shell were also turned down. Burmah Oil's Charles Greenway made much of a suspected Standard-Asiatic concert over India, writing to the Indian government in July 1903 that there was 'little room for doubt that in some form or other the Standard Oil Company do control the operations of the (Asiatic) Combine'.[226] In fact, any cooperation that did exist between Standard and Asiatic was simply one of the oil industry's many temporary, tactical market-sharing truces; indeed, Burmah itself was about to enter into just such an agreement with Standard, and would soon enter into a more long-standing agreement with Asiatic.[227] Perhaps Shell's sales of fuel oil to the German Navy and the Hamburg-Amerika Line were an issue; or its association with Deutsche Bank in the German oil market; or the long-standing connection between Shell's head, Sir Marcus Samuel, and Baron Alphonse de Rothschild.[228] In July 1904 Greenway pushed his argument further, writing to the government that Shell, 'though nominally a British Company, is really cosmopolitan, more than seven-tenths of the capital being held by Jews and foreigners'.[229]

At Shell's 1904 annual meeting, Samuel, having courted the Admiralty so assiduously, expressed his fury: '[W]e have been treated by the Government authorities of Burma in a manner which I do not hesitate to describe as scandalous.'[230] Because of the duty on kerosene imports into Burma which, he argued, effectively subsidized Burmah Oil,

> We, consequently, resolved upon sharing in the production of oil in Burma, and to this end applied to the authorities for ordinary prospecting licenses. Those have been refused to us. Disgraceful as this is, much worse has happened, because certain prospectors who had obtained licenses were threatened with their cancellation if they sold them to this company!

Samuel dismissed several accusations levelled at Shell by its opponents:

> The first is that we are not a *bona-fide* British public company. Our reply is that ... as to nine-tenths, the shareholders are entirely British ... The next obstacle cited is that we are allied with the Standard Oil Company ... [But] it is somewhat notorious that we have pursued our own policy, and maintained our independence consistently, and we are not, and never have been, allied with the Standard Oil Company. Then it is contended that we should only take up these concessions for the sake of injuring the Burma Oil Company, and not with the *bona-fide* intention of increasing the production of Burma oil ... It is quite manifest that the Burma Oil company has great local influence in Burma, and has effectually employed it to retain its monopoly ...

Nevertheless, in March 1905 the Secretary of State for India wrote to the Viceroy explaining his decision to exclude the Asiatic Petroleum Co. as well as its British member, Shell, 'on the ground that there is proof of their foreign character, or that in their constitution, policy and methods [they] are international Trust companies similar to the Standard Oil Corporation'.[231]

Samuel later claimed that it was the British government's exclusion of his company from the oilfields of Burma that forced him to amalgamate with Royal Dutch, as the junior partner.[232] At his company's annual meeting in 1907 he said of the amalgamation,

The occasion is a somewhat painful one to me, because by the capitulation of our rights in controlling [Shell], the one territory [Dutch Borneo][233] capable, in my opinion, of providing supplies of liquid fuel sufficient to meet the naval requirements of this country has passed from British hands. No one would realise the repeated efforts that I made to avert what I look upon as a great blow to British prestige, but I failed to convince the many eminent officials and Ministers with whom I frequently had to communicate of the vast importance of this undertaking to Great Britain. We should have been willing to make a very considerable pecuniary sacrifice to have retained these properties had we been able to obtain any assurance from our own Government that we should have had their support in case of any dispute with the Dutch authorities. Not only was such assurance not forthcoming, but, I venture to say, never in the annals of British trade has so gross a wrong been done to any company as that inflicted by the Indian Government, instigated and supported by the Admiralty, in classifying the 'Shell' Company as a foreign corporation, and refusing them permission to participate in the development of the Indian fields ... Gentlemen, 'great events from little causes spring', and I shall be greatly mistaken if, in the future, the folly – nay, I will say the crime – of compelling a British company to part with property of vital import in the future of naval warfare is not bitterly regretted, and for this folly history must fix the blame on the right shoulders, but this I can assure you, that neither your chairman nor directors have been to blame.[234]

In November 1905 a long-term contract between the Admiralty and the Burmah Oil Co. was finally signed, committing the company to investing, over the next few years, in facilities capable of producing up to 80,000 tons of fuel oil per year and holding a reserve of 20,000 tons. The overall terms of the contract were quite favourable to the Admiralty, but the quid pro quo was that the company was being shielded from foreign competitors that might otherwise have ruined it; as the Admiralty explained, 'The set-off is the assistance the Admiralty have been able – and may possibly hereafter be able – to give the company in their contest with foreign competition.'[235] In late 1906 Burmah Oil decided to implement a proposal, under consideration since 1901, to lay a 300-mile pipeline from the Yenangyaung oilfields to its Rangoon refinery in order to bypass the Irrawaddy Flotilla Co.'s expensive river transit. The Scottish engineering firm G. & J. Weir would begin work on the £700,000 project the following July.[236]

In the 1905 statement on Britain's naval budget, the First Lord of the Admiralty announced, 'it is now quite certain that oil has taken its place as part of the fuel of the Navy, and every arrangement is being made for its supply, storage and distribution.'[237] The Royal Navy began to acquire a tanker fleet and to erect oil storage depots at home ports and on Gibraltar and Malta. Ironically, the navies of oil-rich Russia and the United States fell behind in adopting fuel oil. A Russian oil expert, A. Konshin, lamented, 'We risk the leaving of our Baltic fleet to the mercy of those who have the control of English coal supplies. We do this at a time when there is a possibility of complications with England . . .'[238] The Texas-based *Oil Investors Journal* criticized the US Navy for not converting to fuel oil, with headlines such as 'When Will the United States Navy Department Wake Up!', 'Liquid Fuel for the English Navy: Why Not for Our Own?' and 'The English Navy Sets the Pace'.[239]

Meanwhile, the British Navy, by contrast, now cautiously committed to fuel oil for its advantages, suffered from a serious insecurity of supply, and it monitored developments in the oil industry closely. For example, in July 1905 the Union Oil Co. informed the Admiralty that it was planning to lay a pipeline 50 miles across the Isthmus of Panama; this would give Californian crude a cheap route to the Atlantic, and the new competition would hopefully reduce the price of Texas crude.[240] However, the lack of transatlantic tanker capacity also kept the price of fuel oil in Britain high. According to the editor of *Petroleum World*, 'The question of transport is at the very root of the great trouble of the high price of liquid fuel.'[241]

In 1905 the *Sydney Morning Herald* suddenly alerted its readers to the strategic importance of fuel oil, triggered by suspicions that Germany might be negotiating for a coaling station on Timor,

> a few hours' steam of the Australian coast and within easy reach from Port Darwin and Thursday Island. Quite recently some valuable petroleum discoveries were made there, within Portuguese territory . . . [Perhaps] something more than a coaling station is wanted, and . . . if

Germany could succeed in persuading Portugal to part with its rights in Eastern Timor for a tempting consideration the possession of this oil-bearing territory within a day's steam of the Australian coast may at some time in the future turn out to be a serious thing for the security of the Commonwealth ... The immunity of the Australian coasts from invasion in time of war must largely depend, as it always has depended, on our remoteness. That protective element is being reduced steadily year by year. We have powerful neighbours in the Pacific already, but none of them are so close as Germany would be at Timor, nor so well equipped for effective naval operations in these waters as the ships of that Power would be if they were in a position to use oil instead of coal, and to get unlimited supplies of oil from the petroleum wells of what is now Portuguese Timor. If the territory is to be sold, therefore, it would be a measure of common prudence on the part of Great Britain and of this Commonwealth to become the purchasers themselves. Anything would be better than to allow so tempting a foothold to fall into the possession of a Power whose ambitions at sea are growing into direct rivalry with British interests ... The story of German annexation in New Guinea is still fresh in mind.[242]

Meanwhile, in a notable speculative development in west Africa, a British pioneer in the Galician oilfields, John Bergheim, managed to persuade Britain's Colonial Office to provide him with even more government assistance than Burmah Oil had received, in order that his Nigeria Bitumen Corporation could take on the risk of drilling for oil in Southern Nigeria. Bergheim's chosen location was about 70 miles from Lagos in Ondo state, east of Lekki Lagoon on the westernmost edge of the Niger Delta region.[243] From the seventeenth century British merchants, in concert with indigenous elites, had profited hugely from the trade in human slaves, then, as the slave trade was phased out, from the trade in palm oil and kernels, and later in ivory, cocoa, peanuts and rubber. From the mid-nineteenth century the British began to extend their rule from the coast inland, initially over the island of Lagos. From 1885 the British government, fearing possible French and German encroachments, issued a charter to the Royal Niger Company. This gave the company a mandate to govern – frequently by force, with its private army – the Niger Delta region, named the Oil Rivers

Protectorate for its lucrative palm oil trade. In 1892 the Ijebu, who inhabited the region just north of Lekki Lagoon, were heavily defeated by a British expedition armed with machine guns and artillery in what was a pivotal moment in the so-called 'pacification' of the wider Yoruba population.[244]

In 1902 the pseudonymous 'White Ant' reported in the *Journal of the Royal African Society*, 'A seam of Bitumen runs for about 60 miles East and West in the Eastern District [of Yorubaland]; it is not inflammable. The natives use this for repairing canoes. There might be mineral oil underneath.'[245] Bergheim was the most prominent of a number of prospectors drawn to west Africa at this time by reports of petroleum. Attempts were made to drill for commercial quantities of oil in the Gold Coast (Ghana),[246] and oil was found in the German colony of Cameroon by a plantation company which, finding it to be 'of high value and very similar to Roumanian petroleum', attempted to exploit it commercially.[247] A British syndicate was attempting to commercialize several oil concessions in Portuguese Angola,[248] while Bergheim himself was the chairman of the British-incorporated Société Française de Pétrole Ltd., formed in early 1907 to exploit and search for oil in the French Ivory Coast, 'on the Northern shore of the Tano Lagoon'. In early 1906 an agent for an associated company, the Tano Syndicate Ltd., had reported on 'the Oil deposits, which I am confident will be found by boring, the existing deposits of Bitumen being merely the residues of former outbursts of Petroleum Oil ... The natives ... pointed out to me a place in the Lagoon where there is sometimes a violent outburst of gases.'[249]

In 1906, as formal British rule was consolidated over the Southern Nigeria Protectorate, the Secretary of State for the Colonies, Lord Elgin, responded to Bergheim's lobbying for assistance by declaring that 'if oil exists, this is a very important matter for Nigeria.'[250] Bergheim was given a prospecting monopoly, and the following year a special Oil Ordinance was enacted, incorporating many of his suggestions. The *Lagos Standard* commented that this law 'will give to the Government and the European prospector the kernel of the oil mining business, leaving to the Native the shell in other words giving him the doubtful privilege of working to enrich the white man'.[251] Bergheim was also granted a £25,000 loan for new drilling rigs, a Colonial Office official arguing that if exploration were abandoned it would mean 'the neglect of the possibilities of the oil fields for a generation'.[252] The governor of Southern Nigeria, Sir Walter Egerton, even proposed the purchase of a 20 per cent stake in the Nigerian Bitumen Corporation, although this was rejected by the Colonial Office on the grounds that it would be showing favouritism towards one company.[253] Indeed, the *Financial Times* subsequently went to some length to rebut precisely such a charge, arguing that the Oil Ordinance and the loan

> makes for acceleration of production, the Government's intention in the matter being made all the more clear by the fact that it reserves pre-emption rights to itself – obviously with a view to securing a supply of oil fuel at fair rates for the Navy . . . Here, we think, the Government is to be congratulated . . . rather than carped at as extending undue favour to one particular company.[254]

Over several years of drilling near Lekki Lagoon, Bergheim's team made several oil strikes. The local manager wrote of one in November 1908:

> We got No 5 Well on oil again and it started to flow at the rate of about 2000 bbls per day but after it had flowed a few hours at this rate it came on hot water and we are only getting about 50 bbls of oil from it now. I expect there was great excitement in London when they got the news that Suiogu was flowing so much oil. The lagoon is at present all covered with oil . . . and there was so much oil at our wharf here that the Doctor got all covered last night when he went in swimming, which he does every evening.[255]

There were high hopes in some quarters, expressed the following year in a *Financial Times* editorial:

> It may be stated without much exaggeration that expectation is acute in certain circles at the present moment with regard to news from the oil belt of Southern Nigeria. Boring for oil is a long and expensive operation and needs a great deal of patience as well as capital, so that the space of two and a-half years that has elapsed since regular borings for petroleum commenced in this colony is by no means an extravagantly long waiting period. But the near prospect of a first-class oilfield

coming into British possession has sharpened our sense of the defi-
ciency to which our vast Empire is still subject in the almost total
absence of one of the chief raw materials of commerce ... It is becoming more and more a desideratum for use in our Navy as further
extensions are made to the fleet, and an exclusively British oiling station on the West African coast would be a most useful addition to the
coaling stations we possess at other points of the globe.[256]

Meanwhile, a British consular report of 1909 averred to difficulties
encountered by an Anglo-Portuguese oil syndicate in Angola, suggesting that

> owing no doubt to the hostility of the natives no attempt was hitherto
> made to develop this industry.
>
> Three concessions have now, however, been registered by a British
> engineer, who managed to reach the deposits notwithstanding the
> threatening attitude of the natives. Borings have been made about 8
> miles east of Catumbo, near Barro do Dande.[257]

However, none of these west African enterprises struck oil in large
commercial quantities, although in late 1909 the *Financial Times* informed its readers of 'an interesting report on the petroleum resources
of Angola ... recently laid before the Lisbon Geographical Society by
Señor Joao Carlos da Costa ... [T]he conclusion he arrives at is that
Angola is destined to become one day "a second America as regards
the petroleum industry." '[258]

As for Bergheim's Nigerian Bitumen Corporation, its exploratory
drilling was challenging due to the lack of infrastructure, the swampy
terrain, the intrusion of water into the wells, and the reluctance of the
locals to reveal the whereabouts of oil deposits; and the company's
work would eventually come to an abrupt end following Bergheim's
death in a car crash in 1912.[259]

PETROLEUM SPIRIT

As heavy fuel oil began to be taken up as an alternative to coal for powering steam engines and heating furnaces, some of the lighter petroleum

fractions increasingly found a new role as fuel for internal combustion engines, introduced originally to provide a convenient source of stationary mechanical power in workshops and on farms, and for generating electricity. A growing number of smaller, lightweight versions of these engines – developed particularly for land vehicles, small boats and, even, the aeroplane – required a more volatile liquid fuel, variously known as petroleum spirit, motor spirit, benzine, petrol and gasoline, leading to recurrent fears of shortages in its supply.

For over a decade from the mid-1890s, France – facilitated by the high quality of its road network – set the pace in the manufacture and export of motor vehicles, and in 1900 there were about 14,000 petrol vehicles in the country compared to about 1,500 in both Britain and the United States. Demand for these new 'pleasure motors' or 'touring cars' initially came largely from the wealthiest in society, whose jaunts and expeditions in these novel vehicles could be newsworthy. As the *Daily Telegraph* reported in 1900, for example,

> King Leopold of Belgium, who will probably leave Paris to-morrow, is taking with him three crack automobile carriages. These are a four-place car of twenty-horse power, in which his Majesty has made frequent excursions here, a voiturette for two persons, which he intends, it is said, to ship in his yacht when cruising, and to use for exploring the countryside at places where he may land; and, thirdly, a large carriage for six passengers, with a coachman's box-seat, in which he will drive in Brussels. All three vehicles are worked by petroleum motors.[260]

As a cheaper option, motorcycles, often fitted with sidecars, were popular, and in 1903 half of the 40,000 petrol vehicles in France were motorcycles. The United States, which in the late 1880s had been ahead of Europe in introducing electric tram systems on a large scale, was slow to adopt gasoline-powered vehicles, due in part to the poor condition of most of its roads and to an early enthusiasm for steam- and electric-powered vehicles (see Fig. 15). In 1901 the *Scientific American* conveyed the diversity of vehicles:

> Since all the fashionable world has taken to automobiling, and this sport is no longer a fad, the inventors of the country seem to have turned their attention to bringing out improvements in motors, carriages and

other parts. The number of patent applications that are being received . . .
is so great that it has been found necessary to have five special examin-
ers on this work . . . One division handles electric motors, another steam
motors, another gas and acetylene motors and another looks out for the
compressed-air motors.[261]

Due to the relative reliability and familiarity of electric- and steam-
powered vehicles they continued to feature disproportionately among
the small but growing number of motor vehicles put into service as
taxis, buses and delivery vans; but from 1905, amidst a storm of
vehicle patent disputes, ownership of gasoline-powered automobiles in
the United States began to surge ahead and by 1907 their production –
concentrated in Detroit, now home to the Olds Motor Works, Cadillac
and Ford – overtook that of France.[262] In 1905, Thornycroft & Co.
rolled out a fleet of 8 mile-per-hour, petrol-driven delivery vehicles for
Standard Oil's British subsidiary, the Anglo-American Oil Co., upon
which the *Petroleum Review* commented,

> there appears every reason to assume that ere long the now common
> horsed conveyances will disappear from our streets. The petrol-driven
> motor has come to stay, for now it is not only adapted for the purposes
> of pleasure, but week by week it is entering more into our daily life. The
> ordinary two-horse road tank wagons, which now form such familiar
> sights in our streets, would also appear to be doomed.[263]

By 1908 about a third of London's approximately 3,000 omnibuses –
traditionally horse-drawn – were petrol-powered. Three years later,
two-thirds were motorized (see Fig. 17).[264]

In 1912 Walter C. Baker, founder of electric car manufacturer the
Baker Motor Vehicle Co. based in Cleveland, Ohio, would estimate
that about 5,000 electric cars were being produced in the United
States per year. But, he said,

> Of course, the storage battery is the drawback in any electric car. I
> believe that if it were possible to get a storage battery much lighter for
> the same power, the electric car would be *the* car. We all know that
> from the electric tramcar. If it were ever possible to get a storage battery
> from which we could obtain the same power as we can from the gas-
> olene machine of the same weight, there is no doubt as to which would

be the accepted automobile. Mr. Edison held out the prospect that the Edison battery was going to give us per pound weight nearly twenty-five times the capacity that we had with the ordinary lead battery. That was probably five years ago, and at that time I looked forward to the electric automobile to displace all others, but since then we have found that the power in the Edison battery is probably no greater than that in the lead battery, and so we have had to fall back on the petrol motor.[265]

Just as steam traction engines had been drafted into military service since the 1870s and were used by the British Army during the Boer War, so military applications were soon found for the internal combustion engine. As early as 1881 a submarine built in New York harbour by John P. Holland was powered by a 15h.p. Brayton petrol engine. In 1901 the British Navy ordered five A-class submarines from Vickers, each powered by a 500h.p. Wolseley petrol engine.[266] At this experimental stage, many mariners' lives were lost; but, as Admiral Sir John Fisher commented when one submarine went down as the result of a petrol explosion, 'you can't have an omelette without breaking any eggs.'[267] Armoured, petrol-powered vehicles were first used from 1902 by the French Army, one of which was bought by the Russian government and deployed in 1905 against protesters demonstrating against the Tsar's authoritarian regime (see Fig. 16).[268] In 1906 a Los Angeles gold-mining company operating in Mexico was reported to have ordered an armoured car, fitted with a Gatling gun, 'to provide protection against the attacks of Yaqui Indians'.[269] That year, as British companies started to produce military motor vehicles, the War Office began considering wartime requisition, by 'registering heavy motor wagons, such as light tractors, trucks and motor omnibuses, for the service of the country in time of war, the same as horses are now being registered'.[270]

The internal combustion engine was also taking to the air. As early as 1872 Paul Haenlein, a German engineer, propelled an airship using a Lenoir internal combustion gas engine fuelled by gas from the balloon.[271] Flight by craft heavier than air drew on centuries of experiments with gliders, and in 1896 an American scientist, Samuel P. Langley, achieved one of the first, albeit tentative, powered flights with a model plane, its propeller driven by small steam engine. With the American declaration of war against Spain in 1898 the War Department, and President William McKinley himself, became interested in developing a full-size, person-carrying aeroplane and granted Langley $50,000 for the endeavour.[272] An internal combustion-powered model plane of his flew falteringly in August 1903, but the War Department was disappointed by the failure of his full-sized, piloted plane and concluded, 'we are still far from the ultimate goal, and it would seem as if years of constant work and study by experts, together with the expenditure of thousands of dollars, would still be necessary before we can hope to produce an apparatus of practical utility on these lines'.[273]

Four months later, however, two bicycle manufacturers, the brothers Wilbur and Orville Wright, began piloting petrol-driven planes in increasingly sustained and controlled flights, having built their own especially lightweight internal combustion engine.[274] While their breakthrough was ignored for several years in the United States, some senior British military figures showed great interest, and the invention sparked the launch of an aviation industry in Europe, particularly in France. In late 1904 Colonel John E. Capper, Superintendent of the Government Balloon Factory in Britain, visited the US and met with the Wrights, reporting to the War Office in December,

> The work they are doing is of very great importance, as it means that if carried to a successful line, we may shortly have as accessories of warfare, scouting machines which will go at a great pace, and be independent of obstacles of ground, whilst offering from their elevated position unrivalled opportunities of ascertaining what is occurring in the heart of an enemy's country.[275]

In November 1906 the owner of *The Times* and the mass-circulation *Daily Mail*, Alfred Harmsworth (later Lord Northcliffe), launched a concerted and sustained campaign for the urgent development of the aeroplane for the defence of the realm; due to the invention of the aeroplane, predicted a *Daily Mail* leader article,

> New difficulties of every kind will arise, not the least serious being the military problem caused by the virtual annihilation of frontiers and the acquisition of the power to pass readily through the air above the sea.

The isolation of the United Kingdom may disappear ... They are not mere dreamers who hold that the time is at hand when air power will be an even more important thing than sea power.[276]

For several years the Wright brothers attempted to sell their invention to any government around the world for the highest price offered, and they took on a prominent American financier and arms dealer, Charles R. Flint, as their representative.[277] However, in 1907 Samuel F. Cody and Colonel Capper secretly began building *British Army Aeroplane No.1* at Farnborough, using an Antoinette engine from an airship. A year later, as the US Army also began to take an interest in the Wrights' work, Major Baden F.S. Baden-Powell wrote, 'that Wilbur Wright is in possession of a power which controls the fate of nations is beyond dispute.'[278] When Louis Blériot, in 1909, made the first flight across the Channel from Calais to Dover, H.G. Wells proclaimed in the *Daily Mail*, 'in spite of our fleet, this is no longer, from the military point of view, an inaccessible island.'[279]

In 1900, well over 100,000 barrels of the lightest petroleum distillates, collectively known as 'petroleum spirit', were already being imported into Britain, mainly from the United States. It was used as an illuminant, for enriching coal gas, as a solvent, as a fuel for stoves and as an anaesthetic.[280] Now, however, with its increasing use as an engine fuel, the supply chain came under pressure to meet the rising demand. The number of outlets in London licensed to sell motor spirit leapt from 15 in 1900 to over 150 two years later.[281] In the US, Standard Oil began to import gasoline from Royal Dutch and Shell; this was due partly to the distance of the West Coast from the regions east of the Mississippi that produced lighter grades of crude, but more to the difficulties the eastern industry was beginning to experience in meeting growing demand. *Petroleum World* reported,

> Quite a sensation has been caused in the American part of the oil world by the arrival of tank steamers with full cargoes of benzine from the Far Eastern fields. American oil men never expected to see the day when other countries would send oil into New York and Philadelphia. It is like the proverbial 'carrying coal to Newcastle' to discharge foreign petroleum products at New York.[282]

From 1900 Shell had the production and refining capacity on Borneo and Sumatra to manufacture plenty of petrol, and the company set its sights on the growing European market for motor-spirit. At its AGM in June 1900, Sir Marcus Samuel told shareholders,

> We are producing very large quantities of benzine ... [B]y the end of the year we should be able to market the enormous quantity of about 150,000 tons per annum ... [T]he supply has been wholly unequal to the demand. We should realize a large profit ... because the consumption for motor-cars is not only large as it is, but if a supply of thoroughly good benzine can be placed upon the market ... a great impetus will be given to the trade in motor-cars, because benzine is the most economical power yet devised ...[283]

However, Shell was only just beginning to build up the capacity to store and distribute the product; four months earlier, the company's Borneo managers had written, 'We must destroy the benzine or shut down the refinery. We are burning it ... Benzine, for which you ask, we unfortunately have no means of storing this, and consequently we have been destroying it, for the safety of the enterprise.'[284] Fred Lane told of how 'it was pumped into a lake which was constantly alight',[285] Royal Dutch's Henri Deterding recalled seeing the 'vast masses of smoke belching up into the sky',[286] and for years the conflagration served as landmark at night for shipping in the Strait of Malacca.[287] Thus, while Shell was burning off huge quantities of petrol in the Dutch East Indies, increases in the price of motor-spirit, particularly in Britain between 1902 and 1904, prompted fears for its future supply and the championing of kerosene and alcohol as alternative fuels. In the summer of 1903 the *Manchester Guardian* warned,

> Automobilists are more threatened than they imagine by the difficulties, or alleged difficulties, surrounding the question of petrol supply, the consumption of which, with the ever-increasing number of automobiles, is going up by leaps and bounds. Few realise that so far as this country is concerned we are almost entirely in the hands of the Standard Oil Trust of America, or what is known here as the Anglo-American Oil Company ... On the Saturday before Whitsuntide and again last weekend, London was suffering from a great shortage of petrol ... Without cheap alcohol, or a

perfected heavy oil carburetter, we have only the petrol string to our bow, and that seems straining almost to the breaking point.[288]

A year later, it was argued in the newspaper that the difficulties encountered in designing engines that could run on other fuels were 'but molehills to the mountainous *impasse* which shortage, failure, or cornering of petrol would set up before us all'.[289] The editors of *Petroleum Review*, predicting a rise in demand for petrol as motoring spread, admitted, 'Whether the supply will be equal to this unknown though certainly great demand, we cannot say.'[290] In mid-1905 the American consul in Berlin reported,

> Several years ago, when the motor vehicle for military and industrial purposes began to assume a new and extraordinary importance, the German Government became impressed with the necessity of building motors which could be operated with some liquid fuel that could be produced in Germany, in the event that through the chances of war or other cause the supply of imported benzine and other petroleum products should be cut off. Alcohol offered the solution to this problem, and all the influence of the Government was exerted to encourage its production and its more extended use for motor purposes ... The net result ... has been to extend so rapidly the use of alcohol for heating, lighting and chemical manufacturing purposes that when the drought of last summer reduced somewhat seriously the output of potato alcohol, the previous surplus was exhausted and the price advanced until alcohol became too costly for economical use as fuel for motors.[291]

In the United States the editors of *The Horseless Age* favoured the availability of alternative fuels; for car manufactures, 'the fact that the great industry in which they are engaged in building up is entirely dependent upon a fuel the supply of which is in the hands of a monopoly, and is also limited by natural conditions, must be somewhat disquieting.'[292]

A campaign to repeal the tax on alcohol as fuel was lobbied against by Standard Oil but, given the corporation's negative public image as a market-abusing behemoth, this only served to strengthen the pro-repeal movement. 'This fuel,' hailed an engineer in the *The Horseless Age*, 'far from being controlled by a monopoly, is the product of the tillers of the soil.'[293] However, although Congress repealed the tax in May 1906, by the time the law became effective the production and distribution of gasoline and gasoline-powered automobiles had reached a scale at which a fledgling, parallel alcohol-based industry could not compete.[294]

Kerosene, as a lamp oil was, by contrast, already easily available through a vast and long-established supply chain. However, being a significantly less volatile oil fraction than gasoline, it was unsuitable for fuelling engines that needed to be lightweight, high-speed, and reasonably easy to start. Nevertheless, this heavier grade of oil was suitable for the newly emerging diesel engine, developed during the 1890s simultaneously by Herbert Ackroyd Stuart in Britain and Rudolph Diesel in Germany. Its design was so accommodating of fuel types that Diesel even experimented with coal dust as fuel – an idea perhaps inspired by the recent discovery that some serious explosions in coal mines were caused by airborne coal dust.[295] This type of engine became effective for heavy-duty work where engine weight was less of an issue, and it began to be introduced, for example, for powering drilling rigs in the Russian oilfields. As well as for stationary power, diesel engines began to be used for powering ships and submarines.[296]

As the number of motor vehicles rose to 150,000 in Britain and 200,000 in the United States by 1908 – the year Ford brought out its Model T – even assuming that the production of light crude were to continue at high levels, would the oil industry have the capacity to refine and distribute that much petrol?[297] In the UK, which was entirely dependent on imports, the price of petrol jumped sharply in 1907, prompting the fuels committee of the Motor Union of Great Britain and Ireland to warn that 'a famine in petrol appears to be inevitable in the near future, owing to the demand increasing at a much greater rate than the supply'.[298] Yet at the same time, the newly merged Royal Dutch-Shell was burning off 30 million gallons of petrol a year at its sites in the Dutch East Indies.[299] However, with the merger, supplies from Sumatra and Borneo surged, assisted by the decision of the Suez Canal authorities to allow the passage of gasoline in bulk, while further supplies arrived from Romania, and from Grozny in Russia's North Caucasus.[300] Similarly, fears over petrol supplies in the US began to be assuaged, from 1905, by the growing production of high-grade crude from the new Mid-Continent field, which extended from Kansas southwestwards into Oklahoma Territory and Indian Territory.[301]

'THE DESTRUCTION OF THE COMMUNISTIC SYSTEM'

Indian Territory had been the last expansive refuge for numerous native tribes who had been herded there from all over the United States under varying degrees of duress throughout the nineteenth century, most notoriously in the deadly Trail of Tears during the early 1830s.[302] As the populations of the surrounding states of Kansas, Arkansas, Missouri and Texas increased, the original solemn promises given by the US government that Indian Territory would be inviolable were gradually forgotten. Cattle ranchers crossed the territory for grazing, many thousands of white settlers entered illegally, and the clamour increased for opening up Indian Territory for railways, coal mining and drilling for oil and gas. C.J. Hillyer, attorney for the Atlantic & Pacific Railroad, argued in 1871, 'There is no such sacredness in a treaty stipulation made years ago with an Indian tribe as to require it to permit it to obstruct the national growth.'[303] The head-on clash of cultures was eloquently conveyed in the same year by Cheyenne Chief Stone Calf:

> What use have we for railroads in our country? What have we to transport from other nations? Nothing. We are living wild, really living on the prairies as we have in former times.[304]

Congress pre-emptively gave incentives to the railway companies to construct lines through the territory, to which access was facilitated by mixed-blood Indian citizens, through intermarriage, who were culturally 'white'. Full-blood Indians were fast becoming a minority in their own country and increasingly seen by the US government as an obstacle to 'progress'. A delegation of four tribes declared that the encouragement that was being given to the railway companies by the government seemed designed to provoke a crisis, one that 'would lead to a war upon our poor and weak people which would result in their ruin, their utter destruction, and the possession of their lands by railroad speculators and political adventurers – who care as little for their country's honor and good name as they do for an Indian's rights'.[305]

In 1872, after hearings conducted by the House Committee on Territories into one in a constant stream of bills to create 'the Territory of Oklahoma', Congressman H.E. McKee conveyed in a minority report the view of the great majority of the Cherokee:

> The real root of this movement [for territorial bills] springs from the fact that Congress, in an unwise moment [in 1866], granted millions of acres belonging to the Indians to railroad corporations contingent upon the extinction of Indian title. And now these soulless corporations hover, like greedy cormorants over this Territory and incite Congress to remove all restraints and allow them to sweep down and swallow over 23 million acres of the land ... And why must we do this? In order that corporations may be enriched and railroad stocks advanced on Wall Street.[306]

Congress unilaterally decided to no longer treat Indian tribes on an equal basis as independent sovereign nations but instead as wards over which it had guardianship.[307] In a bill sponsored by Senator Henry L. Dawes, it was further proposed to dismantle the tribal structure by breaking down land title into individual allotments: from past experience, bargaining with isolated individual Indians made it easier to gain rights to their land. In 1880 the Committee on Indian Affairs recommended enacting the bill, but a minority report by a group of dissenting senators argued that the legislation had been

> devised by those who judge [the Indian] exclusively from *their* standpoint instead of from *his*. From the time of the discovery of America, and for centuries probably before that, the North American Indian has been a communist ... in the sense of holding property in common. The tribal system has kept bands and tribes together as families ... The real aim of this bill is to get at the Indian lands and open them up to settlement. The provisions for the apparent benefit of the Indian are but the pretext to get at his lands and occupy them. With that accomplished, we have securely paved the way for the extermination of the Indian races upon this part of the continent. If this were done in the name of Greed, it would be bad enough; but to do it in the name of Humanity ... is infinitely worse.[308]

The majority report claimed that Indians living in traditional, communal ways were generally destitute, and they cited uncritically a comment made by an Indian Agent in 1855 on the Kansas Indians: '[I]n my opinion, they must soon become extinct, and the sooner they arrive at this period the better it will be for the rest of mankind.'[309] In 1884 Washington Gladden articulated in *Century* magazine the world view behind the bill:

> The savage has few wants; the fully developed Christian has many; the progress of the savage from barbarism to Christian civilization consists largely in the multiplication of his wants ... Christianity always had the effect to develop faculties that require for their exercise the possession of property, and to awaken desires that can be gratified only by ... wealth.[310]

In 1886 Senator Dawes argued that even though the Cherokee generally seemed quite content as they were, 'They have got as far as they can go, because they hold their land in common ... [T]here is no enterprise to make your home any better than that of your neighbors. There is no selfishness, which is at the bottom of civilization.'[311] The senator's daughter wrote in *Harper's Magazine*,

> Those ... who have discovered for themselves individual wants, and who have learned to supply them each man for himself, have found the meaning of progress and civilization. They have done away with the much-talked-of equality, and ... achieved the destruction of the communistic system. They, and they alone, have developed anything further than the brute beast ... Not because its people are Indians, but because of their communism, is civilization but a partial success ... in the Indian Territory.[312]

The Dawes Act, passed into law in 1887, was later hailed as a great success by President Theodore Roosevelt, who, in his First Annual Message to Congress in 1901, described it as 'a mighty pulverizing engine to break up the tribal mass'.[313]

Signs of oil had been noticed in Indian Territory since before the Civil War, but the lack of transport access, the initial qualms of the federal government, and investors' concerns about the legal status of Indian leases held up exploration until the 1880s. In the 1890s the extension of Standard Oil's pipelines southwestwards into Kansas to take the state's new production of high-grade crude stimulated speculative exploration south of the border: here, oil companies began to acquire sizeable drilling leases in the Cherokee and Creek Nations and a huge lease in the Osage Nation. The first significant oil strikes were made on either side of the Osage-Cherokee boundary near Bartlesville in 1896–7, but commercial production had to wait for the arrival of nearby railway, pipeline and storage facilities. The Nellie Johnstone No. 1 discovery well, opened up with a shot of nitroglycerine by the Cudahy Oil Co., financed by Michael Cudahy, the wealthy owner of a meat-packing company, therefore had to be capped. However, it was not properly sealed and a stream of oil ran into the nearby Caney River. When locals lit a fire on the riverbank during the winter, flames spread back up the oil stream and engulfed the derrick, destroying it.[314]

As oil speculators poured into Bartlesville, even those Cherokee who were in favour of allowing in the oil industry suddenly felt out of their depth; as one explained,

> It is a fact that I am illiterate as compared with the representatives of Government of the United States . . . I cannot understand all these questions and cannot readily grasp the situation and master it, in matters as great as this. I am being shoved along too fast, and there is not time given to master these questions and reach what I think is just for my place.[315]

In 1898 the Cherokee negotiators backtracked and tried to halt oil development, as the Keetoowah, representing full-blood Cherokees, protested that it was not for their benefit 'that the powerful syndicates and corporations are turning heaven and earth to get control of [our] land'.[316] As calls in Congress grew louder to absorb Indian Territory into a new state – Oklahoma – the Keetoowah and the Creek Snakes engaged in sporadic rebellion in defence of their tribal sovereignty and traditional, communal way of life.[317] Nevertheless, the oil industry set down its roots in their land. In mid-1899 the Atchinson, Topeka and Santa Fe Railroad opened a service to Bartlesville and the following year, after the construction of pipelines from the rail depot to the oil wells, oil began to be shipped out of the Osage Nation bound for Standard Oil's Neodesha refinery in Kansas.[318] Then, in June 1901, a big oil strike just within the Creek Nation at Red Fork, four miles southwest of Tulsa, triggered a great oil rush into the region.[319]

By mid-1903 the Prairie Oil & Gas Co., a Standard Oil subsidiary, had installed 35,000 barrels of storage capacity at Bartlesville, and after intense lobbying its application for rights of way to lay a trunk pipeline through the Cherokee Nation out of Indian Territory was accepted by Congress in early 1904. This enabled Standard to connect Indian Territory to its Kansas pipeline system that, in turn, was connected in 1905 to its eastern system by a 460-mile pipeline from Humboldt, Kansas to its huge refinery at Whiting, Indiana. The new supplies of high-grade, Mid-Continent crude could now, if necessary, be pumped all the way to the East Coast. As the number of motor vehicles registered in the United States passed 75,000, the demand for gasoline certainly existed. But the burgeoning crude production, particularly after another big oil strike in the Creek Nation at Glenn Pool, exceeded the capacity of the mainly Standard Oil pipelines and refineries to keep pace, and storage tanks and earthen reservoirs across the Mid-Continent oilfields overflowed.[320] According to one report,

> The waste of oil is large, extending onto the Verdigris and Caney rivers, which now carry a thick scum of oil to the Arkansas River. This affects fish, water fowl, and also transportation where small streams are forded.[321]

The consequently plummeting price of Mid-Continent crude was a salvation for Gulf Oil and the Texas Co., whose initial boom in the production of heavy crude in Texas had not lasted. In 1906 they turned to the profitable refining of this new crude into gasoline, each laying a trunk pipeline over 400 miles northwards 'so as to obtain a supply of Indian Territory oil', the *Petroleum Review* reported, 'which is a much higher grade and more suitable for refining purposes than Texas oil, [and] being especially desirable for the manufacture of gasolene and other light products'.[322] From 1907 the Texas companies began to take about a quarter of the Mid-Continent crude production, Standard Oil having previously been almost the sole purchaser – while the new state of Oklahoma absorbed the last major remnants of Indian nationhood.[323]

7

A Volatile Mix

THE 'ASPHALT WAR'

In 1901 the *New York Times* reported from 'the scene of the so-called "Asphalt War", now raging' over Pitch Lake in the state of Sucre, northeastern Venezuela:[1]

> The asphalt from the wonderful lake is the finest and purest in the world . . . It is a huge black sea of wealth, stretching out as far as the eye can reach, and dig and dig all the year round, the excavations fill up as rapidly as the workmen leave them.[2]

In 1883 an American, Horatio Hamilton, had married a Venezuelan woman whose family was close to President Blanco Guzmán, and as a wedding gift the President awarded the groom a concession for exploiting asphalt and other resources from an area around Pitch Lake. Hamilton returned to the United States to form the New York & Bermudez Co. (NY&B) for operating the concession, but it was not successfully run and was taken over in 1893 by the Trinidad Asphalt Co.[3] This was a subsidiary of the Barber Asphalt Paving Co., a major player in a merger movement that culminated, in 1903, in the formation of the General Asphalt Co. of New Jersey – a movement described by US judge William Calhoun as being aimed at controlling 'all the principal sources of asphalt production'.[4] In response, a large street-paving company outside the General Asphalt 'trust', the Warner-Quinlan Asphalt Co., purchased rights to exploit the La Felicidad area of Pitch Lake, an area that the NY&B, however, claimed was included under its original Hamilton concession. The asphalt trust had powerful friends in the US, and the American chargé d'affaires in

Caracas was instructed by the State Department to support the NY&B in the ensuing legal battle.[5]

Since most of General Asphalt's production came from the huge asphalt lake in nearby Trinidad through its British-registered subsidiary, the New Trinidad Lake Asphalt Co., the asphalt trust also received the support of British officials. If a competitor to General Asphalt were allowed to exploit asphalt from an area claimed by the NY&B in Venezuela, asphalt production there would most likely increase, causing a decrease in the price of asphalt regionally and hence in the profitability of the Trinidad-based asphalt monopoly.[6] Lower revenues for Trinidad would be the consequence, argued the colony's acting governor, if 'the control of the Venezuelan "pitch lake" were to pass into the hands of an independent Company which would be a formidable rival to the New Trinidad Lake Asphalt Company'.[7]

Indeed, the NY&B had not given the impression of wanting to increase its production in Venezuela significantly – particularly when compared with the rate of increase from asphalt deposits at Inciarte in Zulia state, worked by another American company, the United States-Venezuela Co. – and this gave the government of Cipriano Castro cause for taking a dim view of the NY&B's case against Warner-Quinlan.[8]

In January 1901 – amid a confusion of conflicting claims, legal proceedings and corruption – a US gunboat arrived on the scene after the Venezuelan authorities attempted to confiscate a contested area of Pitch Lake from NY&B.[9] The *New York Times* reported, 'The property of the Asphalt Trust has been attacked by an armed force of Venezuelans. The trust has a private army of its own on the scene ... The orders to Commander Sargent direct him to resist any attempt to dispossess the New York and Bermudez Company.'[10] Although this particular incident blew over, it was but one of many disputes between Castro and a host of foreign companies and individuals who felt they had been treated unfairly by the Castro government, whose inner circle were accused of personally benefiting unduly from their economic nationalism. At the same time, while Venezuelan officials were often corrupt, neither was the behaviour American companies always above board, as the US minister in Caracas, Herbert Bowen, acknowledged: 'It is now time for our American Companies in Venezuela and in all the other republics of South, and Central, America to change their policy. I have never known one of them to conduct relations with this Government in an honest way.'[11] Many of the aggrieved foreign parties – including both the NY&B and the United States-Venezuela Co. – supported a rebellion led by Manuel Antonio Matos and, after its eventual failure, continued to be involved in anti-Castro activity, particularly among the exile communities on British Trinidad and Dutch Curaçao.[12]

By the end of 1901 Germany was seeking American assent to despatch a joint Anglo-German naval mission to enforce European claims against Venezuela. President Roosevelt gave his tacit approval, replying to the German ambassador, 'if any South American State misbehaves toward any European country, let the European country spank it.'[13] During the subsequent Venezuelan Crisis of December 1902, the German Navy sank two Venezuelan gunboats and bombarded a fort at the mouth of Lake Maracaibo; the British bombarded Puerto Cabello; and Italian warships joined the show of force. This generated some disquiet back in Britain for this attack on a relatively defenceless country without apparent good reason, while many felt affronted that the Royal Navy was appearing to play second fiddle to the Germany Navy.[14] Furthermore, the British chargé d'affaires reported to Foreign Secretary Lord Lansdowne rumours that 'a serviceable harbour has been discovered in Margarita which the German Government will lease for forty years as a coaling station and that it is proposed to exploit the coal mines at Guramichate ... for the use of the German navy in these waters'.[15]

This supposition was later conveyed by the *New York Times*,[16] and American fears of European encroachment into the region led Roosevelt into making increasingly assertive declarations of US power over Latin America – though often couched in the self-righteous language of moral duties and responsibilities. Since 1823, under the Monroe Doctrine the US had pronounced the region to be no longer subject to European power. In 1895, during a boundary dispute between Venezuela and British Guiana, US Secretary of State Richard Olney had sent an ultimatum to Britain, declaring that 'today the United States is practically sovereign on this continent, and its fiat is law upon the subjects to which it confines its interposition.'[17] Now, in 1904 – in what became known as the Roosevelt Corollary to the

Monroe Doctrine – Roosevelt formally announced that the US had the right to actively intervene in Latin American countries where it perceived its own interests to be at stake.[18]

As President Castro manoeuvred to eject both the NY&B and its adversary Warner-Quinlan from Pitch Lake, Roosevelt increasingly felt the urge to take military action, 'to take the initiative and give Castro a sharp lesson,' he wrote: 'it will show those Dagos that they will have to behave decently.'[19] However, Judge Calhoun, Roosevelt's own special commissioner tasked with assessing the NY&B case, concluded that it was by no means clear that the Venezuelan authorities had acted improperly, and the US opted instead for international arbitration.[20]

As the NY&B continued to conspire with Castro's opponents to bring him down, he granted other asphalt and oil concessions to allies and friends. In early to mid-1907 General Antonio Aranguren was given a concession covering the Maracaibo and Bolívar districts of Zulia state; Andrés J. Vigas was given a concession covering the Colón district of Zulia; F. Jiménez Arráiz was allotted the Acosta and Zamora districts of Falcón and the Silva district of Lara; and to Bernabé Planas went the Buchivacoa district of Falcón.[21] The concessions ranged from half a million to two million hectares in size, were of fifty years' duration and came with relatively low taxes. However, the concessionaires lacked the kind of capital necessary to start up large-scale asphalt or oil production, and it was only by attracting foreign investment that they could hope to reap rich rewards. But since foreign capital was increasingly averse to the Castro government, by granting these concessions to his allies Castro risked inadvertently turning them into enemies; indeed, it was rival *caudillos* of the economic elite who were the most prominent players in the Venezuelan political arena, one of whom, Manuel (Mocho) Hernández, tried to solicit financial support from American donors in return for pledges of favourable concessions to US companies.[22]

Castro's relationship with one of his once close allies, Vice-President Juan Vicente Gómez, deteriorated and Gómez began to manoeuvre for a coup, signalling that he would smooth over Venezuela's foreign relations and welcome foreign capital.[23] As the Dutch Navy, with the acquiescence of the United States, seized Venezuelan warships, in

December 1908 Gómez took advantage of Castro's absence from the country to launch the coup.[24]

Gómez had barely taken power when US Special Commissioner William I. Buchanan arrived, accompanied by several warships, and won a settlement of claims on much better terms for the American companies concerned, prominent among which was the NY&B. Since many potential rivals to Gómez were supporting him precisely out of a fear that the United States would use its disagreements with Castro as an excuse to intervene and reduce Venezuela to a virtual protectorate – like Panama or Cuba – the new Venezuelan government encountered some early disapproval for bowing to US demands and abdicating its sovereignty.[25] Sir Vincent Corbett, British minister in Caracas, reported that 'the fear of eventual absorption by the United States is undoubtedly almost universal here, and it must be confessed that it is not entirely without cause.'[26]

Gómez held on to power and quickly embarked on a programme of 'modernization' that prioritized stable government and investment in transportation infrastructure in order to attract foreign investment.[27] In an address to the Venezuelan Congress in October 1911, Gómez called for

> capital, brawn, science and experience for our development; and as we do not have such essential factors, it is necessary to receive them from the foreigner who offers it to us in good faith. We have natural wealth that remains unexploited; mostly uncultivated fertile land, immense empty territories. Our industries lack the necessary resources for their development. Our mining treasures lie at the heart of the earth because there are insufficient funds to bring them to the surface.[28]

The anti-American feeling prevalent in Venezuela may go some way to explaining why oil concessions would be taken up almost entirely by Dutch and British companies. However, the country's oil production potential was also, as yet, unproven – speculative territory that Standard Oil, for one, typically considered unnecessarily risky to enter.[29]

REVOLUTION IN RUSSIA – A 'DRESS REHEARSAL'

Russian crude-oil production reached a peak in 1901 at 85 million barrels, having exceeded American production for several years.[30] However, the oil industry of the Transcaucasus rested on unstable economic and political foundations.

Firstly, the programme of rapid industrial expansion, forced through by Finance Minister Sergei Witte in a race to catch up with western European economies, sat uneasily with a traditional, still largely feudal society ruled by an autocratic monarchy. Many, from the landowning elite to the peasant turned factory worker, felt dislocated by the pace of industrialization. Across the political spectrum, a growing nationalism arose out of fears of loss of sovereignty due to foreign indebtedness, foreign ownership of industry, and the government's submission to foreign pressure to lower its import tariff barriers. From the left, Lenin bemoaned the influence of French and German banking houses over the Russian government, exclaiming, 'The economy of the "great Russian state" under the control of the henchmen of Rothschild and Bleichröder: this is the bright future you open before us, Mr. Witte!'[31] There was little political unity, however. Some on the left saw in foreign investment and industrialization a force for wider modernization and progressive change that the landed nobility would be unable to hold in check. Others rejected capitalist industrial enterprise as a new form of exploitation of peasant workers, while yet others opposed factory labour in any form as dehumanizing. However, the government brutally suppressed any expression of dissent, yet while all independent labour organizations were banned Witte encouraged business leaders, by contrast, to organize and articulate their needs and concerns.[32] Labour strikes occurred nevertheless, and even a government report admitted, 'Only the removal of basic causes for industrial unrest can pacify the masses.'[33] Similarly, at universities – expanded to increase the study of science and technology needed for industrialization – the government's crackdown on any signs of political activity led to radicalization among the students.[34]

Secondly, after many decades of aggressive territorial expansion, the

Russian Empire was becoming overstretched. To the east, Russia was expanding in pursuit of an ambitious plan to draw much of the sea trade between Europe and the Far East onto its projected 4,500-mile Trans-Siberian Railway.[35] To the west and south, the government was carrying out an aggressive programme of 'Russification', by systematically suppressing – politically, religiously and culturally – the ethnic groups indigenous to the colonized regions and promoting the settlement of Russian Orthodox peasants. Only a restricted number of Muslims were allowed official representation, for example, and the Armenian Church had its property confiscated. Russification was pursued particularly vigorously in the Caucasus, from 1896 to 1904, under the administration of Prince Grigory Golitsyn who, furthermore, tried to deflect criticism of the government by playing off one ethnic group against another. Consequently, even the hard-line chief of police blamed him for much of the mounting civil strife: 'Golitsyn . . . carried out policies that are at the root of the disorders in the Caucasus during the past decade', by his attempts to 'Russify the Caucasus not through moral and spiritual authority, but through violence and the police apparatus'.[36]

Baku, Tiflis and Batum – linked together by the Trans-Caucasus Railway and to central Russia and the wider world by the oil industry's radiating distribution channels – became hubs of political activity. In 1900 the combination of surging oil production and an economic downturn caused oil prices to fall dramatically, pushing the Caucasus oil industry into a depression. The fallout impacted most heavily the impoverished workers who were now also cut off from the traditional social safety net of family and village. Such was the pressure that had built up – due to the illegality of independent trade unions, strikes and demonstrations – that even relatively minor disputes could trigger protests at which wider political grievances began to be articulated. Labourers walked out on strike in the oilfields – where working conditions were particularly bad – and at the refineries, on the railways and at the ports' packing and loading facilities.[37]

One voice in the cacophony of various socialist and nationalist protest movements in the Caucasus was that of the young Joseph Dzhugashvili – later 'Stalin'. As a child he would have seen the first trains of oil tank cars passing through his village of Gori on their way from Baku to

Batum. In 1890, as the Nobels were blasting their way through the rock a few miles down the railway line at the Suram Pass bottleneck, the teenage Stalin was subjected to 'Russification' when most of his teachers at the Orthodox seminary he was attending were replaced by Russians and his native Georgian became officially a foreign language. Stalin became animated by the nationalist and revolutionary ideas secretly discussed among the students at the seminary, and upon leaving he became a radical political activist among the oil workers of Baku where he helped to clandestinely print and distribute the Georgian Marxist *Brdzola* (The Struggle) and Lenin's *Iskra* (The Spark).[38]

Despite the efforts of the secret police, a strike begun in July 1903 by the Baku oil workers spread to other major industrial centres, leading Tsar Nicholas II to dismiss his long-time Finance Minister Witte. Then, in December 1904, a prolonged strike in Baku achieved, for the first time in Russia, a collective agreement between workers and employers.[39] The protests were becoming increasingly violent, but oil industrialists – notably the producers – remained relatively sanguine. Due to the prevailing overproduction, 'in some quarters the strike is looked upon as a blessing in disguise', reported the *Petroleum World*.[40] Also, it was accepted that many oil companies had failed to keep the promises they had made to their workers after a strike two years earlier. Nevertheless, a Rothschild agent charged that 'the bad spirit ... reigns among the port workers, who never fail to look for a pretext for making complaints, the one more ridiculous than the other'; and *Petroleum World* urged the oil companies to present a stronger, united front against their employees: 'In this contest with labour it has been made apparent that the masters' organisation is deplorably weak, and ... working as unorganised entities, they have shown themselves incapable of acting with firmness ... The masters ... should organise to protect the industry from the political firebrand.'[41]

By this time Russia was embroiled in a war with Japan which, with imperial ambitions of its own, was attempting to halt Russia's expansion into Manchuria and Korea. Britain and the United States backed Japan, while France was allied with Russia.[42] British, Dutch and American oil companies all supported Japan due to their fears that the sizeable kerosene market of northern China might be lost to Russian oil, as *Petroleum World* explained,

A Russian oil company was constructing tanks in the principal towns [of Manchuria], and intended to carry the kerosene tank cars to all towns along the railway line ... This would give the Russian oil a great advantage in the trade, Russian oil having been sold throughout Manchuria at a figure less than United States oil. In the northern part of Manchuria, with Harbin as a centre, Russian oil was fast supplanting the United States product.[43]

At Shell's annual meeting of 1904, Sir Marcus Samuel – after chiding Russia for having imposed unreasonable regulations on tanker shipments from Batum – told his audience,

had Russia not been most justly tackled by Japan, Manchuria would have been lost to European trade, and particularly to our company, because all kerosene but Russian would certainly have been excluded by prohibitive tariffs. Manchuria, however, is a natural market for our Borneo and Sumatra oils, which can be laid down, owing to the geographical position, at much lower prices than is possible for either Russian or American oil ... [W]ith the progressive and able administration of the Japanese, trade, both in Korea and Manchuria, will advance far more rapidly ...[44]

Samuel's trading company provided Japan with many of its war supplies, while Standard Oil's John D. Rockefeller bought up many of the Japanese war bonds issued by Wall Street and the City of London.[45] As Russia's Baltic fleet steamed all the way to the Yellow Sea – some by the Cape route and others via the Suez Canal – its lack of global coaling stations meant that even with supplies provided by its ally, France, and by Germany, it continually ran low on fuel.[46] This gave Samuel an opportunity to make the case for fuel oil: '[I]t is not a little remarkable that Russia, with the largest supplies in the world, should still be burning coal in her own navy. The whole of her difficulties in getting the Baltic fleet out would have been easily overcome had the fleet been capable of steaming on liquid fuel.'[47]

Throughout 1904 the Russian army and navy suffered a series of humiliating defeats to the Japanese, which included the loss of most of its Pacific battle fleet and of its only warm-water port, the Far Eastern naval base at Port Arthur that it had occupied since 1897. The spark came on 22 January 1905, on what became known as Bloody Sunday, when troops of the Russian Imperial Guard, alongside Cossack cavalry, opened fire on peaceful marchers delivering a petition to the Tsar's Winter Palace in St Petersburg. This was followed, in May, by the calamitous destruction of Russia's fleet, after its epic journey from the Baltic, by Japan's British-built navy at the Battle of Tsushima. After these two episodes in particular, in the eyes of many Russians the Tsar had now, finally, lost his aura of benevolence and power, and a wave of rebellion broke out among industrial workers and peasants – particularly across the arc of colonization from the Baltic region to the Caucasus – that the army, having been heavily defeated in Manchuria, was too weak to contain. The Tsar attempted to stem the revolt by making some minimal concessions towards political participation; but, as Edmond de Rothschild wrote, 'the Tsar has so much the habit of doing everything too late, so that one can think that when he decides to make peace, it will be too late to check the revolutionary movement of Russia.'[48]

When this failed to satisfy the political demands, the Tsar supported a conservative backlash of vitriolic propaganda and extreme violence, and the scapegoating of ethnic minorities, particularly Jews, intensified, all of which was designed to divide and confuse the already disparate opposition. The state's secret police and bands of thugs known as the Black Hundreds were, for example, the first to disseminate widely *The Protocols of the Elders of Zion*, a document – denounced as a fabrication – purporting to expose a Jewish-Freemason conspiracy to take over Christian Russia, Europe and then the world by spreading liberalism and capitalism.[49]

Baku and Batum descended into a maelstrom of violence and murder that continued for many months and included clashes between local Muslims and Armenians, attacks on both company bosses and striking workers, and the large-scale destruction of the oil industry's infrastructure. Reports were widespread that the violence seemed to have been incited and instigated by the authorities and the Black Hundreds. Oil producers complained that the government had ignored clear warnings that trouble was brewing in the oil region and that it had been extremely slow to respond once it broke out; and they complained that the government had refused to allow the oil companies to organize their own security forces.[50]

In a despatch of 2 February 1905 the British consul in Batum, Patrick Stevens, reported to the Foreign Secretary back in London, the Marquess of Lansdowne:

[T]he situation at this port is daily becoming more and more critical. For some time past the working classes have completely got out of hand and the Administration appears to be unable, or unwilling to deal with them. The workmen, organized and led by the members of the Committee of the Russian Social Democratic Labour Party . . . became somewhat unruly in so far as their work in connection with the loading and discharging of steamers and the manufacture of cans and cases for petroleum was concerned . . . [M]atters have daily drifted from bad to worse until they have today reached the verge of complete anarchy which the Authorities appear powerless to suppress. The present excited condition of the working classes, throughout the Empire, have lead to the spread of all manner of wild rumours in this town and it is possible that we are on the eve of very serious troubles. With examples before them such as recent events at Baku and again in other principal labour employing cities of the Empire, the men here, being chiefly composed of hot-tempered and dissatisfied eastern elements may break out into open revolt at any time, more especially so, as the existing excitement is being fanned by the distribution of inflammatory proclamations of a Political character. The trade of the port is practically paralysed, work on steamers has been completely stopped and several murders and attempted murders have taken place in the town. The population is in a state of terror with but little hope of having the protection of the authorities extended to them in the hour of need in view of the inefficient Police force at the disposal of the Administration.[51]

Six days later Stevens reported that the striking workers on the oil shipping facilities and the railway had now been joined by many other workers: 'Everything is at a complete stand-still and Cossacks and Infantry continue to patrol the town.' At the end of March, Stevens wrote, 'the political and agrarian rising which has been undertaken by the peasantry of Georgia, is now becoming general, and is fast spreading to all parts of the Trans-Caucasus, the movement increasing in intensity as it gains a larger number of adherents.' The populace was demanding 'the introduction of reforms on constitutional principles' and was in a state of ... revolt against Russian methods of Government, and in many cases the rural administrative offices and other Government institutions and schools have either been burnt or otherwise destroyed ... Acts of violence of every description both against the authorities and landed proprietors in the country are of hourly occurrence notwithstanding the state of siege that has been proclaimed ... [N]otwithstanding the stringent measures which have been adopted consequent on the introduction of a State of Siege in the Trans-Caucasus by the Acting Governor at Tiflis and the newly appointed Governor-Generals of Baku and Kutais, the peasantry and working classes of Georgia ... continue to show a decided disinclination to peacefully settle down to their respective avocations. Incendiarism, disobedience to persons in authority, conflicts with the Police and other armed forces of the Government, robberies, individual cases of assassination, intimidation, attacks upon banks and post offices, and every other imaginable act of violence and terrorism continues to be rife throughout the country and the rural districts are simply reeking with disaffection which is only prevented from breaking out into open rebellion by the fear, or the use, by the troops quartered in the various districts, of exceedingly repressive measures against the offending parties ... As the troops stationed at various places have been found insufficient for the maintenance of law and order in the country, large bodies of Cossacks are being drafted from the northern Caucasus to the disaffected localities ... I am told ... that in all about 16,000 men are to be transferred ... to the Trans-Caucasus to assist the authorities in bringing the population to unconditional submission. Arrests, en masse, are being daily effected and the prisons are so full of criminal and political prisoners that in the fortified towns, the casements of forts have had to be converted into temporary prisons. In consequence of the attitude taken up by the factory workmen and port labourers at Batoum, the trade of the place is completely ruined, at any rate for the time being ... Merchants have recently endeavoured to Charter steamers in London to load case oil at this port, but have completely failed in their attempts. British owners are unwilling to allow their vessels to come to Batoum under any conditions ... Work and traffic on the Trans-Caucasus railway has also recently been thrown much out of order ... Clandestine meetings of the working men are daily taking place, at which inflammatory and seditious speeches against Autocracy and the existing regime are freely delivered.[52]

Towards the end of May Stevens reported that shipping from Batum had 'dwindled down to almost nil . . . [and] the American [oil] trade . . . has taken advantage of the existing unsettled condition of affairs in the Caucasus'. Total economic losses were 'reckoned to amount to 22,000,000 roubles, or somewhat in excess of 2,200,000 pounds' and, as he saw the situation, 'current events do not point to any inclination on the part of the peasantry to surrender to the authority of their Russian rulers'.[53] At the end of May he reported 'the assassination of Prince Nakashidze, Governor of Baku, a further general strike at the oil wells at Balakhani and other parts of the oil territory, a massacre between Armenians and Mahomedans . . . and a fresh outbreak of insurrectionary movement by the peasantry'.[54]

In early August Stevens updated Lansdowne, 'after a somewhat prolonged interval of comparative tranquillity . . . an outburst of disorder at Novorossisk, and strikes of workmen on the Vladikavkaz railway and several of the principal petroleum producing properties at Baku, have to be reported'. At Novorossiysk, an export outlet for oil from the Grozny field, workers who had taken over the railway 'were charged by detachments of Cossacks who used their whips freely'. Shooting then broke out, leading to the deaths of one Cossack and at least thirty-six strikers.[55] On 8 September Stevens telegrammed Lansdowne, 'Situation of affairs at Baku very serious. It is rumoured that the whole of Balakhani, Bibieibat and Blacktown have been rased to the ground. Massacre extended to town. Troops inclusive of Cavalry and Artillery firing into rebels.'[56] Four days later Stevens reported, 'All the oilfields have been literally burnt out and the general opinion seems to be that the Trans-Caucasian Mineral Oil Industry is ruined, seeing that most of the petroleum producers cannot resume work without the grant of a loan, or unless they are indemnified for the colossal losses they have sustained' (see Figs. 18 and 19). There had been 'fighting between Armenians and Tartars' in Baku and the surrounding oilfields:

> The sacrifice of life has been terrible and up to the 8th . . . it is stated, 275 Tartars, 150 Russians, 95 Armenians and about 100 soldiers have been killed . . . [But] it is more than probable that the figures given

represent but a small proportion of the actual number of persons killed and wounded . . . I am glad to be able to report that so far as the information received at this Consulate goes, there has been no loss of life amongst the members of the British Community, seeing that they sought refuge on board a steamer under British management called the 'PADDY' on which vessel, I believe, they have remained ever since the disturbances commenced. It is further reported here, that the Tartars of Baku have made up their minds to continue the struggle until they succeed in driving the Armenian petroleum producers out of the locality and thereby remove a rival race which, owing to its superiority of intellect and its frugality has been able to work up a competition which the Tartars can never aspire to cope with any degree of success.

There was 'considerable revolutionary activity' in the region: 'They hold daily mass meetings [and] march through the streets in large crowds . . . defying the authorities and troops, singing the Marseillaise and shouting "Down with the Tzar".'[57] At the end of October, the Acting Vice-Consul at Baku reported,

The situation in this town appears to have been practically the same as that which has been ruling over the greater part of the Russian Empire during the last fifteen days. Here, however, the antipathy shown towards the Jews has been to a certain extent overshadowed by the antipathy felt towards the Armenians . . . undoubtedly brought about by the recent massacres of the Armenians by the Mahommedans, and the fact that the Armenians, as a community, have shown themselves to be a strong revolutionary agent.[58]

An executive for the Rothschilds' Bnito oil company, Arnold Feigl, wrote,

Whole villages have been massacred, plundered, and burned. These human slaughters have been immediately reflected in the population of Baku, since one is here as on a barrel filled with explosives . . .

Since the bloody days of October, the population is under the terror of anarchy. One fears 'la bande noire', a military riot is feared, and one never knows when lying down what tomorrow will bring. The population is armed and the least rustle provokes a cannonade . . . To all this one must also add the military terror which currently prevails almost everywhere in Russia. The soldiers meet by the thousands, holding speeches, elaborating and presenting their demands. In certain military centres the blood runs with the shooting ordered by the commanders . . . This is a miserable situation without any exit![59]

One witness who visited the oilfields recorded,

the spectacle presented simply defies description . . . Out of the 200 derricks of Bibi Eybat, 188 had been destroyed, and the majority of the other buildings were in heaps of black ruins. The whole area was covered with debris and wreckage, thick iron bars snapped asunder like sticks, or twisted by the fire . . . broken machinery, blackened beams, fragments of cogged wheels, pistons, burst boilers, miles of steel wire ropes . . . The whole atmosphere was charged with the smell of oil.[60]

The Times was particularly concerned for British investment in the Russian oil industry:

The oilfields of Baku represent an undertaking in which a large number of Englishmen are commercially interested. The issued capital of the six most important English companies engaged in it, as quoted in the Stock Exchange Official List, amounts alone to close upon five millions sterling; and this large total, it must be remembered, is exclusive of the private enterprises which do not figure in that list. The warfare – for it is nothing less – will thus cost numerous investors in this country dear . . . [61]

In early October Stevens informed Lansdowne,

the number of wells and other property destroyed was not so great as the earlier reports from the scene of the great catastrophe led one to suppose . . . [I]t has been brought to light, that about 58 per cent of the wells have been completely wiped out whereas about 42 per cent remain intact, so that a considerable production of the raw material is ensured whenever work may be resumed.[62]

Nevertheless, oil production, refining and distribution were all severely disrupted by the combined effects of the labour disputes, the rioting and destruction, and by the government's diversion of resources to the war with Japan. And this was the first time in history that oil operations and the global oil market were disrupted on a major scale due to political turmoil.

In a circular to shareholders towards the end of 1905 the British-registered Bibi-Eybat Petroleum Co. summarized its 'very trying' year. Although some periods of oil production had been possible,

> The severe strike in December, 1904, was followed by terrible snow-storms, which caused a cessation of work for about ten days. Then came the massacres of February, and the political strike in the month of May, succeeded by the widespread incendiarism and anarchy which prevailed in Baku at the end of August and all through September last ... After the 20th August ... all work was stopped at Baku; panic seized the workmen, and Russians, Armenians, Tartars and Persians left Baku in large numbers ... [T]here is a great scarcity of all kinds of labour – more especially of skilled labour – so that it is still difficult to work the properties fully.[63]

Stevens reported to the Foreign Secretary,

> The total losses sustained by the mineral oil industry of the Trans-Caucasus is estimated at nearly three million Sterling, but I am inclined to think that these figures have been somewhat under estimated and that the actual losses are nearer forty million roubles, or four and a quarter million Sterling.[64]

In October the main British oil companies operating in the Caucasus submitted claims to Lansdowne for him to forward to the Russian government, arguing that it was liable for the financial losses they had incurred. Sir Charles Hardinge, British ambassador at St Petersburg, duly wrote – on behalf of the Baku Russian Petroleum Co., the Russian Petroleum and Liquid Fuel Co. and the European Petroleum Co. – to Russia's Foreign Minister, Count Vladimir Lamsdorf, stating that

> the local and provincial Authorities at Baku and Tiflis were appealed to for assistance at the commencement of the year but took no notice of the warning. Two days before the outbreak of August last a deputation waited on the Governor General of Baku and warned him of the imminence of a renewed massacre. This warning was treated lightly by General Fadeieff and apparently disregarded since one

thousand men of a total contingent of seventeen hundred who were quartered at Baku and on the oil-fields were withdrawn on the very morning of the outbreak . . . It was not until a considerable time after the commencement of the massacre that reinforcements began to arrive, and when they did arrive at least a week elapsed before they were dispatched to the oil-fields for the purpose of maintaining order. Even then the dispositions taken were so inadequate that although the massacres began on the 19th of August it was not till a week ago, or after a delay of fully five weeks, that regular work could be resumed on the oil-fields. Under these circumstances it is evident that the Companies have a prima facie case in maintaining that their losses are directly due to the neglect of the local and provincial Authorities.

Total losses sustained by these three companies, wrote Hardinge, were over 2 million roubles as the result of destruction of property and 3 million roubles for the loss of production. He continued,

One Company alone has paid in royalties and taxes in the past four years nearly two million Roubles although no return on their investments has been made to the share-holders since September 1901. Owing to recent events the operations of the Petroleum Companies are seriously crippled and the confidence of English investors has been so shaken that there is now no possibility of raising capital in London for investment in the Russian oil business.[65]

Routed through Hardinge, the Foreign Office received the reply of the Minister of Finance, Count Vladimir Kokovtsov, who retorted that to grant such levels of compensation in the current circumstances 'would be like placing the Russian Government on the brink of an unfathomable abyss'. In any case, he argued, it was not international custom for a government to be so liable for outbreaks of disorder, and he was rejecting similar requests from other countries. Furthermore, he was of the view that these companies had made 'enormous profits . . . and that these gains had been taken from the pockets of the Russian people'. Nevertheless, he was 'ready to give them a loan on easy terms in order to facilitate the renewal of the oil industry' and would 'grant indirect compensation . . . by facilitating the acquisition by them of new oil-lands now in the possession of the Government'.[66] Two months later, the acting ambassador to Russia, Cecil Spring-Rice, wrote to Lansdowne that three British-owned oil companies – Baku Russian Petroleum Co., the Russian Petroleum and Liquid Fuel Co. and the Schibaieff Co. – had settled with the Finance Minister for a Russian loan of 'twenty million roubles (two million sterling)' at 5 per cent interest over about four years, adding, 'In view of the agitation with regard to the supposed favours accorded to foreign shareholders ... I do not think it advisable to approach the Russian Government, at any rate for the present, with a view to obtaining discriminatory treatment for British capital.'[67]

Domestically, much of Russia's industry, which had come to rely on plentiful supplies of cheap fuel oil from the Caucasus, returned to burning coal for which there now existed a domestic source in the Donetz region. Exports of kerosene plummeted to less than half their previous levels and Russian lamp oil began to lose its central role in world markets. The largest oil companies in Russia were, however, able to protect themselves by means of cartel agreements.[68] The Grozny oilfield in the North Caucasus – where around half of the investment was British – was relatively unaffected by the unrest, and oil companies operating there looked forward to profiting from the increase in oil prices resulting from the sudden decline in output from Baku; however, while there had been no loss of life at Grozny due to political or ethnic violence, six labourers died when a gusher struck by the Spies Petroleum Co. caught fire.[69]

The world's other oil exporting countries all stood to benefit from the reduced competition, particularly Romania, whose oil industry was seeing an influx of American and German investment.[70] 'Russia's Weakness the Standard's Opportunity', proclaimed *Petroleum World*,[71] which was soon reporting, in consequence, an 'increased and almost sensational demand for American oil which has arisen in various parts of Europe, and even in Russia itself'.[72]

Russia had entered a period of recurrent political and economic instability, and, while American oil production in particular continued to soar ever higher, it would be several decades before Russian production regained even its 1900 levels. As Karl Hagelin of Nobel Brothers told a meeting of Baku producers,

Were it only a matter of a repetition of the Armenian-Tartar outbreak we would probably have restarted our oilfields. We, however, foresee trouble from another source. What we fear are those starving masses, without distinction of race, who at times of anarchy like the present fall easy prey to the teachings of revolutionary agitators. That attempts to start a revolution are being made is an open secret.[73]

Indeed, twenty years later Lenin would look back on the upheavals of 1905 – which in June included the mutiny on the Black Sea battleship *Potemkin* – as Russia's revolutionary 'dress rehearsal'.[74]

REVOLUTION IN PERSIA, AND THE PLAIN OF OIL

The political ferment in the Caucasus resonated among Persians to the south, thousands of whom had for years travelled north – across the border bisecting the once unified province of Azerbaijan – to find work in Baku's oil industry, a migration that included the movement of political activists and the distribution of political literature.[75] Both populations were becoming increasingly disenchanted with their corrupt, bankrupt and autocratic rulers, but the situation in Persia was further complicated by the intense geopolitical tussle between Britain and Russia for influence over the country, which was pivotal in their wider imperial contest for control over the Persian Gulf and the western approaches to India.[76]

To strengthen its influence in the south, Britain worked to spread its economic presence from the head of the Gulf at Mohammerah, up the Karun River and on to Isfahan. This entailed building up relationships with the tribes of the southwest – most importantly the Bakhtiari – who operated almost autonomously of the weak Persian government, to the extent that the region became virtually a British protectorate. In late 1905, following the disappointing results of drilling at Chiah Surkh in Zohab province, the Burmah Oil Co. shifted its explorations southwards to the Bakhtiaris' winter pasture area near the head of the Karun River in Khuzistan, southeast of Shuster and about 150 miles north of the coast (see Fig. 25); one of the drilling sites was near the temple ruins of Masjid i Suleiman in the Maidan i Naftun, or 'Plain of Oil', in the vicinity of well-known bitumen seeps.[77]

In mid-October the consul general at Isfahan, J.R. Preece, conducted negotiations with the Bakhtiari chiefs on behalf of Burmah's local subsidiary, the Concessions Syndicate Ltd., in a village in the foothills of the Zagros Mountains. The Bakhtiari demanded that they should become partners in the enterprise and receive 20 per cent of the profits, Preece recording that they had

> great faith in the capacity of their oil-bearing country; they think that Baku will be nothing to it in comparison. They describe the oil-belt as being about 100 miles long by about 8 miles to 12 miles broad ... There is one place known as Maidan-i-Man Naftan (the plain of much naphtha), which is full of oil according to them.[78]

At the insistence of William D'Arcy, the originator of the Persian oil enterprise, Preece negotiated the Bakhtiari down to a 3 per cent share, although their chief negotiator, Ali Quli Khan Sardar Asad, warned the other khans, 'If you want to sign this Agreement do so, but you are creating a very dangerous situation for yourselves and the tribes.' Preece recorded his surprise that they had settled for 'next to nothing', while acknowledging that they 'have but the very crudest ideas of Companies, shares, and such like things'.[79] Although the central government in Tehran vehemently opposed these independent negotiations as a violation of its sovereignty, the British nevertheless accorded the Bakhtiari khans a 3 per cent share in the Concessions Syndicate plus £2,000 per year, in return for the rights to drill, build roads and lay pipelines, and for the protection of the company's personnel and installations.[80]

Just as Sardar Asad had predicted, the Bakhtiari Agreement soon began to falter amid payment disputes and feuding between the khans. Furthermore, the Agreement had sown discord among the Bakhtiari population more generally. Since only the Haft Lang khans of the Zagros foothills were party to the Agreement, the other Bakhtiari tribes were aggrieved that rights over their traditional communal lands had been signed away for private profit. The khans who had signed the Agreement were also resented by many of their own tribal subjects for having merely enriched themselves and sold out to the

British; meanwhile, those tribesmen who became lowly labourers for the company protested at the dangerous working conditions and poor wages.[81] In December 1905 several of the company's senior staff, including chief engineer George Reynolds, were harassed and shot at; Reynolds wrote to one of the khans, 'You will see that it is impossible to work among the wild men of these parts without a proper authority from you, and as the delay in communicating with you is very great, I must beg that you will forthwith send such orders in writing as being respected will absolutely put an end to all behaviour such as I have described.'[82]

In early 1906 Sardar Asad led a delegation to the British legation in Tehran to protest that he and his fellow khans had been misled and cheated by the company. However, in Preece's opinion the 'agreement was made by the Chiefs with their eyes wide open' and their complaints were an expression of the 'greed, dishonesty, and cupidity of the Bakhtiari'.[83] The sometimes violent altercations on the ground continued, periodically interrupting the company's operations, and the dispute escalated when, in October, the British consul at Ahwaz, Captain David Lorimer, disregarded the khans' instruction that new drilling be stopped until an agreement on the payments for the guards they were providing had been renegotiated (see Fig. 20). Lorimer ordered Reynolds to continue drilling at Maidan i Naftun and 'to engage for himself a few guards', as 'the Bakhtiaris must not be allowed to think they can stop the work of the Company.'[84] The Foreign Office became concerned that the Russians might exploit these rifts: Preece informed the Foreign Secretary, Sir Edward Grey, that it was 'the payment of the Guards which apparently is the real crux of the matter', and a subsequent additional payment of £500 appeared to settle the matter. D'Arcy was now optimistic that the issues had been resolved:

> As unreliable as the Bakhtiaris are, they are not such fools as to attempt to kill the goose that lays the golden egg, and the increased amount that they have agreed to take and are presumably satisfied with, will, let us hope cause them to continue in this frame of mind until Reynolds gets oil – and I do not think this is very far off ...[85]

Nevertheless, disputes and security concerns continued. In late 1907 Lorimer arrested two tribesmen suspected of attacking a company driller and 'had them tied to the carts of the Oil Syndicate, and well beaten in public', and had them 'ejected from the village with their families', in the hope that this would have what he called a 'moral effect' on the local population.[86] Based on Lorimer's advice, the Foreign Office now felt it necessary – in the face of protests from the Persian government and despite the risk of provoking a Russian reaction – to deploy a gunboat with twenty Bengal Lancers up the Karun, 'with the object of affording some assistance to . . . the Oil Syndicate in view of the importance attached by His Majesty's Government to the maintenance of British enterprise in South West Persia'.[87]

Simultaneously, from December 1905 Persia had entered a period of revolutionary turmoil. After mounting protests, Mozaffar al-Din Shah conceded the formation of an elected assembly, the *majlis*. This had become an immediate cause of anxiety for D'Arcy as one of its first initiatives was to establish a special commission to investigate the highly secretive oil concession. Given the deep unpopularity of the monarchy and its history of enriching itself by selling large concession rights to foreigners, the financial and operating terms of D'Arcy's concession might be exposed, and even its validity brought into question – which may have been a factor in Burmah Oil's decision at this time to renegotiate its contract with D'Arcy. The existence of the *majlis* itself, however, was put in jeopardy soon after its formation, following the death of Mozaffar al-Din in January 1907 and the accession to the throne of his more reactionary son, Mohammad Ali.[88]

Since both Russia and Britain were experiencing crises of imperial overstretch, they each supported a political compromise in Persia that would reduce the chances that the 'Great Game' would escalate into war. With the Anglo-Russian Convention of 1907, the two empires divided Persia into an agreed 'Russian' zone of influence in the north, a 'British' zone in the southeast, and a neutral or common zone in between (see Fig. 25), the British intention being to keep Russian influence away from the Afghan border and from the Persian Gulf – in other words, from India and the Gulf approach to it. There were protests in Britain against this alliance with the violently autocratic Tsar – *The Times* performed an abrupt U-turn in this respect – and the British minister in Tehran, Sir Cecil Spring-Rice, cautioned the Foreign Secretary,

You will be judged by your friends and associates; and if Russia, as is the case, is notoriously hostile to the patriotic movement in Persia, and if you make an agreement with Russia, the simple people here [in Persia] will take for granted that in your heart you think as Russia does . . . [Y]ou must be prepared to pay the cost and as far as I can judge part of the price is a great loss of popularity here which may react unfavourably in your position in other Mohamedan countries.[89]

Spring-Rice subsequently wrote to Grey, 'We are worse off than the Russians because we are not feared as they are, and because we are regarded as having betrayed the Persian people.'[90] He sought assurances from Grey that British interests in Persia would not be sacrificed:

I assume that the amendments desired by you are carried into effect and that Russia is debarred from eventually exercising control over the coast-line outside the Gulf and the frontier to Afghanistan and Baluchistan, and also that British Concessions, as the D'Arcy Oil Concession, the Kanikin telegraph line, and the Sultanabad-Dizful road . . . will be fully safeguarded . . .[91]

Grey would also be vulnerable, Spring-Rice warned, to attack from Liberal supporters of Persia's independence and its emerging Constitutionalist Movement: 'There would be a row at home . . . An anti-popular policy here [in Persia], in alliance with Russia, will not be a popular policy in England, unless the English are gone quite mad over the Russian Entente.'[92] Indeed, the developments prompted a Liberal MP, Henry Lynch, to form the Persia Committee to conduct intensive lobbying in Britain in support of the constitutionalists. Lynch's family firm, Lynch Bros., ran most of the Karun River trade in alliance with the Bakhtiari, with a subsidy from the British government, and D'Arcy's Concessions Syndicate relied on Lynch Bros. for its transport.[93]

However, Sir Arthur Nicolson at the Foreign Office was in favour of the Anglo-Russian Convention:

[T]he chief advantage . . . is that we keep Russia at a distance from our land frontiers . . . This is important, as on land she might conceivably be stronger than we are, and could cause us serious embarrassments. But as regards the Gulf . . . our position is thoroughly assured so long as we retain our sea supremacy: and if we lose our sea supremacy we lose our Empire.[94]

For Grey, the Convention was aimed at improving relations with Russia generally:

> The policy of the agreement is to begin an understanding with Russia, which may gradually lead to good relations in European questions also and remove from her policy designs upon the Indian frontier either as an end in themselves or as a means of bringing pressure to bear upon us to overcome our opposition to her elsewhere.[95]

The Convention was also motivated by the shared desire of the British and Russian governments to contain the rising power of Germany which was, with some success, presenting itself abroad as a friendly alternative and counterforce to the established European colonial powers. German shipping and trade was beginning to break into Britain's monopoly in the Persian Gulf; a German syndicate had been granted rights by Turkey to construct a railway to Baghdad, with extensions eastwards to the Persian border and southwards to the Gulf, and including oil drilling rights along the way, a development that was now being seen as a major emerging geopolitical threat to both British and Russian interests in the region; and there were concerns at the Foreign Office that increasing German influence over the Persian government might enable the Germans to muscle in on D'Arcy's oil concession.[96] In his eagerness, therefore, to curry favour with Russia, Grey even signalled that Britain might consider relaxing its long-held policy of denying the Russian Navy passage from the Black Sea, through the Turkish Straits, to the Mediterranean.[97]

D'Arcy seemed optimistic that oil would soon be struck near the Karun River in the Convention's neutral zone; the commander of the detachment of Bengal Lancers, Lieutenant Arnold Wilson, recalled,

> Two rigs pounded night and day with percussion tools that are regarded to-day [1941] as curiosities: everything had to come from England, India, or America to Mohammerah, thence by river steamer to Ahwaz, thence after transhipment by tramway, to a point on the Upper Karun by a little sternwheeler, the *Shushan*, owned by Lynch

Brothers, and thence by mule back or, after about June 1908, by a cart-road over the Tulkhaiyat (Tailor's Peak) to Masjid-i-Sulaiman. As many as 900 mules were in use in the following year – motors had not yet been acclimatized.[98]

However, Burmah Oil became increasingly reluctant to continue funding the drilling. It already had ample supplies in Burma and it seemed folly to burden itself with a possible excess of crude, especially since there appeared to have been a change of heart at the Admiralty towards fuel oil: Burmah's deputy chairman, Charles W. Wallace, wrote, 'the professional engineers at the Admiralty are advising the head not to commit themselves too deeply to Fuel Oil, as in their opinion the next few years will probably see a development in internal combustion engines worked by coal gas.'[99]

When, in late April 1908, Lieutenant Wilson heard rumours that Burmah Oil was about to give up drilling at Masjid i Suleiman, he immediately wrote to Britain's Political Resident in the Gulf, Lieutenant Colonel Sir Percy Cox, that this seemed

a short-sighted decision, which may involve the cancellation of the D'Arcy Concession. The search for oil will quite certainly be continued, but by the Germans or by one of Rockefeller's Companies. In neither case will any difficulty be found in raising capital. It amazes me that the directors of the Concession Syndicate Ltd. should be in a position to risk the complete loss of a concession covering all oil deposits over the greater part of Persia, without consultation with the F.O. and without telling you or the Minister or the Government of India. This is the 'neutral' zone. What is to stop a Russian controlled oil company from getting a new concession from Persia? What is to prevent C.S. Ltd. from selling D'Arcy's rights to an American or German Company? The directors of C.S. Ltd. are Scotsmen of the hard-headed, short-sighted sort who would not hesitate to do so if they saw a profit, and there is nothing to prevent them. Reynolds thinks they are trying to do so, though the Burma Oil Company took a financial interest in C.S. Ltd. because they did not want a rival company within reach of India, where they have something like a monopoly at present. It may be too late to prevent a stoppage of work here, but not too late to stop the dismantling of plant and the dispersal of trained staff. I am tired of working here for these stay-at-home business men who in all the years they have had the concession have never once come near it: — and — [other British firms operating in the Persian Gulf] are not much better. They have all the vices of absentee landlords.

Cannot Government be moved to prevent these faint-hearted merchants, masquerading in top hats as pioneers of Empire, from losing what may be a great asset? I know that the Government of India regard the prospects of oil in SW. Persia as very poor, because they once sent one geologist to Persia (but not here) for a few weeks. Such obiter dicta should not carry weight ... Lord Curzon, who mentioned this place [Masjid-i-Sulaiman] as petroliferous in his book, should know what is afoot. He, at least, would understand.[100]

Then, on 26 May, Wilson hurried to Shuster to cable a cryptic message to Captain Lorimer, referring to two lines from the Psalms, that Wilson later rendered as 'That he may bring out of the earth oil to make him a cheerful countenance' and 'the flint stone into a springing well': 'As I had no Telegraph Code I wired to Lorimer: "See Psalm 104 verse 15 third sentence and Psalm 114 verse 8 second sentence". This told him the news and in the circumstances was quite as effective as a cipher.'[101] The Concessions Syndicate had struck oil at a depth of 1,180 feet, flowing at nearly 300 barrels per day (see Fig. 21). 'It rose 50 feet or so above the top of the rig,' Wilson recalled, 'smothering the drillers and their devoted Persian staffs who were nearly suffocated by the accompanying gas'.[102] He wrote to his father,

> It is a great event: it remains to be seen whether the output will justify a pipe line to the coast, without which the field cannot be developed. It will provide all our ships east of Suez with fuel: it will strengthen British influence in these parts. It will make us less dependent on foreign-owned oilfields: it will be some reward to those who have ventured such great sums as have been spent. I hope it will mean some financial reward to the Engineers who have persevered so long, in spite of their wretched top-hatted directors in Glasgow, in this inhospitable climate.[103]

This was the first major industrial-scale oil discovery in the Middle East; but to the British Foreign Office its significance was still primarily geostrategic. It was 'excellent news for our interests in south-western

Persia'.[104] As Captain Lorimer argued, 'the greater the aggregation of our interests in this quarter, the more possible it is for us to exert pressures on the Persian Government.'[105] The Foreign Office added, 'it is most desirable from the point of view of British interests, both political and commercial' that Burmah Oil should develop the concession, otherwise 'there is every likelihood that the business will be secured and worked by subjects of a foreign Power.'[106] D'Arcy was now able to use the British government's interest in his enterprise as a bargaining chip. When the Foreign Office suggested that he ought to pay for the protection provided by the Indian Army, he replied that his enterprise 'should be looked upon with favour, if not indulgence, by the British Government, since it may in the near future become a source of valuable fuel oil for our navy'.[107]

When, a year later in April 1909, Burmah Oil formed the Bakhtiari Oil Co. and the Anglo-Persian Oil Co. (forerunner of British Petroleum, or BP) to exploit the region's oilfield, British officials were anxious to ensure that their political objectives would not be jeopardized. Anglo-Persian was concealing from the Bakhtiari the true value of its concession, Wallace having been told by Burmah's main petroleum consultant, the prominent Boverton Redwood, that Persia would become 'in all human probability the biggest oil producer in the world – bar none';[108] Lorimer feared that Anglo-Persian might deprive the Bakhtiari of their due return by shifting profits between subsidiaries; and D'Arcy was being highly secretive and evasive in his dealings also with the Persian government. In late 1909 one Foreign Office official was 'still in some doubt whether the Company meant to swindle the Khans or not', and another argued that if the Bakhtiari were to be cheated, the British government had 'not only the right but the duty to intervene for their protection, not less on moral grounds than on political grounds, since their friendship is important to us from a general point of view'.[109]

Indeed, by mid-1909 the Bakhtiari had become even more significant for British interests in Persia. Soon after the Masjid i Suleiman oil strike, the new Shah launched a coup with the help of the Russian-officered Persian Cossacks against the Constitutional Movement that, in the volatile conditions, ignited civil war.[110] D'Arcy wrote to Preece in Isfahan, 'Things at Teheran look bad and will not I should imagine be better until the Shah is bombed or dethroned.'[111] The British-allied Bakhtiari now saw a chance to participate in toppling the central government and to assert their power on the national scene. However, the British officially stood on the sidelines. Explicit British support for the constitutionalists might undermine their nationalist credentials, as many Persians saw the Anglo-Russian Convention as an imperialist carve-up of their country. More importantly, supporting the constitutionalists risked harming relations with Russia – which the Foreign Office under Grey's leadership was keen to avoid at almost any cost – and might give the Russians an excuse to intervene directly on behalf of the Shah. Furthermore, the British did not wish to appear to be condoning incipient nationalist-constitutionalist movements in regions more fully under British control such as Egypt, or in India itself. As Sir Cecil Spring-Rice put it,

> The attitude of Russia towards the popular party here is as well grounded as our own toward similar movements in Egypt and India. They threaten our interests and we naturally object. If the popular party succeeds here it will be quite impossible for Russia to maintain her control, which she exercises through a certain number of priests and statesmen and especially the Shah ... If we take part with the popular party against Russia we shall be doing what the Germans are accused of doing in Egypt and Morocco. We shall be also giving a very bad example to other nations who would be justified in playing the same game in our gardens.

On the other hand, he continued,

> if we take sides against the popular party we cannot hope that the Mussulmans all over the world who are watching affairs here with the greatest interest will fail to take note. An association of England and France in Morocco and England and Russia in Persia both directed against a popular Mussulman movement; can we hope that this will pass unnoticed? ... It seems therefore that our cooperation with Russia should *not* include active cooperation against the patriotic movement.[112]

Nevertheless, in July 1909, as the Persian government began to protest strongly at what they considered to be D'Arcy's duplicitous and illegal behaviour over the oil concession, his hopes regarding political developments in the country were realized. Bakhtiari forces – funded both directly and indirectly by Anglo-Persian and Lynch Bros. – seized control of Isfahan and then, with other groups from the north, took over Tehran. The Shah abdicated and parliamentary government was re-established, in which the Bakhtiari would play a leading role.[113]

Burmah's initial draft of Anglo-Persian's prospectus emphasized its strong relationship with the Admiralty, but Redwood, also an adviser to them, conveyed their 'strong objections' to this being made public and that they 'would go so far as to deny the statement' and claim that they had no record of this; the wording was toned down, advertising that Anglo-Persian's heavy crude would 'be of immense benefit to the British Navy, and substantial contracts for Fuel Oil may be confidently looked for from the Admiralty'.[114] Its share issue in April 1909 was over-subscribed, enabling it to construct a pipeline and pumping stations to send the crude oil 140 miles over two ranges of hills and along the Karun River to its proposed refinery on the island of Abadan, south of Mohammerah (see Fig. 25). The company employed around a thousand nomadic tribesmen who were provided with miserable accommodation, subjected to abusive and often violent treatment from their European employers, and were expected to continue their heavy manual labour regardless of the season and through the fierce summer heat (see Figs. 22 and 23). This was in stark contrast to their traditional lifestyle, as described by an engineer who made a study of Anglo-Persian's operations in the region some years later:

> The Bakhtiari follow the grass. As summer approaches to scorch the grass on the desert plains, they migrate with their flocks and herds, over the foothills, across the bridgeless rivers and over the snow-clad mountain ranges, to the high lands where grass is then to be found. In the late autumn, when the grass dies down on the high plateau or in the mountain vales, they trek back to the low-lying plains, where, in the mild winter, they can find enough grass on the desert to feed their flocks. They live in tents or in rude shelters constructed of sun-baked mud or of the loose sandstone and shale of the hills. Their wants are few and the hardships of such a life are to them scarcely irksome. Of money they have little need and such exchange as they require is done largely by barter . . . The nomadic instinct is not easily extinguished.[115]

As reported in *The Times*, however,

> Recent times have brought to the Bakhtiaris another source of revenue . . .
> Eight wells have been drilled to a depth varying between 300ft. and
> 1,600ft., and in some of them the pressure is so great that the oil spurts
> up between the ground and the iron casing. A large quantity now runs
> to waste, and in time of flood passes along the local streams and floats
> down the Karun in ugly brown floes.[116]

The newly emboldened Bakhtiari khans demanded £40,000 from
Anglo-Persian for rights of way for the pipeline while, for the refinery
site, the company paid £1,000 per year in rent to the Sheikh of
Mohammerah, the most powerful Arab ruler of the region, who controlled the mouth of the Karun at the Shatt-al-Arab. And the British
were now more deeply entrenched in southwest Persia.[117]

BRITISH SUPREMACY OVER BAGHDAD

In early 1908 the Persia correspondent for *The Times* told its readers,

> The Germans, as you know, are extremely clever at coming on the
> stage when the audience does not expect them, and at adding life and
> movement to the play by giving scenes not on the original program,
> such as the Morocco Crisis, Adventures in China, and Wanderings in
> Mesopotamia.[118]

For almost a century Britain had proclaimed its military and political
supremacy over the Persian Gulf and its coastal periphery in order to
protect this channel of trade and communications with India. However, from the early 1860s parts of Persia and the Ottoman Empire
extending further inland had suddenly taken on greater importance.
During both the Anglo-Persian War of 1856–7 – in which Britain fought
to secure Afghanistan as a buffer state against Russian encroachment
onto India's north-western frontier – and the Indian Mutiny that immediately followed, the British had been stymied by the extreme slowness
of communications with its distant, imperial theatres of war. Consequently, Britain pushed for the construction of unbroken telegraphic
communications with India by extending the existing European

network eastwards from Constantinople. After the Ottoman government, in 1861, had completed the line to Baghdad, from here two separate cables were laid to India, as the British considered both vulnerable to sabotage. One cable ran directly eastwards through Persia, via Tehran; the other, completed in 1865, took a southwards turn through lower Mesopotamia and via Basra. The Ottoman government hoped that the Mesopotamian cable would help to coordinate its attempts to impose its sovereignty over the myriad tribes of this outlying region. By contrast, the British, whose primary aim here was the protection of the telegraph line, were open to winning the allegiance of these tribes regardless of the nature of their relationship with Constantinople.[119]

Thus, in addition to the strategic importance to Britain of the headwaters of the Persian Gulf and its associated, long-established, trading interests up the Tigris and Euphrates into Mesopotamia and beyond, the British Empire now relied on a telegraph network running through the region. The Foreign Secretary, Lord Salisbury, declared in 1878, 'whatever Party be in power, we feel convinced, knowing what the spirit of the people of the country is, that they will never tolerate that Russian influence shall be supreme in the valleys of the Euphrates and the Tigris.'[120]

Similarly, in 1892 the future Viceroy of India, George Curzon, considered Russia to be Britain's primary rival in the Persian Gulf:

> A Russian port in the Persian Gulf, that dear dream of so many a patriot from the Neva or the Volga, would, even in times of peace ... shake the delicate equilibrium so laboriously established ... I should regard the concession of a port upon the Persian Gulf to Russia by any power as a deliberate insult to Great Britain, as a wanton rupture of the status quo, and as an intentional provocation to war ...[121]

Curzon argued, further, that it was just as important to deny the Russians any influence at Baghdad as around the Persian Gulf itself: 'Baghdad, in fine, falls under the category of the Gulf ports, and must be included in the zone of indisputable British supremacy.'[122]

However, from the turn of the century the British government began to regard Germany as being as much of a threat in the region as Russia. This development had its roots not only in Germany's rapid industrialization and in the increasing size of its navy, but also in the attempt by the crumbling Ottoman regime to shore up its sovereignty against the economic and political incursions of Britain and France by seeking alternative partners in trade and investment. Turkey had periodically lost territory to an expanding Russia; it resented Britain's declared de facto sovereignty over parts of Turkish Arabia and its occupation of Cyprus in 1878 and Egypt in 1882; and it felt strangled by the Anglo-French control over its national debt and state budget. When, therefore, from the late 1880s Turkey began to develop its railway network eastwards – with the aims of stimulating large-scale economic development and extending its administrative control and projection of military power – it turned increasingly to German capital and expertise.[123]

In 1846 the influential German economist Friedrich List had envisaged a railway system stretching from Ostende, via Munich, Constantinople, Mesopotamia and the Persian Gulf, to Bombay; German industrialists had funded a trip made by List to London to propose to British parliamentarians that the scheme be implemented as a joint Anglo-German project, but the initiative came to nothing.[124] Now, in 1888, the Anatolian Railway Co., controlled by the German Deutsche Bank, gained a preliminary concession for the construction of a railway from Constantinople to Ankara – with the speculative, distant prospect of extending the main line all the way to the Persian Gulf.[125]

The immediate commercial viability of the Anatolian Railway system was linked largely to the agricultural potential of the region – particularly if facilitated by irrigation works – and to the export route it would open up for European manufactured goods. It also came with onerous 'kilometric guarantees' from the Turkish state, guaranteeing a minimum income to the Anatolian Railway Co. from completed sections of the line regardless of traffic, partly because of the railway's military value to Turkey for the transport of troops across the empire.[126] As was typical for railway concessions of the time, Deutsche Bank's concession also came with promises that it would be given first consideration regarding future mineral exploitation rights along the route.[127]

Natural bitumen pits and petroleum springs in Mesopotamia had been known about, and made use of, since early antiquity – as a waterproofing agent, as cement or glue, for road surfacing, as a medicine, and as a source of heat and light when burned – and during the early nineteenth century European travellers had written of its continued use.[128] In 1836, for example, the East India Company's Resident at Baghdad, Claudius Rich, wrote of the commercial exploitation of naphtha springs near Kifri and mentioned 'Baba Goorgoor ... a spot three miles from Kerkook, where, in a little circular plain, white with naphtha, flames of fire issue from many places'.[129] More detailed accounts of the region's petroleum resources began with a geological study by William Loftus published in 1855. As a member of an Anglo-Russian-Turkish commission attempting to define the Turkish-Persian border, Loftus travelled from Mohammerah, at the head of the Persian Gulf, 600 miles northwards to Mount Ararat: 'Our route,' he wrote, 'lay through the midst of those wild mountain tribes who from time immemorial have never acknowledged law or subjection, and who regarded our movements with infinite distrust.'[130] In 1871 a French consular report described the gathering of crude oil from springs in the vicinity of Mendeli near the Persian border, how it was conveyed in leather bags by mules to the town for simple refining, and how the refined oil was then transported in copper and tin vessels to Baghdad, about 80 miles to the west (see Fig. 26).[131]

There were also several known signs of petroleum in northeastern Turkey in the region of the headwaters of the Euphrates River. In 1882, for example, the British consul for Kurdistan reported an oil spring near Pulk, about 15 miles east of Erzingan (Erzincan): 'This spring is held in great veneration by the inhabitants, both Kurdish and Armenian. Once a year they assemble around it and sacrifice lambs to the protecting saint, using the oil for curing rheumatism.'[132]

During the early and mid-nineteenth century in Mesopotamia, some of the oil wells being worked were owned by a village, some privately by families, and some by the state.[133] However, in 1888 the Turkish government began to indicate that it would give favourable consideration to Deutsche Bank when granting future oil concessions, and the Minister of the Civil List, or 'Privy Purse', advised Sultan Abdul Hamid II to issue decrees to bring the oil exploitation rights and revenues in the provinces of Mosul and Baghdad under his personal control.[134] As the Sultan received an effusive report on the petroleum potential of the Mosul region from one of his officials, the British and the Germans requested his permission to allow teams of archaeologists to investigate the region – which, he soon learned, they were both using as cover to carry out oil prospecting.[135] The Minister of the Civil List was a close friend of the prominent Gulbenkian family, who were steeped in the Baku oil trade; and in 1892 he asked the young Calouste Gulbenkian, who had studied engineering and the petroleum industry at university in London, to produce a report on the prospects for a Mesopotamian petroleum industry, for which Gulbenkian drew much of his evidence from engineers of the Anatolian Railway Co. who had been in Mesopotamia.[136] That same year, the American chargé d'affaires in Istanbul wrote to John D. Rockefeller suggesting that by drawing on Mesopotamian oil it might be possible for Standard Oil to displace Russian oil from the eastern markets.[137]

Further studies of Mesopotamia's petroleum were conducted in the 1890s, including an extensive one by Captain F.R. Maunsell, published in the British *Geographical Journal* in 1897.[138] In Baghdad, Maunsell reported, domestically produced lamp oil had been displaced by higher-quality American and Russian oil; but, he concluded, Mesopotamia appeared to be fertile ground for a lucrative, modern oil industry:

> [T]he most important section of the Mesopotamian petroleum field extends from Mosul to Mendali, a distance of 220 miles, with a breadth of about 60 miles. The Tigris navigation offers a natural outlet towards the Persian gulf, and is now regularly used for traffic all the year round between Busra and Baghdad. A line of railway from Baghdad through Kifri, Tuz Khurmatli, and Kirkuk to Mosul could bring the produce from these places to the river. If such a line were extended from Mosul to the Mediterranean, communication towards both seas would be complete. Possibly one result of the present political troubles of Turkey may be a greater facility in obtaining concessions to develop some of these remarkable mineral riches on modern lines, and, if properly explored, there is no doubt but that the Mesopotamian petroleum field might be made [to] yield results of the greatest commercial importance.[139]

In late 1900 the Sultan sent a Civil List mining engineer, Paul Grosskopf, as part of a broader technical commission, to assess the petroleum prospects of Mosul, Kirkuk and Baghdad, and in the commission's report of October 1901 Grosskopf wrote,

> There are rich oil resources around 15 kilometers north of Kirkuk. The quantity and quality of petroleum, which is extracted and cleaned in primitive forms, is not less than Baku oil resources. It is possible to make effective use of these resources with the railways, which will be built ... I have visited several oil wells all over the world, both before and after exploitation ... but none of these have proved to be so rich ... I have never seen the like as yet ... I respectfully submit my opinion with high precision that the petroleum mines [fields] on the banks of the Euphrates and Tigris will be one of the most prolific sources in the world.[140]

Meanwhile, the Anatolian Railway Co. – financed by German, French and some British capital – had by 1893 completed its line to Ankara; by 1896 the company had extended it to Konya, and in 1898 it applied for a concession to extend the line all the way to the Persian Gulf.[141] The German ambassador at Constantinople wrote in a despatch to the Chancellor,

> one thing we must claim for ourselves, and that is the connecting up of the present sphere of interests of the Anatolian Railway with the river districts of the Tigris and Euphrates, and so on to the Persian Gulf ... Whether or not the Sultan's wish to extend the Anatolian Railway to Baghdad is 'music of the future' ... no one else should get in front of us here.[142]

'Without question,' Kaiser Wilhelm II noted in the margin.[143] Diplomatic ties between Germany and Turkey had strengthened since the Kaiser had visited Istanbul in 1889. Now, in 1898, with great pomp and ceremony, he visited the Sultan again – despite the latter's poor reputation in Europe for the massacre of Christian Armenians – and the following year a preliminary concession was signed to extend the Anatolian Railway to the Persian Gulf, including mineral rights along the route, defeating British and French proposals for railways linking the Gulf with the Mediterranean.[144]

Although Russia was still seen as Britain's primary threat around the Persian Gulf, Lord Curzon, now Viceroy of India, began to warn also of German economic and political incursions into Mesopotamia and Persia.[145] Curzon therefore moved, in early 1899, to avert the possibility that any other nation might obtain rights to the best location for a port at the head of the Gulf – and thus a possible terminus for the Baghdad Railway – by striking a secret, exclusive agreement with the Sheikh of Kuwait, paying him 15,000 Indian rupees in return for rendering his territory a virtual British protectorate.[146] Turkey, however, claimed sovereignty over Kuwait and regarded the sheikh as merely a local chieftain; and it viewed Britain's relationship with him as yet another infringement of its sovereignty, similar to Britain's agreements with other tribal leaders of the Gulf region.[147] As Curzon had written of Oman, for example, it 'may, indeed, be justifiably regarded as a British dependency. We subsidise its ruler; we dictate its policy; we should tolerate no alien interference.'[148] When Germany, indeed, approached Sheikh Mubarak for a Kuwaiti rail terminus and port, Turkish attempts to impose sovereignty over the sheikh were repelled by British naval forces.[149] As *The Times* would interpret these events a decade later, 'Germany has been, in effect, warned off.'[150] In the Persian Gulf generally, wrote Foreign Secretary Lord Lansdowne in 1902, 'we shall resist to the utmost all attempts by other Powers to obtain a foothold on its shores for naval or military purposes. This, I take it, is the "bed rock" of our policy in the Gulf.'[151]

Meanwhile, in 1900 the British ambassador in Constantinople, Sir Nicholas O'Conor, noted one particular aspect of the Baghdad Railway project, that 'Discoveries of bitumen and naphtha would greatly increase the productiveness of the line,'[152] while a German report of 1901 mentioned 'innumerable natural petroleum springs', those at Gyara, about 15 miles down the Tigris from Mosul, creating pools and even lakes of petroleum (see Fig. 26).[153] In 1902 the influential German writer Paul Rohrbach produced a pamphlet, *Die Bagdadbahn*, promoting the Baghdad Railway as Germany's main route to economic expansion, and this was extensively quoted in several British Foreign Office memos in 1903.[154] Among the many economic opportunities accruing from the scheme, wrote Rohrbach,

> we ought to attach the greatest importance to the circumstance that the Baghdad Railway will pass close to the petroleum districts. The only

thing to be feared is that foreign gold and foreign speculators should succeed in securing a preferential right in the exploitation of Mesopotamian naphtha before any effective German initiative has been taken.[155]

As a correspondent for the *Morning Post* elaborated,

> Having fixed on the shortest and cheapest route for the main line, it was a simple matter to arrange branches to various spots away from the trunk railway by which the grain-growing districts might be tapped, and the right to develop the petroleum fields of Mesopotamia secured. It will be remarked that branches will eventually go to Erbil and Kifri, east of the Tigris, and to Hit on the Euphrates, all places where the surface indications of petroleum are most pronounced, while the branch to Khanikin gives the syndicate control over the Bagdad end of the great trade route of Western Persia. Too much importance can hardly be attached to this point, for it is plain that if oil can be produced in large quantities in Mesopotamia, where it can be led along the level floor of the great Mesopotamian valley to the Persian Gulf, or carried in tanks down the Tigris River, the possibilities in store for the German syndicate are enormous. And the expense of the exploitation is comparatively small. The various branches of the railway running to the oil centres can be constructed very cheaply because the conditions concerning the gauge and strength of the railway apply to the main line alone. In fact, the branches need not be built at all unless oil is first proved to exist in paying quantities. The mere tracing of the line on paper secures the mining rights to the syndicate.[156]

William D'Arcy, who had gained the huge oil concession in Persia in 1901 for British investors, now looked to Mesopotamia for an adjacent concession – the whole tract of oil, after all, extended either side of the Turkish-Persian border – and he sent his representative, Alfred Marriott, to Istanbul armed with thousands of pounds for the purpose.[157] Deutsche Bank now faced British competition for Mesopotamian oil concessions.

In March 1903, following several years of organizational and financial obstacles, the Baghdad Railway Convention was finally signed, incorporating the Baghdad Railway Co. under the auspices of the

Deutsche Bank and the Imperial Ottoman Bank, the latter dominated by French investment. Included in the concession were the right to build port facilities at Basra, the construction of a branch line from Baghdad to Khanikin on the Persian border, and navigation rights on the Tigris for the transport of materials; and it also formalized the earlier promise to Deutsche Bank of preferential petroleum rights along the route, extending 20 kilometres either side of the line.[158] Although the project was presented as an international undertaking, the French government was opposed. It feared for its economic and political position in Syria and its close ally, Russia, did not want to see Turkey becoming militarily stronger nearer the Caucasus, even though Turkey had already agreed to shift the proposed route southwards for just this reason.[159] The British government, by contrast, had assented to British investment in the Baghdad railway project. So long as Britain could control any section of the railway that reached the Persian Gulf itself, facilitating the movement of Ottoman troops by rail to Mesopotamia would deter a possible Russian military advance into the region; or, as Lansdowne put it diplomatically, internationalization of the project seemed 'best calculated to remove the international rivalries to which the construction of such a line was sure to give rise'.[160] However, the Germans appeared to want to retain majority control of the project, and in April 1903 the British government joined the French and Russians in opposition, following an outcry in the press against the prospect of British capital being invested in a project that might advance Germany's commercial position in Turkey at the expense of Britain's.[161]

In 1904 Deutsche Bank was granted the right to survey the oilfields of Mesopotamia with the option of commencing production within one year.[162] D'Arcy replaced the unsuccessful Marriott in Constantinople with H.E. Nichols, who proved more successful, and the British government became more actively interested in this aspect of the Baghdad railway project. Lansdowne instructed his ambassador to pass to Nichols a copy of Grosskopf's report on the petroleum indications along the projected railway's route that Maunsell – now the military attaché at the British Embassy in Istanbul – had paid his draughtsman to steal from the Baghdad Railway Co. Included in the report was the suggestion that even before the construction of the railway some of the oil might be conveyed out of the area by pipeline.[163] As Deutsche Bank prepared to send out a team to conduct a more thorough survey of the region's oil potential, Ambassador O'Conor at Constantinople forwarded to the British Foreign Office another report, *The Petroliferous Districts of Mesopotamia*, compiled by an embassy attaché, Captain Mark Sykes. Appearing to draw partly on Grosskopf's report, Sykes also proposed that while many oil deposits lay near the future route of the Baghdad railway, some could indeed be exploited by pipeline before its completion.[164] The railway, however, tended to be seen as a precondition for effective oil production; and a US consular report noted at the time that the dependency might go both ways: 'The importance of such a source of fuel to the railroad in a country entirely destitute of other fuel . . . is sufficiently apparent.'[165] In fact, in 1909 the Deutsche Bank-controlled Steaua Română would start supplying Romanian fuel oil to the Anatolian Railway – although the railway also relied on supplies from Standard Oil.[166]

British officials began to view the future of railway development, Tigris navigation rights, irrigation schemes and oil concessions in southern Mesopotamia as interdependent, and all as crucial to Britain's strategic presence in the area.[167] With the First Morocco Crisis of 1905–6, Germany became overtly singled out as a threat on the imperial stage and – since Britain, France and Russia all had large Muslim subject populations – as being inconveniently, and cynically, pro-Islamic.[168] The Committee of Imperial Defence concluded, 'it is important that England should have a share in the control of the extension of the Baghdad Railway to the Persian Gulf, with a view to insuring the effective neutralisation of the terminus.'[169] If Britain did not participate in at least the southern section of the Baghdad railway, O'Conor argued, it would lose the 'chance of developing the petroleum fields of Mesopotamia'.[170] *The Times* conveyed to its readers an ominous report in a French newspaper:

> [T]he Germans estimate that Anatolia, Syria, and Mesopotamia would furnish more corn than all Russia, and that Kerkuk would yield ten times as much naphtha as Baku . . . and cotton grown in Mesopotamia would take the place of American cotton in Germany . . . [It] would open up a magnificent field for German expansion when . . . direct railway communications is established between Hamburg and the Persian

Gulf [and] would furthermore give Germany a large share in the Indian market.[171]

However, Deutsche Bank's prospectors were equivocal on the economic viability of Mesopotamian petroleum. With the Baghdad Railway Co. facing great financial uncertainties, engineering challenges and political obstacles, Deutsche Bank prevaricated for so long over commencing exploratory drilling that in January 1907 Turkey's Minister of the Civil List declared that the terms of this aspect of the agreement had been broken, and the Germans seemed close to forfeiting their oil concession.[172] At the same time, the British authorities began to provide official support to D'Arcy's efforts to acquire oil rights, O'Conor writing to Foreign Secretary Edward Grey, 'there is every ground for believing that this concession, especially as Mesopotamia is developed by the extension of railway communications and irrigation works, will prove exceedingly valuable, and ... the creation of such important British interests in that country will greatly enhance our influence and general position.'[173]

The Baghdad Railway Committee, established by Grey, concluded in March that 'From a political point of view, the possibility of the shortest route to India being exclusively under the auspices of Germany gives rise to serious anxiety.'[174] Britain's strategic position at the head of the Gulf was seen to be in jeopardy, and the government formally acquired a lease on the Kuwait shore to thwart German designs on a terminus and port there, particularly since the German Hamburg-Amerika Line had just opened up a direct service to the Gulf.[175] Regarding a German port terminus at Kuwait, a specially convened defence committee concluded, 'This outlet, though at first commercial, might eventually be transformed into a strategic base, but by steps so gradual and clandestine as to render protests difficult or impracticable.'[176] In the analysis of the Assistant Quartermaster-General,

So long as we depend entirely upon sea-power for the preservation of our Empire, and remain ... a negligible quantity on land, our policy throughout the East must ... be merely the maintenance of the *status quo*. It may not be a very heroic or enterprising policy, but it is nevertheless one which is forced upon us by our comparative impotence on land.[177]

He voiced also the prevalent British fear that Germany might wield the balance of power by allying with Russia. The time was close, he thought, 'when Russia and Germany must gravitate towards each other'.[178] Consideration was given to establishing a British naval base in the area, perhaps at Basra; but the dangers of an attack from inland, of the harsh climate, and of overstretched military resources led Grey to discount the option, writing that it would involve the expenditure of 'millions a year in keeping a new frontier in a state of defence'.[179] However, the Anglo-Russian Convention, signed in August, signalled a reduced threat from Russia, at least, to British hegemony in and around the Persian Gulf; strategic policy at the head of the Gulf, therefore, would be based on the principle, as Grey put it, that 'commercial enterprise and political influence have gone together in these regions'.[180]

The intricate play of political forces was further complicated by the Young Turk revolution of 1908, motivated in large part by a common perception among Turks that their decadent monarchy was complicit in the dismemberment and exploitation of their empire by the European powers.[181] Britain's new ambassador in Constantinople, Sir Gerard Lowther, saw in the revolt a German-Jewish-Freemason conspiracy, writing to Sir Charles Hardinge at the Foreign Office,

[The Jew] seems to have entangled the pre-economic-minded Turk in his toils, and as Turkey happens to contain the places sacred to Israel, it is but natural that the Jew should strive to maintain a position of exclusive influence and utilize it for the furtherance of his ideals, viz. the ultimate creation of an autonomous Jewish state in Palestine or Babylonia ... Mesopotamia and Palestine are only ... the ultimate goals of the Jews. The immediate end for which they are working is the practically exclusive economic capture of Turkey and new enterprises in that country ... The Jew hates Russia and its Government, and the fact that England is now friendly to Russia has the effect of making the Jew to a certain extent anti-British in Turkey and Persia – a consideration to which the Germans are, I think, alive. The Jew can help the Young Turk with brains, business enterprise, his enormous influence in the press of Europe, and money in return for economic advantages and

the eventual realisation of the ideals of Israel, while the Young Turk
wants to regain and assert his national independence and get rid of the
tutelage of Europe, as part of a general Asiatic revival . . .[182]

In addition, new contenders for railway and oil concessions also
appeared on the scene.[183] The US State Department supported an
attempt by Admiral Colby M. Chester to acquire an ambitious concession for railway construction and oil production across eastern
Turkey.[184] The head of Deutsche Bank, Arthur von Gwinner, complained to the American ambassador in Berlin that the project would
compete with the Anatolian and Baghdad railways; furthermore,
Standard Oil was rumoured to be behind the project and, according to
the ambassador, Gwinner protested that this was a proposal not 'for
bona fide railroad development but a scheme for controlling certain
undeveloped oil fields in order to keep their product out of the market'.[185] This may simply have been a useful rumour for Deutsche Bank,
although a senior Standard Oil geologist did happen to be visiting the
region looking for possible foreign sources of oil.[186] Chester's project
eventually fell by the wayside, however, due to German opposition and
a distinct absence of British support.

At the same time, the Royal Dutch-Shell group began to manoeuvre
towards acquiring oil concessions in Turkey; and the French Rothschilds, major shareholders in the group, also expressed an interest.[187] A
number of French reports, commissioned by Ottoman officials,
expressed enthusiasm for greatly expanding on the current, small-scale
commercial exploitation of the oil in Mosul province. In 1908 Professor L.C. Tassart of France's *École des Mines* focused on the areas
around Qayyarah, Zakho and Kirkuk, and concluded that there 'certainly appear to be signs indicating a vast oil-bearing region . . . to
permit the hope particularly rich oilfields will be discovered there',
although their exploitation would depend on adequate transportation
access. The manager of the Rothschilds' oil interests now wrote to
Royal Dutch-Shell's negotiator Fred Lane, 'You know that the Mesopotamian matter has always been of keen interest to my chiefs, and in
their opinion this is now the time to move forward with this question.'[188] Lane and Sir Marcus Samuel, the head of Shell, approached the
British Foreign Office for its support, but their request was rejected on

the grounds that Royal Dutch-Shell was not British enough, and that it was 'a huge international combine for the control of the petroleum markets of Turkey and the world'.[189] By contrast, in early 1910 the Foreign Office was instructing Lowther to give his 'strong support' to D'Arcy's application and to tell the Turkish government that the British 'attach much importance' to the Mesopotamian oil concession;[190] the previous autumn, Hardinge had written to Sir Arthur Nicolson at the Foreign Office that Mesopotamia 'has always been the sphere of British influence and we intend to keep it so'.[191]

REVOLUTION IN MEXICO – 'POLICE THE SURROUNDING PREMISES'

Simultaneously with the 1908 Persian oil strike, British industrialist Weetman Pearson struck oil in Mexico in immensely larger quantities. Pearson benefited greatly from concerns felt throughout Mexico, and shared by President Porfirio Díaz, at the extent to which key sectors of the economy were run by large American corporations. As part of Díaz's move to counterbalance this with European investment, Pearson's family business, S. Pearson & Son, became a favoured contractor for major engineering projects in Mexico, and Pearson benefited from his close relationship with the authoritarian Díaz regime. In 1906 Pearson had been given a huge fifty-year oil concession covering much of the state of Veracruz in return for 3 per cent of the profits for Veracruz state and 7 per cent for the central government, with the proviso that the operating company should be incorporated in Mexico. American business interests became resentful at this perceived Díaz–Pearson conspiracy, particularly when, a year later, Díaz began to nationalize the largely US-run railways that were seen as having the potential – as they had in the United States – to control the wider Mexican economy. The railways were also important, as purchasers of fuel oil and lubricants, for the Mexican Petroleum Co., owned by the American Edward Doheny, and for the Standard Oil-affiliated Waters-Pierce Co. Pearson, in addition, was breaking into the American monopoly over the sale of other oil products in the country, although – due to the inadequate level of production from his San Cristóbal oilfield on the

Isthmus of Tehuantepec – only by following Waters-Pierce's practice of importing cheap Texas crude for refining in Mexico.[192]

However, in July 1908 Pearson's drillers struck a huge flow of relatively light oil near where Doheny was also exploring, in northern Veracruz at San Diego de la Mar, opening up a region of prolific oilfields that became known as the Golden Lane.[193] At a depth of 1,830 feet the Dos Bocas well exploded into a 1,000-foot fountain of high-pressure oil at temperatures approaching 70 °C. It was soon ignited by the steam boiler that powered the drilling rig, turning into 'a monster with a dark mane crowned by a wide and shiny headdress of fire' that could be seen from ships 200 miles out at sea, recorded a Mexican engineer. Pools of flaming oil floated across the nearby Tamiahua Lagoon 'like phantom ships' and reached the Gulf of Mexico via the Carbajal River.[194] Blisteringly sulphurous black rain was carried by the wind up to 70 miles away as the fire raged uncontrollably for fifty-seven days until the torrent of around 100,000 barrels per day subsided; Pearson estimated the total spillage at 10 million barrels. The initial 8-inch-diameter hole had become a 3-hectare crater, leaving a toxic environmental legacy that remains to this day.[195] A geologist recorded,

> What had been lush monte was now a gaunt spectre of dead trees. The air stunk with the smell of rotten eggs. There was no sign or sound of animal, bird or insect life. Nothing stirred in the breeze. The silence was appalling. It was eerie and frightening ... It smelled and looked like I imagined hell might look and smell.[196]

Pearson's £5 million investment in Mexican oil – which, though somewhat speculative, was protected by his relationship with the Díaz government – could now begin to pay off. He formed the *Compañía de Petróleo 'El Aguila', S.A.*, or Mexican Eagle, and included many of Díaz's inner circle on its board of directors. Oil explorers now poured into the region, many from Oklahoma and Texas, and in late 1910 two further big oil strikes – south of the Dos Bocas well and again causing environmental catastrophes – set off a Mexican oil boom. Meanwhile, Doheny had been so confident of the region's potential that as soon as his Huasteca Petroleum Co. had acquired property at Juan Casiano he began constructing twelve 55,000-barrel steel storage tanks and an

8-inch-diameter pipeline with ten pumping stations, 125 miles northwards to the deep-water port of Tampico; President Díaz intervened personally to get one landowner to allow the pipeline to pass 7 miles through his hacienda. In the autumn of 1910 Doheny's drillers struck an immense gusher on the scale of Pearson's Dos Bocas strike, justifying the completion of the pipeline, the construction of a refinery at Tampico, and an expanded tanker fleet. In December, Pearson's Mexican Eagle made a second uncontrollable 100,000 barrel-per-day oil strike at Potrero del Llano, 40 miles northwest of the port of Tuxpan, to where a pipeline was laid, followed by another to Tampico. Pearson and Doheny paid a local landowner and labour contractor, Manuel Peláez, to corral the area's mestizo ranchers and the Huastecos, or Téenek, indigenous population into combating the massive oil spills and vast conflagrations; for much of the indigenous population – vast areas of whose ancestral land had been utterly devastated – the discovery of this subterranean wealth now entailed being terrorized into obedience to the oil companies.[197]

As the drillers began to learn how to control such powerful oil flows – the greatest ever yet encountered – the only limit on crude exports seemed to be the available pipeline and tanker capacity, and some wondered whether Mexico might contain the world's largest oil reserves. Crude production surged from 3.5 million barrels in 1910 to 12.5 million in 1911 – a figure itself far exceeded in the following years – ranking it as the world's third-largest producing country after the United States and Russia, with over 4 per cent of total world output, just ahead of the Dutch East Indies. Much of the crude was taken up by refineries on the Gulf Coast of Texas, where production had recently plummeted from 28 million barrels in 1905 to 9 million in 1910, and Texas oil companies consequently began to look to Mexico for new production opportunities.[198]

The oil strikes of late 1910 coincided, however, with the eruption of popular discontent and the overthrow of the repressive and corrupt Díaz regime, seen as having enriched itself by allowing Mexico to be exploited by foreign – particularly American – business corporations. Thus, added to the shifting dynamics between various factions in the erupting Mexican Revolution, a highly complex web of intrigue was woven by the conflicting interests of foreign governments and competing business concerns. In the light of attempts by President Díaz to limit US influence over Mexico, many believed that Standard Oil, Waters-Pierce and the American government had backed Mexico's new leader, Francisco Madero.[199] In turn, former President Díaz's favoured oil industrialist Pearson – now Lord Cowdray – was suspected of supporting the overthrow of Madero in early 1913 by Victoriano Huerta, whose presidency was soon recognized by Britain and other European powers, but not by the US.[200] In June 1912 Lord Cowdray had written to the First Lord of the Admiralty, Winston Churchill, requesting to meet him, as

> ... we are being pressed to sell the control of the Mexican Eagle Oil Company, Limited to one of the big existing Oil Companies. Should we do so the fuel oil supplies suitable for Admiralty purposes would thereby become controlled by a foreign Company ... The Mexican Eagle ... unquestionably owns the finest deposits of crude oil suitable for fuel purposes that today are controlled by any one Company.[201]

Some eight months later, in February 1913, Cowdray successfully lobbied Foreign Office officials for the early recognition of the Huerta government, Foreign Secretary Sir Edward Grey concluding, 'our interests in Mexico are so big that I think we should take our own line without making it dependent upon that of other Governments.'[202]

A newspaper owned by Cowdray's main rival in Mexico, Henry Clay Pierce, clamoured that 'Cowdray has taken more out of Mexico than any man since Cortez', and within President Woodrow Wilson's administration it was widely believed that the Mexican conflict had been ignited by the oil companies and that Britain's policy towards Mexico was driven primarily by its oil interests.[203] In April 1913 the US Navy Secretary Josephus Daniels recorded in his diary, 'The general opinion in the Cabinet was that the chief cause of this whole situation in Mexico was a contest between English and American Oil Companies to see which would control; that these people were ready to foment trouble and it was largely due to the English Company that England was willing to recognize Mexico before we did.'[204] Indeed, in July, Churchill announced the imminent signing of a contract for 200,000 tons of naval fuel oil from – he told Parliament – Mexican Eagle, 'the greatest British controlled oil company in the world . . . The Mexican supplies of oil are abundant, and cannot be neglected by the Admiralty . . . They come to us over an ocean route which we can easily and effectively control.'[205]

Furthermore, Britain appointed as its new ambassador to Mexico Sir Lionel Carden, who – as well as having a personal financial stake in Mexican concerns – not only became a relentless proponent of Lord Cowdray's Mexican oil interests but was also fiercely resistant to the emergence of the United States as a rival imperial power to Britain.[206] Under the Roosevelt Corollary to the Monroe Doctrine, the US had, since 1904, claimed not only that Latin American countries should no longer be subject to European power but, furthermore, that the US now had the right actively to intervene in the region in order to uphold its own interests. Before leaving for Mexico, Carden clarified his mission as he saw it. Over recent decades,

> the intervention of the United States Government in the domestic affairs of their weaker neighbours has only been effected by force of arms, whether by open war as in the case of Cuba, or by promoting or aiding revolutions, as in Panama, Nicaragua, Honduras and Mexico. In all these cases British interests have suffered severely . . . [W]e should leave ourselves free to afford effective protection to the great interests we have at stake which are being constantly imperilled by the ill considered or interested action of the United States.[207]

The US had come to regard Central America, the Caribbean and the northern parts of South America as strategically key due to its expanding commercial and security interests in the region. During the Spanish-American War, the press had sent the public a clear message of how the Central American isthmus presented a barrier to inter-oceanic communications, and its reports closely followed the sixty-six-day journey of the battleship USS *Oregon* as it steamed from America's West Coast all the way round Cape Horn in order to reach Cuba.[208] In 1902 the US government had authorized the construction of the Panama Canal and supported a rebellion in Colombia to split off the new, US-compliant nation of Panama; 'I took the Isthmus,' Theodore Roosevelt later said.[209] Given its strategic significance, the canal's associated sea lanes and coastal regions – from Mexico to Ecuador and Venezuela, including the Gulf of Mexico and the Caribbean Sea – would, in turn, be brought under US purview, to the exclusion of Britain and the rising naval forces of Japan and, particularly, Germany.[210] In 1900 the prominent naval strategist Captain Alfred T. Mahan explained to the public that the expanding interests of the US

> make it necessary to connect our Atlantic and Gulf seaboard with the Pacific by a canal across the Central American isthmus . . . The isthmus and its immediate surroundings thus become the greatest of our external interests . . . [W]e must be prepared to resist, forcibly if need be, any attempt to obtain adjacent territory or ports which may serve as stations for a navy hostile to ourselves.[211]

The Secretary of War, Elihu Root, wrote in 1902, 'American public policy in the large sense . . . must certainly bring the West Indies . . . under the political and naval control of the United States.'[212] The US Army Corps of Engineers began constructing the canal in 1908, and it would open in August 1914; by 1912 a US General Staff Report was declaring,

> Upon its completion, the Panama Canal will be our most important strategic position. By our control of this highway between the two oceans the effectiveness of our fleet and our general military power will be enormously increased. It is therefore obvious that the unquestioned security of the canal is our most important military problem.[213]

It was observed in Britain's leading defence journal that 'The canal was pressed for by President Roosevelt as a protection for the Navy, and the increase of the Navy is now urged as a protection for the canal.'[214] As Root had written in 1905, 'The inevitable effect of our building the Canal must be to require us to police the surrounding premises.'[215]

The even broader context was that the United States pinned great economic hopes on China becoming a huge export market – in which it was hoped American lamp oil would figure largely – that would require naval bases and coaling stations from the Panama Canal across the Pacific, for which the US had gone to war against Spain in 1898.[216] This, then, was the US's extensive geostrategic programme that unified its Monroe Doctrine in Latin America with its Open Door trade policy towards China. Now, in 1913, Britain was testing the limits of the Monroe Doctrine in Mexico by recognizing President Huerta, who was relying on the support of countries such as Britain, Germany and Japan.[217] The international political currents, however, were as complex, conflicting and shifting as those of the Mexican Civil War, and all interested parties were manoeuvring at the limits of their capabilities. Some business figures and politicians argued for full US military intervention – Lord Cowdray, himself, told the American ambassador to London, 'in my opinion, the country must be ruled by a strong hand, or a semi-constitutional one, supported by foreign troops'[218] – whereas others argued that this was the one thing that would unite the rival forces against the US, and in all likelihood lead to the destruction of the oil infrastructure.[219] The American consul in the port of Tampico warned Admiral Henry Mayo,

> A war of American intervention would be a great calamity. All other nations will stand to reap all the advantages; whatever the result might be, our country would bear all the expense and reap all the crop of resulting hatred and vengeance. Americans will be unable for many years to come to work in the outlying districts in the oil fields and other parts of Mexico.[220]

It was feared that Britain, Japan and Germany might take the opportunity of a US–Mexican war to gain influence, and even seize territory, in Latin America, the Caribbean and the Pacific, particularly strategic locations for deep-water ports and coaling stations;[221] and Carden himself was not averse to seeing the US becoming embroiled in such a protracted, costly conflict.[222]

However, in November 1913 Britain backed down by withdrawing its support for Huerta, thereby implicitly acknowledging the Monroe Doctrine: such was the perceived threat now from Germany that it did not seem worth alienating the United States, a potential ally.[223] One factor in this respect, perhaps, was the British Navy's eventual assessment that Mexican Eagle's fuel oil failed to meet its standards, being too heavy, sulphurous and smoky; consequently, it would continue to depend a great deal on oil from the US, and on US-owned companies, in time of war.[224] America agreed to take over the role of upholding current British oil concessions in Mexico; but it pressured the government of Colombia into refusing Cowdray oil concessions

that would have included associated port facilities of strategic significance that, it was feared, might even provide the basis for a British
alternative to the Panama Canal.[225] As the industry journal *Oil* put it,
'the building of a harbour on the Gulf of Darien and the enjoyment of
certain rights in this harbour has excited particular attention, in view
of the principle ... that the acquisition by a foreign company of a
base of political or military value near the Panama Canal would be
considered an infraction of the Monroe doctrine.'[226]

However, the United States conceded British demands for equality
of treatment in Panama Canal transit fees.[227] 'In the scramble that is
going on for oil wells in the regions adjacent to the Panama Canal',
wrote the *New York Times,*

> Great Britain is showing great commercial activity. Since the adoption
> of oil fuel in the British Navy every possible effort has been put forth
> to acquire all the oil to be had in Mexico, the West Indies, and Central
> America. At this time concessions are pending in both Ecuador and
> Colombia, in which the Pearson syndicate is endeavoring to secure a
> monopoly ... With the completion of the Panama Canal and the gen-
> eral adoption of oil fuel in the merchant marine these resources will
> become immensely valuable. Competition in the carrying trade will
> turn more or less upon the control of oil at points from which it can be
> cheaply transported to the terminals of the canal.[228]

Throughout this period of revolutionary upheaval, the oil industry in
Mexico actually flourished in contrast to the near collapse of the rest
of the domestic economy, since with a few exceptions the warfare was
conducted outside the oil regions.[229] The Golden Lane's coastal ports
were held by alternating factions that levied oil export taxes as a
source of revenue; and the oilfields themselves were mostly controlled
by local *caudillo* Manuel Peláez, whose private army took protection
money from the oil companies, in a volatile but enduring symbiotic
relationship.[230] As Cowdray had predicted in April 1912, 'My own
feeling is that the oil industry will not be seriously interfered with by
either revolutionists or bandits or instability of the Government. The
interests that are involved are too great: they are owned by foreigners,
and if they were stopped it would immediately stop the distribution
of a great deal of money.'[231]

Nevertheless, over these years many foreign oil workers were attacked and killed: American, British and other European gunboats patrolled the Pamuco River oil installations and their respective warships maintained a looming presence in a tense stand-off – except for a brief invasion of Veracruz by US marines in April 1914, apparently in order to intercept a German arms shipment to Huerta.[232]

AMERICA 'AT THE THRESHOLD OF A REVOLUTION?'

Following in the slipstream of the great railroad corporations, the rise of large industrial corporations had dramatically transformed economic life in the United States. These corporations typically replicated the financing pattern of the railroads by being funded, controlled and even owned by Wall Street banks. From around 1890, Wall Street intensified this process of centralization by orchestrating a corporate merger movement, peaking between 1898 and 1904, in what was 'literally a revolution instead of an evolution', as Theodore Roosevelt described it in 1900.[233] Wall Street's explicit aim in this merger drive was to rein in the excessive production capacity of industry and to reduce the intensity of competition between rival firms that was seen as chaotic, wasteful, and responsible for pushing profits down to unsustainably low levels.[234]

The largest of the vanguard industrial corporations, Standard Oil, was something of an exception in having become self-financing in the early 1880s. As Rockefeller then began to diversify into other industries through his burgeoning stock portfolio, and as Standard's cash reserves continued to pile up at National City Bank (forerunner of Citibank), the Rockefellers became a major force on Wall Street in their own right. John D. Rockefeller's brother, William – president of Standard Oil of New York – took up his position on the board of National City, which became dubbed the 'Oil Bank'.[235] Meanwhile, the investment bank J.P. Morgan & Co. had grown and diversified in the other direction, from banking into owning railroad and industrial corporations.[236] By 1905, Rockefeller and Morgan interests had become dominant on Wall Street and hence over the wider economic and political landscape. Some years later, the Pujo Committee investigation into monopoly power in banking identified Morgan and Rockefeller interests as the main 'agents of concentration', and concluded, 'there is an established and well-defined identity and community of interest between a few leaders of finance . . . [T]he situation is fraught with too great peril to our institutions to be tolerated.'[237] The *Bankers' Magazine*, no less, had already argued in 1901 that

> as the business of the country has learned the secret of combination, it is gradually subverting the power of the politician and rendering him subservient to its purposes. More and more the legislatures and the executive powers of the Government are compelled to listen to the demands of organized business interests. That they are not entirely controlled by these interests is due to the fact that business organization has not reached full perfection.[238]

The editorial had endorsed this development so long as it did not evolve into a form of socialism, as was envisaged by Edward Bellamy and his utopian socialist followers.[239] Bellamy, a best-selling writer of the 1890s, prophesied social harmony if the advance of incorporation and corporate merging were to continue to its logical conclusion. Commercial competition and class conflict would be distant memories, once all economic activity had become perfectly rationally and efficiently organized – as one of Bellamy's fictional characters put it – by being 'entrusted to a single syndicate representing the people, to be conducted in the common interest for the common profit. The nation, that is to say, organized as the one great business corporation in which all other corporations were absorbed.'[240]

Along with economic depression and labour unrest – which Secretary of State Walter Gresham saw as 'symptoms of revolution' – the 1890s saw the rise of 'muckraking' investigative journalism that sought to reveal to popular view the more troubling social and political aspects of modern industrial capitalism. One of the most prominent of these exposés was by the journalist Ida Tarbell, who had grown up, in the 1860s, in the heart of the Oil Region surrounded by the pollution, explosions and flames of Rouseville. Her father had applied his carpentry skills to make sought-after wooden oil-storage tanks, and she remembered how he would rail against the encroaching Standard Oil monopoly.[241] From 1902 to 1904 *McClure's Magazine* published a series of Tarbell's articles that made available to the general reader her in-depth and highly critical study of the rise to dominance of Standard Oil, now the country's largest corporate behemoth alongside U.S. Steel Corp., itself recently consolidated jointly by Rockefeller and J.P. Morgan.[242] At the same time, antitrust legal actions taken against Standard subsidiaries in Texas, Missouri, Kansas and in other states caught the attention of the national press.[243] 'It is funny how we have all found the octopus; an animal whose very existence we denied ten or a dozen years ago,' wrote a conservative Kansas newspaper editor in 1905.[244]

In Kansas, the explosive growth of crude production in 1903–5 from this region of the new Mid-Continent field led to a collapse in prices offered by Standard Oil's local subsidiary, the Prairie Oil & Gas Co.[245] In apparent incomprehension, its boss, Daniel O'Day, wrote, 'The oil producers seem determined to flood the country with crude oil regardless of how it is to be cared for and of its price. It is hard to account for this disposition on their part, as in every other line of business the law of supply and demand prevails and an overproduction must be regulated.'[246] O'Day can hardly have been unaware that it was entirely normal for a multitude of uncoordinated, small producers to descend upon a new oil patch and for each frantically to extract as much crude as possible ahead of others, under the rule of capture. Ironically, it was precisely this type of 'market failure' – the less than optimal outcome of a laissez-faire 'free' market – that was given as the justification for Standard Oil's absorption of competing refineries and for the recent wave of corporate mergers generally: to control prices through the cooperative coordination of output across an industry. By contrast, it was only very rarely that oil producers managed collectively to coordinate output, given the fast-changing conditions in the oilfields and the large number of operators involved. Furthermore, if producers had organized as a cooperative 'cartel', rather than as a single corporation, they were likely to fall foul of antitrust laws.

To compound their difficulties, Kansas producers found their access to independent refineries impeded by a conspiracy between Prairie Oil & Gas and the Santa Fe Railroad; accordingly, they accused Standard Oil of taking advantage of their predicament and abusing its monopoly power. The state's attorney general filed proceedings against Prairie

Oil & Gas, and the Kansas Oil Producers' Association pushed anti-discrimination bills through the state legislature. Since, furthermore, Prairie Oil & Gas was piping Kansas crude out of state to be refined by Standard Oil refineries in Missouri and Indiana, some independent producers attempted to organize a $50 million integrated oil company that would include refining, but this proved beyond their private means. Consequently, Samuel M. Porter, a Republican senator, introduced a bill for the construction of a state-run refinery.[247] In January 1905 the state's new Republican governor, Edward W. Hoch, told the legislature,

> Our oil interests are in jeopardy . . . Rather, therefore, than permit the great monopolies to rob us of the benefits of the vast reservoirs of oil which have been stored by the Creator beneath our soil, I am inclined to waive my objection to the socialistic phase of this subject and recommend the establishment of an oil refinery of our own in our own state for the preservation of our wealth and the protection of our people.[248]

After Prairie Oil & Gas, in apparent retaliation for its treatment in Kansas, halted all of its new work on pipelines and storage facilities, the state accelerated its oil legislation programme.[249] The Cleveland *Plain Dealer* denounced Kansas's plans as 'bizarre economic experiments',[250] and Governor Hoch had to defend himself against widespread accusations of socialism:[251]

> I wish to emphasize and re-emphasize that the State refinery method of protecting State oil interests is not socialism. It is not the spirit of socialism, but the very reverse of it. It may have the semblance of socialism, but its soul is that of competition. Socialism is a heresy . . . the fallacy of which I am more than ever convinced . . . It is not the possession and exercise of property rights which the people of this nation object to; it is the abuse of property rights to which objection is made. [The state refinery plan is] to restore equality of opportunity in the oil industry . . . It is an attempt to encourage competition, not to destroy competition, as socialism does . . . [Independent refining] is now impossible on account of the greatest socialistic corporation now doing business on earth, the Standard Oil Company . . . No greater question confronts the American people than the control of the great aggregations of capital,

all of them socialistic in character, and which are antagonistic to the essential element of all national progress, the competitive system.[252]

In this bold counter-attack, Hoch was credibly able to denounce Standard Oil itself for being 'socialistic', as his audience would have been familiar with Bellamy's corporate-utopian socialism that had been influential in Kansas politics during the 1890s.[253] Hoch saw the state refinery bill passed by the legislature, but in July the Kansas Supreme Court ruled that it violated the state's constitution, whose provisions guarded against

> the public disaster that history shows would follow its engaging in a purely private business enterprise. It has been the policy of our government to exalt the individual rather than the state, and this has contributed more largely to our rapid national development than any other single cause ... [I]f we now intend to reverse this policy, and to enter the state as a competitor against the individual in all lines of trade and commerce, we must amend our constitution and adopt an entirely different system of government.[254]

A faith that the public interest is best served by almost exclusively private economic activity had been written into the state's very constitution. Perhaps it had not been imagined that a day might come when a single private corporation would have the power to siphon Kansas's resources and associated profits right out of the state. Now that it had, citizens lacked the democratic powers equal to the challenge. Michael Cudahy – whose Cudahy Oil Co. had been instrumental in opening up the oilfields of Indian Territory for exploitation – wrote to Standard Oil's Daniel O'Day, 'I never thought that I would see such rampant socialism in America.'[255]

Following journalists' exposés of corporate power and the dramatic events in Kansas and in many other states, President Roosevelt was keen that something should be seen to be done, especially after the confirmation of suspicions that business corporations, including Standard Oil, were involved in large-scale secret funding of political campaigns.[256] Between 1905 and 1906 the federal government ordered several investigations – by the Bureau of Corporations and the Interstate Commerce Commission – into Standard's monopoly practices and abuses of its economic power. On the basis of these investigations, in November 1906 the government launched an antitrust lawsuit against the Standard Oil of New Jersey holding company, under the Sherman Antitrust Act.[257] As William Taft later wrote – after having served a term as President in 1909–13 – this holding company 'was indeed an octopus that held the trade in its tentacles, and the few actual independent concerns that kept alive were allowed to exist by sufferance merely to maintain appearance of competition rather than anything substantial'.[258]

From the perspective of Standard Oil's control over American crude production – which had fallen from 92 per cent in 1880 to 59 per cent in 1906 and 56 per cent in 1911 – it could appear that Taft was making a gross overstatement. However, it was also the company's control over pipelines, refinery capacity – 70 per cent in 1906 and 64 per cent in 1911 – and the marketing of oil – 59 per cent between 1906 and 1911 – that enabled it to dominate the industry as a whole.[259] As the *Wall Street Journal* quoted from the second report of the Commissioner of Corporations:

> The Standard has repeatedly claimed that it has reduced the price of oil; that it has been a benefit to the consumer; and that only a great combination like the Standard could have furnished oil at the prices that have prevailed ... [However,] [i]t now appears conclusively that the result of this domination has been to increase the prices of oil paid by the consumer, and correspondingly to increase the profits received by the Standard ...
>
> [H]ad the industry followed the normal course of development and had no great combination arisen to exercise substantial control therein, prices to the consumer would have been much less than they actually are or have been ...
>
> Its domination has not been acquired or maintained by its superior efficiency, but rather by unfair competition and by methods economically and morally unjustifiable ...
>
> The profits of the Standard Oil Co. are enormous, both in absolute amount and in proportion to the investment of the company ... It is clear that the domestic consumer has been compelled to pay an exorbitant tribute to the oil monopoly.[260]

Evidence was presented that Standard Oil's business practices routinely transgressed the bounds of fair competition. The most common methods adopted by Standard to stifle competition were, firstly, making it difficult for rivals to transport or refine their oil; and, secondly, engaging in 'predatory pricing' – in other words, undercutting rivals in a particular region by selling its oil products below cost, and subsidizing this by charging higher prices in regions where it already monopolized the market. Standard had run affiliated companies, known as 'blind tigers', that passed themselves off as being independent, and the *Wall Street Journal* found a statement of its accounts for 1906 to be highly suspicious:

> [T]he statement is compiled in such a way as to conceal many facts and figures to which the stockholders are entitled ... The key to this extraordinary manipulation of figures is probably held by a few officials and large stockholders of the company ... [T]he statement of the Standard Oil Co. is vague and incomplete ...[261]

It was of some significance, however, that the government reports concentrated on a rather narrow economic analysis of the Standard Oil phenomenon. Almost no consideration was given to the wider social and political implications of the emergence of such large business corporations – which had been an important factor behind the framing of the Sherman Act – and it is likely that this limited the scope of the courts' deliberations, thus influencing the Supreme Court's eventual ruling.[262]

In a speech given in 1907 by Frank B. Kellogg, the head of the most comprehensive investigation into Standard Oil's practices and the former legal counsel for U.S. Steel, he warned against the rapid concentration of economic power:

> A few men by means of such ingenious corporate organizations, may build up a great financial institution, may perpetuate their power, and reduce the great majority of the people of this country from progressive independent business men to mere dependent employes [*sic*] ... A system which makes a few men enormously rich – with the power which goes with wealth – and reduces the balance to the subservient, menial

position of employe [*sic*], must in the end have a disastrous effect upon our civilization.

But, it is said that combination is an economic principle, against which it is useless to combat. I deny that many of the great corporations and trusts of this country are the result of economic principles. They are rather the result of the genius and cupidity of men who love wealth and power, and who have ever been abusive of power, and who will not stop in their grasp for power short of the industrial enslavement of their fellow men . . .

[T]here is nothing to prevent a few men, by the device of holding companies, from controlling all of the industries of the country . . .[263]

In the popular imagination, America was exceptional in the degree of freedom its citizens enjoyed from the distorting and oppressive imbalances of economic and political power characteristic of other nations. On this unusually level playing field, it was believed, the natural forces of free competition between individuals and small enterprises would benefit all. However, during the last half-century the rules of the game, the legal landscape, had gone through seismic changes. With free incorporation, the business corporation had broken free from its origin as an agent of the sovereign power – in the US, the notionally democratic legislature – such that small entrepreneurs were now up against huge corporations commanding vast financial, logistical and legal resources. Whereas small businesses would circulate wealth within a local community, these large corporations tended to concentrate wealth in the hands of relatively few shareholders and Wall Street banks. This was widely seen as an unnatural invasion of the economic and political order, requiring measures, such as antitrust laws, to redress the imbalance. However, when New Jersey pioneered the availability of the holding company device, corporations were legally able to circumvent these measures – by transforming into one corporation what would have been an illegal cartel of separate corporations – and thus were able to expand to mammoth proportions, often primarily in the interests of Wall Street finance capital.[264] As a House of Representatives committee report on the Clayton Bill put it in 1914, the 'holding company' was a company whose *primary* purpose is to hold stocks in other companies ... As thus defined a 'holding company' is an abomination and in our judgment is a mere incorporated form of the old-fashioned trust ...

The concentration of wealth, money and property in the United States under the control and in the hands of a few individuals or great corporations has grown to such an enormous extent that unless checked it will ultimately threaten the perpetuity of our institutions.[265]

Thus, the increasing size of private business corporations was at least as much an artefact of the legal environment and the interests of the financial sector as it was a 'natural' consequence of modern technology – as financiers, industrialists and most economists preferred to see it. Kellogg was just one of many who expressed doubts as to whether there was a genuine economic basis for the seemingly endless combination and merging. Beyond a certain size, a business may become less efficient – in which case, perhaps the advantages of size were more those of market dominance, concentration of wealth and political influence.[266]

Furthermore, argued Senator Albert B. Cummins during a Congressional debate on the Federal Trade Commission Bill in 1914, even real increases in efficiency might come at a greater cost:

We often do wrong, I believe, in assuming that because a great corporation, a vast aggregation of wealth, can produce a given commodity more cheaply than can a smaller concern, therefore it is for the welfare and the interest of the people of the country that the commodity shall be produced at the lower cost. I do not accept that article of economic faith. I think we can purchase cheapness at altogether too high a price, if it involves the surrender of the individual, the subjugation of the great mass of people to a single master mind.[267]

The political implications were, indeed, experienced as being equally troubling. In the first ruling of the drawn-out federal legal case against Standard Oil, Justice Archelaus Woodson predicted, in December 1908, that if monopoly abuses continued, 'It would be only a question of time until they would sap the strength and patriotism from the very foundations of our government, overturn the republic, destroy our free institutions, and substitute in lieu thereof some other form of government.'[268]

Woodrow Wilson, in a speech of April 1911 as governor of New Jersey, railed at the degree to which big business had corrupted politics, thereby undermining the principles commonly believed to be behind the Declaration of Independence:

> By privilege, as we now fight it, we mean control of the law, of legislation and adjudication, by organizations which do not represent the people, by means which are private and selfish and worthy of all condemnation. We mean specifically the conduct of our affairs and the shaping of our legislation in the interest of special bodies of capital and those who organize their use ... There is not enough debate of [legislation] in open house, in most cases, to discover the real meaning of the proposals made. Clauses lie quietly undiscovered in our statutes which contain the whole gist and purpose of the act; qualifying phrases which escape the public attention and casual definitions which do not attract attention, classification so technical as not to be generally understood and which everyone most intimately concerned is careful not to explain or expound, contain the whole purpose of the law. Only after it has been enacted and has come to adjudication in the courts is its scheme as a whole divulged. The beneficiaries are then safe behind their bulwarks.[269]

William Taft regarded the Sherman Antitrust Act as having being directed against business practices that 'had resulted in the building of great and powerful corporations which had, many of them, intervened in politics and ... threatened us with plutocracy'.[270] According to Senator Robert La Follette, even public debate was being stifled as a result of corporate control over much of the press, either directly, or indirectly via the constraints that resulted from being dependent on advertising revenue: 'One would think that in a democracy like ours, seeking for instruction, able to read and understand, the press would be [the people's] eager and willing instructors ... But what do we find has occurred in the past few years since the money power has gained control of our industry and government? It controls the newspaper press.'[271]

In early 1911 the national press reported a speech delivered by George Perkins – a recently retired partner at J.P. Morgan – in which he likened the previous decade of political agitation in America to the run-up to the Civil War.[272] The following year Wilson, while on his campaign trail, asked,

> Why are we in the presence, why are we at the threshold, of a revolution? ... We have been dreading all along the time when the combined power of high finance would be greater than the power of the government. Have we come to a time when the President of the United States or any man who wishes to be the President must doff his cap in the presence of this high finance, and say, 'You are our inevitable master, but we will see how we can make the best of it?' ...
>
> [We are] no longer a government by free opinion, no longer a government by conviction and the vote of the majority, but a government by the opinion and the duress of small groups of dominant men ...
>
> What a prize it would be to capture! How unassailable would be the majesty and the tyranny of monopoly if it could thus get sanction of law and the authority of government! By what means, except open revolt, could we ever break the crust of our life again and become free men, breathing an air of our own, living lives that we wrought out for ourselves?[273]

America, it turned out, was far from replaying its own bloody political history and descending into the kind of violent turmoil engulfing its immediate neighbour to the south. Nevertheless, the nation was clearly experiencing an acute identity crisis.

'A PAPER VICTORY FOR THE PEOPLE'

During the presidential contest of 1912 Woodrow Wilson had deployed some strident language condemning the economic and political injustices associated with the rising power of big business. He was responding to the increasing electoral support for socialist policies: in 1910 Victor Berger had become the first member of the Socialist Party of America to be elected to Congress, and in 1912 the party's presidential candidate, Eugene Debs, polled 6 per cent of the vote. Over this period the citizens of nearly seventy cities had elected a socialist as mayor.[274]

Wilson's electoral victory thus alarmed leading Wall Street figures. In early 1913 – two days before the release of the Pujo Commission report on the monopolistic power of high finance – Wilson's close confidant, 'Col.' Edward M. House, met with bankers to address their fears. House was struck by their apparent naivety regarding electioneering rhetoric and reassured them: 'I said, his utterances while idealistic were intended to raise the moral stamina of the nation, and were not for the purpose of writing such sentiment into law.'[275] Indeed, as public concerns at the escalating wealth and power of corporations intensified, the primary response of the nation's economic and political elite was, in reality, to become correspondingly more adept at creating the impression that they were taking these concerns seriously while actually furthering the advance of corporate power. The fate of a presidential candidacy, after all, was already largely determined by the level of corporate backing.[276]

The attitude of the economic and political elite was summed up in a *Wall Street Journal* editorial of 1907, 'No Lynching of Corporations', that appeared to concede that there were some real problems:[277]

> For a number of years the people have been smarting under the crimes of corporations, which, regardless of law and justice have been seeking their own profit without any thought of public rights or public interest . . . The States have been made the creatures of the very corporations which have been incorporated under their laws, and there have been such a multitude of abuses and exactions, large and small, as at last to drive the people to the point of frenzy.

However, this concession was then rhetorically self-negated. The views of 'the people' ought now to be dismissed as the irrational and dangerous reactions of a frenzied mob:

> What has happened is exactly the same as has taken place in the South in the lynching of negroes and in the city of New York in the lawless attacks on men suspected of outrageous crimes. In other words, the people have taken to lynching corporations. The lynching is in the acts of legislatures enacting confiscatory laws.

Corporate leaders, joined by the most prominent academic economists, saw the growth of corporations in an essentially positive light.

The rise of big business was a natural evolutionary advance in organizational, technological and economic efficiency, one that was necessary, beneficial and ultimately inevitable. Indeed, it was the progress of civilization itself, representing an improvement at the material and even moral and spiritual levels. This view was shared by all four presidents of the era – McKinley, Roosevelt, Taft and Wilson – contrary to the impression created by their occasional critical rhetoric.[278] In 1899, two years before becoming president, Roosevelt encapsulated the stance of the political establishment in a letter to one of his close advisers – all of whom took, at the very least, a fundamentally benign view of large corporations:[279] 'I have been growing exceedingly alarmed at the growth of popular unrest and popular distrust on this question. It is largely aimless and baseless, but . . . what I fear is if we do not have some consistent policy to advocate then the multitudes will follow the crank who advocates an absurd policy.'[280] These politicians essentially agreed with Rockefeller when he wrote,

> combinations of capital are bound to continue and to grow, and this need not alarm even the most timid if the corporation, or the series of corporations, is properly conducted with due regard for the rights of others . . . The chief advantages from industrial combinations are those which can be derived from a coöperation of persons and aggregation of capital . . . It is too late to argue about the advantages of industrial combinations. They are a necessity.[281]

Yet many people clearly found plenty to argue with. Standard Oil had initially tried to remain silent to avoid attracting further attention, but from the 1880s it began to manage its public image. It acquired stock in two Cleveland newspapers, the *Herald* and the *Leader*, that put a consistently positive spin on the activities of Standard Oil; it maintained financial relationships with other newspapers, and it cultivated the friendship of editors. Via the services of the Malcolm Jennings News Bureau and Advertising Agency it paid newspapers and magazines, either directly or indirectly, to publish pro-Standard articles and notices. Readers were instructed, for example, that 'Monopoly and octopus, combines and trusts are haughty words, but the best goods at lower prices are beneficial things. It is much easier to say harsh words than it is to make things cheap.'[282]

By the mid-1890s leading corporate figures had begun mobilizing collectively to counter possible political threats to their interests, and in 1900 they established the National Civic Federation (NCF) – presenting itself as a 'broadly representative', grass-roots organization – to mediate in labour disputes, formulate policy, lobby government and conduct public relations; and it was not averse, also, to adopting darker tactics.[283] The first president of the NCF, Marcus A. Hanna, had known Rockefeller since their childhood in Cleveland and had, as William McKinley's Republican presidential campaign manager in 1896, received over $250,000 in secret contributions from Standard Oil.[284] Hanna now wrote, 'The menace of today, as I view it, is the spread of a spirit of socialism.'[285]

The NCF welcomed conservative union leaders, in other words those interested only in bargaining on behalf of their members and not calling for wider social and political change, and to the Socialist Party leader Eugene Debs it seemed that the NCF merely feigned friendship with labour in order to 'take it by the hand and guide it into harmless channels'.[286] At the same time, the NCF worked to instil some flexibility in the many business leaders who, as one of its founders put it, 'would smash every union in the country if they could. In fact, our enemies are the Socialists among the labor people and the anarchists among the capitalists.'[287] To the latter, Roosevelt made the case for some form of regulation of big business:

> Against all such increase of Government regulation the argument is raised that it would amount to a form of Socialism ... [T]he 'conservatives' will do well to remember that ... unfair and iniquitous methods by great masters of corporate capital have done more to cause popular discontent with the propertied classes than all the orations of all the Socialist orators in the country put together.[288]

A privately circulated memorandum indicated the evolving sophistication of the NCF's strategy: there needed to be 'a carefully planned and wisely directed effort to instruct public opinion as to the real meaning of socialism', but in a manner that would 'not give unjust offense' or create 'a false impression of partisanship'. The memo advised, 'Socialism has the great sentimental advantage of being based upon a desire to benefit all human beings [so] in opposing socialism this same sentimental advantage must be claimed and held for the anti-socialist view.' It recommended a readiness to accept proposals that socialists 'would naturally favor [but are] not necessarily in conflict with the underlying principles of the existing industrial order'.[289]

The overwhelming majority of American socialists were committed to the ballot box and to the health of representative democracy, and their policies were not hugely radical: for example, improved labour rights, a modest redistribution of wealth to meet basic social needs and to promote equality of opportunity, and the nationalization of public utilities and perhaps of natural industrial monopolies.[290] Nevertheless, their policies were routinely caricatured as totalitarian communism, and according to the NCF socialism had to be combated 'if our American political system and its underlying economic institutions are to be preserved'.[291] Frank Kellogg, the Bureau of Corporations' chief investigator into Standard Oil, told an NCF conference that socialists would 'compel the government to take possession of and manage the commerce, the manufactures, all the industries, the cultivation of the soil for the benefit of all'.[292] Newspaper readers were instructed by the American Bankers' Association that socialists 'all agree upon the fundamental proposition that the government – that is, the community as a whole – should own all the real estate, all manufacturing enterprises, all banks, all transportation companies – in short, all the money-making utilities'.[293]

For President Taft in 1910, the most important issue of the day was not the threat posed by private corporate power, but was

> the question of the preservation of our institution of private property or its destruction and the substitution of a certain kind of co-operative joint enjoyment of everything, which is the ideal of Socialism. The institution of private property, in my judgment, has done more to bring about modern progress and civilization than any of our institutions except that of personal liberty ... [Socialism] will be a failure, and the result will be a substitution of a tyranny of governing committees ... instead of a self-acting system in which the industrious and the prudent and the far-sighted are rewarded and the lazy and the inattentive fall behind.[294]

Lest anyone should mistake Wall Street investment banks and corporate boards for 'a tyranny of governing committees', Rockefeller enlightened them about shareholder democracy:

> Just see how the list of stockholders in the great corporations is increasing by leaps and bounds. This means that all these people are becoming partners in great businesses. It is a good thing – it will bring a feeling of increased responsibility to the managers of the corporations and will make the people who have their interests involved study the facts impartially before condemning or attacking them.[295]

For over a decade before the 1906 launch of the federal legal case against Standard Oil, successive administrations – including those of Roosevelt – had artfully dodged calls for taking real action.[296] The much-trumpeted enforcement of antitrust law that the federal government did carry out – that against Standard Oil being the most prominent – was primarily a political performance, intended to mollify critical sections of the public.[297] As Roosevelt, would-be 'trustbuster', explained, 'The machinery of modern business is so vast and complicated that great caution must be exercised in introducing radical changes for fear the unforeseen effects may take the shape of widespread disaster. Moreover, much that is complained about is not really the abuse so much as the inevitable development of our modern industrial life.'[298]

Ironically, it was largely the radical removal of legal constraints on corporate business charters and the innovations of free incorporation and the holding company, rather than any natural inevitability, that had led to the 'unforeseen effects' of such a 'vast and complicated' economic system – which many regarded as a 'widespread disaster' for the body politic. Indeed, in May 1911, four and a half years after the launch of the federal antitrust case against Standard Oil, the Supreme Court finally ruled that it was precisely the ability of the New Jersey holding company to own the stock of a large number of subsidiary companies that enabled Standard artificially to inhibit independent competition in the oil industry.[299] The control that Standard Oil wielded therefore arose 'not as a result of normal methods of industrial development, but by new means of combination, which were resorted to in order that greater power might be added than would otherwise have arisen had normal methods been followed'.[300] However, the consequent decree dissolving the Standard Oil of New Jersey holding company was quite inert – almost as if by design – as it apportioned the shares of the Standard subsidiaries pro rata to the shareholders in the holding company, thus preserving the original profile of ownership. This enabled the major shareholders to maintain the same control over the Standard group, only on an informal basis, just as they had done in the period between the break-up of the original Standard Oil trust and Standard's reincorporation as a holding company in New Jersey; both arrangements were technically illegal, but very difficult to challenge and prove.[301] Indeed, it was the assessment of the head of Standard's legal team that 'We will be able to continue business as before', and, as Standard New Jersey's share price rose significantly following the judgement, the *Chicago Tribune* declared, 'Like the chameleon, the more Standard changes the more it is the same thing.'[302] Roosevelt, no longer president, was now free to ridicule President Taft and Wilson for their ineffectual antitrust policies:

> Under the decree of the court the Standard Oil Company was split up into a lot of smaller companies ... What has been the actual result? All the companies are still under the same control, or at least working in such close alliance that the effect is precisely the same ... If we had our way, there would be an administrative body to deal radically and thoroughly with such a case as the Standard Oil Company. We would make any split-up of the company that was necessary real and not nominal.[303]

The British government soon came to a similar view, and the *New York Herald* concluded that the break-up of Standard Oil was illusory, merely 'a paper victory for the people. The real victory, it is now admitted by Mr. Rockefeller, was won by the oil company.'[304] (The main present-day descendants of this dissolution are Amoco, now merged into BP; Esso/Exxon and Mobil, now merged as ExxonMobil; Chevron; and Marathon.)

In a supplement to the ruling against Standard Oil, Chief Justice Edward White declared that judges should follow the 'rule of reason' in antitrust cases. Since, in the majority view of the panel of judges, corporate combination was beneficial and inevitable, the activities of big business should be deemed to violate antitrust law only if they

amounted to an 'unreasonable' interference with the economic freedom of others in a manner that harmed the public interest.[305] This
qualified reading of the Sherman Act was strongly denounced by Justice John Harlan in his dissenting opinion. Several attempts – promoted
by the NCF[306] – to amend the Act along precisely these lines had
failed in the Senate, so White's interpretation was clearly a 'usurpation
by the judicial branch of the government of the functions of the legislative department', Harlan argued.[307] The application of the Sherman Act
was now reduced to the arbitrary decisions of judges who were mostly
former corporate lawyers who shared the values, views and social life
of the wealthy establishment. Senator La Follette told reporters, 'I fear
that the court has done just what the trusts have wanted it to do and
what Congress has refused steadfastly to do.'[308]

To the side-stepping of the democratic process by judicial discretion
was added the circumscribing of democratic debate by expert commission.[309] Government commissions and inquiries would, in the first
instance, make it look as if something were being done. In 1907 Ralph
Easley, one of the NCF's founders, wrote that its National Conference on
Trusts and Combinations – which had the support of the White House –
would help 'in mollifying the radical sentiment', and 'at a time when the
public mind is so hysterical, it might have a very calming effect'.[310] Easley
wrote to Nelson Aldrich – an influential, pro-corporate senator – that the
conference should be the precursor to creating 'a high class, representative commission' that would 'take the matter out of politics'; in addition,
it would provide the attorney general with 'good grounds' to 'slack up a
little on its prosecutions'.[311] The evidence of carefully selected 'experts'
and the contents of their reports would then limit the parameters of
debate by avoiding any serious suggestion of measures that might be considered politically radical. The series of reports on the monopolization of
the oil industry, for example, may have been critical of Standard Oil, but
they stopped short of the kind of analysis required to offer the public and
the legislature a politically diverse range of policy options to debate and
choose from. This was most apparent regarding the ownership and use
of pipelines, which was key to the control of the oil industry; furthermore, the relevant report was not even published until a year after
pipeline legislation – in the form of the Lodge Amendment to the Hep-
burn Bill of 1906 – was actually debated in Congress.[312]

Similarly, investigations into the banking panic of 1907 ensured that the inner workings of Wall Street – dominated by Morgan and Rockefeller interests – would remain a mystery both to the voting public and to senators tasked with formulating legislation. Wall Street interests were well served by the ideas that money must be based on gold – the basis, at that time, for international finance – and that it should be issued primarily by the private banking sector in the form of debt. Alternative, eminently reasonable views – and arguably more socially equitable and democratic – were marginalized, ridiculed or simply ignored.[313] Perversely, Frank Vanderlip of the National City Bank, speaking for the Rockefeller camp, wrote, 'The subject is technical. Opinions formed without a grasp of fundamental principles and conditions are without value. The verdict of the uninformed majority gives no promise of being correct.'[314] As one Wall Street banker wrote to Senator Aldrich, who was to head the National Monetary Commission, 'reform can only be brought about by educating the people up to it, and such education must necessarily take much time. In no other way can such education be effected more thoroughly and rapidly than by means of a commission . . . [to] present an exhaustive report, which could be made the basis for an intelligent agitation.'[315] On forming the Commission, Aldrich wrote privately, 'My idea is, of course, that everything shall be done in the most quiet manner possible, and without any public announcement.'[316] Indeed, Aldrich convened a top-secret conference of the core trio of Wall Street banking interests – one being the Rockefellers – at which the Federal Reserve Act was drafted.[317] This was regulation of a kind that these bankers actually wanted. It was primarily designed to create the legal infrastructure, modelled on the banking systems of Britain, Germany and France, required to promote the dollar to the status of an international currency and enabling Wall Street to enter the hugely lucrative world of international finance. One pillar of this was to legalize a cartel run by Wall Street's core group, a de facto federal reserve that had evolved informally but on legally dubious foundations.[318] The bill was carried through Congress on the back of a concerted Wall Street sales pitch masquerading as the supposedly objective work of the Monetary Commission; when it came to the vote, on Christmas Eve 1913, senators seemed seriously to misunderstand what the Act really meant.[319] Sold largely on the grounds that it would bring domestic economic stability and prosperity, it turned out that this would primarily be for Wall Street bankers and the associated privileged few. Fifteen years later, the banking system would collapse, the wreckage falling almost entirely on the average citizen.

Throughout the Progressive Era, the increased federal regulation of big business was portrayed to the public as restraining corporate power in the common interest. As Roosevelt put it, 'In order to meet the inevitable increase in the power of corporations produced by modern industrial conditions, it would be necessary to increase in like fashion the activity of the sovereign power which alone could control such corporations.'[320] In reality, the regulatory programme was more about merging corporate and government power; such regulation was actually desired by big business, which ensured that it was tailored and applied in a way that was respectful to its interests. As early as 1899, Rockefeller told the Industrial Commission that he would like to see 'First. Federal regulation under which corporations may be created and regulated, if that be possible. Second. In lieu thereof, State legislation as nearly uniform as possible encouraging combinations of persons and capital . . .'[321]

State legislatures – in closer contact with the popular mandate – displayed greater tendencies to check the power of big business. By contrast, it would be easier for the corporate community to influence one central, federal authority – at a further remove from the public and more amenable to focused lobbying – to fashion corporate-friendly legislation.[322] This implied a regularized partnership, rather than regulatory restraint. As Herbert Knox Smith, commissioner of the Bureau of Corporations, reported in 1907, 'The time is ripe' for a system of regulation that 'will afford a permanent and practical ground for contact and cooperation between the Government officials charged with this work on the one hand and corporate managers on the other'.[323] In 1910 George Perkins, about to retire his partnership at J.P. Morgan, wrote that regulation should be written in 'cooperation between our statesmen and our businessmen'[324] and he exhorted citizens, in the name of economic efficiency, to unite with big business and to head off socialism:

> We need mutualization between labor and capital. Now as never before, an industrial campaign is on for the markets of the world.

Disagreements, discords and strikes will lose it for us ... Co-operation and not competition is the only method of attaining efficiency. The nation that first realises this will lead in the world's commerce ... We must have co-operation between owners and managers and employees and the public.[325]

Towards the end of the decade, this nationalistic vision of a solid unity between the state and private capital, along with dutifully subservient workers, would begin to crystallize as a formal political ideology, in Italy, under the name of 'Fascism', according to which political protest, and even basic human rights and democracy, came to be seen as threats to the smooth operation of the nation's capitalist economy.[326]

ECONOMIC WARFARE

In 1906, as US federal government agencies began to produce a series of reports highly critical of Standard Oil in preparation for legal action against it, Frederick D. Asche of Standard's export department issued a public defence of the company on the grounds of its importance for international commerce. Since the United States was exporting about 60 per cent of its petroleum products, Asche argued, everyone directly or indirectly involved in the US oil industry therefore had

> interests in maintaining the supremacy of American petroleum in the competitive markets of the world, [and] cannot fail to view with deep concern and alarm the damage and menace to our foreign commerce, traceable to the persistent attacks upon the Standard Oil Co ...
>
> The formidable competitors to American petroleum, Russian, German, Roumanian, Galician and Oriental, are very naturally utilising these press attacks to their own great advantage and to our great detriment. And this at a time when the petroleum of this country needs an increased outlet on account of the largely accumulating stocks above ground, due to the immense new oil fields of Kansas, Oklahoma and Illinois ...
>
> [W]e find the competitive obstacles greater from year to year because of the rapidly developing petroleum industries of other countries. These

have the marked advantages of proximity to large consuming centers, to pauper labor employed in production, and the stimulation afforded by the support of their respective Governments and active encouragement extended by their local press.[327]

A few months later, the New York *Sun* published a further defence of the scale and scope of Standard Oil by its head of foreign operations, William H. Libby: 'Concurrently with attempts of the Government to disintegrate the ... Standard Oil Company and other companies, and thus to paralyze their great commerce at home and abroad, it is ... pertinent to note the policy and progress of their most formidable foreign rivals.'[328] Libby enumerated the various European-owned oil conglomerates that had arisen and 'become international in their scope'. They had evolved broadly in accordance with the English legal principle that

> everyone has a right to form a combination, and that everybody else has an equal right to combine against such combination. This absolute liberty of commercial action ... apparently crystalizes the wisdom of the most experienced commercial nation in history ... [Combinations,] so far from receiving the opposition of governments, press or communities, so far from being regarded 'conspiracies in restraint of trade' or as ingenious subterfuges of a trade autocracy, are regarded abroad as being the natural pathway of legitimate, economic and progressive commerce, and are especially commended when the motive is emphasised of eliminating the American product from the competitive markets. Against this array of formidable elements and innumerable other opposing factors the Standard Oil Company is fighting in the world markets for the continued supremacy of American petroleum.

Foreign oil concerns, Libby pointed out, were unconstrained by antitrust law; Royal Dutch and Shell, for instance, had recently combined to control the oil industry of the Dutch East Indies, so they were now in a stronger position 'in competing for the vast petroleum commerce of the Oriental countries, containing more than two-thirds of the world's population'.

Libby seemed to have a point, insofar as European countries had not enacted antitrust laws in the way the United States had.[329] However, the

US Supreme Court gave antitrust law such a weak interpretation that it hardly checked the growth of big business at all, while the eventual dissolution of Standard Oil would be more apparent than real: Libby's demand that Standard be spared from harsh antitrust measures at home was largely met. The legal and political environments for big business in the US and in other countries were not, in practice, that different; in both, official policy was, in one form or another, friendly towards industrial combination, which tended to be organized with the close involvement of a country's major banks.[330]

A more significant factor behind the initial emergence of Standard Oil's foreign competitors was, as Asche put it, 'the stimulation afforded by the support of their respective Governments'. As oil regions outside the United States began to be exploited, the local authorities typically promoted their fledgling domestic oil industries and manoeuvred to protect them from being taken over by established foreign companies, most obviously Standard. From the 1880s the colonial government of the Dutch East Indies had facilitated oil exploration by its nationals by providing military and other practical support, which the Royal Dutch Co. continued to receive after its incorporation by royal charter in 1890. Similarly, the Rangoon Oil Co. benefited from Britain's seizure, in 1885, of Upper Burma – which included the Yenangyaung oilfields – and led to the formation of the Burmah Oil Co.[331] From the 1890s, Standard found itself shut out of the Caucasus due to Russian fears that the American giant would take control of its oil industry.[332] In 1898 the Dutch government blocked the sale to Standard Oil of the Moeara Enim oil company in the Dutch East Indies; Royal Dutch was officially shielded from being taken over by foreigners, specifically by Standard Oil and by the recently formed Shell; and the Dutch blocked continued attempts by Standard to obtain oil concessions in their Far Eastern territories.[333] From 1900, the Romanian government made it difficult for Standard to enter its oil industry due to fears of foreign takeover.[334] From 1902, Standard Oil, as well as Royal Dutch and even Shell, were refused entry into Burma, as Britain adopted a policy of permitting only purely British oil companies to operate oilfields within the British Empire, with the aim of securing domestic control over supplies of fuel oil for the Royal Navy.[335] This principle was extended to territory regarded as being within British spheres of influence, when Britain arranged for Burmah Oil to invest in William Knox D'Arcy's exploratory drilling in Persia to the exclusion of Standard Oil and the Rothschilds. The Colonial Office also granted a British company exclusive rights to drill for oil in Nigeria, and even extended a loan to the company.[336]

At the turn of the century international competition in the oil industry was intensifying as the once commanding position of Standard Oil was challenged not only by the Russian industry but also by oil from the Dutch East Indies, Burma and eastern Europe, and then from Mexico. Across the industry there was a clear tendency towards the emergence of large oil companies, for several interrelated reasons. Firstly, it was very capital-intensive at the transportation and distribution phases – particularly as pipelines, tankers, and storage facilities increased in size – and also at the refining stage, especially as this became more technically sophisticated. Secondly, just as Standard's refineries commanded a virtually guaranteed supply of crude due to the company's extensive pipeline and storage network across the United States, so the rising non-US oil companies sought to guarantee their crude supplies by seeking rights to large production areas, as crude production from any one oilfield could be highly unpredictable. Thirdly, oil companies tended to combine forces in order to reduce competition between themselves and to unite against common rivals. At one end of the spectrum, competitors would enter a temporary truce, usually by apportioning between themselves exclusive sales regions or percentages of product sales within a region. In a stronger form of cooperation, companies would join forces in a more integrated, active, longer-term association. At the furthest end of this spectrum, companies would fully merge or be bought up under common ownership, although the conglomerate would usually operate as a plethora of subsidiary companies dedicated to particular regions, activities and product ranges. Combination was often focused at the transportation, refining and marketing stages – most notably in the US – though elsewhere crude production also came under more combined control. The degree of state orchestration varied widely, from official acquiescence in the US – which could itself be seen as a form of government facilitation – to much more active state involvement.

Of particular significance for Standard Oil was the formation in

1902 of the Asiatic Petroleum Co. Under the umbrella of Asiatic, the Rothschilds' Bnito – one of the two most dominant concerns in the Russian oil industry – joined forces with Royal Dutch and Shell to coordinate their efforts in the eastern markets against Standard Oil and Burmah Oil; and Asiatic went on to become a major player on the international oil scene for many years.[337] Also from 1902, Deutsche Bank, with the backing of the German government, began to coordinate the activities of German oil companies in Romania in order to combine their export potential.[338] Then, in 1904 two large German banks, Disconto-Gesellschaft and Bankhaus S. Bleichröder, in alliance with the Austro-Hungarian government, formed a petroleum trade association – commonly known as OLEX – to pool the marketing of Galician oil products in Europe.[339] In 1905 Russia's military defeat by Japan and the Russian revolution of that year set back the Russian oil industry, which strengthened the positions of Standard Oil, Royal Dutch and Shell in the Far East. Consequently, the Rothschilds and the Nobels joined with Deutsche Bank in 1905 to form the European Petroleum Union (EPU) as a counterforce to the extensive European marketing operations of Standard, and to the emergence of those of Shell; in fact, the EPU's pooling arrangements – obscure or simply secret – were merely the most prominent of many similar agreements between oil companies.[340] One trade journal surmised that the EPU had the backing of the German government, from its 'concern about providing the German navy with deliveries of liquid fuel'.[341] In 1907 a huge amalgamation was finalized between Royal Dutch and Shell, forming the Royal Dutch-Shell group. This was designed to strengthen their position primarily against Standard, and also against the rising presence of German banks in the oil industry;[342] but the head of Shell, Sir Marcus Samuel, blamed an absence of support from the British government – in particular its exclusion from the oilfields of Burma – for forcing him to amalgamate, as the junior partner, with Royal-Dutch.[343] The head of Royal Dutch-Shell, Henri Deterding, hailed corporate combination in the language of socialist and nationalist rhetoric: 'In the long run, selfishness never pays. God so made the world that we are all of us interdependent . . . "Eendracht maacht macht", runs an old Dutch proverb, meaning "Co-operation gives power".'[344]

In addition to Standard Oil's outright exclusion from some countries and the tendency of its competitors to join forces against it, most foreign governments also nurtured their oil industries by placing tariff barriers and other obstacles in the way of petroleum imports. Most European countries imposed high tariffs on the import of petroleum products, particularly Russia, France, Spain, Austria-Hungary and Romania, in order to protect their domestic refining industries, whether or not they were crude producers. In France, for example, the refining of vegetable oils and alcohols, plus a small amount of domestically produced shale oil, was a long-established industry which now also drew on semi-refined US crude that escaped the import tariffs on finished products.[345] Although Britain allowed almost tariff-free imports into the UK, it protected its imperial oil interests in Burma by erecting an import tariff barrier around India.[346] Standard Oil repeatedly complained that these tariffs were unfair; however, the United States itself had imposed a protective tariff on kerosene imports from 1861, and it was only with the 1909 Payne–Aldrich Tariff Act that import duties were removed for all crude or refined oils, except residuum and tar.[347] Indeed, since the era of Alexander Hamilton in the early nineteenth century, the US had followed the practice of other countries by regarding import tariffs as a normal and necessary tool of economic development. As famously described by the nineteenth-century political economist Friedrich List, tariffs and other protections were routinely deployed to shield 'infant industries' until they could compete on an equal footing with their foreign counterparts.[348] One American commentator wrote, 'I presume Russia is charging such a high import duty because it knows that if it were not for this, the Standard Oil Company would first ruin all the Russian oil industries by offering oil at cost if necessary, and afterwards charging extravagant prices to the Russian consumer.'[349]

From 1908, sudden production booms in Galicia and Romania, and a collapse in crude prices there, presented Standard Oil with an opportunity to take greater control over these countries' oil industries. However, both respective governments – accusing Standard of unfair, overbearing business practices – took measures to avert this. The Austro-Hungarian government – as well as arranging for the state railways to adopt fuel oil and building a state-run refinery at

Drohobycz – discriminated against Standard when conferring pipe-line construction rights and setting rail freight rates.[350] When Standard, along with a large French oil company, protested at the discrimin-atory freight rates, *The Times* noted the irony. Austria had deployed

> methods of administrative chicanery and railway discrimination strik-ingly similar to those which made the name of the Standard Oil Company a byword in the United States. The tactics of the Austrian authorities are as indefensible, or as defensible, as are those of the Stand-ard Oil Company; but, as a sense of humour is not always an asset in business and diplomacy both the United States and the French Govern-ments made diplomatic representations to the Austro-Hungarian Foreign Office on behalf of the American and French interests affected by Austrian flattery of Mr. Rockefeller.[351]

Similarly, just as Standard Oil was excluded by Britain from crude production in Burma – and now also in Trinidad and Egypt[352] – as well as by Russia and by the Dutch in the East Indies, so was the British-owned Mexican Eagle Co. in 1913 excluded from Colombia under political pressure from the US government.[353]

National security considerations were often behind these exclu-sionary policies and, in the end, economic and military power were often seen as being inseparable – as the *United States Investor* had put it in 1898 at the outbreak of the Spanish-American War: 'Back of every political complication of today are trade considerations. Wars, treaties, colonization schemes and all the intricacies of diplomacy are really, in the final analysis, but means to one great end – the extension of trade, and the enlargement of the means of acquiring wealth.'[354] In his 1899 annual message President William McKinley conveyed a vision of global economic combat:

> In this age of keen rivalry among nations for mastery in commerce, the doctrine of evolution and the rule of the survival of the fittest must be as inexorable in their operation as they are positive in the results they bring about. The place won in the struggle by an industrial people can only be held by unrelaxed endeavor and constant advance in achievement.[355]

In 1907, as Standard Oil executives fumed at the mounting challenges to their foreign operations, the future American president Woodrow

Wilson – then president of Princeton University – made a virtual declaration of war on behalf of corporate interests:

> Since trade ignores national boundaries and the manufacturer insists on having the world as a market, the flag of his nation must follow him, and the doors of the nations which are closed against him must be battered down. Concessions obtained by financiers must be safeguarded by ministers of state, even if the sovereignty of unwilling nations be outraged in the process. Colonies must be obtained or planted, in order that no useful corner of the world may be overlooked or left unused.[356]

The terminology of warfare permeated business discourse.[357] In 1902 Frank Vanderlip, of the Rockefeller-allied National City Bank, wrote an effusive series of articles entitled 'The American "Commercial Invasion" of Europe', in which he argued that US exports would ultimately prevail over any European import tariffs, and that the US export surplus would translate inevitably into increased American foreign investment and ownership:

> The industrial world as yet is by no means prostrate at our feet. We have before us a long campaign . . . [along] the road to the commercial domination of the world . . . [But] [o]f all nations the United States has the most unbounded wealth of natural resources . . . the most abundant and the cheapest raw materials and supplies of fuel in the world . . . and the resources of no other great power are for one moment to be compared to them.[358]

This sentiment was mirrored in Deterding's opposingly imperialistic vision of Royal Dutch-Shell's destiny:

> In the years of clash and conflict which followed from 1903, it used to be said that I had taken as our motto: 'Our field is the world.' . . . [W]e had no other alternative but to expand, expand, expand. Had we restricted our trading purely within certain areas, our competitors could easily have smashed us by relying on the profits they were making in other countries to undercut us in price. So, to hold our own, we had to invade other countries, too.[359]

An account in *The Times* of an outbreak of intensive price-cutting competition between Standard Oil and its rivals in 1910 was entitled

'The Petroleum "War"', which the *Washington Post* saw as a global face-off between the Rothschilds and the Rockefellers:

> The Rothschilds, in the guise of the Asiatic Petroleum Company, between whom and the Rockefellers a sharp rivalry exists with respect to the control of the Far East trade, plan to extend the hostilities to America ... The invasion would appear to be in retaliation for the Standard's policy abroad, where its tactics have brought it into conflict with oil interests generally. A trade war already of great magnitude thus is now spread over practically the entire world.[360]

Indeed, in 1910, once Standard Oil had weaned itself off Sumatran gasoline for supplying its West Coast market, the Asiatic-Royal Dutch-Shell combine established its own marketing infrastructure there to sell its gasoline directly.[361] The *Washington Post*, among other newspapers, termed this an 'invasion',[362] and Honolulu's *Evening Bulletin* asked, 'Is Hawaii nei [beloved Hawaii] soon to be invaded by the Royal Dutch? ... [Its] plan is to fight the Standard in its own territory and the battle will be one to the last ditch.'[363] As Deterding put it,

> we obviously had to dig ourselves in as traders on American soil; otherwise we would have lost our foothold everywhere else. Until we started trading in America, our American competitors controlled world prices – because ... they could always charge up their losses in underselling us in other countries against business at home where they had a monopoly ... I decided that we must take America, with her vast resources both for production and for trading, into our general working plan.[364]

Spurred by Standard Oil's attempt in early 1912 to acquire oil concessions in Sumatra via a Dutch subsidiary, Royal Dutch-Shell started buying producing properties in Oklahoma and consolidated them under the Roxana Petroleum Co. The following year it began to purchase production sites in California and to establish its own oil transport and refining capacity such that by 1914 Royal Dutch-Shell had become, by one measure, the largest single foreign industrial enterprise in the United States – although, as the *Oil & Gas Journal* pointed out, its investment of $25 million was still only half that of the larger domestic companies such as the Texas Co. and the Gulf Oil Co., and just a tiny fraction of the thirty-four Standard Oil companies' $1.5 billion capitalization.[365]

This was just one component of a huge global expansion drive in which, from 1907 to 1914, Royal Dutch-Shell increased its combined asset value by over two and a half times.[366] Its Astra Română subsidiary became one of the largest oil companies in Romania;[367] in 1911 it took over its one remaining rival in the Dutch East Indies;[368] and in 1912 it acquired the Rothschilds' Russian oil interests, for which the Rothschilds received a large shareholding in the group, giving the Anglo-Dutch combine a roughly equal production capacity in the East and West.[369] In addition, it expanded its oil production operations into Sarawak – a British protectorate in north Borneo – into Egypt, Trinidad, Mexico and even New Zealand, where J.D. Henry, a prolific writer on oil matters, claimed, 'the oilfields ... are certain to be large producers in the very near future.'[370] The company bought a huge oil concession in Venezuela and it began contending for concessions in Mesopotamia in eastern Turkey.[371] Lord Cowdray, however, turned down an approach by Royal Dutch-Shell to enter an alliance with his Mexican Eagle Oil Co., as this would most likely end in its absorption; as he explained to his vice-president,

> The Royal Dutch will be most formidable competitors, and if they came into business they might spoil the position we shall otherwise have, as no other company could do. By that, I mean that they are very energetic, enterprising, with endless resources and are great transport people. I know their ambition is to get into Mexico, secure a big supply of crude, and with it their fields in Roumania and the East Indies, to circle the globe with a chain of fuel oil stations.[372]

'Germany Opens War on Standard Oil' declared a front-page news item in the *New York Times* in 1910.[373] German oil traders had long resisted the growing monopoly power of Standard Oil, which now controlled about 80 per cent of the German market, and the few remaining independent traders were campaigning for government intervention to loosen Standard's grip and to facilitate the import of oil from Galicia and Romania where German oil companies predominated.[374] In the words of the editor of Germany's main oil trade journal, *Petroleum*, 'The Standard has taken the map of Germany and staked the entire empire out as its own particular claim. No hamlet is too small to merit the octopus's attention. The result has inevitably been the destruction of the small dealer.'[375]

By late 1912 the Reichstag was considering a bill to virtually nationalize the industry: Standard Oil would be forced to sell its German assets and only purely German firms would be allowed to trade, under government supervision. In the view of an American observer,

> There is no doubt that the Kaiser's Government is proceeding on the theory that the United States Government would not raise a finger to defend the interests of the much-maligned 'Rockefeller octopus' ... that, quite apart from the Standard's unpopularity at home, no American Administration would risk declaring a tariff war with Germany on Mr. Rockefeller's behalf.[376]

The head of Deutsche Bank, Arthur von Gwinner, backed the monopoly proposal with the argument that Germany could thereby become less dependent on Standard-controlled sources of petroleum. However, given that around 75 per cent of Germany's petroleum came from the United States, how could this be achieved? When Gwinner attempted to secure oil from independent American companies he was outmanoeuvred by Standard's fixer in Germany, Walter Teagle, who signed contracts for those supplies ahead of Deutsche Bank.[377] Teagle gloated, 'This would seem to leave comparatively small quantities of oil available during 1914 to be bought up if the German monopoly should be created.'[378] Kaiser Wilhelm II himself became involved in attempts to persuade the Russian government of the benefits of a German, Austro-Hungarian, Romanian and Russian oil union, the question being, as he put it, 'How can European states which possess oil resources (including Russia) create by their own efforts a special organization, whose goal is to provide European states with liquid fuel and distilled products of crude oil, but without relying upon transoceanic markets?'[379]

However, Russia counted Austria-Hungary as a geopolitical adversary, while the Russian Navy by this time considered Germany to be a more serious rival than Britain. Indeed, one of Germany's motives behind its monopoly scheme seemed to be to gain greater control over supplies of fuel oil for its navy, particularly in time of war.[380] The *New York Times* reported,

German military and naval circles are expressing misgivings in regard to the effect of the proposed State petroleum monopoly on the defensive establishment of the empire.

The naval men especially are said to be somewhat alarmed owing to the increasing use they expect to make of oil fuel in the fleet. They declare that inasmuch as Germany, under even the best circumstances, will need to buy large quantities of oil from the Standard Oil Company, it is very necessary to contract for such supplies for a long term of years ahead. Count von Bernstorff, German Ambassador to Washington, during his recent visit to Berlin, is said to have made strong representations in this direction and called attention to the fact that favorable contracts with the Standard were much in doubt if a controversy arose.

The situation ... is all the more interesting in view of the fact that Herr [von Gwinner], Managing Director of the Deutsche Bank, who won the Kaiser over to the monopoly project, has made special use of the argument that the German Navy could not afford to be dependent for all its fuel on the grasping American monopoly.[381]

Thus, once these pressing practical matters were considered, support in the Reichstag for the state monopoly bill waned. Perhaps this was also partly because the head of Standard Oil's German subsidiary was one of the main funders of the Roman Catholic Centre Party, the *Frankfurter Zeitung* declaring, 'Rockefeller now rules, too, in the German Reichstag.' Teagle returned to New York – on the steamship *Lusitania* – victorious.[382] From 1912 to the summer of 1914 about 75 per cent of Germany's petroleum imports came from the United States. Although, over this period, it shifted its sources of gasoline away from the US (20 per cent) and towards Romania (28 per cent), the Dutch and British East Indies (21 per cent) and Russia (20 per cent), with Austria-Hungary supplying 12 per cent, the US remained the source of 78 per cent of Germany's staple kerosene illuminant imports and 42 per cent of its petroleum-based lubricants.[383]

Indeed, in the decade or so leading up to 1914, although the geographical spread of oil production shifted significantly, the United States would remain by far the largest producer, its crude output surging to 248 million barrels in 1913, accounting for 65 per cent of global production. Meanwhile, although Russian production had briefly overtaken that of the US at the turn of the century, since the revolutionary upheavals of 1905 its output had fallen to about 80 per cent of its previous levels and had remained roughly constant, so that by 1913 its 63-million-barrel output accounted for only 17 per cent of worldwide production. Mexican output, by contrast, jumped from zero in 1903 to 26 million barrels in 1913, 7 per cent of global production. The rising outputs of Romania, the Dutch East Indies and Burma would account for a further 4 per cent, 3 per cent and 2 per cent respectively, while Galicia's declining output, down to 8 million barrels, represented another 2 per cent.[384]

In the United States, production from the original Appalachian fields declined only slightly from their 1890s levels, while falling production from the Lima-Indiana fields was offset by new production from Illinois. But the country saw a major shift in production with an explosion of new output from the Mid-Continent fields (primarily Oklahoma, plus northern Texas and Louisiana, and Kansas) and from California, along with a significant, though smaller, increase from the Gulf fields (coastal Texas and Louisiana). California retained its position, since 1903, as the top-producing state with a crude output in 1913 of 98 million barrels, followed by Oklahoma's 64 million barrels, representing 39 per cent and 26 per cent of US production respectively.[385] With the surge in production came colossal levels of waste and pollution. The largest of California's many out-of-control oil strikes was made on 15 March 1910 at the Union Oil Co.'s Lakeview well in the prolific Midway field in San Joaquin Valley, south of Buena Vista Hills in Kern County. For a year and a half it gushed tens of thousands of barrels per day, totalling an estimated 9 million barrels, of which around 5 million were lost (see Fig.14).[386] From Oklahoma's huge Cushing field it was reported in early 1914, 'Every gallon of storage in the field is filled and the tanks are slopping over. From 3,000 to 5,000 bbls. [barrels] are going to waste daily.'[387] Millions of barrels of crude evaporated from their tanks and, according to the US Bureau of Mines, in Oklahoma 'many tanks are lost by fire caused by lightning ... and it is a common occurrence in the oil fields during a thunder shower to see a great tank containing thousands of barrels of oil rapidly burning'. Simultaneously, vast quantities of associated gas were released into the atmosphere, and it was estimated that in 1913 100 billion cubic feet of gas was wasted in this way from the Cushing field alone, which 'would have met the wants of nearly 1,000,000 families for 1 year . . . Not only was the gas allowed to waste, but such tremendous volumes of this inflammable material hung over the oil fields that . . . in many cases disastrous fires were started, resulting in the loss of life and property.'[388]

Following a similar pattern to developments in California, the inability of Standard Oil's Prairie Oil & Gas Co., alongside Gulf Oil and the Texas Co., to build the storage, pipeline and refinery capacity to keep pace with the torrent of crude in the Mid-Continent field, continued to open up opportunities for the formation of rival integrated oil companies such as Magnolia Petroleum and the Sinclair Oil & Refining Co., as well as Royal Dutch-Shell's Roxana Petroleum. Yet the consequent surge in output of refined petroleum products by the American oil industry was barely able to meet the even more rapidly rising domestic and foreign demand – particularly for gasoline and fuel oil – and prices often rose steeply. Whereas the United States was exporting about 40 per cent of its oil products around the turn of the century, this would fall to 25 per cent by 1914.[389]

Meanwhile, as increasing domestic demand in Russia absorbed even more of the country's post-1905 stagnant output of oil products, its global market share plummeted. Very little Russian fuel oil was exported and, due to its high price in 1913, the Trans-Caucasus Railway – Baku's long-standing export outlet, along with the parallel pipeline – actually imported cheaper consignments from Mexico: 'A regular case of carrying coal to Newcastle,' commented the British consul in Batum.[390] Crude production at Baku was also impacted by the declining output of existing fields, by the use of old drilling technology and the continued working of many hand-dug wells, and by the stalled leasing of new oil territory by the government and its high royalty rates. However, high domestic oil prices attracted an infusion of foreign capital from 1910 that propped up the ailing industry, over which Nobel Bros. would nevertheless remain dominant. British capital in particular was channelled through a new combine, the Russian General Oil Corporation, and into Royal Dutch-Shell subsidiaries that would control nearly a fifth of Russian production by 1914. Much of this investment went into opening up new oilfields that somewhat offset the impaired output from Baku: in the north Caucasus at Grozny in the Terek district and at

Maikop in the Kuban region; in the Ural-Emba region just north of the Caspian Sea; and at Cheleken on the eastern shore of the Caspian. However, oil operations were significantly impacted by labour strikes across the Caucasus from mid-1913.[391]

Thus, with the falling away of Russian exports, oil-importing countries became almost singularly dependent on the United States as by far the world's leading source of supply. For example, by 1903 Russia had risen to join the US as Britain's two almost exclusive sources of petroleum products; thereafter, imports from Russia declined dramatically while those from the US increased until, between 1909 and 1910, they were supplying approximately just 10 per cent and nearly 60 per cent of Britain's petroleum respectively. Then, until 1913, as Russian imports increased slightly, these were now far outstripped by further rising imports from the US, were now roughly equalled by imports from the Dutch East Indies, and were exceeded by those from Romania. From 1911, Britain's fuel oil imports suddenly surged; but, in an exception to the pattern of primary dependency on the US, these came mostly from Mexico and Romania, on which note the *Petroleum Review* concluded its summary of Britain's petroleum trade for 1913:

> In this respect Mexico stands in the pre-eminent position . . . Roumania, however, is gaining headway, while increasingly larger quantities are coming from the United States. If one adds to these figures the enormous amounts which the British Admiralty consumes, it will be seen that the liquid fuel aspect of our petroleum import trade is to-day one of the most important.[392]

'A VOLLEY OF ARROWS'

Since 1885, the assertion of European sovereignty over the island of Papua, just north of Australia across the Torres Strait, had been divided between Holland in the west, Germany in the northeast and Britain in the southeast. As the British Foreign Office would later put it, 'the natives in general accepted the new arrangements quietly; and the murder of Europeans, which had been continuous since 1845, the vendetta, head-hunting, and cannibalism, common in the territory, virtually ceased as the authority of government was extended.'[393]

Towards the turn of the century a number of syndicates began to establish large-scale plantations and in 1905 British New Guinea, renamed the Territory of Papua, came under the administration of Australia as its formal dependency.[394] In 1911 two former employees of the British New Guinea Development Co., Lewis Lett and Garnet Thomas, set out to establish their own coconut plantation near Kiri Creek, a tributary of the Vailala River 15 miles from its mouth on the Gulf of Papua. That August they were told by two gold prospectors that, following up the reports of a local native, they had smelt kerosene in the jungle near the creek. Excited by the prospect of a new commercial possibility, the two planters investigated further and found a bubbling gas spring, or vent, in the vicinity of the nearby village of Opa. Many similar vents were soon discovered in the area along with indications of oil which Lett and Thomas, the two gold prospectors, and the British New Guinea Development Co. registered with the colonial authorities. The government, however, asserted its own claim over this likely oil patch as Crown property because of fears that it might be taken over by Standard Oil.[395] Several official expeditions through the jungle waterways of the delta recorded numerous oil and gas springs between the Vailala and Purari rivers. In December the Acting Resident Magistrate of the Gulf Division wrote of being shown the oil and gas seeps near the Vailala, a few miles up from the village of Akaunda:

> The froth caused by the bubbling had a very decided odour of kerosene. In this vicinity there were also a number of places where highly inflammable gas was issuing from small fissures in the dry earth. The noise of the gas forcing its way upward could be heard from some distance. Once lighted, these jets of gas would burn brightly for an indefinite time. One remained alight for well over two hours, and was still going strong, after having boiled a billy-can full of water in quick time.[396]

In May 1912 the government geologist, J.E. Carne, reporting on a circuitous expedition through this new potential oil region, wrote,

> It speaks eloquently for the Government policy of patrolling and general oversight, that this populous region – in the not distant past a

dangerous one – can now be traversed ... without danger, a result achieved with little or no bloodshed. Punishment of offenders has been secured, weaker villages protected, and general confidence instilled in the native mind, by firm, but just and humane treatment, the culprits returning to their villages impressed with the power and fairness of the 'Governor'.[397]

Carne's overall conclusion was that 'the possibilities of a valuable national asset in oil in Papua are great', with the qualification that 'however favourable the natural indications may be, upon the drill alone depends proof of workable oil-fields'.[398] Another concern, which affected the Papuan colonial economy generally – primarily gold mining and plantation agriculture – was the shortage of manual labourers. Not only had the indigenous population been impacted by European diseases, but the materialism and work ethic of the relatively small number of white settlers were alien to them. According to one report, 'They detest work. They laugh at the white man, and would infinitely prefer to fritter away their lives in their own villages than to toil and hew.'[399] Nevertheless, 'in some districts,' wrote Papua's Lieutenant Governor, Judge John Hubert Murray, 'the natives have become so used to the white man's luxuries, which can only be obtained by money, that they have come to look upon labour for the European as one of the normal features of their existence.'[400] At the same time, Murray was concerned at how the colonists' frustrations at indigenous Papuans would manifest in coercion and abuse, and in early 1913 he wrote to his brother Gilbert, professor of Greek at Oxford University, that he was 'coming to the conclusion that any white community left with absolute power over "natives" would resort to slavery within three generations – I can feel the tendency myself'.[401]

As the settlers' attention moved to the Vailala River, in the spring of 1912 a team of gold prospectors ventured up towards its headwaters and came into conflict with some men of the Kukukuku (Anga) tribe standing on a large rock. In the words of the team leader,

> one of them fired an arrow and struck me on the left breast, about two inches from the heart. The arrow travelled in the direction of the left kidney, penetrating almost through me. I snatched the arrow out, then fired at the native. Priddle and the boys also fired. The native who

fired at me I shot and killed . . . Mr. Priddle had a stretcher made, and got the boys to carry me back to the depot . . . I am of the opinion that had these men been on the ground where I could have walked up to them the attack would not have been made, but that on the rock they considered themselves safe from attack and took our rifles for clubs. I consider the shooting of the men absolutely necessary for the safety of the entire party.[402]

High hopes that Papua was on the verge of an oil boom and that Papuan oil would fuel the Australian Navy's new oil-fired destroyers and cruisers were disappointed, however, as the government drilling team, led by an American petroleum engineer, failed over the next few years to strike oil around the Valaila in commercial quantities.[403]

Meanwhile, in Venezuela President Juan Vicente Gómez, in October 1911, called for foreign investment in order to modernize the country: 'We have natural wealth that remains unexploited; mostly uncultivated fertile land, immense empty territories.'[404] However, many of these territories were not 'empty'.

In 1908 the newly inaugurated President Gómez had added to his predecessor's grants of millions of hectares of petroleum concessions by granting another covering the Benitez district of Sucre state and a second, huge concession covering 27 million hectares that encompassed twelve states. These new concessions were soon taken over by, respectively, the Bermudez Co. and the Caribbean Petroleum Co., subsidiaries of the General Asphalt Co., a US company with a politically fraught history in Venezuela. They began surveying and exploratory drilling in 1912, but due to their lack of experience with oil and the huge size of the larger lease, after spending nearly $500,000 on exploration they had little to show for it, at which point General Asphalt sold a controlling stake in Caribbean Petroleum to Royal Dutch-Shell for $10 million, with the involvement of the French Rothschilds. Sir Henri Deterding, head of Royal Dutch-Shell at the time, later wrote that the takeover of Caribbean Petroleum was 'universally reckoned to be our most colossal deal [in which] I made, perhaps, the most speculative venture of my life'.[405] One of the Castro-era concessions was taken up by another General Asphalt subsidiary, the Colón Development Co., and another by a British company, Venezuelan Oil Concessions Ltd., both of which were also taken over by Royal Dutch-Shell a few years later.[406] Standard Oil, by contrast, was averse to going into upstream production in South America at this time; as one Standard executive wrote of Peru, 'It will be more profitable to buy production than to hunt for it, and to buy oil than to run the risks in this territory for producing it.'[407]

As Royal Dutch-Shell's team of geologists surveyed the 'immense empty territories' presided over by Gómez, the local inhabitants were their first valuable resource:

> It may be said that many of the seepages in remote or forested regions were shown to our men by observant inhabitants of the country ... We found the people in Venezuela good observers and reliable guides, and much credit should go to them for finding many important prospective areas ... In parts of our work we had to find a Guarauno Indian to act as a guide to show us where there were asphalt deposits in the area. At first it was exceedingly hard to get one, as they would run into the jungle and hide. This was due to the Venezuelans having rounded up numbers of them and forced them to labor as slaves gathering rubber in the jungle. After we had employed one or two and given them canned foods and cigarettes as well as paying for their labor, we were over-whelmed by them.[408]

The indigenous populations of Venezuela had suffered from the violent incursions of European and mestizo farmers and ranchers for centuries, but since the mid-nineteenth century some had actually managed to regain lost territory.[409] In 1894 the American consul in Maracaibo had advised Washington on how best to conduct relations with the local Barí – or 'Motilones', as they had been named by the Spanish invaders – west of Lake Maracaibo: 'A well-organized raid upon these [Barí], giving no quarter, without regard to age or sex, is, in the opinion of those most competent to advise, the only method to pursue. Burning all houses and settlements and uprooting all plantations would naturally follow ... [T]he Motilones tribe can only be treated as beasts of prey.'[410]

This region was covered by Royal Dutch-Shell's Caribbean Petroleum concessionary area that the geologists now, in 1913, intended to survey. One wrote,

The boundary of the Indian country is roughly the Rio Negro about ten miles south of the town of Machiques. The ranches along the northern bank of the river are in a constant state of defense and men are occasionally picked off ... Aside from a natural and apparently justifiable desire for revenge on the Venezuelans, the object in attacking the ranches is to get machetes and salt ... These Indians show signs of not being low savages. Twenty or thirty years ago they were at peace with the Venezuelans and had communications with them, but the barbarous treatment they have received, according to the stories which are told, has driven them into their present state of perpetual hostility. They seem to be more than holding their own in the conflict as they have been slowly driving the Venezuelans from their ranches.[411]

The team employed Venezuelans to cut through the jungle with machetes, and all were armed: '[O]ne with a rifle went ahead of the machete men, another with a rifle stayed ... in the rear. We continually saw signs of Indians, small palm leaf huts, footprints, trails, and places where they had made fires ... Often the sentinels would shoot at something they took to be Indians in the dark.'[412] Indeed, the Venezuelans often seemed to be rather trigger-happy: '[T]hey saw something which they took to be Indians and began to shoot. Some of the men rushed out from the camp and started shooting in every direction.'[413] On another occasion, when travelling by canoe,

We turned the bend and about 150 yards ahead of us were four or five Indians on a raft ... I did not get a very good view of them as I was busy trying to keep the two Venezuelans with me from shooting at them ...

At a sharp bend in the river we came on a bunch of Indians building a raft. I counted ten, including one child ... They did not see us until we were within 60 yards of them when I had to shout to stop one of my men from shooting.[414]

When deep in Barí territory, however, the geologists came under attack themselves on numerous occasions. On the Oro River in the Colón district southwest of Lake Maracaibo, for example,

Just as we were getting into the canoes, about twenty Indians stepped out of the brush on the opposite side, fifty yards distant, and let fly a volley of arrows. We stepped back, as far as we could on the sand bar, and began to shoot. I saw two of the Indians drop and the rest disappeared ... Our men calculated that we hit at least twelve. I have to hand it to them, the Motilones are brave men. Their real last attack, when they came at us from both sides, was just as vicious as the others, in spite of the fact that many of them must have been killed. I do not believe that there were more than forty Indians altogether, who took part in driving us out.[415]

Given the hostile conditions, the survey team wondered how future oil operations might be made possible:

Captain Dame tells me that in Colombia there is a mission school where Motilones are being educated. It would be worthwhile to go to a very large expense, which might not be required, to get some of these Motilones and try to establish communications with those of the [Rio] Ariguaisa. Aside from the danger, difficulty and very large expense of going up into this country and cleaning them out, I do not like the idea of destroying a whole community of men, women and children. But this would be the only thing to do unless peace is made. After three centuries of continuous warfare between the Venezuelans and the Motilones, opening up peaceful relations is a matter of no small difficulty, but it can be done if the problem is attacked in a serious manner. If oil is found up the Lora, peaceful relations with these Indians would be worth several hundred thousand dollars to the company.[416]

Meanwhile, two oil prospectors searching in the region on behalf of an American company 'were attacked by a band of hostile Indians and had a narrow escape from death or capture', reported the *Boston Globe*. One of the men 'had been severely wounded by arrows during the skirmish, but ... the attacking party had finally been driven off by a hot fire from the Americans' automatic pistols after two members of the hostile band had been killed'.[417]

Royal Dutch-Shell's explorations around Lake Maracaibo at this time led to the discovery of a number of significant oilfields, such as Mene Grande, Las Cruces and the Bolívar coastal field;[418] and Venezuela's minister of development, in his 1913 report to Congress, looked forward to 'a new source of income that shortly will be of major importance. Oil . . . that sought-after source of fuel . . . has ceased to be a treasure hidden in the bowels of the Venezuelan earth.'[419]

In early 1914, at a Venezuelan Oil Concessions general meeting, its chairman told of the 'vast potentialities' of the company's concession. The shareholders, he declared:

> . . . stand to reap immense profits. It is estimated that over £80,000,000 of British money has been expended in one way or another in connection with the oil industry in Europe, Russia, Roumania, Galicia, &c., and the adjacent countries. In the event of a great European conflagration, the whole of the results being obtained by that vast expenditure would cease within a few hours of declaration of war, and apart from the financial loss, the very results of that expenditure could be used against this Empire with an effect that would be ludicrous, were it not a calamity. In such an event one can now only turn at the present time to two other parts of the world where regular supplies can be obtained, the East and America. The immense concession we are now handling is situated in a unique and strategic position. A great chain of the Andes guards it on three sides . . . and on the other side the way is practically open to England by deep water.[420]

8

Oil for War

OIL 'FOR A GREAT WAR'

Around the turn of the twentieth century the world's navies had begun adopting fuel oil for its many advantages over coal as a raiser of steam – and the British Navy, in order to maintain its huge naval supremacy, was determined to lead the way.[1] From 1905 auxiliary oil-firing, for sudden acceleration and a greater top speed, had become standard for the Royal Navy's ships, including its main battleships. Since then, an increasing number of its smaller warships were built to be solely oil-fired, although fears over oil supplies in the event of war prompted a temporary reversal in the 1908–9 naval programme when sixteen coal-fired destroyers were ordered instead of the previous year's twelve oil-only destroyers. By 1911 the Royal Navy had fifty-six oil-only destroyers either built or under construction, and from 1912 most of its cruisers would be built to be solely oil-fired so they could keep up with the oil-fired destroyers.[2] By February 1914 it would have a total of 166 oil-only ships completed or under construction: five battleships (*Queen Elizabeth*, *Barham*, *Warspite*, *Valiant* and *Malaya*), sixteen light cruisers, 109 destroyers and thirty-six torpedo boats.[3] Most other navies were adopting oil-firing by 1912, though often only as an auxiliary to coal.[4]

'As soon as I knew for certain I was to go to the Admiralty I sent for Fisher,' Winston Churchill recalled of his being appointed First Lord of the Admiralty in October 1911.[5] Admiral Sir John Fisher, who had been First Sea Lord until the previous year, had long been called the 'Oil Maniac' for his fiery enthusiasm for oil, and Churchill shared his passion for modernizing the Royal Navy and for adopting any new technologies that might help to maintain its lead in the global naval arms race, particularly relative to Germany.[6] In December, Fisher urged Churchill on by conjuring up the prospect that not only might the US Navy launch oil-only, steam-powered battleships ahead of the Royal Navy but also that the German Navy was on its way to developing the first internal combustion, diesel-powered battleship:

> Your old women will have a nice time of it when the new American Battleships are at sea burning oil alone and a German Motor Battleship is cocking a snook at our 'Tortoises'![7]

Unfortunately for Churchill, in early 1912 a committee chaired by Fourth Sea Lord Captain William Pakenham rejected a proposal to extend oil-only propulsion from the navy's smaller warships to its battleships; in addition, it calculated that a much larger oil reserve – for a projected twelve months of wartime consumption – was urgently needed, concluding, 'it is essential that construction for storage of a further 1,000,000 tons of oil at once [be] commenced.'[8]

Churchill was unwilling to accept the huge financial burden of these high oil-storage recommendations, and the soaring costs of the navy was the big issue in his March 1912 speech on the Navy Estimates:

> The second source of certain and uncontrollable increase lies in the consumption of fuel, coal and oil ... due to the rapidly increasing horse-power of the Fleet ... The adoption and supply of oil as a motive power raises anxious and perplexing problems. In fact, I think, they are among the most difficult with which the Admiralty have ever been confronted.[9]

Neither, however, could Churchill accept the Pakenham committee's cautious approach to the adoption of oil; so his response was to set up a royal commission to re-examine the subject – to be chaired by the fervently pro-oil Lord Fisher. On 31 July 1912 King George V formally instituted the Royal Commission on Fuel and Engines, to inquire into 'the means of supply and storage of liquid fuel in peace and war and its application to warship engines, whether indirectly or by internal combustion'; and, according to the Commission, these were matters 'of pressing interest gravely affecting the great question of our Naval supremacy'.[10] The inquiry was to be conducted in utmost secrecy: 'There can be no question as to the Report of the Commission being made public, nor as to the evidence taken before it,' Churchill told Parliament a year later.[11] Some of the Commission's deliberations were not even printed in the secret report, and one Commission member commented during the proceedings, 'Everything here is absolutely private; I do not think all the Cabinet Ministers even have seen the papers.'[12] Under this cloak of secrecy Churchill, furthermore, even pre-empted the work of the Commission. While it was still only being formed he decided that the five new *Queen Elizabeth* class of dreadnought battleships, to be the pride of the fleet, would be solely oil-fuelled.[13] Churchill wrote to Fisher,

> the matters wh we are concerned are too serious (I'm sure) for anything but plain language. This liquid fuel problem has got to be solved, & the natural, inherent, unavoidable difficulties are such that they require the drive & enthusiasm of a big man ... You have got to find the oil: to show how it can be stored cheaply: how it can be purchased regularly & cheaply in peace, and with absolute certainty in war.[14]

From 1902 to 1906 the Royal Navy had sourced all of its fuel oil from Texas, rising from 1,200 tons per year to 25,000 tons. From 1907, as its total purchases began to increase quite dramatically, it took a regular consignment of 7,000 tons per year from the Burmah Oil Co. In 1910, when it bought a total of 128,000 tons, its purchases from the United States began to decline, replaced by mostly Romanian oil plus significant quantities of Scottish shale oil. In order to stimulate sources of supply under British sovereignty, the Admiralty appears to have purchased both the Burmese and the Scottish oil at higher than the prevailing market prices, while a small amount of oil was sourced from Borneo for its Far East fleet. In 1912 it bought a total of 278,000 tons, just under 200,000 of which was used that year by the navy; supplies from the US suddenly quadrupled on the previous year to 180,000 tons, and it now took 30,000 tons from Persia. Thus, with its soaring demand, the Royal Navy became a major customer in the world fuel-oil market.[15] In early 1913 Churchill told Parliament that the Admiralty generally endeavoured to keep its fuel-oil contracts secret, revealing only that 'the bulk of the contracts have usually fallen to American and Roumanian contractors', among whom

Standard Oil's subsidiaries – the Anglo-American Oil Co. and Româno-Americană – turned out to be prominent.[16] Nevertheless, in July 1913 he did reveal that the Admiralty was signing a major contract with the British-owned company Mexican Eagle for the supply of 200,000 tons of fuel oil. However, this was cancelled six months later – either because the navy found that the oil fell below its specifications, or, as the Additional Civil Lord of the Admiralty, Sir Francis Hopwood, wrote to Churchill, because 'Fisher had taken fright at the paragraph in the *Times* hinting at "scandals" with regard to a Mexican oil contract, and is now suggesting that we ought to do without it.'[17] At the same time, by 1913 imports of Mexican fuel oil for Britain's growing civilian market had surged to overtake those from Romania.[18] Production from Trinidad, Egypt and Borneo was relatively small, but each was contributing to the navy's supplies of fuel oil.[19] Over these years, neither the Admiralty nor the civilian economy imported any fuel oil from Russia, where all of it was used domestically, and from 1912 continuing labour strife led Russia even to import some fuel oil, from Mexico, Romania and Egypt.[20]

Freight charges were a major component in the price of fuel oil, comprising about 45 per cent of the cost of Texas oil and about 60 per cent of Burmese. The greater use of fuel oil led to a tanker construction boom after 1900, and in 1908 an average of two oil tankers were being launched from British shipyards every week. This kept freight rates reasonably low until 1911, however, when freight rates rose dramatically as the rising demand for fuel oil outstripped the available tanker capacity. One striking illustration of this increasing demand was the construction of a 250-mile, 4-inch-diameter pipeline from the port of Matadi to Léopoldville (Kinshasa) in the Belgian Congo, initiated in 1910 and completed in 1913, to supply the steamers of the Upper Congo with fuel oil. At this time the Royal Navy only had two rather inadequate tankers of its own, *Kharki* and *Petroleum*, plus the former collier *Isla* used for supplying fuel to submarines. Such was the demand now for tankers that the Admiralty was unable to acquire them commercially, so in 1912 it embarked on its own tanker construction programme, as did the German Navy in late 1913.[21]

The Royal Navy's supply of fuel oil was thus largely dependent on the vagaries of the global oil market, to which it was particularly exposed due to its minimal strategic oil reserve. In the period up to 1909 in particular, its strategic oil storage consisted largely of whatever was held by private oil companies – as if its strategic reserve of this new, imported fuel could be modelled on its traditional, more secure reliance on domestic coal producers. Even the oil-storage construction programme that the navy subsequently initiated was judged by the Pakenham committee to be woefully inadequate, providing for only 205,000 tons compared to the one year's wartime consumption of 1.5 million tons it thought necessary.[22]

The Royal Commission's calculation of the size of the strategic oil reserve required by the navy would be dependent primarily on its analysis of where, in the event of war, oil production and transit routes might be interrupted. A detailed study was made of the world's current and potential sources of production, and the Commission was faced with the perennial issue that, vast as the British Empire was, it did not possess a single major oil-producing region. Burma's output was moderate but came nowhere near to meeting the navy's demands. This fundamental predicament was widely appreciated, as the *Economist* had put it in 1910, with a hint of desperation:

> The man in the street may have omitted to notice or remember that a number of destroyers and cruisers and other crafts built of late years have had their furnaces adapted for the burning of oil as well as or in place of coal . . . [B]ut the difficulty remains that . . . we have to depend on foreign sources of supply for our fighting ships. And that is a weakness to which we cannot submit permanently . . . We need to develop the oil resources of the Empire . . . We cannot possibly leave our naval developments absolutely dependent on supplies of fuel that may be cut off at any moment – or at all events at the most critical moment when it will most be wanted.[23]

Despite patriotic hopes for the oil potentials of Trinidad, Egypt, Canada, Newfoundland, British Borneo, Nigeria, South Africa, Australia, New Zealand and Tasmania, their actual prospects appeared to be only moderate, if that.[24] In Australia, there was some brief excitement at the prospect of fuelling the two oil-fired warships – the *Parramatta* and the *Yarra* – of the new Royal Australian Navy with oil from Papua and from Portuguese Timor. In April 1912, according to the Melbourne *Age*, the Minister of Defence was considering whether:

> ... the Defence department should establish a national refinery in Papua, and obtain there the oil required for the fleet unit. The department's bill for oil fuel will be, when the fleet unit is completed next year, at least £52,000 a year, and an immense saving would he achieved if the Papuan petroleum could be used. The destroyers and cruisers will use 14,000 tons of oil fuel a year, and the naval board has been paying from 60/ to 65/ a ton. Another and even more important aspect of the find is that the fleet will be independent of imported fuel should the Papuan deposits prove as good as expected.[25]

Three months later, the Sydney *Sun* hoped that, following the discovery of oil at Vessoro in Timor, the Timor Petroleum Concessions Ltd., headed by Sir Joseph Carruthers, former premier of New South Wales, would 'wring Timor dry' for oil for the Australian Navy; and, the newspaper asked, 'Will Timor become a Commonwealth possession?'[26] However, a native rebellion in Timor was not an incentive to investment and early attempts at oil production both there and in Papua floundered.[27] Neither did a few minor oil strikes in Britain itself – such as at Willesden, northwest of London, and at Kelham in Nottinghamshire – turn out to be the discovery wells of significant oil reserves, although 'One often hears the conviction expressed that "somewhere under England" there is a payable oil field,' wrote the *Petroleum World* optimistically.[28] Churchill told Members of Parliament in 1913, 'It is calculated that Scottish shales alone, if developed to their fullest capacity, would yield between 400,000 and 500,000 tons for 150 years – at a price. Immense deposits of kimmeridge clay, containing the oil-bearing bands or seams, stretch across England from Dorsetshire to Lincolnshire. There are extensive shale beds in Nova Scotia and New Brunswick.'[29]

However, although the present level of Scottish shale-oil production was reliably significant, it was dwarfed by the navy's consumption. The use, as fuel oil, of the liquid by-products of coal gasification in Britain's gasworks was considered quite feasible, and even *Oil News* exhorted Britain to emulate Germany here: 'The more we in England learn about what they are doing in Germany in this respect, the nearer we shall come to solving the great question of the supply of liquid fuel. While the supremacy of petroleum in this respect cannot be challenged, a very useful auxiliary supply of liquid fuel for certain purposes can be obtained from coal.'[30] It was calculated that this would not be commercially viable, however, such was the demand from US railroad companies for the increasingly lucrative creosote: 'Owing to their eating up their forests so quickly they are having to creosote their sleepers now, and that is taking a very large quantity of creosote from this country.'[31]

It seemed, therefore, that Britain was going to be dependent on foreign sources of oil for the foreseeable future. In addition to output from the United States and Romania, large increases were expected from Mexico and the Dutch East Indies, along with anticipated major new production from Persia and Venezuela.[32] But oil production had repeatedly proved unsustainable or disappointing, even across parts of the US, and Galician production, for example, had entered a decline that would turn out to be terminal. As the head of Royal Dutch-Shell, Henri Deterding, put it to the Royal Commission, 'we know exactly how rich an oilfield has been after it is exhausted ... [F]rom my experience there is not anybody who can be certain of his supply.'[33] Many British investors had recently been burned in the speculative excitement over the Maikop field in the north Caucasus, as *Petroleum World* reported at the end of 1911: 'One of the earliest notable events of the year was the completion of the first Maikop pipeline ... [A]t last the pipe was opened and the stored-up oil found an outlet into the world beyond. The importance of this event has since been somewhat dwarfed by the fact that the production of this field has not come up to expectations.'[34] As the highly experienced oil prospector, William McGarvey, told the Admiralty, even if 'all the conditions which belong to a petroleum field exist ... the drill and the drill alone can tell if petroleum exists in large and paying quantities'; and as Shell's Sir Marcus Samuel told the Commission, 'it is only the drill that proves'.[35]

Where oil production was high, it was often nevertheless taken up largely by local demand, as was fuel oil in Russia and, to a lesser extent, in California and Romania, while demand in the United States was raising the cost of Mexican oil.[36] According to *The Times*, 'Practically at present our oil-fuel supply can be drawn only from those sources which have a surplus over the home consumption; these are getting fewer year by year, and when they fail we shall have only our own Colonies to fall back upon.'[37] In addition, a lack of storage and transit infrastructure in oil-producing regions often created a wide disparity between the local production and the quantities available for export; this was particularly the case for Mexico.[38] Romania's oil companies struggled to handle their post-1910 production boom, after which the government moved to address the long-standing rail-transit limitations by building three export pipelines; however, their construction and the country's oil exports were obstructed by the Balkan Wars of 1912–13. In the autumn of 1913 it was eventually reported, 'The laying of the State pipe lines for oil from Ploeshti to Constantza [sic], which was interrupted by the war, has been resumed.'[39] Similarly, due to the lack of transportation infrastructure from Austria-Hungary's oil region in Galicia to its main Adriatic port, Trieste, and to its naval base at Pola, its navy actually had to import most of its oil from Romania, for which it commissioned a tanker to run the three-week round trip from Constanța.[40]

However, even when oil was available for export, was there the tanker capacity available to deliver it? When in 1911 the rising demand for fuel oil began to run up against the limited availability of tank steamers, the competition over chartering them led to a sharp rise in oil freight rates. As the Director of Navy Contracts, Sir Frederick Black, explained to the Royal Commission, 'a shortage of a very few steamers rapidly affects prices.'[41] In February 1913 the Admiralty reported a disappointing response to its oil tenders, and its Director of Stores warned, 'if not quite impracticable, extreme difficulty would certainly be experienced in obtaining and freighting the quantities [of oil] required annually for consumption and for reserves, even for light cruisers and destroyers only, apart from battleships.'[42] As Churchill told the House of Commons during that year's Navy Estimates speech in July, 'The first and greatest of all new features in the oil market in the last two years has been the great upward movement in prices. Oil, which in 1911–12 could practically compete on favourable terms with coal, has now almost doubled in cost. At the same time, owing to a temporary scarcity of oil transport, freights have risen by 60 or 70 per cent.'[43]

Thus, the oil industry's production and distribution capacity was lagging behind the rising demand for fuel oil. As Fred Lane, the

Rothschilds' pivotal oil specialist, summed up the situation, 'It is sel-dom that there is any surplus of oil in the world, and if at any time there be a surplus, it is simply because there has not been time to cre-ate the necessary facilities for sale and distribution.'[44] The cost to the Royal Navy was escalating, and the *Economist* argued, 'surely, with oil at these exorbitant prices, there is much to be said for the cautious policy of the German Admiralty, which frees a great deal of money for other purposes.'[45]

In the case of war, Black took an especially pessimistic view, imply-ing that both Atlantic and Mediterranean freight would be in great peril: 'It is essential to bear in mind that the only large sources of sup-ply open at the moment within Admiralty specification and reasonable cost of oil, viz., Texas and Roumania, are not likely to be available in time of war.'[46] The Burmah Oil Co. could supply 100,000 tons of fuel oil per year, but 'The freighting of this oil to Europe would involve delay, especially if the Cape route had to be followed', in other words, if shipment via the Suez Canal were to be impeded, which would also incur the additional cost of freight.[47] As a cautionary tale, during the Turkish-Italian War of 1912, the transit route for the Admiralty's sup-plies of Romanian fuel oil was cut off for some time when Turkey closed the Dardanelles Strait by laying mines. Cargo ships, 'principally laden with cereals, flour, and petroleum', were now unable to make the passage from the Black Sea to the Mediterranean, crowding the Bos-phorus, and Romanian oil bound for Germany went instead, for a while, by rail.[48] Later that year, the Royal Navy's tanker *Petroleum* was on a Romanian trip when it passed close to a mine still in the Straits; the *Petroleum World* warned, 'We need not enlarge upon the importance of the Dardanelles to the trade of Europe in general and to the oil trade with Roumania and Russia in particular . . . the only last-ing preventative is to forbid any Power to carry on war in the Dardanelles and on that understanding to forbid Turkey to place mines in this narrow passage.'[49]

Given the foreseeable limited tanker capacity and the vulnerability of tankers at sea, Black concluded,

> It is clear, therefore, that unless war brings a very serious curtailment of the world's oil trade, all possible pressure on owners to obtain the

steamers will be required for feeding the Fleet, and it will be extremely difficult to obtain, in addition, the steamers required for replenishing tanks. Everything, therefore, points to the necessity of our being independent of foreign supplies during war time.[50]

The Royal Commission concurred. The need for a larger strategic fuel oil reserve was 'a matter ... of primary importance and great urgency'; and its chairman, Lord Fisher, concluded, 'the removal of all cause for anxiety will only be completely achieved by getting sufficiently large stocks of oil accumulated within the next few years, and to this end there should be a steady progressive programme of Government storage to provide fully for a great war.'[51]

By early 1913, however, the navy had only built a reserve capacity of 250,000 tons of fuel oil in Britain and its overseas bases, as part of a construction programme for merely around half the storage capacity recommended by the Pakenham committee. Its six tankers under construction were, similarly, less than half the number it was thought would be required in the event of war.[52] The French Navy – Britain's indispensable ally for maintaining the security of the Mediterranean – was, meanwhile, experiencing an acute fuel-oil shortage. According to Oil News, its warships at Toulon recently had no more than two days' supply: 'The gravity of the position in which our allies were thus placed is hardly capable of exaggeration.'[53]

Three months before the October 1913 launch of HMS Queen Elizabeth – Britain's first oil-only battleship – a parliamentary debate on the 1913–14 Navy Estimates became heated over the subject of the provision of fuel oil to the navy.[54] Churchill was charged with pursuing a reckless policy. Admiral Lord Charles Beresford, former commander of the Channel Fleet, warned that, in addition to the problem of ensuring access to oil production, the questions of its transport, storage and distribution to the fleet had not been adequately addressed. Churchill, he complained, had allowed the navy to develop an acute oil dependency

> before taking the ordinary business precaution of getting the transports from the source of origin, the necessary storage and the transports for distribution ... [T]he right hon. Gentleman has not put up his store tanks in sufficient numbers, and he ought to have put up

a great many more of them before he started building these oil-burning ships. There will be 304 vessels burning oil. At the present moment there are 234 burning oil and coal, and some are burning all oil, and therefore we ought to have these reserves, and the storage should be provided as early as possible ... In three days two flotillas of destroyers burn nearly 5,000 tons of oil. That was merely at manoeuvres, and if you expend oil like that at manoeuvres, just imagine what it means in war time! My point is that the Admiralty ought never to have agreed to the laying down of those ships for oil until they had provided everything that was necessary to keep them going if strained relations suddenly occurred ... You have not nearly enough [tankers] to supply the 1,000,000 tons which are required, and those vessels cannot be built by the time you have finished your five big oil-burning ships ... They ought to have their storage full with at least 1,000,000 tons, and they ought to have ships for distributing it among the Fleet.[55]

Indeed, the 'R' class battleships ordered in the 1913 naval programme reverted to dual coal- and oil-firing, due to the increasing cost of oil and its storage, and to the warnings both of Pakenham and of a senior member of the Royal Commission.[56] As Churchill told Parliament, 'It is therefore possible to use coal as their main motive power, and this, it must be admitted, is convenient in view of the very high prices now ruling for oil.'[57]

Even the US Navy had been hesitant to adopt fuel oil due to fears over the sustainability of its domestic oil supplies and over the future price of fuel oil in the face of growing global demand. In 1911 the Chief of the Bureau of Steam Engineering had written to the Secretary of the Navy,

The probability of an eventual demand for petroleum greatly exceeding the supply, together with the fact that we produce the greater part of the world's supply, should give us a distinct military advantage over other nations. The control of our exports of oil might limit the extent of the adoption of the oil engine by our possible enemies. Under the present law, however, any foreign nation can draw on the production of the oil fields of the United States to an unlimited extent, and can thus accumulate in time of peace a reserve supply.[58]

In 1913 the US Navy finally decided that its new battleships would be oil-only, beginning with the USS *Nevada*, only after extensive measures had been taken to safeguard its provision of fuel oil. This included setting aside public oil lands as a strategic naval reserve, at Elk Hills and Buena Vista Hills in California, followed later by Teapot Dome in Wyoming, along with the construction of oil-storage facilities at naval bases, the largest of which were at Guantanamo Bay, Cuba, and Pearl Harbor, Hawaii.[59] The German Navy, the Royal Navy's main rival, became even more cautious towards oil after adopting auxiliary oil-firing for its smaller warships. Only in 1912, eight years after the British, did it eventually opt for oil-only torpedo boats; and in the same year, as the Royal Navy was ordering five oil-only dreadnoughts, the German Navy was only just incorporating auxiliary oil-firing in its three new *Derfflinger* class battlecruisers.[60] Admiral Alfred von Tirpitz recalled,

> As we didn't possess any oil wells near enough, we were compelled to maintain a supply for war. This was, however, next to impossible in the case of the big ships. The expenditure of hundreds of millions would have been required, but even with the torpedo boats we had to exercise economy. In the year 1912 we were, however, obliged to adopt oil as the sole fuel in the torpedo boats because we could no longer cope with the heavy claims on transport which coal demanded.[61]

Given its meagre domestic oil production, Germany was, like Britain, hugely reliant on imports: these came predominantly from the United States, and Standard Oil held a near monopoly over the German market. About 15 per cent of its imports came from Romania – either by sea or via the Danube – and about 5 per cent from Galicia, via the Elbe. Although Germany, primarily through Deutsche Bank, had built up a large stake in the Romanian oil industry, since 1906 it had been importing far less oil from Romania than had Britain.[62] Thus, there was considerable anxiety in Germany over the vulnerability of the country's oil supplies – particularly in the event of war – in view of its acute reliance on imports from overseas, mostly from across the Atlantic. The German authorities encouraged the domestic manufacture of petroleum alternatives: plant-based alcohol and coal-based tar oil; and an appreciable quantity of the latter was used to fuel the navy's oil-burning warships.[63] The government also moved to create a national petroleum monopoly that was intended to reduce Germany's dependency on the US, in particular on Standard Oil, and towards sourcing more of its oil from Romania, Galicia and Russia. This failed, however, when it proved impossible to source enough non-Standard and non-US oil to meet Germany's burgeoning demand.[64] Indeed, by 1914 even the Russian Navy was becoming concerned about the limited domestic production and increasing cost of fuel oil.[65]

Between 1904 and 1910 a German company made oil discoveries in Cameroon,[66] and in early 1914 *The Times* reported that 'the German naval authorities were deeply interested' in a new oil find in German New Guinea, for which its committee for the colonies had assigned £25,000 for preliminary geological borings; according to the *Petroleum Review*, the New Guinea finds 'have been of sufficient merit to warrant the formation of several large companies ... The obvious importance of such oil deposits for Germany's own needs is such that the Government cannot treat the matter lightly. For its navy it must have large quantities of liquid fuel if it is to keep abreast of the times, and if these can be secured under her own flag so much the better.'[67]

However, nothing came of these somewhat desperate and highly speculative German attempts at sourcing oil from the country's overseas colonies. Nevertheless, Britain's Royal Commission was given the impression that Germany's oil supplies were particularly secure due to its relative proximity to Romania and Galicia – even though these were the source of only about 20 per cent of its imports. The Commission was told, 'it is an open secret that the German Government have their arrangements in case of war for a supply of petroleum from Galicia and Roumania'; and 'we have been told that Germany has ear-marked certain portions of the territory as a source of oil for the German Navy in case of need ... [T]he oil could be brought up the Danube and distributed, via the Upper Danube, into Germany, and in that way they could get a supply of German-controlled oil, which would render them independent of American oil.'[68] Special prominence was given, in its evidence, to an article from the *Financial News* of 14 November 1912, entitled 'Will Germany Control the Oil Supply for Our Navy?': 'It is very suggestive ... to note the elaborate system of transcontinental water transport which has been created in order to render German buyers of oil quite independent of the various chances of ocean navigation of tank steamers, which might otherwise have had to pass from the Black Sea into the Mediterranean, and thence through the English Channel to North Sea ports.'[69]

In the July 1913 Navy Estimates debate, Winston Churchill attempted to soothe British anxieties over oil security by explaining that since, indeed, no single source of fuel oil could be relied upon with certainty, the Admiralty's policy was based on diversifying its sources of supply and on the cultivation of multiple sources so that the sum total of at least several of these would suffice at any one time. He explained to Parliament,

> three governing principles have been observed: First, a wide geographical distribution, to guard against local failure of supplies and to avoid undue reliance on any particular source so as to preserve as much security and as much expansive power or elasticity in regard to each source as possible; secondly, to keep alive independent competitive sources of supply, so as to safeguard the Admiralty from becoming dependent on any single combination; and, thirdly, to draw our oil supply, so far as possible, from sources under British control or British influence, and along those sea or ocean routes which the Navy can most easily and most surely protect ... [I]t is not a case of choosing between alternatives – it is not a case of choosing this course against that. On no one quality, on no one process, on no one country, on no one company, and no one route, and on no one oil field must we be dependent. Safety and certainty in oil lie in variety, and in variety alone.[70]

Lord Beresford replied, however,

> The right hon. Gentleman made a great point about getting oil from other countries. Look at the danger of that! We get in this country about 100,000 tons of oil from Roumania. She has just declared war against Bulgaria, and therefore the whole of our contracts with Roumania for the supply of oil are off at this moment. Something has been said about getting oil from Mexico. I have the greatest objection to that ... [W]hat has happened in Roumania may happen in Mexico, and a rebellion there might stop our supply. The Eagle Company has

got enormous contracts with the Mexican railways and the Standard Oil Company take precedence over us. We may get a new Mexican Government, and they may repudiate the contracts we have got, and object to the methods by which they were obtained. All those are dangers which are happening, or will happen, with regard to obtaining oil from Mexico.[71]

In the House of Lords, the Earl of Selborne – who, as First Lord of the Admiralty from 1900 to 1905, had overseen the Royal Navy's first adoption of fuel oil – also remained unpersuaded: '[T]o lay down five great ships last year and design them so that they could only be propelled by oil was a rashness which I would characterise as almost unpardonable.'[72] Meanwhile, the *Observer* pondered the world-historic shift in play: 'When we began to win naval supremacy the island by its western position had a unique advantage from the winds. In the age of steam it enjoyed another unique advantage in the extent, situation and character of its coal deposits. In changing to oil we forfeit all favour from Nature, and depend upon conditions relatively more artificial and precarious.'[73]

Even with Churchill's ally Lord Fisher at the helm of the Royal Commission, such was the concern of its members over the security of oil supplies that it recommended a strategic naval reserve of four years' peacetime consumption – which Churchill still regarded as being beyond his budgetary powers. In order to engineer a lower reserve standard, he turned to his newly formed Naval War Staff for a fresh study, this time based on the assumptions that, during wartime, Britain would pay any price for oil and would manage to retain absolute command of the seas.[74] Churchill justified this latter assumption by portraying the survival of the British Empire in terms of its all-or-nothing naval supremacy:

> Our power to obtain additional supplies of oil fuel in time of war in excess of those which are being stored in this country, depends upon our preserving the command of the sea. If that is endangered, oil is not the only commodity which will be excluded from these Islands. If we cannot get oil, we cannot get corn, we cannot get cotton, and we cannot get a thousand and one commodities necessary for the preservation of the economic energies of Great Britain.[75]

As Major Henry Guest, MP, had argued in Parliament a year earlier, in Churchill's presence, 'the whole fate of our Empire is based on the principle of naval defence ... At some future date, for all we can tell, we may be vitally dependent on the supply of oil coming from the Black Sea or the Persian Gulf.'[76] Lord Beresford argued, 'we must remember that if a foreign country captured our oil supply we should have to demobilise our Fleet at once. If once our supply is cut off we have not enough oil in store, and we are not likely to have. In this matter the Admiralty have made a mistake.'[77]

By mid-1913 Fisher had come to the view that submarines could jeopardize Britain's oil supplies – 'nothing can stand against them' and they would 'sweep all surface warships from the seas' – and he even supported the idea of a Channel Tunnel for 'getting our oil in war as well as food'.[78] But in 1912 he had shared Churchill's outlook:

> Oil tankers are in profusion on every sea and as England commands the Ocean (*she must command the Ocean to live!!*) she has peripatetic re-fuelling stations on every sea ... Before very long there will be a million tons of oil on the various oceans in hundreds of oil tankers. The bulk of these would be at our disposal in time of war. Few or none could reach Germany.[79]

This assumed the maintenance of Britain's traditional naval supremacy that, ironically, fuelled calls for expanding the Admiralty's budget even further – possibly a calculated gamble on Churchill's part. As the *Observer* argued,

> If we are to make the stupendous change from coal to oil ... we shall require a definitely increased standard of sea-power in order to ensure an adequate supply of the new fuel ... [B]ut with an optimism to tempt the gods, Mr. Churchill proclaims the age of oil. And at the same time he announces a naval programme falling below his own statements of the bare minimum required ... It is a thing which concerns every citizen in the land and will influence for ever the destinies of Britain and the Empire ... Let us face the truth. If we are to depend on oil, the whole conditions of its supply dictate for us, whose sea-power is our all, a standard of naval strength definitely increased, and, once established, to be maintained at any cost. Otherwise the new departure, in

spite of all the unquestioned technical advantages of oil, would be the beginning of our end . . .

More than ever, it is certain that the British Empire cannot be maintained on the cheap. In a moral sense, Mr. Churchill will find that he cannot be a great man on the cheap. So far, in respect of the standard of naval strength, the main matter of his business, he has been weak and he has failed.[80]

In early 1914 reports of a reduction in the Royal Navy's fleet activity due to a shortage of fuel oil turned out to be well-founded. The previous September, Churchill had written to the Naval War Staff, 'During a year when we are making every effort to accumulate rapidly our oil fuel reserve . . . special efforts are needed to reduce by every possible means the great consumption of oil by flotillas . . . It may be necessary to omit the manoeuvres altogether next year . . . or in any case to leave the oil burning vessels out of them.'[81]

Lord Beresford now confirmed in the House of Commons that sixty-two new 'L' class torpedo-boat destroyers were out of action for just this reason:

I repeat what I said last year on the oil question, that the building of these five 'Queen Elizabeth's' and of an extra number of oil-burning ships is a dangerous experiment and gigantic gamble that might bring grave danger to this country in a sudden emergency. We have to provide a large amount of oil, and we cannot do it except by stopping the manoeuvres. Remember more has to come. There are sixty-two ships to be provided for, and I forgot also the 'L' class of T.B.D.'s. They have all got their new torpedoes and their new guns, but we have not been able to try them because we have not the oil.[82]

The navy complained to Churchill about the dire straits it was in, and he had to admit to his Cabinet colleagues that the German fleet was, as a consequence, now spending more time at sea than the Royal Navy.[83]

INTERNAL COMBUSTION

As the world's navies struggled and competed to secure the oil needed to power their oil-fuelled, steam-driven warships, there came into view on the distant horizon a potential saviour, a technology that promised to reduce dramatically the quantities of oil required for marine propulsion.[84] As one witness before the Royal Commission put it,

> the utilisation of the oil ... in the simple burning of the crude oil or proper fuel-oil under boilers (thus indirectly producing power), is the most wasteful that can possibly be ... [T]he enormous amount wasted will compel people gradually to adopt different methods: that is to say the internal combustion engine will replace the wasteful system of using fuel-oil under boilers.[85]

Rudolph Diesel's new type of internal combustion engine was more of a workhorse than the greyhound-like petrol engine, and diesel engines began to be used for their reliable output of power for mechanical work and electricity generation. While petrol engines were good for propelling small motorboats, the heavy-weight diesel engine appeared more suited for larger marine vessels. Diesel engines were fuelled by heavier grades of oil than the more volatile and explosive petrol, so they began to be adopted for powering submarines, the early, petrol-driven versions having been prone to catastrophic engine room explosions. During the same period, in 1903 Nobel Bros. launched *Vandal*, the first of a fleet of diesel-powered oil tank barges plying the Volga, and in 1908 they launched the first of several diesel tankers, *Dyelo*. In 1909 Royal Dutch-Shell ordered perhaps the first truly ocean-going diesel vessel, the tanker *Vulcanus*, which came into service in early 1911.[86] A number of diesel-powered merchant ships began to appear on the high seas, while many more were under construction, and in mid-1912 the *Toiler* became the first to cross the Atlantic.[87] The Royal Commission duly noted a recent observation by the US government's chief petroleum geologist in his 1912 annual report: 'In view of the inability of the producers to furnish fuel oil in the desired quantity, it is most fortunate that it is no longer necessary to convert oil into power by burning it under boilers.'[88] The chairman of the Royal

Commission, Lord Fisher, wrote to Churchill, 'When a cargo steamer can save 78 percent in fuel and gain 30 percent in cargo space by the adoption of internal combustion propulsion and practically get rid of stokers and engineers – it is obvious what a prodigious change is at our doors with oil!'[89] After asking the navy's engineer-in-chief how long it took its steam engines to heat up before they could reach full speed, Fisher wrote again to Churchill, 'He said about 8 hours but it could be less. With the oil engines you start full speed at once – *see the fighting advantage!!* and yet there are bloody fools who like Canute want to stem the tide of internal combustion!'[90]

By now both the German and British navies were at work developing the diesel engine. In late 1912 the Royal Navy launched its first experimental diesel-powered destroyer, the *Hardy*, while it had under construction a diesel-powered tanker, *Trefoil*.[91] Fisher, however, was anxious. Towards the end of 1911 he wrote to Shell's Sir Marcus Samuel, 'had I remained one more year at the Admiralty we should be now constructing a motor Battleship ... No funnels to discover us to the enemy ... We know the German Admiralty are designing an armoured cruiser that will go round the world without requiring to replenish her fuel.'[92] Indeed, early the previous year the German Navy had drawn up plans for its new battlecruiser *Derfflinger* to have a large diesel engine installed on the centre shaft of a three-shaft arrangement, and the battleship *Prinzregent Luitpold* was to be similarly constructed.[93] In Admiral von Tirpitz, Germany now had its own 'oil maniac': '[T]he construction of the 1911 ships with diesel engines is the most important of all [design issues],' he declared; 'all others fade into the background', since the diesel engine might give Germany the chance 'for a real leap ahead of other navies'.[94]

Throughout 1911 Fisher fulminated at the Royal Navy's apparent lack of initiative, writing to Samuel, '[F]or the first time in Sea History the British Admiralty is not leading the way but following the Germans ... *The present Admiralty officials are simply d**** fools!*'[95] In a memorandum of 1912 he exhorted the navy to action:

[P]lease imagine the blow to British prestige if a German warship with Internal Combustion Propulsion is at sea before us and capable of going round the World without re-fuelling! What an Alabama!!! What an upset to the tremblers on the brink who are hesitating to make the plunge for Motor Battleships . . .

[I]t must be admitted that the burning of oil to raise steam is a roundabout way of getting power! The motor car and the aeroplane take little drops of oil and explode them in cylinders and get all the power required without being bothered with furnaces or boilers or steam engines, so we say to the marine engineers, 'Go and do thou likewise!'[96]

Shell's Samuel was of a similar view and Fisher wrote to Lord Reginald Esher, King Edward VII's closest adviser, 'I've got enthusiastic colleagues in the oil business! They're all bitten! Internal Combustion Engine Rabies!'[97] Fisher told Churchill that Samuel had asked him if he would join the Shell board, but that he had declined: 'I'm a pauper and I am deuced glad of it! but if I wanted to be rich I would go in for oil!' Fisher had, indeed, invested heavily in Shell but assured Churchill that he had sold all his shares before taking up the chairmanship of the Royal Commission.[98] Fisher wrote again to Esher,

My idea now is to raise a syndicate to build the 'Non-Pareil'! A few millionaires would suffice, and I know sufficient of them to do it. All the drawings and designs are quite ready. The one *all pervading, all absorbing* thought is to get in first with motor ships before the Germans! Owing to our apathy during the last two years they are ahead with internal combustion engines! *They have killed 15 men in experiments with oil engines and we have not killed one!*[99]

Fisher seemed hysterical, and he certainly made sure that the Royal Commission was aware of the German Navy's hopes for diesel power.[100] He even threatened to resign his chairmanship unless Churchill agreed to the construction of a diesel-powered battleship.[101] Fred Lane, the Rothschilds' intermediary with Royal Dutch-Shell, told the Commission,

When I consider what an enormous advantage Germany would possess in the event of the successful development of the Diesel engine from the fact that she has behind her as fuel for this purpose the supplies of Galicia, Russia and Roumania, always available by land, in addition to the small supplies she herself possesses, I feel that she will press forward to the utmost and in the most rapid manner the development of this engine.[102]

But the Commission soon established that the overwhelming consensus of marine engineers was that diesel engines of the size required for large ships were still very much at the experimental stage and that years of development would be needed before these huge internal combustion engines would achieve the reliability of the tried and tested steam engine. Even many of the relatively small diesel-powered vessels had been plagued with technical problems.[103] Indeed, the German Navy had recognized almost immediately the practical impossibility, in the foreseeable future, of Tirpitz's dream of a diesel-powered 'leap ahead of other navies'.[104] By early 1913 Churchill was telling Parliament rather more soberly, 'We are not very far away – we cannot tell how far – from some form of internal combustion engines for warships of all kinds, and the indirect and wasteful use of oil to generate steam will, in the future, give place to the direct employment of its own explosive force. That position is not, however, reached at present.'[105]

Meanwhile, the runaway success of the petrol engine was creating its own problems. Motor transport was becoming prevalent in civilian life, and air flight was now another popular, petrol-powered pastime of the wealthy. Britain's petrol consumption had increased from 18 million gallons in 1905 to nearly 100 million by 1914 – 60 per cent of which came from the Dutch East Indies and 20 per cent from the United States. However, since the number of motor vehicles was rising at an even greater rate than was the supply of petrol, particularly in the US, a petrol 'famine' was feared.[106] As *Petroleum World* put it, 'Motor spirit is no longer a luxury; it is a necessity that all the world wants.'[107] To the *Daily Telegraph* it seemed that 'There is reason for believing we are approaching a period which will be known as the Oil Age.'[108]

Petrol-powered traction engines began to replace their steam-powered antecedents for agricultural and engineering work, and following the British Army's use of armoured steam traction engines for haulage during the Boer War, armies around the world now began to experiment with internal combustion-powered haulage. One new development was the caterpillar-track vehicle. The first working version was a steam-powered log hauler, built in the United States in 1900 by the Alvin O. Lombard Traction Co., and from 1903 they became widely used in Maine and New Hampshire.[109] In Britain, steam- and internal combustion-powered tracked vehicles were developed by Hornsby & Sons from 1904; several were trialled by the army for artillery haulage, one of which the *Illustrated London News* in 1909 called 'the machine that walks over hedges and ditches'.[110] However, the following year the military correspondent for *The Times* turned his guns on one of these vehicles for contributing to a seven-hour delay of a marching column of troops during military manoeuvres: '[T]he leading gun of the heavies was being dragged by the tractor monstrosity known as the "caterpillar", a nightmare machine which travels half a mile an hour uphill when in good humour, and eight miles an hour downhill when its fuel holds out. The man who placed this machine in the C. column of march deserves hanging.'[111] The California-based Holt Manufacturing Co. commercialized caterpillar-track vehicles on a much larger scale internationally from 1906, mainly for agricultural uses.[112]

Ordinary wheeled motor vehicles began to be incorporated into plans for the rapid movement of troops. In 1909 the London General Omnibus Co. made twenty-four of its motor buses available for testing their use for troop mobilization, and the following year the Automobile Association organized the movement of a battalion from London to Hastings in 316 privately owned cars.[113] In the spring of 1914, as civil war in Ireland appeared imminent, the Ulster Volunteer Force one night organized convoys of hundreds of cars and lorries to rapidly distribute tens of thousands of rifles and ammunition, smuggled through ports near Belfast, to British loyalists across Northern Ireland; *The Times* effused, 'The whole operation was conducted in a manner unparalleled in the history of the United Kingdom, and the daring and audacity, as well as the completeness of the scheme and its undoubted success, have astonished even those in Belfast.'[114]

The first significant deployment of armoured motor vehicles in combat occurred in 1907 when the French Army used a converted Panhard & Levassor touring car mounted with a machine gun against rebels in Morocco. In 1909 the Ottoman Sultan ordered four Hotchkiss armoured cars from France, which were then captured by the Young Turks, who used them in the overthrow of the Sultan's regime. During the Italian-Turkish War of 1911–12, the Italian Army used Isotta-Fraschini and Fiat armoured cars during its invasion of the north African Ottoman province of Tripolitania, leading to the creation of Italian Libya.[115] Several inventors designed caterpillar-tracked versions of armoured vehicles, and in 1912 an Australian, Lancelot de Mole, submitted his design to the British War Office; it was rejected, but seven years later it would be recognized as having been ahead of its time.[116]

As for aviation, by the end of 1912 *Petroleum World* was announcing, 'The aeroplane has now been used in three wars : – 1. The Mexican revolution. 2. The Turco-Italian war. 3. The Balkan war.'[117] Aeroplanes – often naval seaplanes – were seen primarily as a powerful new reconnaissance tool, although they also began to be used more offensively.[118] In the autumn of 1911, soon after Italy's invasion of north Africa, Lieutenant Giulio Gavotti wrote back to his father, 'Today two boxes full of bombs arrived. We are expected to throw them from our planes . . . It will be very interesting to try them on the Turks . . . It is the first time that we will try this and if I succeed, I will be really pleased to be the first person to do it.'[119] On 1 November he secured his place in history by flying over Arab allies of the Ottoman army at Ain Zahra, east of Tripoli, and dropping several bombs on them.[120] '[I]t is already clear that no nation can afford to go to war with a marked inferiority in aerial strength,' commented the correspondent for *The Times*.[121] Aeroplanes were used several years into the Mexican Revolution, most notably by the US Navy for reconnaissance purposes during its occupation of Veracruz in early 1914.[122] In December 1912, during the First Balkan War, Lieutenant Michael Moutoussis of the Greek Army flew over Janina, where 'he threw down bombs, creating a veritable panic amongst the Turkish troops. Many hostile bullets tore the fabric, but the machine continued its flight unaffected.'[123]

In early 1913, Major Frederick H. Sykes of the Royal Flying Corps told an audience of the Aeronautical Society at the Royal United Services Institution,

> There can be no doubt that unless one side definitely obtains command
> of the air . . . the cards will be more openly displayed for both . . . [T]he
> 'fog of war', the 'hill' behind which Wellington could not see, will, to a

certain extent, be quietly and quickly removed . . . I dream, in the not
far distant future, of scouting aeroplanes of 120 miles an hour; fighters
to carry pilot and assistant, gunner and observer at a speed of 100
miles; weight-carriers to transport troops, rations and equipment ten
or twelve at a time a distance of 30 miles and make five trips a day.
Four hundred of these and some twenty to twenty-four thousand men
are landed a double march ahead, with no weariness of the flesh, but
rather physically and mentally braced up by a pleasant journey.[124]

However, the government was criticized for failing to build up Britain's air power in line with that of other countries and was accused of
providing figures that misleadingly inflated the size of its air fleet,
while the press warned that Germany was capable of unleashing an
invasion of Zeppelin airships.[125] *Oil News* wrote,

any of the Powers that count on the continent . . . can add to their military and naval strength aerial fleets far superior to the incongruous
collection of toy airships and foreign-built aeroplanes of every make
which the War Office and the Admiralty can at the present time put
together. One is, accordingly, doubly grateful to the 'Review of Reviews',
the 'Sphere', the 'Daily Mail', and the Navy League for their efforts in
waking up England to the air peril for which she is so deplorably
unprepared . . . What is more, it will give a much-needed impetus to the
general trade in aircraft in this country, and to the oil trade.[126]

Flight magazine urged its readers to 'insist that the safety of the country shall take precedence of schemes of so-called social reform, which
no one wants and which are frankly designed to catch the votes of the
unthinking populace'.[127] In a speech at the Lord Mayor's Banquet in
November 1913, Churchill conceded the criticism to a degree:

[I]t is not only in naval aeroplanes that we must have superiority . . .
[T]he enduring safety of this country will not be maintained by force
of arms unless over the whole field of aerial development we are able
to make ourselves the first nation. That will be a task of long duration,
and many difficulties have to be overcome. Other countries have
started sooner. The native genius of France, the indomitable persever-
ance of Germany, have produced results which we at present cannot
equal.[128]

However, the oil industry was hardly able to keep up with the sky-rocketing demand for petrol, which was almost doubling each year. Sir Marcus Samuel told Shell shareholders,

> so great is the economy of distributing our petrol by means of motor traction, that we have decided to scrap the whole of our horse transport as quickly as we can get the new motors delivered ... [T]he days of steam or of oil fuel for generating it are numbered, and the internal combustion engine will wipe out the steamer as certainly as that in its time superseded the old sailing ship.[129]

As an MP noted in Parliament, 'You have only to look at the streets of London to see how horses have disappeared and how petroleum is doing their work. Motor boats are now being driven through the water by petroleum, and to-day petroleum is doing not a little towards the conquest of the air.'[130]

The oil companies were profiting greatly from the escalating demand and rising prices, and Standard Oil and Shell were accused of unfair profiteering. Calls went out for the development of alternative plant- and coal-based motor fuels, and the *Daily Mail* suggested that motorists could become independent of the oil companies by forming a petrol co-operative: 'It might not be so difficult for a great co-operative body to obtain its own supplies. Even if it meant going into production, refining, transport, and distribution, is not the great body of motorists rich enough to do this for its own profit?'[131] The famous racing driver Charles Jarrott warned of Britain's increasing dependence on oil in a letter to *The Times*:

> The oil supply is practically monopolised, and although we have several companies outside the ring, nevertheless we are entirely dependent upon foreign countries for our supply, and the result of this in time of war would probably be disastrous. Aeroplanes and motors of every description are going to play a very important part in the next great war, and in the course of time there will be many thousands of them used for purposes of warfare.[132]

The *Observer* exclaimed, '[I]t is a matter of the most vital national importance to produce in this country at least a material proportion of liquid fuel increasingly essential to our needs instead of allowing us to remain, as we are to-day, wholly dependent on consignments from overseas, which consignments might be interrupted at any time in the event of an an outbreak of war, thereby rendering our motor vehicles useless as means of transport, and, further, rendering our Navy also liable to become useless.'[133]

This raised the question of Britain's strategic oil reserves, the accumulation of which was threatening to sink the Admiralty budget. In early 1913 the MP Basil Peto suggested that Parliament provide an incentive for the creation of a domestic oil-refining industry – and thereby a national, private-sector oil reserve – by levying higher taxes on imported petrol:

> I consider that if [the Chancellor of the Exchequer] made a reasonable difference between the tax charged upon petrol refined in this country and the tax charged upon petrol refined abroad, a very large part of the £1,000,000 capital expenditure for oil reserves would be wholly unnecessary, even in view of the fact that the Navy is to become a large user of oil in the near future. If the tax upon oil refined in this country were half that upon oil refined abroad, there would grow up here an immense oil-refining industry, and automatically without paying a half-penny, let alone a million sterling for capital expenditure, we should have an immense reserve of oil in the oil refineries of this country.[134]

Indeed, as Samuel told the Royal Commission, 'at a certain moment, owing to the loss of a ship off the coast, there was not seven days' supply of petrol in England ... for the submarines, aeroplanes and motor transport.'[135]

SPHERES OF INTEREST

As the European powers tugged at the fraying fabric of the Persian and Ottoman empires, the period 1907–14 saw the division of the Middle East more clearly into spheres of interest. Under the Anglo-Russian Convention of 1907, Persia was formally divided into a 'Russian' zone in the north and a 'British' zone in the south and east, with a 'neutral' zone in between (see Fig. 25). Meanwhile, Asiatic Turkey progressively underwent a tentative, informal division into three

regions of influence: a 'French' Syria; a 'German' Anatolia and northern Mesopotamia; and a 'British' southern Mesopotamia.[136] The
proto-partition of Turkey was conducted by diplomats largely in
secrecy; as Charles Hardinge at the British Foreign Office observed, 'I
cannot conceive that the Turks in their present frame of mind would
agree to any scheme which contained a proposal to create spheres of
influence for Foreign Powers.'[137] Sir Louis Mallet, while British ambassador at Constantinople, wrote to Foreign Secretary Sir Edward Grey,
'All the Powers, including ourselves, are trying hard to get what they
can out of Turkey. They all profess to wish the maintenance of Turkey's integrity but no one ever thinks of this in practice.'[138] The
associated agreements over Turkey – formal and informal – between
the European powers were uneasy and unstable: they were based on
the frustrating limits of their respective projection of power in the
region and on the shifting patterns of antagonism and cooperation
between them more widely.

The oil strike in Persia's 'neutral' zone in 1908 by William D'Arcy's
exploratory syndicate, with the subsequent formation of the Anglo-
Persian Oil Co., became a valuable new chess piece in the political
and commercial struggle for strategic supremacy around the head of
the Persian Gulf, and the British Foreign Office was adamant that
the Persian oil concession should not be 'secured and worked by
subjects of a foreign Power'.[139] Following Britain and Russia's carve-
up of Persia, the previous year, into spheres of influence under the
Anglo-Russian Convention, there were signs that the financially
strapped Persian government was, like the Turks, turning to Germany for support. In December 1909 the Foreign Office informed
Anglo-Persian that a British subject was attempting to purchase the
Persian government's shares in the company, with the attendant risk
that these could then pass into foreign hands. This prompted D'Arcy
to suggest that the British government buy some of these shares,
which 'would be a great coup for the Govt., almost a Suez Canal
over again'.[140] As several other investors made moves to acquire the
Persian shares, Britain's consul general at Isfahan responded to the
perceived danger by insisting that Anglo-Persian follow to the letter
the concession terms agreed with the Persian government, since 'if
they can annul the Concession for any cause they may get a lump

sum from the Germans.'[141] Britain was beginning to fear the spread
of German influence at the head of the Persian Gulf. This was due to
the arrival of some German commercial competition in the area; to
the advance of the German-controlled Baghdad Railway and its
planned branch lines; and to the approaching end of a ten-year
Anglo-Russian moratorium on railway construction in Persia, along
with signs that Russia was becoming less hostile to German railway
construction in the region. A British General Staff report of 1910
warned of the dangers of any railway construction in Persia at all;
the question was

> one of supreme importance . . . [which] must vitally affect the defence of
> our Indian Empire and possibly the entire situation in Asia . . .
> Geographically, Persia lies between India and Russia. So long as it
> remains devoid of the means of rapid communication, the country inter-
> poses a practically insuperable military obstacle between the two great
> neighbouring States. The introduction of railways will at once remove
> the barrier and render possible, if not actually provoke, an aggressive
> policy.[142]

While a Russian consortium – led by an oil industrialist – promoted
the construction of a Trans-Persian Railway from the 'Russian' zone
in northern Persia to the Gulf, a British consortium, the Persian Railways Syndicate – chaired by Anglo-Persian's managing director,
Charles Greenway – promoted complementary railway schemes. The
Persian Railways Syndicate sought construction rights for railways
radiating from Bunder Abas and Bushire in the 'British' zone, and for
a line running from Mohammerah in the 'neutral' zone of southwest
Persia northwards to Khoremabad near the 'Russian' zone, via the
Karun valley, Anglo-Persian's field of operations. At the same time,
the British applied to the Turkish government for a Tigris valley railway concession in southern Mesopotamia which, alongside the Karun
valley railway concession, formed part of Britain's strategy to thwart
German encroachment at the head of the Gulf.[143] As one Foreign
Office memo of early 1909 described the situation, 'The Germans are
the invaders, and HMG are on the defence. The proposed Tigris Rly.
is a weapon of defence.'[144] The *Daily Telegraph* noted the agricultural
potential of both the Tigris and Karun valleys, adding,

Much more important than this, however, has been the discovery among the Bakhtiari Mountains, thirty miles east of Shustar, of one of the first-class oil-fields of the world. The future of this part of Persia has, in consequence, assumed a sudden importance, which would have seemed incredible five years ago. Unfortunately, this valuable enclave lies wholly within the neutral sphere, and close to that part of the western frontier of Persia against which Germany has inaugurated a silent but persistent campaign of official encroachment ... Russia and England are slowly realising that the net is being spread, not so much to catch the fish as to catch the sea itself.[145]

Overall, British policy was – in the spirit of the Anglo-Russian Convention of 1907 – that while Britain and Russia would each formulate their railway schemes to promote their own interests, they should compromise and coordinate them with a view to denying Germany any significant inroads: Hardinge hoped that 'As regards Persia, if we and the Russians present a solid front ... I think we shall in the end defeat the Germans.'[146]

Regarding the Tigris valley railway concession in Turkey, Hardinge wrote to the British ambassador at Constantinople, 'Upon your success depends largely our future position in Mesopotamia.'[147] However, Germany held more sway in Turkey, where government officials baulked, for example, at Hardinge's refusal – via the French- and British-controlled Ottoman Public Debt Administration – to allow them to increase customs duties 'until we get all we want, and that is the full control and construction of the Baghdad end of the Baghdad Railway, and possibly a promise of a concession for a railway from Baghdad to the Mediterranean'.[148]

Attempts to preserve the shipping monopoly of Lynch Bros. on the Tigris led to anti-British riots in Baghdad, and one official at the British Embassy warned Grey that 'the designs of England for territorial aggrandisement in the regions about the Turkish [Persian] Gulf' had created 'a widespread, if vague distrust of our policy'.[149] In December 1910 Grey was concerned that if Britain were to 'cold shoulder' Turkey, it could have 'precipitated them into the arms of the Triple Alliance and into a Pan Islamic policy'.[150] Similarly, Churchill later warned Grey of 'the consequences of throwing Turkey ... into the arms of Germany'.[151] This danger, the Foreign Office concluded, meant that Turkey was actually 'a Power to be reckoned with'.[152] In September 1910 Churchill even wrote to Grey in apparent exasperation, that the 'only view I have formed about this part of the world of ruined civilizations and systems, and harshly jumbled races is this – why can't England & Germany come together in strong action and for general advantage?'[153] As *The Times* saw it, in an editorial of January 1911,

> the only concrete question in which any definite antagonism may be argued to exist between British and German interests is that of the Baghdad Railway, and mainly in connexion with the later stages of development ... The nearer the Baghdad Railway approaches to the last section, between Baghdad and the Persian Gulf, in which British interests are most closely concerned, the more we shall ... insist that an arrangement is desirable and even necessary ... Sooner or later, whenever the section from Baghdad to the Persian Gulf is reached, the question will have to be discussed between us and the Germans, for the Baghdad Railway will then approach a region in which the vital importance of British interests has been affirmed by successive British Governments.[154]

The bottom line for Britain, *The Times* declared two months later, was this: 'The continuance of the *status quo* in the Persian Gulf is a condition essential to the stability of British rule in India.'[155]

In November 1910, Turkey had secured a new loan from Germany, rather than from France or Britain.[156] The latter were now close allies: '[T]he maintenance of the *entente* ... [is] ... the bedrock of our policy,' Sir Arthur Nicolson wrote from the Foreign Office to British bankers who were potentially in competition with their French counterparts.[157] Turkey then refused to grant Britain's request for a Tigris valley railway monopoly while, in March 1911, it granted Germany the right to participate in railway construction south of Baghdad.[158] In response, Lord Curzon, former Viceroy of India, reiterated in the House of Lords his view of the strategic importance to Britain of southern Mesopotamia: 'It would be a mistake to suppose that our political interests are confined to the Gulf. They are not confined to the Gulf; they are not confined to the region between the Gulf and Busra; they are not confined to the region between Busra and Baghdad; they extend over the whole region right away up to Baghdad.'[159] In the House of Commons, the Earl of Ronaldshay – who had worked as a close aide to Lord Curzon when he was Viceroy – argued that Britain must

> counteract the advance of German influence [in] ... the regions south of Baghdad ... [and] for political and strategic reasons the task upon which we should concentrate our efforts is that of securing control of any railway which may be built in future from Baghdad to the Persian Gulf.[160]

He supported his argument by quoting from a passage of a recent edition of *Die Bagdadbahn* by the popular German writer Paul Rohrbach:

> A direct attack upon England across the North Sea is out of the question ... It is necessary to discover another combination in order to hit England in a vulnerable spot – and here we come to the point where the relationship of Germany to Turkey and the conditions prevailing in Turkey become of decisive importance for German foreign policy, based as it now is upon watchfulness in the direction of England ... England can be attacked and mortally wounded by land from Europe only in one place – Egypt. The loss of Egypt would mean for England not only the end of her dominion over the Suez Canal, and of her connections with India and the Far East, but would probably entail the loss of her possessions in Central and East Africa. The conquest of Egypt by a Mohammedan Power, like Turkey, would also imperil England's hold over her sixty million Mohammedan subjects in India, besides prejudicing her relations with Afghanistan and Persia. Turkey, however, can never dream of recovering Egypt until she is mistress of a developed railway system in Asia Minor and Syria, and until, through the progress of the Anatolian Railway to Baghdad, she is in a position to withstand an attack by England upon Mesopotamia. The Turkish army must be increased and improved, and progress must be made in her economic and financial position ... The stronger Turkey grows, the more dangerous does she become for England ... Egypt is a prize which for Turkey would be well worth the risk of taking sides with Germany in a war with England. The policy of protecting Turkey, which is now pursued by Germany, has no other object but the desire to effect an insurance against the danger of a war with England.[161]

In early August the British ambassador in Constantinople, Sir Gerard Lowther, and Grey became concerned that Germany and France seemed to be gaining control over all the railways across Turkey – although, in fact, all sides tended to consider their own position to be perilous. The French chargé d'affaires in Constantinople, for example, warned, 'If we are not careful . . . we shall find our place taken when the time arrives for the liquidation of Asiatic Turkey . . . Our whole future in Syria depends on our economic and industrial activity. If we do not modify our methods, our position will be lost and in several years Syria will be English, German and Italian.'[162] In October the British consul at Erzurum wrote to Lowther that 'as an indirect consequence of the railway schemes now under consideration for establishing communications between Van, Erzeroum, and the sea, the possibilities of working the petroleum deposits in these districts seem now to be receiving some attention from both British and French capitalists.'[163] A British syndicate was investigating the possibilities of drilling for Turkish oil in Van province at Korzut, near the northeastern tip of Lake Van, and in Erzurum province at Pulk and Chymaghyl, while a French company was also investigating in the latter province at Kiskim and Terjan. Meanwhile, Grey lobbied the Turkish authorities on behalf of the British-owned Syrian Exploration Co., which would start drilling for oil in Palestine, near Makarim on the Hejaz Railway, in 1913.[164]

Regarding Mesopotamia, the Foreign Office, feeling its position under threat, decided that Britain should make sure of obtaining all irrigation and oil concessions there. The latter – in the provinces of Mosul and Baghdad, through which the Baghdad Railway was projected to run – were being sought by William D'Arcy, founder of the Anglo-Persian Oil Co.[165] It had become clear, in negotiations that would lead to the Potsdam Agreement of 19 August 1911, that Germany and Russia were moving close to a mutual recognition of their spheres of interest in Turkey and Persia. This involved Russian acquiescence to the continued construction of the Baghdad Railway and an agreement to extend a branch line from Baghdad, via Khanikin, to Tehran, which led one Foreign Office clerk to predict that 'we shall have what will shortly become a Turco-German hegemony at Teheran.'[166] One particular aspect of this, Lowther wrote to Grey, would be that 'we must be prepared to see the petroliferous fields around

Suleimanieh fall into German hands.'[167] Was this the 'solid front' that Hardinge had hoped Britain and Russia would present to Germany? Coming on top of the German loan to Turkey, Nicolson wrote to Lowther,

> [Russia] seems to me to be giving everything away and receiving nothing in return ... Personally I am not particularly keen on seeing the present Turkish regime too well provided with funds. This would only assist towards the creation of a power which, I think, in the not far distant future – should it become thoroughly consolidated and established – would be a very serious menace to us and also to Russia ... and there is the additional danger that it would be able to utilise the enormous Mussulman populations under the rule of Christian countries. I think that this Pan-Islamic movement ... is one of our greatest dangers in the future, and is, indeed, far more of a menace than the 'Yellow Peril' which apparently produces such misgivings in the mind of the German Emperor. Germany is fortunate in being able to view with comparative indifference the growth of the great Mussulman military, she having no Mussulman subjects herself, and a union between her and Turkey would be one of the gravest dangers to the equilibrium of Europe and Asia.[168]

The Times believed that the agreement between Russia and Germany was unlikely to fracture the Triple Entente between Britain, France and Russia, particularly considering that Russia had joined Britain in supporting France over its dispute with Germany in the ongoing Agadir, or Second Morocco, Crisis. Nevertheless, Britain's plan for a railway from Mohammerah/Khor Musa to Khoramabad was felt to be a necessary precaution against Russo-German plans for a Baghdad-Khanikin-Tehran line:

> We are inclined to think that a good deal of water will run down the Tigris before the main line reaches Baghdad, still more before the branch to Khanikin is constructed; but it none the less behoves British and Russian diplomacy to be prepared for what is not only a possible but a probable development ... There are many advantages to [the Khor Musa-Khoramabad] scheme. It would serve the promising oil industry that is now being developed by British enterprise in the Karun

district, and would supply an excellent route into Persia for British and other trade. The right to create such a route would enable us to contemplate with even greater equanimity the possibility that the Baghdad Railway might be completed without British participation.[169]

Charles Greenway, Anglo-Persian's managing director, added that 'The oilfields in Zohab will, when the Baghdad and other contemplated railways in Mesopotamia and Northern Persia are operating, be of considerable value, as oil will be the only cheap fuel available for these railways.'[170]

However, in addition to the other setbacks for the British Foreign Office in Mesopotamia, the Sheikh of Kuwait came under renewed pressure to submit to Turkish sovereignty, and it was suggested that a military force be sent to the head of the Gulf to 'make a stand against Turkish encroachment', as Sir Louis Mallet saw it from the Foreign Office, or to 'show the Turks we mean business', in the words of Hardinge, now Viceroy of India.[171] In October, Lord Crewe, Secretary of State for India, made contingency plans for an expeditionary force to Mesopotamia, but on reflection it was decided that Britain was already militarily stretched to the limit. It could spare neither the naval nor ground forces necessary to achieve anything more than a temporary occupation of Kuwait and Basra, particularly since this carried the risks of an uncontrollable escalation of the conflict and, it was thought, of a wider uprising of Muslim populations against British rule.[172] The costs of a full-blown partition of Turkey – which, unified, afforded protection against the Russian threat to India's western approaches – would most likely outweigh the benefits; Mallet now wrote, 'a division of the Asiatic provinces into spheres of interest could not benefit us ... and might bring about a European war.'[173] The India Office calculated that a military occupation at the head of the Gulf would put 'an intolerable strain upon our already overburdened military resources', and Mallet was forced to conclude that 'we had better not have a sphere of interest which we could not defend.'[174]

Following Persia's Constitutional Revolution of 1905–11, the Bakhtiari khans – on whose land lay Anglo-Persian's producing wells and a stretch of its intended pipeline route – had become leaders of the new Persian government and they were therefore in a strong position when it came to negotiating with the oil company over land prices.[175] In March 1910 Greenway wrote to the Foreign Office,

> the Bakhtiari khans have demanded extortionate compensation from the Anglo-Persian Oil Company for the cultivated land required by the latter for their pipe line and works, and they threaten to stop the operations of the company's workmen unless these demands are satisfied, and [the Company has] asked that a telegram might be sent to His Majesty's Minister at Tehran, instructing him to urge on the khans' representatives that reasonable compensation should be accepted, and to insist that the company's work should on no account be interfered with.[176]

In July, at Anglo-Persian's first Ordinary General Meeting in London, the company's chairman, Lord Strathcona, emphasized the crucial importance of the pipeline and the associated construction work:

> [T]he pipe line ... is the keystone of our whole undertaking, since until it is completed it is no use producing oil at our wells and our refinery cannot obtain the oil necessary to keep it running. Consequently, we are doing our utmost to get this part of our work completed by the earliest possible date ... [W]e have had to put our own launches and barges on the lower and upper parts of the River Karun, to build a short tramline connecting the lower and upper rivers, make roads and bridges from the upper river to the field, send out tractors for heavy hauling work, and to procure from Mesopotamia many hundreds of mules to carry the pipes from various points of the river to the places where they have to be laid, and up the mountain passes to our oil fields.[177]

Eventually, in April the following year, a price of £22,000 was agreed for the land purchases – which Greenway regarded as too high – that included oil-bearing territory and a one-foot-wide strip of land 17 miles long for the pipeline from the wells at Masjid i Suleiman to the limits of the Bakhtiari's territory. A thousand local labourers worked for a year and a half under thirty-seven European foremen, often in hazardous conditions, to construct the 6-inch-diameter pipeline, 140 miles long and rising up to 1,400 feet above sea level, that was completed in July 1911 (see Figs. 22, 23 and 25). One Anglo-Persian manager became tired of inspecting the pipeline on horseback and brought the first aeroplane to Persia for the purpose.[178]

Due to the Bakhtiari's strong negotiating position, Anglo-Persian decided to conduct exploratory drilling at an alternative site at Ahwaz, on land owned by the Sheikh of Mohammerah.[179] The Sheikh was a particularly important ally of the British as he controlled coastal territory around the Shatt al-Arab that, like Kuwait, had potential locations for a port at the head of the Gulf. Britain's Political Resident in the Gulf, Lieutenant Colonel Sir Percy Cox, therefore ensured that Anglo-Persian's negotiations with the Sheikh over the rental of land for their refinery on Abadan Island were conducted respectfully and to mutual advantage. In a manner similar to Kuwait's in relation to Turkish claims of sovereignty, this southwestern region of Persia – 'Arabistan', as it was known, controlled by the Bakhtiari khans and the Sheikh of Mohammerah – became virtually a British protectorate.[180] A few weeks after the pipeline's completion, *The Times* commented on Turkey's independent assertiveness since its revolution, and on its claims to control navigation on the Shatt al-Arab:

> For some fifty miles up the river, the left bank of the Shatt is Persian, and it is on that bank that the Karun empties itself into the Shatt, the Persian town of Muhammerah lying at the point of confluence. The interests of British trade and shipping at Muhammerah and on the Karun River are already of no mean importance, especially since trade has been driven away from Bushire by the insecurity of the roads from that port into the interior, and they are expected to grow by leaps and bounds as soon as the Anglo-Persian Oil Company has begun to operate its concession and oil flows down its pipes – not later, it is expected, than next October – from the wells in the Ahwaz district to a refinery on the Persian bank of the Shatt.[181]

Lord Strathcona had told Anglo-Persian's shareholders that the refinery at Abadan would 'be favourably situated, not only for carrying on the internal trade of Southern Persia and Asiatic Turkey, but also for supplying all the markets east of Suez'; and, he had stated optimistically, it would 'be completed in good time to deal with the first oil that passes through our pipe line'.[182] Almost as soon as the oil did start flowing, Ottoman officials in Basra reported to Istanbul that the oil was polluting the Shatt al-Arab, the source of drinking water for the local population and of irrigation for the Dawasir date plantations. Meanwhile, the refinery would not begin to manufacture viable products until mid-1913; the type of crude being produced from the Persian wells presented a steep learning curve for the Burmah Oil Co.'s refining engineers sent from Rangoon, and for Anglo-Persian's management team, one of whose directors warned, 'If the refinery is not properly organized and run, it may be considered a scrap heap as far as results are concerned.'[183]

Without tankers and a marketing system of its own, and in the context of a sudden rise in charter tanker freight rates, the company lacked the resources to sell its first rudimentary products or even its unrefined crude oil; and after years of investment sunk in exploration and infrastructure, the lack of sales revenue took the company to the verge of financial collapse. In 1912 it was forced to turn to Royal Dutch-Shell's distribution subsidiary, the Asiatic Petroleum Co., with which Anglo-Persian signed a ten-year contract for the marketing of its kerosene and benzine and for the purchase of its crude, which was shipped by Royal Dutch to one of its refineries on Sumatra.[184] Sir Percy Cox dutifully kept the Foreign Secretary informed of the latter's tanker movements in the Persian Gulf, 'in connection with the apprehensions expressed ... by the Anglo-Persian ... as to possible coercion by the Shell', while one of Asiatic's directors, Robert Waley Cohen, wrote of his great relief on signing the agreement with Anglo-Persian: 'I think the situation of these people, apparently with very large supplies, made them rather a serious menace in the East.'[185]

Anglo-Persian resisted including fuel oil in the contract with Asiatic, however, and began to pin its hopes on selling fuel oil independently of Royal Dutch-Shell.[186] On the possible Mohammerah-Khoramabad railway through southwest Persia, Lord Strathcona remarked,

> Such a railway would be of great benefit to this company, not only because it would open up the central markets of Persia to our products, but also because it would perforce require to burn oil fuel, and it is therefore, perhaps, needless for me to say that the directors are giving this project their utmost support.[187]

The British Admiralty was well aware of this potential source of fuel for its warships and wrote to the India Office in early 1912, 'The APOC urge that as fuel oil cannot be remuneratively shipped from the Persian Gulf to markets west of the Suez Canal in competition with oil produced from Russia and Rumanian oilfields, the only likely outlet for Persian oil, other than the Admiralty, is with the Indian Railways.'[188] In May, Anglo-Persian received a formal request from the Admiralty for information 'on the suitability of Persian Gulf Oil for use as a Fuel'.[189]

Given that Anglo-Persian's primary importance to the British government was for its contribution to excluding the commerce of Germany, Russia and other countries from the head of the Persian Gulf, it was somewhat ironic that the company bid for a fuel oil contract with the Hamburg-Amerika Line.[190] Then Lynch Bros. – supported by the Foreign Office for serving British strategic interests – reached an agreement with Deutsche Bank to ship materials from the head of the Gulf up the Tigris so that railway construction could begin from Baghdad before the main line of the Baghdad Railway reached the city from the west.[191] Simultaneously, the Foreign Office was surprised to learn from Hugo Baring that the National Bank of Turkey, of which he was manager, had joined with Royal Dutch-Shell and Deutsche Bank to attempt to secure Mesopotamian oil concessions, in a syndicate organized by Calouste Gulbenkian.[192]

The National Bank of Turkey had been created in 1908 by a group of British and Turkish financiers, including the hugely wealthy Sir Ernest Cassel, a dual German-British national who had established the National Bank of Egypt in 1898. The new British-controlled bank was supported by Foreign Secretary Grey in the hope that it might counter-balance the French-dominated Imperial Ottoman Bank, and it soon included a range of British notables such as Lord Revelstoke of Barings Bank and Sir Henry Babington Smith, along with the Armenian-Turkish Gulbenkian, who had taken up British citizenship in 1902.[193] By early 1911, however, the bank had fallen out of favour with the Foreign Office; Sir Arthur Nicolson wrote to Lowther,

> We cannot rely with certainty on any of these financiers being animated by disinterested and patriotic motives. They look solely and simply at

the profits . . . and leave entirely on one side the political character of the questions with which they have to deal. It is a matter of perfect indifference to them . . . whether the ends which they pursue are or are not in harmony with the interests of this country.[194]

Now, in August 1912, the Foreign Office, on learning of this new international syndicate manoeuvring for Mesopotamian oil, initially considered whether Anglo-Persian might join it; the Foreign Office offered its diplomatic support 'in the event of British interests being adequately represented in the Syndicate' and asked Greenway for his opinion.[195] He appeared to be open to the possibility of an association; but, he wrote to a Burmah Oil director, 'an amalgamation would doubtless mean the early exploitation of the Mesopotamian Oil fields which would not suit our interests at all.'[196] However, outright competition would be even more dangerous since, if the new syndicate were to gain exclusive control over Mesopotamian oil, it 'will have achieved its object and the A.P.O.C. will be obliged to come to terms because the Mesopotamian oil supply will be overwhelming'.[197] Meanwhile, Babington Smith asked the Foreign Office for its endorsement of his new syndicate, which would be a partially Anglo-German partnership.[198]

During September, Grey was persuaded by Greenway's arguments. These were that Anglo-Persian would not survive in competition with Mesopotamian oil being produced right on its doorstep; that the Persian oilfields had been developed with the Admiralty's requirements for fuel oil in mind; and that foreign domination of oil supplies might endanger Britain's naval supremacy in time of war. Grey decided, therefore, to oppose the new syndicate unless the wholly British Anglo-Persian were to replace Royal Dutch-Shell entirely.[199] Furthermore, Sir Charles Marling wrote from Constantinople that 'This financial alliance with German interests . . . [was] . . . most disquieting', and that 'the only possible issue seems to me to throw Cassel over . . . [for] . . . his bank has a disagreeable political flavour through its relations with the Committee [of Union and Progress]', that is, with the new independent-minded Turkish government.[200]

Anglo-Persian recognized that the recent turn of events might move the British government to provide it with greatly increased support, and Greenway saw a 'unique opportunity for effecting a great "coup"'.[201]

One director wrote to him that a government subsidy would be 'quite ideal', though it was an option that 'the Government could not contemplate for there was no precedent ... [A]lthough a policy that has no precedent is almost essential to a successful progressive business, it is anathema nowadays in the eyes of British Governments.'[202]

On 25 September the new syndicate was formally organized as the Turkish Petroleum Company, in which the National Bank of Turkey held a 50 per cent share; Deutsche Bank received 25 per cent in return for its claimed Mesopotamian oil rights associated with the Baghdad Railway concession; and Royal Dutch-Shell received the remaining 25 per cent.[203] Three days after its creation, the Foreign Office, however, told Babington Smith that it could not support the Turkish Petroleum Co. in its current form.[204] Nicolson wrote to Marling in Constantinople,

> After consulting with the Admiralty, who do not at all like the idea of a foreign syndicate having control of large oil supplies in Mesopotamia, we have ... told the National Bank that we are unable to support their request for a concession. Of course our main reason is that we have already advocated and supported the demands of the Anglo-Persian Oil Co., and it would certainly be unfair to the latter to encourage a competitor who is largely backed up by foreign capital. The reason which we have given to the National Bank is simply that we have already pledged ourselves to support Mr D'Arcy and company and that it is therefore impossible at a late hour to encourage another competitor, but we are going to verbally tell Babington Smith confidentially that after considering the whole matter our Government do not like the idea of entrusting to a syndicate which, though nominally British, is in reality composed of two very powerful foreign syndicates, the control of so large a supply of oil fuel.[205]

'WILL GERMANY CONTROL THE OIL SUPPLY OF OUR NAVY?'

The German Navy – with almost no domestic production – had adopted oil-firing merely as an auxiliary to coal in its battleships, although in 1912 it began constructing torpedo boats that were solely oil-fired. In that year, by contrast, the British Navy – even though it could draw on only very modest levels of oil production across its empire – adopted sole oil-firing in its new battleships, and its torpedo boats had been solely oil-fuelled since 1906.[206] Accordingly, Britain embarked on a drive to increase oil production from territory under its control, the *Economist* exhorting investors in 1910, 'We need to develop the oil resources of the Empire ... We cannot possibly leave our naval developments absolutely dependent on supplies of fuel that may be cut off at any moment – or at all events at the most critical moment when it will most be wanted.'[207]

Yet the potential for oil production within the British Empire seemed limited. Even the Burmah Oil Co. – the only entirely British-owned oil company operating within British territory that was also capable of delivering on an Admiralty fuel oil contract – was now faced with declining production and no new oil strikes.[208] By contrast, the prospects of the fully British-owned Anglo-Persian Oil Co. seemed much more encouraging; and while the region where it was operating did not come under formal British sovereignty, the British government considered it to be one where British predominance was strategically vital. However, both of these strategic imperatives appeared to come under threat when, in mid-1912, the Turkish Petroleum Co. – a consortium of the Deutsche Bank, the National Bank of Turkey and Royal Dutch-Shell – was formed with the intention of exploiting Mesopotamian oilfields, which were contiguous with Anglo-Persian's oilfields across the Turkish-Persian border: Anglo-Persian's managing director, Charles Greenway, told the Foreign Secretary, Sir Edward Grey, that it was the intention of Royal Dutch-Shell 'to be in a position to attack us on the flank'.[209]

In October 1912 Greenway submitted to the Admiralty a formal request for significant government support, assuring the company's founder, William D'Arcy, 'that the whole question has been thoroughly discussed with the Foreign Office and the Admiralty ... and that they are unanimously agreed that it is absolutely essential to maintain British control over the Persian Oilfields and that to ensure this British control must be secured over the Mesopotamian Oilfields'.[210] Greenway wrote at the same time to Cargill, another fellow director, that 'unless the Government can come to some arrangement with this company to assist not only in the matter of contracts for supplies, but also some form of subsidy to aid development and active support in respect of applications for concessions in Mesopotamia, it will probably be impossible for the APOC to preserve its independence.'[211] In his submission to the Admiralty, Greenway argued that Anglo-Persian would need, among other things, £2 million of investment, along with the concession for Mesopotamian oil, in order to 'ensure for all time British control over the Oil produced from the Persian Oilfields ... [and] that supplies from this important Oilfield, probably the largest by far in any British or British controlled country, would not in time of war be subject to the restraints which might occur were this Company forced into a commercial alliance with a foreign Combine'.[212]

Without a distribution and marketing network of its own, Anglo-Persian had turned to Royal Dutch-Shell's marketing subsidiary, the Asiatic Petroleum Co., to sell its first products, but had resisted including fuel oil in the arrangement, which it hoped it could sell independently. Greenway forwarded the correspondence of these negotiations to the Foreign Office, where Sir Louis Mallet concluded,

> It is clear ... that the Shell group are aiming at the extinction of [Anglo-Persian] as a competitor. One of their objects being to control the price of liquid fuel for the British Navy ... [Shell wants] to control the supply to the Admiralty ... which would be of most serious import to this country ... I think that we should go to every length in supporting the independence of [Anglo-Persian] and subsidise them if necessary, as well as supporting their Mesopotamian concession claims. On commercial grounds alone we should be seriously criticised for assisting in the formation of such a gigantic and powerful ring – and those are the grounds on which I would base my opposition.[213]

Sir Edward Grey concurred, recording, 'Evidently what we must do is to secure under British control a sufficient oil field for the British Navy.'[214]

Meanwhile, Greenway failed – unsurprisingly, as the head of a small, struggling company – to negotiate for both a 50 per cent controlling stake in the Turkish Petroleum Co. and the exclusion of

471

Deutsche Bank, preconditions that Grey demanded before the Foreign Office would give the consortium its blessing.[215] Much was then made of an unsigned article in the *Financial News* on 14 November 1912, 'Will Germany Control the Oil Supply for Our Navy?', which was reproduced in full by the Royal Commission in its first report.[216] The article suggested that proposals for a German oil monopoly, currently being debated in the Reichstag, would see the Deutsche Bank ally with Royal Dutch-Shell to enable Germany to become independent of Standard Oil for its oil imports, which would be made up by supplies from Romania, Galicia and Russia. In fact, the scheme was clearly over-ambitious and was already on the point of collapse;[217] nevertheless, the article painted a terrifying picture of

> the extraordinary ramifications of German capital and influence in the oil world ... British-owned oilfields lack all approach to organisation which could make their production available for national purposes, while our Government is powerless, compared with that of Germany, which is actually entering into a trading alliance with its banks and their subsidiary oil companies. It is scarcely open to doubt that, very few years hence, all modern naval and mercantile vessels will be using oil fuel, and the advantages now conferred upon us by the possession of Welsh smokeless coal will then be lost. The German petroleum monopoly should be a reminder that we, too, must secure ample and regular supplies of the more modern fuel if we wish to keep our position at the head of maritime nations.[218]

In evidence to the Royal Commission, Greenway endorsed the contents of the article; furthermore, he added, the monopoly 'has been engineered by the ... Royal Dutch Company', while the Royal Dutch-Shell combine 'it seems likely will shortly be under the control of the German Government itself'.[219] The secretary to the Royal Commission, Captain Philip Dumas, noted in his diary of Royal Dutch-Shell, 'As they are Dutch and again controlled by the Deutsche Bank our condition in time of war can better be imagined than described unless we possess ample reserve storage in England.'[220] Given Anglo-Persian's overstretched finances, however, it might have no alternative, Greenway told the Commission, but to enter into a fuller partnership with Royal Dutch-Shell; therefore, he continued, we have got to make up our minds whether we are to attach ourselves to the Admiralty, or rather whether the Admiralty are to exercise the control that we thought they would desire to exercise, or whether we should join hands with the Shell Company ... [which] will result in their securing a practical monopoly of oil in the Eastern Hemisphere, if not the whole world. They are working in the direction of securing a monopoly of Oil Fuel for the whole world ...

If we join with the Shell Company in a combine in any shape or form ... it seems possible that we should become, through the Royal Dutch Shell Company, under the control of the German Government itself. That, of course, from a political point of view is a very serious situation, because it is not to the interests of this country that Germany should have any control or voice at all in any part of Persia, nor that that country should be in a position to control the supplies of oil from Persia ... If we join hands with the Shell we shall make a very large sum of money and be a very prosperous concern; and if we refrain from doing that we want a *quid pro quo* in some shape or form – we want guarantees and a contract that will at any rate give us a moderate return on our capital.[221]

While Greenway claimed that Anglo-Persian's directors were moved by patriotic ideals, he made it abundantly clear that the board was duty-bound to maximize profits for its shareholders, an argument he repeated the following day in a meeting at the Foreign Office.[222] When he told the Foreign Office that Royal Dutch-Shell had attempted to buy Burmah Oil's holding in Anglo-Persian and that the German government had requested a quote for a contract, Mallet was not alone in wondering whether he was engaging in blackmail: 'I do not like the attitude of the Anglo-Persian Oil Company who have hitherto posed as ultra-imperialist ... Greenway now threatens complete absorption with the Shell unless the Admiralty give him a contract.'[223]

On 5 February 1913 Mallet conveyed Grey's riposte to Greenway, in which the Foreign Secretary included a measure of counter-bluff of his own. Despite the Admiralty's strong support for Anglo-Persian,

There would, however, be great objections to ... a policy of finding Two [*sic*] million pounds of capital for a British Company operating in territory which although presumably largely subject to British influence is nevertheless foreign territory ... The Admiralty doubt whether such a policy could be recommended to Parliament on naval grounds, and if such a policy could be advocated from that point of view alone there are other oil fields in which the arguments for similar action might prove to be equally or even more cogent especially if the fields were in actual British territory and within a shorter open sea voyage from the United Kingdom than the route from the Persian Gulf either via the Suez Canal or the Cape ... The Admiralty regret any extension of influence which may tend to a greater monopoly in the oil fuel market by powerful combinations, but they have to consider from the purely Admiralty point of view in each case whether Admiralty interests would be so jeopardised as to necessitate the consideration of the large question of subsidizing commercial companies ... and they consider that it is at present very difficult to say in regard to any oilfield, however important in certain eventualities it might become, whether it is likely to stand in any absolutely pre-eminent or exclusive position from the Admiralty point of view ...

I am to observe that the support which His Majesty's Government have given to your Company in the past, both in obtaining their concession in Persia and in other ways, was given on the understanding that the enterprise would remain British and it would be a matter of great surprise and regret if your Company made any arrangement whereby a syndicate predominantly foreign obtained control of their interests in that country.

I am to point out that, in such a contingency, your Company could not of course hope to get from His Majesty's Government the same support as in the past.

I am to add that while, in the contrary case, your Company would continue to enjoy in Persia the treatment hitherto accorded to them by His Majesty's Government, there is some reason to suppose that the German claims to the Mesopotamian concession may rest on a more substantial foundation than was at first imagined.[224]

The Foreign Office, nevertheless, was in fact persuaded by Greenway's argument that both significant financial support and the acquisition of the Mesopotamian oilfields would be required to preserve Anglo-Persian's independence.[225] From the geostrategic point of view, however, the India Office argued that this was the responsibility of the Foreign Office, since Anglo-Persian was not operating in Persia's 'British' zone – as defined by the Anglo-Russian Convention – but in the 'neutral' zone and that 'the vital interests of India are sufficiently safeguarded so long as the British sphere [in Persia] remains intact and British power is supreme at sea and controls the entrance to the Gulf' (see Fig. 25).[226] There was frustration felt in the Foreign Office at the other departments' apparent failure to appreciate the importance of Anglo-Persian's operations to Britain's strategic presence at the head of the Gulf; Sir Arthur Nicolson – who had received an extensive letter on the subject from Captain Dumas, along with a copy of the *Financial News* article – perceived significant danger in there being 'some risk of the oil fields in Mesopotamia and the whole of Persia being under foreign and largely German control'.[227] Mallet reminded the India Office of

the paramount importance of maintaining our existing influence on the shores of the Persian Gulf; and emphasis has been laid upon the fact that our political position is largely the result of our commercial predominance. Sir Edward Grey cannot but apprehend that our position both commercial and political will be seriously jeopardised if the most important British concession in Persia, the Anglo-Persian Oil Company, is allowed to pass under foreign control by absorption in the Shell Company.[228]

Meanwhile, Shell's Sir Marcus Samuel, who had been Lord Mayor of London in 1902–3, kept up a spirited defence of his company's patriotic motives, telling the Royal Commission, 'The Admiralty know perfectly well they have only got to ask me for anything I could give them, and I should do it immediately, notwithstanding ... that we have been very badly treated, shamefully treated.'[229] He complained that Shell had combined with Royal Dutch in 1907 as 'the direct result' of having been denied access to the oilfields of Burma by the British Indian government in 1904, just when it was experiencing a shortfall in its oil production;[230] he had warned at the time that Britain might thereby be sacrificing its only British-owned oil company of any significant standing. The Indian authorities, instead, had granted a near monopoly to the Burmah Oil Co. after one of its executives, a certain Charles Greenway, had gained the support of the Secretary of

State for India, telling him that Shell, 'though nominally a British Company, is really cosmopolitan, more than seven-tenths of the capital being held by Jews and foreigners'.[231] Since then, Shell had pursued oil prospects within the British Empire in the hope, according to Shell executive Robert Waley Cohen, that this would 'strengthen our hands enormously in our dealings with the British Admiralty and also our dealings with the Government of India, who always insist on treating us in a very unfriendly way as foreigners'.[232]

Shell struck oil in Egypt in 1909 and in north Borneo in 1910, at Miri on the border between the British protectorates of Sarawak and Brunei, and it began investing in Trinidad in 1911. It tried drilling for commercial quantities of oil in New Zealand and even attempted to develop Dorsetshire shale oil.[233] Samuel explained to the Royal Commission that Royal Dutch-Shell would be a reliable supplier of fuel oil as it was producing from a wide range of oilfields around the world, from Romania to California to Sumatra, and he argued that 'the sole fields which could give an undoubted output over a course of years for the supply of the British Navy are the oilfields of Borneo.'[234] Samuel also complained about the Foreign Office's refusal in 1909 to support Royal Dutch-Shell's application for a Mesopotamian oil concession.[235] However, as Shell was part of a conglomerate that was 60 per cent controlled by Royal Dutch, it was an uphill task for Samuel to convince the Commission that his company's oil would definitely be available even in time of war. Sir Boverton Redwood – adviser to Burmah Oil, Anglo-Persian and the Admiralty, and a member of the Commission – argued, 'I take it that, the Shell Company being under Dutch control, pressure could be put upon Holland by Germany, and that even if Sir Marcus Samuel were willing to let the Admiralty have oil, Germany might say to Holland, "No".'[236]

The British authorities in fact displayed an ambivalent attitude towards Royal Dutch-Shell. In 1910 the Admiralty's attention had been drawn to the British colony of Trinidad, which was seeing a mini oil boom, with the recent flotation of Trinidad Oilfields Ltd. being oversubscribed in a speculative fever for cashing in on sales of fuel oil to the Royal Navy. Randolph Rust, in London to promote his own Trinidadian oil company, declared, 'Trinidad, England's most valuable possession in the West Indies, being as it is one of the keys to the Panama Canal, now rapidly approaching completion, might herself one day be one of the chief sources of supply of oil fuel, and thanks to that and her unique position, might become one of our most important naval bases.'[237] Here, the British and Foreign Oil and Rubber Trust – a syndicate of British oil companies including Burmah Oil and Anglo-Persian, and chaired by Greenway – requested a long-term lease, a large acreage for prospecting and a pipeline monopoly on the island. At a conference with the Governor of Trinidad, representatives of the Colonial Office and the Admiralty, Greenway warned that if his Trust were to commence oil production 'it might be worth the while of a powerful opposing company (that is, of course, the Standard) to lay lines even at a loss merely to interfere with them'.[238] It was taken as a given within the industry that Standard Oil's rise to power had been achieved via its control over the transit and refining of oil; as the *Economist* put it, 'the Standard Oil Company ... not only exercises complete sway in its own country, but largely controls the sale of refined oil all over the world. Though not itself a producer, its monopoly of marketing and refining place it in a position to control the whole industry.'[239]

Greenway's call for exclusive access to Trinidad's oil was supported by Redwood, who suggested that 'a *de facto* monopoly such as that which the Burmah Oil Company secured through their agreement with the Secretary of State for India, might be arranged.'[240] On learning, in early 1911, that Burmah Oil was in favour of Royal Dutch-Shell and the Rothschilds having a large share in the Trust, the Admiralty insisted that it should be a purely British enterprise. However, as an interdepartmental committee reviewed the prospects for petroleum developments across the empire – in the West Indies, Canada, Newfoundland, Egypt, New Zealand, Brunei and Somaliland – the Admiralty backed down once it realized that the Trust, on its own, would not have the financial resources to make Trinidadian oil economically viable.[241] As a Colonial Office official put it, 'I wish we could have had a purely British oil company, but ... we cannot do without either the Shell or Standard Oil. Other companies may be able to put up sufficient money for development, but Trinidad oil can hardly hope to get on the market without one of these two corporations to sell it.'[242]

In 1912 the Trust was reorganized as the United British West Indies

Petroleum Syndicate, with the inclusion of Royal Dutch-Shell. However, the Governor of Trinidad warned the Royal Commission of a strong German commercial presence in neighbouring Venezuela and, he added, he was 'quite certain that if there ever was the slightest row between us and Germany, Trinidad is one of the first places they would go for'.[243]

Simultaneously, the Admiralty's attention turned to Egypt. In 1909, Royal Dutch-Shell had struck oil, rich in fuel oil, on the Gemsah Peninsula; as its manager in Egypt pointed out, 'A very large proportion of the shipping of the world passes within two miles of our wells, which are situated within a few hundred yards of deep water.'[244] This led to the 1911 formation of a subsidiary, Anglo-Egyptian Oilfields Ltd., with the involvement of the National Bank of Egypt, and the Admiralty initially accepted – as in Trinidad – the need for non-British investment and expertise in Egyptian oil production.[245] However, in 1912, with the appointments of Lord Kitchener as Consul General for Egypt and Winston Churchill as First Lord of the Admiralty, British policy became more nationalistic, Churchill writing to the Additional Civil Lord of the Admiralty, Sir Francis Hopwood, and agreeing with Kitchener's view 'that it would be much better for the Egyptian oil to be worked as a government monopoly for the Egyptian Government and for the Admiralty, we making an unlimited forward contract for what he can supply. This is much better than trading it off to the Shell Company.'[246]

One concern of the British was that Royal Dutch-Shell might not move quickly enough to develop Egyptian oil production. The Financial Adviser to the Egyptian government wrote to Anglo-Egyptian, 'Your company, in its natural desire to establish a virtual monopoly, is anxious to control large areas of undeveloped land, and possibly to restrict temporarily the production of petroleum. The Government, on the other hand, is anxious to develop its oil-bearing territory.'[247] From the Admiralty, Hopwood wrote to Foreign Secretary Grey requesting that a small, speculative British oil company, the Cairo Syndicate, be supported due to the 'dangers lying ahead if the Dutch influences are allowed to absorb the majority of the new fields'.[248] After all, as the Syndicate's chairman told the Royal Commission, 'The German Government have, as is common knowledge, been making a petroleum monopoly in their own country to prevent, ostensibly, the undue inflation of prices, but probably really to be able to commandeer as much oil as they require for their own purposes at fair prices.'[249] But how hostile could the Admiralty afford to be towards a company as well-resourced as Royal Dutch-Shell, upon which it might well have to depend for a large proportion of its supply of fuel oil?[250] The evidence given to the Royal Commission by the group's head, Henri Deterding, made quite an impression on Lord Fisher, chairing the Commission, who, in typically bombastic style, now wrote to Churchill, 'The greatest mistake you will ever have made ... will be to quarrel with Deterding. He is Napoleon and Cromwell rolled into one. *He is the greatest man I have ever met!* Placate him, don't threaten him! Make a contract with him for his fleet of 64 oil tankers in case of war. Don't abuse the Shell Company or any other oil company. *You want to get oil from everyone everywhere.*'[251] A few weeks later, he reiterated, '*Get a lien on Deterding's 64 Oil Tankers for War.* Reward! Make Deterding a Knight! Don't forget! ... He has a son at Rugby or Eton and has bought a big property in Norfolk and building a castle! Bind him to the land of his adoption! ... He has just bought up Sir Robert Balfour and all his vast store of Californian Oil. He isolated him and then swallowed him! Napoleonic in his audacity: Cromwellian in his thoroughness!'[252]

The Financial Adviser in Egypt wrote to Royal Dutch-Shell that 'the simplest solution would be for the Government to become a shareholder in [Anglo-Egyptian]', with continuous working clauses included in its leases.[253] In September 1913 an agreement was signed by which the Egyptian government took a 10 per cent share in Anglo-Egyptian, had the right to nominate one director on its board, and set a maximum price for fuel oil and kerosene sold in Egypt.[254] As *The Times* later commented, the government was right to require British oil companies operating in British-controlled territory to be in a position to provide oil for the navy:

[T]he need of British oil for the British Navy is one very practical aspect of the importance of 'developing the Imperial estate' on business-like lines. Egypt ... is not the only part of the world under British control where oil is likely to be found by private enterprises, and . . . a business-like arrangement is best made beforehand, with an eye on the combined interest of both parties.[255]

Royal Dutch-Shell was thus simultaneously seen as an indispensable asset and as a potential threat. In October 1913, Fisher wrote to Churchill, 'I had a chance of urging Sir E. Grey and Lord Crewe to push on decision about Persian Oil and Mesopotamian oil concession. Otherwise Deterding will have it.'[256]

BRITAIN'S STATE OIL COMPANY

When the Royal Navy embarked on the use of fuel oil, the British government began protecting British oil enterprises operating within the empire or in regions considered to be subject to its exclusive influence. From 1902 it shielded the Burmah Oil Co. from foreign competition by denying production licences in Burma to foreign, or even partly foreign-owned, oil companies and by erecting a tariff barrier against the import of petroleum products to India; and in 1905 the Admiralty signed a long-term fuel oil contract with the company. In 1904 a Joint Committee, convened by the Admiralty and the Colonial Office, elaborated a general policy designed to stimulate the production of fuel oil for the Royal Navy by British companies operating within the empire and, crucially, to protect them from foreign competition.[257]

In 1905 the Admiralty engineered the rescue, by Burmah Oil, of a British syndicate – now the Anglo-Persian Oil Co. – drilling for oil in western Persia, which, although not within the British Empire, was deemed to be a crucial British sphere of interest around the head of the Persian Gulf. As Winston Churchill subsequently put it in 1914, as First Lord of the Admiralty, 'For many years it has been the policy of the Foreign Office, the Admiralty, and the Indian Government to preserve the independent British oil interests of the Persian oil-field, to help that field to develop as well as we could and, above all, to prevent it being swallowed up by the Shell or by any foreign or cosmopolitan companies.'[258]

In 1906 the Colonial Office extended a £25,000 loan to the Nigerian Bitumen Corp., operating in the British colony of Southern Nigeria, in the hope that it would find commercial quantities of oil.[259] However, the Colonial Office rejected the further proposal by the colony's governor that the government should purchase a 20 per cent stake in the company, as it would be 'an invidious thing for Government to hold shares in an industrial undertaking. There would always be an inclination to view the affairs of such an undertaking too favourably, and ... there would always be room for competing concerns to raise the vexatious and embarrassing cries of favouritism.'[260]

In 1907 the Foreign Office sent a military detachment to Persia to protect the oil prospectors working there, and after their oil strike of 1908 the newly named Anglo-Persian Oil Co. became especially prized by the Foreign Office – more for its contribution to Britain's strategic presence around the head of the Persian Gulf than for its oil production, which was as yet an unknown quantity.[261] In 1910 Anglo-Persian's managing director, Charles Greenway, led a syndicate with ambitions to become the leading, all-British oil company in the colony of Trinidad, under state protection, and in this he was supported by the Admiralty. However, when it became apparent that the syndicate would not have the necessary distribution infrastructure to become commercially viable, Royal Dutch-Shell, which was at least partly British, was reluctantly allowed to join in the formation, in 1912, of the United British West Indies Petroleum Syndicate.[262]

By this time, the Nigerian Bitumen Corp. had sunk fifteen wells in Southern Nigeria, with encouraging results, and in September 1912 it went to the Admiralty to apply for more funding from the Colonial Office. The renowned oil prospector William McGarvey, a long-time associate of the company's chairman, John Bergheim, considered that 'it would be almost a crime not to continue the drilling ... and I hope, in the interest of the British Empire, such a calamity will not be allowed to come to pass.'[263] From the Admiralty, Sir Hopwood wrote to the Colonial Office, 'You can well imagine how satisfactory it will be to secure a good supply of oil for the Navy from a British Company in a British colony only about twelve days steaming from home.'[264] However, the Colonial Office threw the ball back into the Admiralty's court, an official noting on Hopwood's letter, 'If the success of this Company is important to the Navy, the Admiralty ought to assist.'[265] At just this time, Bergheim was killed in a car crash, the impact of which was also terminal for his company.[266]

In Egypt, a de facto British protectorate, the authorities decided, as in Trinidad, that it would only be possible to extract the country's oil reserves on a large scale through Royal Dutch-Shell's local subsidiary, the Anglo-Egyptian Oil Co. In early 1913 it was suggested that the government could gain a measure of control over the company by taking a 10 per cent share in it along with the right to appoint one director to the board; an agreement to this effect was signed in September and Hugo Baring, a director of the Turkish Petroleum Co., was appointed to the board.[267]

In November 1912 the Royal Commission, tasked with inquiring into the security of the Royal Navy's oil supplies, took the view that it was not qualified 'to be able to discriminate with exactitude as to the rival claims of, say, the Anglo-Persian Oil Company, the Nigerian Bitumen Corporation, the Trinidad Oil-fields, Sir H. Blake's Newfoundland venture, Mr. Mackenzie's Mount Sinai claims, and many more, all desirous of Government aid'.[268] However, by early 1913, after the price of fuel oil had increased sharply and its availability had diminished for well over a year, the Admiralty began to consider the navy's oil supplies to be perilously uncertain. The ensuing deliberations and parliamentary debates over the matter were, therefore, among the most significant in the history of the British Empire. They addressed the urgent problem of how Britain might secure the medium- and long-term supplies of fuel oil upon which the Royal Navy – the lynchpin of the empire – had become existentially dependent; and, by voting ultimately for British state ownership of oilfields in southwest Persia, Britain's MPs would increase their country's imperial interests in the Middle East, extending these beyond protecting the routes to India to include gaining control over the region's oil reserves. Moreover, since these oilfields were located in Persia's 'neutral zone' as defined under the 1907 Anglo-Russian Convention, this decision would render Britain even more diplomatically dependent on Russia – with potentially momentous consequences (see Fig. 25).

Officials became persuaded by Greenway's argument that Royal Dutch-Shell was close to absorbing Anglo-Persian and, furthermore, that the Anglo-Dutch group was likely to come under German control. This was seen as a major threat from Germany both to the navy's future supplies of fuel oil and to Britain's strategic position at the head of the Persian Gulf – long regarded as crucial to the security of India but seen as being vulnerable to Russian, and now German, incursion. In March the Royal Commission decided, after all, that financial support should be given to Anglo-Persian; and on 16 June Churchill presented his case for this to the Cabinet in a pivotal memorandum entitled 'Oil Fuel Supply for His Majesty's Navy'.[269] His central argument was that

> In the last few months, since the proposals of the Anglo-Persian Oil Company were first received by the Admiralty, the trend of prices, as well as the results of experiments with other descriptions of oil, and the conclusions arrived at by the Royal Commission have all contributed to render the Persian supply relatively more important than was originally anticipated, without altering the Admiralty view that too much reliance should not be placed on any one source of supply . . .
>
> So far the British Admiralty has adhered to the system of annual contracts. To continue longer on such a system is to make sure of being mercilessly fleeced at every purchase, and to run a very great risk of not being able to secure on any particular occasion supplies of a fuel which will be as vital to the Navy as ammunition itself.[270]

On 9 July Churchill told the Foreign Secretary, Sir Edward Grey,

> The Admiralty are proposing to make a contract with the Anglo-Persian for the supply of a large quantity of oil in consideration of an advanced payment on account of the oil by the Admiralty of £100,000 a year for 20 years . . . The Admiralty regard the development of the Anglo-Persian oil supplies as indispensable to the solution of the liquid fuel problem . . . It is hoped that India will come to our assistance. If she does not, we must go forward alone by the more expensive method. It appears to me that there can be no doubt that the Anglo-Persian Oil Company will receive substantial financial aid in one form or another from the British or Indian Governments.[271]

At a Cabinet meeting two days later it was formally agreed that 'in view of the vital necessity to the Navy of a continuous and independent supply of oil in the future it was desirable that the Government should acquire a controlling interest in trustworthy sources of supply, both at home and abroad.'[272] On 17 July Churchill, during his annual Navy Estimates speech to the House of Commons, introduced his plans for a state-owned oil company to safeguard the Royal Navy's supplies of fuel oil. The Admiralty's 'interim policy' was to sign several forward contracts, of about five years' length, and to build up its fuel reserves. However, he told MPs, the Admiralty had decided on a course of 'moving towards that position of independence outside the oil market which it is our ultimate policy to secure'. Although Churchill admitted 'that to a very large extent the causes which raise prices and create stringency are natural and automatic' – in other words, the usual result of market forces – he charged that the growing demand for fuel oil and petrol 'has given rise to vast and formidable schemes on the part of a comparatively small number of wealthy combinations to control the oil market and raise and maintain prices'. The oil market was being subjected to 'artificial manipulation', and 'the open market is becoming an open mockery'.[273]

The Admiralty's 'ultimate policy' of taking up a 'position of independence outside the oil market' would be achieved in three ways, Churchill continued. Firstly, by having the necessary storage and transport capacity it would be able to 'override price fluctuations' by buying up large quantities of oil, either crude or refined, when it was cheap; this would put the Admiralty in a stronger bargaining position. As Shell's Sir Marcus Samuel had told the Royal Commission, 'You cannot have too much storage ... because there are times when there are gluts of oil, and then Mr. Black [Frederick Black, the Navy's Director of Contracts] can buy very cheaply ... but he cannot buy if he has got nowhere to put it.'[274] Churchill continued,

> Th[e] second aspect of our ultimate policy involves the Admiralty being able to ... refine ... crude oil of various kinds until it reaches the quality required for naval use. This again leads us into having to dispose of the surplus products – another great problem – but I do not myself see any reason why we should shrink, if necessary, from entering this field of State enterprise. We are already making our own cordite, which is a most complex and difficult operation. We already keep our great system

of the dockyards in full activity in order to provide a check on private constructors, and I see no reason, nor do my advisers, why we should shrink from making this further extension of the vast and various businesses of the Admiralty.

'The third aspect of the ultimate policy,' Churchill then declared, 'is that we must become the owners, or at any rate, the controllers at the source of at least a proportion of the supply of natural oil which we require.'[275]

Churchill did not detail how his policy would be implemented, although by now the Cabinet, except for the Chancellor of the Exchequer David Lloyd George, was close to agreeing that it was likely to involve extending significant financial support to Anglo-Persian as part of a wider policy of securing British control over the oilfields of both western Persia and Mesopotamia. The company's wells appeared capable of producing large volumes of crude; however, production was being 'shut in' due to the limited capacity of the pipeline, the continuing inadequacy of its refinery, and the near unavailability of tankers. In fact, the company was on the verge of bankruptcy. Greenway had been requesting £2 million of government investment since the previous autumn, and the Admiralty had proposed a twenty-year forward contract for just this amount.[276]

In order to give an official stamp of approval to Anglo-Persian, and to rebut critics, Churchill sent a commission to Persia to report back on the company's prospects. Headed by Vice-Admiral Edmond Slade, the commission arrived at Mohammerah in southwest Persia on 23 October, a few days after Greenway had written to the navy's Director of Contracts, Frederick Black, hoping for a 'closer relationship than that of temporary independent buyers and sellers'.[277] In early November Slade privately communicated to Churchill, 'It seems to be a thoroughly sound concession, which may be developed to a gigantic extent with a large expenditure of capital. It would put us into a perfectly safe position as regards the supply of oil for naval purposes *if we had the control of the company* and at very reasonable cost.'[278] Meanwhile, Churchill sought to persuade sceptics at the Treasury that assistance could be extended to Anglo-Persian in a manner that avoided putting the government in the position of 'having borne a large part of the financial burden at a crucial time ... then leaving private capitalists to reap the profit if the affairs of the Company take a highly prosperous turn'.[279]

At just this time, Lord Cowdray offered the Chancellor a major contract whereby his Mexican Eagle Oil Co. would supply the Royal Navy with fuel oil in return for £5 million of government investment. He would, he said, 'specifically dedicate ... the Mexican Eagle Company to the primary purpose of supplying the Admiralty with fuel oil ... If it were desired that the control of the Company should remain in British hands for say seven years, this can be arranged.'[280] However, the government had effectively already decided to place its investment in Anglo-Persian; Mexico was deemed too politically unstable, and Britain had agreed to respect the United States' Monroe Doctrine by not pressing for more extensive oil concessions in Latin America. The Admiralty also claimed that Mexican Eagle's fuel oil failed to meet its specifications – although so did that of Anglo-Persian.[281]

The Slade Commission returned to England at the end of January 1914 and officially reported that the Persian concession was 'an extremely valuable one ... [W]e strongly recommend that control of the Company should be secured by the Admiralty [and] that it would be a national disaster if the concession were allowed to pass into foreign hands ...'[282] After lengthy discussions, on 18 February the Cabinet agreed that 'a contract should be entered into with the Anglo-Persian Oil Company, whereby the Government would acquire for about £2 000 000 a controlling interest in that undertaking'.[283] This second state rescue of the ailing company, on a far greater scale than the first, would not only keep it alive as a going concern under British control. It would also put it in a position, exactly one month later, to take a controlling stake in the Turkish Petroleum Company, under a compromise agreement between the British and German governments.[284]

In a paper presented to Parliament in mid-May, Churchill argued,

> foreign navies are on the road to become large consumers of oil, and ... in the course of time they will be entering the market for increased quantities, and competing still more keenly with His Majesty's Government for prompt supplies. This will probably have the effect of inflating prices or reducing available supplies, and for this

reason alone it is necessary to make provision now to safeguard the position.

At the present time the production and distribution of petroleum oil is fairly well distributed among a number of Companies. There is, however, an evident tendency for the oilfields of the world and the marketing of their products to fall more and more under the control of a comparatively few large concerns. To a great extent this is probably unavoidable. The policy of amalgamation is founded on considerations of expediency similar to those which induce any large oil consumer to distribute orders as widely as possible, viz., greater security of supply, in order that a possible shortage in one direction may be counterbalanced by abundance elsewhere.

A large consumer like the Admiralty, with the national interest of naval defence in its keeping, cannot, however, place itself in a position of dependence for vital supplies upon a few large Companies whose interests are necessarily cosmopolitan and financial, although the Admiralty may properly place a reasonable portion of its contracts in the hands of such organisations, admittedly efficient, and possessing great resources. It is important and essential in naval interests to secure that at least one large British Oil Company shall be maintained, having independent control of considerable supplies of natural petroleum, and bound to the Government by financial and contractual obligations. Such an arrangement can hardly fail to have its effect not merely on the supplies directly so obtained, but on the terms and conditions on which the whole of the Admiralty's requirements will be met, owing to the greater independence and bargaining power thus obtained.[285]

It took nearly three months to work out the details of the government's purchase of the company and of the associated Admiralty contract, but these were finally approved and the contracts signed in mid-May 1914. All that was now left to do – after the fact – was to sell the scheme to the public.[286] On 25 May *The Times* told its readers,

Since Lord Beaconsfield, with the financial cooperation of the Rothschilds, purchased on behalf of the British Government a controlling interest in the Suez Canal, there has been no similar transaction comparable in importance to the oil deal announced on Saturday ... The policy underlying the new announcement, considered apart from its

details, is one with which most people will agree. It is of the first importance that there should be available for the use of our battleships a
sufficient supply of oil, obtainable from territory under British influence, outside the danger of market manipulation, and uncontrolled by
any group or combination of financiers. In default of such an independent supply the country might in a time of national crisis find its usual
sources closed by foreign interference, or be confronted with a combination of oil interests taking advantage of the conditions to force up
prices. It is essential of any such arrangement that the source of the
supply should be exempt from liability to foreign attack or internal
disorder. Mexico has lately furnished abundant illustration of the dangers to which oilfields are exposed, and the first question that suggests
itself in connexion with the scheme relates to the prospect of similar
difficulties occurring in Persia ... [T]he main consideration for the
Admiralty is to secure an adequate supply of oil fuel virtually under
their own control. If that end is reasonably assured, too much stress
need not be laid upon the purely financial objections that may be taken
in some quarters. The efficiency of the Navy in time of war is the first
object to be secured.[287]

However, two days later the newspaper acknowledged that 'the
agreement ... raises certain important considerations of foreign policy and defence which require careful scrutiny'. For instance, as
defined by the Anglo-Russian Convention of 1907, Anglo-Persian's
producing wells were located not in the 'British' zone of Persia but in
the 'neutral' zone (see Fig. 25),

> a region where Great Britain not only has no preferential control nor
> influence, but has expressly divested herself of such influence as many
> people once considered her to possess. It is a region notoriously turbu-
> lent, and even Admiral Slade, who has considerable knowledge of
> Southern Persia, admits that though the operations of the company
> have had a tranquillizing effect, there may be a relapse into lawlessness.
> The very frontier between Persia and Turkey in this neighbourhood has
> been for fifty years the subject of a dispute not yet finally adjusted. The
> oil-wells are, further, situated in a region where sovereignty of any kind
> is growing feeble, and where international influences which may be
> conflicting tend to expand. The Middle East has been called the real

cockpit of the world, and it may become so again ... A great trunk railway [the Baghdad Railway] will soon be at work within striking distance across the Turkish frontier, and an extremely vulnerable pipeline will continuously be at the mercy of swarms of fickle tribesmen in a land where internal authority has almost lapsed. At any time the Admiralty property may require protection, and the duty will assuredly devolve, not upon the Navy, but upon the Army of India. In no spirit of carping criticism, but with a very real sense of anxiety, we desire to point out that it was precisely to avoid this kind of responsibility that the spheres of influence in Persia were drawn in the manner finally settled upon.[288]

On 17 June 1914 the bill for the government's acquisition of a controlling stake in the Anglo-Persian Oil Co. was debated in Parliament.[289]

'SURROUNDED BY MATERIAL FAR MORE INFLAMMABLE THAN THE OIL'

In the midsummer of 1914 the First Lord of the Admiralty, Winston Churchill, had to stand up in the House of Commons and explain to Members of Parliament why the government – supposedly committed to private enterprise as a matter of principle – was investing £2.2 million in a small, new oil company operating in Persia, a distant land that, as one MP put it, the public 'only know through the Press as a country of constant disturbance'.[290]

The essence of Churchill's argument was that the Admiralty was irreversibly committed to adopting fuel oil in the Royal Navy's steamships. However, Britain's home production of fuel oil, from Scottish shale, was quite limited, as was production from the wider British Empire. Consequently, the navy had become dependent upon fuel oil from foreign sources, and supplied mainly by foreign oil companies. Furthermore, the Admiralty's ability to diversify its sources of supply was limited, as the two huge conglomerates – Standard Oil and Royal Dutch-Shell – were rapidly gaining control over the world's oilfields and over refining and distribution. In particular, Churchill argued, the recent dramatic rise in the price of fuel oil had in large part been engineered by these oil companies as they wielded their growing monopoly power. The Admiralty needed somehow to check the increasing market dominance of foreign oil companies and to secure some independent control over supplies of a fuel upon which the navy was vitally dependent:

> [W]e must not let ourselves be deprived of oil. We must not be forced to content ourselves with less efficient war machines because of the difficulties, perfectly superable difficulties, attendant on obtaining a supply of oil ... We must continue to make efforts to give our sailors the finest and most suitable weapons of war which science can devise or money can buy. We cannot allow these great advantages to pass to other nations, to whom naval strength is of so much less consequence than it is to us.[291]

The only solution appeared to be for the Admiralty to have its own oil company. Churchill concluded that none of the British oil companies working within the British Empire would be capable, any time soon at least, of reliable, long-term fuel-oil production on a scale that would meet a significant proportion of the navy's projected requirements. As Foreign Secretary Sir Edward Grey expressed the predicament to Parliament a few weeks later,

> I would very much rather, I fully admit, that the Admiralty had been able to make the arrangements inside the British Dominions, but they could not do it. The British Empire was never planned, and the importance of oil was never foreseen; so, even if it had been planned, I doubt whether this omission to secure a first-rate supply of oil in the British Empire would have been remedied.[292]

By contrast, the Anglo-Persian Oil Co. – operating in the southwest of Persia's 'neutral' zone as defined by the 1907 Anglo-Russian Convention (see Fig. 25) – appeared to have much greater potential. Churchill told MPs, 'The extent and variety of the ... oozings of oil to the surface over the whole of that region is fully described in [Admiral Slade's] Report. They abound, not only in the neutral, but over the British zone. Some of them are not inland, but close to the sea. Others are close to the Indian border.'[293]

Although southern Persia was clearly foreign territory, it was a basic tenet of British foreign policy that it should remain firmly under British influence. The government should therefore buy a controlling stake in Anglo-Persian and provide the investment needed for it to fulfil its great potential. Otherwise, 'there is no doubt whatever in my mind ... that amalgamation or merger of the Anglo-Persian with the Shell ... would probably take place in a very little while.' Churchill emphasized that this was a measure intended primarily to address the price of fuel oil in peacetime and that the Admiralty would continue to diversify its sources of supply widely so as not to be dependent on just a few, in case any were cut off: '[T]here are no grounds, in the opinion of the Admiralty, for misgiving on the score of an oil famine in this country in time of war, and I hope we shall not hear in this Debate any of the nonsense I have read in some of the newspapers, about the British Empire becoming dependent upon a slender pipe line running through 150 miles of mountainous country and barbarous tribesmen.'[294] Churchill acknowledged that although his policy of major state intervention might appear to be a radical departure from the prevailing market ideology,

> The general principle of partial State ownership has not been impugned and I do not expect that it will be impugned in this discussion. After all, it is only what we do in regard to the shipbuilding trade of the country by the competition of Royal dockyards, the general manufacture of cordite, which is a very complicated operation, and the Whitehead torpedoes, and in regard to the general purposes of the Navy ... I do not think that there is any question of principle at stake.[295]

In an editorial published the same day, the *Daily Telegraph* agreed that 'while every right-minded person would revolt against any action by the State calculated to injure unnecessarily private enterprise, national interests come above all others ... If the Admiralty prove the need, we do not believe that members of the House of Commons will shrink from the onus of intervention in the interest of security and economy.'[296]

In response to Churchill's first public announcement in Parliament a year earlier that the Admiralty would have to go into the oil business, the chairman of the parliamentary Labour group, Ramsay MacDonald, had praised Churchill, tongue-in-cheek, for the magnificent Socialist position which the right hon. Gentleman has taken up in reference to oil. I congratulate him on his progress ... He has discovered that the ordinary operations of capitalism produce monopoly. He has discovered that in the supply of oil, which is the food of his ships, two or three gentlemen sitting in some back parlour or other, can manipulate his prices, collar his oil fuel, and make him very accurately describe the open market as something that requires a great deal of imagination to conceive. I congratulate him. He is perfectly right in his description and in his conclusions. The Admiralty ... must protect itself against the ordinary operations of capitalism and monopoly, by establishing its own supplies to secure itself against these monopolists. What is true of food for the ships is true of food for the people but that is a field into which the Chairman will not allow me to go.[297]

The *Economist* had ridiculed the idea of a state-owned oil company from a different angle:

We can understand Mr Keir Hardie, an ardent Socialist, who represents the steam coal area, forgetting the interest of the Welsh coalminers in his joy over the Admiralty's new Socialistic policy of purchasing oil fields. The monopolisation of the means of distribution and production by the State is the great end and aim of Socialism. But it is amusing to see the *Daily Telegraph* and the *Daily Chronicle* joining hands with Mr Keir Hardie ... [W]e are amazed to think that [Churchill] and his Board, not one of whom would profess to be a business man, should think of looking round the world for suitable oil fields in competition with Mr John Rockefeller. We would put it to any person of City experience whether they think that Mr Churchill, with the help of three or four admirals and two or three Civil Servants, is likely to outwit Mr Rockefeller or Lord Cowdray, or any other oil magnate, in the purchase of oil fields.[298]

The *Observer*, however, agreed with Churchill that

the financial undertaking is promising and as legitimate ... as holding shares in the Suez Canal or building our own warships. It must be carefully noted that the British Government is not the pioneer in this form of State enterprise. The German Government, whose difficulties of all kinds in connection with the oil question are greater than our own, is trying to establish a national petroleum monopoly. The American

Government, as Mr. Churchill points out, is laying down a pipe-line of its own and starting refineries.[299]

Whereas the German government's state oil plan had, in fact, long since failed to materialize, the United States government had indeed been taking measures to safeguard its navy's supplies of fuel oil, despite the country's own vast oil production. In 1912 it had set aside two oilfields in California for the navy, and at the end of 1913 the US Secretary of the Navy, Josephus Daniels, like his British counterpart, blamed the recent sharp rise in the price of fuel oil on the private oil companies, and he decided that the navy should therefore enter the oil business:

> I desire to recommend to Congress the immediate consideration of providing fuel oil for the Navy at reasonable rates, and the passage of legislation that will enable the department to refine its own oil from its own oil wells and thus relieve itself of the necessity of purchasing what seems fair to become the principal fuel of the Navy in the future, at exorbitant and ever-increasing prices, from the private companies that now completely control the supply ... [I]n Mexico and elsewhere English firms are reaching out for oil fields from which to supply the demands of the English Navy. On the other hand, the price of oil is steadily creeping upward, so that to-day the United States Navy is paying over twice as much for its oil as it did in 1911 ... [H]ow can the United States ... escape the charge of willful waste of public money if it continues to purchase oil at prices which may fatten the pockets of a few oil companies, but which increases the burden of the taxpayer? ... This proposed step is no new departure, for the Navy now builds some of its own ships, maintains large industrial Navy yards, a gun and a clothing factory – all of which are indispensable to the supply of superior articles for the Navy and for the control of prices from commercial concerns furnishing similar articles.[300]

In his report for 1914 Daniels again put the case for the navy's role in the manufacture of arms and supplies: 'It has been suspected in some countries that makers of armor and powder, guns and fighting craft have promoted steadily increasing equipment for giant navies and large armies for their personal enrichment ... The incentive of personal aggrandizement by preparations for war should not be permitted to exist in the United States.'[301] In January 1914 the Senate ordered a study into the 'feasibility, expense and desirability' of a government pipeline from oilfields of the Osage Indians in Oklahoma to the Gulf of Mexico, 'with a view to providing and conserving at all times an adequate and available supply of oil for the use of the Navy'.[302] However, the proposals ultimately went no further, due to a lack of support in Congress and a belief that state regulation of the oil industry would suffice.[303]

During the Commons debate, Churchill was scathingly critical of Shell and its chairman, Sir Marcus Samuel; the Conservative MP Watson Rutherford called it 'a great pity' that Churchill had felt it necessary 'to raise the question of monopoly and to do a little bit of Jew-baiting', referring to the fact that Sir Marcus, as well as Shell's major shareholders, the Paris Rothschilds, were Jewish.[304] Samuel Samuel, another Conservative MP, put up a defence on his brother's behalf. He denied that a monopoly existed in the oil market, as there were numerous oil companies operating around the world, many of them British; and he criticized the Admiralty for insisting on such secrecy regarding its oil contracts that he was unable to prove that Shell had offered it fair prices for fuel oil.[305] Sir Marcus himself responded publicly a few days later at a Shell AGM: 'In 1913 of the total quantity of oil imported by the Admiralty our company supplied less than one-eighth. The remaining seven-eighths were derived from nine separate vendors, not one having the most remote connexion nor alliance nor agreement with the Shell Company, nor among themselves.'[306] Indeed, Churchill admitted during the debate, 'It is quite true that these price movements arose largely out of an increased demand by the world which is eager to use such an extraordinarily convenient fuel.'[307]

Sir Samuel argued not only that was there was no oil monopoly, but also that the Admiralty ought to be able to rely on Shell for secure supplies of fuel oil, particularly from its oilfields under British control in Sarawak, Egypt and Trinidad. The company's policy had been

the development of fields all over the world, especially including British territory or places under British jurisdiction. The First Lord of the Admiralty has misconstrued this action, and has assumed that our purpose was to raise prices. The motive that prompted us is to obtain large, constant, and widespread supplies from as many sources as possible. This obviates dependence upon any one field, and enables us to undertake business and to give contracts of supply practically without a *force majeure* clause ... It is in the best interests of consumers that certainty of supply should be established.[308]

Behind the scenes, Samuel had put it to the Admiralty that it would be counter-productive for the government to invest heavily in Anglo-Persian as this would 'so discourage the commercial production of fuel oil in other parts of the world that the tendency will be to throw the Admiralty back purely upon Persia as its source of supply'.[309] As to the accusation that Shell was subservient to its Dutch partner in the Royal Dutch-Shell group, 'The reproach that we might suffer from a foreign alliance would possibly have held good if it were with a country having conflicting interests with Great Britain and having a fleet requiring large supplies of liquid fuel. Neither of these conditions applies to Holland.'[310]

However, British officials had already decided that in the event of war with Germany, its proximity to Holland might enable it to put adverse pressure on Royal Dutch-Shell.[311] Similarly, Mexico, where the production of Lord Cowdray's Mexican Eagle Oil Co. was now prolific, was in a state of civil war and it was uncertain what the foreign policy of a future Mexican government might be. Furthermore, the United States had reasserted its Monroe Doctrine over Latin America by preventing Lord Cowdray from gaining additional, large-scale British control over oil reserves in the region.[312] Sir Edward Grey even argued that there might be some uncertainty over supplies from the US. Britain's view of the international law of wartime contraband was that neutral countries should allow the export of oil; but, Grey asked, 'Can you be quite sure that a Government such as that of the United States would take that view?'[313]

But how did these risks compare with those that came with the Persian oilfields and the associated transit routes? The 500,000 square miles of Anglo-Persian's 1901 oil concession covered the whole of Persia except for the five northernmost provinces bordering Russia.[314] The signing, six years later, of the Anglo-Russian Convention – designed to reduce tensions between the two powers across Central Asia from Persia, through Afghanistan, to Tibet – had designated northern Persia as within the Russian sphere of influence. This actually covered the most northern parts of Anglo-Persian's concessionary area, which included Kermanshah and the capital, Tehran (see Fig. 25). Due to the weakness of the Persian government – rocked by revolutionary political upheavals – and the lack of law and order, 12,000 Russian troops were now brutally occupying the northern zone under the pretext of protecting trade routes and Russian commercial interests; as the British consul at Kermanshah had written in 1908, 'with the Russians political influence and commerce are, in Persia, interchangeable terms.'[315] A late 1913 cartoon in *Punch* magazine depicted a British lion glaring at a Russian bear sitting on the tail of a Persian cat, and Grey came in for sustained, intense criticism: for abandoning Persia to Russia, for failing to support Persia's nascent Constitutionalist movement, and for honouring the country's independence and sovereignty in words only. One member of the Persia Committee lamented, during the Commons debate,

> It is almost amusing the way the great Powers, when discussing a matter of this sort, consider that they are conferring an untold benefit on the country in question, and the interests of that country so far as its population is concerned are entirely disregarded. I suppose that the Persian Government has been consulted, although I daresay that would be considered an unnecessary formality. It has been the policy of the British Government too often to concentrate attention on the material development of a country without sufficient regard to the welfare and the liberties of the inhabitants to whom that country belongs. That has been too often our policy. It is a matter of small moment to any hon. Member in this House how the Persians will fare. We think in our arrogance that, of course, British capital and British enterprise can do nothing but confer an immediate benefit and advantage on a country in such a backward state as the Persian Empire . . . This unfortunate country has had very little chance of asserting itself, of recovering, and of getting a good Government established, because there has been perpetual interference and intervention from these two Powers . . . I only say that in establishing ourselves in 500,000 square miles of territory in

another country we might just give a passing thought to the interests of the inhabitants of that country.[316]

Relative to Russia, which could easily send troops into Persia, Britain felt unable to deploy troops to protect its harassed commercial interests in the south and was instead relying on cultivating close ties with local tribes – primarily the Bakhtiari khans and the Sheikh of Mohammerah – to safeguard Anglo-Persian's operations.[317] In mid-1913 Grey was warned by his minister in Tehran that Russia might extend its influence into the 'neutral' zone:

> Russia's word is law in Isfahan, and ... unless we speedily do something to prove to the Bakhtiaris and the rest of Persia that we are faithful to our ancient friendship for the tribesmen, that we can furnish the backbone of a strong natural barrier between North Persia and that part of the country in which British interests are still predominant, we shall find the wedge of Russian influence driven down to the Karun by peaceful penetration under arms.[318]

By March 1914 Grey was suggesting to his ambassador in Russia, Sir George Buchanan, that a partition of the neutral zone might be negotiated: 'The chief point on which we shall have to ask for a modification of the Anglo-Russian Agreement is in connection with the concessions ... The weakness of our position in Persia is that the Russians are prepared to occupy Persia, and we are not. We wish Persia to be a neutral buffer state; they are willing to partition it.'[319] Grey feared that Persia might, indeed, be destined for partition, writing to Lord Crewe at the India Office in mid-May, 'It looks as if the expected but dreaded break-down in Persia is coming.'[320] At the end of the month, after the public announcement of the government's £2.2 million takeover of Anglo-Persian and the attendant press criticism, Grey anxiously enquired of Churchill,

> Is it not the case that some of the oil wells of importance are in the British sphere in Persia? And as to the rest in the Neutral Sphere what distance are they from the coast? Perhaps Admiral Slade would send me a memo of these facts ... The real point is that S. Persia near the coast is more controllable by us than other centres of oil production in the world, which are entirely out of our reach.[321]

Four days later, Grey's colleague Sir Eyre Crowe urged a more assertive renegotiation of the Convention: 'There is only one possible way
now of keeping Russia out of southern Persia: we must establish our
own exclusive authority there ourselves. If we take the line that we
cannot afford to incur expenditure for such Imperial interests, it will
be like a declaration of imperial bankruptcy in respect of those
regions.'[322] In mid-June, a few days before the debate on the government's takeover of Anglo-Persian, Sir Arthur Hirtzel at the India
Office wondered how the company's oilfields might be defended in
the event of war:

> It may be assumed that if the relations of the British empire with Russia
> or Turkey became strained, the oil-field would be a probable objective
> for an enemy and its security in advance would be a very pressing mat-
> ter. The Indian Government from its geographical position would be
> naturally looked to for troops. But it is clear . . . that in the event of war
> the Indian Government will require every available soldier to provide
> for the internal defence of India and to place on the north-west frontier
> a field army of the necessary strength.[323]

Now George Lloyd, a Conservative MP who had travelled and studied the region extensively, stood up and told fellow MPs,

> I happen to know the districts very well . . . [The oilfields are] in a coun-
> try which is surrounded by war-like tribes such as the Bakhtiari and
> others whom we have never been able regularly to control in spite of
> having treaties and other friendly relations with them, and which is in
> the hands of turbulent tribesmen whose influence is proportionate
> locally for their capacity to terrorise and raid, and whose policy is
> directed by no respect for foreign undertakings or treaties . . . In the
> West what have you got? You have got a Turkish Army Corps, the 6th
> Army Corps, a Regular Army Corps, regularly resident in Bagdad and
> Irak, and on the East of these wells and quite close to them you have got
> the fierce and turbulent Bakhtiari tribes, favourable at times and very
> formidable at other times, and very difficult to deal with, and not at all
> the kind of gentlemen the right hon. Gentleman alludes to in his very
> easy way. You have got on the North a force of Russian Cossacks who
> are likely to remain there some time, and Russian control, and you have

14. Some of the first offshore oil wells, at Summerland beach, Santa Barbara County, California, in 1902.

15. 1902 adverts for electric and steam cars, illustrating their continuing popularity in the US around turn of the 20th century.

16. One of several illustrations accompanying a prophetic March 1906 article by the Marquis Jules-Albert de Dion, co-founder of the world-leading French De Dion-Bouton car manufacturing firm, in which he predicts how motor vehicles will have changed the world in forty years' time. The article was published in English in several American newspapers.

17. Traffic and road works at Piccadilly Circus, London, in May 1910, showing an equal mix of horse-drawn and motor vehicles.

8. A Cossack patrol in the Balakhany oilfield near Baku in September 1905 during the Russian revolution of that year.

9. The remains of oil well machinery destroyed by fire during the insurrection.

20. A 1908 photo of a group of Bakhtiari tribesmen, paid by the Anglo-Persian Oil Co. to guard its operations in southwest Persia.

21. In May 1908 the Burmah Oil Co.'s subsidiary the Concessions Syndicate Ltd., drilling at Masjid i-Suleiman in southwest Persia, made the first major oil strike in the Middle East. This would lead to Burmah Oil forming the Anglo-Persian Oil Co. – later the Anglo-Iranian Oil Co., British Petroleum and ultimately BP.

22. Horse-drawn transport of sections of the pipeline being constructed by the Anglo-Persian Oil Co. 140 miles southwards from its oil wells at Masjid i Suleiman to its refinery on Abadan Island at the head of the Persian Gulf, circa 1911.

23. Bakhtiari tribesmen, circa 1910, laying the Anglo-Persian Oil Co.'s pipeline, which was completed in July 1911.

24. The Union Oil Co.'s Lakeview gusher of 1910 on the prolific Midway oilfield in Kern County, California. Reflected in its own lake of oil, which threatened to flow into Buena Vista Lake.

25. A June 1914 map in *The Times* showing the Anglo-Persian Oil Co.'s operations near the head of the Persian Gulf, the extent of its oil concession, and the Slade Commission's reports of 'oil shows', in the context of the zones of influence as defined by the 1907 Anglo-Russian Convention.

REFERENCE.
British Sphere
Russian Sphere
Oil Wells - Proved Areas
Reported Oil Shows
Concession Boundary
International "

MILES
50 100 200

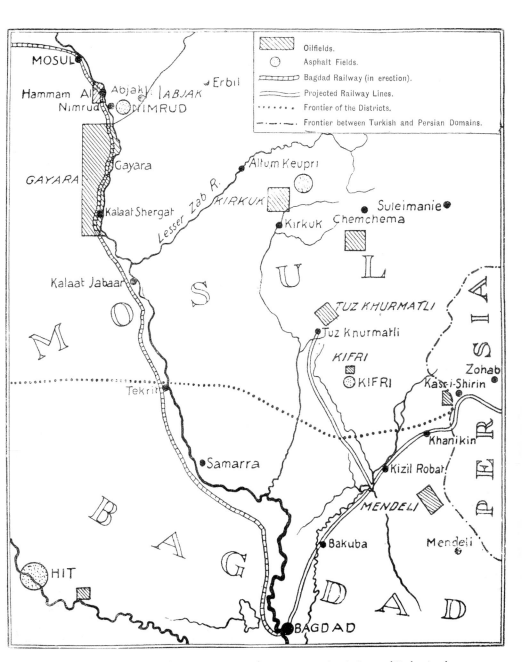

26. Map featured in 'The Petroleum Deposits of Mesopotamia: A Second Baku in the Making', *Petroleum Review* (23 May 1914), including part of the projected route of the German-financed Baghdad Railway.

27. During the years prior to the outbreak of WWI, Royal Dutch-Shell had come under fire from the British government for being foreign-controlled, and during the war Shell placed many adverts in British newspapers and magazines proclaiming its patriotic dedication to the Allied cause, this full front-page advert in the *Daily Mail* in November 1914 being the most prominent.

got these wells within range of the most turbulent of all the Arab tribes, who do not regularly cross into Persia, but when they do everybody knows about it – the Munte[fik] Arabs ... Nobody denies that the properties are valuable, but ... the whole of your properties is surrounded by material which is far more inflammable than the oil which you seek ... If you cannot defend your property, you are offering a leverage to many other interests, foreign interests ... and which you will find it very hard to recover unless you have a scheme, and unless you mean clearly to defend the properties you are acquiring. Is it going to be by the Navy, which you are going to send round through the Mediterranean, which you have recently abandoned, or largely abandoned, for the control of France? ... If not, are you going to disembark a large landing party to protect this oil industry, or who is going to protect it?[324]

Churchill reassured MPs that Anglo-Persian's operations were quite secure:

We are told that the tribesmen are wild and that the Persian Government are weak. The investments of capital, the development of roads, railways, and industries in which the tribesmen and the Persian Government are both interested, and from which both profit, ought to make the Persian Government strong and the tribesmen tame ... [It is] a region where we already have great interests ... where our relations with great Powers are already regulated by agreement, where it is our interest and policy to sustain the native inhabitants and the native Government, and where neither the native Government nor the native inhabitants are capable of pursuing a prolonged or formidable policy of hostility towards us.[325]

In a letter to *The Times* a few days later, on 24 June, the company's vice-chairman, Charles Wallace, backed up Churchill's positive view of the security arrangements:

Besides paying for the cost of upkeep of armed Bakhtiari guards, we formed subsidiary companies to produce the oil [in which] we gave to the chiefs in whose territories different wells ... are situated shares which will of course give them dividends only if they assist us to work in peace and security. Turbulent and impatient of restraint by the Northern Government, as it is locally called, these men may be, but they are

neither savages nor ignorant. Their chiefs are highly educated and pol-
ished gentlemen quite able to see where their own interest lies, and most
unlikely after pursuing it steadily for so many years to wreck the value
of their own shareholdings ... [A]nother great factor which has served
to build up the excellent relations subsisting between our staff in Persia
and the natives with whom they come into contact – viz., the fine work
done among Persians and tribesmen by our medical staff. Dr. D.M.
Young and his coadjutors have been and are worth more than a military
police force could have been, for by their successful surgical and medical
treatment of all and sundry who come to them from far and near they
have turned hundreds or thousands of potential enemies into firm
friends.[326]

British views of the Bakhtiari were as ambivalent as the Bakhtiari's
views of the British. Having spearheaded the armed uprising against
the Shah, the Bakhtiari now dominated the Constitutionalist government, and during the Commons debate a member of the Persia
Committee praised 'the Bakhtiaris, for whom I have a great admiration, as their gallant efforts contributed largely to the securing of the
liberties of Persia'; another MP agreed that 'The Bakhtiari tribes are
a turbulent and troublesome lot ... I know them personally, and
although years ago they may not have been above cutting anybody's
throat for a few krans ... I know that they have, too, gone through a
new experience since they have virtually governed Persia for no in-
considerable period after the abdication of the last Kajar Shah.'[327]
However, British liberal supporters of the Constitutional movement
had been troubled when, in December 1911, the Bakhtiari-dominated
Cabinet in Tehran had carried out a military putsch and closed down
the *majlis*, or parliament.[328] Similarly, the relationship between the
Bakhtiari and Anglo-Persian had long been one of mutual suspicion,
and in 1912 Dr Young had complained, 'It has been our chief trouble
during the past two years that there was no proper authority here to
whom matters could be referred for immediate attention. Since the
senior khans have gone to Tehran and other towns, they have regarded
the Governorship of their own country as of minor importance.'[329]

The transit routes required for Persian oil to reach Britain raised
another security issue, and the *Observer* argued, 'Submarines will

render impossible the Mediterranean route in time of war. The contract is another factor compelling even in peace times the reinforcement of our naval squadrons in that sea and the strengthening by every available means of our position and influence both in the Near East and the Middle East.'[330] However, Churchill expressed confidence that the Royal Navy would reign supreme and that even if the Suez route were to become inaccessible the longer Cape route would be sufficient:

> Any difficulties we may experience on our trade routes will occur at the very beginning of a war, particularly if it began by surprise. Every day the war continues we shall become stronger on every trade route, and our opponents will become weaker ... Along the Cape route we find it would be exceptionally easy to protect trade because of the disposition of our squadrons and bases, and the enemy's vessels would find it exceptionally difficult to maintain themselves on that route for the purpose of attacking our trade.[331]

In a subsequent debate the Earl of Ronaldshay agreed that, 'So long as South Africa remains British, the route round the Cape ... would be available', while another MP added, regarding the Cape route, 'If we are not in a position to keep that part of the sea open, it is no use talking about oil or the Empire or anything else, because the game is up.'[332] Several MPs argued that, overall, the eventual military cost to Britain of defending Anglo-Persian's operations would quite likely outweigh any savings on the price of fuel oil. John Dillon, a member of the Persia Committee, said that he might have supported the government's bill

> if they could defend that possession without incurring other and still greater risks and burdens which might more than counterbalance the advantages obtained from the possession of the oil-field ... I am convinced that ... by an irresistible chain of cause and effect, you will see that British troops and the British flag will follow the enterprise, and that all public opinion in this country will be compelled to recognise the necessity of protecting this great enterprise. What will be the result? You will be an active party in partitioning this ancient empire of Persia.[333]

The day after the debate, *The Times* also expressed serious scepticism:

The real issue is whether the Government have, by this new enterprise, entered upon a fresh and dangerous policy in the Middle East, which may in the end lead them into responsibilities of a character which Ministers still seem unable to comprehend . . . The fact is that Mr. Churchill's bargain may bring the Navy a fresh supply of oil, but it may also at any moment place the whole of British policy in the Middle East in the melting-pot. It may impose new and heavy demands upon the military resources of India, demands which in our opinion India would be unable to bear unless her forces are increased . . . We want the Navy to have oil, but we do not want to run the risk of fresh embroilment anywhere in the Middle East; and it is for this reason that we fear the country may come to regret an impetuous and careless undertaking.[334]

Churchill had concluded his speech thus:

[I]t would be much easier and pleasanter for us simply to sit still, and loll with supine ease, while we watched the absorption of every independent oilfield, to sit still and observe the whole world being woven into one or two great combinations, to treat those combinations with the utmost consideration, to buy from hand to mouth in the so-called open market what we wanted from time to time, to pay the great oil trusts what they would consider an encouraging price, and to present the bill to the Treasury and the House of Commons year by year . . . It is for Parliament to balance the trouble and mental exertion unquestionably required from its Members in the process of securing an independent oil supply against the extortion of which the taxpayer would otherwise be the victim. We are confident that Parliament, in facing the difficulty and making the exertion, will only be doing its duty to the State.[335]

When it came to the vote, the bill for the government takeover of Anglo-Persian was passed by 254 votes to 18; and when Greenway asked Churchill, 'how did you manage to carry the House with you so successfully?', he replied that 'it was his attack on monopolies and trusts that did it.'[336] The *Daily Telegraph* agreed: 'What came out in the debate yesterday more clearly than anything else was that it was the operations of the great Oil Trusts which had driven the Government to take this step.'[337] Indeed, during the rather cursory debate,

Sir Edward Grey had brazenly denied 'that this matter comes before the House in its present advanced stage because it is connected with foreign politics. That is not so at all ... Foreign policy is merely incidental to it.'[338] This blatantly contradicted the fact that since 1901 the Foreign Office had given its firm backing to Anglo-Persian precisely as a means of entrenching the British presence in Persia. Similarly, although in public Churchill and Grey were claiming that their government's acquisition of Anglo-Persian would in no way jeopardize relations with Russia, Grey in fact did fear that the Russians might consider this to be a violation of the Anglo-Russian Convention. Indeed, the maintenance of good Anglo-Russian relations was such a bedrock principle of British foreign policy under Grey that, in March, Buchanan had proposed to offer the Russians a share in Anglo-Persian's more northern oilfield at Qasr-ishirin on the Turkish border (see Fig. 25).[339]

During a follow-up debate in the Commons on 29 June, the Conservative MP Sir Mark Sykes warned,

> there may be internal troubles where our oil supply is, in which case
> inevitably we have to have occupation, and if we have occupation,
> there is the possible hypothesis ... of a complete break-up of the Otto-
> man Empire ... [T]hat will provide us with a German frontier in
> Mesopotamia. But, on the other hand, we have the possible hypothesis
> of a break-up of Persia, which would give us a Russian frontier. Great
> Britain will then be like a stranded whale on a mud bank, with a river
> hippopotamus on one side and a rhinoceros charging down from the
> hills straight in front.[340]

A week later in Parliament, the Conservative MP Bertram Falle asked Grey why the Admiralty seemed to have ignored oil on the other side of the Persian Gulf – presumably referring to indications of oil recently discovered in Kuwait and Bahrain, both of which were effectively British protectorates and whose sheikhs had pledged to defer to the British government before permitting any exploitation of oil there. The Slade Commission, sent to the Gulf in 1913 to assess possible sources of fuel oil for the Admiralty, had, in fact, visited these sites in addition to the Persian oilfields (see Fig. 25). Falle criticized the Admiralty for not considering oil from these 'places that belong to us, and which must be – probably are – full of oil. Instead of that we are going to invest our money in Persia to get a cheap and certain supply, which will, I think, be neither cheap nor certain. This adventure may cost this country a great war . . .'[341] Others, however, such as the Earl of Ronaldhsay, welcomed a more 'forward' policy in southern Persia:

> [We] have seen . . . countries like Persia and Turkey wholly unable to resist the policy of peaceful penetration adopted towards them by strong Continental Powers. [We] have seen these Powers acquiring rights here and concessions there until they have come to occupy positions of paramount influence similar to the position which we now see Russia occupying in Northern Persia. [We] have come to the conclusion . . . that the best way in the long run, and the cheapest way, of preventing these possible hostile agents from penetrating too far towards the Persian Gulf was for us to occupy the ground ourselves by obtaining concessions and strengthening our hands in that quarter, in order that, having interests of our own to protect, we should be in a position to offer a legitimate protest against the encroachment of others.[342]

The *Observer* was in accord with this sentiment: 'As an Imperial people we are not yet retiring from business. The contract is in the spirit that made the Empire. Let us resist the more passive habit that tends to creep up on us, put pessimism aside, and go out in the old paths of constructive adventure.'[343]

THE TURKISH PETROLEUM COMPANY

From 1911 Britain had retreated somewhat from its strongly assertive policy in southern Mesopotamia in recognition of the fact that it lacked the necessary financial, military and diplomatic strength to exclude all other European powers from any significant commercial activity in the area. The more 'forward' policy had risked further alienating the Turkish government, which was moving steadily closer to Germany in order to counterbalance the power wielded by Britain and France; it risked worsening the already strained relations with Germany; and it risked triggering a partition of Turkey by the European powers, leading perhaps to a wider European war and an insurrection of aggrieved Muslim populations across the British Empire.[344] In March 1911 the Foreign Secretary Sir Edward Grey told the House of Commons that, although he would not reveal the details of the delicate negotiations currently under way with Germany over the projected southern Mesopotamian sections of the Baghdad Railway, 'I frankly want to see an agreement, because, if an agreement is come to which is satisfactory to Turkey, to ourselves, and to Germany . . . then one great apprehension and one source of possible friction will be removed.'[345] In mid-1912 Grey wrote to his ambassador in Berlin that if Britain and Germany could resolve various conflicting interests of theirs around the world, 'the effect on the political atmosphere would in the course of two or three years be very ameliorating'.[346]

In the case of Mesopotamia, Grey told Parliament, Britain would focus on its top priority of ensuring 'that there should be no possible risk of there being in the hands of another Power a fortified position on the Persian Gulf which might be used on the flank of our communications with India'.[347] While the Foreign Office, therefore, withdrew its demand for a majority British stake in the entire section of railway running from Baghdad to the Gulf, and while it acknowledged the fact that Turkey had granted its approval to the Deutsche Bank-led Baghdad Railway Co. for the construction of most of this section, it insisted that any section of railway running south of Basra to a Gulf port must come under British control. Grey wrote, 'a settlement recognizing our *de facto* position in the Persian Gulf and on the Shatt-el-Arab [is] the only way in which we can now safeguard our great interests in Mesopotamia.'[348]

In fact, Britain's concession to Germany was not so very great, given the extremely slow and faltering progress of construction along most sections of the Baghdad Railway, which the Foreign Office now thought would merely facilitate local transport rather than become a new route to India. Nevertheless, the Foreign Office demanded in compensation that Turkey should grant all other commercial concessions in southern Mesopotamia – irrigation, navigation and petroleum – to British concerns, and it continued to paint a picture of a rapid increase in German commercial activity in the region as a looming threat to the security of the British Empire.[349] In November 1912 Sir Arthur Nicolson at the Foreign Office wrote to Grey, 'The Germans already control the Baghdad Railway, they are believed, through the National Bank of Turkey, to be seeking to control navigation on the Mesopotamian rivers, and if they also get the oil concessions in Mesopotamia and Persia, they cannot fail to acquire enormous political influence at British expense, in regions which are of supreme importance to India.'[350]

Grey had by this time been persuaded by Charles Greenway, managing director of the fledgling Anglo-Persian Oil Co., that the company could not survive the competition and remain a purely British concern if the Mesopotamian oil concession were to be secured by the Turkish Petroleum Co.: this was 50 per cent owned by the National Bank of Turkey, an Anglo-Turkish consortium; 25 per cent by the Royal Dutch-Shell group; and 25 per cent by Deutsche Bank. The National Bank was seen as being too close to the nationalist Turkish government, Royal Dutch-Shell was deemed not British enough, and Deutsche Bank was seen as the financial force behind Germany's geopolitical ambitions. Consideration was given to the possibility of Anglo-Persian joining the Turkish Petroleum Co., but Anglo-Persian and the Foreign Office concluded that only actual British control over a Mesopotamian oil company would suffice to protect both Britain's geopolitical and petroleum interests.[351] In March 1913 Grey telegrammed Ambassador Lowther at Constantinople, 'Our object is to maintain the independence of the Anglo-Persian Oil Company in order to keep competition alive with the Shell Company. Amalgamation of the various interests would not therefore suit us at all, nor would it be to the general interest in view of the great importance of oil in the future.'[352]

Although Deutsche Bank claimed to hold the large Mesopotamian oil concession granted under the Baghdad Railway Convention of 1903, its failure to meet the contractual terms for exploratory work put this in doubt; in the view of one of Turkish Petroleum's own lawyers, it was 'not what I should call a "strong case"'.[353] Equally uncertain, however, was the claim of Anglo-Persian's founder, William D'Arcy, that he had secured the promise of a concession from the Turkish government. He was ready to bribe the relevant officials, though his agent in Constantinople wrote back in early 1913 warning him, 'Officially, of course, neither the Ambassador nor the Foreign Office must know of any such arrangement.'[354] Nevertheless, the British and German governments each argued that they were certain of their own claims; Alwyn Parker, at the Foreign Office, told Richard von Kühlmann, the Counsellor to the German Embassy in London negotiating on behalf of Deutsche Bank, that 'in the absence of an agreement acceptable both to Germany and to Great Britain, there would be no satisfactory settlement and a good deal of ill-feeling which it was most desirable to avoid.'[355]

Under equal and opposite diplomatic pressures from Britain and Germany, the Turkish authorities prevaricated. The Foreign Office threatened – via the French- and British-controlled Ottoman Public Debt Administration – to withhold approval of the Turkish government's planned revenue-raising increases in customs duties, unless it made 'without delay arrangements concerning the Mesopotamian oil-wells, which will ensure British control'.[356] However Turkey, likewise, had the power to withhold the oil concession, and Lowther warned Grey that the Grand Vizier's

> urgent advocacy of the amalgamation of the British and German groups for working the oilfields, and his declaration to me that he was prepared to see that the preponderating share in such a combine should be British, seems however to indicate the recognition of certain moral rights acquired by Mr. D'Arcy, coupled with the intention of keeping faith with the Germans. It seems to me that the negotiations on behalf of Mr. D'Arcy have now reached a point not far from a deadlock, and I am apprehensive lest the pressure which may be exerted in favour of the rival groups by ourselves and the Germans may result in determining the Government not to give exclusive rights to either groups.[357]

Indeed, other groups were actively seeking Mesopotamian oil concessions. Thomas de Ward was lobbying on behalf of an Anglo-German group, the Central Mining and Investment Corp. It included S. Pearson and Son, headed by Lord Cowdray – founder of the Mexican Eagle Petroleum Co. which was currently producing Mexican oil in vast quantities – and his associate Lord Murray of Elibank, and it had offered the Turkish Ministry of Mines an advance of £500,000 for the concession. Then, in June, the *Frankfurter Zeitung* reported that a Rothschild bank had offered an advance of the same amount; and in September Grey was cabled from his Constantinople Embassy telling him that the Russian ambassador was lobbying the Turks. So too was the US ambassador there, John D. Rockefeller having offered an advance of 10 million francs, and Standard Oil would soon receive licences to drill for oil in northern Anatolia and in Palestine, south of Beersheba near Kurnub. Here, in January 1914, two Standard Oil prospectors posing as tourists would find themselves being shadowed by a British intelligence agent using his archaeological activities as cover – Thomas E. Lawrence, later known as 'Lawrence of Arabia', who had written to his parents that he and his colleague were 'obviously only meant as red herrings to give an archaeological colour to a political job'.[358] As the India Office put it regarding the Ottoman oil concessions, 'Other, and independent, eagles have gathered about the carcase.'[359] Lowther informed Grey that the Turkish Ministry of Mines was, in fact, considering 'putting all known petroleum deposits up to auction . . . which appears for the present to be blocked owing to the action of the Grand Vizier, no doubt in consequence of the pressure which is being exerted on him by the British and German embassies . . .'[360]

Ironically, a US trade report described how petroleum at Mosul was already being commercialized indigenously, albeit on a relatively small scale (see Fig. 26):

> There are wells at Harbol, in the casa of Sakho, 60 miles north of Mosul; at Gayara, 40 miles south of Mosul; and at Kerkouk . . . There is a very small still at Gayara, which has an annual output of from 2,000 to 2,500 4-gallon tins of oil. This oil is not properly refined and therefore not suited for lighting, but it is used instead for oil engines. For lighting the better class of houses American and Russian oil is used. It is imported through Bagdad and carried from Bagdad to Mosul by caravan, which brings up the price to almost double that prevailing at Bagdad. The native Gayara oil sells at Mosul for about 10 cents a gallon retail, while the American oil retails at 35 to 45 cents a gallon.[361]

Simultaneously with the proliferating foreign interest in Mesopotamian oil, the rapid rise in demand for fuel oil around the world saw its price rise sharply and its availability plummet, particularly as tankers came into short supply and freight rates escalated. This development – coming just after the Royal Navy had taken the plunge from dual coal- and oil-firing to sole oil-firing in its new dreadnoughts – sent shock waves through the Admiralty. Given its assessment of the meagre oil supplies within the British Empire that were under purely British control – in stark contrast to its vast domestic resources of high-quality steam coal – it suddenly appeared to the Admiralty much more important to preserve Anglo-Persian as an independent, British company – operating, as it was, in a region over which the Foreign Office also considered it crucial that British influence should prevail.[362] Since Greenway had persuaded the British government that securing British control over future Mesopotamian oil production would be the only way to guarantee that Anglo-Persian would remain a purely British enterprise, this development therefore raised the stakes in Mesopotamia. Churchill explained, 'The future of the oil market is so uncertain and the present prices are so unfavourable ... [A]ction is urgent as the future oil supplies are being increasingly bought up, the absorption of the smaller independent producers is imminent or is actually proceeding, and the oil market is being rapidly contracted both from natural and artificial causes.'[363]

Anglo-German negotiations concerning the Baghdad Railway – in which Germany would agree, further, not to establish a port or rail terminus on the Gulf coast – were currently in progress, and *The Times* reported on a study delivered at the Institute of Mining and Metallurgy indicating the interdependence of the construction of railways and the exploitation of minerals in the Ottoman Empire: 'That little mining had been done in the country was owing to the lack of communication and absence of facilities of transport; it was doubtful if there was a country so full of virgin mineral wealth which required only railway communication for its development.'[364] Richard von Kühlmann now told Alwyn Parker at the Foreign Office that his government was very keen to reach a solution to the oil question, and he even indicated that so long as Germany were to be guaranteed a fair share of fuel oil and other oil products, his government would be willing to cede a good deal of control over the Turkish Petroleum Co. to the British.[365] Sir Edward Grey, spurred by the Admiralty's oil-supply crisis, was similarly anxious to strike a deal, which Lowther described to him as 'a matter of supreme importance to British interests and the vital issue of the naval defence of the Empire'.[366] Grey told Ibrahim Hakki Pasha, Turkey's chief negotiator in London, 'that His Majesty's Government attach very great importance to a satisfactory arrangement in regard to the Mesopotamian oil concession', and Sir Louis

Mallet wrote from the Foreign Office to the Board of Trade, 'Sir Edward Grey is so much impressed by the importance of reaching an agreement that he considers a further effort in this direction should be made, and he would be prepared to bring such pressure as may be possible to bear on the representatives of the National Bank group to agree to [a] compromise if the Board of Trade consider that it would adequately secure British control.'[367]

Similarly, Sir Eyre Crowe wrote to the Board of Trade that Grey was 'for political reasons anxious that a settlement of this difficult question should, if possible, be reached by consent', while another Foreign Office official wrote that the wider negotiations were intended to 'remove the last outstanding cause of possible friction between British and German interests in Asiatic Turkey'.[368] At the Board of Trade, Hubert Llewellyn Smith now saw the oil question 'not as an isolated transaction but as part of a general settling up of questions in Mesopotamia', and Mallet subsequently wrote to Grey that the over-all aim of the various agreements regarding Turkey had been that 'a mingling of European interests all over Turkey will diminish the danger which is threatened by each country having a sphere of influence to itself'.[369]

For the British government, the British element of control over the Mesopotamian oil concessions meant that of Anglo-Persian, the official view being that neither the National Bank nor Royal Dutch-Shell was British enough. The former, Parker minuted, 'behaved in a very unpatriotic manner', while the latter, wrote Smith, was 'a very powerful foreign combination with which it is quite impossible to discuss frankly the strategical needs of the Empire'.[370] Both the British and German governments were keen to access oil independently of the two global oil giants – Standard Oil and Royal Dutch-Shell – and Sir Arthur Hirtzel at the India Office even saw in the Anglo-German negotiations evidence of 'a limited kind of partnership' between the two countries; Mallet later wrote to Grey that by joining with the Germans 'we shall be in a better position to fight . . . Standard Oil'.[371]

By early August 1913 the Board of Trade and the Foreign Office together had worked out the basic principles and minimum requirements for a deal with the Germans. As set out by Smith, these began with,

1. The desire of the Admiralty to secure supplies of fuel oil from the Mesopotamian oil-fields, with the minimum risk that supply may fail them in time of war or strained relations.

2. The desire of the Admiralty for other reasons to preserve the independence of the Anglo-Persian Oil Company, and with that object to prevent any company connected with the Shell combination from using the supplies of oil obtained from the Mesopotamian oil-fields to under-sell the Anglo-Persian Oil Company in the Middle Eastern markets, which it has made peculiarly its own, thus forcing the company to choose between ruin and absorption into their system.[372]

Elaborating on point 1, Smith added,

in ordinary times of peace, the object of the Admiralty would be sufficiently attained by an option on fuel oil up to a sufficient amount, or proportion of the total output. In time of war or of strained relations, however, such an option might become valueless, and in these circumstances the Board consider that the only possible security that the fuel oil shall reach the navy in time of strained relations is the predominance on the board of the company of British directors, who in such times will, in the last resort, obey the instructions of His Majesty's Government on matters of vital importance to Imperial safety and defence. It also seems necessary that this ... should be coupled with such marketing arrangements as will ensure that the oil destined for the use of the British navy will actually be available for that purpose, and not diverted by middlemen to other destinations.[373]

The British should have a controlling interest of at least 50 per cent, and the associated directors should be approved by the government and pledge loyalty to Britain in the event of conflict; Anglo-Persian should control 50 per cent of the marketing of products and have exclusive marketing rights in key regions; and the British government should have an option to purchase up to 50 per cent of the total fuel oil produced, the German government up to 40 per cent and the Turkish up to 10 per cent.[374]

Both the British and German authorities were eager to reduce Royal Dutch-Shell's share as much as possible.[375] However, contractual obligations made it difficult for Deutsche Bank to distance itself from the group, while the fuel-oil supply scare highlighted the importance of maintaining a working relationship with it.[376] Mallet questioned the wisdom of alienating 'so powerful a Syndicate which controls such large supplies of oil [and] which is never tired of professing itself to be British'; and the Foreign Office wrote to the Board of Trade that 'to exclude a company which commands such large supplies might possibly have regrettable effects from a naval point of view.'[377] When the Board of Trade asked the Admiralty if there were any means 'by which the Anglo-Saxon Company could be eliminated' – the Anglo-Saxon Petroleum Co. being the transportation subsidiary of Royal Dutch-Shell – Sir Francis Hopwood replied from the Admiralty, 'We lead a very stormy life in our interviews with the "Shell". I always try to keep on good terms with them myself, because I feel we want their stuff.'[378]

However, as the negotiations between the Board of Trade, Greenway and Deutsche Bank's chief oil adviser, Emil von Stauss, drew close to a compromise solution, relations with both Royal Dutch-Shell and the National Bank of Turkey soured further. The Foreign Office suddenly learned that Calouste Gulbenkian – the key figure behind the formation of the Turkish Petroleum Co. and an intimate associate of, and investor in, Royal Dutch-Shell – would use his 30 per cent shareholding in the National Bank to block the sale of its 50 per cent share in Turkish Petroleum to Anglo-Persian. Gulbenkian threatened legal action if the Foreign Office attempted to force them to sell; he later wrote, 'It was for *me* to decide with the Royal Dutch Shell group what we should do.'[379]

Simultaneously, the plan for turning the Mesopotamian oil concession over to British control came in for such heavy criticism in the German press and in the Reichstag that the German negotiators could not concede, further, Greenway's demand that it should always be his group that would nominate the chairman of Turkish Petroleum.[380] On 1 December 1913 the *Daily Telegraph* ran a report stating, 'the British Government is granted a concession of all the oil wells situated in Arabia, Mesopotamia, and ... Syria.'[381] This prompted the leader of the National Liberal Party in the Reichstag, Ernst Bassermann, to put to the Foreign Minister, Gottlieb von Jagow, as reported by a Reuters correspondent, 'what the German Government intended to do to maintain freedom of access to these oil supplies for the German Navy'. Jagow replied that 'Negotiations were in progress between British and German groups with regard to the acquisition of considerable petroleum concessions, especially in Mesopotamia. German naval interests would be guarded. The government would support all German undertakings aimed at securing for Germany a proper share in the petroleum output of the world.'[382] Indeed, a few weeks later Deutsche Bank's chairman, Arthur von Gwinner, delivered a lecture in Berlin that he concluded with his prediction that 'in future the armies and navies of the world would have more need of petroleum.'[383]

Nevertheless, Gulbenkian's refusal to move aside united the British and Germans – albeit ambivalently – against him. Together, they had the power to impose their scheme for the exploitation of Mesopotamian oil and to block any other. Denied loans and the permission to increase taxes, the Turkish government would face bankruptcy. As Alwyn Parker put it, 'neither the Anglo-Saxon [Royal Dutch-Shell] nor Mr Gulbenkian could attach much importance to being shareholders in a paralytic company.'[384] Grey cabled Mallet in Constantinople, instructing the ambassador, 'You may state categorically that if concession for these fields is given to any company in which D'Arcy group does not receive 50 per cent of the whole, I shall be compelled to break off all negotiations with Hakki Pasha and to reconsider terms on which HMG could consent to customs increase and monopolies.'[385]

The two governments even initiated plans to create a purely Anglo-German enterprise to take up the oil concession, with the backing of the most senior figures involved, including Jagow and the British Admiralty's Director of Contracts, Frederick Black. If this was a bluff to intensify the pressure on Gulbenkian and Royal Dutch-Shell, it only half worked, however. While Gulbenkian was now persuaded to relinquish control and to hold, instead, a 5 per cent beneficial interest in Turkish Petroleum, Churchill's immediate and acute concern for the navy's oil supplies moved him to mend relations with the head of Royal Dutch-Shell, Henri Deterding, who had been as furious as Gulbenkian over the way they had been treated by the British. Churchill told Parker of the need 'to smooth him over . . . as he might be of great service to the Government in the future'.[386] Given, also, that Deutsche Bank was vulnerable to legal action from Royal Dutch-Shell, the group was in a strong enough position to retain its share in Turkish Petroleum.[387]

In mid-February 1914 the British Cabinet agreed to a state rescue and takeover of Anglo-Persian, which would put the now almost bankrupt company in a position actually to take up the controlling stake in Turkish Petroleum that the government had negotiated for it.[388] On 19 March all the parties concerned gathered at the Foreign Office to sign the so-called 'Foreign Office Agreement', one component of a larger package of secretive Anglo-German agreements over Turkey and the Baghdad Railway. Under this agreement, the National Bank of Turkey departed the Turkish Petroleum Co.; a 50 per cent share went to the British D'Arcy group, 25 per cent to Deutsche Bank and 25 per cent to Royal Dutch-Shell; Gulbenkian's 5 per cent beneficial holding was split between stakes in the holdings of the D'Arcy group and of Royal Dutch-Shell; and all parties agreed to engage in petroleum operations in the Ottoman Empire only through Turkish Petroleum, with a few exceptions. At this time, the question of Mesopotamian oil in relation to Britain's strategic interests at the head of the Persian Gulf overlapped very explicitly when the Foreign Office challenged the Board of Trade's lack of interest in evidence of oil at Kuwait and Basra, both potential locations for a Baghdad Railway port terminus, although in a further agreement with the Foreign Office three months later, the German government would commit to Basra as the railway's terminus and renounce the right to a direct transit connection between the port and the Persian Gulf without the prior agreement of the British. Regarding the agreement over Turkish Petroleum, crucial issues, however, were left unresolved, such as the protection of Anglo-Persian's markets and the three governments' rights to fuel-oil purchases. The latter appeared to be addressed by a supplementary agreement that each should have the right to purchase a third of any fuel oil available for export;[389] Kaiser Wilhelm II was told,

> Along with the agreements on railway and shipping questions a thorough understanding has been achieved on action with regard to the production of petroleum in Asia Minor ... It will probably be possible to obtain an arrangement, whereby a third of the total production will

be earmarked each for the German and British navies, and only a third will be on sale in the open market.[390]

However, the oil concession itself had still not formally been secured, partly because the revised Turkish Mining Law prohibited monopolies; and negotiations continued over the Turkish government's royalties, its share of the profits and the geographical scope of the concession. Thus, the diplomatic and legal wrangling continued through the summer of 1914.[391] Nevertheless, despite the many uncertainties the *Petroleum Review* subtitled a comprehensive article on 'The Petroleum Deposits of Mesopotamia' as 'a Second Baku in the Making'. A map indicated the route of the Baghdad Railway, projected to run through the largest oilfield near 'Nimrud', south of Mosul (see Fig. 26), and the article concluded, 'There is no doubt that in the very near future a forest of modern derricks will appear in the fields of ancient Assyria, Babylon and Chaldæa, and in a short time an extensive flourishing petroleum industry will arise there, thus becoming a second Baku, which will have an extraordinary influence on the world's petroleum trade.'[392]

A further political complication was that the Foreign Office, in particular, was adamant that the D'Arcy group, which had now occupied the driver's seat of the Turkish Petroleum Co., should be officially portrayed as being quite separate from Anglo-Persian, even though the directors were almost identical and the Foreign Office itself, in its official discussions, had described the group as 'Anglo-Persian'. If the British government were seen to be participating in the control of Turkish Petroleum via its control of Anglo-Persian, the Secretary of State for India, Lord Crewe, noted, 'The transaction may be represented – quite untruly – as a colonial acquisition of interest by HMG.'[393] The Admiralty acknowledged that this would make it 'impossible to resist any claim to similar direct participation by the German Government'.[394] Sir Eyre Crowe, at the Foreign Office, emphasized that Grey

had understood that if His Majesty's Government acquired a controlling interest in the Anglo-Persian Oil Company, one condition, and in his opinion a very essential one, would be that the company should be kept as far as possible distinct from any connection with the Mesopotamian concession, and . . . that there would be grave objections of foreign policy to His Majesty's Government becoming interested, through a

company which they are to control, in the Mesopotamian oil concessions. Such a consummation would almost inevitably entail a demand
for a corresponding Russian Government control of oil concessions in
Armenia, and of French Government control of oil concessions in Syria,
and it is impossible to view with indifference the possibility of such political interests being created in different portions of the Ottoman
Empire, or to disregard their possible consequences upon the maintenance of the integrity of that Empire.[395]

The British government, having applied so much pressure to the Turkish government over the oil, irrigation and navigation concessions,
simultaneously endeavoured to counter the perception that it was
attempting to exert de facto sovereignty over southern Mesopotamia –
and thus risk an international conflict that might rapidly spiral out of
control.[396] Many of those living in the region itself had long struggled
for independence from the Ottoman Empire and saw, in British moves
to take control over so much economic activity, a new imperial power
beginning to replace the old. The Baghdad *Al-Misbah* warned readers
that, as in India, British companies were 'harbingers of a vast colony
here', and that they should 'arm themselves to fight the pioneers of the
colonizing army'.[397]

At the same time, the Foreign Office was caught in a delicate balan-
cing act on the Gulf coast precisely by allying with Arab tribal leaders
in their resistance to Turkish sovereignty. In 1911, for example, Sir
Louis Mallet had written, 'we must not at the moment enter upon
negotiations with Ibn Saud, but keep on friendly terms with him in
the event of not coming to an understanding with the Turks.'[398] After
'serious' unrest along the lower Tigris in 1912, the British Resident in
Turkish Arabia had argued that 'the support of the Baghdad province
will be lost to any party which does not concede to the Arabs political
equality with the Turks.'[399] In early 1913, after Britain had tempered
its relations with tribal leaders in an attempt to placate the Turkish
government, Mallet wrote,

> It would be a pity to alienate finally three powerful sheikhs, who will
> turn to some other country. We are always talking of our prestige and
> interests in Mesopotamia, we do not know what may happen in that
> part of the world during the next few years – there has been talk of an

Arab revolt and declarations of independence. It is a question of whether without committing ourselves we might not find some way of showing the Arabs that we are not entirely indifferent to their interests.[400]

'THE SANCTITY OF BRITISH SUPREMACY'

As Germany's economy had industrialized and expanded, its growing overseas trade and merchant fleet were seen as inevitably requiring naval protection. According to the naval historian Captain Alfred T. Mahan – whose theories were highly influential at this time in naval and political circles around the world – this implied a dynamic of imperialism that simply followed centuries of precedent.[401] In 1897 the German Chancellor Bernhard von Bülow had told the Reichstag, 'we don't want to put anyone in the shade, but we too demand our place in the sun'; and as Admiral Alfred von Tirpitz would later put it,

> Without sea-power Germany's position in the world resembled a mollusc without a shell. The flag had to follow trade, as other older states had realized long before it began to dawn upon us. As *The Fortnightly Review* put it, both tersely and correctly, in 1893: 'Commerce either engenders a navy which is strong enough to protect it, or else it passes into the hands of foreign merchants, who already enjoy such protection.'[402]

However, the established European powers baulked at the prospect of having to share further their international economic and political power, and from the late 1890s Russia, France and Britain – despite their differences, especially between Russia and Britain – improved their mutual relations and began manoeuvring against this European newcomer to the global stage. This dynamic was explicit in the Franco-Russian Alliance, in the *Entente Cordiale* between France and Britain, and in the 1907 Anglo-Russian Convention. From 1905 secret Anglo-French agreements and military planning – to which only a select few ministers and officials were privy – evidenced intentions to prepare for a conflict that would give Britain, alongside France and Russia, the opportunity to subjugate Germany.[403] Already in 1901, Admiral Sir John Fisher, now Britain's First Sea Lord, had written,

'The Germans are our natural enemies everywhere! We ought to unite with France and Russia!'[404] As Theobald von Bethmann Hollweg, the German Chancellor from 1909, would reflect some years later, the British generally 'looked upon a Germany that kept growing as an unwanted and troublesome intruder on the sanctity of British supremacy over the commerce and oceans of the world'.[405]

Following secret Anglo-French military and naval discussions, conducted just after the height of the First Morocco Crisis of 1905–6, Fisher wrote to Foreign Secretary Lord Lansdowne, 'This seems a golden opportunity for fighting the Germans in alliance with the French, so I earnestly hope you may be able to bring this about . . . All I hope is that you will send a telegram to Paris that the English and French Fleets are *one*. We could have the German Fleet, the Kiel Canal, and Schleswig-Holstein within a fortnight.'[406] In January 1906 the new Foreign Secretary, Sir Edward Grey, wrote to the Secretary of State for War, Richard Haldane, 'Fisher says he is ready, by which I take it he means that his ships are so placed that he can drive the German Fleet off the sea and into shelter at any time.'[407] For German naval commanders, following Mahanian doctrine, this was precisely the rationale behind their own fleet, that at virtually a moment's notice the Royal Navy had the capacity to entirely choke off Germany's seaborne imports and exports via the North and Baltic seas.[408]

Later that year, amid concerns over the appearance of German commercial and political activity in the Middle East, the Foreign Office was negotiating the terms of the Anglo-Russian Convention, Viceroy of India Charles Hardinge hoping that 'As regards Persia, if we and the Russians present a solid front . . . I think we shall in the end defeat the Germans.'[409] The Secretary of State for India, Lord Crewe, recorded Grey's view that 'the key to German diplomacy is to prevent anything like a triple entente of England, France, Russia. Her policy is expanded commerce. Her triumphant rival in this field is England. To weaken England, she will use the Russian and the Turk wherever she can. She increases her fleet, to give courage to her merchants.'[410] While the German government may have regarded its fleet as primarily defensive, the British government saw it as primarily offensive, since it might eventually become capable of interrupting the lifelines of Britain's sea trade.[411]

An influential memorandum drawn up for Grey by Sir Eyre Crowe in January 1907 signalled Britain's tough line:

> The general character of England's foreign policy is determined by the immutable conditions of her geographical situation on the ocean flank of Europe as an island State with vast overseas colonies and dependencies, whose existence and survival as an independent community are inseparably bound up with the possession of preponderant sea power. The tremendous influence of such preponderance has been described in the classical pages of Captain Mahan. No one now disputes it ... [I]t [has] to be assumed that Germany is deliberately following a policy which is essentially opposed to vital British interests and that an armed conflict cannot in the long run be averted, except by England either sacrificing those interests, with the result that she would lose her position as an independent Great Power, or by making herself too strong to give Germany the chance of succeeding in a war.[412]

Diehard British supremacist journalists such as James L. Garvin urged readers to give no quarter:

> [Germany] has challenged the naval supremacy which is the life of our race. That is precisely why we have been so urgently moved to settle our outstanding differences with the rest of the world. That is why we have been brought in the last seven years to view in a totally altered light our relations with the Third Republic and the Empire of the Tsars. That is why we have made the real but sensible sacrifice of minor interests to major interests.[413]

Echoing the uncompromising imperialists at the Foreign Office regarding the German Navy, Fisher wrote to King Edward VII, 'that we have eventually to fight Germany is just as sure as anything human can be, solely because she can't expand commercially without it.'[414] Fisher reassured the king that the British Navy was four times stronger than that of the Germans, 'but we don't want to parade all this to the world at large', as this would make it harder to argue the case for the Admiralty's huge fleet-construction programme.[415] In order to justify this the British public was, instead, subjected to a litany of German invasion scare stories led by newspapers owned by the press baron Lord Northcliffe, which included *The Times* and *Daily Mail*.[416] Lord

Reginald Esher – the king's close adviser and a permanent member of the top-secret Committee of Imperial Defence – wrote to Fisher, 'A nation that believes itself secure, all history teaches is doomed. Anxiety, not a sense of security, lies at the root of readiness for war. An invasion scare is the mill of God which grinds you a Navy of Dreadnoughts, and keeps the British people war-like in spirit.'[417] Where one side saw its armaments increases as defensive, the other saw them as offensive, and vice versa, creating a vicious cycle that stoked an arms race. In 1908 Albert Ballin, head of the Hamburg-Amerika Line, had warned Chancellor Bülow, 'We just cannot afford a race in dreadnoughts against the much wealthier British.'[418] Similarly, in Britain the following year, amid a high tide of anti-German sentiment in the press and urgent demands for a larger navy, some began to wonder whether mass hysteria had taken over. The *Daily News* argued, 'The appetite of this monster of armaments grows by what it feeds on. Give it four Dreadnoughts and it asks for eight, eight and it asks for sixteen, sixteen and it would still be unsatisfied. It is an appetite without relation to needs or facts. It is the creation of irrational hates and craven fears. It can only be stopped by the determination of the democracies of the two countries that this wild rivalry shall end.'[419] Even Grey cautioned Parliament, 'half the national revenue of the great countries in Europe is being spent on what is, after all, preparations to kill each other. Surely the extent to which this expenditure has grown really becomes a satire, and a reflection on civilization. Not in our generation, perhaps, but if it goes on at the rate at which it has recently increased, sooner or later I believe it will submerge that civilization.'[420]

Attempts to bring about a freeze in battleship construction were seen from Germany as an obvious ruse by Britain to maintain the status quo of its naval supremacy to enforce German subservience globally. Admiral Lord Charles Beresford – now Conservative MP for Portsmouth – told a meeting of the conservative Primrose League that 'The feeling in Germany had originated through our own fault. We began by threatening Germany with the Dreadnought, saying that that ship could sink her whole fleet, and much of what had happened and been said in Germany was due to our attitude towards that country.'[421]

In early 1912, as Haldane was negotiating with the German government for a possible pause in fleet construction, Winston Churchill gave a speech – while visiting Glasgow's shipyards as First Lord of the Admiralty – almost designed to scupper the 'Haldane mission': '[T]he German Navy is to them more in the nature of a luxury. Our naval power involves British existence. It is existence to us; it is expansion to them.'[422] Churchill would later echo this sentiment when he argued in Parliament in June 1914, 'we must not let ourselves be deprived of oil ... We must continue to make efforts to give our sailors the finest and most suitable weapons of war which science can devise or money can buy. We cannot allow these great advantages to pass to other nations, to whom naval strength is of so much less consequence than it is to us.'[423]

In June 1912 *The Times* reported on a series of articles in a German newspaper in which prominent Germans replied that their navy was not intended as a threat to Britain and that the two nations might profitably work together: 'Herr von Gwinner, director of the Deutsche Bank and president of the Baghdad Railway,' had written,

> English public opinion has been persuaded that the building of Germany's fleet is only for the purpose of an attack upon England. That sea supremacy is, for Great Britain, a life-and-death question is understood and appreciated in Germany ... Can public opinion on the other side of the North Sea not be convinced likewise that the possession of a strong fleet is for Germany, if not to such a complete extent, still to an important degree also a question of vital importance? So long as the German Empire was an agricultural State it could do without a fleet, but since it has become industrial it requires, just as the British Empire does, a Navy for the protection of its world-wide commerce.[424]

However, the British government was set on containing German naval expansion, a policy that extended to ensuring that Germany should also be contained on land. Britain would lose its crucial naval alliance with France if the latter were to be invaded by Germany; and if Germany were to supplement its currently inadequate ports with those on the Dutch, Belgian or French coasts, British sea trade could be greatly imperilled. Hence the British Army's secret discussions for an expeditionary force to be sent to assist their French and Belgian counterparts in the event of a war against Germany – discussions that were kept secret from even most of the British Cabinet until late 1911.[425] In July

1912 Grey told the Committee of Imperial Defence, 'if a European conflict ... arose ... our concern in seeing that there did not arise a supremacy in Europe which entailed a combination that would deprive us of the command of the sea would be such that we might have to take part in that European war. That is why the naval position underlies our European policy.'[426]

In late 1912 the Royal Commission on the British Navy's oil supplies became concerned that the Royal Dutch-Shell group, based in Holland, might come under German influence in time of war. This would likely impact the Shell oil company – the junior partner in the group – which supplied about an eighth of the Admiralty's fuel-oil purchases by 1913; and as the Royal Navy hit a fuel-oil supply crisis, naval officials predicted that they would become more reliant on Shell in future: 'we want their stuff,' wrote Sir Francis Hopwood from the Admiralty.[427] By early 1914 the Royal Navy had a total of 166 oil-only ships, either built or under construction, including five battleships. However, at just this time, between 1912 and 1914, the availability of fuel oil had fallen and its price had risen dramatically as global demand escalated. As a consequence, from late 1913 the Admiralty ordered its fleet to conserve fuel and many of its most advanced, oil-fuelled ships were taken out of action entirely.[428] In just one decade the Royal Navy's security of fuel supplies had turned from a near certain availability of practically unlimited reserves of British steam coal to an existentially perilous dependency on highly uncertain sources of fuel oil – while, thereby, introducing yet another cause for geopolitical conflict.

Of greater immediate significance for the Admiralty than the future status of Shell was the fact that the Royal Navy was sourcing over a third of its fuel oil from Romania – 49 per cent in 1911, 23 per cent in 1912 and 40 per cent in 1913 – although due to the secrecy surrounding its oil contracts this was not generally known outside the Admiralty.[429] Romanian fuel oil – the output of which was expected to increase significantly in the coming years – was delivered to the Royal Navy, as well as to civilian consumers, by tanker from the Black Sea via the Bosphorus and Dardanelles Strait, and thence via the Aegean and Mediterranean seas. In the summer of 1914 a *Times* correspondent warned Britons that 'the mere closing of the Dardanelles would be sufficient to cut you off from this source of supply', which had indeed happened twice during recent conflict in the Balkans and the Aegean.[430] As Lord Beresford had reminded the House of Commons in March, 'I pointed out last year we had got 100,000 tons of oil from Roumania, but directly the Balkan war broke out that supply was stopped immediately.'[431]

Even though the German Navy, likewise, was dependent on this Mediterranean route for a significant proportion of its fuel oil, concerns were expressed during the Royal Commission's hearings that Germany might become independent of this route by bringing its Romanian oil supplies directly westwards across Europe by river and rail – although this would not alter the fact that Germany was far more reliant on oil from the United States.[432] From 1913, however, the security of British merchant shipping in the Mediterranean became dependent upon France after the two countries reached an agreement whereby the French Navy would largely take over responsibility for the Mediterranean, allowing the Royal Navy to concentrate its fleet in British home waters.[433] Thus, if France were to come under the control of Germany, the Royal Navy might lose access to a third of its oil supplies – just at a time when fuel-oil availability was tight as a result of high global demand. In this imagined scenario, all of the sea trade upon which Britain was utterly dependent could be at risk, including its fuel-oil and petrol imports. Additionally, although the Admiralty had endeavoured to remain independent of Royal Dutch-Shell, Britain as a whole – which included its growing military and aviation requirements – was dependent on this oil company for almost 60 per cent of its yearly imports of petrol, approaching a total of 100 million gallons per year by 1914.[434] Thus, the prospect of losing France as an ally also raised the possibility that the normal route taken by Shell's tankers – from the Dutch East Indies via the Suez Canal and the Mediterranean – might be rendered impassable. Although the company's tankers could take the alternative, much longer Cape route, they – along with all of Britain's Atlantic trade – might still be vulnerable in the Atlantic. It may also not have gone unnoticed that the French Rothschilds were major investors in Royal Dutch-Shell, which, in addition, controlled much of Romania's oil production through its subsidiary Astra Română.[435]

The Dardanelles was of further importance in several respects. It was the outlet not only for Romanian oil but also for a great deal of Russia's exports, on which its economy had become increasingly dependent. Almost all of Russia's oil exports reached world markets via the Dardanelles, and although its output of fuel oil, specifically, was currently being taken up entirely by domestic use, this might not always remain so.[436] A major proportion of Russia's huge grain exports were also shipped by this route and the recent interruptions of trade through the Dardanelles due to military conflict in the region had significantly impacted Russia's export-oriented economy, as Tsar Nicholas II lamented to British ambassador George Buchanan in April 1914.[437] A year later a British newspaper would be commenting on the closure of the Dardanelles, including its importance for Russia's oil exports – at this time primarily lamp oil, gasoline and lubricants:

> Corn and oil ... The wheat of South Russia shipped from the Black Sea ports ... is always enormous ... Add two and a quarter millions of gallons of petroleum products exported from the port of Batoum, or close on 850,000 tons ... Remember also [other commodities], and you realise that it is not only for the passage of munitions of war and the moral effects of the occupation of Constantinople that the Allies' shells are dropping on the forts of the Bosphorus and the Dardanelles.[438]

And as the *Financial Times* would summarize Royal Dutch's report for 1914,

> When the war broke out the company's business was affected by a shortage of tonnage, which became still more accentuated when the Dardanelles were closed, export from Russia and Roumania through the Black Sea became impossible, and that part of the demand which had always been satisfied from this quarter had to be supplied from other oil-producing countries.[439]

Since the Dardanelles was also the route via which Russia's Black Sea fleet might access the Mediterranean, the Turkish Straits had long been regarded as strategically key – and thus a continual source of dispute. The Straits had been at the root of the Crimean War; and for Russia, Britain was still the primary obstacle to its gaining free access through the Straits for its navy. However, given Russian perceptions of Germany's influence at Constantinople – which the Germans themselves, ironically, felt was decidedly on the wane and of decreasing value – the fact that the Bulgarian and Serb armies had advanced as far as Adrianople in November 1912 prompted Foreign Minister Sergey Sazonov to ask the Tsar to 'imagine what would happen if, instead of Turkey, the Straits were to go to a state which would be able to resist Russian demands'.[440] It was indicative both of the immense strategic complexity and ramifications of the Balkan wars and of the strength of the Franco-Russian alliance against Germany that, following signs of Austrian troop movements, the Russian ambassador to Paris, Alexander Izvolsky, was reassured on 17 November by the new French Prime Minister, Raymond Poincaré, that 'If Russia goes to war, France will also, as we know that in this question Germany is behind Austria.'[441] Two weeks later, the German ambassador to London, Prince Karl Lichnowsky, reported being warned by Grey that

If . . . a European war were to arise through Austria's attacking Serbia, and Russia, compelled by public opinion, were to march into Galicia rather than again put up with a humiliation like that of 1909, thus forcing Germany to come to the aid of Austria, France would inevitably be drawn in *and no one could foretell what further developments might follow*. This is the second time that Sir Edward Grey has given me this hint, a hint that cannot be misunderstood . . . [F]or, despite the fact that there were no secret agreements with France, it was for England of vital necessity *to prevent that country from being crushed by Germany*; England . . . would have no alternative but to come to the aid of France should Germany, as is expected here, prove victorious over the French.[442]

One manifestation of the international jockeying for influence over Turkey – where France held over 60 per cent of the national debt, with Britain and Germany holding the rest about equally – was in the training of its navy by British officers and the training of its army by German officers.[443] In December 1913, General Otto Liman von Sanders arrived in Constantinople to head a new military mission that would give him command over the Ottoman First Army Corps, which was responsible for the defence of the Straits and of Constantinople itself. The Russians protested vehemently at what they again saw, as they had in late 1912, as a fundamental shift in the balance of power regarding the Straits, and they considered taking military action. The Russian minister to Romania wrote in a widely circulated report that 'the possession of the Dardanelles represents for Russia a question of its very existence', and Sazonov warned the Tsar, 'the state which possesses the Straits will hold in its hands not only the key of the Black Sea and Mediterranean, but also that of penetration into Asia Minor and the sure means of hegemony in the Balkans.'[444]

It was decided – given the intricate interdependencies between the Straits, European Turkey, the Balkans and the balance of power in Europe – that this would inevitably involve a general European war, in which most of the Russian Army would have to be directed against Germany and Austria. The Russians calculated that even with France and Britain on its side, victory for the Triple Entente might not lead to Russia gaining control over the Straits since, as a Russian Navy spokesman later warned at a conference on the question, 'others might seize them while we are fighting on our western front'.[445] Therefore, Sazonov, 'with greater seriousness and openness than on any other occasion', told the British chargé d'affaires in St Petersburg that the Liman mission was 'the first question seriously involving Russian interests' where he was seeking British support and that the strength of Britain's support against German involvement in the security of the Straits would be 'a test of the value of the Triple Entente' to Russia.[446] Even as the Germans began to conciliate, Ambassador Buchanan was warned by Sazonov that inadequate British support would mean a breakdown in Anglo-Russian ties; this, the chargé d'affaires was told, would lead Turkey to see 'that the strength lies on the side of the Triple Alliance' of Germany, Austria-Hungary and Italy.[447] Sazonov wrote to Izvolsky, 'the further strengthening and development of the so-called "Triple Entente" and its conversion (so far as is possible) into a new alliance, presents itself to me as an urgent task.'[448]

The Tsar spoke with Buchanan about the benefits of a full Anglo-Russian alliance, and talks were begun regarding a possible naval convention; he told Buchanan that 'He had reason to believe that Germany was aiming at acquiring such a position at Constantinople as would enable her to shut in Russia altogether in the Black Sea. Should she attempt to carry out this policy He would have to resist it with all His power, even should war be the only alternative.'[449]

From the Foreign Office, Sir Arthur Nicolson signalled to Buchanan that he was in favour of negotiations over a naval agreement with Russia: 'I do not know how much longer we shall be able to follow our present policy of dancing on a tight rope, and not be compelled to take up some definite line or other. I am also haunted by the same fear as you – lest Russia should become tired of us and strike a bargain with Germany.'[450]

In early 1914, as the Germans conceded to transferring Liman to another posting, Sazonov nevertheless looked again into Russia's military preparedness to seize the Straits, while he acknowledged to Buchanan, ominously, that 'separate action by Russia would inevitably cause war into which we [Britain] should be dragged in the end.'[451] By midsummer another closure of the Straits by Turkey loomed, as a likely defensive measure against an anticipated Greek pre-emptive assault on the Turkish Navy's new dreadnoughts. In that eventuality, Grey proposed that Britain might intervene to keep the Straits open, since their closure caused 'no end of trouble to and loss to British shipping' – which, he might have added, included the transit of a crucial, not easily replaceable, third of the Royal Navy's supplies of fuel oil from Romania.[452]

'WE HAVE LOST CONTROL AND THE LANDSLIDE HAS BEGUN'

In 1910 Count Franz von Harrach, an officer in the Austrian Army's transport unit, had bought himself a new, Viennese-made Gräf & Stift Double Phaeton, a luxury six-seater touring car. Now, on a summer's Sunday on 28 June 1914, he was sitting in one of its backwards-facing rear seats opposite the heir to the Austro-Hungarian throne, Archduke Franz Ferdinand, and his wife Sophie, Duchess of Hohenberg. Harrach had lent the use of his car for the royal couple's tour of Sarajevo, the capital of Bosnia-Herzegovina, a former province of the Ottoman Empire, annexed by Austria-Hungary in 1908 but coveted by Serbia. Warnings that trouble might be afoot from violently nationalistic Bosnian Serbs proved justified after an assailant threw a bomb at the car, only for it to bounce off the folded-down soft top and explode under the car behind them in the convoy. Nevertheless, after stoically attending an event at the city hall, the royal couple returned to the car to continue their tour – although it was decided that they should divert from the original, publicly advertised route. This created some confusion: the driver took a wrong turn, and as he attempted to reverse the lumbering vehicle another assassin, Gavrilo Princip, took this unexpected opportunity to shoot the archduke and his wife at nearly point-blank range. Both died within a few minutes.[453]

The crumbling edges of the tottering Ottoman Empire had been falling away for decades; and wherever subjects within its peripheral provinces – from east to west – began to wrest autonomy from Constantinople, control and influence over the spoils was fought over by competing local factions, by neighbouring countries and by other interested European powers. On the empire's westernmost European fringes in the Balkans, territorial control was contested by local nationalist groups and by Austria-Hungary, Italy and Greece, while Russia, France, Britain and Germany endeavoured to assert their power and influence.[454] It had been feared that the Italian-Turkish War of 1911–12, the First Balkan War of 1912–13 and the Second Balkan War of 1913, each a crushing blow to the Ottoman Empire, might escalate into a wider European war. That they did not – apparently due to the success of international diplomacy – seemed a reason for optimism, particularly also in the light of the Anglo-Russian Convention of 1907 and the resolution of the Agadir Crisis in 1911.[455] At the same time, the frequency of such crises – coupled with an industrialized arms race – led many to believe that an immense European war was eventually inevitable.

In Austria-Hungary and Serbia, news of the assassination of the archduke was received with alarm; but elsewhere across Europe and the wider world it generally elicited merely a sadness tinged with boredom – would internecine conflict in the Balkans never end? – and follow-up reports were relegated to the newspapers' inside pages.[456] This was partly due to the sheer complexity of the region. The international power politics being played out through the Balkans was something like a combination of chess and poker, played at an advanced level and involving ten or more players. In any case, the conflicts that had broken out there in recent years had remained localized. The major European powers – France, Russia and Britain of the Triple Entente, and Germany, Austria-Hungary and Italy of the Triple Alliance – had not come to blows directly. However, just as in a game of chess or poker, it may not be apparent quite how finely balanced the complex pattern of opposing forces are – and just how rapidly and dramatically the direction and pace of the game might suddenly change.

By late 1913 the so-called Concert of Europe began to break apart following British Foreign Secretary Sir Edward Grey's insistence that Turkey yield several strategic northern Aegean islands to Greece, and his shift away from accommodating the wishes of Germany and Austria-Hungary and more towards accommodating those of Russia.[457] This was largely due to the latter's increasing leverage over Britain regarding Persia. As a result of the Royal Navy's fuel-oil supply crisis of 1913, through the winter of 1913–14 the British government moved ever closer to formally deciding, in mid-February, to rescue the Anglo-Persian Oil Co. by purchasing a controlling stake in it. After the announcement of this decision in late May, and during subsequent parliamentary debates in June, Grey and the First Lord of the Admiralty, Winston Churchill, claimed in public that the decision had been taken on purely commercial grounds and that it had no great geopolitical and security implications. In private, however, the Foreign Office shared the serious concerns of many critics who argued that the Admiralty would now be heavily dependent on oil supplies from a particularly volatile region. The Anglo-Russian Convention of 1907 had been agreed precisely to stabilize tensions between Britain and Russia over the western and northern approaches to India: The two powers had agreed to a 'neutral' zone in Persia to serve as a buffer between a 'Russian' sphere of influence in the north and a 'British' sphere in the south and east. Yet by turning Anglo-Persian into a state-owned company, the British government was thereby implicitly asserting a formal colonial claim over the neutral zone of southwest Persia, where Anglo-Persian was operating (see Fig. 25). The Foreign Office understood that their Russian counterparts would be likely to view this as a violation of the Anglo-Russian Convention. However, while Russia had sent over 10,000 troops across its land border into the northern zone of Persia, the

Foreign Office had calculated that the British Army was not capable of deploying troops effectively in southern Persia. Thus, from this position of relative vulnerability and weakness, by June 1914 the British government found itself urgently having to renegotiate the terms of the faltering Anglo-Russian Convention – now to achieve not only 'repose' around the approaches to India, but also the security of a major proportion of the Admiralty's anticipated future oil supplies.[458]

Three weeks before the assassination of Archduke Franz Ferdinand, Grey wrote to Russian Foreign Minister Sergey Sazonov that a close Anglo-Russian relationship was pivotal to British foreign policy, that 'Unless the situation is remedied, and that soon, the whole policy of Anglo-Russian friendship, on which H.M. Government had built, and which was the corner-stone of their foreign relations, would come to a disastrous end.'[459] Grey complained about 'a military occupation by Russia of the north of Persia and the establishment of a political protectorate there'.[460] But in late June a Foreign Office official suggested that the status of the neutral zone would have to be 'radically changed' by being incorporated into the British zone for 'commercial purposes, Russia receiving as a *quid pro quo* full elbow room in the north'. Another official, Sir George Clerk, agreed, arguing that the location of the Admiralty's oil operations in Persia 'makes it more than ever imperative for us to define in concert with Russia what the future position of both countries in the neutral zone is to be'.[461]

On 6 July Germany – fearful of harming its relationship with its one strong, close ally and losing its strategic foothold in the Near East – gave Austria-Hungary what became known as 'the blank cheque' to punish Serbia for the suspected complicity of the latter's government agents in the assassination of the archduke. Yet the German Chancellor, Bethmann Hollweg, was nervous, as 'an attack on Serbia can lead to world war'.[462] The British Foreign Office, however, appeared unconcerned, Sir Arthur Nicolson writing to their ambassador in Vienna, 'We are ... chiefly busying ourselves with endeavouring to arrange matters with Russia in regard to Persia, and, in a secondary degree, Tibet.'[463] Two days later, Grey tasked Clerk with drawing up a memorandum on British policy towards Russia regarding Persia; and Clerk suggested that he incorporate the views of the Admiralty and the India Office 'on Sir G. Buchanan's proposal to offer the Russians a share in the Kasr-i-Shirin oil-fields. I would then . . . prepare a memorandum to shew the desiderata of H.M.G. and the basis of a possible negotiation with Russia, on which everything else hangs.'[464] Following further criticism that the Admiralty's Persian oil operations might damage relations with the Russians while giving them a powerful bargaining chip, on 14 July Grey wrote to Buchanan, his ambassador in St Petersburg, that he was 'urging the Admiralty to see if they cannot agree that the development of oil-wells further north than the one already working should be leased to a Russian, or Anglo-Russian, subsidiary Company'.[465]

On 21 July Clerk completed the Foreign Office memorandum, which stated,

> After seven years His Majesty's Government are faced with the urgent necessity of taking stock of their position in Persia, for the incapacity of the Persians and the steady advance of Russia have together created a situation which cannot be allowed to drift any longer without the most serious danger to those British interests whose maintenance constitutes one of the most cardinal principles of Imperial policy . . . What we are seeking to do is to restrain the steady trend of Russia southwards to the Gulf . . . Once Russia reaches the Gulf, our secular position collapses like a house of cards, unless we are ready to defend it by force of arms.[466]

The memorandum emphasized that the point behind the Anglo-Russian Convention could not be 'more succinctly put' than it had been in a minute written by Grey in 1907; it was 'to begin an understanding with Russia which may gradually lead to good relations in European questions also, and remove from her policy designs upon the Indian frontier, either as an end in themselves or as a means of bringing pressure to bear upon us to overcome our opposition elsewhere'.[467] By continually protesting to Russia over its activities in Persia, however, Clerk's policy memorandum continued,

> We are . . . in the process alienating the friendship, and keeping alive the suspicions, of the one Power with whom it is our paramount duty to cultivate the most cordial relations . . . [T]he first principle of our foreign policy must be genuinely good relations with Russia, and founded on the belief that if we do not make relatively small sacrifices, and alter

our policy, in Persia now, we shall both endanger our friendship with Russia and find in a comparatively near future that we have sacrificed our whole position in the Persian Gulf, and are faced in consequence with a situation where our very existence as an Empire will be at stake.[468]

The memorandum dismissed the possibility that Britain might 'call in Germany to help us stay Russia's advance. Is it unreasonable to say that such a solution would be suicidal, and that it has only to be stated to be ruled out? It would be changing King Log for King Stork with a vengeance.'[469]

Since 1900, the Foreign Office had come to regard Germany as a more menacing rival in the Middle East than Russia. Under the Baghdad Railway Convention of 1903, Turkey had granted Deutsche Bank its concession to lay a railway from Anatolia to Baghdad, and possibly all the way to the Persian Gulf. Despite the extremely slow and faltering construction of the line, the British Foreign Office became increasingly concerned as it inched its way closer to the Gulf, over which Britain claimed unquestioned hegemony, and to which end it won a treaty agreement that Germany would not build a Gulf coast port terminus without British approval. In parallel, the associated oil rights granted to Deutsche Bank under the Convention presented a particular problem, increasingly so from 1909 when Standard Oil and Royal Dutch-Shell began vying with the Germans and the British for Mesopotamian oil concessions. The Anglo-Persian Oil Co. – which had struck commercial quantities of oil in Persia in 1908 – persuaded the Foreign Office that in order to continue its Persian oil operations as an independent, purely British concern, it would be necessary to secure British control also over Mesopotamian oil. However, by the spring of 1914 the best that the Foreign Office had been able to negotiate for was a 50 per cent controlling interest in the Turkish Petroleum Co., alongside Deutsche Bank and Royal Dutch-Shell with 25 per cent each. Curiously, given the intense Anglo-German naval rivalry, the deal gave each navy rights to a third of the fuel oil available for export for thirty years, which was reaffirmed in the secret Anglo-German Baghdad Railway Convention of 15 June 1914. It was only after both the British and German governments reminded the Turkish government that permission to implement its planned tax increases was conditional on it actually granting them the oil concession that, on 28 June – the day of the assassination of the Austro-Hungarian archduke and his wife – Constantinople finally moved closer to formally issuing the grant.[470] For the Foreign Office and the Admiralty this was a tricky predicament, anxious as they were to maintain British control over the Persian Gulf and to protect their newly acquired Middle Eastern strategic oil reserves.

The solution that the Clerk memorandum recommended for Persia was to propose to Russia that the neutral zone be partitioned along economic lines. In return for securing its commercial presence in southwest Persia and its relations with the Bakhtiari and the Sheikh of Mohammerah, Britain would relinquish the more northern sectors of Anglo-Persian's oil concession and of another mining concession to Russian interests, and would suggest a realignment of the planned Trans-Persian Railway.[471] In the Foreign Office, Crowe and Nicolson agreed, the former re-emphasizing 'that the principal thing we must keep before us is the necessity of placing our relations with Russia on a satisfactory and, so far as possible, lasting foundation'.[472]

The following day, on 22 July, Grey explained to Russia's ambassador in London, Count Benckendorff, in a note circulated also to Prime Minister Herbert H. Asquith and to Churchill,

We were certainly not going to develop at present oil wells north of the one already in use, on the contrary we were going to spend money on developing the very promising prospects of oil in the islands, the British sphere, or near the south coast ... [B]y giving up development far inland in the north, and handing it over on terms of reasonable profit to another Company, the Admiralty were positively avoiding the risk of having to send troops into the interior of Persia, which had been urged in Parliament as one of the great drawbacks of the Concession.[473]

As Clerk was compiling the memorandum on Anglo-Russian relations, and as Austria-Hungary hesitated for two weeks before taking action against Serbia, some became concerned at how events in the Balkans might escalate. In the French Parliament the Socialist leader Jean Jaurès – himself assassinated on 31 July – voiced his suspicions as to what President Poincaré's and Prime Minister René Viviani's real agenda might be for their upcoming trip to Russia: 'We find it inadmissible . . . that France should become involved in wild Balkan adventures because of treaties of which she knows neither the text, nor the sense, nor the limits, nor the consequences.'[474] Most of Russia's top officials and its press – fearing, since the Liman Affair, that Germany was manoeuvring towards taking control of the Turkish Straits – appeared ready to prosecute a long-anticipated war against Germany; and in St Petersburg the two French leaders – as Jaurès suspected – proceeded secretly to give a 'blank cheque' of their own to Russia, their closest ally.[475] Even before this trip, the German ambassador to London, Prince Lichnowsky, reported being warned by Grey that, 'Should . . . violent excitement arise in Russia in consequence of Austria's military measures, he would be quite unable to put a curb on Russian policy and, in view of the feeling of dissatisfaction which at present prevails in Russia against England . . . he will, he says, have to humour Russian sensibilities.'[476]

On 23 July Austria-Hungary issued an ultimatum to Serbia, to which the latter replied with a dissimulating apparent acceptance that was, in practice, a rejection. In this, Serbia was emboldened by signals of support from Russia with hints that the latter was about to begin mobilizing its army. Given the vast distances over which Russian troops had to move, this was to be a mobilization by stealth, as had been set out in Russian directives of November 1912 and repeated in March 1913; namely, 'to complete concentration without beginning hostilities, in order not to deprive the enemy irrevocably of the hope that war can still be avoided. Our measures for this must be masked by clever diplomatic negotiations, in order to lull to sleep as much as possible the enemy's fears.'[477]

Much of Germany's advantage lay precisely in its ability to mobilize more rapidly than Russia, but if it did so it might appear in public opinion that Germany had initiated hostilities, particularly since – as was widely known – Germany would have to make the most of its time advantage by first invading France, Russia's assumed ally in war, under the Schlieffen Plan, most likely via Belgium.[478] As Chancellor Bethmann Hollweg saw the dilemma, according to his personal secretary, 'Should war break out, it will result from Russian mobilization *ab irato*, before possible negotiations. In that case we could hardly sit and talk any longer, because we have to strike immediately in order to have any chance of winning at all.'[479] Although Bethmann Hollweg was hoping that the conflict over Serbia could be localized and that Grey might 'use his influence in Petersburg in this direction',[480] Grey instead told Lichnowsky on the 24 July that he could no longer 'endeavour to exercise a moderating influence at St Petersburg', explaining, 'I felt quite helpless as far as Russia was concerned.'[481] Presumably, this was because of Britain's vulnerable position in Persia, with all that meant, firstly for the security of India and the routes to it and, secondly, for the Admiralty's recently acquired oil insurance policy. When Sazonov told Buchanan that 'Russia would at any rate have to mobilise', both Buchanan and the French ambassador agreed.[482] France confirmed its 'blank cheque' to the Russian government which, in turn, handed a 'blank cheque' to Serbia.[483] A European-wide war – with Russia, France and most probably Britain, set against Germany and Austria-Hungary – was being set in motion secretly by diplomats; and yet it was only now that the crisis began to receive extensive press coverage.[484]

On 25 July Buchanan, after meeting with Sazonov, cabled to Grey 'For ourselves position is a most perilous one, and we shall have to choose between giving Russia our active support or renouncing her friendship. If we fail her now we cannot hope to maintain that friendly cooperation with her in Asia that is of such vital importance to us.'[485] On 26 July, as German officials began to receive the first reports of possible Russian mobilization, Bethmann Hollweg hoped that Britain might exert a pacifying influence on Russia: his secretary recorded him as thinking, 'Everything depends upon Petersburg, will it mobilize immediately and be encouraged or discouraged by the West?'[486] Instead, on 27 July Russia received encouragement when Churchill and Grey, without even consulting the Cabinet, ordered the Royal Navy to mobilize its home fleet, which was announced publicly the next day.[487] Lichnowsky reported back from London,

> The impression is constantly gaining force here – and I noticed it plainly at my interview with Sir Edward Grey – that the whole Serbian question has devolved into a test of strength between the Triple Alliance and the Triple Entente. Therefore, should Austria's intention of using the present opportunity to overthrow Serbia ('to crush Serbia,' as Sir E. Grey expressed it) become more and more apparent, England, I am

certain, would place herself unconditionally by the side of France and of Russia, in order to show that she is not willing to permit a moral, or perhaps a military, defeat of her group. If it comes to war under these circumstances, we shall have England against us.[488]

On 28 July Nicolson wrote to Buchanan telling him that Britain would have to go to war alongside Russia since, 'I foresaw as well as you did that this crisis might be taken by Russia as a test of our friendship, and that were we to disappoint her all hope of a friendly and permanent understanding with her would disappear.'[489] Prime Minister Asquith told King George V,

> As far as this country is concerned, the position may thus be described. Germany says to us, 'if you will say at St. Petersburgh that in no conditions will you come in and help, Russia will draw back and there will be no war'. On the other hand, Russia says to us 'if you won't say you are ready to side with us now, your friendship is valueless, and we shall act on that assumption in the future'.[490]

On 29 July, as the Russians considered ordering full mobilization – after at least three days of partial mobilization – and as Austria-Hungary declared war on Serbia, Bethmann Hollweg cabled his ambassador in Russia, 'Kindly call Mr Sazonov's serious attention to the fact that further continuation of Russian mobilization measures would force us to mobilise, and in that case a European war could scarcely be prevented.'[491] However, in the absence of calls for restraint from Britain or France – and with cries of support for Russia from the Northcliffe press such as The Times and the Daily Mail, plus other newspapers such as the Daily Telegraph – on 30 July the Russians proceeded to full mobilization; and it became ever clearer to German officials that the Triple Entente was united against them.[492] In Germany, decision-making power shifted to the military, meaning, as Bethmann Hollweg saw it, 'we have lost control and the landslide has begun'.[493]

The next day Admiral Sir John Fisher, who had recently headed the Royal Commission on the Royal Navy's oil supplies, wrote to Churchill reassuring him that Royal Dutch-Shell would be on Britain's side: 'I have just received a most patriotic letter from Deterding to say he means you shan't wait for oil or tankers in case of war – Good old

Deterding! How these Dutchmen do hate the Germans! Knight him when you get the chance'[494] (see Fig. 27).

From this time onwards the course of events was, perhaps, unstoppable. Russia's full mobilization meant that the German authorities felt forced to set in train their war plans. On 31 July Germany declared war on Russia, after which Grey pledged naval support to France and Churchill ordered the full, empire-wide mobilization of the Royal Navy. On 2 August, Germany sent an ultimatum to Belgium to allow its army free passage to invade France; but, with British backing, Belgium refused, upon which Germany invaded on 4 August. This finally handed Britain's interventionists the pretext that they knew would enable them to reach the threshold of political support needed to draw their country into the war alongside Russia and France. By their masterful deployment of deception, rhetorical trickery and threats, Grey and his associates could now blame Germany for having started the war. Furthermore, they suggested to the many waverers that Britain's role in it would be only a minimal naval involvement.[495] Churchill told the Chancellor of the Exchequer, David Lloyd George, 'the naval war will be cheap', and Grey told the House of Commons, 'For us, with a powerful Fleet . . . we shall suffer but little more than we shall suffer even if we stand aside.'[496]

The Russian authorities had decided that the time was right to go to war against Germany and Austria-Hungary. Primarily, this was in order to avert the danger, as the Russians saw it, that European Turkey, and the strategically and economically crucial Turkish Straits, might come under German control; and the assertion of Russian control over the Straits would hopefully rule out their closure, which had caused so much interruption to Russia's trade over the previous few years. French leaders grasped at the opportunity to join their main ally in crushing their main rival. Furthermore, Britain's foreign policy under Foreign Secretary Grey had become closely tied to that of both France and Russia, the result of more or less secret military and naval agreements with France and of the threat that Russia posed to the approaches to India. By 1914, however, an additional factor was the serious difficulty that the British Admiralty faced in securing adequate and reliable supplies of fuel oil. Firstly, the Turkish Straits was the channel not only for much of Britain's merchant shipping but also for a third of the Royal

Navy's oil, originating from Romania; if the Straits were to come under Russian influence, Russia had better be cultivated as an ally. And secondly, the Admiralty's takeover of the Anglo-Persian Oil Co. in June 1914 put the Royal Navy's long-term oil insurance policy at the mercy of the Russian Army, now occupying northern Persia with over 10,000 troops; here, again, Britain could ill afford to risk alienating Russia. Thus, while in the years, months and weeks leading up to the outbreak of war between the major European powers, Grey continually asserted that Britain retained a 'free hand', in reality his hands were bound tightly behind his back by France and, in particular, by Russia. Blindfolded, the people of Europe and the wider world were made to walk the plank of this foreign policy.

Little could the MP Bertram Falle have imagined quite how portentous his criticism of the British government's purchase of Anglo-Persian was when he had warned, in the House of Commons just three weeks earlier,

> we are going to invest our money in Persia to get a cheap and certain supply, which will, I think, be neither cheap nor certain. This adventure may cost this country a great war . . .[497]

Epilogue

The First World War greatly accelerated our advance into the petroleum age. When the British Expeditionary Force was despatched to France in August 1914 it had, in addition to 60,000 horses, about 1,000 motor transport vehicles, the majority of which had been requisitioned from civilians. By the end of the war on the Western Front in November 1918, the British, French and Americans had around 120,000, 70,000 and 45,000 transport vehicles respectively, along with over 6,000 armoured vehicles and tanks. Germany, by contrast, had only 40,000 trucks and forty-five tanks, although it also had thirty-seven surviving diesel-fuelled 'U-boat' submarines. The Allies began the war with about 200 aeroplanes, mostly French, and the Germans with about 180, but during the war aeroplane production soared: Britain manufactured 55,000, France 68,000, Italy 20,000 and Germany 48,000; nevertheless, in 1918 the Allies' 4,500 and Germany's 3,500 aeroplanes on the Western Front were, in large part, a reflection of the rapid obsolescence of machines and horrific attrition of pilots. The Royal Navy, with an ever-increasing number of oil-fired warships, gorged on 90 per cent of Britain's fuel-oil imports as the Admiralty's monthly fuel-oil consumption surged from 59,000 tons in late 1914 to 309,000 tons in 1918, and Britain's imports of petroleum products doubled over this period. However, with the closure of the Dardanelles Strait the Allies were cut off from Russian and Romanian oil, and they became even more dependent on the United States, which now supplied over 80 per cent of their petroleum. The rest, supplementing the US's overstretched crude production, came from Mexico – supplying 8 per cent of the Royal Navy's fuel oil in 1917 – Borneo, Sumatra, Burma, Persia and Trinidad, along with some

Scottish shale oil. After it became apparent at the end of 1914 that gasworks were unable to supply enough toluol (a by-product of coal gasification) demanded for the increasing production of TNT high explosives, Royal Dutch-Shell supplied the Allies with plentiful quantities from its Borneo crude, for which in January 1915 it even dismantled its Rotterdam refinery to be secretly shipped by the Royal Navy to England and reassembled (see Fig. 27).[1]

At the beginning of the war only Russia and Austria-Hungary were self-sufficient in oil, as were Romania and the United States when they subsequently entered the conflict in August 1916 and April 1917 respectively. The other main combatants, by contrast – Britain, France and Germany – were so dependent on external sources of oil that this became one of the major determinants of the war's progress and outcome. A key factor here was Britain's drive to minimize its great dependence on US oil by safeguarding sources of supply currently under British control and by seeking to expand its control over foreign sources for the future. Thus, although only about 10 per cent of the Royal Navy's wartime supply of fuel oil was sourced from Persia, this was of far greater significance to the British than its relatively small contribution might suggest. With Turkey's declaration of war on the side of Germany in November 1914, Anglo-Persian's oil pipeline and its refinery on the Gulf coast at Abadan Island, near the Persian-Turkish border, became targets (see Fig. 25). Attacks by Turkish and German forces were repelled by a British-Indian expeditionary force, which then occupied Basra 20 miles further up the Shatt al-Arab river; but the pipeline was successfully sabotaged the following February by a team of German agents and Turkish troops, who blew it up at several places along a 12-mile stretch northeast of Ahwaz. They were assisted by a number of disaffected Bakhtiari, whose tribal leaders were being paid by Anglo-Persian for the security of their oil installations; in response, Britain's Political Resident in the Persian Gulf, Sir Percy Cox, cemented the Bakhtiaris' allegiance by paying them more money and giving them stronger assurances of political support in the future. The security of Anglo-Persian's operations was restored, as was the status of southwest Persia as a virtual British protectorate. Ironically, in this context, at the company's next annual general meeting its chairman, Charles Greenway, assured investors that the steps taken would ere long reduce these lawless Germans to a proper recognition of their obligations as residents in a neutral country, and thus enable Persia not only to maintain its neutrality, but also its independence, which at present is gravely imperilled by the actions of Prince Reuss [the German ambassador to Tehran] and his tag-rag [*sic*] and bobtail following of military adventurers, and of irresponsible Persian rebels, who care nothing for the fate of their country so long as their cupidity is fed with German gold.[2]

Greenway added,

The amount of excess oil which had to be burnt, owing to lack of sufficient storage tanks and to avoid pollution of rivers, in the 4½ months during which we suffered from this enforced shut-down was, I am sorry to say, about 36,000,000 gallons, or 144,000 tons, [which was] in excess of the large quantity which we ordinarily have to dispose of in the same manner, owing to our present pipeline not being of sufficient capacity to carry the full quantity of oil produced from our flowing wells.[3]

The *Financial Times* commented, 'point is lent by the event to the criticism of those who, when the Government acquired control of the Anglo-Persian Company, dwelt specially on the vulnerability of the oilfields in Persia to attack by hostile forces. A pipeline 150 miles in length is good only so long as it is unbreached.'[4]

Meanwhile, over the winter of 1914–15 some saw military conflict between the Russian and Turkish armies in the southern Caucasus border region as being motivated at least in part by Turkish and German designs on Baku's oil reserves – the *Wall Street Journal* entitling an article 'Germans Through Turks Seek the Oil Fields' – until the Turks were crushingly defeated at the Battle of Sarikamish.[5]

The Allies achieved a near total sea blockade on Germany, which responded with a devastating campaign of U-boat attacks on Allied merchant shipping. Indeed, the acute oil shortage experienced by the Allies, which became critical in 1917, was due far more to the shortage of tankers and their loss at sea than to insufficient global oil production. As French president Georges Clemenceau wrote to US president Woodrow Wilson in December 1917, insisting that US oil companies divert more tankers from the Pacific to the Atlantic:

> At the decisive moment of this War, when the year 1918 will see military operations of the first importance begun on the French front, the French army must not be exposed for a single moment to a scarcity of the petrol necessary for its motor lorries, aeroplanes, and the transport of its artillery. A failure in the supply of petrol would cause the immediate paralysis of our armies, and might compel us to a peace unfavourable to the Allies ... If the Allies do not wish to lose the War, then, at the moment of the great German offensive, they must not let France lack the petrol which is as necessary as blood in the battles of to-morrow.[6]

Since the Royal Navy was receiving a significant proportion of its fuel oil from Mexico, the Germans considered targeting the country's oilfields, although the proposals were either deemed unfeasible or never came to fruition. In 1915, for example, Franz von Papen, the German military attaché to the United States and Mexico, suggested that 'In view of the great importance of the Tampico (Mexico) oil wells for the English fleet and the large English investments there, I have sent Herr v. Petersdorf there in order to create the greatest possible damage through extensive sabotage of tanks and pipelines.'[7]

Yet the Central Powers were in a far more vulnerable position than the Allies. Before the war, Germany had imported about 75 per cent of its oil from the United States, alongside a much smaller, though significant, percentage from Russia; but as the naval blockade tightened it was only able to draw on imports from Romania and Austro-Hungarian Galicia, with just 8 per cent of its now diminished consumption supplied by some domestic production from near Hanover and from Alsace-Lorraine. However, the Galician oilfields, now supplying 60 per cent of the Central Powers' petroleum supplies, were captured by the Russians in mid-September 1914. The province was retaken in May 1915, the oil wells and refineries having sustained less damage from the retreating Russians than had been expected; but, following years of overproduction, output from Galicia's oilfields was now in terminal decline. Hence, on the entry of Romania into the war alongside the Allies in August 1916, as Germany's chief military strategist General Erich Ludendorff later recalled, 'I now saw quite clearly, we should not have been able to exist, much less carry on the war, without Rumania's corn and oil, even though we had saved the Galician oil-fields at

Drohobycz from the Russians ... The stocks of oil were uncommonly low; it was urgently necessary to increase the supply from Rumania.'[8]

By October, advancing German and Austro-Hungarian forces had captured large stocks of Romanian petroleum products, including gasoline, and the oilfields were in sight, at which point – against some resistance from the Romanian government – the Allies decided to destroy as much of the country's oil industry as possible to deny it to the Germans. About 800,000 tons of oil went up in flames and installations across the oil region were dynamited and wrecked, putting most production and refining out of action. It took about five months until German engineers were able to revive oil operations – increasing from a third to four-fifths of the previous output from late 1917 to 1918 – and to lay new export pipelines to the Danube. As Ludendorff recounted,

> the Rumanian oil was of decisive importance. But even when we had this source the question of rolling stock remained very serious, and made both the carrying on of the war and life at home very difficult ... [P]roduction of oil in Rumania had increased to the limits of the possible, [and] this could not make good the whole shortage. It now seemed possible to supplement it from Trans-Caucasian sources, and in particular from Baku, if transport facilities could be provided ... In 1918 the supplies in the Caucasus promised better times.[9]

With the collapse of the tsarist regime in 1917 and the end of war with Russia, the new Soviet government agreed to allow the Germans access to Baku's oil. However, the Caucasus was in turmoil, with a recurrence of inter-ethnic violence between Muslims and Christian Armenians and Georgians, which on the part of the Turks had been genocidal. The violence was fuelled by Russian-Turkish conflict earlier in the war, and now by Turkey's designs on long-lost territory in the region as German-Turkish relations began to break down. Furthermore, the local Baku Commune refused to cooperate with the Germans. 'We could only expect to get oil from Baku if we helped ourselves,' Ludendorff later wrote.[10] The Germans leapt at the Georgians' request for support, as Georgia lay on the Baku-Tiflis(Tbilisi)-Batum oil transit route westward, while the British sent a small specialist force from Baghdad via northern Persia, under the leadership of General Lionel Dunsterville, to prevent Baku from being taken by either the Germans or the Turks, or to sabotage the oil installations if either appeared imminent. But the chaotic tangle of violence and proxy warfare became a near stalemate, and while the Germans clung to a desperate optimism that they could reach Baku's prized oil, the final battles on the Western Front and in Palestine drew the war to a bloody close. The British 'Dunsterforce' that arrived in Baku in August 1918 was soon ejected by Turkish forces and Azerbaijani militia – who then massacred local Armenians in revenge for an Armenian massacre of Azerbaijanis a few months earlier.[11]

In late July 1918 a detailed and forthright report by Admiral Sir Edmond Slade, the long-standing authority on oil supplies for the Royal Navy, had strengthened the position of those in Britain's War Cabinet – including Lord Curzon, with special responsibility for the Middle East – who were pushing for British forces to take possession of northern Mesopotamia (Iraq). Slade wrote,

> In Persia and Mesopotamia lie the largest undeveloped resources at present known in the world ... If this estimate is anywhere near the truth, then it is evident that the Power that controls the oil lands of Persia and Mesopotamia will control the source of supply of the majority of the liquid fuel in the future ... We must therefore at all costs retain our hold on the Persian and Mesopotamian oil fields ... and we must not allow the intrusion in any form of any foreign interests, however much disguised they may be. We shall then be in a position of paramount control ... and enjoy all the advantages that this will give us if we find ourselves forced into another war. These advantages are very great and we cannot expect to enjoy them without making some sacrifices for them and we must be prepared to defend our claim against everybody. Conventions and Treaties are only paper and can be torn up and are not sufficient safeguard. We must have absolute security in this matter ... The case ... is sufficiently alarming to make the question one for immediate decision else we may find ourselves out-manoeuvred not only by the enemy but by neutrals or even by our present Allies.[12]

The Chief of the Air Staff, among others, was in full agreement with what the Secretary of the War Cabinet, Sir Maurice Hankey, called this 'most vitally important paper'. Hankey wrote to the First Lord of the Admiralty, 'As regards the future campaign, it would appear desirable that before we come to discuss peace, we should obtain possession of all the oil-bearing regions in Mesopotamia and Southern Persia, wherever they may be.'[13] Meanwhile, to the Prime Minister, David Lloyd George, he wrote that while 'there is no *military* advantage in pushing forward in Mesopotamia . . . there may be reasons other than purely military for pushing on. Would it not be an advantage before the end of the war, to secure the valuable oil wells in Mesopotamia?'[14] In mid-August the War Cabinet was persuaded that, as Hankey put it, 'The only big potential supply that we can get under British control is the Persian and Mesopotamian supply . . . [and] that the control over these oil supplies becomes a first-class British war aim.'[15] These deliberations clearly became a major reason why, several days after the armistice with Turkey on 31 October, the British Army was sent northwards to occupy Mosul.[16]

Ten days after the European Armistice of 11 November, Lord Curzon presided over a celebratory gathering of the Inter-Allied Petroleum Conference in London at which he declared,

Even before the War, oil was regarded as one of the most important national industries and assets. It was being increasingly used for economical and transport purposes but with the commencement of the War oil and its products began to rank as among the principal agents by which [the Allies] would conduct and by which they could win it . . . Without oil how could they have procured the mobility of the Fleet, the transport of their troops, or the manufacture of several explosives? . . . All products of oil fuel, gas oil, aviation spirit, motor transport spirit, lubricating oil, etc., played an equally important part in the War; in fact, [we] might say the Allies floated to victory on a wave of oil.[17]

Notes

1. EARTH OIL

1. 'Bitumen – Its Uses', *Scientific American* (25 October 1856); Halleck, *Bitumen*; Cooke, 'Naphtha'; Owen, *Trek*, ch. 1.
2. Boëda et al., 'New Evidence'; Cârciumaru et al., 'New Evidence'.
3. Connan and Van de Velde, 'Overview'; Schwartz and Hollander, 'Annealing'; Connan, 'Use'; Brown et al., 'Sourcing'; Wendt and Cyphers, 'How the Olmec'; Forbes, *Ancient Technology*, vol. 1, 56–109; Moorey, *Ancient*, 331–5, 356–7; Clark et al., 'Significance'.
4. Connan and Van de Velde, 'Overview'; Schwartz and Hollander, 'Uruk'; Oron et al., 'Early'.
5. Finkel, *Ark*, 118–19, 172–82, 307–8, 351–6; Woolley, 'Excavations', 635–6; Hamilton, *Book*, 349, 352.
6. Ovid, *Metamorphoses*, bk. 9, ll. 659–60; McDonald, 'Georgius'.
7. Herodotus, *Histories*, bk. I, 179; Rich, *Memoir*, 23–4, 31, 55–6, 62–3; Forbes, *Ancient Technology*, vol. 1, 56–83.
8. Herodotus, ibid., bk. VI, 119; Lockhart, 'Iranian', 4–5, 7–8; Forbes, ibid., vol. 1, 40–1; Alizadeh, *Ancient*, 273–81; Sorkhabi, 'Pre-Modern', 155.
9. Vitruvius, *Ten Books*, VIII.III.
10. Forbes, *Ancient Technology*, vol. 1, 44–5.
11. Strabo, *Geography*, vol. 7, 16.2.42.
12. Ibid., 16.1.15; Toutain, *Economic*, 124; Forbes, *Ancient Technology*, vol. 1, 83–4; Partington, *History*, 3–4.
13. Strabo, ibid., 16.1.15.
14. Pliny, *Natural History*, vol. 1, II.108–10; ibid., vol. 2, VI.26 and VII.13; ibid., vol. 5 (1856), XXXI.39; ibid., vol. 6 (1857), XXXV.51; Sorkhabi, 'Pre-Modern', 162; Partington, *History*, 3–4.
15. Ammianus Marcellinus, *Ammianus*, bk. 23, ch. 6, 15–16.
16. Plutarch, *Plutarch*, vol. 7, XXXV, 1–2 and 3–5 respectively.

17. Forbes, *More Studies*, 70–6; Partington, *History*, 1–3.
18. Pliny, *Natural History*, vol. 1, II.108; Forbes, ibid., 77; Partington, ibid., 28–32; Day, 'Sir Gawain', 13–14.
19. Dio, *Roman History*, vol. 3, bk. XXXVI, 1.
20. Ibid., vol. 9, Epitome of bk. LXXVI, 12.
21. Procopius, *Procopius*, vol. 5, VIII, xi, 33–40; Lockhart, 'Iranian', 8.
22. Constantine Porphyrogenitus, *De Administrando*, 13.73–103.
23. Forbes, *More Studies*, 78–90; Partington, *History*, 2–5, 10–22, 27–32; Pryor and Jeffreys, *Age*, 26, 31–2, 46, 61–2, 72–3, 86, 189, 203, 378–9, 607–31; Haldon, '"Greek Fire"'; Day, 'Sir Gawain', 14–16; Roland, 'Secrecy'; Constantine Porphyrogenitus, ibid., 53.493–511; Shepard, 'Closer', 24–6.
24. Burger et al., 'Identification'.
25. Hitti, *History*, 291–2, 348–9.
26. Shaban, *Islamic*, vol. 2, 56–9.
27. Fishbein, *History*, 136.
28. Forbes, *Studies*, 154; Lockhart, 'Iranian', 10.
29. Mas'ūdī, *Historical*, vol. 1, 421–2; Lockhart, ibid.; Sorkhabi, 'Pre-Modern', 163.
30. Minorsky, *History*, 27, 120; Minorsky, *Abū-Dulaf*, 35; Lockhart, 'Iranian', 10; Sorkhabi, ibid., 164; Jackson, *From Constantinople*, 28–9.
31. Le Strange, *Lands*, 289; Lockhart, ibid., 15; Sorkhabi, ibid., 162–3.
32. Ouseley, *Oriental*, 57, 77, 133–4, 250, 272; Sorkhabi, ibid., 162–3.
33. Bosworth, *History*, 64, 68, 71; Hitti, *History*, 327; Minorsky, *Studies*, 65.
34. Partington, *History*, 21–8; Ayalon, *Gunpowder*, 9–15; Ayalon, 'Reply'.
35. Quoted in Partington, ibid., 24–5; Day, 'Sir Gawain', 13.
36. Joinville, *Histoire*, 113–17 (my translation); Partington, ibid., 25–6.
37. Shaban, *Islamic*, vol. 2, 58.
38. al-Hassan and Hill, *Islamic*, 106–11, 144–6.
39. Quoted in Forbes, *Studies*, 149; Forbes, *Ancient Technology*, vol. 1, 47–8.
40. Partington, *History*, 200–1; Hoffmeyer, *Military*, 149–50; Forbes, *Ancient Technology*, vol. 1, 48.
41. Quoted in Forbes, *Short*, 53.
42. Raymond, *Cairo*, 75–7; Lev, *Saladin*, 60–1.
43. Lockhart, 'Iranian', 11; Forbes, *Studies*, 155.
44. Le Strange, *Lands*, 63; Schwarz, 'Khanikin', 901.
45. Polo, *Book*, vol. 1, 46.
46. Quoted in Kerr, *General*, vol. 7, 146–7; Forbes, *Ancient Technology*, vol. 1, 40.
47. Morgan and Coote, eds., *Early*, vol. 2, app. IX, 439–40.

48. Olearius, *Voyages*, 402, 356.
49. Quoted in Forbes, *Studies*, 158–9; Sorkhabi, 'Pre-Modern', 164–5.
50. Quoted in Alekperov, *Oil*, 12.
51. Quoted in Forbes, *Studies*, 30–1.
52. Fryer, *New*, 318; Sorkhabi, 'Pre-Modern', 163.
53. Matveichuk, 'Peter'; Alekperov, *Oil*, 13–17; Lockhart, 'Iranian', 14; Vassiliou, *Historical*, 388–9; Forbes, *Studies*, 161–2.
54. Hanway, *Historical*, vol. 1, 381–3; Redwood, *Petroleum*, vol. 1, 4–5.
55. Rich, *Memoir*, 63–4.
56. Kinneir, *Geographical*, 359–60; Wright, *English*, 150.
57. Kinneir, ibid., 38–9.
58. Ibid., 298.
59. Lynch, 'Tigris', 476; Ross, 'Notes', 449, 443.
60. Rawlinson, 'Notes', 79, 94.
61. Layard, *Early*, vol. 2, 32.
62. Layard, 'Description', vol. 16 (1846), 81–2, 73 and 49 respectively; Khazeni, *Tribes*, 115.
63. Layard, *Nineveh*, vol. 2, 381.
64. Layard, *Discoveries*, 109.
65. Ibid., 171–2. Layard added in a footnote, 'In a few hours the pits are sufficiently filled to take fire again.'
66. Lopes de Lima, *Ensaios*, 29; Domingues da Silva, 'Atlantic'.
67. Quoted in Simon, *Scientific*, 159 (translated by Esmé Carter); ibid., 82.
68. Ibid., 80, 94–7; Vansina, 'Ambaca', 1–2.
69. Livingstone, 'XVI – Explorations', 233; Livingstone, *Missionary*, 418, 421; Petrusic, 'Violence'; Domingues da Silva, 'Atlantic', 117–19; Ferreira, 'Conquest'.
70. Quoted in Needham, *Science*, vol. 5, pt. 7, 75; ibid., vol. 3, 608–9; Yinke, *Ancient*, 40; Vogel, 'Bitumen', 160–1.
71. Aston, *Nihongi*, bk. XXVII, 289; Redwood, *Petroleum*, vol. 1, 74.
72. Quoted in Deng, *Ancient*, 40; Needham, *Science*, vol. 5, pt. 7, 75.
73. Needham, ibid., 80–94; Feng et al., *Chinese*, 2.
74. Quoted in Needham, ibid., vol. 3, 609; Deng, *Ancient*, 40.
75. Quoted in Vogel, 'Bitumen', 160–1.
76. Quoted in Vogel, '"That which soaks"', 483 and 478 respectively; Feng et al., *Chinese*, 3.
77. Quoted in Vogel, ibid.
78. Vogel, 'Types', 463–9; Vogel, '"That which soaks"', 480–8, 494–9; Jung and Fang, 'Account', 232–3.
79. Dion, 'Sumatra', 139, 142; Reid, *Southeast*, vol. 1, 75.

80. Burnell, ed., *Voyage*, vol. 1, 109; Poley, *Eroica*, 6.
81. Mills, 'Eredia', app. II, 238.
82. Quoted in Forbes, *Studies*, 172; Beaulieu, *De rampspoedige*, 135.
83. Quoted in Forbes, ibid., 112; Forbes, *More Studies*, 47–8.
84. Bontius, *Account*, 4–5; Forbes, *Studies*, 173.
85. Baker, 'Short', 172; Longmuir, *Oil*, 8–9; Scott, *Gazetteer*, pt. 1, vol. 2, 256.
86. Symes, *Account*, 261–2; Noetling, 'Occurrence', 7–14.
87. Symes, ibid., 441–3.
88. Cox, 'Account', 130–5. Lord Archibald Cochrane, 9th Earl of Dundonald, had patented and commercialized a process for distilling tar products from coal. (Clow and Clow, 'Lord Dundonald'.)
89. Moriyama, *Crossing*, 51, 217; Bolitho, 'Echigo', 261; Forbes, *Studies*, 180–1.
90. Quoted in Woodman, *Making*, 73–4.
91. Crawfurd, *Journal*, 53–5, 427, 445.
92. Quoted in Huc, *Journey*, vol. 1, 300–6; Forbes, *Studies*, 175; Vogel, '"That which soaks"', 494.
93. Wilcox, 'Memoir', 415.
94. 'Proceedings of the Asiatic Society', 169.
95. Saikia, 'Imperialism', 49–50.
96. Groen, 'Colonial', 280–1.
97. Quoted in Poley, *Eroïca*, 38, 37 and 38–9 respectively; ibid., 43, 54.
98. Yule, *Narrative*, 20–1; Pollak, 'Origins'.
99. De La Rue and Müller, 'Chemical', 222; Noetling, 'Occurrence', 143–4, 156.
100. Corley, *History*, 7–10.
101. Quoted in Poley, *Eroïca*, 41–2.
102. Cârciumaru et al., 'New Evidence'.
103. Nardella et al., 'Chemical'.
104. Herodotus, *Histories*, bk. IV, 195; Forbes, *Ancient Technology*, vol. 1, 43.
105. Vitruvius, *Ten Books*, VIII.III.
106. Forbes, *Ancient Technology*, vol. 1, 43; Pliny, *Natural History*, vol. 5, XXIV.25; ibid., vol. 6, XXXV.51; Faraco et al., 'Bitumen'.
107. Dioscorides, *De materia*, I.73.1.
108. Ibid.; Forbes, *Ancient Technology*, vol. 1, 43–4; Pliny, *Natural History*, vol. 1, II.110; ibid., vol. 6, XXXV.51; Partington, *History*, 4.
109. Santoro, 'Crafts', 271–3.
110. Adler, *Itinerary*, 8.
111. Stapleton, ed., *Magni*, vol. 2, xx; McDonald, 'Georgius', 353.
112. Lopez and Raymond, trans., *Medieval*, 109–110, 112.

113. Gerali and Lipparini, 'Maiella', 277.

114. Quoted in Forbes, *Studies*, 107; ibid., 37–40, chs. 7–8; Brévart, 'Between', 21–5.

115. Quoted in Forbes, ibid., 94.

116. Gerali, 'Scientific', 91, 96–8; Gerali, 'Development', 173–7; Gerali, 'Oil Research', 202–3.

117. Poggi, 'Account', 323–4; Saussure, 'Experiments'; Del Curto and Landi, 'Gas-Light', 11–12.

118. Holland, *Travels*, 18–19, 517–23.

119. Forbes, *Studies*, ch. 5, 108–12.

120. Quoted in Forbes, *Ancient Technology*, vol. 1, 46, and Forbes, *Studies*, 46, respectively; McDonald, 'Georgius', 352.

121. Forbes, *Studies*, 47–55.

122. Quoted in ibid., 3.

123. Ibid., 70–7.

124. Quoted in ibid., 14.

125. Ibid., 21–7, 78–82, 183–6; Butt, 'Technical', 513; 'Asphaltic Mastic', 383–6; 'French Asphalt'; 'The Asphalts'; Lay, *Ways*, 208–17, 229–30.

126. 'Petroleum in Alsace', *Scientific American* (24 August 1872).

127. Camden, *Britannia*, vol. 2, 971; Selley, 'U.K. Shale Gas', 105.

128. Eele et al., 'Manufacture'.

129. Butt, 'Technical', 512; Luter, '"British Oil"'; Craig et al., 'History', 4–5; Forbes, *Studies*, 52–3.

130. Luckombe, *Beauties*, 294, also 288.

131. Craig et al., 'History', 5; Trinder, *Industrial*, 90; Huxley, *Britain*, 28–31.

132. Quoted in 'Coalbrookdale', 338.

133. Pennant, *Tour*, 421; Lewis, *Topographical*, vol. 1, 'Flintshire'.

134. Clow and Clow, 'Lord Dundonald'; Sugden, 'Archibald', 8–16; Sugden, 'Lord Cochrane', 14–20, 166–8, 304, 332–3; Luter, 'Archibald'.

135. Lea, 'Derbyshire'.

136. Butt, 'Technical'; Dean, 'Scottish'; McKay, *Scotland*, ch. 1.

137. Wołkowicz et al., 'History'; Sozański et al., 'How the Modern', 812.

138. Sozański et al., ibid., 812–15; Mikucki, *Nafta*, 3–9; Schatzker and Hirszhaut, *Jewish*, 5–12; Forbes, *More Studies*, 93–6.

139. Brice, 'Abraham'; Butt, 'Technical', 514–15; Forbes, *Studies*, ch. 13; Forbes, *More Studies*, 91–9; Schatzker and Hirszhaut, ibid., 24–41; Frank, *Oil*, 48–58.

140. Schatzker and Hirszhaut, ibid., 36–7.

141. Buzatu, *History*, vol. 1, 33–7; Forbes, *More Studies*, 99–104; Sell, 'Statistics', vol. 1, 22–3; Stoicescu and Ionescu, 'Romanian', 134–6.

142. See section 'Middle East and Central Asia' above on Emperor Constantine.
143. Galkin et al., 'Oil'; Vassiliou, *Dictionary*, 507.
144. Alekperov, *Oil*, 13–17.
145. Ibid., 19–23.
146. Ibid., 25–7, 49–54, 56–60.
147. Ibid., 40–2.
148. Vassiliou, *Dictionary*, 530–1; Alekperov, ibid., 43.
149. Boxt et al., 'Isla Alor', 64–71, 75–7, 79; Wendt and Cyphers, 'How the Olmec'.
150. Boxt et al., ibid., 77–9; Mathews, *Chicle*, 6–11.
151. Sahagún, *Florentine*, bk. 10, pt. XI, 88–9; Forbes, *Studies*, 146–8.
152. Mathews, *Chicle*, 9–10.
153. Oviedo, *Natural History*, 20.
154. Humboldt and Bonpland, *Personal*, vol. 7, 57; Mazadiego Martínez et al., 'Information', 242.
155. Oviedo, *Natural History*, 20.
156. Oviedo, *Historia General*, vol. 1, 214–15, 501–2 (translations by Dr Enrique Zapata-Bravo); Mazadiego Martínez et al., 'Information', 239–43.
157. Francis, *Invading*; Moses, *Spanish*, vol. 1, chs. 1–4; Romero, 'Death and Taxes'.
158. Oviedo, *Historia General*, vol. 1, 501.
159. Ibid., 591, 593.
160. Ibid., vol. 2, pt. 1, 301; Moses, *Spanish*, vol. 1, ch. 4.
161. Oviedo, ibid., 370; Mazadiego Martínez et al., 'Information', 240; Avellaneda, *Conquerors*, 40–3; Van Isschot, *Social*, 22–3.
162. Oviedo, ibid., vol. 1, 214, 502.
163. Quoted in Bergman, 'Medical', 12 (n. 41); Mazadiego Martínez et al., 'Information', 240; Morón, *Orígenes*, vol. 1, 192–3.
164. Quoted in Mazadiego Martínez et al., ibid., 241; Forbes, *Studies*, 142.
165. Acosta, *Naturall*, 173; Forbes, ibid., 142.
166. Carletti, *My Voyage*, 37.
167. Quoted in Boxt et al., 'Isla Alor', 78; Scholes and Roys, *Maya*, 16.
168. Raleigh, *Discovery*, 13; Higgins, *History*, 3–5; Spielmann, 'Who Discovered'; Forbes, *Studies*, 143–4.
169. Raleigh, ibid., 186–7.
170. Barba, *Arte*, 21.
171. Masefield, ed., *Dampier*, vol. 2, 224.
172. Ibid., vol. 1, 158.

173. Lossio, 'Del copey', 2–17; Bosworth, *Geology*, 341–4; Larkin, *Report*, 10–11; Pinelo, 'Nationalization', 2–6; Kuczynski, *Peruvian*, 111–12.

174. Riguzzi and Gerali, 'Veneros', 751–2; Rippy, 'United States', 20.

175. Smith, *Generall*, vol. 2, 275.

176. Hughes, *Natural*, 50–1; Forbes, *Studies*, 144–5.

177. 'Barbadoes Mineral Oil, or Green Naphtha', 394–5.

178. Anderson, 'Account', 67.

179. Nugent, 'Account', 67–9.

180. Ibid., 68; Higgins, *History*, 12.

181. Lyon, *Journal*, 44–5.

182. Garay and Moro, *Survey*, 121.

183. 'Asfalto y sal gema', 168 (translations by Dr Enrique Zapata-Bravo).

184. Ibid., 169.

185. Ibid., 172.

186. Williams, *Isthmus*, 11; Townsend, *Malintzin*, 13–17, 150–1; Bassie-Sweet et al., 'History', 9–11.

187. Williams, ibid., 159, 230.

188. Gerali and Riguzzi, 'Inicios', 66; Middleton, *Industrial*, 48–9.

189. Taylor and Clemson, 'Notice', 166.

190. Sugden, *Lord Cochrane*, 168–73; see section 'Europe' above on Earl of Dundonald.

191. Quoted in Thomas and Bourne, *Life*, vol. 2, 318.

192. Higgins, *History*, 6–7, 12–14, 37–8; Daintith, *Finders*, 126; Beaton, 'Dr. Gesner', 35–7; Lucier, *Scientists*, 38, 43, 46, 157; Sugden, *Lord Cochrane*, 166–8; Hughes, *Energy*, 51–2.

193. Manross, 'Notice', 160.

194. Quoted in Hughes, *Energy*, 53.

195. 'Money-Market and City Intelligence', *The Times* (6, 13 and 18 March 1856).

196. Sugden, *Lord Cochrane*, 279; Higgins, *History*, 13–14, 38–41; Wall and Sawkins, 'Report', 94–7.

197. Quoted in Higgins, ibid., 38.

198. Brown et al., 'Sourcing', 66–7.

199. Warner, *Texas*, 18; Clayton et al., *De Soto*, vol. 1, 162

200. Brown, 'Asphaltum'; Wärmländer et al., 'Could the Health'.

201. Thwaites, ed., *Jesuit*, vol. 43, 261; Finley, *French*, 375–6; Ash, 'Friar'.

202. Coyne, *Exploration*, pt. 1, 29, 81; Williamson and Daum, *Illumination*, 10.

203. Hudson's Bay Company Archives, 'Instructions from James Knight'; Yerbury, 'Protohistoric', 21.

204. Quoted in Fitzgerald, *Black*, 11.

205. Charlevoix, *Voyage*, vol. 1, 196; Severance, *Story*, 126–7.

206. Pritchard and Taliaferro, *Degrees*, 21–2, 172–5; Black, *Petrolia*, 23–4; Barnhart, *American*, 63; Day, *Historical*, 637–8.

207. Hulbert and Schwarze, eds., 'David', 52–3.

208. Loskiel, *History*, pt. 1, 117–18.

209. Crespí, *Description*, 341–3; Hodgson, 'California'.

210. Crespí, ibid., 343, 355.

211. Ibid., 407, 699.

212. Ibid., 477, 717; Fages and Priestley, 'Historical', 89.

213. Crespí, ibid., 703.

214. Fages and Priestley, 'Historical', 75; Hodgson, 'California', 48–53.

215. Quoted in Cook, *Washington*, 65; Jefferson, *Notes*, 36–7; Glanville and Mays, 'William'; Pendelton, 258–68.

216. Lincoln, 'Account', 375–6.

217. Jefferson, *Notes*, 36–7.

218. Sparks, ed., *Writings*, vol. 1, 584.

219. Kovarsky, *True*, 2, 38–9; University of Virginia, 'Notes on the State of Virginia'.

220. Davidson, *North West*, 262, 264; Lewis, 'Recognition', 32–7.

221. Tyrrell, ed., *Journals*, 386; Ruggles, *Country*, 52–4.

222. Lamb, ed., *Journals*, 129.

223. Cruikshank, ed., *Correspondence*, vol. 1, 290; Robertson, ed., *Diary*, 155; Thames River Background Study Research Team, *Thames*, 58–9, 63, 71–3, 94–5.

224. Vancouver, *Voyage*, vol. 2, 449; Yerkes et al., 'Petroleum', 14.

225. 'From the Ontario Repository', *The Enquirer* (Richmond, VA) (30 June 1804).

226. Harris, *Journal*, 46.

227. Ashe, *Travels*, 46, 82–4.

228. Cuming, *Sketches*, 84; Griffenhagen and Harvey, *Old*, 157–64.

229. Rosenberg and Helfand, 'Every Man', 49 (n. 22); Ritter, *History*, 247–8; Miller, ed., *This Was Early*, 1.

230. Eaton, *Petroleum*, 57–8; Giddens, *Birth*, 30–1.

231. Hildreth, 'Observations', 61–2; Babcock, *Venango*, vol. 1, 43, 57–9; Waples, *Natural Gas*, 11–13; *Derrick's Hand-Book*, 8–11.

232. Hildreth, ibid., 64–5.

233. Clinton, *Introductory*, 25.

234. 'Valuable Discovery', *Richmond Enquirer* (Richmond, VA) (29 December 1818).

235. Nuttall, 'Of Kentucky', 69–71; McLaurin, *Sketches*, 34–8.

236. Atwater, 'Facts', 5.

237. Alden, 'Antiquities', 310–11.

238. Franklin, *Narrative*, 516; ibid., 137, 507, 514, 519.

239. Howison, *Sketches*, 194; National Historic Sites Directorate, *To Confirm*.

240. Thomas, *Manufactured*, 23–56; Tomory, 'Building'; Binder, 'Gas Light', 359–62; *Mineral Resources of the U.S., 1885*, 169; Waples, *Natural Gas*, 12–13.

241. 'A Village Lighted by Natural Gas', 398–9.

242. Waples, *Natural Gas*, 13–14.

243. Quoted in *Derrick's Hand-Book*, 9.

244. Hodgson, 'California'; Brown et al., 'Sourcing'.

245. Flint, ed., *Personal*, 214–15.

246. 'Geological Phenomena', *Recorder* (Hillsborough, NC) (2 September 1829); Nuttall, 'Of Kentucky', 71–3.

247. Pickering, *Inquiries*, 122.

248. Silliman, 'Notice', 98–100; Giddens, *Pennsylvania*, 7, 32; Giddens, *Birth*, 15–16; Williamson and Daum, *Illumination*, 13, 17.

249. Zallen, *American*, 214–28, 236.

250. Quoted in Atkinson, *History*, 234.

251. Quoted in ibid., 235.

252. Quoted in Howe, *Historical*, 346; Atkinson, ibid., 234–6; Waples, *Natural Gas*, 13–16.

253. Lee, *Burning*, 11–13; White, 'Petroleum', 140–5.

254. Zallen, *American*, chs. 1–4, 236.

255. Black, *Crude*, 31–3; Bone, *Petroleum*, 20–1; Brice, 'Samuel', 78.

256. Giddens, *Pennsylvania*, 14–15; ibid., 7, 15–17, 33–4, 69, 127; Giddens, *Birth*, 16–17.

257. Williamson and Daum, *Illumination*, 17–23; Brice, 'Samuel', 79–81; Giddens, *Pennsylvania*, 10–19; Daintith, *Finders*, 42–3; 'Read! Read! What Kier's Petroleum, Or Rock Oil, Has Been Doing', *Mountain Sentinel* (Ebensburg, PA) (20 May 1852); 'Kier's Petroleum, Or Rock Oil', *Hillsdale Standard* (Hillsdale, MI) (17 January 1854); 'A Case of Total Blindness Cured', *New Orleans Daily Crescent* (21 April 1854).

258. 'Discovery of Rock Oil', *Meigs County Telegraph* (Pomeroy, OH) (19 June 1851); Williamson and Daum, *Illumination*, 64–5.

259. Brice, 'Samuel', 82–5; Giddens, *Birth*, 23–5; Giddens, *Pennsylvania*, xii, 21–3; Miller, *This Was Early*, 5–6.

260. Lucier, *Scientists*, chs. 5–6; Williamson and Daum, *Illumination*, 27–56; Beaton, 'Dr. Gesner'; Zallen, *American*, 228–30; see Polish oil lamps in section 'Europe' above.
261. 'Oils from the Breckenridge Coal', *New York Times* (17 November 1856).
262. Blake, 'Preliminary', 433; Lucier, *Scientists*, 280.
263. Silliman, *Professor*, 20–1, 24, 26–7.
264. California State Mining Bureau, *Third Annual*, 294; Miller, 'North', 317–18; Bancroft, *History*, vol. 5, 403–5, 422; Testa, 'Los Angeles', 80; Yerkes et al., 'Petroleum', 13–14; White, *Formative*, 2–4; Franks and Lambert, *Early*, 4–5.
265. Burr, *Canada*, 62.
266. Ibid., 38–9, 47–9, 64–6, 87–8; May, *Hard*, 28–39; Taylor, *Imperial*, 22–4.

2. SENECA OIL

1. Maybee, *Railroad*, 3–7; Bell, ed., *History*, 222–37.
2. Brooks, *Frontier*, 71–3; Wilhelm, 'Wheeler'; Babcock, *Venango*, vol. 1, 57–9; Bates, *Our County*, 300, 304, 463–4, 607, 620, 650, 733–4, 793; see Allegheny Valley use of petroleum in ch. 1, section 'North America'.
3. Giddens, *Pennsylvania*, xii–xiii, 45–55, 69; Lucier, *Scientists*, 189–97; Williamson and Daum, *Illumination*, 64–9; see Silliman, Sr in ch. 1, section 'North America'.
4. Quoted in Giddens, ibid., 129; Lucier, ibid., 195–200; Giddens, *Birth*, 26–40.
5. Quoted in Lucier, ibid., 200.
6. Lucier, ibid., 201–3; Giddens, *Birth*, 41–4.
7. 'Oil from Water', *Boston Evening Transcript* (14 May 1855); 'Oil from Water', *New England Farmer* (19 May 1855); 'Oil from Water', *North Carolina Standard* (30 May 1855).
8. Quoted in Giddens, *Beginnings*, 65–6; Lucier, *Scientists*, 201–3.
9. Giddens, *Pennsylvania*, xii, 3, 35–42.
10. Lucier, *Scientists*, 203–4; Giddens, *Birth*, 47–53; Brice, 'Edwin', 11–17.
11. Giddens, ibid., 49–50.
12. Yergin, *Prize*, 20, 29.
13. Painter, 'Oil'; Hugill, *Transition*.
14. Calloway, *Scratch*, 4–11; Ward, *Breaking*.
15. Jennings, *Empire*, 50–1.
16. Ibid., 125–31; Higonnet, 'Origins'; Clayton, 'Duke'.

17. Jennings, ibid., 10–13, 118–19, 240–1; Calloway, *Indian*, 45–51; Nester, *Great*, 32–3; Ward, *Breaking*, 26–8; Egnal, 'Origins', 410–12; Livermore, *Early*, 75–82; Abernethy, *Western*, 5–9; Richter, *Before*, 376–7.

18. Richter, 'Onas', 141; Richter, *Before*, 373, 383–5; Nester, *Great*, 36–8; McConnell, *Country*, 89–98.

19. Quoted in Jennings, *Empire*, 39–40.

20. Jennings, *Ambiguous*, 360–2; Anderson, *Crucible*, 23; Nester, *Great*, 31–2.

21. Jennings, ibid., 325, 332–9, 388–97; Jennings, *Empire*, 388–97; Nester, ibid., 36–8; Ward, *Breaking*, 25–9.

22. Quoted in Egnal, 'Origins', 412.

23. Jennings, *Empire*, 52–3; Nester, *Great*, 38–9; Ward, *Breaking*, 29–30; McConnell, *Country*, 101–2; Reynolds, 'Venango'.

24. Jennings, ibid., 118–20; McConnell, ibid., 107–8; Nester, ibid., 41–5; Ward, ibid., 31–2; Higonnet, 'Origins', 74.

25. Jennings, ibid., 153–68, 187–206, 276; Richter, *Before*, 386–95; Nester, ibid., chs. 4–5; Ward, ibid., 33–178; McConnell, ibid., 108–30; Anderson, *Crucible*, 163–4; Calloway, *Scratch*, 48–50.

26. See Evans in ch. 1, section 'North America'.

27. Evans, *Geographical*, 15; Pritchard and Taliaferro, *Degrees*, 172.

28. Nuttall, 'Of Kentucky', 69; Howe, *Historical*, 352.

29. Anderson, *Crucible*, 232–9; Nester, *Great*, ch. 3.

30. Anderson, ibid., 268–71.

31. Quoted in ibid., 271.

32. Ibid., 269–80, 455; Ward, *Breaking*, 178–82; Richter, *Before*, 395–9.

33. Quoted in Jennings, *Empire*, 409; Ward, ibid., 182–5.

34. See Allegheny logging in section 'Drilling in Ohio Country' above.

35. Anderson, *Crucible*, 472–5, 524–37; Richter, *Before*, 399–403.

36. Jennings, *Empire*, 442; Calloway, *Scratch*, 16, 73–81.

37. Wallace, *Death and Rebirth*, 120; Anderson, *Crucible*, 535–7; McConnell, *Country*, ch. 8; Ward, *Breaking*, ch. 7; Richter, *Before*, 403–6.

38. Anderson, ibid., 538–41; Wallace, ibid., 115–16; Ward, ibid., ch. 8; Calloway, *Scratch*, 29, 66–72.

39. Quoted in Richter, *Before*, 411–12.

40. Quoted in Lehman, *Bloodshed*, 49 and 82 respectively; Anderson, *Crucible*, 540–1; Wallace, *Death and Rebirth*, 115–18.

41. Quoted in Parmenter, 'Pontiac's', 628, and Calloway, *Scratch*, 73, respectively.

42. Quoted in Jennings, *Empire*, 441.

43. Quoted in Middleton, *Pontiac's*, 110.
44. Quoted in Calloway, *Scratch*, 73.
45. Quoted in ibid., 87; Jennings, *Empire*, 447–8.
46. Anderson, *Crucible*, 560–71; Calloway, ibid., 92–8; Richter, *Before*, 406–9; Parmenter, 'Pontiac's', 628; Blaakman, 'Speculation', 382–5, 392–3.
47. Anderson, ibid., 617–26.
48. Calloway, *Indian*, 181–212; Holton, *Forced*, 3–38; Calloway, *Scratch*, 56–65, 98–100; Richter, *Before*, 409–12; Chandler, *Land*; Anderson, ibid., 525–7, 740; Hutson, 'Benjamin', 433–4; Billington, *Westward*, 134–7; Alvord, 'Virginia', 21–5; Abernethy, *Western*, 14–39; Livermore, *Early*, 113, 215.
49. Livermore, ibid., 111–15; Holton, 'Ohio', 457–68; Wallace, *Death and Rebirth*, 122; Billington, ibid., 148–57; Calloway, *Scratch*, 28, 60–5; McConnell, *Country*, ch. 10; Del Papa, 'Royal'.
50. Hauptman, *Tribes*, 27–38; Calloway, ibid., 11–14, 111; Richter, *Before*, 412–14; Davidson, *Propaganda*; Dion, 'Natural'.
51. Fruchtman Jr, *Thomas*; Davidson, 'Whig', 442–53.
52. Quoted in Holton, *Forced*, 36; Morris, *American*, 37–9; Alvord, 'Virginia', 25–8; Onuf, 'Toward', 353–66; Egnal, 'Origins', 415–18; Curtis, 'Land', 217–22; Miller, *Origins*, 264–7; Furstenberg, 'Significance', 654–5; Calloway, *Scratch*, 122.
53. Winsor, 'Virginia', 436–43; Holton, ibid., 34; Billington, *Westward*, 146, 152, 166–8; Livermore, *Early*, 91, 108; Ellis, 'Ohio', 130; Abernethy, *Western*, 5–6, 105–6; McConnell, *Country*, ch. 11; Calloway, *Indian*, 206–12.
54. Jennings, *Creation*, 228, 242–51; Taylor, *Divided*, 5–6, 81, 83, 217; Wallace, *Death and Rebirth*, 125–48; Billington, ibid., 174–95.
55. Sioussat, 'Chevalier', 391–418; Onuf, 'Toward', 353–74.
56. Bakeless, *Lewis*, 29.
57. Jennings, *Creation*, 275; Billington, *Westward*, 181.
58. Boyd, ed., *Papers*, vol. 3, 259; Ostler, '"To Extirpate"'.
59. Quoted in Mintz, *Seeds*, 4, 76; Abler, *Cornplanter*, 81; Calloway, *Indian*, ch. 11; Taylor, *Divided*, 4–6, 98–102; Billington, *Westward*, 185; Stone, 'Sinnontouan', 212–18.
60. Fitzpatrick, ed., *Writings*, vol. 15, 189–92. Transcribed by Alexander Hamilton, phrase in brackets added by Washington.
61. Graymont, *Iroquois*, 204, 214–15, 218; Mintz, *Seeds*, 79, 116.
62. Quoted in Moore, *Diary*, vol. 2, 218–19.

63. Quoted in Richter, 'Onas', 136; Blaakman, 'Speculation', 385–90.

64. Fitzpatrick, ed., *Writings*, vol. 27, 140.

65. Billington, *Westward*, 215–17; Stagg, 'Between', 394, 413; Jennings, *Creation*, 308–9; Furstenberg, 'Significance', 655–68; Calloway, *Victory*, 17–19; Blaakman, 'Speculation', 393–6, 409.

66. Fitzpatrick, ed., *Writings*, vol. 13, 20; Livermore, *Early*, 254–5.

67. Boyd, ed., *Papers*, vol. 4, 237–8.

68. Fitzpatrick, ed., *Writings*, vol. 27, 488–90; Furstenberg, 'Significance', 665.

69. Moore and Moore, *Thomas*, 156–7; Rolt, *From Sea*, 2, 40, 111, 178.

70. Hauptman, *Conspiracy*, 18–21, 29–31, 65, 121–43.

71. Quoted in Troup, *Vindication*, app., 4.

72. Quoted in Hauptman, *Conspiracy*, 13.

73. Stagg, 'Between', 388, 405; Richter, 'Onas', 133–4, 150–1; Hauptman, ibid., 69, 102–6; Blaakman, 'Speculation'.

74. Quoted in Taylor, *Divided*, 298.

75. Quoted in ibid., 159.

76. Richter, 'Onas', 141; Horsman, 'American', 35–53; Lehman, 'End', 523–47; Campisi and Starna, 'On the Road', 468–71; Jennings, *Creation*, 275–85; Blaakman, 'Speculation', 387–8.

77. Quoted in Campisi and Starna, ibid., 474.

78. Quoted in Lehman, 'End', 528.

79. Richter, 'Onas', 149.

80. Ibid., 148–59; Campisi and Starna, 'On the Road', 477–9; Calloway, *Victory*, 93–105; Abler, *Cornplanter*, 2, 69–75; Blaakman, 'Speculation', 225–7, 390–1, 403–4, 457–8; Bell, *History*, 62–5; Mt. Pleasant, 'Independence'; Huston, 'Land', 324–32.

81. Blaakman, ibid., 215–300, 396–402, 409–15, 437–74; Calloway, ibid., ch. 2.

82. Quoted in Abler, *Cornplanter*, 81–2.

83. Ibid., 83–4, 135–6; Swatzler, *Friend*, 132–3; Deardorff, 'Cornplanter'.

84. Ford, ed., *Works*, vol. 6, 242.

85. Calloway, *Victory*, chs. 3–5.

86. Bergmann, '"Commercial"', 157; Ward, *Breaking*, 76–7.

87. Abler, *Cornplanter*, 64–110; Gallo, 'Improving'; Ilisevich, 'Early', 291–4; Arbuckle, 'John Nicholson', 353–64.

88. Quoted in Kent and Deardorff, 'John Adlum', 310.

89. Quoted in Knopf, ed., *Anthony*, 252.

90. Quoted in ibid., 354 and 357; Calloway, *Victory*, 139–52.

91. Quoted in Hauptman, *Conspiracy*, 13.

92. Abler, *Cornplanter*, ch. 6; Hauptman, ibid., 88–95, 144–5; Arbuckle, 'John Nicholson', 364–85; Swatzler, *Friend*, xiii, 103–4, 134–44; Wallace, *Death and Rebirth*, 159–83.
93. Quoted in Ganter, ed., *Collected*, 79
94. Quoted in Wallace, ed., 'Halliday', 128–9, 132.
95. Wallace, *Death and Rebirth*, 63–4, 184–92; Wonderley, *At the Font*, ch. 4; Abler, *Cornplanter*, ch. 7.
96. Giddens, *Pennsylvania*, 5; Swatzler, *Friend*, 30, 48, 133.
97. Bates, *Our County*, 8–9, 300, 481, 528–9, 597, 607; Kerrigan, *Johnny*, 44–7.
98. Campisi and Starna, 'On the Road', 479–87; Hauptman, *Conspiracy*, 18–21, 68–9, 76–85.
99. Richardson, ed., *Compilation*, vol. 1, 340; Owens, 'Jeffersonian', 405–35; Horsman, 'American', 47–53.
100. Peterson, ed., *Writings*, 1,117–20.
101. Esarey, ed., *Messages*, vol. 1, 25; Huston, 'Land', 331–6.
102. Billington, *Westward*, 268–89; Friedenberg, *Life*, 298–310.
103. Joy, *American*, 34–8.
104. Stagg, 'Between'; Hauptman, *Conspiracy*, 82–146.
105. Quoted in Hauptman, ibid., 20.
106. Richardson, ed., *Compilation*, vol. 2, 585.
107. Haines, *Considerations*, 7.
108. Ibid., 10, 15, 27, 29.
109. Ibid., 15–16.
110. Ibid., 20.
111. Ibid., 21.
112. Ibid., 17, 30, 32, 35.
113. Ibid., 24.
114. Hauptman, *Conspiracy*, 17, 21, 32, 77–8, 123; Jakle, 'Salt', 702–9; Rezneck, 'Coal', 58–9.
115. Haines, *Considerations*, 24.
116. Callender, 'Early', 111–62; Hauptman, *Conspiracy*, 1–3.
117. Abler, *Cornplanter*, 178–85.
118. Wilkins, *History*, 94–104; Hidy and Hidy, 'Anglo-American', 150–69; Billington, *Westward*, 329–48, 387–402; Emerson, 'Geographic'; Veenendaal, Jr, *Slow*, 49–55.
119. Cave, 'Abuse'; Satz, *American*; Huston, 'Land', 336–41.
120. Richardson, ed., *Compilation*, vol. 2, 520–1.
121. 'Artificial Illumination – Burning Fluids', *Scientific American* (2 January 1858); Lucier, *Scientists*, chs. 5–6; Williamson and Daum, *Illumination*, 49–60; Zallen, *American*, 229–34.

122. Brice, 'Edwin', 16–25; Black, *Petrolia*, 29–32; Vassiliou, *Dictionary*, 536; Lucier, ibid., 204–5; Giddens, *Birth*, 53–9; Williamson and Daum, ibid., 77–81.

3. OILDOM

1. 'Discovery of Subterranean Fountain of Oil', *New-York Daily Tribune* (13 September 1859); Black, *Petrolia*, 37–8, 126–30; Giddens, *Birth*, 60–3.
2. E.g. 'Discovery of an Oil Fountain', *Daily Herald* (Wilmington, NC) (19 September 1859); 'Discovery of a Subterranean Fountain of Oil', *Belmont Chronicle* (Saint Clairsville, OH) (29 September 1859).
3. Giddens, *Birth*, 60–75, 83–91; Clark, *Oil*, 29–30, 42–5; Williamson and Daum, *Illumination*, 100–14.
4. 'Doings in Oildom', *Raftsman's Journal* (Clearfield, PA) (25 April 1860).
5. Miller, 'Fountain', 33–4.
6. 'Explosion of an Oil Well', *Plymouth Weekly Democrat* (Plymouth, IN) (2 May 1861); Giddens, *Pennsylvania*, 213–16; Giddens, *Birth*, 76–8.
7. Miller, *This Was Early*, 19–21, 48–52; Giddens, *Birth*, 76–84; Williamson and Daum, *Illumination*, 113–14.
8. Quoted in Giddens, *Pennsylvania*, 206–7; Sabin, '"A Dive"', 482.
9. Giddens, ibid., 80–3; Clark, *Oil*, 47–8, 101.
10. *Derrick's Hand-Book*, 24.
11. Daintith, *Finders*, chs. 1–3; Black, *Petrolia*, 39–44; Clark, *Oil*, 46, 84–99; Williamson and Daum, *Illumination*, 758–66; Giddens, *Birth*, 63.
12. Sulman, 'Short'; Williamson and Daum, ibid., 83–6, 103–11, 164–79; Giddens, ibid., 101–11; Johnson, *Development*, 2–4.
13. McLaurin, *Sketches*, 314–15; Giddens, ibid., 102–3.
14. Daum, 'Petroleum in Search', 28; Williamson and Daum, *Illumination*, 111–12.
15. Gale, *Wonder*, 54.
16. Wrigley, *Special*, 72.
17. 'The Petroleum Region – The Rock Oil Business', *Scientific American* (22 February 1862).
18. Bell, *History*, 29–30, 71–5, 91–2, 121, 316, 432–3; McLaurin, *Sketches*, 152–3; Cone and Johns, *Petrolia*, 492; Giddens, *Birth*, 169–70.
19. 'Petroleum: The Oil Regions of Pennsylvania', *Morning Post* (London) (24 January 1865).
20. 'Oil City and Pond Freshets', *Venango Spectator* (21 May 1862).
21. 'Fires in the Oil Regions', *New York Times* (8 August 1865).
22. 'Oil', *Chicago Tribune* (15 March 1865).

23. Williamson and Daum, *Illumination*, 106–7, 120–1; Giddens, *Birth*, 83–7; Clark, *Oil*, 70.

24. Giddens, ibid., 91–3; Williamson and Daum, ibid., 109–11, 288–92; Daum, 'Petroleum in Search', 28–30; Hauptman, *Iroquois*, 118; Maybee, *Railroad*, 5–7; Simmons, *Atlantic*, 4.

25. Quoted in Giddens, ibid., 96.

26. 'Atlantic and Great Western Railway', *American Railroad Journal* (25 May 1861); Maybee, *Railroad*, 11–12.

27. *Commercial Relations of the U.S., 1861*, 176; Zallen, *American*, 242–50; Giddens, *Birth*, 194–5; Williamson and Daum, *Illumination*, 309–22; Clark, *Oil*, 106–9.

28. 'Oil Wells in Upper Canada', *Watertown Republican* (Watertown, WI) (30 August 1861); May, *Hard*, 46.

29. Burr, *Canada*, 28–53, 66–74, 87–96; May, ibid., chs. 3–4.

30. 'Money-Market & City Intelligence', *Times* (18 December 1861); Henry, *Thirty-Five*, 5; Williamson and Daum, *Illumination*, 324.

31. 'Money-Market & City Intelligence', *Times* (3 July 1862); Williamson and Daum, ibid., 329–30; Maybee, *Railroad*, 245–9; Hidy and Hidy, *Pioneering*, 129–30; Hidy, 'Government', 82, 90; Martell, 'On the Carriage', 6–7; Butt, 'Scottish'.

32. Lee, *Burning Springs*, chs. 2–4; White, 'Petroleum', 146–7; Whiteshot, *Oil-Well*, 67–8; Nuttall, 'Of Kentucky', 74–8; Zallen, *American*, 236–42.

33. Scott, ed., *War of the Rebellion*, ser. 1, vol. 25, pt. 1, 120.

34. Maybee, *Railroad*, 11.

35. Ibid., 5–17; Wilkins, *History*, 105–6; Giddens, *Birth*, 111–13; Williamson and Daum, *Illumination*, 170–8; Simmons, *Atlantic*, 4.

36. Quoted in Morris, *Derrick*, 50.

37. Williamson and Daum, *Illumination*, 322–31; Zallen, *American*, 242–55; Daum, 'Petroleum in Search', 32–4; Black, *Crude*, 31–3; Dolson, *They Struck*, 85–9; Giddens, *Birth*, 98; Miller, *This Was Early*, 57.

38. 'A Good Time Coming for Whales', *Littell's Living Age* 66 (September 1860), 810–12; York, 'Why Petroleum'.

39. 'Petroleum: The Oil Regions of Pennsylvania', *Morning Post* (London) (24 January 1865).

40. Quoted in Giddens, *Pennsylvania*, 223.

41. Quoted in Giddens, *Birth*, 99.

42. *Commercial Relations of the U.S., 1865*, 430.

43. Ibid., 122; ibid., *1866*, 223.

44. Ibid., *1866*, 407.

45. Ibid., 422–3.

46. Quoted in Clark, *Oil*, 84–5.
47. Gale, *Wonder*, 44–5.
48. Giddens, *Birth*, 96–100.
49. Quoted in Smith, *Life*, vol. 2, 822; Giddens, ibid., 121–6; Peskin, *Garfield*, 248–50, 267–9.
50. 'How Oil was Discovered', *Morning Herald* (Titusville, PA) (27 July 1866).
51. Peto, *Resources*, 205–8; Giddens, *Birth*, 195–6.
52. Wlasiuk, *Refining*, 14–15; Whitten, *Emergence*, 19; Chernow, *Titan*, 45, 71, 111; Bogart, 'Early', 56.
53. Quoted in Chernow, ibid., 47.
54. Ibid., 6–26, 60, 68–72; 'Co-Partnership Notice', *Daily Leader* (Cleveland, OH) (18 March 1859).
55. Sulman, 'Short', 55–69; Maybee, *Railroad*, 18–22; Clark, *Oil*, 69.
56. Maybee, ibid., 223–4; Clark, ibid., 110–12; Chernow, *Titan*, 77–8, 83–8, 111–12; Williamson and Daum, *Illumination*, 172–4, 291–300, 321; Johnson, *Development*, 15; 'Excelsior Oil Works', *Daily Leader* (Cleveland, OH) (21 February 1865); 'Rockefeller Tells of Oil Trust Start', *New York Times* (19 November 1908).
57. Chernow, ibid., 78–9, 100–5, 150, 161, 168–9, 202–3; Montague, 'Rise', 267–8, 277–9; Williamson and Daum, *Illumination*, 178–80, 344–5; Hidy and Hidy, *Pioneering*, 127; Rockefeller, *Random*, 17–20.
58. Williamson and Daum, ibid., 117–20; Black, *Petrolia*, 113–15, 148–9; Clark, *Oil*, 70.
59. Zheng and Palmer, 'Bamboo'; Jung and Fang, 'Account', 230–2.
60. Williamson and Daum, *Illumination*, 170–2, 183–4; Johnson, *Development*, 4–6; Clark, *Oil*, 78–81; Maybee, *Railroad*, 175–8.
61. Quoted in Johnson, ibid., 6.
62. Williamson and Daum, *Illumination*, 122–3; Black, *Petrolia*, 149–50, 163–6.
63. Peto, *Resources*, 203; Johnson, *Development*, 2.
64. Johnson, ibid., 6–8; Williamson and Daum, *Illumination*, 140–1, 184–5; Giddens, *Birth*, 143–7; Redwood, *Petroleum*, vol. 2, 656.
65. Quoted in Johnson, ibid., 8.
66. Ibid., 9–10; Clark, *Oil*, 79–81; Chernow, *Titan*, 110; 'Incendiary Mob at Shaffer', *Morning Herald* (Titusville, PA) (21 April 1866).
67. Giddens, *Birth*, 145; Johnson, *Development*, 9–10; Williamson and Daum, *Illumination*, 185.
68. Johnson, ibid., 10–12; Williamson and Daum, ibid., 185–9; Clark, *Oil*, 119–20; Black, *Petrolia*, 166.

69. *Derrick's Hand-Book*, 84; Giddens, *Birth*, 141, 151-2; Maybee, *Railroad*, 240; Williamson and Daum, ibid., 180-2; Clark, ibid., 83.

70. 'A Week on Oil Creek', *Scientific American* (1 September 1866).

71. 'Petroleum', *New York Times* (28 October 1865).

72. 'The Oil Wells of Pennsylvania', *Morning Leader* (Cleveland, OH) (11 October 1864).

73. 'The Oil Trade – The Price and Supply to be Regulated', *Pittsburgh Gazette* (28 November 1861).

74. Daintith, *Finders*, chs. 1-3; Black, *Petrolia*, 39-44; Clark, *Oil*, 46, 73-8, 84-99; Williamson and Daum, *Illumination*, 160-3, 351, 357-9, 758-66; Chernow, *Titan*, 129-30.

75. 'Destruction of Oil Boats', *Scientific American* (27 December 1862).

76. Black, *Petrolia*, 87-91, 102-3, 122-3; Giddens, *Birth*, 103-9; Cone and Johns, *Petrolia*, 528-39.

77. 'A Week on Oil Creek', *Scientific American* (1 September 1866).

78. 'Fires in the Oil Regions', *New York Times* (8 August 1865).

79. Black, *Petrolia*, 102-3, 121-3, 167; McLaurin, *Sketches*, 437-41; Williamson and Daum, *Illumination*, 192-3; 'The Great Oil Fires of 1865-66', *Buffalo Daily Courier* (14 November 1866).

80. Peto, *Resources*, 194; Black, *Petrolia*, 160.

81. *Derrick's Hand-Book*, 376-7.

82. 'Another Nitro-Glycerine Horror', *Titusville Morning Herald* (20 May 1871); Black, *Petrolia*, 77-8; Williamson and Daum, *Illumination*, 149-56; Henry, *Early*, 251-4; McLaurin, *Sketches*, ch. 17; May, *Hard*, 86-92.

83. Clark, *Oil*, 81-3; Williamson and Daum, ibid., 182-3, 192-4; Wlasiuk, *Refining*, 28-35.

84. Quoted in Chernow, *Titan*, 101; Wlasiuk, ibid., 7-9, 36-9.

85. Quoted in Chernow, ibid.; Wlasiuk, ibid., 8, ch. 3.

86. Quoted in Wlasiuk, ibid., 62.

87. Ibid., chs. 3-4.

88. 'Awful Conflagration', *Philadelphia Inquirer* (9 February 1865).

89. 'Great Fire in Philadelphia', *American Citizen* (Butler, PA) (15 February 1865).

90. 'The Great Fire in Philadelphia – Further Particulars', *Buffalo Courier* (16 February 1865).

91. 'The Petroleum Disaster', *Philadelphia Inquirer* (9 February 1865).

92. 'Awful Petroleum Fire in Antwerp', *Cleveland Daily Leader* (30 August 1866).

93. 'The Dangers in Storing Petroleum', *Scientific American* (15 December 1866).

94. Williamson and Daum, *Illumination*, 124–5.
95. 'Petroleum', *Titusville Morning Herald* (2 May 1866).
96. Clark, *Oil*, 54, 69–77, 95–101; Williamson and Daum, *Illumination*, 121–9; Giddens, *Birth*, 78–80, 85–7, 183; Black, *Petrolia*, 173–4.
97. 'Petroleum: The Oil Regions of Pennsylvania', *Morning Post* (London) (24 January 1865).
98. Quoted in Peto, *Resources*, 195.
99. 'Monthly Petroleum Report', *Titusville Morning Herald* (12 July 1870).
100. Williamson and Daum, *Illumination*, 386–93, 411; Brown and Partridge, 'Death of a Market', 571–2; Johnson, *Development*, 138.
101. *Derrick's Hand-Book*, 346; Williamson and Daum, ibid., 390–3; 'Oil Scouts', *Daily Argus* (Rock Island, IL) (19 October 1882).
102. Giddens, *Birth*, 183–7; Hornsell, *Oil*, 226–8.
103. Williamson and Daum, *Illumination*, 102–3; Giddens, ibid., 182–91; Maybee, *Railroad*, 206; Henry, *Early*, 279–82.
104. 'Oil City as It Was and Is', *Titusville Morning Herald* (11 July 1870); Clark, *Oil*, 106–7.
105. Wrigley, *Special*, 72; Williamson and Daum, *Illumination*, 488–9.

4. CORPORATE CONTROL

1. Roy, *Socializing*, ch. 4; Chandler, Jr, *Scale*, 51–62.
2. Roy, ibid., 129–40; Whitten, *Emergence*, 11–56; Maybee, *Railroad*, 103–11, 286; Porter, *Rise*, 27–54; Wilkins, *History*, 94–123; Chernow, *Titan*, 113–17; Clark, Jr, *Railroads*; Ward, *Railroads*, 41–55, 136–70; Montague, 'Rise', 276–9.
3. Quoted in Nicolay, *Personal*, 380–1.
4. Peterson, *Lincoln*, 160.
5. Bergeron, ed., *Papers*, vol. 12, 111, 113.
6. Quoted in Miller, *New History*, 293.
7. Berthoff, *Republic*, 160–8, 197–215; Roy, *Socializing*, 110–14; McCurdy, 'Justice'; Purcell, Jr, 'Ideas'; Destler, 'Opposition'.
8. Williamson and Daum, *Illumination*, 194–201, 297–346; Maybee, *Railroad*, 402–4; Chernow, *Titan*, 113–17, 135; Giddens, *Birth*, 153; Montague, 'Rise', 268–71; Clark, *Oil*, 112.
9. Quoted in Maybee, ibid., 141.
10. Williamson and Daum, *Illumination*, 173, 198; Maybee, ibid., 36–7; Johnson, *Development*, 3–14; Giddens, *Birth*, 145–9.
11. Giddens, ibid., 147; Williamson and Daum, ibid., 173–8, 189; Maybee, ibid., 120, 155, 179, 223–4; Johnson, ibid., 16, 28.

12. Johnson, ibid., 14–16; Williamson and Daum, ibid., 300–5; Chernow, *Titan*, 113.

13. Quoted in Maybee, *Railroad*, 263; ibid., 221–4, 263–6.

14. Quoted in Williamson and Daum, *Illumination*, 376; ibid., 296, 364–6; Maybee, ibid., 155, 179–81, 353, 375; Johnson, *Development*, 15–16; Montague, 'Rise', 273.

15. Quoted in Chernow, *Titan*, 130–2, 148–9; Williamson and Daum, ibid., 302–3, 367.

16. Quoted in Williamson and Daum, ibid., 354.

17. Granitz and Klein, 'Monopolization', 1–27; Klein, '"Hub-and-Spoke"'; Crane, 'Were Standard'; Priest, 'Rethinking'; Maybee, *Railroad*, 286–357; Johnson, *Development*, 17–25, 192–4; Williamson and Daum, ibid., 346–68; Chernow, *Titan*, 129–68; Helfman, 'Twenty-Nine'.

18. Quoted in Chernow, ibid., 130.

19. Quoted in Williamson and Daum, *Illumination*, 403; ibid., 431–3; Johnson, *Development*, 82–6, 113–14; Chernow, ibid., 208–14.

20. Quoted in Williamson and Daum, ibid., 371; ibid., 131–5.

21. Quoted in Johnson, *Development*, 53.

22. Johnson, ibid., 26–31; Williamson and Daum, *Illumination*, 396–405.

23. Johnson, ibid., 31–55; Williamson and Daum, ibid., 405–12.

24. Johnson, ibid., 49–58; Williamson and Daum, ibid., 412–21.

25. Quoted in Williamson and Daum, ibid., 423.

26. Ibid., 423–9; Johnson, *Development*, 62–5.

27. Quoted in Chernow, *Titan*, 203.

28. Quoted in Johnson, *Development*, 66.

29. Williamson and Daum, *Illumination*, 463–6; Hidy and Hidy, *Pioneering*, 205, 213; Chernow, *Titan*, 207–9, 261–2.

30. Quoted in Williamson and Daum, ibid., 383.

31. Ibid., 383–90, 432.

32. Ibid., 577, 599, 619–20; Hidy and Hidy, *Pioneering*, 87–8, 278–82; Brown and Partridge, 'Death of a Market', 572; 'Oil Producers to Unite', *New York Times* (12 February 1890).

33. Quoted in Chernow, *Titan*, 207.

34. Williamson and Daum, *Illumination*, 438–40; Johnson, *Development*, 70–4; Montague, 'Later', 294–5.

35. Williamson and Daum, ibid., 431–3, 440–1; Johnson, ibid., 76–81.

36. Quoted in Chernow, *Titan*, 209.

37. Williamson and Daum, *Illumination*, 438–44; Johnson, *Development*, 74–6, 111; Chernow, ibid., 207–12.

38. Quoted in Johnson, ibid., 77.

39. Ibid., 114–21.
40. 'The Free Pipe Line', *The New Era* (Lancaster, PA) (21 April 1883).
41. 'Truth and Oil Both Leaking Out', *The New Era* (Lancaster, PA) (9 June 1883).
42. U.S. Congress: *Hearings Before the Industrial Commission*, 660; Johnson, *Development*, 114.
43. Quoted in Johnson, ibid., 96.
44. Quoted in ibid.
45. Quoted in ibid., 95.
46. Quoted in Chernow, *Titan*, 211.
47. Johnson, *Development*, 100–7, 123–6; Williamson and Daum, *Illumination*, 448–51; 'The Salamanca Pipe Line', *Buffalo Express* (26 July 1880); 'The Oil Monopoly's Hand', *New York Times* (24 January 1882).
48. Quoted in Johnson, ibid., 111–12.
49. Quoted in ibid., 97.
50. Quoted in U.S. Congress: *Trusts*, 355–6; Clark, *Oil*, 116–19.
51. 'The Oil Octopus', *The Daily Graphic* (New York) (4 February 1879).
52. Lloyd, 'Story'.
53. Quoted in Hidy and Hidy, *Pioneering*, 213.
54. Quoted in Chernow, *Titan*, 262.
55. Ibid., 261.
56. Huston, 'American'.
57. Roy, *Socializing*, chs. 4–5.
58. Ibid., 51–5; Maier, 'Revolutionary'; Barkan, *Corporate*.
59. Quoted in Bilder, 'Corporate', 522.
60. Quoted in ibid., 514 and 530 respectively.
61. Lamoreaux and Novak, 'Corporations', 6–7.
62. Sayles, *Medieval*, 181–7, 439–42; Morton, *People*, 62, 68–74; Harrison, *Common*, 56–64; Platt, *English*, ch. 5.
63. Dent, '"Generally"'.
64. Harris, 'Trading'; Jha, 'Financial'; Brenner, *Merchants*.
65. Stern, *Company*.
66. Davis, *Essays*, vol. 1, Essay I; Klein, 'Voluntary'; Roy, *Socializing*, ch. 3; Maier, 'Revolutionary'; Davis, Jr, 'Corporate', 605–10; Seavoy, *Origins*, chs. 1–3; 'Incorporating the Republic'; Wilentz, 'America'; Nelson, *Royalist*.
67. Reid, 'America'.
68. Quoted in Davis, 'Corporate', 605.
69. Quoted in 'Incorporating the Republic', 1,898.

70. Griswold, 'Agrarian', 668–72, 681; Seavoy, *Origins*, 53–68, 73–6; Davis, 'Corporate', 607–12; Dickinson, 'Partners', 571–4; Kessler, 'Statistical'.

71. Quoted in Davis, ibid., 608–9.

72. Seavoy, *Origins*, 56–76; Blumberg, 'Limited', 590–4; Dodd, *American*, 84–93, ch. 5.

73. Ford, *Works*, vol. 12, 44.

74. Denham, Jr, 'Historical'; McLaughlin, 'Court'.

75. Quoted in Grossman and Adams, 'Taking', 64.

76. Quoted in Wilgus, 'Need', 364.

77. Roy, *Socializing*, chs. 4–5.

78. Quoted in Seligman, 'Brief', 259.

79. Quoted in 'Incorporating the Republic', 1,901.

80. Quoted in ibid., 1,896.

81. Agg, *Proceedings*, vol. 1, 58.

82. Quoted in Maier, 'Revolutionary', 59–60.

83. Quoted in Seligman, 'Brief', 257; Thompson, 'Radical'.

84. Hamill, 'From Special', 97–105; Davis, 'Corporate', 610–15; Hovenkamp, 'Classical', 1,634–40; Lamoreaux and Novak, 'Corporations', 10–13; Horwitz, 'Santa Clara', 20–1; Taylor, *Creating*.

85. Hinsdale, ed., *Works*, vol. 2, 61.

86. Roy, *Socializing*, 83–97.

87. Quoted in Johnson, *Development*, 117; Williamson and Daum, *Illumination*, 439.

88. Williamson and Daum, ibid., 466–70; Freyer, *Regulating*, 84–5; Montague, 'Later', 298–9; Hidy and Hidy, *Pioneering*, 40–9.

89. Lloyd, 'Story'; Johnson, *Development*, 112.

90. Quoted in Johnson, ibid., 114.

91. Horwitz, 'Santa Clara'; Bloch and Lamoreaux, 'Corporations'; Hartmann, *Unequal*, 95–119; Arnold, *Folklore*, chs. 8–9.

92. Quoted in Wilgus, 'Need', 371.

93. Hovenkamp, 'Classical', 1,672–81.

94. Quoted in Johnson, *Development*, 195.

95. Quoted in Chernow, *Titan*, 149–50.

96. Prettyman, 'Gilded'; Quint, *Forging*, ch. 3.

97. Roy, *Socializing*.

98. Bringhurst, *Antitrust*, 102–12; Johnson, *Development*, 93, 121–2, 140–62, 219–42; Morris, 'Sheep', 99–103; Purcell, Jr, 'Ideas'.

99. *Congressional Record*, 21 March 1890, 2457, 2460. Briareus being a giant, in Greek mythology, with a hundred arms.

100. Peritz, '"Rule of Reason"', 292–7; Hazlett, 'Legislative'; May, 'Antitrust', 262–309.
101. *Congressional Record*, 21 March 1890, 2460.
102. Ibid., 25 March 1890, 2598; Dickson and Wells, 'Dubious'; Gordon, 'Attitudes'; Letwin, 'Congress', 250.
103. Quoted in Chernow, *Titan*, 298.
104. Seligman, 'Brief', 264; Bringhurst, *Antitrust*, 12–17.
105. Quoted in Bringhurst, ibid., 16.
106. Roy, *Socializing*, 15–16, 148–54, 198–203, 213, 220, 262; Grandy, 'New Jersey'; Davis, 'Corporate', 603, 615–19; Winkler, '"Other People's Money"', 906–12; Johnson, *Development*, 184, 196; Hidy and Hidy, *Pioneering*, 201–32, 305–13, 323–32; Bringhurst, *Antitrust*, 32–3, 108–9; Chernow, *Titan*, 332–3.
107. Steffens, 'New Jersey'.
108. Quoted in Hidy and Hidy, *Pioneering*, 214, and in Chernow, *Titan*, 295.
109. *Congressional Record*, 21 March 1890, 2461.
110. Roy, *Socializing*, 210–20; McCurdy, 'Knight'.
111. Quoted in Roy, ibid., 220.
112. Freyer, 'Sherman'; Bringhurst, *Antitrust*, 157–79; Weinstein, *Corporate*, 62–70.
113. Roy, *Socializing*, 100–14, chs. 5–9; North, 'Life'; Winkler, '"Other People's"'.
114. Quoted in Thomas, 'Fifty', 271.
115. Hidy and Hidy, *Pioneering*, 77–89.
116. Quoted in ibid., 84.
117. Ibid., 155–68; Williamson and Daum, *Illumination*, ch. 22.
118. Gale, op. cit., ch. 3, section 'Wartime Demand and Destruction'.
119. Quoted in Williamson and Daum, *Illumination*, 602.
120. Ibid., 558–62; Johnson, *Development*, 130–7, 164–70.
121. Quoted in Johnson, ibid., 137; Nevins, *John D.*, vol. 2, 110–11; Hidy and Hidy, *Pioneering*, 205, 213.
122. Quoted in Williamson and Daum, *Illumination*, 562.
123. Quoted in ibid., 564–5; Johnson, *Development*, 137–40; Hidy and Hidy, *Pioneering*, 176–88; 'Oil Producers Combining', *The Sun* (New York) (2 August 1887).
124. Williamson and Daum, *Illumination*, 596–629; Johnson, ibid., 162–3; Johnson, *Petroleum*, 4–14.
125. Johnson, *Development*, 173; Williamson and Daum, ibid., 569–71.
126. Williamson and Daum, ibid., 571–3.
127. Quoted in Johnson, *Development*, 175.

128. Williamson and Daum, *Illumination*, 583–5.
129. Quoted in ibid., 176; Hidy and Hidy, *Pioneering*, 169–88, 281–2.
130. Yergin, *Prize*, 53–4, 249–69, 402–3, 512–40, 567–8, 578–652.
131. Williamson and Daum, *Illumination*, 569–76; Johnson, *Development*, 173–83; Hidy and Hidy, *Pioneering*, 278–82.
132. Williamson and Daum, ibid., 576–81.
133. Ibid., 471, 491–2, 614, 663, 689–90, 740–57; Hidy and Hidy, ibid., 254–7, 282–302; Cochran, *Encountering*, 12–24.
134. Quoted in Hidy and Hidy, ibid., 122–3.

5. A GLOBAL INDUSTRY

1. White, *Formative*, 2–4.
2. Cleland, 'Early'; Ruiz, 'American'.
3. Madley, *American*; Lindsay, *Murder*.
4. Quoted Madley, ibid., 173, 353.
5. Lucier, *Scientists*, 278–87.
6. Quoted in ibid., 284–5.
7. Quoted in ibid., 287.
8. Ibid., 290–311; White, *Formative*, 4–22.
9. Larkin, *Report*, 7, 11–13.
10. Ibid., 13–17; Pinelo, 'Nationalization', 7–9; 'Petroleum Found in Peru', *New York Times* (30 September 1879); Henry, *Early*, 143, 350; Thorp and Bertram, *Peru*, 95–9; Bosworth, *Geology*, 345–7; Lossio, 'Del copey', 19–20.
11. Gerali, 'Brief', 245; Gerali and Riguzzi, 'Los inicios', 66–7.
12. Murphy, *Petroleum*, 3–4, also 24; Ministerio de Fomento, *Memoria*, 355–9; Gerali and Riguzzi, 'Entender'; Miller, 'Arms'.
13. Murphy, ibid., 10, 12.
14. Ibid., 21.
15. Ibid., 10–11, 22; Gerali and Riguzzi, 'Entender', 4–5.
16. Quoted in Schoonover, 'Dollars', 39.
17. Quoted in Garner, *British*, 138 and 164 (n. 1).
18. U.S. Congress: *Evacuation of Mexico*, 47; Chynoweth, *Fall*, 54–5.
19. Gerali and Riguzzi, 'Entender', 13.
20. 'Mexican Petroleum', *New York Times* (11 August 1867).
21. *Commercial Relations of the U.S., 1868*, 652–3; Schoonover, 'Dollars', 40; Murphy, *Petroleum*, p. 15.
22. Gerali, 'Brief', 248–50; Ryan, Jr, 'History', 48–50.

23. Murphy, *Petroleum*, 28; Brown, *Oil*, 17–18; Williamson and Daum, *Illumination*, 543; Hidy and Hidy, *Pioneering*, 128.

24. Quoted in Higgins, *History*, 46.

25. Quoted in Hughes, *Energy*, 56; Higgins, ibid., 41–5.

26. 'Trinidad Petroleum', 764.

27. Kingsley, *At Last*, 235.

28. Schatzker and Hirszhaut, *Jewish*, 37–8, 45, 49.

29. Quoted in ibid., 37–8.

30. Ibid., chs. 4–5; Frank, *Oil*, 1–108; Forbes, *More Studies*, 94–9; Daintith, *Finders*, 142–5; Henry, *Early*, 148–52.

31. Forbes, *More Studies*, 99–107; Pearton, *Oil*, 7–9, 16–17; Buzatu, *History*, 37–9; Daintith, ibid., 149–53; Gerali and Gregory, 'Understanding'; *Commercial Relations of the U.S., 1865*, 477.

32. Gerali, 'Oil Research', 202–9; Gerali, 'Scientific', 99–104.

33. Fairman, *Treatise*, v–vi.

34. 'The Zante Petroleum Company', *Times* (29 April 1865).

35. Monteiro, *Angola*, vol. 1, 150–1; ibid., vol. 2, 64.

36. Ibid., vol. 2, 8–13.

37. Quoted in Noetling, 'Occurrence', 25; Corley, *History*, 10–15.

38. Longmuir, *Oil*, 50–109; Henry, *Early*, 154–60; Noetling, ibid., 25, 216–18; Hereward Holland, 'In Myanmar, China's Scramble for Energy Threatens Livelihoods of Villagers', *National Geographic* (5 September 2014); Saikia, 'Imperialism', 50–1.

39. Dodd, *Journal*, 209–10; Redwood, *Petroleum*, vol. 1, 74; ibid., vol. 2, 365–6.

40. Davidson, *Island*, 493–5; Giddens, ed., 'China'.

41. Giddens, ibid., 33–4.

42. Quoted in Gerretson, *History*, vol. 1, 29, 31, 38; Poley, *Eroïca*, 34–5, 52–3.

43. Quoted in Gerretson, ibid., 38–9; Poley, ibid., 53, 62.

44. Gerretson, ibid., 43–6; Poley, ibid., 63, 65–76; Burr, *Canada*, 158–9.

45. Quoted in Matveichuk, 'Peter'; Alekperov, *Oil*, 16–17; Henry, *Baku*, 21; Marvin, *Region*, 164; Thompson, *Oil Fields*, 397–9; Vassiliou, *Dictionary*, 554.

46. Hanway, op. cit., ch. 1, section 'Middle East and Central Asia'; Searight, *British*, 59–61.

47. Allen and Muratoff, *Caucasian*, 17–21; Atkin, *Russia*, chs. 3–10; Swietochowski, *Russia*, 1–5; Mostashari, *On the Religious*, 12–18; Sahni, *Crucifying*, 37–9; Ingram, *Britain*, 74–6.

48. Quoted in Atkin, ibid., 75–6.
49. Henry, *Baku*, 24; Forbes, *Studies*, 162.
50. Quoted in Wright, *English*, 150; Ingram, *Britain*, 2–3, 15–16, 23–41.
51. Kinneir, *Geographical*, 359–60.
52. Kelly and Kano, 'Crude Oil', 309, 315–17; McKay, 'Entrepreneurship', 49; Alekperov, *Oil*, 29–32.
53. Alekperov, ibid., 37–9; Schaefer, *Insurgency*, 58–61.
54. Quoted in Sahni, *Crucifying*, 41; Seton-Watson, *Russian*, 183.
55. Quoted in Gammer, 'Russian', 47.
56. Bennigsen Broxup, 'Introduction', 4; Henze, 'Circassian', 65; Gammer, ibid., 49; Zelkina, 'Jihad'.
57. Quoted in Zelkina, ibid., 254.
58. Quoted in Sahni, *Crucifying*, 38; Seely, *Russo-Chechen*, 19–40.
59. Atkin, *Russia*, 157–60; Mostashari, *On the Religious*, 23–5; Swieto-chowski, *Russia*, 5–7; Seton-Watson, *Russian*, 289–92; Hopkirk, *Great*; Clayton, *Britain*, 54–8; Anderson, *Eastern*, 72–3.
60. Crawley, 'Anglo-Russian'; Al-Sayyid Marsot, *History*, 74–5; Clayton, ibid., 66–76; Hopkirk, ibid., 69–76, 86–7, 149–61, 502–7; Hopkins, 'Growth', 172; Kelly, *Britain*; Bailey, 'Economics'; Greaves, *Persia*, ch. 2; Ingram, *Britain*, 15–26, 39–41; Atkin, ibid., 32–44; Yapp, 'British', 647–65; Marvin, *Region*, 112–56; Ingram, 'Great Britain', 160–71.
61. Alekperov, *Oil*, 32–7.
62. Quoted in Henze, 'Circassian', 78; Ingram, *Britain*, 75.
63. Ditson, *Circassia*, 275.
64. Seely, *Russo-Chechen*, 36–52.
65. Quoted in Alekperov, *Oil*, 39.
66. Quoted in Gammer, 'Russian', 54; Seton-Watson, *Russian*, 292–3.
67. Alekperov, *Oil*, 43–4, 54.
68. Rawlinson, *England*, 69; Seely, *Russo-Chechen*, 52–9.
69. Anderson, *Eastern*, 128–36; Clayton, *Britain*, 105–11; Alekperov, *Oil*, 43.
70. Quoted in Henze, 'Circassian', 81.
71. Henze, ibid., 87–97; Seton-Watson, *Russian*, 319–31; Anderson, *Eastern*, 139–41; Clayton, *Britain*, 112–13, 126; Brock, 'Fall', 406; Clark, *Sleepwalkers*, 136.
72. Henze, ibid., 91–9; Mostashari, *On the Religious*, 57–8.
73. Quoted in Henze, ibid., 102.
74. Mostashari, *On the Religious*, 37–46; Henze, ibid., 99–104; Atkin, *Russia*, 149–50, 158; Swietochowski, *Russia*, 10–12.

75. Henze, ibid., 104–5; Clayton, *Britain*, 139–48; Crawley, 'Anglo-Russian', 67; Williams, 'Approach', 216–35; Issawi, 'Tabriz', 22–4.

76. Kelly and Kano, 'Crude Oil', 309–17; McKay, 'Entrepreneurship', 50–2; Kelly, 'Crisis', 292–5; see Mining Corps in ch. 1, section 'Europe' above; Alekperov, *Oil*, 46–9, 74–6; Tolf, *Russian*, 44; Henry, *Baku*, 32–3.

77. Westwood, *History*, 45, 61; Henry, ibid., 112–13; McKay, ibid., 51; Alekperov, *Oil*, 44–6.

78. Quoted in 'Russian Oil Industry Celebrates 150th Anniversary!', *МАСЛА@ЛУКОЙЛ* (*Masla@LUKOIL*) (September 2016), 3; Alekperov, *Oil*, 61–3.

79. *Commercial Relations of the U.S., 1866*, 299.

80. Quoted in Alekperov, *Oil*, 63.

81. Ibid., 65–70.

82. Stewart, 'Account', 311–12; Shapira, 'Karaite', 105; Marvin, *Region*, 160–6.

83. McKay, 'Entrepreneurship', 51–5; McKay, 'Baku', 606–7; McKay, 'Restructuring', 87; Alekperov, *Oil*, 54–6, 74–9, 81–2; Tolf, *Russian*, 44–51; Henry, *Baku*, 38, 72.

84. Thompson, *Oil Fields*, 303.

85. General Staff, War Office, 'Military Report on Trans-Caucasia' (1907), in Burdett, ed., *Oil*, vol. 1, 229.

86. Quoted in McKay, 'Entrepreneurship', 56; Alekperov, *Oil*, 79–81, 101–3; McKay, 'Restructuring', 87; McKay, 'Baku Oil', 607.

87. Arnold, *Through*, vol. 1, 130.

88. Ibid., 133–4.

89. Quoted in McKay, 'Entrepreneurship', 58.

90. Henry, *Baku*, 56–7, 62, 71–8; McKay, ibid., 57–8; 'Russia's Naphtha Industry', *Economist*, Monthly Trade Supplement (15 February 1890), 7–8; Alekperov, *Oil*, 89–93.

91. Henry, *Baku*, 37.

92. McKay, 'Entrepreneurship', 58–68; Tolf, *Russian*, 53, 69–71; Alekperov, *Oil*, 89–95; Hidy, 'Government', 82–3; Marvin, *Region*, 251–74.

93. Tolf, ibid., 36, 45–6; Alekperov, ibid., 84–6; McKay, ibid., 63–4; Marvin, ibid., 280–1; Fursenko, *Battle*, 12–14; Fursenko, 'Oil', 445–8.

94. McKay, ibid., 64–83; McKay, 'Baku Oil', 610–23; Alekperov, ibid., 86–7; Marvin, ibid., 197–202; Henry, *Baku*, 72–4, 88–9.

95. Quoted in McKay, 'Entrepreneurship', 67; Tolf, *Russian*, 79.

96. Martell, 'On the Carriage', 5–8; Williamson and Daum, *Illumination*, 329–31, 637–8; Henry, *Thirty-Five*, 6–8; Redwood, *Petroleum*, vol. 2,

679–80; Henry, *Baku*, 224–5; Gerretson, *History*, vol. 1, 207–9; Hidy and Hidy, *Pioneering*, 145.

97. McKay, 'Entrepreneurship', 68–72; McKay, 'Baku Oil', 608; McKay, 'Restructuring', 88; Tolf, *Russian*, 54–67, 79; Kelly and Kano, 'Crude Oil', 312–13; Williamson and Daum, ibid., 329–30, 511–18, 637; Henry, *Baku*, 78, 224–5; Marvin, *Region*, 288–90, 299–306; Martell, 'On the Carriage', 16.

98. McKay, 'Entrepreneurship', 73–7; Tolf, ibid., 98; Williamson and Daum, ibid., 517.

99. Stewart, 'Account', 312. The temple was found to be in a similar condition and use over twenty-five years later (Jackson, *From Constantinople*, vii, 56–7).

100. Marvin, *Region*, 196; Thompson, *Oil Fields*, 385; Hidy and Hidy, *Pioneering*, 132–3.

101. Osmaston, *Old*, 249–50; McKay, 'Baku Oil', 609; Marvin, ibid., 191–2, 218–20; Tolf, *Russian*, 100.

102. Marvin, ibid., 203; McKay, 'Restructuring', 89; Williamson and Daum, *Illumination*, 512–13.

103. Tolf, *Russian*, 57–9, 98; McKay, 'Baku Oil', 622–3.

104. Marvin, *Region*, 139, 305–8; McKay, 'Entrepreneurship', 60; McKay, 'Baku Oil', 609–10; McKay, 'Restructuring', 92; Williamson and Daum, *Illumination*, 635–6.

105. Tolf, *Russian*, 84–7; Alekperov, *Oil*, 110–11; Gerretson, *History*, vol. 1, 213; ibid., vol. 3, 39; Henry, *Baku*, 113–14; Hidy and Hidy, *Pioneering*, 131; Williamson and Daum, ibid., 636; Henriques, *Marcus*, 73.

106. Quoted in Peckham, *Report*, 153; Williamson and Daum, ibid., 518–19.

107. Marvin, *Region*, ix, 209, 246.

108. McKay, 'Entrepreneurship', 78–86; Henry, *Baku*, 113–14; Tolf, *Russian*, 96; Fursenko, *Battle*, 17–21, 25–6; Fursenko, 'Oil', 448–9.

109. Quoted in Alekperov, *Oil*, 111.

110. McKay, 'Baku Oil', 610; McKay, 'Entrepreneurship', 78–86.

111. Quoted in McKay, 'Baku Oil', 611; Hidy and Hidy, *Pioneering*, 144; Williamson and Daum, *Illumination*, 518.

112. Marvin, *Region*, 318; McKay, ibid., 613–16; McKay, 'Restructuring', 91; Tolf, *Russian*, 97.

113. Quoted in McKay, 'Baku Oil', 613.

114. Quoted in ibid., 611.

115. Ibid.; Gerretson, *History*, vol. 3, 272; Hidy and Hidy, *Pioneering*, 144.

116. Quoted in McKay, ibid., 612.

117. Quoted in ibid., 616.
118. Martell, 'On the Carriage', 18.
119. McKay, 'Baku Oil', 616–18; McKay, 'Restructuring', 90–2; Marvin, *Region*, 339–44; Henry, *Baku*, 116–17; 'The Russian Kerosene Pipe Line', *New York Times*, 9 July 1889; Curzon, *Persia*, vol. 1, 64–5; Tolf, *Russian*, 96–7; Williamson and Daum, *Illumination*, 635; Foreign Office, *Diplomatic and Consular Reports: Batoum, 1889*, 6.
120. Quoted in McKay, 'Baku Oil', 617–18; Hidy and Hidy, *Pioneering*, 137–8.
121. Williamson and Daum, *Illumination*, 635–7; Gerretson, *History*, vol. 1, 213–14; ibid., vol. 2, 29–35; Fursenko, *Battle*, 44.
122. Martell, 'On the Carriage', 6–15; Henry, *Baku*, 225–7, 232–4; Williamson and Daum, ibid., 641, 648, 658, 662; Tolf, *Russian*, 57–9, 93, 105; Gerretson, ibid., vol. 2, 103–4, 113–14; Henriques, *Marcus*, 75–6; Hidy and Hidy, *Pioneering*, 144; Foreign Office, *Diplomatic and Consular Reports: Batoum, 1889*, 5; ibid., *1892*, 2–3.
123. 'Terrible Explosion at Calais', *Times* (18 October 1888); 'Fatal Explosion', *The Times* (27 December 1886); Henry, *Thirty-Five*, 11, 15, ch. 10; Henry, *Baku*, 226–7; Redwood, *Petroleum*, vol. 2, 696–710; Henriques, *Marcus*, 76; *Report from the Select Committee on Petroleum*, apps. 5–6; Little, *Marine*, 19–24.
124. Williamson and Daum, *Illumination*, 633; Hidy and Hidy, *Pioneering*, 124.
125. Martell, 'On the Carriage', 11.
126. Williamson and Daum, *Illumination*, 638–43; Tolf, *Russian*, 59; Martell, ibid., 6–15; Hidy and Hidy, *Pioneering*, 145–6.
127. Gerretson, *History*, vol. 2, 103.
128. Williamson and Daum, *Illumination*, 493–4, 642, 648–9, 651–9, 690–6; Hidy and Hidy, *Pioneering*, 123–52, 236–54; Reader, 'Oil'; Tolf, *Russian*, 92–5; Jones, *State*, 19–21; Schenk and Boon, 'Trading', 16–21.
129. 'Raw Materials: Oils and Oilseeds: Petroleum', *Economist* supplement (21 February 1891).
130. Quoted in Hidy and Hidy, *Pioneering*, 136; Tolf, *Russian*, 104–5; Henriques, *Marcus*, 148–9, 158, 179–82, 189–91, 269.
131. Butt, 'Technical'; Dean, 'Scottish'; McKay, *Scotland*, ch. 1.
132. Schatzker and Hirszhaut, *Jewish*, 45–94; Frank, *Oil Empire*, 79–101; May, *Hard*, 102, 126, 135–49, 165; Burr, *Canada*, 163–4; Daintith, *Finders*, 144–7; Gerretson, *History*, vol. 3, 66–72; Bielen, 'Nationalism', 5–9.
133. *Reports from the Consuls of the United States*, vol. 19, 155–6.
134. Pearton, *Oil*, 8–9, 17, 22; Redwood, *Petroleum*, vol. 1, 26; *Report of the Royal Commission*, vol. 2, 32; Forbes, *More Studies*, 99–107; Tulucan et al., 'History', 193.

135. Williamson and Daum, *Illumination*, 663.
136. Longmuir, *Oil*, 46–8, 61–76; Corley, *History*, 8–24; Noetling, 'Occurrence', 27–8, 163, 216–25; Webster, 'Business', 1,003–7, 1,012–13; Thant Myint-U, *Making*, 178–85.
137. Quoted in Longmuir, ibid., 80–1.
138. Ibid., 76–86; Webster, 'Business'; Turrell, 'Conquest'; Christian, 'Anglo-French'; Fieldhouse, *Economics*, 384–93; Thant Myint-U, *Making*, 186–9.
139. Quoted in Longmuir, ibid., 81; Webster, ibid., 1,014–25.
140. Quoted in Turrell, 'Conquest', 157.
141. Quoted in Woodman, *Making*, 233.
142. Quoted in Thant Myint-U, *Making*, 190–1; Aung, *Lord Randolph*, chs. 1–13; Hussain, 'Resistance', 19–22.
143. Quoted in Longmuir, *Oil*, 84; Thant Myint-U, ibid, 189–93; Boyer, 'Picturing'.
144. Quoted in Longmuir, ibid., 84; Corley, *History*, 26–7.
145. Corley, ibid., 30–1; Longmuir, ibid., 111, 117.
146. 'Petroleum is such an important article', *Times of India* (21 June 1886).
147. Marvin, *England*, v.
148. 'England as a Petroleum Power', *Times of India* (4 May 1887).
149. Yergin, *Prize*, 141.
150. Fielding Hall, *Soul*, 53, 65; Hussain, 'Resistance', 28–30.
151. Quoted in Thant Myint-U, *Making*, 198.
152. Ni Ni Myint, *Burma*, 81–90, 127, 131, 144, 147–8, 150, 152–3; Hussain, 'Resistance', 64–6, 69–71, 80, 105, 107, 109, 113, 116–19, 121, 136–41; Thant Myint-U, *Making*, 198–207; Aung, *Stricken*, 94–5; Woodman, *Making*, chs. 15–17.
153. Geary, *Burma*, 248, 236–7.
154. 'Paper read by Colonel Stewart', *Times of India* (17 July 1886).
155. Quoted in Longmuir, *Oil*, 119.
156. Crosthwaite, *Pacification*, 129–30; Thant Myint-U, *Making*, 207–8; Longmuir, ibid., 122–30.
157. Marvin, *England*, vi, 19.
158. 'Russian Petroleum in India', *Financial Times* (13 March 1888).
159. Visvanath, *Hundred*, 6–9.
160. Quoted in Poley, *Eroïca*, 81.
161. Ibid., 81–4.
162. Quoted in Gerretson, *History*, vol. 1, 77; Poley, ibid., 84–9.
163. Gerretson, ibid., 77–87, 110–13; Poley, ibid., 89–92, 103–5.
164. Gerretson, ibid., 82–3; Poley, ibid., 89–91; Reid, *Contest*, 30–40.

165. Quoted in Reid, ibid., 57.

166. Ibid., 52-78; Farnie, *East and West*, 174-6.

167. Quoted in Reid, ibid., 61.

168. Quoted in ibid., 76-7; Ricklefs, *History*, 174-6.

169. Bakker, 'Aceh War', 56-8; Kreike, 'Genocide'; Groen, 'Colonial', 284-5; Gerretson, *History*, vol. 2, 127; Reid, 'Colonial, 98-103; Ricklefs, ibid., 176-7.

170. Quoted in Poley, *Eroïca*, 93; Gerretson, *History*, vol. 1, 100-2; Beaton, *Enterprise*, 23-4.

171. Gerretson, ibid., 130.

172. Quoted in ibid., 133; ibid., 110-33; Poley, ibid., 93-7; Beaton, *Enterprise*, 24-7; Doran, *Breaking*, 100-6.

173. Reid, *Contest*, 271; Gerretson, ibid., 165-7; Beaton, ibid., 29; Henriques, *Marcus*, 148.

174. Quoted in Gerretson, ibid., 167-8; ibid., 262-3; Hidy and Hidy, *Pioneering*, 263.

175. Yergin, *Prize*, 75, 118.

176. Gerretson, *History*, vol. 1, 168-85; Poley, *Eroïca*, 97-119.

177. Williamson and Daum, *Illumination*, 492, 666, 675, 743; Hidy and Hidy, *Pioneering*, 359; Farnie, *East and West*, 440-3; Dasgupta, *Oil*, 14-15; Gerretson, ibid., 215.

178. Fletcher, 'Suez'; Marlowe, *Spoiling*, 24-35.

179. Marlowe, ibid., ch. 2; Searight, *British*, ch. 11; Furber, 'Overland'; Hoskins, *British*, chs. 1-11; Hoskins, 'First'; Hoskins, 'Growth'; Harris, 'Persian'; Shahvar, 'Tribes'; Shahvar, 'Concession'; Onley, 'Britain'; Ingram, *Beginning*, ch. 6; Bailey, 'Economics'; Farnie, *East and West*, ch. 2.

180. Pudney, *Suez*, ch. 4; Hoskins, *British*, ch. 12; Farnie, ibid., 29-31.

181. Quoted in Harrison, *Gladstone*, 45.

182. Quoted in Farnie, *East and West*, 73; Pudney, *Suez*, ch. 8; Brown, 'Who Abolished'.

183. 'The Suez Canal', *Economist* (20 November 1869); Fletcher, 'Suez', 564.

184. Coffin, *Our New Way*, 507.

185. Quoted in Pudney, *Suez*, 182; Harrison, *Gladstone*, 51-3; Marlowe, *Anglo-Egyptian*, 90-8; Farnie, *East and West*, ch. 14; Chamberlain, *Scramble*, 37-9.

186. Pudney, ibid., 74; Marlowe, *Spoiling*, 55-6; Farnie, ibid., 38, 45-7; Hoskins, 'British'.

187. 'British Interests', *Times* (25 June 1877); Farnie, ibid., 260-1.

188. Quoted in Farnie, ibid., 270.

189. Quoted in Roberts, *Salisbury*, 229.
190. Harrison, *Gladstone*, 49–59, 83–7; Newsinger, 'Liberal', 57–9; Marlowe, *Anglo-Egyptian*, 112–16; Hopkins, 'Victorians', 380–2; Mowat, 'From Liberalism', 112–14; Al-Sayyid Marsot, *History*, 82–6.
191. Biagini, 'Exporting', 211–16; al-Sayyid, *Egypt*, 5–13.
192. Quoted in Kudsi-Zadeh, 'Afghānī', 33, 34; Harrison, *Gladstone*, 54–9, 85–7; Steele, 'Britain'.
193. Harrison, ibid., 83–5.
194. 'A Plea for the Egyptians', *New York Times* (9 July 1882).
195. Matthew, *Gladstone*, 383–4; Harrison, *Gladstone*, 75; Newsinger, 'Liberal', 59–62.
196. Quoted in al-Sayyid, *Egypt*, 13.
197. Quoted in Harrison, *Gladstone*, 59.
198. Lucas, *Divided*, 182; Steele, 'Britain', 2; Shaw, *Eden*.
199. Abrahamian, *Coup*; Owen Jones, 'Even the Crisis in Afghanistan Can't Break the Spell of Britain's Delusional Foreign Policy', *Guardian* (19 August 2021); George Monbiot, 'Who's to Blame for the Afghanistan Chaos? Remember the War's Cheerleaders', ibid. (25 August 2021); Miller, ed., *Tell Me Lies*; Robinson et al., *Pockets*; David Cromwell and David Edwards, 'The Great Iraq War Fraud', *Media Lens* (13 July 2016); David Edwards, 'The Great Libya War Fraud', ibid. (3 October 2016).
200. Schölch, '"Men on the Spot"'; Matthew, *Gladstone*, 387–9; Cain and Hopkins, *British*, 339–43.
201. Harrison, *Gladstone*, 77–8, 113–15.
202. Quoted in ibid., 72; Farnie, *East and West*, 362; Hopkins, 'Victorians', 373–4.
203. Quoted in Hopkins, ibid., 383; Huffaker, 'Representations'; Pinfari, 'Unmaking'; Harrison, ibid., 74–5.
204. Quoted in Galbraith and al-Sayyid Marsot, 'British', 476.
205. 'Patriotism in the East – From the London Spectator', *New York Times* (4 July 1882).
206. Quoted in Harrison, *Gladstone*, 91.
207. Quoted in al-Sayyid Marsot, *Egypt*, 16.
208. Harrison, *Gladstone*, 77, 91; Newsinger, 'Liberal', 55–6.
209. Quoted in Harrison, ibid., 92.
210. Ibid., 93.
211. Ibid., 92–5; Chamberlain, 'Alexandria'.
212. 'Our Policy in Egypt', *Economist* (1 July 1882).
213. 'Leader – The Eastern Question', *Illustrated London News* (3 June 1882).

214. Quoted in Farnie, *East and West*, 292–3.
215. 'Arabi Pasha's Appeal to England', *Times* (24 July 1882).
216. Quoted in Galbraith and al-Sayyid Marsot, 'British', 486; James, *Empires*, 80–2; Holland, *Blue-Water*, 112–14.
217. HC Deb, 12 July 1882, vol. 272, c. 169.
218. Quoted in Galbraith and al-Sayyid Marsot, 'British', 487; Harrison, *Gladstone*, 96–126; Newsinger, 'Liberal', 65–6; al-Sayyid Marsot, *Egypt*, 25.
219. Quoted in Harrison, ibid., 110.
220. Quoted in ibid., 110; Newsinger, 'Liberal', 67–8.
221. 'A Holy War', *New York Times* (4 August 1882).
222. Quoted in Galbraith and al-Sayyid Marsot, 'British', 478; Harrison, *Gladstone*, ch. 6; Newsinger, 'Liberal', 68–70; Hunter, 'Tourism', 39.
223. Harrison, ibid., ch. 7; Galbraith, 'Trial'; Newsinger, 'Liberal', 70–3.
224. *Correspondence Respecting the Passage of Petroleum*, 51, 75; Williamson and Daum, *Illumination*, 675; Cochran, *Encountering*, 22–8.
225. Jones, *State*, 19–21; Henriques, *Marcus*, 50, 66–84; Doran, *Breaking*, 47–56, 62–9.
226. Henriques, ibid., 76–86, ch. 3; Farnie, *East and West*, 442–9; Gerretson, *History*, vol. 1, 214–17; ibid., vol. 2, 144–6; Doran, ibid., 69–70, 75–9, 82–94; Hidy and Hidy, *Pioneering*, 259; Williamson and Daum, *Illumination*, 664–8.
227. Quoted in Henriques, ibid., 109–10.
228. Ibid., 119–20; Gerretson, *History*, vol. 1, 233–4; Doran, *Breaking*, 94–6.
229. Samuel, 'Liquid Fuel', 386; Henriques, ibid., 136–7.
230. Williamson and Daum, *Illumination*, 633, 660–8, 742–3.
231. Quoted in Issawi, ed., *Economic*, 328–9.
232. Foreign Office, *Diplomatic and Consular Reports: Batoum, 1894*, 11–12; Issawi, ibid.
233. McKay, 'Baku Oil', 614, 621–2; McKay, 'Restructuring', 92–103; Gerretson, *History*, vol. 2, 105–11, 148–50; Hidy and Hidy, *Pioneering*, 132, 236; Foreign Office, *Diplomatic and Consular Reports: Batoum, 1895*, 3–8, 22–3; Thompson, *Oil Fields*, 29–32; Alekperov, *Oil*, 119–22; Henriques, *Marcus*, 162–3.
234. Tolf, *Russian*, 114–17; Hewins, *Mr Five*, 24–7; Foreign Office, *Diplomatic and Consular Reports: Batoum, 1894*, 2–3; ibid., *1895*, 2–13; Gerretson, ibid., 108–11, 115, 146–51; Fursenko, *Battle*, 15–34, 71–3.
235. Quoted in Nowell, *Mercantile*, 50–1.

236. Quoted in Williamson and Daum, *Illumination*, 650; Fursenko, *Battle*, 23-34.

237. Leader: 'It is interesting to learn . . .', *New York Times* (23 September 1891), 4; Fursenko, ibid., 67-75.

238. Quoted in Chernow, *Titan*, 248.

239. Quoted in Williamson and Daum, *Illumination*, 650.

240. 'Petroleum Trade Division', *New York Times* (27 July 1895); Henriques, *Marcus*, 265-8.

241. Williamson and Daum, *Illumination*, 577, 599, 619-20; Hidy and Hidy, *Pioneering*, 87-8, 278-82; Brown and Partridge, 'Death of a Market', 572; 'Oil Producers to Unite', *New York Times* (12 February 1890); 'Collapse of an Oil Exchange', *Washington Post* (22 September 1890); 'The Standard Fixes Oil Prices', *New York Times* (23 January 1895); 'A Further Sharp Advance in Oil', *New York Times* (17 April 1895); 'Panic in the Oil Market', *New York Times* (18 April 1895).

242. Williamson and Daum, ibid., 650-3, 662; Hidy and Hidy, ibid., 238-45, 253; Tolf, *Russian*, 116; Henriques, *Marcus*, 126-8, 140-1, 163-5.

243. Henriques, ibid. 140-3, 150; McKay, 'Restructuring', 103; Gerretson, *History*, vol. 2, 147-51; Fursenko, *Battle*, 23-4, 29-34, 43, 54-62; Chernow, *Titan*, 248.

244. Gerretson, ibid., 39-40, 64, 75, 179, 283; Hidy and Hidy, *Pioneering*, 259-64.

245. Curzon, *Persia*, vol. 1, vii, 1-4, 59-60.

246. Ibid., 61.

247. Ibid., 63-5.

248. Ibid., 66-7.

249. McKay, 'Restructuring', 92-103; Fursenko, 'Oil', 451-4; Gerretson, *History*, vol. 2, 110-11, 149; Foreign Office, *Diplomatic and Consular Reports: Batoum, 1894*, 7-9.

250. Jones, *State*, 56-62.

251. Thompson, *Oil Fields*, 7, 16-19, 126-32; Alekperov, *Oil*, 113-15; Jones, *State*, 55; Gerretson, *History*, vol. 3, 272.

252. Thompson, *Oil Pioneer*, 69-70; Thompson, *Oil Fields*, 305.

253. Thompson, *Oil Fields*, 307, 317-8, 353; 'Great Petroleum Fire at Baku', *Observer* (18 July 1897).

254. Thompson, ibid., 308.

255. Ibid., 382; Alekperov, *Oil*, 138; 'Hand-Dug Wells Forbidden', *Oil News* (29 March 1913).

256. Thompson, ibid., in the earlier 1904 edition.

257. Foreign Office, *Diplomatic and Consular Reports: Batoum, 1893*, 7–8, 21–2; ibid., *1894*, 29; ibid., *1895*, 5–6, 22; Alekperov, *Oil*, 116.

258. Fursenko, *Battle*, 39, 44–74, 92–9; Jones, *State*, 54–8; McKay, 'Restructuring', 103; Gerretson, *History*, vol. 2, 98–9, 102, 111–17; ibid., vol. 3, 271–3; ibid., vol. 4, 160; Fursenko, 'Oil', 452–4; Thompson, *Oil Fields*, 7, ch. 4; 'The European Petroleum Company', *Times* (18 May 1896); 'Russian Petroleum and Liquid Fuel Company', ibid. (1 November 1897); 'Baku Russian Petroleum Company', ibid. (21 June 1898); 'Bibi-Eybat Petroleum Company', ibid. (29 October 1900); 'Russian Petroleum and Liquid Fuel', *Financial Times* (7 December 1905); 'Baku Russian Petroleum', ibid. (13 December 1905).

259. 'The Importation of Petroleum', *Manchester Guardian* (14 June 1899).

260. Thompson, *Oil Pioneer*, 296.

261. Quoted in Fursenko, *Battle*, 54.

262. 'Russian Petroleum Companies', *Financial Times* (13 December 1898).

263. 'Petroleum Craze at Baku', ibid. (12 May 1898).

264. Thompson, *Oil Fields*, 7 (as stated also in the earlier 1904 edition); Jones, *State*, 56–8.

265. 'Russian Petroleum Companies', op. cit.

266. Quoted in Fursenko, *Battle*, 56; Thompson, *Oil Pioneer*, 67.

267. Quoted in Fursenko, ibid., 60.

268. Thompson, *Oil Pioneer*, 53–4, 67.

269. Von Laue, 'Russian'.

270. Quoted in Fursenko, *Battle*, 57.

271. Quoted in Von Laue, *Sergei*, 3.

272. Gerretson, *History*, vol. 1, 280–3; ibid., vol. 2, 39–40, 55, 75, 130–1, 226–34, 243–8; Hidy and Hidy, *Pioneering*, 260–5, 516; White, *Formative*, 196; Alekperov, *Oil*, 117–18.

273. Quoted in Gerretson, ibid., vol. 2, 42.

274. Deterding, *International*, 48.

275. Quoted in Gerretson, *History*, vol. 1, 260–1; ibid., 171–4, 217–35, 240–1, 251–3, 282–6; ibid., vol. 2, 92; Williamson and Daum, *Illumination*, 672–6; Beaton, *Enterprise*, 30; Deterding, ibid., 50–3, 55–6; Poley, *Eroïca*, 137–8; Cochran, *Encountering*, 22–8; 'Baku Petroleum Producers' Association and the Oil Markets of the Far East', *Petroleum World* (4 November 1904), 705–6; 'American Case Oil for the Orient', ibid. (14 January 1905), 110; Hidy and Hidy, *Pioneering*, 532.

276. Gerretson, ibid., vol. 2, 49, 309; Locher-Scholten, *Sumatran*, 194–5.

277. Gerretson, ibid., 50.

278. Ibid., 53.

279. Ibid., 46–60.
280. Ibid., 61–73; ibid., vol. 1, 222–4, 260–70, 285–6; Henriques, *Marcus*, 221–3, 234; Fursenko, *Battle*, 82–4.
281. Gerretson, ibid., vol. 2, 66, 68.
282. Ibid., 68.
283. Deterding, *International*, 45–6.
284. Quoted in Gerretson, *History*, vol. 2, 72.
285. Ibid., 71–4; Fursenko, *Battle*, 83–5.
286. Gerretson, ibid., 67–9, 76–86; Beaton, *Enterprise*, 31; Hidy and Hidy, *Pioneering*, 266–7.
287. Henriques, *Marcus*, 162–5; Gerretson, ibid., 99, 151; Williamson and Daum, *Illumination*, 674.
288. Black, 'The "Lastposten"', 284–8; Gerretson, ibid., 155–7.
289. Gerretson, ibid., 40, 168, 206–20; Henriques, *Marcus*, 184–220; Jonker and Zanden, *History*, vol. 1, 35, 56–9; Locher-Scholten, *Sumatran*, 208–9.
290. Quoted in Henriques, ibid., 184; ibid., 224–6.
291. Gerretson, *History*, vol. 2, 91–5, 121–6; Deterding, *International*, 56–7; Doran, *Breaking*, 111–13.
292. Gerretson, ibid., 127.
293. Quoted in ibid., 131.
294. Ibid., 132.
295. Quoted in Doel, 'Military', 63; Kitzen, 'Course', 242–8; Reid, *Contest*, ch. 8; Ricklefs, *History*, 177–8.
296. Doran, *Breaking*, 120–1; Ricklefs, ibid., 178; Locher-Scholten, *Sumatran*, 191–2, 200–1; Lindblad, 'Economic', 11–12; Bakker, 'Economisch', 55–61; Reid, ibid., 270–80.
297. Quoted in Doel, 'Military', 72; Zondergeld, *Goed*, 85–119; Groen, 'Colonial', 289–90; Vickers, *History*, 10–14.
298. Quoted in Langeveld, *Dit leven*, vol. 1, 59 (trans. Roos Geraedts); Jonker and Zanden, *History*, vol. 1, 154; Haan, 'Knowing', 783–8; 'Dutch Reassess "hero" behind Bali Killings', *Times* (12 May 1998).
299. Groen, 'Colonial', 284, 278; Kreike, 'Genocide', 299.
300. Quoted in Henriques, *Marcus*, 293.
301. Gerretson, *History*, vol. 2, 135–8; Jonker and Zanden, *History*, vol. 1, 52–3; Doran, *Breaking*, 121–5.
302. Quoted in Gerretson, ibid., 139–40.
303. Ibid., 138–42.
304. Quoted in Locher-Scholten, *Sumatran*, 204 and 211; ibid., 212–17.

305. Henriques, *Marcus*, 237–8; Fursenko, *Battle*, 86–94; Gerretson, *History*, vol. 2, 101–19; Deterding, *International*, 57.

306. Quoted in Fursenko, ibid., 87; Henriques, ibid., 235; Gerretson, ibid., 82–6, 169–71.

307. Gerretson, ibid., 168; Henriques, ibid., 227–30, 271–88; Samuel, 'Liquid Fuel', 384–5; Fursenko, ibid., 87; Jones, *State*, 10.

6. OIL FOR POWER

1. 'Petroleum Fuel for Steamships', *New York Times* (15 June 1867); Williamson and Daum, *Illumination*, 240–2; Williamson et al., *Energy*, 178–9; Ross, *Air as Fuel*.

2. Quoted in Jones, 'Oil-Fuel', 135; Brown, 'Royal Navy', 42–9; *Second Report on Coal Supplies*, vol. 2, 214–18.

3. Williamson and Daum, *Illumination*, 595–613; Jones, ibid., 133–5; Jones, *State*, 4–10; North, *Oil*, chs. 5–6; Boyd, *Petroleum*, ch. 8; Tolf, *Russian*, 69–71; Hidy and Hidy, *Pioneering*, 162–3; Henriques, *Marcus*, 271–2; 'Oil as Fuel for Ships', *New York Times* (12 September 1891).

4. White, *Formative*, chs. 2–5; Franks and Lambert, *Early*, 3–7, 40–2, 54–7; Watts, *Oil*.

5. Quoted in Patrick Lee, 'Field in Trouble', *Los Angeles Times* (22 September 1989).

6. Franks and Lambert, *Early*, 20, 28; White, *Formative*, 150–2; 'Oil from Peru', *Morning Call* (San Francisco) (17 November 1893); *Mineral Resources of the U.S., 1893*, 516–24; Williamson and Daum, *Illumination*, 661.

7. Moreno, *Petroleum*, 8–10, 20–9, 34–9; Caraher, 'Construction', 41.

8. Quoted in White, *Formative*, 151.

9. Quoted in ibid.

10. Testa, 'Los Angeles', 81–3; Lee, 'Field in Trouble', op. cit.; Davis, *Dark*, ch. 2.

11. 'Oil Wells in the Sea', *Boston Globe* (30 April 1900); Rintoul, *Spudding*, ch. 14; Brantly, *History*, 1364–8; White, *Formative*, 152–61.

12. Olien and Olien, *Oil*, 1–13; Rundell, Jr, *Early*, 17–33; Spellman, *Spindletop*, 9–15.

13. 'The Crisis in Petroleum', *Financial Times* (14 April 1888).

14. Williamson and Daum, *Illumination*, 678, 683; Williamson et al., *Energy*, 174–9; Hidy and Hidy, *Pioneering*, 299–301; Gibson, '"Oil Fuel"', 110–12; Gröner, *German*, vol. 1, 10–13; 'Naval and Military Notes', *United Service* (March 1896), 267; 'Naval and Military Notes', ibid. (April

1896), 360, 362; 'Naval and Military Notes', ibid. (May 1896), 448; 'Liquid Fuel for German Warships', *Daily Picayune* (16 February 1896); 'Liquid Fuel in the German Navy', ibid. (5 October 1896); 'About Masut', ibid. (14 August 1897); DeNovo, 'Petroleum', 641; Samuel, 'Liquid Fuel', 386–7; Henriques, *Marcus*, 274; Brown, 'Royal Navy', 49–50; 'Liquid Fuel for Steamships', *Washington Post* (23 February 1894); 'Petroleum as Navy Fuel', *New York Times* (18 November 1890); 'Test to be Made of Liquid Fuel', ibid. (18 November 1895); 'Liquid Fuel for the Navy', ibid. (25 September 1896); 'Oil Fuel on Torpedo Boats', ibid. (2 November 1898); 'German Torpedo Boats', *Manchester Guardian* (25 July 1898).

15. 'Liquid Fuel for Steamers', *Times* (14 September 1898).

16. Henriques, *Marcus*, 271–80; Samuel, 'Liquid Fuel', 386–9; Brown, 'Royal Navy', 44–7.

17. Bartholomew, *Early*, 3; 'Fighting Motor Car', *Horseless Age* 2, no. 1 (November 1896); 'Motor Vehicles in Warfare', ibid. 2, no. 9 (July 1897); Eckermann, *World*, 37–9; Lieckfeld, *Oil Motors*, 140–1.

18. Donkin, Jr, *Gas*; Graffigny, *Gas*, 1–11; Lockert, *Petroleum*, 18–21; Eckerman, ibid., 10–13, 19–23; Bryant, 'Silent', 184–94; 'A Gunpowder Motor', *Horseless Age* 2, no. 3 (January 1897), 10.

19. Eckermann, ibid., 14–19; 'Gasolene vs. Gas Engines', *Horseless Age* 2, no. 12 (October 1897); White, Jr, '"Perfect Light"', 64–9; Dowson and Larter, *Producer*; Webber, *Town*; Williamson and Daum, *Illumination*, 234–8; Williamson et al., *Energy*, 179; Donkin, ibid., 51, 288–94; Lieckfeld, *Oil Motors*, 18–22, ch. 2; Hutton, *Gas-Engine*, chs. 6–8, 10; Lockert, ibid., 35–63.

20. 'Scientific Results of the Month', *Illustrated London News* (4 July 1874).

21. Donkin, *Gas*, 55–74, 179, 197–200, 296–303; Eckermann, *World*, 25–32; Lieckfeld, *Oil Motors*, 24–7, 139, 180–4; Lockert, *Petroleum*, 64–6, 89–96, 122–5; Redwood, 'Motor', 512–13; Williamson et al., *Energy*, 184–5; 'The Flying Machine', *Horseless Age* 2, no. 7 (May 1897).

22. 'Petroleum', *Financial Times* (10 October 1888).

23. 'The New Motor Law in England', *Horseless Age* 2, no. 1 (November 1896).

24. 'The Oil Famine Bugaboo', ibid. 2, no. 3 (January 1897).

25. Estes, *Our History*, 16–18, 128–9; Utley, *Indian*, ch. 9; Andrist, *Long*, 1–2, 330–54; Campbell, *Special*, 1–2.

26. Campbell, ibid.; McCormick, *China*, ch. 1; LaFeber, *New*, chs. 2–8; Conant, 'Economic'; Healy, *US Expansionism*, chs. 9–13; Holmes, 'Mahan', 28–37; Merk, *Manifest*, ch. 11; Fordham, 'Domestic'; Williams, 'United States'; Tuason, 'Ideology'.

27. Quoted in Campbell, *Special*, 5.

28. Ince, 'Friedrich'; Gibson, *Wealth*, chs. 1–2.

29. Palen, 'Protection'; Campbell, *Special*, 5–9; Fordham, 'Protectionist'.

30. Quoted in McCormick, *China*, 66.

31. Quoted in Hidy and Hidy, *Pioneering*, 264; ibid., 49–52, 63, 126, 228–31, 259–68; Cochran, *Encountering*, 22–31; Hunt, 'Americans', 281–4; Henriques, *Marcus*, 223–6, 320, 322–3; *Mineral Resources of the U.S., 1899*, 277–82.

32. McCormick, *China*, chs. 1–3; LaFeber, *New*, ch. 4, 300–25, 352–62; Campbell, *Special*, chs. 2–6; Fordham, 'Protectionist'; O'Brien and Pigman, 'Free', 94–5, 103; Holmes, 'Mahan', 44; Lee, 'China', 55–6; 'The Break-up of China and our Interest in it', *Atlantic Monthly* 84, no. 502 (August 1899).

33. Merk, *Manifest*, 231–7; LaFeber, ibid., ch. 1; Schoultz, *Beneath*, ch. 5; Healy, *US Expansionism*, chs. 1–2.

34. Healy, *James*, ch. 3; Wheeler and Grosvenor, 'Our Duty', 631–3; Campbell, *Special*, 13–18; Shulman, *Coal*, 157–63; Fordham, 'Domestic'; McCormick, *China*, ch. 4; Fisher, 'Destruction'; Zinn, *People*, ch. 12.

35. 'The Philippines', *The Sun* (New York) (19 May 1898); Bradley, *Imperial*, ch. 4.

36. 'Keep the Philippines', *The Sun* (New York) (9 September 1898).

37. Quoted in Sklar, *Corporate*, 397.

38. *Mineral Resources of the U.S., 1899*, 278; also Hidy and Hidy, *Pioneering*, 262.

39. McCormick, *China*, chs. 6–7; Campbell, *Special*, chs. 7–10; Vevier, 'Open'; Casserly, *Land*, 288–91; Scott, *China*, 143–55.

40. *Mineral Resources of the U.S., 1899*, 279.

41. Cochran, *Encountering*, 31–2; Hidy and Hidy, *Pioneering*, 532, 547–53.

42. White, *Formative*, 6, 45, 219–21, 263, 271–2, 276, 281–4; Henriques, *Marcus*, 326.

43. Quoted in Challener, *Admirals*, 22; Hunt, 'Americans', 289–95.

44. Bradley, *Imperial*; Miller, *Bankrupting*; Jim Lobe, 'Iraq: Chalabi, Garner Provide New Clues to War', *Inter Press Service* (20 February 2004).

45. Williamson and Daum, *Illumination*, 633; *Mineral Resources of the U.S., 1899*, 191–8; ibid., *1900*, 594–9; U.S. consular report: 'Russian Petroleum Trade', 195–9; Foster, 'Russian'; Alekperov, *Oil*, 121–2; Thompson, *Oil Pioneer*, 59.

46. Johnson, *Petroleum*, 3–4, 14; Williamson et al., *Energy*, 72–4, 170–4, 242–57; Hidy and Hidy, *Pioneering*, ch. 19.

47. Quoted in Hidy and Hidy, ibid., 510; 'Big Oil Strike in Texas', *New York Times* (13 January 1901); 'Texas Oil Discovery', ibid. (17 January 1901); 'The Russian and Texas Oil Fields Compared', *Petroleum Review* (March 1904), 146.

48. *Mineral Resources of the U.S., 1900*, 579.

49. Ibid., *1904*, 712; Williamson et al., *Energy*, 19-22, 74-90; Olien and Olien, *Oil*, chs. 1-2; Spellman, *Spindletop*, chs. 1-3; Rundell, *Early*, 35-9.

50. Bringhurst, *Antitrust*, ch. 2, 190-1; Pratt, 'Petroleum'; Singer, *Broken*; Hidy and Hidy, *Pioneering*, 118; Chernow, *Titan*, 254-5.

51. Williamson et al., *Energy*, 75-88; Johnson, 'Early', 516-22; Henriques, *Marcus*, 336-9.

52. *Mineral Resources of the U.S., 1904*, 685, 716.

53. Jonker and Zanden, *History*, vol. 1, 67; Doran, *Breaking*, 151-4.

54. Quoted in Jones, 'Oil-Fuel', 136.

55. Quoted in Jones, 'State', 364; Brown, 'Royal Navy', 43-5.

56. *Second Report on Coal Supplies*, vol. 2, 197; Corley, *History*, 39; Jones, 'Oil-Fuel', 137-42; Jones, *State*, 97-9; Sorkhabi, 'Sir Thomas'.

57. 'From a Lamp Oil to a By-Products Age', *Petroleum Times* 53, no. 1353 (17 June 1949); Jones, *State*, 12.

58. Quoted in Chernow, *Titan*, 431.

59. Henriques, *Marcus*, 290, 326, 354, 366.

60. Quoted in Beaton, *Enterprise*, 53; Henriques, ibid., 245, 258, 315.

61. Henriques, ibid., 366-7; Hidy and Hidy, *Pioneering*, 504; Gerretson, *History*, vol. 4, 195-6.

62. Henriques, ibid., 326, 356-80, 389-90; Hidy and Hidy, ibid., 343; Jonker and Zanden, *History*, vol. 1, 68-72.

63. Quoted in Hidy and Hidy, ibid., 509-10.

64. Fursenko, *Battle*, 43, 92-9.

65. Hidy and Hidy, *Pioneering*, 498, 548; *Mineral Resources of the U.S., 1904*, 749-51.

66. Quoted in Fursenko, *Battle*, 108; Pearton, *Oil*, 22-34, 45; Gerretson, *History*, vol. 3, 78-83.

67. Arnold-Forster, 'Notes', 134-5.

68. Hidy and Hidy, *Pioneering*, 515-17; *Mineral Resources of the U.S., 1900*, 611-13; ibid., *1904*, 740-1; Fursenko, *Battle*, 109; Pearton, *Oil*, 30.

69. Hidy and Hidy, ibid., 512-15; Frank, 'Petroleum', 20-4; Frank, *Oil*, 150-71; Frank, 'Environmental', 178-80, 186-9; *Mineral Resources of the U.S., 1900*, 608-10; ibid., *1904*, 736-9.

70. Quoted in Jones, *State*, 99; Noetling, 'Occurrence'; Hidy and Hidy, ibid., 499–501; Sorkhabi, 'Sir Thomas', 439–40.

71. Quoted in Jones, 'State', 361; Jones, *State*, 91–3; Hidy and Hidy, ibid., 499–501; Corley, *History*, 65–8; Longmuir, *Oil*, 164–72, 266–7; Dasgupta, *Oil*, 14–23.

72. Quoted in Longmuir, ibid., 161.

73. Quoted in Gerretson, *History*, vol. 1, 341.

74. Quoted in Jones, *State*, 89.

75. *Second Report on Coal Supplies*, vol. 2, 210.

76. Quoted in Henriques, *Marcus*, 364; Jones, *State*, 19–21, 93–5; Jonker and Zanden, *History*, vol. 1, 72–9; Conlin, *Mr Five*, 67–71; Doran, *Breaking*, ch. 11; Jonker and Zanden, 'Searching', 21–3; Fursenko, *Battle*, 89–92; Nowell, *Mercantile*, 52–3, 57–8; Hidy and Hidy, *Pioneering*, 553, 563–72; Beaton, *Enterprise*, 49–50; Deterding, *International*, 64–72.

77. Henriques, ibid., 361–494; Corley, *History*, 68–71; Longmuir, *Oil*, ch. 10; Hidy and Hidy, ibid., 504–6, 547–53; Dasgupta, *Oil*, 20–5; Beaton, ibid., 46–57; Jones, 'State', 101–3; Cochran, *Encountering*, 20–43; 'Baku Petroleum Producers' Association and the Oil Markets of the Far East', *Petroleum World* (4 November 1904), 705–6; 'American Case Oil for the Orient', ibid. (14 January 1905), 110; *Mineral Resources of the U.S., 1899*, 277; Williamson et al., *Energy*, 258–9.

78. Beaton, ibid., 56–7.

79. *Mineral Resources of the U.S., 1904*, 675–724.

80. 'Boryslaw, the Oil Field of 1903', *Petroleum World* (December 1903).

81. 'An English Syndicate's Success in Alaska', *Petroleum World* (February 1904); White, *Formative*, 196, 625 (n. 5); *Second Report on Coal Supplies*, vol. 2, 196; *Mineral Resources of the U.S., 1899*, 167; ibid., *1900*, 587; ibid., *1904*, 724–6.

82. *Mineral Resources of the U.S., 1899*, 179–80; ibid., *1900*, 590–1; ibid., *1904*, 731; Redwood, *Petroleum*, vol. 1, 151–3; ibid., vol. 3, 1,090; 'Peruvian Petroleum', *Petroleum World* (May 1904); 'Peruvian Oil in 1903', ibid. (4 November 1904).

83. *Mineral Resources of the U.S., 1899*, 167–79; ibid., *1900*, 587–9; ibid., *1904*, 728–9; Redwood, *Petroleum*, vol. 3, 1,092; Williamson and Daum, *Illumination*, 689–90.

84. Towler, 'Pitch Lake'; McBeth, *Gunboats*, 42–4; Holley, Jr, 'Blacktop', 706–13; Higgins, *History*, 14–16, 40; Daintith, *Finders*, 123–38; Redwood, ibid., 1,125.

85. Higgins, ibid., 51–64; Jones, *State*, 106–7; J.D. Henry, *Oil Fields*, 3–157; 'Principal Petroleum Resources' (30 June 1904), 107, 175–82.

86. McBeth, *Gunboats*, 42–4, 58–9, 105, 113; Tinker Salas, *Enduring*, 41–2; 'The Asphalt Lake in Venezuela', *New York Times* (17 February 1901); Gerretson, *History*, vol. 4, 268–9.

87. *Mineral Resources of the U.S., 1900*, 593.

88. Rafael Arráiz Lucca, 'Táchira's Petrolia: First Private Oil Company in Venezuela', *Caracas Chronicles* (31 July 2018); Tinker Salas, *Enduring*, 40–1; Arnold et al., *First*, 36–7, 99; Gerretson, *History*, vol. 4, 273; U.S. consular report, 'Petroleum and Kerosene Oil', 454–9.

89. Notestein et al., 1,167; Beckerman and Lizarralde, *Ecology*, 74.

90. Jones, *State*, 105–6; *Second Report on Coal Supplies*, vol. 2, 195; 'Principal Petroleum Resources' (29 September 1904), 182–4; Gerretson, *History*, vol. 4, 269–70.

91. Cote, *Oil*, 1–5, 12–13.

92. Quoted in Ayemoti Guasu, 'God Man', 240; Langer, *Expecting*, 186–9; Gustafson, *New*, 33–6; Langer, *Economic*, chs. 3, 6.

93. Gustafson, ibid., 36–8; Langer, *Expecting*, 189–92.

94. Quoted in Langer, *Economic*, 133.

95. Cote, *Oil*, 4–12.

96. Oliveira and Figueirôa, 'History', 15–19.

97. 'Cuban Mining and Development', *Financial Times* (30 April 1903); 'Cuban Petroleum Company', ibid. (27 March 1906); Brown, *Report*, 82–3.

98. 'Mexican Petroleum', *New York Times* (11 August 1867).

99. U.S. consular report, 'Mexican Asphaltum'.

100. Munch, 'Anglo-Dutch', 135–45, 150–61; Santiago, 'Culture', 63–5; Santiago, *Ecology*, 64–75; Gerretson, *History*, vol. 4, 254–8; Jones, *State*, 63–5.

101. Quoted in Santiago, *Ecology*, 65; Brown, *Oil*, 25–53; Brown, 'Domestic', 387–406; Davis, *Dark*, 34–59; Redwood, *Petroleum*, vol. 1, 132–3; ibid., vol. 3, 1,088, 1,123; *Mineral Resources of the U.S., 1899*, 181–2; ibid., *1900*, 589–90; ibid., *1904*, 730; 'Mexican Oil Development', *Petroleum World* (12 August 1904), 465; Garner, *British*, 139–49.

102. Lisa Bud-Frierman et al., 'Weetman', 279–86; Redwood, ibid., vol. 1, 137.

103. Quoted in Santiago, *Ecology*, 66.

104. *Mineral Resources of the U.S., 1899*, 184; ibid., *1900*, 591–3; Redwood, *Petroleum*, vol. 1, 147–51; Gerali and Riguzzi, 'Gushers', 413–14.

105. Redwood, ibid., 19–48; *Mineral Resources of the U.S., 1899*, 218–44; ibid., *1900*, 613–17; ibid., *1904*, 742–6; Nowell, *Mercantile*, 46–8, 72; Williamson et al., *Energy*, 248; McKay, *Scotland*, 50–62; Gerali and

Lipparini, 'Maiella', 290–6; Gerali et al., 'Historical', 313–14; 'Galicia and Romania', *Petroleum World* (21 October 1905); 'The Oil Fields of Germany and Italy', ibid.; 'Natural Gas in England', ibid.; 'Principal Petroleum Resources' (30 June 1904), 107.

106. *Mineral Resources of the U.S., 1904,* 733–4; Redwood, ibid., 11–19.

107. 'Saghalin Island', *Petroleum World* (2 December 1904).

108. Visvanath, *Hundred,* 5–16; Redwood, *Petroleum,* vol. 1, 55–61; ibid., vol. 3, 1,087–9; *Mineral Resources of the U.S., 1904,* 754–6; 'Principal Petroleum Resources' (30 June 1904), 101–2.

109. *Mineral Resources of the U.S., 1899,* 263–78; ibid., *1904,* 749–53; 'Petroleum in Japan', *Economic Geology* 13, no. 7 (November 1918), 512–15; 'Japanese Fields', *Petroleum World* (21 October 1905), 716.

110. Bradley, *Imperial,* ch. 4; McCormick, *China.*

111. Quoted in *Mineral Resources of the U.S., 1899,* 260.

112. 'The Petroleum Industry in the Philippine Islands', *Petroleum Review* (21 January 1905), 38.

113. Quoted in *Mineral Resources of the U.S., 1899,* 263; ibid., *1900,* 622; ibid., *1904,* 749.

114. Hidy and Hidy, *Pioneering,* 262–3; *Mineral Resources of the U.S., 1899,* 278; Gerretson, *History,* vol. 3, 190–4; Lee, 'China', 55–8; 'Great Chinese Concessions', *New York Times* (2 February 1899).

115. Redwood, *Petroleum,* vol. 1, 155–6; *Mineral Resources of the U.S., 1899,* 291–2; ibid., *1904,* 757; *Second Report on Coal Supplies,* vol. 2, 196; *Report of the Royal Commission,* vol. 2 (1913), 58–60, 109–17.

116. 'Australian Maritime History – No. VI', *Australian Star* (Sydney) (11 April 1891).

117. 'Eastern Notes', *Northern Territory Times and Gazette* (Darwin) (22 May 1891).

118. 'A Promising Venture', *Northern Territory Times and Gazette* (Darwin) (10 May 1901); Rau, *Mineral,* 108–9; 'Petroleum Mining in Timor', *Daily Telegraph* (Sydney) (26 October 1905).

119. *Second Report on Coal Supplies,* vol. 2, 196; Kesse, 'Search', 3.

120. Macdonald, *Gold,* 127.

121. *Second Report on Coal Supplies,* vol. 2, 196; Owen, *Trek,* 429, 1451; 'An American Consul and the Petroleum Fields of Algeria', *Petroleum World* (12 August 1904); Redwood, *Petroleum,* vol. 1, 48.

122. Jones, *State,* 113–14; Redwood, ibid., 49–51; Ardagh, 'Red Sea', 502–7; U.S. consular report, 'Petroleum Fields of Egypt', 423–5; Stewart, *Report.*

123. See ch. 7, section 'British Supremacy over Baghdad', and section '"The Virgin Oil Fields of Persia"' below; *Second Report on Coal Supplies*, vol. 2, 196.
124. 'The World's Supply of Oil', *Petroleum World* (May 1904), 241.
125. Loftus, 'On the Geology', 269–70.
126. Siegel, *Endgame*, 1–15; Wright, *English*; Ingram, *Britain*; Kelly, *Britain*.
127. Quoted in Greaves, *Persia*, 25.
128. Wright, *English*, 128–34; Shahvar, 'Concession', 181–4; Shahvar, 'Iron'; Shahvar, 'Tribes'; Searight, *British*, 168–70.
129. Quoted in Kazemzadeh, *Russia*, 151.
130. Quoted in Malley, 'Layard', 633; Kazemzadeh, ibid., 148–65; Greaves, *Persia*, 161–6; Greaves, 'British Policy – I', 46; Shahnavaz, *Britain*.
131. Kelly, *Britain*, 475–84, 499.
132. Quoted in Greaves, *Persia*, 162.
133. Ibid., 161–2; Khazeni, *Tribes*, ch. 3; Garthwaite, 'Bakhtiyari', 30; Kazemzadeh, *Russia*, 165–8; Kelly, *Britain*, 482–4, 496; Greaves, 'British Policy – I', 42; Shahnavaz, *Britain*, 163.
134. Curzon, *Persia*, vol. 1, 480; Keddie, *Religion*, 4; Galbraith, 'British Railway', 483–8.
135. Curzon, ibid., 484; Keddie, ibid., 8.
136. Curzon, ibid., 3–6.
137. Quoted in Greaves, 'Sistan', 93.
138. Quoted in Ferrier, *History*, vol. 1, 27; Issawi, *Economic*, 328.
139. Curzon, 'Karun', 526–7; Curzon, *Persia*, vol. 2, 519–21; Kazemzadeh, *Russia*, 164–215; Brockway, 'Britain', 41–2; Wright, *English*, 103–4; Khazeni, *Tribes*, 115–16.
140. Quoted in Ferrier, *History*, vol. 1, 26.
141. Bayat, *Iran*, 3–33; Keddie, *Religion*; Lambton, 'Tobacco'; Wright, *English*, 106–7.
142. Quoted in Keddie, ibid., 99.
143. Quoted in ibid., 78–9.
144. Quoted in Lambton, 'Tobacco', 142.
145. Quoted in ibid., 149.
146. Quoted in ibid., 139–40.
147. Quoted in Greaves, 'British Policy – I', 36.
148. Ibid., 43–4; Ferrier, *History*, vol. 1, 26.
149. Quoted in Greaves, ibid., 44.
150. Quoted in ibid., 41.
151. Quoted in Issawi, *Economic*, 312; Kazemzadeh, *Russia*, 358.

152. Gordon, *Persia*, 101.

153. Ferrier, *History*, vol. 1, 29; Davoudi, *Persian*, 11–13; Keddie, *Religion*, 46–7.

154. Goldsmid, 'De Morgan', 477; Khazeni, *Tribes*, 116.

155. Maunsell, 'Mesopotamian', 529.

156. Ferrier, *History*, vol. 1, 29–43; Jones, *State*, 97–9; Sorkhabi, 'Sir Thomas', 441; Greaves, 'British Policy – II', 295–6; *Report of the Royal Commission*, vol. 1 (November 1912), 315–17.

157. Quoted in Ferrier, ibid., 35.

158. Quoted in Davoudi, *Persian*, 66.

159. Ferrier, *History*, vol. 1, 38–43, 54; Davoudi, ibid., 13–22, 78; Kazemzadeh, *Russia*, 354–6.

160. Quoted in Greaves, 'British Policy – II', 297; Jones, *State*, 129–30.

161. Quoted in Ferrier, *History*, vol. 1, 43.

162. Quoted in ibid., 641; Kazemzadeh, *Russia*, 357; Shafiee, *Machineries*, 21, 26–8.

163. Issawi, *Economic*, 328–9; Fursenko, *Battle*, 130–1.

164. Quoted in Issawi, ibid., 330; Kazemzadeh, *Russia*, 359–60; Ferrier, *History*, vol. 1, 43–5.

165. Quoted in Greaves, 'British Policy – II', 301–2.

166. Quoted in Issawi, *Economic*, 331; Kazemzadeh, *Russia*, 328, 361–2.

167. Kazemzadeh, ibid., 383–5; Ferrier, *History*, vol. 1, 46; Davoudi, *Persian*, 31–2; Issawi, ibid., 333–4; Fursenko, *Battle*, 131–3.

168. Marder, ed., *Fear God*, vol. 1, 275; ibid., vol. 2, 235; Fisher, *Records*, 202.

169. Greaves, 'British Policy – II', 296–307; Ferrier, *History*, vol. 1, 60.

170. Issawi, *Economic*, 317–18; *Report of the Royal Commission*, vol. 1, 316, 554–5, 566–7; Ateş, *Ottoman*.

171. Ferrier, *History*, vol. 1, 54–67; Davoudi, *Persian*, 35–41; Corley, *History*, 96, 100.

172. Quoted in Ferrier, *History*, vol. 1, 60; Corley, ibid., 98–100; Mitchell, *Carbon*, 50.

173. Jones, 'Oil-Fuel', 137–42; Jones, *State*, 40–2; Fletcher, 'From Coal', 1–5; Snyder, 'Petroleum', 145–6; Jonker and Zanden, *History*, vol. 1, 67–8, 78, 81; 'Liquid Fuel Prospects in this Country', *Petroleum Review* (30 March 1907), 181–2.

174. Quoted in Snyder, 'Petroleum', 70.

175. Ibid., chs. 2–3; O'Brien, 'Titan', 146–54; Shulman, *Coal*, 157–63.

176. Snyder, ibid., 54.

177. Quoted in Mahajan, *British*, 129.

178. Quoted in O'Brien, 'Titan', 146.

179. Holmes, 'Mahan', 49–50; Clark, *Sleepwalkers*, 138–40, 158; Neilson, '"Greatly Exaggerated"', 695–720; Docherty and MacGregor, *Hidden*, chs. 3–5.

180. Melville, *Report*, 56; Snyder, 'Petroleum', chs. 4–5, 152–4, 157–8; Madureira, 'Oil', 83–9.

181. Quoted in Jones, 'Oil-Fuel', 144; Gibson, '"Oil Fuel"', 112–13; Snyder, ibid., 62–4, 126–44; Brown, 'Royal Navy', 49–50; Fletcher, 'From Coal', 6–7; Madureira, 'Oil', 88; Sullivan, 'Italian', 7–8.

182. Marder, ed., *Fear God*, vol. 1, 185, 220.

183. Quoted in Jones, 'Oil-Fuel', 144; Marder, ibid., 228–9.

184. Brown, 'Royal Navy', 43–9; Gröner, *German*, vol. 1, 10–14; 'German Torpedo Boats', *Manchester Guardian* (25 July 1898); 'World's Naval News', *San Francisco Call* (14 January 1901); 'Oil as Fuel for German Warships', *Washington Post* (12 September 1904); 'The Growing Importance of Liquid Fuel', *Petroleum World* (7 October 1904).

185. Marder, ed., *Fear God*, vol. 1, 235.

186. Melville, *Report*, 55; DeNovo, 'Petroleum', 643; Snyder, 'Petroleum', 113–15.

187. Snyder, ibid., ch. 5, 188–9; Brown, 'Royal Navy', 49–52, 59–60; Brown, *Grand Fleet*, 21–3; Henriques, *Marcus*, 386–7, 399–402; Longmuir, *Oil*, 203; 'Liquid Fuel in the British and American Navies', *Petroleum World* (1 July 1904); 'Oil Fuel in the Navy', *Petroleum World* (11 March 1905); 'British Ships to Try Oil', *Houston Daily Post* (15 February 1903); 'British Are Not Expert in The Use of Liquid Fuel', *San Francisco Call* (13 April 1903); 'The Naval War', *Daily Telegraph* (2 July 1906).

188. Roberts, *Battleship*, 25; Massie, *Dreadnought*, ch. 26.

189. Sondhaus, *Naval*, 211 (n. 39); Pearton, *Oil*, 49.

190. Jones, *State*, 40–5, 78.

191. Henriques, *Marcus*, 288, 355.

192. Jones, *State*, 24, 44; Snyder, 'Petroleum', 114–15, 128, 142; Henriques, ibid., 400–2; Nowell, *Mercantile*, 56; Melville, *Report*, 62.

193. Melville, ibid., 389; ibid., 62, 389–90, 394–5.

194. Jones, *State*, 43–5, 106; DeNovo, 'Petroleum', 642.

195. Dasgupta, *Oil*, 17–24; Corley, *History*, 68–71.

196. Quoted in Longmuir, *Oil*, 202.

197. Quoted in Snyder, 'Petroleum', 156.

198. Quoted in ibid., 157.

199. Ibid., 184–7; Williamson et al., *Energy*, 255; Pearton, *Oil*, 40–1; Heidbrink, 'Petroleum', 58; Flanigan, 'Some Origins', 121; Gerretson, *History*, vol. 3, 63–4, 78; ibid., vol. 4, 178; Hidy and Hidy, *Pioneering*, 516.

200. 'Liquid Fuel and Coal', *Petroleum World* (10 March 1906), 232.

201. Quoted in Snyder, 'Petroleum', 157; ibid., 184–7; Miller, *Straits*, 423.

202. Jones, *State*, 11–14; Jones, 'Oil-Fuel', 144–7; Snyder, ibid., 180–4; Corley, *History*, 80–1; Longmuir, *Oil*, 200–2; Brown, 'Royal Navy', 53–4.

203. Jones, ibid., 97, 100; Ferrier, *History*, vol. 1, 68; Corley, ibid., 81–5; Longmuir, ibid., 203–8; Snyder, ibid., 159–77.

204. Quoted in Snyder, ibid., 157.

205. Quoted in Jones, *State*, 99.

206. Quoted in ibid.

207. Quoted in ibid.

208. Quoted in Ferrier, *History*, vol. 1, 68.

209. Quoted in Jones, *State*, 108.

210. Ibid., 105–8; Higgins, *History*, 64–70, 79–81.

211. Quoted in Jones, ibid., 107–8.

212. Ibid., 108.

213. Quoted in ibid., 133.

214. See section '"The Virgin Oil Fields of Persia"' above; Ferrier, *History*, vol. 1, 44–6, 60–2; Corley, *History*, 98–104; Davoudi, *Persian*, 45–51; Nowell, *Mercantile*, 50–4, 61; Mitchell, *Carbon*, 53–4.

215. Quoted in Ferrier, ibid., 61.

216. Quoted in Corley, *History*, 101; HC Deb, 17 June 1914, vol. 63, cc. 1139, 1189–91; 'Company Meetings – Anglo-Persian Oil Company', *Times* (30 May 1914); Davoudi, *Persian*, 50, 103–8, 126–7.

217. Ferrier, *History*, vol. 1, 59–62, 67–72; *Report of the Royal Commission*, vol. 1, 335, 339, 342–3; Jones, *State*, 132–5; Mitchell, *Carbon*, 53–4.

218. Quoted in Longmuir, *Oil*, 207; Snyder, 'Petroleum', 174.

219. Quoted in Longmuir, ibid., 200.

220. Quoted in ibid., 208.

221. Quoted in ibid.; ibid., 212; Snyder, 'Petroleum', 174.

222. Jones, *State*, 109.

223. Quoted in ibid., 101.

224. Quoted in Longmuir, *Oil*, 166–7.

225. Quoted in ibid., 214; Jones, *State*, 91–3; Snyder, 'Petroleum', 162–3, 175–6; Hidy and Hidy, *Pioneering*, 501; Corley, *History*, 124.

226. Quoted in Jones, ibid., 96; Corley, ibid., 54–6.

227. Jones, ibid., 96–7; Corley, ibid., 88–93; Henriques, *Marcus*, 326, 356–80, 389–90; Jonker and Zanden, *History*, vol. 1, 68–72; Williamson et al., *Energy*, 253–9.

228. Henriques, ibid., 67, 81–2, 355, 360, 375, 490–1; Jonker and Zanden, ibid., 67, 79, 81–2; Pearton, *Oil*, 42–3.

229. Quoted in Jones, *State*, 96.
230. 'Meetings – "Shell" Transport and Trading Company, Limited', *Economist* (24 September 1904).
231. Quoted in Jones, *State*, 95; Longmuir, *Oil*, 210–12; Corley, *History*, ch. 6.
232. *Report of the Royal Commission*, vol. 1, 365–7.
233. Ibid., 338.
234. 'Oil Combine', *Daily Telegraph* (30 July 1907); Henriques, *Marcus*, 490–2; Jonker and Zanden, *History*, vol. 1, 81–2; Jones, *State*, 21–3; Beaton, *Enterprise*, 51–2.
235. Quoted in Jones, ibid., 102; Corley, *History*, 93; Brown, 'Royal Navy', 54, 58.
236. Corley, ibid., 60–2, 74–5, 125–6.
237. *Statement of the First Lord of the Admiralty*, 7.
238. Quoted in 'A Russian View of Liquid Fuel', *Petroleum World* (23 September 1904).
239. DeNovo, 'Petroleum', 642–4.
240. Snyder, 'Petroleum', 186–9; 'Fight Standard Oil by Way of the Isthmus', *New York Times* (1 March 1906); 'The Panama Pipe Line', *Petroleum World* (10 March 1906); 'Liquid Fuel Prospects in This Country', *Petroleum Review* (30 March 1907); 'The Union Oil Company's Panama Pipe Line', ibid. (27 April 1907); 'The Trans-Isthmian Pipe Line', ibid. (8 May 1909).
241. J.D. Henry, 'Oil Fields of the Empire', *Times* (15 January 1908); 'The Recent Crisis in the Bulk Oil-Carrying Trade', *Petroleum Review* (29 August 1908).
242. 'The Question of Timor', *Sydney Morning Herald* (11 January 1905).
243. Steyn, 'Oil', 252–3; Carland, 'Enterprise', 192–8; *Report of the Royal Commission*, vol. 1, 8, 158, 543; 'Petroleum Prospects in Nigeria', *Petroleum Review* (13 October 1906).
244. Falola, *Colonialism*, ch. 1; Isichei, *History*, ch. 15; Crowder, *Story*, ch. 12.
245. White Ant, 'Practical'.
246. 'Gold Coast Exploration and Trading Company Limited', *Economist* (19 July 1902); 'Colonial Oil Syndicate', *Financial Times* (25 January 1907); Kesse, *Search*, 3; Redwood, *Petroleum*, vol. 1, 159.
247. 'Petroleum in West Africa', *Financial Times* (27 June 1904); Rudin, *Germans*, 275–6, 284, 289; 'Is There Petroleum in Cameroon?', *Petroleum Review* (12 May 1906).
248. Foreign Office, *Diplomatic and Consular Report: Portugal, 1905*, 7.
249. 'Société Française de Pétrole, Limited', *Times* (18 February 1907); Thompson, *Oil Pioneer*, 308–12; Steyn, 'Oil', 255.

250. Quoted in Carland, 'Enterprise', 195.

251. Quoted in Steyn, 'Oil', 256–7; Carland, ibid., 195–8.

252. Quoted in Carland, ibid., 198; *Report of the Royal Commission*, vol. 1, 545–6.

253. Carland, ibid., 198–9.

254. 'The Nigerian Oilfield', *Financial Times* (3 August 1909).

255. Quoted in Steyn, 'Oil', 253–4.

256. 'The Nigerian Oilfield', op. cit.

257. Foreign Office, *Diplomatic and Consular Report: Portugal, 1908*, 47; Couceiro, *Angola*, 334–6.

258. 'Oil Share Market Notes', *Financial Times* (27 November 1909); Costa, *A riqueza*.

259. Steyn, 'Oil', 254; Carland, 'Enterprise', 200–3; *Report of the Royal Commission*, vol. 1, p. 158.

260. 'Paris Day by Day', *Daily Telegraph* (12 November 1900).

261. 'Big Rush for Automobile Patents', *Scientific American* (27 July 1901).

262. Barker, 'German', 1–25; Barker and Gerhold, *Rise*, 74–84; Barker, 'Spread', 149–67; Williamson et al., *Energy*, 186–93; U.S. consular report, 'Motor Machines', 9–16; Mom and Kirsch, 'Technologies'; Black, *Internal Combustion*, chs. 4–7; Black, *Crude*, 106–16.

263. 'Petrol Driven Road Tank Waggons', *Petroleum Review* (4 March 1905).

264. Barker and Robbins, *History*, vol. 2, 118–36, 166–70.

265. Coffin, 'American', 43.

266. Dash, 'British', 67, 98, 123–4, 210; Jones, 'Oil-Fuel', 148.

267. Quoted in Dash, ibid., 156.

268. Bartholomew, *Early*, 5; White, *Tanks*, 112.

269. 'New Armored Car', *Horseless Age* (18 April 1906); 'New Warfare Against Yaquis', *St. Louis Post-Dispatch* (22 April 1906).

270. 'Requisition of Motor Trucks for Military Purposes', *Horseless Age* (14 March 1906); 'British Military Gasoline Motor Tractors', ibid. (23 May 1906); White, *Tanks*, 113.

271. Taylor, *Aircraft*, 8.

272. Gibbs-Smith, *Aviation*, 1–64; Gollin, *No Longer*, 22–3; Taylor, ibid., 8–9.

273. Quoted in Gibbs-Smith, ibid., 67.

274. Ibid., chs. 11–12; Taylor, *Aircraft*, 9–11.

275. Quoted in Gollin, *No Longer*, 70; ibid., chs. 1–5, 12; Gibbs-Smith, *Aviation*, chs. 11–12.

276. Quoted in Gollin, ibid., 194.

277. Ibid., 111–314.

278. Quoted in ibid., 381; Gibbs-Smith, *Aviation*, 121–37.

279. Quoted in Gibbs-Smith, ibid., 148; Gollin, ibid., 281–3, chs. 13–14; Gollin, *Impact*, chs. 1–4.

280. *Second Report on Coal Supplies*, vol. 2, 203–5; Williamson et al., *Energy*, 172–3; 'Gasoline Stoves and How to use Them', *Scientific American* (19 May 1888).

281. 'Petroleum and the Motor-Car Industry', *Times* (23 August 1902).

282. 'Tank Steamers in 1905', *Petroleum World* (27 January 1906); Beaton, *Enterprise*, 56–7; Deterding, *International*, 84–7; 'The Kansas and Indian Territory Oil Fields', *Petroleum Review* (19 November 1904).

283. Quoted in Henriques, *Marcus*, 316.

284. Quoted in ibid., 309.

285. *Report of the Royal Commission*, vol. 1, 329.

286. Deterding, *International*, 41; *Report of the Royal Commission*, vol. 2, 70.

287. Jonker and Zanden, *History*, vol. 1, 27.

288. 'Automobiles', *Manchester Guardian* (10 June 1903); 'Alcohol Fuel Again', *Horseless Age* (17 June 1903).

289. 'Automobiles', *Manchester Guardian* (6 April 1904).

290. 'The Era of Bye-Products', *Petroleum Review* (6 August 1904).

291. U.S. consular report, 'Alcohol in Germany', 201–2; *Second Report on Coal Supplies*, vol. 2, 210–2; 'Oil Combine', *Daily Telegraph* (30 July 1907).

292. Quoted in McCarthy, 'Coming'; 'Gasoline Getting Scarce', *New York Times* (18 January 1906).

293. Quoted in McCarthy, ibid., 51.

294. Ibid., 50–3; 'Alcohol Bill Passes Senate', *Horseless Age* (30 May 1906).

295. Bryant, 'Development', 432–46; 'The Diesel Oil Engine', *Horseless Age* (September 1897); Lytle, 'Introduction', 115–23; Cummins Jr, *Diesel*, 198–9; 'Explosions in Coal-Mines', *Science* 9, no. 222 (6 May 1887).

296. Williamson et al., *Energy*, 185; Tolf, *Russian*, 172–5; Thompson, *Oil Fields*, 230–9; Jones, 'Oil-Fuel', 142–3; Jonker and Zanden, *History*, vol. 1, 132.

297. 'From a Lamp Oil to a By-Products Age', *Petroleum Times* (17 June 1949); Barker, 'German', 48.

298. U.S. consular report, 'Impending Petrol Famine', 61.

299. 'Petroleum Spirit from the Far East', *Petroleum World* (27 January 1906); Jones, *State*, 42.

300. Henriques, *Marcus*, 489–99; 'The Suez Canal and the Carriage of Benzine', *Petroleum Review* (27 April 1907); 'Another Red Letter Day for

Bulk Benzine', ibid. (10 October 1908); 'The Petroleum Trade of England during 1908', ibid. (2 January 1909); 'Oil Combine', *Daily Telegraph* (30 July 1907); Deterding, *International*, 84–5; Gerretson, *History*, vol. 3, 143, 273–9.

301. U.S. consular report, 'Consumption of Petrol', 62; Snider, *Oil*, 139–45; 'The Era of Bye-Products', *Petroleum Review* (6 August 1904); 'The Future of Petrol', ibid. (18 August 1906); 'The Kansas and Indian Territory Oil Fields', ibid. (19 November 1904).

302. Andrist, *Long*, 3–10; Thornton, 'Cherokee'; Thorne, *World's Richest*, 21; Finkelstein, 'History', 32–45.

303. Quoted in Miner, *Corporation*, 79; Carroll, 'Shaping', 127–30.

304. Quoted in Miner, ibid., 48.

305. Quoted in ibid., 29.

306. Quoted in McLoughlin, *After*, 272.

307. Thorne, *World's Richest*, 10; Williams, 'United States', 810–31.

308. U.S. Congress, 'Views of the Minority', *Lands in Severalty to Indians*, 8–10.

309. Ibid., 5.

310. Quoted in Miner, 'Cherokee', 65.

311. Quoted in Nash and Strobel, *Daily*, 239.

312. Dawes, 'Unknown'; 'Indians Coming to Terms', *New York Times* (1 January 1897).

313. Richardson, ed., *Compilation*, vol. 15, 6,672; Thorne, *World's Richest*, 8–10; Carroll, 'Shaping', 130.

314. Franks, *Oklahoma*, 3–20; Miner, 'Cherokee', 46–56; Finney, Sr, 'Indian', 153–5; Wilson, *Underground*, 99–104.

315. Quoted in Miner, ibid., 57.

316. Quoted in ibid., 154–7; Miner, 'Cherokee', 46–7.

317. Thorne, *World's Richest*, 30–1.

318. Franks, *Oklahoma*, 23, 36; Miner, *Corporation*, 171–4.

319. Franks, ibid., 26–35; Miner, ibid., 157–8; Beaton, *Enterprise*, 113–14.

320. Williamson et al., *Energy*, 90–1; Johnson, *Petroleum*, 18–19, 36; Johnson, *Development*, 208–12; 'Indian Affairs', pt. 2, 237; Hidy and Hidy, *Pioneering*, 399–400; Franks, *Oklahoma*, 31, 42; 'The Location of the American Oil Fields', *Petroleum Review* (29 April 1905); 'The American Mid-Continent Fields', ibid. (22 December 1906).

321. Condra, 'Opening', 339.

322. 'From Indian Territory to the Mexican Gulf', *Petroleum Review* (8 December 1906); 'Amalgamation of the Interests of the Gulf-Guffey Oil Companies', ibid. (22 December 1906); 'The American Mid-Continent

Fields', ibid. (22 December 1906); Williamson et al., *Energy*, 92–4; Johnson, 'Early', 523–5; J.D. Henry, 'The Manufacture of Petrol', *Times Engineering Supplement* (11 March 1908).

323. Thorne, *World's Richest*; Beaton, *Enterprise*, 113–14, 124–5.

7. A VOLATILE MIX

1. 'The Asphalt Lake in Venezuela', *New York Times* (17 February 1901).
2. Ibid.
3. McBeth, *Gunboats*, 41–4.
4. Quoted in ibid., 46; Holley, 'Blacktop', 709–13.
5. McBeth, ibid., 44–7.
6. Ibid., 111–2, 115, 141.
7. Quoted in ibid., 112.
8. Ibid., 43, 45, 49–50, 58, 113, 125.
9. Ibid., 49.
10. 'Venezuelan Crisis Acute', *New York Times* (17 January 1901).
11. Quoted in Schoultz, *Beneath*, 179.
12. McBeth, *Gunboats*, 41, 49–50, ch. 4., 105–9, 126–7, 137–47, 151, 160, 173–6, 182–4, 189–97.
13. Quoted in Schoultz, *Beneath*, 180; McBeth, ibid., 84.
14. McBeth, ibid., ch. 6; Holmes, 'Mahan', 45.
15. Quoted in McBeth, ibid., 84.
16. 'Germany in Venezuela', *New York Times* (17 August 1902).
17. Quoted in Schoultz, *Beneath*, 115; Fordham, 'Protectionist', 32–7.
18. Maass, 'Catalyst'; Schoultz, ibid., 183–5; LaFeber, *New*, ch. 6.
19. Quoted in Ewell, *Venezuela*, 104; McBeth, *Gunboats*, 110–23.
20. McBeth, ibid., 184–5.
21. Lieuwen, *Petroleum*, 9–11; McBeth, *Juan Vicente*, 11–12; Tinker Salas, *Enduring*, 42–3; Gerretson, *History*, vol. 3, 274.
22. Lieuwen, ibid., 10–13; McBeth, *Gunboats*, 152, 197–8, 215, 223; Conlin, *Mr Five*, 106–7.
23. McBeth, ibid., 151–9, 167, 226–30.
24. Ibid., 191–235.
25. Ibid., 167, 228–9, 246–55.
26. Quoted in McBeth, *Dictatorship*, 29.
27. Ibid., 15–18, 34, 39–45.
28. Quoted in ibid., 40.
29. Gerretson, *History*, vol. 4, 276; Philip, *Oil*, 10.
30. Coons, 'Statistics', 323–4.

31. Quoted in McKay, *Pioneers*, 274; ibid., 286–94; Fursenko, *Battle*, ch. 2; Siegel, 'Russian', 24–6; Tolf, *Russian*, 183–5; Gerretson, *History*, vol. 4, 147.

32. McKay, ibid., 286–90; Von Laue, 'Russian'; Roosa, *Russian*, 2–3.

33. Quoted in Pospielovsky, *Russian*, 27.

34. Pipes, *Russian*, 4–8.

35. Ibid., 12–13; Hopkirk, *Great*, 502–9.

36. Quoted in Mostashari, *On the Religious*, 100; Swietochowski, *Russia*, chs. 2–3; Altstadt, *Azerbaijani*, 15–35; Tolf, *Russian*, 150–1; Thompson, *Oil Fields*, 125–6, 375–6.

37. Tolf, ibid., 183–5; Von Laue, 'Russian', 75–6.

38. Pospielovsky, *Russian*, 12–48; Tolf, ibid., 152–5; Marriott and Minio-Paluello, *Oil Road*, 44.

39. De Jonge, *Stalin*, 21–8; Deutscher, *Stalin*, 19–21, 66; Von Laue, *Sergei*, 256; Tolf, ibid., 155–6; Mostashari, *On the Religious*, 102–3; Siegel 'Russian', 27.

40. 'The Latest from Baku', *Petroleum World* (14 January 1905).

41. Quoted in Siegel, 'Russian', 27, and 'The Bother at Baku', *Petroleum World* (28 January 1905), respectively.

42. Docherty and MacGregor, *Hidden*, 85–91; Chapman, 'Russia'.

43. 'The War and the Oil Trade of Manchuria', *Petroleum World* (21 October 1904).

44. 'Meetings – "Shell" Transport and Trading Company, Limited', *Economist* (24 September 1904).

45. Henriques, *Marcus*, 384, 482–3; Chernow, *Titan*, 373; Docherty and MacGregor, *Hidden*, 92–3.

46. Gray, 'Black', 201–2; Docherty and MacGregor, ibid., 89–91; Chapman, 'Anglo–Japanese', 52.

47. 'Meetings – "Shell"', op. cit.; Gerretson, *History*, vol. 4, 154.

48. Quoted in Siegel, 'Russian', 27.

49. Pipes, *Russian*, 13–50; Harcave, *First*, 63–4, 107, 112–3; Deutscher, *Stalin*, 66–8, 85; Bayat, *Iran*, 76–98; Altstadt, *Azerbaijani*, 36–71; Alekperov, *Oil*, 133–7; Docherty and MacGregor, *Hidden*, 86–7, 91–3; Gerretson, *History*, vol. 3, 131–4; Roosa, *Russian*, 7; 'Russian Oil Companies and the Strike', *Financial Times* (13 January 1905); 'The Baku Massacre', *Times* (1 April 1905); 'The Anarchy at Baku', ibid. (11 September 1905); 'Leader – The Anarchy in the Caucasus', ibid. (23 September 1905); 'The Outbreak in the Caucasus', ibid. (23 September 1905); 'Leader – Disorders in the Caucasus', ibid. (14 October 1905); 'The Outbreak in the Caucasus', *The Speaker*

(23 September 1905); Penslar, 'Tracing'; Bronner, *Rumor*; Tolf, *Russian*, 180–1.

50. Tolf, ibid., 156–62; Mostashari, *On the Religious*, 101–5, 141–2; Siegel, 'Russian', 27–31; Kelly, 'Crisis', 296–9; Deutscher, *Stalin*, 68; Russian Petroleum and Liquid Fuel Co. Ltd., 'Memorandum of evidence of omission on part of Russian Government', in Burdett, ed., *Oil*, vol. 1, 127–9; 'Despatch No. 55 from Mr P. Stevens', in ibid., 192–4; 'The Truth about Baku', *Petroleum World* (11 March 1905); 'The Baku Industry', ibid. (25 March 1905); 'Serious Outlook at Baku', ibid. (15 July 1905); 'Baku in February and September', ibid. (9 September 1905); 'The Rising in the Caucasus', *Times* (14 September 1905).

51. 'Despatch No. 3 from Mr P. Stevens', in Burdett, ed., ibid., 149–51.

52. 'Despatch No. 21 from Mr P. Stevens', in ibid., 152–3, 157–62.

53. 'Despatch No. 30 from Mr P. Stevens', in ibid., 164–5.

54. 'Despatch No. 31 from Mr P. Stevens', in ibid., 167.

55. 'Despatch No. 36 from Mr P. Stevens', in ibid., 173–4.

56. 'Despatch No. 43 from Mr P. Stevens', in ibid., 177.

57. Ibid., 178–80.

58. 'Despatch No. 55 from Mr P. Stevens', in ibid., 192.

59. Quoted in Siegel, 'Russian', 27–8, 31.

60. Quoted in Mostashari, *On the Religious*, 103.

61. 'Leader – The Disorders in the Caucasus', *Times* (9 September 1905).

62. 'Despatch No. 45 from Mr P. Stevens', in Burdett, ed., *Oil*, vol. 1, 182.

63. 'Bibi-Eybat Petroleum', *Financial Times* (9 December 1905).

64. 'Despatch No. 45 from Mr P. Stevens', in Burdett, ed., *Oil*, vol. 1, 182–3.

65. 'Despatch No. 631 from Sir Charles Hardinge', in ibid., 137–40.

66. Ibid., 132–6.

67. 'Despatch No. 44' from Sir Cecil Spring-Rice, in ibid., 144.

68. Kelly, 'Crisis', 299–318; Foreign Office, *Diplomatic and Consular Reports: Batoum, 1905*, 3–18; Gerretson, *History*, vol. 3, 135–55; 'The Oil Famine in Russia', *Petroleum World* (23 September 1905); 'The Russian Liquid Fuel Crisis', ibid. (7 October 1905); 'Baku and the Oil Trade of the World', ibid. (4 November 1905); 'Russian Petroleum', *Financial Times* (22 August 1905); 'Baku Oil Losses', ibid. (30 September 1905); 'The Caucasus', *Times* (8 September 1905); 'The Anarchy at Baku', ibid. (9 September 1905); Thompson, *Oil Fields*, 30; Hidy and Hidy, *Pioneering*, 511. Fursenko, *Battle*, 159–62; Jones, *State*, 58–9; Tolf, *Russian*, 182.

69. 'The Capture of Baku Trade', *Financial Times* (25 September 1905); 'Dividends and Reports', ibid. (4 December 1905); Thompson, *Oil Pioneer*, 301.

70. 'Roumania Oil Industry', *Financial Times* (22 September 1905).

71. 'The "Big Hand" of the Standard Stretched Out in Europe', *Petroleum World* (26 August 1905); Fursenko, *Battle*, 99.

72. 'Texas Opinion of the Russian Troubles', *Petroleum World* (21 October 1905); Alekperov, *Oil*, 131–7.

73. Quoted in 'Mr. Hagelin and the Forces of Disorder', *Petroleum World* (7 October 1905); 'The Outbreak in the Caucasus', *Times* (22 September 1905).

74. Quoted in Ascher, 'Introduction', 2.

75. Chaqueri, *Russo–Caucasian*; Bayat, *Iran*, 98–105.

76. Siegel, *Endgame*, 1–15.

77. Garthwaite, 'Bakhtiyari', 30–3; Shahnavaz, *Britain*, 111–35; Kazemzadeh, *Russia*, 428–47; Khazeni, *Tribes*, ch. 3, 118–20; Ferrier, *History*, vol. 1, 115–6, 121–2; Siegel, ibid., 56.

78. Quoted in Khazeni, ibid., 121.

79. Quoted in ibid., 121–2.

80. Ferrier, *History*, vol. 1, 55–6, 66–7, 75–7; Davoudi, *Persian*, 55–64; Khazeni, ibid., 121–3; Chen, 'British', 59–65; Corley, *History*, 104–5; Garthwaite, *Khans*, 108–11; Kazemzadeh, *Russia*, 677; H. Hoeffer, 'The Petroleum Deposits of Persia', *Petroleum Review* (29 September 1906); *Report of the Royal Commission*, vol. 1, 556–9.

81. Khazeni, ibid., 123–46, 184–5.

82. Quoted in ibid., 131.

83. Quoted in ibid., 124.

84. Quoted in ibid., 140.

85. Quoted in Ferrier, *History*, vol. 1, 81.

86. Quoted in Khazeni, *Tribes*, 145, and Davoudi, *Persian*, 88.

87. Quoted in Ferrier, *History*, vol. 1, 85–6; Davoudi, ibid., 58–64, 70, 74–5, 81–3, 86–91; Khazeni, ibid., 142–6; Chen, 'British', 64–8; Garthwaite, 'Bakhtiyari', 36; Jones, *State*, 135–7.

88. Bonakdarian, *Britain*, 27–9, 49–58, 63–4; Davoudi, *Persian*, 65–83; Chen, ibid., 107–10; Siegel, *Endgame*, 30–2.

89. Quoted in Greaves, 'British Policy – I', 79; ibid., 80–1; Siegel, ibid., 15–24, 35; Stebbins, *British*, chs. 3–4; Davoudi, ibid., 83–7.

90. Quoted in Siegel, ibid., 24.

91. Quoted in Bonakdarian, *Britain*, 64.

92. Quoted in ibid.

93. Ibid., 37–49, 67–70, 74–105, 133–7; Bayat, *Iran*, 243–8; Siegel, *Endgame*, 62–8.
94. Quoted in Otte, *Foreign*, 309; Wilson, 'Grey', 275–9.
95. Quoted in Wilson, ibid., 277–8; Bonakdarian, *Britain*, 49–51.
96. Abrahamian, *History*, 39–50; Siegel, *Endgame*, 14–32, 49; Klein, 'British', 35–40; Greaves, 'British Policy – I', 73–84; Chen, 'British', 90–2; Shahnavaz, *Britain*, 176–8; Feis, *Europe*, 366–71; Wilson, 'Creative', 58–9; Ferrier, *History*, vol. 1, 91, 141–2; Davoudi, *Persian*, 69; Kent, *Moguls*, 5–23; Kent, *Oil*, 16.
97. Ferguson, *Pity*, 60–1; Siegel, ibid., 46–7.
98. Wilson, *SW. Persia*, 38.
99. Quoted in Ferrier, *History*, vol. 1, 82; Davoudi, *Persian*, 92–3; Corley, *History*, 124.
100. Wilson, *SW. Persia*, 40.
101. Ibid., 42; Khazeni, *Tribes*, 147–8; Ferrier, *History*, vol. 1, 88; Davoudi, *Persian*, 95–6; Corley, *History*, 132–8; Jones, *State*, 138–40; Greaves, 'Some Aspects – II', 298.
102. Wilson, ibid., 42.
103. Ibid.
104. Quoted in Ferrier, *History*, vol. 1, 90; Khazeni, *Tribes*, 146–8.
105. Quoted in Ferrier, ibid., 90.
106. Quoted in ibid., 91.
107. Quoted in ibid., 127; Davoudi, *Persian*, 109.
108. Quoted in Davoudi, ibid., 103.
109. Quoted in Jones, *State*, 140; Davoudi, ibid., 35–7, 56–62, 98–105, 111–9, 139–40; *Report of the Royal Commission*, vol. 1, 558; Ferrier, *History*, vol. 1, 89–104; Garthwaite, 'Bakhtiyari', 36; Cronin, 'Politics', 4–5.
110. Khazeni, *Tribes*, 148–51, ch. 5; Kazemzadeh, *Russia*, 510–49; Siegel, *Endgame*, 38–43; Bonakdarian, *Britain*, 106–76; Chen, 'British', 108–30; Davoudi, ibid., 109.
111. Quoted in Ferrier, *History*, vol. 1, 92.
112. Spring Rice to Grey, 28 March 1907 (NA FO 800/70, 100–1); Bonakdarian, *Britain*, 63; Klein, 'British', 735.
113. Davoudi, *Persian*, 112; Khazeni, *Tribes*, ch. 5; Bonakdarian, ibid., 167–201; Ferrier, *History*, vol. 1, 126–9; Klein, ibid., 746–8; Garthwaite, 'Bakhtiyari', 37; Shahnavaz, *Britain*, chs. 6–7, 162–8; Kazemzadeh, *Russia*, 542–5; Bayat, *Iran*, 258–60; Siegel, *Endgame*, 57–60.
114. Quoted in Ferrier, ibid., 106; Davoudi, ibid., 107–111.
115. Williamson, *In a Persian Oil Field*, 126–7; Davoudi, ibid., 119–21.
116. 'The Land of the Bakhtiaris', *Times*, 6 May 1910.

117. Ferrier, *History*, vol. 1, 104–40; Khazeni, *Tribes*, 124–9; Chen, 'British', 68–80; Jones, *State*, 141–2; Garthwaite, 'Bakhtiyari', 36.

118. 'German Influence in Persia', *Times* (30 March 1908); Hoffman, *Great*, 164.

119. Onley, 'Britain's Informal'; Shahvar, 'Concession'.

120. HL Deb, 29 July 1878, vol. 242, cc. 508–9.

121. Curzon, *Persia*, vol. 2, 465.

122. Ibid., 578.

123. Kirk, *Short*, 89–90; Feis, *Europe*, 313–44; Hoffman, *Great*, 139–41.

124. Wendler, *Friedrich*, 214–5, 242–6.

125. Earle, *Turkey*, ch. 3; McMurray, *Distant*, 18–23; Barth and Whitehouse, 'Financial', 116–9.

126. Earle, ibid., chs. 2–3; McMurray, ibid., ch. 1; Feis, *Europe*, 313–48; Barth and Whitehouse, ibid., 117–21; Henderson, 'German'.

127. Kent, *Oil*, 16; Longrigg, *Oil*, 27.

128. See ch. 1, section 'Middle East and Central Asia'; Kinneir, *Geographical*, 38–40, 281, 298; Rich, *Narrative*, vol. 1, 22–3, 27–31; Mignan, *Winter*, vol. 2, 25–6; Loftus, *Travels*, 9–11, 26, 30, 41, 46–8, 70, 128–30, 169–70, 246, 249, 376.

129. Rich, ibid., 31.

130. Loftus, 'On the Geology', 248.

131. 'Petroleum Production, 1871', in Issawi, ed., *Fertile*, 402–3.

132. Foreign Office, 'Report by Major Everett'; Akpınar and Altınbilek, 'Pulk–Balıklı'.

133. Rich, *Narrative*, vol. 1, 31; Issawi, *Fertile*, 402; Longrigg, *Oil*, 14; Dann, 'Report', 287–8.

134. Ediger and Bowlus, 'Greasing', 197–8; Longrigg, ibid., 13–14; Kent, *Oil*, 15; 'The Mesopotamian Oilfields', *Oil Engineering and Finance* 3 (17 February 1923).

135. Ediger and Bowlus, ibid., 197; Hopkirk, *On Secret Service*, 22.

136. Black, *Banking*, 101–4.

137. See Rockefeller, ch. 5, section 'The Markets of the World'.

138. Longrigg, *Oil*, 14; Fitzgerald, 'France', 700–1; Maunsell, 'Mesopotamian'.

139. Maunsell, ibid., 532.

140. Quoted in Yavuz, 'Ottoman', 58–9, and Ediger and Bowlus, 'Greasing', 199; Kent, *Oil*, 17, 217 (n. 20); Hopkirk, *On Secret Service*, 22.

141. Earle, *Turkey*, 32–5; Kumar, 'Records', 70–1; Feis, *Europe*, 344–5; Kirk, *Short*, 92.

142. Quoted in Dugdale, ed., *German*, vol. 2, 467–8.

143. Ibid., 468.
144. McMurray, *Distant*, 29–33; Earle, *Turkey*, 39–69; Feis, *Europe*, 344–6; McMeekin, *Berlin–Baghdad*, 43; Ediger and Bowlus, 'Greasing', 198–9; Shorrock, 'Origin', 135–6.
145. Hoffman, *Great*, 159–61.
146. Busch, *Britain*, ch. 3; Black, *Banking*, 118–9; Cohen, *British*, 5.
147. Onley, 'Britain's Informal', 30–2.
148. Curzon, *Persian Question*, vol. 2, 443.
149. Busch, *Britain*, ch. 7.
150. 'Turkey and Koweit', *Times* (28 Jan 1911).
151. Quoted in Lauterpacht et al., ed., *Kuwait*, 18.
152. Quoted in Kent, *Oil*, 16.
153. Quoted in 'The Mesopotamian Oilfields', *The Near East* (26 October 1917); Earle, *Turkey*, 15.
154. Deren, 'German', 147–78; Cohen, *British*, 64 (n. 7).
155. Quoted in Kent, *Oil*, 16.
156. Whigham, *Persian*, 241.
157. Longrigg, *Oil*, 28; Kent, *Oil*, 50.
158. Earle, *Turkey*, 67–92; Barth and Whitehouse, 'Financial', 122–4; Kent, ibid., 16.
159. Shorrock, 'Origin', 135–40; Feis, *Europe*, 348–52; McMeekin, *Russian*, 14–15.
160. Quoted in Cohen, *British*, 35; Kazemzadeh, *Russia*, 404–5.
161. Cohen, ibid., 27–38; Francis, 'British'; Hoffman, *Great*, 143–50; Busch, *Britain*, 222–4; McMeekin, *Berlin–Baghdad*, ch. 2.
162. Kent, *Oil*, 16–17; Earle, *Turkey*, 15.
163. Longrigg, *Oil*, 28; Kent, ibid., 17, 217 (n. 20); Cohen, *British*, 45, 53; Hamm, 'British', 116, 189–90.
164. 'The Mesopotamian Oil Concessions', *Petroleum World* (25 March 1905); Dann, 'Report'; Black, *Banking*, 122; Kent, ibid., 217 (n. 20); Miller, *Straits*, 23.
165. U.S. consular report: 'Oil Fields in Persia', 261; 'Petroleum in Mesopotamia', *Petroleum World* (29 July 1905); 'Occurrences of Petroleum in Mesopotamia', *Petroleum Review* (15 September 1906).
166. Pearton, *Oil*, 49–50; Gerretson, *History*, vol. 4, 185.
167. Kumar, 'Records', 73–5; Kent, *Moguls*, 69; Henderson, 'German Economic Penetration', 60.
168. Docherty and MacGregor, *Hidden*, ch. 4; McMeekin, *Berlin–Baghdad*, ch. 1.
169. Gooch and Temperley, eds., *British*, vol. 6, 325.

170. Quoted in Cohen, *British*, 45.

171. 'Baghdad Railway', *Times* (13 November 1906).

172. Kent, *Oil*, 18–24, 53; Gerretson, *History*, vol. 3, 232; vol. 4, 187–8; Longrigg, *Oil*, 27–8; Issawi, 'Prospects of Petroleum in Baghdad Vilayet, 1918', in Issawi, *Fertile*, 405; *Report of the Royal Commission*, vol. 1, 347.

173. Quoted in Kent, ibid., 21.

174. Quoted in Neilson, 'Baghdad', 160.

175. Busch, *Britain*, 304–12, 353–7, 373–9.

176. Quoted in Kent, *Moguls*, 12.

177. Quoted in Kumar, 'Records', 78.

178. Quoted in ibid.

179. Quoted in Cohen, *British*, 40.

180. Quoted in ibid., 35; Busch, *Britain*, 357–69.

181. Kaya, 'Western'.

182. Quoted in Kedourie, 'Young', 99–100; Öke, 'Young'.

183. Longrigg, *Oil*, 28–30; Cohen, *British*, 118–19.

184. DeNovo, *American*, ch. 3; Kent, *Oil*, 26–9.

185. Quoted in DeNovo, ibid., 71 (n. 31).

186. Gibb and Knowlton, *History*, 285.

187. Kent, *Oil*, 22–5; Conlin, 'Debt', 525, 536–7; Gerretson, *History*, vol. 3, 244.

188. Quoted in Fursenko, *Battle*, 117; Fitzgerald, 'France', 700–1.

189. Quoted in Kent, *Oil*, 25; *Report of the Royal Commission*, vol. 1, 362, 368.

190. Quoted in Cohen, *British*, 119.

191. Quoted in ibid., 110.

192. Garner, *British*, 139–57; Brown, *Oil*, 17–42, 47–55; Brown, 'Domestic', 388–410; Santiago, *Ecology*, 64–7; Skirius, 'Railroad', 25–8; Jones, *State*, 63–8; Katz, *Secret*, 21–30.

193. Santiago, ibid., 83–4, 133–9; Brown, *Oil*, 60–1, 115–20; Brown, 'Domestic', 406–7.

194. Santiago, ibid., 83, 138.

195. Ibid., 84, 134–44.

196. Quoted in ibid., 139.

197. Garner, *British*, 157–63; Bud-Frierman et al., 'Weetman', 286–8; Brown, *Oil*, 42–3, 63–75, 117, 122, 126–7, 257; Davis, *Dark*, 70–9; Santiago, ibid., 85–90, 98–9, 104–34, 145–6, 154–62; Santiago, 'Culture', 67; Santiago, 'Class', 173–84; Hall and Coerver, 'Oil', 229–36; Munch, 'Anglo-Dutch-American', 163–77; Gerali and Riguzzi, 'Gushers'.

198. Williamson et al., *Energy*, 28–9, 176, 182; *Mineral Resources of the U.S., 1912*, pt. 2, 466–72; *Report of the Royal Commission*, vol. 1, 78, 87–8, 99–108, 290–3, 572–3; Brown, ibid., 103–7, 115, 122–4; Hart, *Revolutionary*, 153–4; Redwood, *Petroleum*, vol. 3, 1066,1088.

199. Katz, *Secret*, 5–41; Skirius, 'Railroad', 25–36; Hart, *Revolutionary*, 237–49; Brown, ibid., 89–100, 171–80; Garner, *British*, 170–4; Berger, *St. Louis*, 94–100; 'Sleuths Trailing Newly-Made Lord', *New York Times* (24 June 1910).

200. Katz, ibid., 156–74; Garner, ibid., 172–6, 183–92; Skirius, ibid., 36–40; Brown, ibid., 180–5; Davis, *Dark*, 90–6; Schoultz, *Beneath*, 239–40; Berger, ibid., 100–1; 'Rival Oil Giants in Mexican Broils', *New York Times* (9 November 1913).

201. Quoted in Churchill, *Winston*, vol. 2, Companion pt. 3, 1,930; Garner, ibid., 179–81; Jones, *State*, 75–6.

202. Quoted in Garner, ibid., 186; Calvert, *Mexican*, 163.

203. Quoted in Middlemas, *Master*, 223; Garner, ibid., 167; Schoultz, *Beneath*, 243–4; Jones, *State*, 72; 'Oil Interests in Mexico', *Times* (25 November 1913).

204. Quoted in Schoultz, ibid., 244.

205. HC Deb, 17 July 1913, vol. 55, c. 1479; Churchill, *Winston*, vol. 2, Companion pt. 3, 1952; Jones, *State*, 69, 72.

206. Katz, *Secret*, 170–81, 188–90; Garner, *British*, 193–6.

207. Quoted in Calvert, *Mexican*, 221–3; Healy, *Drive*, 100–9; Challener, *Admirals*, 21, 81–2.

208. Schoultz, *Beneath*, ch. 8, 158.

209. Quoted in ibid., 173; Major, *Prize*, ch. 1–2; Schoultz, ibid., 158–75; Jones, *Limits*, 407–8; Bucheli, 'Negotiating', 532–3; Stuart, 'Panama', 337.

210. Challener, *Admirals*, 13–45, 81–101, 323–5; Healy, *Drive*, ch. 6; Schoultz, ibid., 224, 231; 'England Fortifying West Indies', *New York Times* (28 September 1902); 'Uncle Sam's Naval Base in the Caribbean Sea', ibid. (5 October 1902); 'Germany in Venezuela', ibid. (17 August 1902).

211. Quoted in 'Monroe Doctrine Demands Big Navy', *New York Times* (12 March 1900); 'The Canal in Time of War', ibid. (10 February 1900); Holmes, 'Mahan', 32–3; Mahan, 'Panama'.

212. Quoted Healy, *Drive*, 100.

213. General Staff Report, *Organization of the Land Forces*, 9; Yerxa, 'United States', 182–4.

214. Colquhoun, 'Strategical', 182.

215. Quoted in LaFeber, *Panama*, 42.

216. See ch. 6, section '"The Teeming Millions of the Middle Kingdom"'.

217. Katz, *Secret*, 156–73.

218. Quoted in Garner, *British*, 185.

219. Brown, *Oil*, 190–7, 201–3, 239–44, 279–82; Katz, *Secret*, 165–7, 191–3, 197–8; Skirius, 'Railroad', 41–2; Fursenko, *Battle*, 154–5; 'His View of Wilson Plan', *New York Times* (22 January 1914); 'Tampico Operators to Ask Wilson's Aid', ibid. (8 May 1914).

220. Quoted in Brown, ibid., 193.

221. Herwig and Archer 'Global'; Mahan, 'Fortify'; 'Hobson Says We Are Defenseless Against Japan', *New York Times* (5 February 1911); 'Fortify Panama Canal or England Will Command It', ibid. (26 February 1911); 'Warns Again of War', *Washington Post* (3 March 1911); 'Our Fleet at Canal Opening', ibid. (3 July 1911); Holmes, 'Mahan', 42.

222. Katz, *Secret*, 177–8.

223. Garner, *British*, 196–7; Katz, ibid., 178; Calvert, *Mexican*, 223–4.

224. Brown, *Oil*, 104; Katz, ibid., 178–9; Brown, 'Royal Navy', 102; Snyder, 'Petroleum', 227–8; Jones, *State*, 166.

225. Garner, *British*, 181–3; Katz, ibid., 178–81, 198, 200–1; Jones, ibid., 72–3; Bucheli, 'Negotiating'; Calvert, 'Murray'; Holmes, 'Mahan', 33, 44–5; 'Says Press Beat Colombia Oil Plan', *New York Times* (28 November 1913); 'We Barred Oil Deal?', ibid. (29 November 1913); 'Lord Cowdray's Oil Interests', *Times* (28 November 1913); 'Oil and Ideals in Latin Lands', *Literary Digest* (6 December 1913).

226. 'The Pearson Contracts in Colombia', *Oil* (27 September 1913).

227. Katz, *Secret*, 178; Jones, *State*, 72–6.

228. 'Wants Oil Competition', *New York Times* (26 November 1913).

229. Garner, *British*, 167, 169–70; Brown, *Oil*, 104–6, 199–206, 213, 251–2, 268; Brown, 'Royal Navy', 102; *Report of the Royal Commission*, vol. 2, 72.

230. Brown, *Oil*, 212–30, 253–306; Santiago, *Ecology*, 93–9; Botz, *Edward*, ch. 4; Bud-Frierman et al., 'Weetman', 295.

231. Quoted in Garner, *British*, 167.

232. Brown, *Oil*, 171–223, 251–306; Garner, ibid., 197–8; 'Annual Report of the Royal Dutch Company', *Oil* (4 July 1914), 4; Katz, *Secret*, 196–7; Schoultz, *Beneath*, 245–7; Baecker, 'Arms'; Small, 'United States', 265–6.

233. 'Gov. Roosevelt's Annual Message', *New York Times* (4 January 1900).

234. Livingston, *Origins*, 57–63, ch. 2; Sklar, *Corporate*, 193–4, 398–401, 411–9, 426–9; Roy, *Socializing*, chs. 7–9; Kolko, *Triumph*, chs. 1–3; Leonard, 'American'; Bringhurst, *Antitrust*, 122–3; Youngman, 'Tendency – II'.

235. Chernow, *Titan*, 336–8, 370–93; Youngman, ibid. and 'Tendency – I'.

236. Youngman, 'Tendency – I', 204–8, and 'Tendency – II'.

237. U.S. Congress, *Report of the Committee Appointed Pursuant to House Resolutions*, pt. 2 – Review of the Evidence, 56, 129–30; 'Analysis of "Money Trust"', *New York Times* (1 March 1913).

238. 'Editorial Comment', *The Bankers' Magazine* 62, no. 4 (April 1901).

239. Ibid.

240. Quoted in Prettyman, 'Gilded', 23.

241. LaFeber, *New*, 172–5; Tarbell, *All in the Day's*, 1–26.

242. Weinberg, *Taking*; Chernow, *Titan*, 389–93, 435–58, 520–1; Johnson, *Development*, 201–8; Jones, *Limits*, 371–2; Kolko, *Triumph*, 15–17.

243. Bringhurst, *Antitrust*, ch. 1–4; Hidy and Hidy, *Pioneering*, 682–3.

244. Quoted in Clanton, 'Populism', 579.

245. Bringhurst, *Antitrust*, 77–9; Hidy and Hidy, *Pioneering*, 671–2.

246. Quoted in Hidy and Hidy, ibid., 399.

247. Bringhurst, *Antitrust*, 77–83.

248. Quoted in ibid., 79.

249. Ibid., 81; Hidy and Hidy, *Pioneering*, 675–6.

250. Quoted in Bringhurst, ibid., 83–4.

251. Ibid., 83–4.

252. Edward Wallace Hoch, 'Kansas and the Standard Oil Company', *The Independent* (New York)(2 March 1905), 463.

253. McHugh, 'Midwestern'.

254. Quoted in Kinzer, 'Kansas Attorney General Opinion'.

255. Quoted in Hidy and Hidy, *Pioneering*, 672.

256. Winkler, '"Other People's"'; McCormick, 'Discovery', 247–74; Chernow, *Titan*, 519–20; Bringhurst, *Antitrust*, 103.

257. Bringhurst, ibid., 129–41; Johnson, *Petroleum*, 54–8; Hidy and Hidy, *Pioneering*, 676–91; Williamson et al., *Energy*, 8–10; Chernow, ibid., 520–2.

258. 'William H. Taft Discusses "The Rule of Reason"', *New York Times* (7 June 1914).

259. Williamson et al., *Energy*, 4–9; Hidy and Hidy, *Pioneering*, 417; Johnson, *Petroleum*, 58–64; Walker, 'Oil', 31–6; Bringhurst, *Antitrust*, 109–12, 182–5.

260. 'Standard Oil Company's Profits Are Unreasonable', *Wall Street Journal* (5 August 1907); Bringhurst, ibid., 89–95, 102–12.

261. 'The Standard Oil Company's Investment Holdings and Cash', *Wall Street Journal* (8 October 1907); Walker, 'Oil'; Leslie, 'Revisiting'; Bringhurst, ibid., 109–12; Chernow, *Titan*, 255–9; see ch. 4, section 'Standard Oil, Producers and Independents', on competition and the Billingsley Bill, and ch. 6, section 'An Industry in Flux', on 'blind tigers', etc.

262. See ch. 4, section 'The Standard Oil Trust', on the Sherman Act; Bringhurst, ibid., 102–4; May, 'Antitrust'; Freyer, 'Sherman', 1007–8; Hazlett, 'Legislative'; Leslie, ibid.,

263. Kellogg, 'Enforcement', 216; ibid., 214–16; Bringhurst, ibid., 132.

264. See ch. 4, section '"Corporations Derive Their Power from the People"'; Roy, *Socializing*; Sklar, *Corporate*; Weinstein, *Corporate*.

265. U.S. Congress, *Antitrust Legislation*, 17, 19.

266. Roy, *Socializing*, 7–10, 21–41, 259–65; Bringhurst, *Antitrust*, 102–12.

267. Quoted in May, 'Antitrust', 296.

268. Quoted in Bringhurst, *Antitrust*, 95.

269. 'Wilson Reaffirms Democratic Creed', *New York Times* (14 April 1911).

270. Taft, *Anti-Trust*, 4.

271. La Follette, *La Follette*, 604; Gibson, *Communication*, 11–28.

272. 'Would Limit Capital in Big Business', *New York Times* (14 January 1911).

273. Wilson, *New*, 28, 200–1, 212–13.

274. Weinstein, *Corporate*, 3–6, 118–20; Weinstein, *Ambiguous*, 1–8; Jones, *Limits*, 370–7.

275. Quoted in Link, *Wilson*, 27.

276. Kolko, *Triumph*; Weinstein, *Corporate*.

277. 'No Lynching of Corporations', *Wall Street Journal* (6 August 1907).

278. Livingston, *Origins*, 57–63, ch. 2; Sklar, *Corporate*, chs. 3–6; Roy, *Socializing*, chs. 7–9; Kolko, *Triumph*, chs. 1–3; Hovenkamp, 'Classical', 1627–81; Leonard, 'American'; Bringhurst, *Antitrust*, 122–3.

279. Sklar, ibid., 185–6, 229, 333–64.

280. Quoted in Kohlsaat, *From McKinley*, 82; Weinstein, *Corporate*, 17.

281. Rockefeller, *Random*, 65–7.

282. Quoted in Lawson, *Truth*, 36; Hidy and Hidy, *Pioneering*, 206–19, 652–63, 673–4, 698–708; Bringhurst, *Antitrust*, 28, 80–1, 181–2; Chernow, *Titan*, 267–8; Wlasiuk, *Refining*, 23–5.

283. Weinstein, *Corporate*, chs. 1–2, 74, 129–30; Livingston, *Origins*, 103–16, 208–12; Kolko, *Triumph*, 146–9; Fronc, *New*, 154–7.

284. Chernow, *Titan*, 39, 101, 147, 388.

285. Hanna, *Mark*, 39, 32.

286. Quoted in Weinstein, *Corporate*, 22.

287. Quoted in ibid., 11.
288. Roosevelt, 'Trusts'.
289. Quoted in Weinstein, *Corporate*, 128–9.
290. Stromquist, *Reinventing*, 71–82; Enyeart, 'Revolution'; Burwood, 'Debsian'; Morgan, '"Red Special"'.
291. Quoted in Weinstein, *Corporate*, 128.
292. Kellogg, 'Enforcement', 217.
293. 'Bankers Renew Attack on Currency Bill', *New York Tribune* (9 October 1913).
294. Quoted in 'Taft Disclaims Panic Prophecy', *New York Times* (4 June 1910).
295. Rockefeller, *Random*, 65–6.
296. Bringhurst, *Antitrust*, ch. 5.
297. Ibid., chs. 5–7; Kolko, *Triumph*; Weinstein, *Corporate*.
298. 'Gov. Roosevelt's Annual Message', *New York Times* (4 January 1900).
299. Bringhurst, *Antitrust*, 156–7.
300. Quoted in ibid., 157.
301. Ibid., 147, 151–7, ch. 7; Chernow, *Titan*, 555–8; Hidy and Hidy, *Pioneering*, 713–7.
302. Quoted in Bringhurst, ibid., 180–1, and Giddens, *Standard*, 127 respectively.
303. Hagedorn, ed., *Works*, 311; Chernow, *Titan*, 557.
304. Quoted in 'Dissolved Trusts Under Scrutiny', *Literary Digest* (15 June 1912); Brown, 'Royal Navy', 92; Bringhurst, *Antitrust*, 190; Williamson et al., *Energy*, 96–7, 103–9, 252.
305. Bringhurst, ibid., 157–79; Freyer, 'Sherman', 1007–11.
306. Weinstein, *Corporate*, 74–88.
307. Quoted in Bringhurst, *Antitrust*, 169; Stucke, 'Does', 1389–93; Ross, *Muted*, 44–6; Hazlett, 'Legislative'.
308. 'Standard Oil Company Must Dissolve in 6 Months; Only Unreasonable Restraint of Trade Forbidden', *New York Times* (16 May 1911); 'The Trusts Have Won', *The Commoner* (Lincoln, NE) (26 May 1911).
309. Leonard, 'American', 124–6.
310. Quoted in Weinstein, *Corporate*, 73–4.
311. Quoted in ibid., 74; Livingston, *Origins*, 182–3.
312. Johnson, *Development*, ch. 10; Johnson, *Petroleum*, ch. 2, 4–6; Williamson et al., *Energy*, 104–9.
313. Zarlenga, *Lost*; U.S. Congress, 'Statement of T. Cushing Daniel, of Virginia', *Hearings before the Committee on Banking*, 1159–74; Kreitner, 'Money'; Livingston, *Origins*, 91–4; Aschheim and Tavlas, 'Academic'.

314. Quoted in Livingston, ibid., 174.
315. Quoted in ibid., 182.
316. Quoted in ibid., 188.
317. Wicker, *Great*, ch. 5; Prins, *All the Presidents'*, ch. 1; Broz, *International*, 176–7.
318. Kolko, *Triumph*, ch. 9; Broz, ibid.; Eichengreen and Flandreau, 'Federal'; McCormick, *China*, 31–2, 35; Gorton, 'Clearinghouses'; Tallman and Moen, 'Private', 7.
319. Timberlake, Jr, *Origins*, 198–206.
320. Roosevelt, 'Trusts'.
321. Quoted in Kolko, *Triumph*, 64; 'Rockefeller on Trusts', *New York Times* (11 January 1900); Rockefeller, *Random*, 68–9.
322. Kolko, ibid., 58–9, 63–4, 69, 77–8, 129, 161–3, 178–80; Weinstein, *Corporate*, 31–2, ch. 3; Winkler, '"Other People's"', 896.
323. Quoted in Sklar, *Corporate*, 193.
324. Quoted in Weinstein, *Corporate*, 153.
325. Quoted in 'Perkins Says Labor Must Share Profits', *New York Times* (21 December 1910).
326. Leonard, 'American', 114–26; Sternhell, *Birth*; Crane, 'Antitrust'.
327. 'Oil Trade Hurt by Bitter Attacks', *San Francisco Chronicle* (30 August 1906).
328. 'In Defense of Standard Oil', *The Sun* (New York) (1 December 1906); 'A Government Against an Industry', *Petroleum Review* (22 December 1906).
329. Freyer, 'Sherman', 991–2, 996–1,003, 1,011–17; Wilson, *British*, ch. 3; Walker, 'Policies'.
330. See section '"A Paper Victory for the People"' above; Roy, *Socializing*, chs. 4–9; Chandler, *Scale*, ch. 7, 10–12.
331. See ch. 5, section 'Colonial Oil'.
332. See ch. 6, section 'An Industry in Flux'.
333. See ch. 5, section 'The Royal Dutch and the Shell'; Jonker and Zanden, *History*, vol. 1, 126.
334. See ch. 6, section 'An Industry in Flux'.
335. See ibid., and section 'Liquid Fuel and Sea Power'.
336. See ibid., section 'Liquid Fuel and Sea Power'.
337. See ibid., section 'An Industry in Flux'.
338. Gerretson, *History*, vol. 3, 81–5; Henriques, *Marcus*, 356, 478, 490–2; Jonker and Zanden, *History*, vol. 1, 67.
339. Hidy and Hidy, *Pioneering*, 513–15; Chandler, *Scale*, 438–9; Frank, 'Petroleum', 20–4.

340. Conlin, *Mr Five*, 71–5; Hidy and Hidy, ibid., 554, 564; Fursenko, *Battle*, 67–75, 107–13; Tolf, *Russian*, 183–5; 'German Trust to Oppose Standard Oil', *New York Times* (11 November 1906); Gerretson, *History*, vol. 3, 88–91, 101–3; ibid., vol. 4, 173–6; Nowell, *Mercantile*, 59–61; Gibb and Knowlton, *History*, 201–3; Chandler, ibid., 437–9; Williamson et al., *Energy*, 252–9.

341. Quoted in Fursenko, *Battle*, 111.

342. Fursenko, ibid., 99–100; Henriques, *Marcus*, 489–501; Jonker and Zanden, *History*, vol. 1, 80–5; Jonker and Zanden, 'Searching', 23–4; Jones, *State*, 22–3; Beaton, *Enterprise*, 52–5; 'Company Meetings – A £10,000,000 Amalgamation', *Daily Telegraph* (16 May 1907); Deterding, *International*, 76–7.

343. See ch. 6, section 'Liquid Fuel and Sea Power'; Doran, *Breaking*, ch. 15.

344. Deterding, *International*, 47.

345. Hidy and Hidy, *Pioneering*, 125, 131–2, 238–45, 513–8, 523–6, 557, 568; Williamson and Daum, *Illumination*, 327, 333–4, 496, 651–2; Frank, 'Petroleum', 21; Nowell, *Mercantile*, 45–8; 'Oil Combine', *Daily Telegraph* (30 July 1907); Redwood, *Petroleum*, vol. 2, 595–600.

346. See ch. 6, section 'An Industry in Flux'.

347. Hidy and Hidy, *Pioneering*, 6, 452–3; 'The Countervailing Duty on Petroleum Products', *Washington Times* (30 March 1909); 'Will Try to Call Back Tariff Bill', *New York Times* (11 April 1909).

348. See ch. 6, section ' "The Teeming Millions of the Middle Kingdom" ', on List.

349. 'How Standard Oil Co. Has Made Millions', *Labor World* (15 February 1908).

350. Frank, *Oil*, 167–71; Frank, 'Petroleum', 21–6; Pearton, *Oil*, ch. 3.

351. 'Hungarian Finance – The Failure of the Loan Negotiations', *Times* (15 September 1910); 'Oil Trade Disputes', *Daily Telegraph* (1 October 1910).

352. Jones, *State*, 106–20.

353. See section 'Revolution in Mexico – "Police the Surrounding Premises" ' above.

354. Quoted in Campbell, *Special*, 4, and Etherington, *Theories*, 68 (n. 4).

355. Richardson, ed., *Compilation*, vol. 10, 143.

356. Thorsen, *Political*, 179.

357. McCormick, *China*, 33–4.

358. Vanderlip, 'American', 3–22, 194–213 and 287–306 respectively.

359. Deterding, *International*, 76.

360. 'The Petroleum "War" ', *Times*, Financial and Commercial Supplement (14 October 1910); Frank, 'Petroleum', 17; 'Rothschilds After

Rockefeller', *Washington Post* (8 July 1911); Gibb and Knowlton, *History*, 79; Williamson et al., *Energy*, 259–60.

361. Beaton, *Enterprise*, 56–81; Fursenko, *Battle*, 101–3; Henriques, *Marcus*, 521–6; Jonker and Zanden, *History*, vol. 1, 124–6; Deterding, *International*, 85–8; Doran, *Breaking*, ch. 18.

362. 'Competition in Gasoline', *Washington Post* (14 September 1912); 'May Invade America', *Boston Evening Transcript* (11 September 1912); 'Standard Oil to Have Dutch Rival', *San Francisco Call* (2 May 1911).

363. 'Invasion of the Royal Dutch into Hawaii Now Predicted', *Evening Bulletin* (Honolulu) (10 May 1912).

364. Deterding, *International*, 87–8.

365. Beaton, *Enterprise*, 69–97, 117–22; Gerretson, *History*, vol. 4, 213–43; Gibb and Knowlton, *History*, 91–2; Wilkins, *History*, 289–92; 'Dutch-Shell in America', *Oil and Gas Journal* (14 May 1914).

366. Jones, *State*, 77–8; 'Annual Report of the Royal Dutch Company', *Oil* (4 July 1914), 3–4; Jonker and Zanden, *History*, vol. 1, 116–45; 'Enormous Resources of Royal Dutch–Shell Combine', *Wall Street Journal* (6 May 1914); Owen, *Trek*, 382–4; Beaton, ibid., 55; Deterding, *International*, 88–90; Conlin, *Mr Five*, 104–9.

367. Gerretson, *History*, vol. 3, 127–30; ibid., vol. 4, 167–73; Pearton, *Oil*, 33, 35, 37; Jonker and Zanden, ibid., 117–21.

368. Jonker and Zanden, ibid., 123–4; Jones, *State*, 77.

369. Gerretson, *History*, vol. 4, 135–7; Fursenko, *Battle*, 101, 162–7; Jonker and Zanden, 'Searching', 24; Owen, *Trek*, 431–2; Tolf, *Russian*, 189–90; Jones, ibid., 61; Alekperov, *Oil*, 147–8; 'The Royal Dutch-Shell Combine', *Petroleum Review* (7 February 1914).

370. 'New Zealand Oil Fields', *Financial Times* (6 October 1911); Gerretson, ibid., vol. 3, 236–42; ibid., vol. 4, 82–91, 270–2; Jonker and Zanden, *History*, vol. 1, 98–9, 127; 'Weekly Financial City Letter', *The Globe* (London) (17 January 1911); 'Singapore as a Naval Base', *Homeward Mail* (1 April 1911); Jones, ibid., 86, 110–20; Owen, ibid., 393–95, 404, 408–11, 423; Higgins, *History*, 180–2; Garner, *British*, 180–1.

371. Deterding, *International*, 89–90, 97–9; McBeth, *Juan Vicente*, 11–12; Martinez, *Chronology*, 42; Lieuwen, *Petroleum*, 14; see ch. 8, section 'Spheres of Interest', on Mesopotamia.

372. Quoted in Garner, *British*, 180.

373. 'Germany Opens War on Standard Oil', *New York Times* (23 February 1910).

374. Munro, 'Proposed', 311–25; Flanigan, 'Some Origins', 111–19; Hidy and Hidy, *Pioneering*, 504–9, 556–71; Gibb and Knowlton, *History*, 203–8;

Nowell, *Mercantile*, 61–3; Fursenko, *Battle*, 114–20; 'Germany to War on Standard Oil', *New York Times* (25 February 1912); 'Study Kaiser's Plan for Oil Monopoly', ibid. (26 February 1912); 'United to Fight Standard', ibid. (15 April 1912); 'Germans in Battle Against Oil Trust', ibid. (21 July 1912).

375. 'Germany May Seize Standard's System', *New York Times* (20 October 1912).

376. 'We May Champion the Standard Oil', ibid. (16 October 1912).

377. Gibb and Knowlton, *History*, 207–17; Hidy and Hidy, *Pioneering*, 571.

378. Quoted in Gibb and Knowlton, ibid., 216; Fursenko, *Battle*, 119–23; Nowell, *Mercantile*, 64.

379. Quoted in Fursenko, *Battle*, 121; Clark, *Sleepwalkers*, 316–17.

380. Fursenko, ibid., 120–2; Flanigan, 'Some Origins', 113, 118, 121; Gibb and Knowlton, *History*, 219–20; Munro, 'Proposed', 322; Hidy and Hidy, *Pioneering*, 571. 'Deutsche Bank's Statement in Regard to Petroleum', *Wall Street Journal* (6 April 1914).

381. 'Fear Loss of Oil for Ships', *New York Times* (26 October 1912).

382. 'Standard Oil Wins Its German Fight', ibid. (15 December 1912); 'The Standard Oil Stays in Germany', ibid. (6 December 1912); 'Defeat Menaces Kaiser's Oil Plan', ibid. (15 November 1912); Nowell, *Mercantile*, 63–4; Flanigan, 'Some Origins', 120–1; Gibb and Knowlton, *History*, 220.

383. 'Imports of Petroleum Products into Germany During 1913', *Petroleum Review* (9 May 1914); 'German Petroleum Trade During the First Quarter', *Petroleum Review* (16 May 1914); 'Position of Oil Industry', *Financial Times* (22 August 1914); U.S. consular report: 'German Trade', 1253.

384. *Mineral Resources of the U.S., 1914*, pt. 2, 901–3; Coons, 'Statistics', 323–9; 'Petroleum', *Times* (16 January 1914).

385. *Mineral Resources of the U.S.*, ibid., 897–900; Coons, ibid., 327–42.

386. Rintoul, *Spudding*, ch. 7.

387. 'Cushing. The World's Great High Grade Oil Field', *Wall Street Journal* (27 April 1914); Beaton, *Enterprise*, 130–1.

388. Manning, *Yearbook*, 129 and 123 respectively.

389. Williamson et al., *Energy*, 97–101, 168, chs. 5–7.

390. Foreign Office, *Diplomatic and Consular Reports: Batoum, 1913*, 10; Tolf, *Russian*, 185–9.

391. Jones, *State*, 50–63; Tolf, ibid., 180–93; Lewery, *Foreign*, 7, 12–22, 24; Foreign Office, ibid., 10, 20–1, 36, 38–9.

392. 'The Petroleum Trade of England During 1913', *Petroleum Review* (3 January 1914); Jones, ibid., 32–4; Williamson et al., *Energy*, 256–7;

'Classified Imports into the United Kingdom up to December 20th,1913', *Petroleum Review* (27 December 1913); Toprani, *Oil*, 36.

393. Foreign Office, *British New Guinea*, 20.

394. Lewis, *Plantation*, chs. 1–7.

395. Carne, 'Notes', 35–50; Lewis, ibid., 66–7; 'Papua's Great Wealth', *The Sun* (Sydney) (25 May 1913); 'Oil Share Market Notes', *Financial Times* (22 June 1912); Senator Clemons, Australian Senate, Debates, 11 September 1913, 5th Parliament, 1st Sess., 1105; Senator Pearce, ibid., 26 September 1913, 5th Parliament, 1st Sess., 1533; Senator Pearce, ibid., 18 December 1913, 5th Parliament, 1st Sess., 4667–8.

396. Carne, ibid., 38.

397. Ibid., 45.

398. Ibid., vii.

399. 'Papua's Principal Problem', *The West Australian* (Perth) (25 March 1912).

400. Parliament of the Commonwealth of Australia, *Papua, 1912*, 7.

401. Quoted in Lewis, *Plantation*, 120–1.

402. 'In Papua', *Daily Telegraph* (Sydney) (2 April 1912); Parliament of the Commonwealth of Australia, *Papua*, 12, 35, 77.

403. 'Oil Fields in Papua', *The Age* (Melbourne) (20 April 1912); 'The Oil Search in Papua', ibid. (10 March 1917); Foreign Office, *British New Guinea*, 52–3.

404. Quoted in McBeth, *Dictatorship*, 40.

405. Deterding, *International*, 90,97.

406. McBeth, *Juan Vicente*, 10–14; Lieuwen, *Petroleum*, 9–15; Owen, *Trek*, 405–11; Gerretson, *History*, vol. 4, 276–81; Martinez, *Chronology*, 42–3; Jonker and Zanden, *History*, vol. 1, 245; Conlin, *Mr Five*, 106–7.

407. Quoted in Hidy and Hidy, *Pioneering*, 529.

408. Arnold et al., *First*, 205, 239.

409. Beckerman and Lizarralde, *Ecology*, 61–73.

410. Quoted in ibid., 72.

411. Arnold et al., *First*, 275–6; Beckerman and Lizarralde, ibid., 73–5.

412. Arnold et al., ibid., 276.

413. Ibid., 275.

414. Ibid., 299, 306.

415. Ibid., 301–2.

416. Ibid., 302.

417. 'Reach New York', *Boston Globe* (12 April 1913); 'Give Indians a Tussle', *Los Angeles Times* (7 March 1913).

418. Owen, *Trek*, 407–11; Tinker Salas, *Enduring*, 45–57.

419. Quoted in Tinker Salas, ibid., 44.

420. 'Venezuelan Oil Concessions', *Daily Telegraph* (20 January 1914).

8. OIL FOR WAR

1. See ch. 6, section 'Liquid Fuel and Sea Power'; Churchill, *World*, 90–1.

2. Brown, 'Royal Navy', 45–6, 52, 66; Gibson, '"Oil Fuel"', 113–15; Henriques, *Marcus*, 514; Snyder, 'Petroleum', 188–9; 'The Recent Manoeuvres and Liquid Fuel', *Petroleum Review* (7 July 1906); 'Oil Fuel Turbine-Driven Torpedo Boats for the British Navy', ibid. (8 June 1907); 'War Expresses of the Sea', *Daily Telegraph* (19 December 1907); 'Oil Fuel for the British Fleet', ibid. (30 July 1913); 'Oil Fuel for British Navy', *New York Times* (8 December 1907); 'Oil Fields of the Empire', *Times* (15 January 1908); 'Coal and Oil', ibid. (16 July 1913); 'Liquid Fuel', *Manchester Guardian* (23 July 1908); 'Oil Fuel for the Navy', *Petroleum World* (February 1910); Churchill, *World*, 91–3; Gibson, 'British', 32–3.

3. HC Deb, 18 February 1914, vol. 58, cc. 961–3; 'Navy and Oil Fuel', *The Globe* (London) (19 February 1914).

4. Ediger and Bowlus, 'Farewell'; Jones, 'Oil-Fuel', 144–6; DeNovo, 'Petroleum', 644–8; Hidy and Hidy, *Pioneering*, 454; Dahl, 'Naval', 54; Gerretson, *History*, vol. 4, 154; 'The German Navy of To-Day', *Scientific American* (8 August 1908); 'California Oil – Our Navy's Future Fuel', *San Francisco Call* (30 August 1911); 'Russian Navy's Oil Fuel', *Oil News* (21 December 1912); 'Oil Fuel for Naval Purposes', *Financial Times* (26 July 1913); 'The World's Fastest Oil-Burning Vessel', *Petroleum Review* (20 December 1913).

5. Churchill, *World*, vol. 1, 67; Jones, *State*, 27; Brown, 'Royal Navy', 65.

6. Fisher, *Records*, 202; Marder, ed., *Fear God*, vol. 2, 235; Gibson, 'British', 33–4; Gough, *Churchill*, 174–80.

7. Quoted in Churchill, *Winston*, vol. 2, Companion pt. 3, 1,927; ibid., 1956.

8. Quoted in Brown, 'Royal Navy', 102; Gibson, '"Oil Fuel"', 115–16.

9. HC Deb, 18 March 1912, vol. 35, c. 1551; Jones, *State*, 13–14.

10. *Report of the Royal Commission*, vol. 1, iv, vi; Brown, 'Royal Navy', 68–70; Kent, *Moguls*, 42.

11. HC Deb, 16 July 1913, vol. 55, c. 1219; ibid., 17 July 1913, vol. 55, c. 1475–6; *Report of the Royal Commission*, vol. 1, 77; Brown, ibid., 69; Churchill, *Winston*, vol. 2, Companion pt. 3, 1929, 1951–2.

12. *Report of the Royal Commission*, vol. 2, 8, 151; ibid., vol. 3, 154.

13. Brown, 'Royal Navy', 66–70, 76, 86–7, 105–6; Marder, *From the Dreadnought*, vol. 1, 268–9; Churchill, *World*, 92–3; HC Deb, 17 July 1913, vol. 55, c. 1468–9; ibid., 18 March 1914, vol. 59, c. 2102; Gough, *Churchill*, 176, 188–9.

14. Quoted in Churchill, *Winston*, vol. 2, Companion pt. 3, 1929.

15. Jones, *State*, 11–12, 23–4; Brown, 'Royal Navy', 57–8, 283–4; Snyder, 'Petroleum', 178–80, 229 (n. 31); *Report of the Royal Commission*, vol. 1, 178; HC Deb, 17 July 1913, vol. 55, c. 1472.

16. HC Deb, 10 February 1913, vol. 48, c. 478; Jones, *State*, 23; Gerretson, *History*, vol. 4, 283–4.

17. Quoted in Churchill, *Winston*, vol. 2, Companion pt. 3, 1943; Davoudi, *Persian*, 133–7.

18. Anglo-Mexican Petroleum Products Co., *Mexican Fuel Oil*, 15.

19. Brown, 'Royal Navy', 101–2.

20. *Report of the Royal Commission*, vol. 1, 70; Jones, 'Oil-Fuel', 139–40; 'Russian Oil Outlook', *Oil* (4 October 1913); 'Mexican Liquid Fuel for Russia', *Petroleum Review* (20 December 1913); 'Serious Strike on the Baku Fields', ibid. (13 June 1914).

21. Brown, 'Royal Navy', 57, 81–2; Jones, ibid., 141, 147; *Report of the Royal Commission*, vol. 1, 75; ibid., vol. 3, 177–9; Churchill, *Winston*, vol. 2, Companion pt. 3, 1932; 'Navy Estimates in the Reichstag', *Times* (26 November 1913); Foreign Office, *Correspondence Respecting the Affairs of the Congo*, 77, and *Manual of Belgian Congo*, 250.

22. Brown, ibid., 55–7, 102–10; Jones, ibid., 141–2; 'Chain of Oil Fuel Depots', *Petroleum World* (November 1910).

23. 'Oil and the Admiralty – Extending Consumption', *Economist* (5 March 1910); Jones, ibid., 86.

24. *Report of the Royal Commission*, vol. 1, 5–27, 65–70, 110–12, 123–9, 137–47, 158–9, 178–99, 217–28, 242–50, 274–90, 313–22, 326–7, 370–86, 543–9; ibid., vol. 3, 194–200; 'Colonial Oil Fields and Liquid Fuel', *Petroleum World* (March 1910); 'Prospects for Oil-Fuel', *Times Financial and Commercial Supplement* (4 March 1910); 'The Petroleum Industry in British North Borneo', *Financial Times* (9 March 1911); 'Oil and the Navy', *Oil News* (11 January 1913); 'Supply – Colonial Office Vote', *Times* (1 August 1913); 'Oil at Sarawak', *Oil* (20 September 1913); HC Deb, 17 June 1914, vol. 63, cc. 1236–7; Gerali and Gregory, 'Harsh'; Perkin, 'Oil'; Jones, *State*, ch. 4.

25. 'Oil Fields in Papua', *The Age* (Melbourne) (20 April 1912); 'Australia's Navy', *Daily Telegraph* (Sydney) (25 October 1910); 'Fuel for the Fleet', *The Argus* (Melbourne) (25 May 1912).

26. 'Island Soaked with Oil', *The Sun* (Sydney) (7 July 1910); 'Struck Oil', *Evening News* (Sydney) (19 August 1911); 'To Wring Timor Dry', *Sydney Morning Herald* (2 December 1911); 'Petroleum in Timor', *Cairns Post* (Queensland) (24 October 1912).

27. 'Timor Rebellion', *The Telegraph* (Brisbane) (31 January 1912).

28. 'Oil Discoveries in England', *Petroleum World* (March 1912); 'Petroleum in England', *Financial Times* (2 March 1912); 'Two Oil Finds in England', ibid.; 'Oil in England', *Oil News* (4 January 1913); 'Oil Spring in a Coal Pit', *Daily Telegraph* (10 June 1911); 'Oil for the Navy', *Observer* (27 July 1913).

29. HC Deb, 17 July 1913, vol. 55, c. 1472.

30. 'Oil and Germany', *Oil News* (24 May 1913).

31. *Report of the Royal Commission*, vol. 1, 133; ibid., 162–7, 424–49.

32. Ibid., 5–27, 65–70, 99–109, 290–7, 334–8, 554–79; ibid., vol. 2, 194–203; Snyder, 'Petroleum', ch. 6; 'Oil Prospects in 1913', *Oil News* (28 December 1912).

33. *Report of the Royal Commission*, vol. 2, 61, 66; Frank, *Oil*, 176.

34. 'Oil in 1911 and 1912', *Petroleum World* (December 1911); Alekperov, *Oil*, 145–7, 149; Jones, *State*, 62; Tolf, *Russian*, 188.

35. *Report of the Royal Commission*, vol. 1, 159 and 353 respectively.

36. Ibid., 62–3, 87–8, 291.

37. 'Future of Fuel', *Times* (1 December 1913).

38. *Report of the Royal Commission*, vol. 1, 104, 108, 293, 572–3; ibid., vol. 2, 70.

39. 'Oil Investments in Roumania', *Oil* (27 September 1913); Pearton, *Oil*, 43–8; *Report of the Royal Commission*, vol. 1, 152–3; ibid., vol. 2, 33; ibid., vol. 3, 196; Gibb and Knowlton, *History*, 81, 202; 'Roumania Becomes Normal', *Oil News* (16 August 1913).

40. Frank, *Oil*, 176–9; Sondhaus, *Naval*, 197.

41. *Report of the Royal Commission*, vol. 1, 65; ibid., 75, 87, 298–9.

42. Quoted in Sumida, 'British', 464; Jones, *State*, 13–14, 166–9; 'Wanted – New Oilfields' and 'Oil Fuel and the Coal Strike', *Petroleum World* (March 1912); 'Oil Fuel Prices and Supply', *Internal Combustion Engineering* (22 April 1914); Brown, 'Royal Navy', 74–5.

43. HC Deb, 17 July 1913, vol. 55, c. 1473.

44. *Report of the Royal Commission*, vol. 1, 329.

45. 'Oil and Coal for the Navy', *Economist* (28 March 1914).

46. *Report of the Royal Commission*, vol. 1, 64.

47. Ibid., 66.

48. 'The Dardanelles', *Times* (23 April 1912); 'Oil Share Market Notes', *Financial Times* (18 May 1912); HC Deb, 18 March 1914, vol. 59, c. 2103; Pearton, *Oil*, 72–3.

49. 'The Dardanelles Danger', *Petroleum World* (August 1912); Foreign Office, *Diplomatic and Consular Reports: Batoum, 1913*, 39; Gerretson, *History*, vol. 4, 164–5.

50. *Report of the Royal Commission*, vol. 1, 75.

51. Ibid., vol. 2, 8, and ibid., vol. 1, 641; 'Naval Oil Fuel – A Warning', *Petroleum World* (October 1912).

52. Jones, 'Oil-Fuel', 147; Sumida, 'British', 464–5; Brown, 'Royal Navy', 78–83, 102–10; Kent, *Moguls*, 45–6; *Report of the Royal Commission*, vol. 1, 75.

53. 'French Navy's Fuel Oil Crisis', *Oil News* (25 January 1913); Nowell, *Mercantile*, 74–5.

54. 'The Great Oil Battleship', *Financial Times* (18 October 1913); 'The First Oil Fuel Battleship', *Cassier's Engineering Monthly* 44, no. 5 (November 1913); 'The Battleship Warspite', *Times* (26 November 1913); Gibson, '"Oil Fuel"', 116–19.

55. HC Deb, 17 July 1913, vol. 55, c. 1521.

56. Brown, 'Royal Navy', 68, 71–2, 105; Ferrier, *History*, vol. 1, 177.

57. HC Deb, 17 July 1913, vol. 55, c. 1470.

58. Quoted in DeNovo, 'Petroleum', 647.

59. Ibid., 645–54; Snyder, 'Petroleum', ch. 7; *Annual Reports of the Navy Dept., 1913*, 14–16; *Mineral Resources of the U.S., 1913*, pt. 2, 955–60; 'United States Fuel Oil Supply Ships', *Petroleum World* (August 1912); 'Consumption of Oil Fuel', *Oil* (11 October 1913).

60. Staff, *German Battlecruisers*, 201–6, 257, 285, 290; Gröner, *German Warships*, vol. 1, 56–7; Herwig, 'Luxury', 66, 82–3; Fursenko, *Battle*, 120–3; 'Oil-Firing in the German Navy', *Times* (18 July 1914).

61. Tirpitz, *My Memoirs*, vol. 2, 400–1; Staff, *German Battlecruisers*, 204; 'Oil Fuel in the German Navy', *Times* (7 February 1914).

62. Pearton, *Oil*, 40–4, 68–9; see near end of ch. 7, section 'Economic Warfare', on Standard Oil's German monopoly; 'Oil Fuel for London', *Petroleum World* (February 1912); 'Roumanian Oil – Important Developments', *Financial Times* (9 August 1913); 'Germany's Oil Trade', *Petroleum World* (May 1912); 'A Great Oil Year', *Oil News* (20 December 1913); see ch. 6, section 'Liquid Fuel and Sea Power', on the Danube.

63. 'Oil and Germany', *Oil News* (24 May 1913); Büchi, 'Modern'; 'Petrol and Other Fuels for Motor Cars', *Oil News* (13 September 1913); 'Petrol

and the Public', *Petroleum World* (December 1912); 'Future of Fuel', *Times* (1 December 1913); *Report of the Royal Commission*, vol. 1, 66, 97; ibid., vol. 2, 102; ibid., vol. 3, 18, 94; Ormandy, 'Britain', ch. 25; 'Industrial Alcohol – How it is Made and How it is Used', *Scientific American* (21 July 1906); see ch. 6, section 'Petroleum Spirit', on German alcohol.

64. Gibb and Knowlton, *History*, 203-20; Fursenko, *Battle*, ch. 4; Fursenko, 'Oil', 460-3; 'Fear Loss of Oil for Ships', *New York Times* (26 October 1912); '"A Place in the Sun"', *Times of India* (11 December 1913); Flanigan, 'Some Origins', 118; Hidy and Hidy, *Pioneering*, 571; Gerretson, *History*, vol. 3, 65-6.

65. 'The Exploitation of Russian State Oil-Lands', *Petroleum Review* (11 April 1914); 'Another Strange Russian Project', ibid.

66. Rudin, *Germans*, 275-6; 'German Cameroons', *Financial Times* (18 January 1910); 'Is There Petroleum in Cameroon?', *Petroleum Review* (12 May 1906).

67. 'Oil in German New Guinea', *Times* (27 February 1914); 'Prospective New Oil Supplies for Germany', *Petroleum Review* (7 March 1914).

68. *Report of the Royal Commission*, vol. 1, 160, and vol. 2, 33-4 respectively; ibid., vol. 1, 151, 160; also ibid., 328. See near end of ch. 7, section 'Economic Warfare', on Germany's petroleum imports.

69. *Report of the Royal Commission*, vol. 1, 404-5.

70. HC Deb, 17 July 1913, vol. 55, cc. 1476-7; Kent, *Moguls*, 45-6, 49-50.

71. HC Deb, 17 July 1913, vol. 55, c. 1519.

72. HL Deb, 5 August 1913, vol. 14, c. 1604.

73. 'The Oil-Age and the Admiralty', *Observer* (20 July 1913).

74. Churchill, *Winston*, vol. 2, Companion pt. 3, 1,932-4, 1,941-3; *Report of the Royal Commission*, vol. 2, 16; Brown, 'Royal Navy', 103-13; Kent, *Moguls*, 45-6.

75. HC Deb, 17 July 1913, vol. 55, c. 1470; Brown, ibid., 108.

76. HC Deb, 22 July 1912, vol. 41, c. 917.

77. HC Deb, 17 July 1913, vol. 55, c. 1518.

78. Quoted in Brown, 'Royal Navy', 104, and Churchill, *Winston*, vol. 2, Companion pt. 3, 1,936, respectively; ibid., 1,955; 'Tunnel or Ferry', *Times* (6 August 1913); Miller, *Millstone*, ch. 15; Lambert, 'British', 600-18; Gough, *Churchill*, 191-9.

79. Fisher, *Records*, 195; Sumida, 'British', 464.

80. 'The Oil-Age and the Admiralty', *Observer* (20 July 1913).

81. Quoted in Churchill, *Winston*, vol. 2, Companion pt. 3, 1954; McBeth, *British*, 11; 'News of the Navy', *Observer* (8 and 22 February 1914).

82. HC Deb, 18 March 1914, vol. 59, c. 2103; McBeth, ibid., 11.

83. Brown, 'Royal Navy', 111–14.

84. *Report of the Royal Commission*, vol. 1, 308; ibid., vol. 3, 186.

85. Ibid., vol. 1, 574.

86. 'Review: Diesel Engines for Land and Marine Work', *Times* (8 May 1912); 'Submarine Explosion', *Times* (9 June 1913); Dash, 'British', 123–4; *Report of the Royal Commission*, vol. 1, 85–6, 580–3; Tolf, *Russian*, 172–5; Jones, 'Oil-Fuel', 142–3; Jonker and Zanden, *History*, vol. 1, 132.

87. 'Oil Fuel for Vessels', *Times* (20 February 1912); 'The Clyde's Third Oil-Engined Vessel', ibid. (3 April 1912); 'The Motor Liner Selandia', ibid. (6 March 1912); 'Increased Demand for Oil Tank Vessels', ibid. (18 October 1912); 'Marine Engineering', ibid. (22 January 1913); 'A Diesel Engine Ship', ibid. (24 July 1912).

88. *Mineral Resources of the U.S.*, 1911, 345; Quoted in *Report of the Royal Commission*, vol. 1, 87.

89. Churchill, *Winston*, vol. 2, Companion pt. 3, 1927.

90. Quoted in ibid., 1938.

91. 'The Internal Combustion Engine and Marine Propulsion', *Times* (31 March 1911); *Report of the Royal Commission*, vol. 1, 85–6; Brown, 'Royal Navy', 82–3; 'An Oil-Engined Destroyer', *Times* (12 October 1912); 'The Heavy-Oil Engine in Marine Service: Its Influence Upon Battleship Design', *Scientific American*, Supplement (22 February 1913); 'Oil Versus Coal', *Daily Telegraph* (25 March 1912).

92. Quoted in Henriques, *Marcus*, 530.

93. Staff, *German Battlecruisers*, 201; Staff, *German Battleships*, 5; Herwig, '*Luxury*', 65; Philbin, *Admiral*, 71; 'May Make German Navy the Finest', *New York Times* (6 February 1912); 'Motor Warships', *Observer* (12 March 1911); 'The World's Navies', *Times of India* (16 September 1910); 'Internal Combustion Engines for Battleships', *Petroleum World* (August 1910).

94. Quoted in Staff, *German Battlecruisers*, 201.

95. Quoted in Henriques, *Marcus*, 531.

96. Fisher, *Records*, 196.

97. Fisher, *Memories*, 217; *Report of the Royal Commission*, vol. 1, 350, 352, 359–61; Docherty and MacGregor, *Hidden*, 21–2, 71–3.

98. Quoted in Churchill, *Winston*, vol. 2, Companion pt. 3, 1927; ibid., 1,938–9; Henriques, *Marcus*, 459; 'The Admiralty and the Oil Market', *Economist* (26 July 1913); Nowell, *Mercantile*, 58.

99. Fisher, *Memories*, 218–19; Churchill, ibid., 1949.

100. *Report of the Royal Commission*, vol. 1, 387–9, 594–5, 599.

101. Miller, *Straits*, 431; Churchill, *Winston*, vol. 2, Companion pt. 3, 1928.

102. *Report of the Royal Commission*, vol. 1, 35; ibid., 594–5.

103. Brown, 'Royal Navy', 64–5; *Report of the Royal Commission*, ibid., 386–7, 508, 582–5; ibid., vol. 3, 123–4, 128; 'High-Power Diesel Engines', *Marine Engineer* (June 1914), 415–16; Henriques, *Marcus*, 516.

104. Staff, *German Battlecruisers*, 203; Staff, *German Battleships*, 22–3.

105. HC Deb, 26 March 1913, vol. 50, c. 1773.

106. Redwood, 'Future'; 'Half-Year's Petroleum Imports', *Financial Times* (11 July 1914); 'Alcohol as a Fuel', *Daily Telegraph* (16 February 1914); 'The Oil Outlook', *Times* (11 September 1913); 'Motor Spirit', ibid. (3 December 1913); 'Petrol and the Public', *Petroleum World* (December 1912); 'Cheaper Petrol', *Daily Telegraph* (9 July 1914); Barker and Gerhold, *Rise*, 77–83; Barker, 'Spread', 153–64; Williamson et al., *Energy*, 192–5; Dixon, 'Petrol', 3–4.

107. 'The Price of Petrol', *Petroleum World* (June 1912).

108. 'Oil Industries', *Daily Telegraph* (23 March 1914).

109. Winton, 'British'; Wik, 'Benjamin', 93–5; Rogers and Scribner, 'Lombard'.

110. 'The Battle of Hastings, 1909: War by Machinery', *Illustrated London News* (20 March 1909); 'The Caterpillar that Draws Big Guns', ibid. (23 May 1908); Robinson, *Lincoln*, 86–9; Kaplan, *Rolling*, 13–14; Haskew, *Tank*, 19–21.

111. 'The Army Manoeuvres', *Times* (27 September 1910).

112. Wik, 'Benjamin', 95–103; Karwatka, *Technology*, 103–5.

113. Winton, 'British', 201–5; Sutton, *Wait*, 53; 'Motor Vehicles in War', *Times* (24 July 1913).

114. 'The Arming of Ulster', *Times* (27 April 1914).

115. Bartholomew, *Early*, 4–11; Surlémont, 'French'; Flaherty, '1909', 51–2; White, *Tanks*, 115–16.

116. Haskew, *Tank*, 19–20; Kaplan, *Rolling*, 18–20; DiNardo and Bay, 'First'; Liddell Hart, *Tanks*, vol. 1, 16.

117. 'Petrol and Aviation', *Petroleum World* (December 1912).

118. Gibbs-Smith, *Aviation*, 161; Gollin, *Impact*, 202–3, 277–8; 'The Seaplane Carrier', *Times* (16 March 1914); 'Seaplanes of To-day', *Marine Engineer* (June 1914); 'Paris Day by Day – Military Aviation', *Daily Telegraph* (25 March 1912).

119. Quoted in 'Libya 1911: How an Italian Pilot Began the Air War Era', *BBC News* (10 May 2011).

120. Hippler, *Bombing*, 1–2, 57–62; 'The Tripoli Campaign', *Times* (6 January 1912).

121. 'The Italian Army and Aviation', *Times* (12 August 1912); Clark, *Sleepwalkers*, 242–3.

122. 'Aeroplanes on Way to Mexico', *Albuquerque Morning Journal* (10 July 1912); *Annual Reports of the Navy Department*, 1914, 12; 'Aeroplanes in War', *Daily Telegraph* (30 July 1913).

123. 'Aviation in War', *Flight* (25 January 1913).

124. Quoted in 'Military Aviation', ibid. (8 March 1913); 'The Air, the Public, and the Future', *Spectator* (7 May 1913).

125. Gollin, *Impact*, ch. 11; 'Editorial – Our Aerial Fleet', *Flight* (1 February 1913); 'Editorial – "The Peril of the Air"', ibid. (22 February 1913); 'Editorial – Our Phantom Air Squadron', ibid. (1 March 1913); 'The Black Shadow of the Airship', ibid.; 'The Struggle for the Control of the Air', ibid. (8 March 1913); 'Editorial – Our Phantom Fleet', ibid. (9 August 1913).

126. 'Oil and the Flying Business', *Oil News* (5 April 1913); 'Oil and the Aeroplane', ibid. (15 March 1913).

127. Gollin, *Impact*, 240.

128. Quoted in 'Editorial – Mr. Churchill and British Aerial Supremacy', *Flight* (15 November 1913); Gough, *Churchill*, 199–201.

129. 'Company Meetings: "Shell" Transport and Trading Co. (Ltd.)', *Times* (23 June 1914).

130. HC Deb, 17 June 1914, vol. 63, c. 1234; 'The Petrol Problem', *Daily Mail* (9 January 1913).

131. 'The Petrol Mystery', *Daily Mail* (11 June 1912); 'The Price of Petrol', *Petroleum World* (June 1912); 'The Truth About Petrol', ibid. (November 1912); 'Petrol and the Public', ibid. (December 1912); 'To Solve the Motor Fuel Problem', *Observer* (8 December 1912); 'The London Aerodrome', *Oil News* (26 April 1913); 'Oil and Germany', ibid. (24 May 1913); '"Not Enough Oil"', ibid. (22 March 1913); 'The Oil Outlook – Petrol and its Substitutes', *Times* (11 September 1913); 'Liquid Fuel', *Manchester Guardian* (23 July 1908).

132. 'An Objection to Oil Fuel', *Times* (2 April 1912).

133. 'To Solve the Motor Fuel Problem', *Observer* (8 December 1912).

134. HC Deb, 29 April 1913, vol. 52, cc. 1142–3.

135. *Report of the Royal Commission*, vol. 1, 363.

136. Ireland, *Iraq*, 55–6; Shorrock, 'Origin', 143–53; Feis, *Europe*, 355–9; Barth and Whitehouse, 'Financial', 134–5.

137. Quoted in Cohen, *British*, 142; ibid., 139, 181; Earle, 'Secret'.

138. Quoted in Kent, 'Constantinople', 155.

139. Quoted in ch. 7, section 'Revolution in Persia, and the Plain of Oil'; Strunk, 'Reign', 151–2.

140. Quoted in Davoudi, *Persian*, 113.
141. Quoted in Ferrier, *History*, vol. 1, 93; Davoudi, ibid., 113–15, 117–19; Wilson, 'Creative', 51–9, 72; Hoffman, *Great*, 164–6; Chen, 'British', 89–91, 149.
142. Quoted in Siegel, *Endgame*, 85.
143. Cohen, *British*, chs. 4–5; Spring, 'Trans-Persian'; Siegel, ibid., 65–8, 84–7, 90–3, 96–100, 114, 138–40, 157–65; Chen, 'British', 163–7; see ch. 5, section 'Across the Caucasus to the World', on the industrialist Bunge; Jones, *History*, vol. 1, 128–31; 'A British Railway in Persia', *Times* (29 April 1911).
144. Quoted in Cohen, ibid., 142.
145. 'New Persian Question', *Daily Telegraph* (22 April 1910).
146. Quoted in Gooch and Temperley, eds., *British*, vol. 6, 468.
147. Quoted in Busch, *Britain*, 317.
148. Quoted in Miller, *Straits*, 78; Cohen, *British*, 123, 155.
149. Quoted in Cohen, ibid., 121; ibid., 119–20.
150. Quoted in ibid., 161.
151. Quoted in ibid.
152. Quoted in ibid.
153. Quoted in Miller, *Straits*, 80; Cohen, ibid., 159–60; Neilson, 'Baghdad to Haifa', 161.
154. 'The Baghdad Railway', *Times* (12 January 1911).
155. 'The Baghdad Railway Question', *Times* (10 March 1911).
156. Kent, *Moguls*, 74–9.
157. Quoted in ibid., 77; ibid., 81–2.
158. Cohen, *British*, 151–5.
159. HL Deb, 22 March 1911, vol. 7 cc. 573–612.
160. HC Deb, 23 March 1911, vol. 23, cc. 628–9.
161. Quoted in Sarolea, *Anglo-German*, 277–8.
162. Quoted in Shorrock, 'Origin', 147.
163. Quoted in Demirel, 'Doğu'.
164. Foreign Office, 'Koweit and Bahrein: Oil Deposits. Water Supply', 292–5; U.S. consular report: 'Drilling for Oil in Palestine'; Dominian, 'Fuel', 250.
165. Kent, *Oil*, 28–9; ibid., 222 (n. 80).
166. Quoted in Cohen, *British*, 153; Head, 'Public'; Spring, 'Trans-Persian'; Kazemzadeh, *Russia*, 591–7; Hoffman, *Great*, 164–8; Siegel, *Endgame*, 90–1, 96–7.
167. Sir G. Lowther to Sir Edward Grey, 13 December 1911, FO 424/229, 169.
168. Quoted in Miller, *Straits*, 81, and Lowe and Dockrill, eds., *Foreign*, vol. 5, pt. 3, 469–70.

169. 'The Russo-German Agreement', *Times* (22 August 1911).

170. Quoted in Ateş, *Ottoman*, 289.

171. Quoted in Cohen, *British*, 156; 'British Interests in the Persian Gulf. IX – Turkish Pretensions', *Times* (1 August 1911).

172. Cohen, ibid., 155–9, 178–80; Ireland, *Iraq*, 44 (n. 1).

173. Quoted in Cohen, ibid., 180.

174. Quoted in ibid., 179.

175. Bonakdarian, *Britain*, 176, 200, 288, 297, 328, 348, 395; Klein, 'British', 747–52; Ferrier, *History*, vol. 1, 139–40.

176. Quoted in Chen, 'British', 68; Ferrier, ibid., 139–40, 145–6; *Report of the Royal Commission*, vol. 1, 558.

177. 'Company Meetings: Anglo-Persian Oil Company (Limited)', *Times* (26 July 1910).

178. Ferrier, *History*, vol. 1, 114–15, 130–3, 145, 155–6; Davoudi, *Persian*, 115–17; *Report of the Royal Commission*, vol. 1, 556–63; 'Persian Oil Deal', *Oil* (6 June 1914).

179. Ferrier, ibid., 143–5; *Report of the Royal Commission*, ibid., 558.

180. Ferrier, ibid., 74–86, 115–16, 119–29; Kazemzadeh, *Russia*, 427–47; Shahnavaz, *Britain*, 111–87; Klein, 'British', 746–8; Garthwaite, 'Bakhtiyari', 24–38.

181. 'British Interests in the Persian Gulf. IX – Turkish Pretensions', *Times* (1 August 1911).

182. 'Company Meetings: Anglo-Persian Oil Company (Limited)', *Times* (26 July 1910).

183. Quoted in Corley, *History*, 193; Kurt, 'Discovery'.

184. Ferrier, *History*, vol. 1, 133–9, 147–53, 160–2; Jones, *State*, 141–53; Gerretson, *History*, vol. 4, 185.

185. Foreign Office, 'Turkey in Asia: Oil Concessions', 554; quoted in Jones, ibid., 144.

186. Ferrier, *History*, vol. 1, 162–5; Jones, ibid., 144.

187. 'Company Meetings – Anglo-Persian Oil Company (Limited)', *Times* (29 October 1912).

188. Quoted in Sluglett, *Britain*, 259 (n. 4).

189. Quoted in Ferrier, *History*, vol. 1, 162.

190. Ibid., 164.

191. Cohen, *British*, 194–5; Spring, 'Trans-Persian', 71–2.

192. Kent, *Oil*, 34, 86; Conlin, 'Debt', 537; Ferrier, *History*, vol. 1, 165; Black, *Banking*, 138–9; Longrigg, *Oil*, 29–30; Gerretson, *History*, vol. 4, 187–8.

193. Kent, *Moguls*, 70–4; Conlin, ibid., 525–30; Kent, *Oil*, 80; Gerretson, ibid., vol. 3, 235; Black, ibid., 128, 138–9; Longrigg, ibid., 29–30.

194. Quoted in Kent, *Moguls*, 84.

195. Quoted in Ferrier, *History*, vol. 1, 166; Kent, *Oil*, 34–5.

196. Quoted in Ferrier, ibid., 166.

197. Quoted in Jones, *State*, 150.

198. Kent, *Oil*, 34.

199. Ferrier, *History*, vol. 1, 167; Kent, ibid., 35–6.

200. Quoted in Kent, ibid., 36; Conlin, 'Debt', 537.

201. Quoted in Ferrier, *History*, vol. 1, 168; Davoudi, *Persian*, 126–7.

202. Quoted in Ferrier, ibid., 196.

203. Kent, *Oil*, 34; Black, *Banking*, 140–5; Ferrier, ibid., 165; Gerretson, *History*, vol. 4, 187–91; Longrigg, *Oil*, 29–30.

204. Ferrier, ibid., 166–7; Kent, ibid., 36.

205. Quoted in Miller, *Straits*, 428.

206. See section "Oil for a Great War" above.

207. 'Oil and the Admiralty – Extending Consumption', *Economist* (5 March 1910).

208. Jones, *State*, 104–5.

209. *Report of the Royal Commission*, vol. 1, 336.

210. Quoted in Ferrier, *History*, vol. 1, 169.

211. Quoted in ibid., 168.

212. Quoted in ibid., 169; Kent, *Oil*, 38–9.

213. Quoted in Ferrier, ibid., 170, and Kent, ibid., 37; Jones, *State*, 150–3.

214. Quoted in Ferrier, ibid.

215. Ibid., 167; Kent, *Oil*, 35–6; *Report of the Royal Commission*, vol. 1, 346–7.

216. *Report of the Royal Commission*, ibid., 336–7, 403–6; Miller, *Straits*, 429; Kent, ibid., 224 (n. 42).

217. See ch. 7, section 'Economic Warfare'.

218. Quoted in *Report of the Royal Commission*, vol. 1, 406.

219. Ibid., 336; Jones, *State*, 151–3.

220. Quoted in Miller, *Straits*, 429.

221. *Report of the Royal Commission*, vol. 1, 335, 337.

222. Ibid., 343–4; Ferrier, *History*, vol. 1, 172.

223. Quoted in Ferrier, ibid., 175; Jones, *State*, 151, 162.

224. Foreign Office, 'Turkey in Asia: Oil Concessions', 534–46; Kent, *Oil*, 43; Ferrier, ibid., 175.

225. Ferrier, ibid., 172–3.

226. Quoted in Kent, *Oil*, 41; Jones, *State*, 161–4; Ferrier, ibid., 173.

227. Quoted in Ferrier, ibid.; Miller, *Straits*, 429.

228. Quoted in Kent, *Oil*, 40.

229. *Report of the Royal Commission*, vol. 1, 361.

230. Ibid., 366.

231. See ch. 6, section 'Liquid Fuel and Sea Power'.

232. Quoted in Jones, *State*, 86.

233. Ibid., 86, 110-11, 114-15; Gerretson, *History*, vol. 4, 87-90.

234. *Report of the Royal Commission*, vol. 1, 351; 'British Oil and the Admiralty', *Times* (29 April 1914).

235. *Report of the Royal Commission*, ibid., 362, 368; see ch. 7, section 'British Supremacy over Baghdad'.

236. *Report of the Royal Commission*, vol. 2, 21.

237. Quoted in Higgins, *History*, 98.

238. Quoted in Jones, *State*, 109; Higgins, ibid., 86-7, 98-9.

239. 'Oil Companies and Foreign Fields', *Economist* (10 December 1910).

240. Quoted in Jones, *State*, 109; Higgins, *History*, 93.

241. Jones, ibid., 110-13, 165.

242. Quoted in ibid., 111; Higgins, *History*, 91-2, 100, 180.

243. *Report of the Royal Commission*, vol. 1, 288.

244. Quoted in Gerretson, *History*, vol. 3, 240.

245. Ibid., 236-42; Jones, *State*, 114-17.

246. Quoted in Jones, ibid., 119.

247. Quoted in ibid., 118; *Report of the Royal Commission*, vol. 1, 362.

248. Quoted in Jones, ibid., 120.

249. *Report of the Royal Commission*, vol. 1, 372.

250. Kent, *Oil*, 69, 78, 90.

251. Quoted in Churchill, *Winston*, vol. 2, Companion pt. 3, 1,949; *Report of the Royal Commission*, vol. 3, 49; Fisher, *Records*, 201-2.

252. Quoted in Churchill, ibid., 1,950-1.

253. Quoted in Jones, *State*, 120.

254. Ibid., 116-21; Longrigg, *Oil*, 23.

255. 'British Oil and the Admiralty', *Times* (24 April 1914).

256. Quoted in Churchill, *Winston*, vol. 2, Companion pt. 3, 1,955.

257. See ch. 6, sections 'An Industry in Flux' and 'Liquid Fuel and Sea Power'.

258. HC Deb, 17 June 1914, vol. 63, c. 1139.

259. See ch. 6, section 'Liquid Fuel and Sea Power'.

260. Quoted in Carland, 'Enterprise', 198-9.

261. See ch. 7, section 'Revolution in Persia, and the Plain of Oil'.

262. See section '"Will Germany Control the Oil Supply of Our Navy?"' above.

263. *Report of the Royal Commission*, vol. 1, 159; Carland, 'Enterprise', 200.

264. Quoted in Carland, ibid., 201.

265. Quoted in ibid.

266. See ch. 6, section 'Liquid Fuel and Sea Power'.

267. See section '"Will Germany Control the Oil Supply of Our Navy?"' above; HC Deb, 17 June 1914, vol. 63, c. 1222; Kent, *Oil*, 200.

268. *Report of the Royal Commission*, vol. 1, 641.

269. See ch. 7, section 'British Supremacy over Baghdad', sections 'Oil "for a Great War"', 'Spheres of Interest' and '"Will Germany Control the Oil Supply of Our Navy?"' above and 'The Turkish Petroleum Company' below; Kent, *Moguls*, 46-7; Brown, 'Royal Navy', 94-6; Davoudi, *Persian*, 134-5.

270. Quoted in Kent, ibid., 47-8; Jones, *State*, 169; Miller, *Straits*, 441-2.

271. Quoted in Churchill, *Winston*, vol. 2, Companion pt. 3, 1,948.

272. Quoted in Kent, *Moguls*, 49.

273. HC Deb, 17 July 1913, vol. 55, c. 1473-5.

274. *Report of the Royal Commission*, vol. 1, 369.

275. HC Deb, 17 July 1913, vol. 55, c. 1475.

276. Kent, *Moguls*, 50; see above and section 'The Turkish Petroleum Company' below; Ferrier, *History*, vol. 1, 175-95; Jones, *State*, 168-9; 'Anglo-Persian Oil Company (Limited)', *Times* (29 October 1913).

277. Quoted in Jones, ibid., 150; ibid., 169-70; Kent, ibid., 51; Ferrier, ibid., 183-4; Brown, 'Royal Navy', 96; 'The Persian Oil Deal', *Oil* (6 June 1914).

278. Quoted in Ferrier, *History*, vol. 1, 195.

279. Quoted in ibid., 195-6.

280. Quoted in Jones, *State*, 76.

281. See ch. 7, section 'Revolution in Mexico – "Police the Surrounding Premises"'; Ferrier, *History*, vol. 1, 199-200; Davoudi, *Persian*, 132-4.

282. Quoted in Ferrier, ibid., 196.

283. Quoted in ibid., 197; Kent, *Moguls*, 52.

284. Kent, *Oil*, 85, 90, 97-100; Miller, *Straits*, 456-7. This was, therefore, more than a 'curious coincidence', as Ferrier describes it (*History*, vol. 1, 197); Nowell, *Mercantile*, 71.

285. 'Agreement with the Anglo-Persian Oil Company (Limited)', 11 May 1914 (NA Cab 37/119/61), 4.

286. Kent, *Moguls*, 52-4; Ferrier, *History*, vol. 1, 199; Jones, *State*, 154-5, 171; Brown, 'Royal Navy', 97; Miller, *Straits*, 460; HC Deb, 17 June 1914, vol. 63, cc. 1178, 1208.

287. 'The State as Shareholder', *Times* (25 May 1914); 'Naval Oil Agreement', *Times* (26 May 1914).

288. 'The Political Aspect of the Persian Oil Agreement', *Times* (27 May 1914); Kent, *Moguls*, 55.

289. HC Deb, 17 June 1914, vol. 63, cc. 1085-252.

290. HC Deb, 17 June 1914, vol. 63, c. 1201; Davoudi, *Persian*, 135–9.

291. HC Deb, 17 June 1914, vol. 63, c. 1148.

292. HC Deb, 29 June 1914, vol. 64, cc. 111–12.

293. HC Deb, 17 June 1914, vol. 63, c. 1141.

294. HC Deb, 17 June 1914, vol. 63, cc. 1131–53; 'The Government's Oil Deal', *Daily Telegraph* (17 June 1914).

295. HC Deb, 17 June 1914, vol. 63, c. 1149.

296. 'The Government's Oil Deal', op. cit.

297. HC Deb, 17 July 1913, vol. 55, c. 1504.

298. 'The Admiralty and the Oil Market', *Economist* (26 July 1913).

299. 'Oil-Power and Sea-Power', *Observer* (21 June 1914).

300. *Annual Reports of the Navy Dept., 1913*, 14–15; 'United States Follows Britain's Naval Oil Policy', *Oil* (27 December 1913); DeNovo, 'Petroleum', 648–51; 'World's Search for Petroleum', *Wichita Daily Times* (14 January 1914).

301. *Annual Reports of the Navy Dept., 1914*, 14–15.

302. Quoted in DeNovo, 'Petroleum', 653; 'Daniels Wants Oil Supplies for Navy', *New York Times* (31 January 1914); 'Government Pipeline Was Favored', *Houston Daily Post* (4 June 1914).

303. DeNovo, ibid., 653–4.

304. HC Deb, 17 June 1914, vol. 63, c. 1229; Cohen, *Churchill*, 41–9.

305. HC Deb, 17 June 1914, vol. 63, cc. 1219–22, also 1210, 1243; Jones, *State*, ch. 3.

306. 'Company Meetings: "Shell" Transport and Trading Co. (Ltd.)', *Times* (23 June 1914); HC Deb, 7 July 1914, vol. 64, c. 1036.

307. HC Deb, 17 June 1914, vol. 63, c. 1138.

308. 'Company Meetings: "Shell"', op. cit.

309. Quoted in Miller, *Straits*, 460–1.

310. 'Company Meetings: "Shell"', op. cit.

311. See section '"Will Germany Control the Oil Supply of Our Navy?"' above.

312. See ch. 7, section 'Revolution in Mexico – "Police the Surrounding Premises"'.

313. HC Deb, 17 June 1914, vol. 63, c. 1186.

314. See ch. 6, section '"The Virgin Oil Fields of Persia"'.

315. Quoted in Siegel, *Endgame*, 55.

316. HC Deb, 17 June 1914, vol. 63, c. 1177; Davoudi, *Persian*, 4, 140; 'As Between Friends', *Punch* (13 December 1911).

317. Kazemzadeh, *Russia*, chs. 8–9; Bonakdarian, *Britain*, chs. 9–10; Greaves, 'Some Aspects – II'; 'The Bakhtiari and the Lynch Road', *Times* (13 July 1912).

318. Quoted in Siegel, *Endgame*, 169; Greaves, ibid., 300–1; 'Anglo-Russian Rivalry in Persia', *Economist* (18 July 1914).

319. Quoted in Siegel, ibid., 178, 187; Greaves, ibid., 300–3; Wilson, 'Grey', 279–80; HC Deb, 29 June 1914, vol. 64, cc. 113–14.

320. Quoted in Wilson, ibid., 279.

321. Churchill, *Winston*, vol. 2, Companion pt. 3, 1964.

322. Quoted in Siegel, *Endgame*, 187.

323. Quoted in ibid., 183.

324. HC Deb, 17 June 1914, vol. 63, cc. 1154–7; Davoudi, *Persian*, 138; 'Anarchy in Mesopotamia', *Times* (28 June 1911); 'The Fighting in Arabia', *Times* (2 June 1910); Miller, *Straits*, 23.

325. HC Deb, 17 June 1914, vol. 63, cc. 1143–5.

326. 'Anglo-Persian Oilfields', *Times* (24 June 1914); Greaves, 'Some Aspects – II', 299–300 (n. 36).

327. HC Deb, 17 June 1914, vol. 63, cc. 1198 and 1244 respectively.

328. Bonakdarian, *Britain*, 286–8, 296–300, 327–8, 346–8.

329. Quoted in Garthwaite, 'Bakhtiyari', 37.

330. 'Oil-Power and Sea-Power', *Observer* (21 June 1914).

331. HC Deb, 17 June 1914, vol. 63, c. 1136.

332. HC Deb, 29 June 1914, vol. 64, cc. 54 and 75 respectively.

333. HC Deb, 17 June 1914, vol. 63, c. 1200, also 1155, 1223, 1228.

334. 'The Navy and Persian Oil', *Times* (18 June 1914).

335. HC Deb, 17 June 1914, vol. 63, cc. 1152–3.

336. Quoted in Ferrier, *History*, vol. 1, 199.

337. '£2,200,000 for Oil for the British Navy', *Daily Telegraph* (18 June 1914).

338. HC Deb, 17 June 1914, vol. 63, c. 1178.

339. Greaves, 'Some Aspects – II', 300–3, 306–7; Wilson, 'Grey', 275–80; Siegel, *Endgame*, 114, 117, 126, 184–95; Miller, *Straits*, 460.

340. HC Deb, 29 June 1914, vol. 64, cc. 66–7.

341. HC Deb, 7 July 1914, vol. 64, c. 1052; Owen, *Trek*, 1,321–2; Abu-Hakima, *Modern*, 197; Wheatcroft, *Life*, 81.

342. HC Deb, 29 June 1914, vol. 64, c. 54; Greaves, 'Some Aspects – II', 307–8.

343. 'Oil-Power and Sea-Power', *Observer* (21 June 1914).

344. Cohen, *British*, 159–62, 171–81.

345. HC Deb, 23 March 1911, vol. 23, c. 648.

346. Quoted in Cohen, *British*, 160.

347. HC Deb, 23 March 1911, vol. 23, c. 648.

348. Quoted in Cohen, *British*, 172–3.

349. Ibid., 171–8, 187–98, 209–10; McMurray, *Distant*, 84–102; Busch, *Britain*, chs. 10–11.

350. Quoted in Cohen, ibid., 191.

351. See section 'Spheres of Interest' above.

352. Quoted in Kent, *Oil*, 52.

353. See ch. 7, section 'British Supremacy over Baghdad'; quoted in Kent, *Oil*, 53; Black, *Banking*, 145.

354. Quoted in Kent, ibid., 50; Foreign Office, 'Turkey in Asia: Oil Concessions', 162, 232–8.

355. Foreign Office, ibid., 63.

356. Quoted in Kent, *Oil*, 57; ibid., 51.

357. Foreign Office, 'Turkey in Asia: Oil Concessions', 326; Kent, ibid., 55.

358. Quoted in Anderson, *Lawrence*, 55; ibid., 9–12; Kent, ibid., 62–3, 74, 84, 228 (n. 91); Haque, 'Conflict', 90, 96–8; Miller, *Straits*, 449–50; Ferrier, *History*, vol. 1, 197; Longrigg, *Oil*, 25; Dugdale, *German*, vol. 4, 253; Shwadran, *Middle East*, 449.

359. Quoted in Kent, *Oil*, 74.

360. Foreign Office, 'Turkey in Asia: Oil Concessions', 330.

361. U.S. consular report, 'Trade of Mesopotamia and Irak', 587.

362. See section 'Oil "for a Great War"' above.

363. Quoted in Miller, *Straits*, 442.

364. 'Minerals in the Turkish Empire', *Times* (24 December 1913).

365. Kent, *Oil*, 65–6; Cohen, *British*, 175, 184 (n. 75); Clark, *Sleepwalkers*, 334–5.

366. Foreign Office, 'Turkey in Asia: Oil Concessions', 432.

367. Ibid., 114; quoted in Kent, *Oil*, 61–2.

368. Quoted in Kent, ibid., 76, and Cohen, *British*, 176, respectively.

369. Quoted in Cohen, ibid., 176 and 181 respectively.

370. Quoted in Kent, *Oil*, 62 and 68 respectively.

371. Quoted in Cohen, *British*, 176, 185 (n. 82); Dugdale, *German*, vol. 4, 244–6, 248–9, 253.

372. Foreign Office, 'Turkey in Asia: Oil Concessions', 50; Kent, *Oil*, 69.

373. Foreign Office, ibid.

374. Kent, *Oil*, 70.

375. Ibid., 62, 66–8.

376. Ibid., 69–71, 78, 89–94.

377. Quoted in ibid., 69.

378. Quoted in ibid., 78.

379. Quoted in Black, *Banking*, 149; Kent, ibid., 76–83; Ferrier, *History*, vol. 1, p. 197; Conlin, *Mr Five*, 90–3.

380. Kent, ibid., 83–4.

381. 'A New Treaty Between England and Turkey', *Daily Telegraph* (1 December 1913).

382. '"A Place in the Sun"', *Times of India* (11 December 1913).

383. 'An Interesting Lecture on Petroleum', *Petroleum Review* (27 December 1913).

384. Quoted in Kent, *Oil*, 86; Miller, *Straits*, 450.

385. Quoted in Black, *Banking*, 154; Kent, ibid., 88.

386. Quoted in Kent, ibid., 90; Conlin, *Mr Five*, 93–4.

387. Kent, ibid., 88–92.

388. See section 'Britain's State Oil Company' above.

389. Kent, *Oil*, 92–7, 170–1; Earle, 'Turkish', 269, 277–9; Cohen, *British*, 191, 208–9; Busch, *Britain*, 336, 346; Shwadran, *Middle East*, 407; Black, *Banking*, 154–8; Miller, *Straits*, 450–2; Earle, 'Secret'.

390. Quoted in Dugdale, *German*, vol. 4, 254–5; Miller, ibid., 462–3.

391. Kent, *Oil*, 103–12; Black, *Banking*, 158–9, 163–4.

392. 'The Petroleum Deposits of Mesopotamia', *Petroleum Review* (23 May 1914); Mitchell, *Carbon*, 51.

393. Quoted in Cohen, *British*, 180; Kent, *Oil*, 97–101; Miller, *Straits*, 456–8.

394. Quoted in Miller, ibid., 457.

395. Eyre Crowe, Foreign Office to Admiralty, 13 May 1914 (NA FO 195/2456/60).

396. Cohen, *British*, chs. 8–9.

397. Quoted in Ireland, *Iraq*, 44 (n. 1).

398. Quoted in Cohen, *British*, 202.

399. Quoted in ibid., 199.

400. Quoted in ibid., 202.

401. Holmes, 'Mahan'; MacMillan, *War*, 87–130.

402. Quoted in Holmes, ibid., 37; Tirpitz, *My Memoirs*, vol. 1, 77.

403. Docherty and MacGregor, *Hidden*, chs. 3–8; McGeoch, 'On the Road'; Neilson, '"Greatly Exaggerated"', 695–720; Coogan and Coogan, 'British'; Wilson, *Policy*, chs. 5–7; Miller, *Millstone*, ch. 4.

404. Marder, ed., *Fear God*, vol. 1, 218.

405. Bethmann Hollweg, *Reflections*, pt. 1, 66.

406. Marder, ed., *Fear God*, vol. 2, 55; Miller, *Millstone*, 29.

407. Quoted in Maurice, *Haldane*, vol. 1, 173; Docherty and MacGregor, *Hidden*, 108; Neilson, '"Greatly Exaggerated"', 704–7.

408. Holmes, 'Mahan', 45–7; Lambert, *Planning*, 1–181; Offer, 'Working', 212–17; Marder, *Dreadnought*, vol. 1, 367–83.

409. Quoted in Gooch and Temperley, eds., *British*, vol. 6, 468; see section 'Spheres of Interest' above.

410. Quoted in Wilson, 'Creative', 53–4.
411. O'Brien, 'Titan'; Offer, 'Working', 204–15; Marder, *Dreadnought*, vol. 1, 358–67.
412. Eyre Crowe, 'Memorandum on the Present State of British Relations with France and Germany, January 1, 1907', in Gooch and Temperley, eds., *British*, vol. 3, 402, 414; MacMillan, *War*, 115–16; Holmes, 'Mahan', 49–50.
413. Quoted in Marder, *Dreadnought*, vol. 1, 135.
414. Marder, ed., *Fear God*, vol. 2, 169; ibid., 122–3; Miller, *Millstone*, 90.
415. Fisher, *Memories*, 16, 237.
416. Docherty and MacGregor, *Hidden*, chs. 9–10; Morris, *Scaremongers*; Gollin, *No Longer*, ch. 11, 13–14; Gollin, *Impact*.
417. Quoted in Steiner and Neilson, *Britain*, 54; McGeoch, 'On the Road', 214; Docherty and MacGregor, ibid., 71–3; Coogan and Coogan, 'British', 110–11.
418. Quoted in Berghahn, *American*, 78; Wawro, *Warfare*, 181–4; Marder, *Dreadnought*, vol. 1, 430.
419. Quoted in Marder, ibid., 215; Morris, *Scaremongers*, 219; Docherty and MacGregor, *Hidden*, ch. 9.
420. HC Deb, 29 March 1909, vol. 3, c. 69; Marder, ibid., 135–71.
421. 'Lord C. Beresford on the Navy and the Crisis', *Manchester Guardian* (23 November 1911); Marder, ibid., 130–5, 171–83, 221–33, 272–87; Docherty and MacGregor, *Hidden*, ch. 14.
422. Quoted in Churchill, *World*, 77; Docherty and MacGregor, ibid., 188–92; Marder, ibid., 272–87; Gough, *Churchill*, 201–6.
423. HC Deb, 17 June 1914, vol. 63, c. 1148; section '"Surrounded by Material Far More Inflammable than the Oil"' above.
424. 'British and German Relations', *Times* (27 June 1912).
425. Marder, *Dreadnought*, vol. 1, 121–2; Docherty and MacGregor, *Hidden*, 106–18, 133–4, 181–90, 197–200; Coogan and Coogan, 'British', 121–6.
426. Quoted in Marder, ibid., 429.
427. Quoted in Kent, *Oil*, 78; section 'The Turkish Petroleum Company' above; see sections '"Surrounded by Material Far More Inflammable than the Oil"' and '"Will Germany Control the Oil Supply of Our Navy?"' above.
428. See section 'Oil "for a Great War"' above.
429. Brown, 'Royal Navy', 284; Jones, *State*, 12; *Diplomatic and Consular Reports: Roumania, 1913*, no. 5326 (July 1914), 9–10.
430. 'Trade of Rumania', *Times* (29 June 1914); 'Roumania's Extending Export Trade', *Petroleum Review* (22 November 1913); 'Roumania's

Crude Oil Production', ibid. (16 May 1914); Redwood, *Petroleum*, vol. 3, 1,084; *Report of the Royal Commission*, vol. 2, 32–4.

431. HC Deb, 18 March 1914, vol. 59, c. 2103; see section 'Oil "for a Great War"' above.

432. See ch. 6, section 'Liquid Fuel and Sea Power' and section 'Oil "for a Great War"' above.

433. Marder, *Dreadnought*, vol. 1, 287–309; Docherty and MacGregor, *Hidden*, 190–3.

434. See section 'Internal Combustion' above.

435. See ch. 7, section 'Economic Warfare'.

436. See section 'Oil "for a Great War"' above.

437. Bodger, 'Russia', 78–80; Clark, *Sleepwalkers*, 340–1; Miller, *Straits*, 163, 203 (n. 30).

438. 'The Black Sea and the War', *The Courier* (Dundee) (5 April 1915); Alekperov, *Oil*, 150; Williamson et al., *Energy*, 262–3.

439. 'Royal Dutch Report', *Financial Times* (16 June 1915).

440. Quoted in Clark, *Sleepwalkers*, 341; ibid., 187–8, 250–76; Neilson, '"Greatly Exaggerated"', 708–14; Miller, *Straits*, 163–5, 207–9.

441. Quoted in Taylor, *Struggle*, 492–3; Turner, 'Russian', 253–6; MacMillan, *War*, 497–8.

442. Lichnowsky, *Heading*, 168; Clark, *Sleepwalkers*, 351–8.

443. Seligmann, 'Keeping'.

444. Quoted in Aksakal, *Ottoman*, 82, and McMeekin, *Russian*, 31, respectively.

445. Quoted in Bodger, 'Russia', 90; Gough, *Churchill*, 328–31.

446. Quoted in Miller, *Straits*, 166–7; Clark, *Sleepwalkers*, 338–49, 484–5; Sazonov, *Fateful*, 117–27.

447. Quoted in Miller, ibid., 169–70.

448. Quoted in McDonald, *United*, 203.

449. Quoted in ibid., 202; Miller, *Straits*, 196.

450. Quoted in Nicolson, *Sir Arthur*, 406; Otte, 'Foreign', 22–3; McDonald, *United*, 190–201.

451. Quoted in Miller, *Straits*, 176.

452. Quoted in ibid., 204; Aksakal, *Ottoman*, 86–90; see above on Romanian fuel oil.

453. 'The Car that Witnessed the Spark of World War I', *New York Times* (10 July 2014); King and Woolmans, *Assassination*, 198–206; Neiberg, *Dance*, 13–17; Clark, *Sleepwalkers*, chs. 1–2, 7; Docherty and MacGregor, *Hidden*, ch. 20.

454. Feis, *Europe*, chs. 12–13.

455. MacMillan, *War*, chs. 13–17; Lynn-Jones, 'Détente', 121–37; McMeekin, *Russian*, 23–7; Neiberg, *Dance*, chs. 1–3.

456. Neiberg, ibid., ch. 3; Clark, *Sleepwalkers*, 376–81.

457. Bridge, 'Sir Edward'; Crampton, 'Decline'.

458. See sections 'Britain's State Oil Company' and '"Surrounded by Material Far More Inflammable than the Oil"' above; Siegel, *Endgame*, ch. 8; Chen, 'British', 96–8; Wilson, 'Grey', 278–9.

459. Quoted in Wilson, ibid., 279.

460. Quoted in Greaves, 'Some Aspects – II', 301.

461. Quoted in ibid., 302.

462. Quoted in Jarausch, 'Illusion', 58; Clark, *Sleepwalkers*, 414–19; Craig, *Germany*, 334–6; Aksakal, *Ottoman*, 93–102.

463. Quoted in Wilson, 'Britain', 185; Bridge, 'Sir Edward', 270–1; Fromkin, *Europe*, 166.

464. Quoted in Greaves, 'Some Aspects – II', 302–3; Wilson, 'Struggle', 290.

465. Quoted in Miller, *Straits*, 460.

466. Quoted in Wilson, 'Struggle', 299, 309.

467. Quoted in ibid., 299.

468. Quoted in ibid., 307, 322.

469. Quoted in ibid., 316–17.

470. See ch. 7, section 'British Supremacy over Baghdad' and sections 'Spheres of Interest' and 'The Turkish Petroleum Company' above; Barth and Whitehouse, 'Financial', 135–6; Kent, *Oil*, 109–12; Earle, *Turkey*, 260–5.

471. Wilson, 'Struggle', 309–16, 321–8; Greaves, 'Some Aspects – II', 306–7.

472. Quoted in Greaves, ibid., 307; ibid., 308.

473. Quoted in ibid., 300.

474. Quoted in MacMillan, *War*, 547.

475. Clark, *Sleepwalkers*, 340–9, 409, ch. 9; MacMillan, ibid., 546–53; McDonald, *United*, 202–7; Newton, *Darkest*, 115–16; McMeekin, *Russian*, 6–59; Seligmann, 'Keeping'.

476. Quoted in Lichnowsky, *Heading*, 375; Newton, ibid., 109–10.

477. Quoted in Fay, *Origins*, vol. 2, 380; Clark, *Sleepwalkers*, 451–68, 471–4; Docherty and MacGregor, *Hidden*, 268–76; McMeekin, *Russian*, 50–62.

478. Turner, 'Significance', 205, 216; Clark, ibid., 306, 354–5, 483–4; MacMillan, *War*, 547; Jarausch, 'Illusion', 66–8; Docherty and MacGregor, ibid., 181.

479. Quoted in Jarausch, ibid., 63.

480. Quoted in ibid., 64.

481. Quoted in Newton, *Darkest*, 110.

482. Quoted in Clark, *Sleepwalkers*, 472.

483. Clark, ibid., 476–84; MacMillan, *War*, 550–2; Docherty and MacGregor, *Hidden*, 267.

484. Docherty and MacGregor, ibid., chs. 22–3; Neiberg, *Dance*, chs. 3–4; Thompson, *Politicians*, 18–21.

485. Quoted in Wilson, 'Grey', 283 (n. 33).

486. Quoted in Jarausch, 'Illusion', 64.

487. Newton, *Darkest*, 27–30, 50–4.

488. Quoted in Miller, *Millstone*, 413; Jarausch, 'Illusion', 65.

489. Quoted in Gooch and Temperley, eds., *British*, vol. 11, 157.

490. Quoted in Miller, *Millstone*, 413; Wilson, *Policy*, 78–84.

491. Quoted in Geiss, ed., *July*, 285; MacMillan, *War*, 568.

492. Jarausch, 'Illusion', 66–9; Newton, *Darkest*, chs. 4–11; Clark, *Sleepwalkers*, 494–8; McDonald, *United*, 206–7; McMeekin, *Russian*, 62–75; Docherty and MacGregor, *Hidden*, ch. 24.

493. Quoted in Jarausch, ibid., 69; ibid., 70–7.

494. Churchill, *Winston*, vol. 2, Companion pt. 3, 1965.

495. Newton, *Darkest*, chs. 6–26.

496. Quoted in ibid., 182 and 224 respectively; ibid., 207–8.

497. HC Deb, 7 July 1914, vol. 64, c. 1052; section '"Surrounded by Material Far More Inflammable than the Oil"'.

EPILOGUE

1. Gibson, *Britain*, ch. 2, 75; Jonker and Zanden, *History*, vol. 1, 157–77; Yergin, *Prize*, 171–8; Winegard, *First*, 5, 101; Toprani, *Oil*, 36–41; Jones, *State*, ch. 7; Brown, 'Royal Navy', chs. 6–7, 176; Sumida, 'British'.

2. 'Anglo-Persian Oil Company (Limited)', *Times* (21 December 1915); Brown, ibid., chs. 6–7; Toprani, ibid., ch. 1; Winegard, ibid., 63–8; Ferrier, *History*, vol. 1, 279–82; Townshend, *When God*, chs. 1–4; Garthwaite, 'Bakhtiyari', 39–42; McMeekin, *Berlin–Baghdad*, 203–4, 277–8; Kent, *Oil*, 118–19, 134.

3. 'Anglo-Persian Oil Company (Limited)', op. cit.

4. 'Anglo-Persian Oil', *Financial Times* (7 December 1915).

5. 'Germans Through Turks Seek the Oil Fields', *Wall Street Journal* (20 November 1914); 'What War Moves Mean', *Washington Times* (14 November 1914); Murray, 'Fortnightly', vol. 1, 87–8; 'The Progress of the War', *Review of Reviews* (March 1916).

6. Quoted in Delaisi, *Oil*, 85–6; Brown, 'Royal Navy', ch. 6; Sumida, 'British', 463–73; Gibson, *Britain*, 53–88; Yergin, *Prize*, 176–9.

7. Quoted in Katz, *Secret*, 343; Brown, ibid., 134–5, 142, 168.

8. Ludendorff, *My War*, vol. 1, 287–8, and vol. 2, 525; see near end of ch. 7 section 'Economic Warfare' on Germany's petroleum imports; Toprani, *Oil*, 6; 'Petrol Shortage in Germany', *Financial Times* (12 October 1914); 'As the Germans See It – The Growing Scarcity of Petroleum', *Guardian* (11 November 1914); Fursenko, *Battle*, 180–1; Frank, *Oil*, ch. 6; Schatzker and Hirszhaut, *Jewish*, ch. 9.

9. Ludendorff, ibid., vol. 1, 287–8, 347, and vol. 2, 659; Pearton, *Oil*, ch. 4; Yergin, *Prize*, 179–82; Winegard, *First*, 93–5.

10. Ludendorff, ibid., vol. 2, 659; Yergin, ibid., 182; Winegard, ibid., 97–184.

11. Winegard, ibid., chs. 5–7; McMeekin, *Berlin–Baghdad*, 332–7; Alekperov, *Oil*, 175–84.

12. Quoted in Winegard, *First*, 112.

13. Quoted in ibid., 111.

14. Quoted in Gibson, *Britain*, 90.

15. Quoted in Rothwell, 'Mesopotamia', 289; Jones, *State*, 195–200.

16. Winegard, *First*, 112–15; Sluglett, *Britain*, 65–77; Rothwell, ibid.; Mejcher, 'Oil'; Kent, *Oil*, ch. 8; Toprani, *Oil*, 26, 40, 43–6.

17. Quoted in Delaisi, *Oil*, 29, 86–7.

Bibliography

BOOKS

Abernethy, Thomas Perkins, *Western Lands and the American Revolution* (New York and London: D. Appleton-Century, 1937).

Abler, Thomas S., *Cornplanter: Chief Warrior of the Allegany Senecas* (Syracuse, NY: Syracuse University Press, 2007).

Abrahamian, Ervand, *A History of Modern Iran* (Cambridge: Cambridge University Press, 2008).

——, *The Coup: 1953, the CIA, and the Roots of Modern US-Iranian Relations* (New York: The New Press, 2013).

Abu-Hakima, Ahmad Mustafa, *The Modern History of Kuwait, 1750–1965* (London: Luzac & Co., 1983).

Acosta, Joseph, *The Naturall and Morall Historie of the East and West Indies*, trans. E.G. (London: Edward Blount and William Aspley, 1604).

Adler, Marcus Nathan, *The Itinerary of Benjamin of Tudela: Critical Text, Translation and Commentary* (London: Henry Frowde, 1907).

Aksakal, Mustafa, *The Ottoman Road to War in 1914: The Ottoman Empire and the First World War* (Cambridge: Cambridge University Press, 2008).

Alekperov, Vagit, *Oil of Russia: Past, Present and Future* (Minneapolis, MN: East View Press, 2011).

Alizadeh, Abbas, *Ancient Settlement Systems and Cultures in the Ram Hormuz Plain, Southwestern Iran: Excavations at Tall-e Geser and Regional Survey of the Ram Hormuz Area* (Chicago, IL: Oriental Institute of the University of Chicago, 2014).

Allen, W. E. D., and P. Muratoff, *Caucasian Battlefields: A History of the Wars on the Turco-Caucasian Border, 1828–1921* (Cambridge: Cambridge University Press, 1953).

Altstadt, Audrey, *The Azerbaijani Turks: Power and Identity under Russian Rule* (Stanford, CA: Hoover Institution Press, 1992).

Ammianus Marcellinus, *Ammianus Marcellinus*, trans. John C. Rolfe (Cambridge, MA: Harvard University Press, 1940).

Anderson, Fred, *Crucible of War: The Seven Years' War and the Fate of Empire in British North America, 1754–1766* (New York: Vintage Books, 2000).

Anderson, M. S., *The Eastern Question, 1773–1923: A Study in International Relations* (London: Macmillan, 1966).

Anderson, Scott, *Lawrence in Arabia: War, Deceit, Imperial Folly and the Making of the Modern Middle East* (New York: Random House, 2013).

Andrist, Ralph K., *The Long Death: The Last Days of the Plains Indians* (New York: Macmillan, 1964).

Arnold, Arthur, *Through Persia by Caravan* (London: Tinsley Brothers, 1877).

Arnold, Ralph, George A. Macready and Thomas W. Barrington, *The First Big Oil Hunt: Venezuela 1911–1916* (New York: Vantage Press, 1960).

Arnold, Thurman W., *The Folklore of Capitalism* (New Haven, CT: Yale University Press, 1937).

Ashe, Thomas, *Travels in America, Performed in 1806* (London: Edmund M. Blunt, 1808).

Aston, W. G., trans., *Nihongi: Chronicles of Japan from the Earliest Times to A.D. 697* (London: Kegan Paul, Trench, Trübner & Co.,1896).

Ateş, Sabri, *Ottoman-Iranian Borderlands: Making a Boundary, 1843–1914* (New York: Cambridge University Press, 2013).

Atkin, Muriel, *Russia and Iran, 1780–1828* (Minneapolis, MN: University of Minnesota Press, 1980).

Atkinson, Geo. W., *History of Kanawha County* (Charleston, VA: West Virginia Journal, 1876).

Aung, Maung Htin, *Lord Randolph Churchill and the Dancing Peacock: British Conquest of Burma, 1885* (New Delhi: Manohar Publications, 1990).

——, *The Stricken Peacock: Anglo-Burmese Relations, 1752–1948* (The Hague: Martinus Nijhoff, 1965).

Avellaneda, José Ignacio, *The Conquerors of the New Kingdom of Granada* (Albuquerque, NM: University of New Mexico Press, 1995).

Ayalon, David, *Gunpowder and Firearms in the Mamluk Kingdom: A Challenge to a Mediaeval Society* (London: Vallentine, Mitchell, 1956).

Babcock, Charles A., *Venango County, Pennsylvania: Her Pioneers and People* (Chicago, IL: J. H. Beers & Co., 1919).

Bakeless, John, *Lewis and Clark: Partners in Discovery* (New York: William Morrow & Co., 1947).

Bancroft, Hubert Howe, *History of California* (San Francisco, CA: The History Company, 1890).

Barba, Alvaro Alonso, *El Arte de Los Metales*, trans. Ross E. Douglass and E. P. Mathewson (New York: John Wiley & Sons, 1923).

Barkan, Joshua, *Corporate Sovereignty: Law and Government under Capitalism* (Minneapolis, MN, and London: University of Minnesota Press, 2013).

Barker, Theo, ed., *The Economic and Social Effects of the Spread of Motor Vehicles: An International Centenary Tribute* (London: Macmillan, 1987).

——, and Dorian Gerhold, *The Rise and Rise of Road Transport, 1700–1990* (London: Macmillan, 1993).

——, and Michael Robbins, *A History of London Transport: Passenger Travel and the Development of the Metropolis* (London: George Allen & Unwin, 1974).

Barnhart, Terry A., *American Antiquities: Revisiting the Origins of American Archaeology* (Lincoln, NE: University of Nebraska Press, 2015).

Bartholomew, E., *Early Armoured Cars* (Princes Risborough: Shire, 1988).

Bassie-Sweet, Karen, ed., *The Ch'ol Maya of Chiapas* (Norman, OK: University of Oklahoma Press, 2015).

Bates, Samuel P., *Our County and Its People: A Historical and Memorial Record of Crawford County, Pennsylvania* (Boston, MA: W. A. Fergusson & Co., 1899).

Bayat, Mangol, *Iran's First Revolution: Shi'ism in the Constitutional Revolution of 1905–1909* (Oxford: Oxford University Press, 1991).

Beaton, Kendall, *Enterprise in Oil: A History of Shell in the United States* (New York: Appleton-Century-Crofts, 1957).

Beaulieu, Augustin de, *De rampspoedige scheepvaart der Franschen naar Oostindien* (Amsterdam: Jan Rieuwertsz and Pieter Arentsz, 1669).

Bebbington, David, and Roger Swift, eds., *Gladstone Centenary Essays* (Liverpool: Liverpool University Press, 2000).

Beckerman, Stephen, and Roberto Lizarralde, *The Ecology of the Barí: Rainforest Horticulturalists of South America* (Austin, TX: University of Texas Press, 2013).

Bell, Herbert C., ed., *History of Venango County, Pennsylvania* (Chicago, IL: Brown, Runk & Co., 1890).

Bennigsen Broxup, Marie, ed., *The North Caucasus Barrier: The Russian Advance towards the Muslim World* (London: Hurst & Co., 1992).

Berger, Henry W., *St. Louis and Empire: 250 Years of Imperial Quest and Urban Crisis* (Carbondale, IL: Southern Illinois University Press, 2015).

Bergeron, Paul H., ed., *The Papers of Andrew Johnson* (Knoxville, TN: University of Tennessee Press, 1995).

Berghahn, Volker R., *American Big Business in Britain and Germany: A Comparative History of Two 'Special Relationships' in the Twentieth Century* (Princeton, NJ: Princeton University Press, 2014).

Berthoff, Rowland, *Republic of the Dispossessed: The Exceptional Old-European Consensus in America* (Columbia, MO, and London: University of Missouri Press, 1997).

Bethmann Hollweg, Th. Von, *Reflections on the World War*, trans. George Young (London: Thornton Butterworth, 1920).

Billington, Ray Allen, *Westward Expansion: A History of the American Frontier*, 2nd edn (New York: Macmillan, 1960).

Black, Brian C., *Crude Reality: Petroleum in World History* (Lanham, MD: Rowman & Littlefield, 2012).

——, *Petrolia: The Landscape of America's First Oil Boom* (Baltimore, MD, and London: Johns Hopkins University Press, 2000).

Black, Edwin, *Banking on Baghdad: Inside Iraq's 7,000-Year History of War, Profit, and Conflict* (Hoboken, NJ: John Wiley & Sons, 2004).

——, *Internal Combustion: How Corporations and Governments Addicted the World to Oil and Derailed the Alternatives* (New York: St. Martin's Press, 2006).

Blouet, Brian W., and Merlin P. Lawson, eds., *Images of the Plains: The Role of Human Nature in Settlement* (Lincoln, NE: University of Nebraska Press, 1975).

Bonakdarian, Mansour, *Britain and the Iranian Constitutional Revolution of 1906–1911: Foreign Policy, Imperialism, and Dissent* (Syracuse, NY: Syracuse University Press, 2006).

Bone, J. H. A., *Petroleum and Petroleum Wells*, 2nd edn (Philadelphia, PA: J. B. Lippincott & Co., 1865).

Bontius, James, *An Account of the Diseases, Natural History and Medicines of the East Indies*, trans. from the Latin (London: T. Noteman, 1769).

Bosworth, C. E., trans., *The History of al-Ṭabarī*, vol. 33: *Storm and Stress along the Northern Frontiers of the Abbāsid Caliphate* (Albany, NY: State University of New York Press, 1991).

Bosworth, T. O., *Geology of the Tertiary and Quaternary Periods in the North-West Part of Peru* (London: Macmillan and Co., 1922).

Boxt, Matthew A., and Brian D. Dillon, eds., *Fanning the Sacred Flame* (Boulder, CO: University Press of Colorado, 2012).

Boyd, Julian P., ed., *The Papers of Thomas Jefferson* (Princeton, NJ: Princeton University Press, 1951).

Boyd, R. Nelson, *Petroleum: Its Development and Uses* (London and New York: Whittaker & Co., 1895).

Boyer, Christopher R., ed., *A Land Between Waters: Environmental Histories of Modern Mexico* (Tucson, AZ: University of Arizona Press, 2012).

Bradley, James, *The Imperial Cruise: A Secret History of Empire and War* (New York: Little, Brown and Co., 2009).

Brantly, J. E., *History of Oil Well Drilling* (Houston, TX: Gulf Publishing Co., 1971).

Brenner, Robert, *Merchants and Revolution: Commercial Change, Political Conflict, and London's Overseas Traders, 1550–1653* (Cambridge: Cambridge University Press, 1993).

Bringhurst, Bruce, *Antitrust and the Oil Monopoly: The Standard Oil Cases, 1890–1911* (Westport, CT: Greenwood Press, 1979).

Bronner, Stephen Eric, *A Rumor About the Jews: Reflections on Anti-Semitism and the Protocols of the Elders of Zion* (New York: St. Martin's Press, 2000).

Brooks, Charles E., *Frontier Settlement and Market Revolution: The Holland Land Purchase* (Ithaca, NY: Cornell University Press, 1996).

Brown, David K., *The Grand Fleet: Warship Design and Development, 1906–1922* (London: Chatham, 1999).

Brown, Harriet Connor, *Report on the Mineral Resources of Cuba in 1901* (Baltimore, MD: Guggenheimer, Weil & Co., 1903).

Brown, Jonathan C., *Oil and Revolution in Mexico* (Berkeley, CA: University of California Press, 1993).

Broz, J. Lawrence, *The International Origins of the Federal Reserve System* (New York: Cornell University Press, 1997).

Burdett, A. L. P., ed., *Oil Resources in Eastern Europe and the Caucasus: British Documents 1885–1978* (Cambridge: Cambridge Archive Editions, 2012).

Burnell, Arthur Coke, ed., *The Voyage of John Huyghen van Linschoten to the East Indies* (London: Hakluyt Society, 1885).

Burr, Christina, *Canada's Victorian Oil Town: The Transformation of Petrolia from a Resource Town into a Victorian Community* (Montreal: McGill-Queen's University Press, 2006).

Busch, Briton Cooper, *Britain and the Persian Gulf, 1898–1914* (Berkeley, CA: University of California Press, 1967).

Buzatu, Gheorghe, *A History of Romanian Oil*, trans. Laura Chistruga-Schneider (Bucharest: Mica Valahie, 2011).

Cain, P. J., and A. G. Hopkins, *British Imperialism, 1688–2015* (London: Routledge, 2016).

Calloway, Colin G., *The Indian World of George Washington: The First President, the First Americans, and the Birth of a Nation* (Oxford: Oxford University Press, 2018).

——, *The Scratch of a Pen: 1763 and the Transformation of North America* (Oxford: Oxford University Press, 2006).

——, *The Victory with No Name: The Native American Defeat of the First American Army* (Oxford: Oxford University Press, 2015).

Calvert, Peter, *The Mexican Revolution, 1910–1914: The Diplomacy of Anglo-American Conflict* (Cambridge: Cambridge University Press, 1968).

Camden, William, *Britannia: Or a Chorographical Description of Great Britain and Ireland*, 2nd edn, rev. Edmund Gibson (London: Printed by M. Matthews, for Awnsham Churchill, 1722).

Cameron, Rondo, and V. I. Bovykin, eds., *International Banking, 1870–1914* (New York: Oxford University Press, 1991).

Campbell, Jr, Charles S., *Special Business Interests and the Open Door Policy* (New Haven, CT: Yale University Press, 1951).

Carletti, Francesco, *My Voyage Around the World*, trans. Herbert Weinstock (London: Methuen & Co., 1965).

Casserly, Gordon, *The Land of the Boxers, or China under the Allies* (London: Longmans, Green, & Co., 1903).

Challener, Richard D., *Admirals, Generals, and American Foreign Policy* (Princeton, NJ: Princeton University Press, 1973).

Chamberlain, M. E., *The Scramble for Africa* (London: Longman, 1974).

Chandler, Alfred N., *Land Title Origins: A Tale of Force and Fraud* (New York: Robert Schalkenbach Foundation, 1945).

Chandler, Jr, Alfred D., *Scale and Scope: The Dynamics of Industrial Capitalism* (Cambridge, MA: Harvard University Press, 1990).

Chaqueri, Cosroe, *The Russo-Caucasian Origins of the Iranian Left: Social Democracy in Modern Iran* (Richmond, VA: Curzon Press, 2001).

Charlevoix, Father, *A Voyage to North America* (Dublin: John Exshaw and James Potts, 1766).

Chernow, Ron, *Titan: The Life of John D. Rockefeller, Sr.* (New York: Vintage Books, 2004).

Churchill, Randolph S., *Winston S. Churchill* (London: Heinemann, 1969).

Churchill, Winston S., *The World Crisis, 1911–1918*, abr. and rev. (London: Thornton Butterworth, 1931).

Chynoweth, W. Harris, *The Fall of Maximilian, Late Emperor of Mexico* (London: self-published, 1872).

Clark, Christopher, *The Sleepwalkers: How Europe Went to War in 1914* (London: Allen Lane, 2012).

Clark, Jr, John E., *Railroads in the Civil War: The Impact of Management on Victory and Defeat* (Baton Rouge, LO: Louisiana State University Press, 2001).

Clark, Stanley, *The Oil Century: From the Drake Well to the Conservation Era* (Norman, OK: University of Oklahoma Press, 1958).

Clayton, G. D., *Britain and the Eastern Question: Missolonghi to Gallipoli* (London: University of London Press, 1971).

Clayton, Lawrence A., Vernon James Knight, Jr and Edward C. Moore, *The De Soto Chronicles: The Expedition of Hernando de Soto to North America in 1539–1543* (Tuscaloosa, AL: University of Alabama Press, 1993).

Clemens, A. H. P., and J. Th. Lindblad, eds., *Het belang van de Buitengewesten: Economische expansie en koloniale staatsvorming in de Buitengewesten van Nederlands-Indië, 1870–1942* (Amsterdam: NEHA, 1989).

Clinton, De Witt, *An Introductory Discourse delivered before the Literary and Philosophical Society of New-York* (New York: David Longworth. 1815).

Cochran, Sherman, *Encountering Chinese Networks: Western, Japanese, and Chinese Corporations in China, 1880–1937* (Berkeley, CA: University of California Press, 2000).

Coffin, Charles Carleton, *Our New Way Round the World* (London: Frederick Warne & Co., 1883).

Cohen, Michael J., *Churchill and the Jews, 1900–1948*, 2nd edn (London: Routledge, 2003).

Cohen, Stuart A., *British Policy in Mesopotamia, 1903–1914* (London: Ithaca Press, 1976).

Cone, Andrew, and Walter R. Johns, *Petrolia: A Brief History of the Pennsylvania Petroleum Region* (New York: D. Appleton and Co., 1870).

Conlin, Jonathan, *Mr Five Per Cent: The Many Lives of Calouste Gulbenkian, the World's Richest Man* (London: Profile Books, 2019).

Constantine Porphyrogenitus, *De Administrando Imperio*, ed. Gy. Moravcsik, trans. R. J. H. Jenkins (Washington, DC: Dumbarton Oaks, 1967).

Cook, Roy Bird, *Washington's Western Lands* (Strasburg, VA: Shenandoah Publishing House, 1930).

Corley, T. A. B., *A History of the Burmah Oil Company, 1886–1924* (London: Heinemann, 1983).

Cote, Stephen C., *Oil and Nation: A History of Bolivia's Petroleum Sector* (Morgantown, VA: West Virginia University Press, 2016).

Couceiro, Henrique de Paiva, *Angola (dois anos de governo, junho 1907–junho 1909)* (Lisbon: Tipografia Portuguesa, 1948).

Coyne, James H., *Exploration of the Great Lakes, 1669–1670, by Dollier de Casson and De Bréhant de Galinée* (Toronto: Ontario Historical Society, 1903).

Craig, Gordon A., *Germany, 1866–1945* (Oxford: Clarendon Press, 1978).

Craig, Jonathan, Francesco Gerali, Fiona MacAulay and Rasoul Sorkhabi, eds., *History of the European Oil and Gas Industry* (Geological Society, London: Special Publications 465, 2018).

Crawfurd, John, *Journal of an Embassy from the Governor-General of India to the Court of Ava, in the Year 1827* (London: Henry Colburn, 1829).

Crespí, Juan, *A Description of Distant Roads: Original Journals of the First Expedition into California, 1769–1770*, ed. and trans. Alan K. Brown (San Diego, CA: San Diego State University Press, 2001).

Cribb, Robert, ed., *The Late Colonial State in Indonesia: Political and Economic Foundations of the Netherlands Indies, 1880–1942* (Leiden: KITLV Press, 1994).

Crosthwaite, Charles, *The Pacification of Burma* (London: Edward Arnold, 1912).

Crowder, Michael, *The Story of Nigeria*, 4th edn (London: Faber and Faber, 1978).

Cruikshank, E. A., ed., *The Correspondence of Lieut. Governor John Graves Simcoe, with Allied Documents Relating to his Administration of the Government of Upper Canada* (Toronto: Ontario Historical Society, 1923).

Cuming, F., *Sketches of a Tour to the Western Country* (Pittsburgh, PA: Cramer, Spear & Eichbaum, 1810).

Cummins, Jr, C. Lyle, *Diesel's Engine*, vol. 1: *From Conception to 1918* (Wilsonville, OR: Carnot Press, 1993).

Curzon, George N., *Persia and the Persian Question* (London: Longmans, Green, and Co, 1892).

Daintith, Terence, *Finders Keepers? How the Law of Capture Shaped the World Oil Industry* (Washington, DC, and London: RFF Press, 2000).

Dasgupta, Biplab, *The Oil Industry in India: Some Economic Aspects* (London: Frank Cass, 1971).

Davidson, Gordon Charles, *The North West Company* (Berkeley, CA: University of California Press, 1918).

Davidson, James W., *The Island of Formosa, Past and Present* (London: Macmillan & Co., 1903).

Davidson, Philip, *Propaganda and the American Revolution, 1763–1783* (Chapel Hill, NC: University of North Carolina Press, 1941).

Davis, Joseph Stancliffe, *Essays in the Earlier History of American Corporations* (Cambridge, MA: Harvard University Press, 1917).

Davis, Margaret Leslie, *Dark Side of Fortune: Triumph and Scandal in the Life of Edward L. Doheny* (Berkeley, CA: University of California Press, 1998).

Davoudi, Leonardo, *Persian Petroleum: Oil, Empire and Revolution in Late Qajar Iran* (London: I.B. Tauris, 2021).

Day, David T., ed., *A Handbook of the Petroleum Industry* (New York: John Wiley & Sons, 1922).

Day, Sherman, *Historical Collections of the State of Pennsylvania* (Philadelphia, PA: George W. Gorton, 1843).

De Jonge, Alex, *Stalin and the Shaping of the Soviet Union* (London: Collins, 1986).

Delaisi, Francis, *Oil: Its Influence on Politics*, trans. C. Leonard Leese (London: Labour Publishing Co. and George Allen and Unwin, 1922).

DeNovo, John A., *American Interests and Policies in the Middle East, 1900–1939* (Minneapolis, MN: University of Minnesota Press, 1963).

Deterding, Henri, *An International Oilman* (London: Ivor Nicholson and Watson Ltd., 1934).

Deutscher, I., *Stalin: A Political Biography*, 2nd edn (London: Oxford University Press, 1967).

Dinnerstein, Leonard, and Kenneth T. Jackson, eds., *American Vistas, 1607–1877*, 3rd edn (New York: Oxford University Press, 1979).

Dio, Cassius, *Dio's Roman History*, trans. Earnest Cary (London: Heinemann, 1914).

Dioscorides of Anazarbus, Pedanius, *De materia medica*, trans. Lily Y. Beck, 2nd edn (Hildesheim: Olms-Weidmann, 2011).

Ditson, George Leighton, *Circassia; or, A Tour to the Caucasus*, new and revised edition (New York: Stringer & Townsend, 1850).

Docherty, Gerry, and Jim MacGregor, *Hidden History: The Secret Origins of the First World War* (Edinburgh and London: Mainstream Publishing, 2013).

Dodd, Edwin Merrick, *American Business Corporations until 1860 with Special Reference to Massachusetts* (Cambridge, MA: Harvard University Press, 1954).

Dodd, John, *Journal of a Blockaded Resident in North Formosa, During the Franco-Chinese War, 1884–5* (Hong Kong: 'Daily Press' Office, 1888).

Dolson, Hildegarde, *They Struck Oil* (London: Hammond, Hammond & Co., 1959).

Donkin, Jr, Bryan, *Gas, Oil and Air Engines; or, Internal Combustion Motors without Boiler* (London: Charles Griffin & Co., 1894).

Doran, Peter B., *Breaking Rockefeller: The Incredible Story of the Ambitious Rivals who Toppled an Oil Empire* (New York: Viking, 2016).

Dowson, J. Emerson, and A. T. Larter, *Producer Gas*, 2nd edn (London: Longmans, Green & Co., 1907).

Dugdale, E. T. S., ed. and trans., *German Diplomatic Documents, 1871–1914* (New York: Harper & Brothers, 1929).

Dundonald, Thomas, Eleventh Earl of, and H. R. Fox Bourne, *The Life of Thomas, Lord Cochrane, Tenth Earl of Dundonald* (London: Richard Bentley, 1869).

Dunstan, A. E., ed., *The Science of Petroleum* (London: Oxford University Press, 1938).

Earle, Edward Mead, *Turkey, the Great Powers, and the Baghdad Railway* (New York: Macmillan, 1924).

Eaton, Rev. S. J. M., *Petroleum: A History of the Oil Region of Venango County, Pennsylvania* (Philadelphia, PA: J. P. Skelly & Co., 1866).

Eckermann, Erik, *World History of the Automobile* (Warrendale: Society of Automotive Engineers, 2001).

Edmondson, Linda, and Peter Waldron, eds., *Economy and Society in Russia and the Soviet Union, 1860–1930: Essays for Olga Crisp* (London: Macmillan, 1992).

Erickson, Ljubica and Mark, eds., *Russia: War, Peace and Diplomacy* (London: Weidenfeld & Nicolson, 2004).

Esarey, Logan, ed., *Messages and Letters of William Henry Harrison* (Indianapolis, IN: Indiana Historical Commission, 1922).

Estes, Nick, *Our History is the Future: Standing Rock versus the Dakota Access Pipeline, and the Long Tradition of Indigenous Resistance* (London: Verso, 2019).

Etherington, Norman, *Theories of Imperialism: War, Conquest and Capital* (Beckenham: Croom Helm, 1984).

Evans, Lewis, *Geographical, Historical, Political, Philosophical and Mechanical Essays* (Philadelphia, PA: B. Franklin and D. Hall, 1755).

Ewell, Judith, *Venezuela and the United States: From Monroe's Hemisphere to Petroleum's Empire* (Athens, GA: University of Georgia Press, 1996).

Fairman, E. St. John, *A Treatise on the Petroleum Zones of Italy* (London: E. & F. N. Spon, 1868).

Falola, Toyin, *Colonialism and Violence in Nigeria* (Bloomington, IN: Indiana University Press, 2009).

Farnie, D. A., *East and West of Suez: The Suez Canal in History, 1854–1956* (Oxford: Clarendon Press, 1969).

Fay, Sidney Bradshaw, *The Origins of the World War* (New York: Macmillan, 1928).

Feis, Herbert, *Europe: The World's Banker, 1870–1914* (New York: W. W. Norton & Co., 1965).

Feng, Lianyong, Yan Hu, Charles A. S. Hall and Jianliang Wang, *The Chinese Oil Industry: History and Future* (New York: Springer, 2013).

Ferguson, Niall, *The Pity of War* (London: Allen Lane, 1998).

Ferrier, R. W., *The History of the British Petroleum Company* (Cambridge: Cambridge University Press, 1982).

Fieldhouse, D. K., *Economics and Empire, 1830–1914* (London: Macmillan, 1984).

Fielding Hall, H., *The Soul of a People*, 4th edn (London: Macmillan, 1902).

Figueirôa, Silvia Fernanda, Gregory A. Good and Drielli Peyerl, eds., *History, Exploration & Exploitation of Oil and Gas* (Cham: Springer, 2019).

Finkel, Irving, *The Ark Before Noah: Decoding the Story of the Flood* (London: Hodder & Stoughton, 2014).

Finley, John, *The French in the Heart of America* (New York: Charles Scribner's Sons, 1918).

Fishbein, Michael, trans., *The History of al-Ṭabarī*, vol. 31: *The War between Brothers* (Albany, NY: State University of New York Press, 1992).

Fisher, John, *Records* (London: Hodder and Stoughton, 1919).

Fitzgerald, J. Joseph, *Black Gold With Grit: The Alberta Oil Sands* (Sidney: Gray's Publishing, 1978).

Fitzpatrick, John C., ed., *The Writings of George Washington* (Washington, DC: GPO, 1936).

Flint, Timothy, ed., *The Personal Narrative of James O. Pattie, of Kentucky* (Cincinnati, OH: John H. Wood, 1831).

Forbes, R. J., *A Short History of the Art of Distillation* (Leiden: Brill, 1948).

——, *More Studies in Early Petroleum History, 1860–1880* (Leiden: Brill, 1959).

——, *Studies in Ancient Technology*, 3rd edn (Leiden and New York: Brill, 1993).

——, *Studies in Early Petroleum History* (Leiden: Brill, 1958).

Ford, Paul Leicester, ed., *The Works of Thomas Jefferson* (New York: G. P. Putnam's Sons, 1904–5).

Francis, J. Michael, *Invading Colombia: Spanish Accounts of the Gonzalo Jiménez de Quesada Expedition of Conquest* (University Park, PA: Pennsylvania State University Press, 2007).

Frank, Alison Fleig, *Oil Empire: Visions of Prosperity in Austrian Galicia* (Cambridge, MA: Harvard University Press, 2005).

Franklin, John, *Narrative of a Journey to the Shores of the Polar Sea, in the Years 1819, 20, 21, and 22* (London: John Murray, 1823).

Franks, Kenny A., *The Oklahoma Petroleum Industry* (Norman, OK: University of Oklahoma Press, 1980).

——, and Paul F. Lambert, *Early California Oil: A Photographic History, 1865–1940* (College Station, TX: Texas A&M University Press, 1985).

Fraser, T. G., ed., *The First World War and Its Aftermath: The Shaping of the Middle East* (London: Gingko Library, 2015).

Freyer, Tony, *Regulating Big Business: Antitrust in Great Britain and America, 1880–1990* (Cambridge: Cambridge University Press, 1992).

Friedenberg, Daniel M., *Life, Liberty and the Pursuit of Land: The Plunder of Early America* (Buffalo, NY: Prometheus Books, 1992).

Fromkin, David, *Europe's Last Summer: Why the World Went to War in 1914* (London: Vintage, 2004).

Fronc, Jennifer, *New York Undercover: Private Surveillance in the Progressive Era* (Chicago, IL: University of Chicago Press, 2009).

Fruchtman, Jr, Jack, *Thomas Paine: Apostle of Freedom* (New York and London: Four Walls Eight Windows, 1994).

Fryer, John, *A New Account of East-India and Persia* (London: R.R. for Ri. Chiswell, 1698).

Fursenko, A. A., *The Battle for Oil: The Economics and Politics of International Corporate Conflict Over Petroleum, 1860–1930* (Greenwich, CT: JAI Press, 1990).

Gale, Thomas A., *The Wonder of the Nineteenth Century! Rock Oil, in Pennsylvania and Elsewhere* (Erie: Sloan & Griffeth, 1860).

Ganter, Granville, ed., *The Collected Speeches of Sagoyewatha, or Red Jacket* (Syracuse, NY: Syracuse University Press, 2006).

Garay, José de, and Gaetano Moro, *Survey of the Isthmus of Tehuantepec* (London: Ackermann and Co., 1844).

Gardner, Walter M., ed., *The British Coal-Tar Industry: Its Origin, Development and Decline* (London: Williams and Norgate, 1915).

Garner, Paul, *British Lions and Mexican Eagles: Business, Politics, and Empire in the Career of Weetman Pearson in Mexico, 1889–1919* (Stanford, CA: Stanford University Press, 2011).

Garthwaite, Gene R., *Khans and Shahs: A Documentary Analysis of the Bakhtiyari in Iran* (Cambridge: Cambridge University Press, 1983).

Geary, Grattan, *Burma, After the Conquest* (London: Sampson Low, Marston, Searle, and Rivington, 1886).

Geiss, Imanuel, ed., *July 1914, the Outbreak of the First World War: Selected Documents* (New York: W. W. Norton & Co., 1967).

Gerretson, F. C., *History of the Royal Dutch* (Leiden: Brill, 1953–7).

Gibb, George Sweet, and Evelyn H. Knowlton, *History of Standard Oil Company (New Jersey): The Resurgent Years, 1911–1927* (New York: Harper & Bros., 1956).

Gibbs-Smith, Charles H., *Aviation: An Historical Survey from its Origins to the End of World War II* (London: HMSO, 1970).

Gibson, Donald, *Communication, Power and Media* (New York: Nova Science Publishers, 2004).

Gibson, Martin, *Britain's Quest for Oil: The First World War and the Peace Conferences* (Solihull: Helion & Co., 2014).

——, *Wealth, Power, and the Crisis of Laissez Faire Capitalism* (New York: Palgrave Macmillan, 2011).

Giddens, Paul H., *Pennsylvania Petroleum 1750–1872: A Documentary History* (Titusville, PA: Pennsylvania Historical and Museum Commission, 1947).

——, *Standard Oil Company (Indiana): Oil Pioneer of the Middle West* (New York: Appleton-Century-Crofts, 1955).

——, *The Beginnings of the Petroleum Industry: Sources and Bibliography* (Harrisburg, PA: Pennsylvania Historical Commission, 1941).

——, *The Birth of the Oil Industry* (New York: Macmillan, 1938).

Gollin, Alfred, *No Longer an Island: Britain and the Wright Brothers, 1902–1909* (London: Heinemann, 1984).

——, *The Impact of Air Power on the British People and their Government, 1909–14* (London: Macmillan, 1989).

Golonka, Jan, and Frank J. Picha, eds., *The Carpathians and their Foreland: Geology and Hydrocarbon Resources* (Tulsa, OK: American Association of Petroleum Geologists, 2006).

Gooch, G. P., and Harold Temperley, eds., *British Documents on the Origins of the War, 1898–1914* (London: HMSO, 1930).

Gordon, General Sir Thomas Edward, *Persia Revisited* (London and New York: E. Arnold, 1896).

Gough, Barry, *Churchill and Fisher: Titans at the Admiralty* (Barnsley: Seaforth Publishing, 2017).

Graffigny, H. de, *Gas and Petroleum Engines*, trans. A. G. Elliott (London and New York: Whittaker & Co., 1898).

Graymont, Barbara, *The Iroquois in the American Revolution* (Syracuse, NY: Syracuse University Press, 1972).

Greaves, Rose Louise, *Persia and the Defence of India, 1884–1892: A Study in the Foreign Policy of the Third Marquis of Salisbury* (London: University of London, Athlone Press, 1959).

Griffenhagen, George B., and James Harvey, *Old English Patent Medicines in America* (Washington, DC: Smithsonian Institution, 1959).

Gröner, Erich, rev. and expanded by Dieter Jung and Martin Maass, *German Warships 1815–1945* (London: Conway Maritime Press, 1990).

Gustafson, Bret, *New Languages of the State: Indigenous Resurgence and the Politics of Knowledge in Bolivia* (Durham, NC: Duke University Press, 2009).

Haas, T. C. A. de, and G. W. Tol, eds., *The Economic Integration of Roman Italy: Rural Communities in a Globalizing World*, Mnemosyne Supplements, vol. 404 (Leiden: Brill, 2017).

Hagedorn, Hermann, ed., *The Works of Theodore Roosevelt* (New York: Charles Scribner's Sons, 1926).

Haines, Charles G., *Considerations on the Great Western Canal, from the Hudson to Lake Erie: With a View of its Expense, Advantages and Progress*, 2nd edn (Brooklyn: Spooner & Worthington, 1818).

Halleck, H. W., *Bitumen: Its Varieties, Properties, and Uses* (Washington, DC: Peter Force, Tenth Street, 1841).

Hamilton, Victor P., *The Book of Genesis, Chapters 1–17* (Grand Rapids, MI: William B. Eerdmans, 1990).

Hanna, Marcus Alonzo, *Mark Hanna: His Book* (Boston, MA: Chapple Publishing Co., 1904).

Hanway, Jonas, *An Historical Account of the British Trade over the Caspian Sea* (London, 1753).

Harcave, Sidney, *First Blood: The Russian Revolution of 1905* (London: Bodley Head, 1964).

Harris, Thaddeus Mason, *The Journal of a Tour into the Territory of Northwest of the Allegheny Mountains* (Boston, MA: Manning & Loring, 1805).

Harrison, J. F. C., *The Common People: A History from the Norman Conquest to the Present* (London: Fontana, 1984).

Harrison, Robert T., *Gladstone's Imperialism in Egypt: Techniques of Domination* (Westport, CT: Greenwood Press, 1995).

Hart, John Mason, *Revolutionary Mexico: The Coming and Process of the Mexican Revolution* (Berkeley, CA: University of California Press, 1987).

Hartmann, Thom, *Unequal Protection: The Rise of Corporate Dominance and the Theft of Human Rights* (Emmaus, PA: Rodale Press, 2002).

Haskew, Michael E., *Tank: 100 Years of the World's Most Important Armored Military Vehicle* (Minneapolis, MN: Zenith Press, 2015).

al-Hassan, Ahmad Y., and Donald R. Hill, *Islamic Technology: An Illustrated History* (Cambridge: Cambridge University Press, 1986).

Hauptman, Laurence M., *Conspiracy of Interests: Iroquois Dispossession and the Rise of New York State* (Syracuse, NY: Syracuse University Press, 1999).

——, *The Iroquois in the Civil War: From Battlefield to Reservation* (Syracuse, NY: Syracuse University Press, 1993).

——, *Tribes and Tribulations: Misconceptions about American Indians and their Histories* (Albuquerque, NM: University of New Mexico Press, 1995).

Healy, David, *Drive to Hegemony: The United States in the Caribbean 1898–1917* (Madison, WI: University of Wisconsin Press, 1988).

——, *James G. Blaine and Latin America* (Columbia, MI: University of Missouri Press, 2001).

——, *US Expansionism: The Imperialist Urge in the 1890s* (Madison, WI: University of Wisconsin Press, 1970).

Henriques, Robert, *Marcus Samuel: First Viscount Bearsted and Founder of the 'Shell' Transport and Trading Company 1853–1927* (London: Barrie and Rockliff, 1960).

Henry, J. D., *Baku: An Eventful History* (London: Archibald Constable & Co., 1905).

——, *Oil Fields of the Empire* (London: 32 Gt. St. Helens, 1910).

——, *Thirty-Five Years of Oil Transport: The Evolution of the Tank Steamer* (London and Tonbridge: Bradbury, Agnew & Co., 1907).

Henry, J. T., *The Early and Later History of Petroleum* (Philadelphia, PA: Jas B. Rodgers Co., 1873).

Herodotus, *The Histories*, trans. Tom Holland (London: Penguin, 2013).

Herwig, Holger H., *'Luxury' Fleet: The Imperial German Navy, 1888–1918* (London: Allen & Unwin, 1980).

Hewins, Ralph, *Mr Five Per Cent: The Biography of Calouste Gulbenkian* (London: Hutchinson, 1957).

Heywood, Anthony J., and Jonathan D. Smele, eds., *The Russian Revolution of 1905: Centenary Perspectives* (London: Routledge, 2005).

Hidy, Ralph W., and Muriel E., *Pioneering in Big Business, 1882–1911: History of Standard Oil Company (New Jersey)* (New York: Harper & Brothers, 1955).

Higgins, George E., *A History of Trinidad Oil* (Port of Spain: Trinidad Express Newspapers Ltd., 1996).

Hinsdale, Burke A., ed., *The Works of James Abram Garfield* (Boston, MA: James R. Osgood and Co., 1883).

Hinsley, F. H., ed., *British Foreign Policy under Sir Edward Grey* (Cambridge: Cambridge University Press, 1977).

Hippler, Thomas, *Bombing the People: Giulio Douhet and the Foundations of Air-Power Strategy, 1884–1939* (Cambridge: Cambridge University Press, 2013).

Hitti, Philip K., *History of the Arabs from the Earliest Times to the Present*, 10th edn (Basingstoke: Macmillan, 1970).

Hoffman, Ross J. S., *Great Britain and the German Trade Rivalry, 1875–1914* (New York: Russell & Russell, 1964).

Hoffmeyer, Ada Bruhn, *Military Equipment in the Byzantine Manuscript of Scylitzes in the Biblioteca Nacional in Madrid* (Granada: Instituto de Estudios sobre Armas Antiguas, 1966).

Holland, Henry, *Travels in the Ionian Isles, Albania, Thessaly, Macedonia, &c. During the Years 1812 and 1813* (London: Longman, Hurst, Rees, Orme, and Brown, 1815).

Holland, Robert, *Blue-Water Empire: The British in the Mediterranean Since 1800* (London: Allen Lane, 2012).

Holton, Woody, *Forced Founders: Indians, Debtors, Slaves, and the Making of the American Revolution in Virginia* (Chapel Hill, NC: University of North Carolina Press, 1999).

Hopkirk, Peter, *On Secret Service East of Constantinople: The Plot to Bring Down the British Empire* (London: John Murray, 1994).

——, *The Great Game: On Secret Service in High Asia* (Oxford: Oxford University Press, 1990).

Hornsell, Paul, *Oil in Asia* (Oxford: Oxford University Press, 1997).

Hoskins, Halford L., *British Routes to India* (New York: Longmans, Green & Co., 1928).

Houtsma, M. Th., A. J. Wensinck, T. W. Arnold, W. Heffening and E. Lévi-Provençal, eds., *First Encyclopaedia of Islam, 1913–1936* (Leiden: Brill, 1993).

Howe, Henry, *Historical Collections of Virginia* (Charleston, VA: Babcock & Co., 1845).

Howison, John, *Sketches of Upper Canada* (Edinburgh: Oliver & Boyd; London: G. & W. B. Wittaker, 1821).

Hoxie, Frederick E., Ronald Hoffman and Peter J. Albert, eds., *Native Americans and the Early Republic* (Charlottesville, VA, and London: University Press of Virginia, 1999).

Huc, M., *A Journey Through the Chinese Empire* (New York: Harper and Brothers, 1855).

Hughes, David McDermott, *Energy Without Conscience: Oil, Climate Change, and Complicity* (Durham, NC, and London: Duke University Press, 2017).

Hughes, Griffith, *The Natural History of Barbados* (London: Printed for the author, 1750).

Hugill, Peter J., *Transition in Power: Technological 'Warfare' and the Shift from British to American Hegemony since 1919* (Lanham, MD: Lexington Books, 2018).

Humboldt, Alexander de, and Aimé Bonpland, *Personal Narrative of Travels to the Equinoctial Regions of the New Continent During the Years 1799–1804*, trans. Helen Maria Williams (London: Longman, Rees, Orme, Brown and Green, 1829).

Hutton, Frederick Remsen, *The Gas-Engine* (New York: John Wiley and Sons; London: Chapman and Hall, 1903).

Huxley, John, *Britain's Onshore Oil Industry* (London: Macmillan, 1983).

Ingram, Edward, *Britain's Persian Connection, 1798–1828: Prelude to the Great Game in Asia* (Oxford: Clarendon Press, 1992).

——, *The Beginning of the Great Game in Asia, 1828–1834* (Oxford: Clarendon Press, 1979).

Ion, A. Hamish, and E. J. Errington, eds., *Great Powers and Little Wars: The Limits of Power* (Westport, CT: Praeger, 1993).

Ireland, Philip Willard, *Iraq: A Study in Political Development* (London: Jonathan Cape, 1937).

Isichei, Elizabeth, *A History of Nigeria* (London: Longman, 1983).

Issawi, Charles, ed., *The Economic History of Iran, 1800–1914* (Chicago, IL: University of Chicago Press, 1971).

——, *The Fertile Crescent, 1800–1914: A Documentary Economic History* (New York and Oxford: Oxford University Press, 1988).

Jackson, A. V. Williams, *From Constantinople to the Home of Omar Khayyam: Travels in Transcaucasia and Northern Persia* (New York: Macmillan Co., 1911).

James, Lawrence, *Empires in the Sun: The Struggle for the Mastery of Africa* (London: Weidenfeld & Nicolson, 2016).

Jefferson, Thomas, *Notes on the State of Virginia* (Richmond, VA: J. W. Randolph, 1853).

Jeffreys, Elizabeth, ed., *Byzantine Style, Religion and Civilization* (Cambridge: Cambridge University Press, 2006).

Jennings, Francis, *Empire of Fortune: Crowns, Colonies and Tribes in the Seven Years War in America* (New York and London: W. W. Norton & Co., 1988).

——, *The Ambiguous Iroquois Empire: The Covenant Chain Confederation of Indian Tribes with English Colonies from its Beginnings to the Lancaster Treaty of 1744* (New York and London: W. W. Norton & Co., 1984).

——, *The Creation of America: Through Revolution to Empire* (Cambridge: Cambridge University Press, 2000).

Johnson, Arthur Menzies, *Petroleum Pipelines and Public Policy, 1906–1959* (Cambridge, MA: Harvard University Press, 1967).

——, *The Development of American Petroleum Pipelines: A Study in Private Enterprise and Public Policy, 1862–1906* (Ithaca, NY: Cornell University Press, 1956).

Joinville, Jean, Sire de, *Histoire de Saint Louis* (Paris: Librairie de Firmin Didot Frères, Fils et Cie, 1874).

Jones, Geoffrey, *The History of the British Bank of the Middle East* (Cambridge: Cambridge University Press, 1986).

——, *The State and the Emergence of the British Oil Industry* (London: Macmillan, 1981).

Jones, Maldwyn A., *The Limits of Liberty: American History, 1607–1992*, 2nd edn (Oxford: Oxford University Press, 1995).

Jonker, Joost, and Jan Luiten van Zanden, *A History of Royal Dutch Shell* (Oxford: Oxford University Press, 2007).

Joy, Mark S., *American Expansionism, 1783–1860: A Manifest Destiny?* (London: Pearson Education, 2003).

Kaplan, Philip, *Rolling Thunder: A Century of Tank Warfare* (Barnsley: Pen & Sword Books, 2012).

Karwatka, Dennis, *Technology's Past: America's Industrial Revolution and the People Who Delivered the Goods* (Ann Arbor, MI: Prakken Publications, 1996).

Katz, Friedrich, *The Secret War in Mexico: Europe, the United States and the Mexican Revolution* (Chicago, IL: University of Chicago Press, 1981).

Kazemzadeh, Firuz, *Russia and Britain in Persia, 1864–1914: A Study in Imperialism* (New Haven, CT: Yale University Press, 1968).

Keddie, Nikki R., *Religion and Rebellion in Iran: The Tobacco Protest of 1891–1892* (London: Frank Cass & Co., 1966).

Kelly, J. B., *Britain and the Persian Gulf, 1795–1880* (Oxford: Clarendon Press, 1968).

Kennedy, Greg, ed., *Imperial Defence: The Old World Order 1856–1956* (London: Routledge, 2008).

Kennedy, Paul M., ed., *The War Plans of the Great Powers, 1880–1914* (London: George Allen & Unwin, 1979).

Kent, Marian, ed., *Moguls and Mandarins: Oil, Imperialism, and the Middle East in British Foreign Policy 1900–1940* (London: Frank Cass, 1993).

——, *Oil and Empire: British Policy and Mesopotamian Oil, 1900–1920* (London: Macmillan, 1976).

——, *The Great Powers and the End of the Ottoman Empire*, 2nd edn (London: Frank Cass, 1996).

Kerr, Robert, *A General History and Collection of Voyages and Travels* (Edinburgh: William Blackwood, 1812).

Kerrigan, William, *Johnny Appleseed and the American Orchard: A Cultural History* (Baltimore, MD: The Johns Hopkins University Press, 2012).

Khazeni, Arash, *Tribes and Empire on the Margins of Nineteenth-Century Iran* (Seattle, WA: University of Washington Press, 2009).

Kindleberger, Charles P., and Guido di Tella, *Economics in the Long View: Essays in Honour of W. W. Rostow*, vol. 1: *Models and Methodology* (London: Macmillan, 1982).

King, Greg, and Sue Woolmans, *The Assassination of the Archduke: Sarajevo 1914 and the Murder that Changed the World* (London: Macmillan, 2013).

Kingsley, Charles, *At Last: A Christmas in the West Indies* (New York: Harper & Brothers, 1871).

Kinneir, John Macdonald, *A Geographical Memoir of the Persian Empire* (London: John Murray, 1813).

Kirk, George E., *A Short History of the Middle East: From the Rise of Islam to Modern Times* (London: Methuen & Co., 1948).

Knopf, Richard C., ed., *Anthony Wayne, A Name in Arms* (Pittsburgh, PA: University of Pittsburgh Press, 1960).

Kohlsaat, H. H., *From McKinley to Harding: Personal Recollections of Our Presidents* (New York: Charles Scribner's Sons, 1923).

Kolko, Gabriel, *The Triumph of Conservatism: A Reinterpretation of American History, 1900–1916* (New York: Free Press of Glencoe, 1963).

Kovarsky, Joel, *The True Geography of Our Country: Jefferson's Cartographic Vision* (Charlottesville, VA: University of Virginia Press, 2014).

Kuczynski, Pedro-Pablo, *Peruvian Democracy under Economic Stress: An Account of the Belaúnde Administration, 1963–1968* (Princeton, NJ: Princeton University Press, 1977).

La Botz, Dan, *Edward L. Doheny: Petroleum, Power and Politics in the United States and Mexico* (London: Praeger, 1991).

La Follette, Robert M., *La Follette's Autobiography: A Personal Narrative of Political Experiences* (Madison, WI: The Robert M. La Folette Co., 1913).

LaFeber, Walter, *The New Empire: An Interpretation of American Expansion, 1860–1898* (Ithaca, NY: Cornell University Press, 1963).

——, *The Panama Canal: The Crisis in Historical Perspective* (New York and Oxford: Oxford University Press, 1989).

Lamb, W. Kaye, ed., *The Journals and Letters of Alexander Mackenzie* (Cambridge: Hakluyt Society, 1970).

Lambert, Nicholas A., *Planning Armageddon: British Economic Warfare and the First World War* (Cambridge, MA: Harvard University Press, 2012).

Lamoreaux, Naomi R., and William J. Novak, eds., *Corporations and American Democracy* (Cambridge, MA: Harvard University Press, 2017).

Langer, Erick D., *Economic Change and Rural Resistance in Southern Bolivia 1880–1930* (Stanford, CA: Stanford University Press, 1989).

——, *Expecting Pears from an Elm Tree: Franciscan Missions on the Chiriguano Frontier in the Heart of South America, 1830–1949* (Durham, NC: Duke University Press, 2009).

Langeveld, Herman, *Dit leven van krachtig handelen. Hendrikus Colijn, 1869–1944*, vol. 1: *1869–1933* (Amsterdam: Uitgeverij Balans, 1998).

Lauterpacht, E., C. J. Greenwood, Marc Weller and Daniel Bethlehem, eds., *The Kuwait Crisis: Basic Documents* (Cambridge: Grotius Publications, 1991).

Lawson, Linda, *Truth in Publishing: Federal Regulation of the Press's Business Practices, 1880–1920* (Carbondale, IL: Southern Illinois University Press, 1993).

Lay, M. G., *Ways of the World: A History of the World's Roads and of the Vehicles That Used Them* (New Brunswick, NJ: Rutgers University Press, 1992).

Layard, Austen Henry, *Discoveries Among the Ruins of Nineveh and Babylon* (New York: Harper & Brothers, 1853).

——, *Early Adventures in Persia, Susiana, and Babylonia* (London: John Murray, 1887).

——, *Nineveh and its Remains* (London: John Murray, 1849).

Le Strange, G., *The Lands of the Eastern Caliphate: Mesopotamia, Persia, and Central Asia from the Moslem Conquest to the time of Timur* (Cambridge: Cambridge University Press, 1905).

Lee, Howard B., *The Burning Springs, and Other Tales of the Little Kanawha* (Morgantown, VA: West Virginia University, 1968).

Lehman, Tim, *Bloodshed at Little Bighorn: Sitting Bull, Custer, and the Destinies of Nations* (Baltimore, MD: Johns Hopkins University Press, 2010).

Lev, Yaacov, *Saladin in Egypt* (Leiden: Brill, 1999).

Lewis, D. C., *The Plantation Dream: Developing British New Guinea and Papua 1884–1942* (Canberra: Journal of Pacific History, 1996).

Lewis, Samuel, *A Topographical Dictionary of Wales*, 2nd edn (London: S. Lewis & Co., 1840).

Lichnowsky, Prince Karl, *Heading for the Abyss: Reminiscences by Prince Lichnowsky*, trans. F. S. Delmer (New York: Payson & Clarke, 1928).

Liddell Hart, B. H., *The Tanks: The History of the Royal Tank Regiment and its Predecessors* (London: Cassel, 1959).

Lieckfeld, G., *Oil Motors: Their Development, Construction and Management* (London: Charles Griffin & Co., 1908).

Lieuwen, Edwin, *Petroleum in Venezuela: A History* (Berkeley, CA: University of California Press, 1954).

Lindsay, Brendan C., *Murder State: California's Native American Genocide, 1846–1873* (Lincoln, NE: University of Nebraska Press, 2012).

Link, Arthur S., *Wilson: The New Freedom* (Princeton, NJ: Princeton University Press, 1956).

Little, George Herbert, *The Marine Transport of Petroleum* (London: E. & F. N. Spon, 1890).

Livermore, Shaw, *Early American Land Companies: Their Influence on Corporate Development* (New York: Commonwealth Book Fund, 1939).

Livingston, James, *Origins of the Federal Reserve System: Money, Class and Corporate Capitalism, 1890–1913* (Ithaca, NJ: Cornell University Press, 1986).

Livingstone, David, *Missionary Travels and Researches in South Africa* (London: John Murray, 1857).

Locher-Scholten, Elsbeth, *Sumatran Sultanate and Colonial State: Jambi and the Rise of Dutch Imperialism, 1830–1907*, trans. Beverley Jackson (Ithaca, NY: Southeast Asia Program, Cornell University, 2003).

LoCicero, Michael, Ross Mahoney and Stuart Mitchell, eds., *A Military Transformed? Adaptation and Innovation in the British Military, 1792–1945* (Solihull: Helion & Co., 2014).

Lockert, Louis, *Petroleum Motor-Cars* (New York: D. Van Nostrand, 1899).

Loftus, William Kennett, *Travels and Researches in Chaldæa and Susiana* (New York: Robert Carter & Brothers, 1857).

Longmuir, Marilyn V., *Oil in Burma: The Extraction of 'Earth-Oil' to 1914* (Bangkok: White Lotus Press, 2001).

Longrigg, Stephen Hemsley, *Oil in the Middle East: Its Discovery and Development*, 2nd edn (London: Oxford University Press, 1961).

Lopes de Lima, José Joaquim, *Ensaios sobre a Statistica das Possessões Portuguezas na Africa Occidental e Oriental* (Lisbon: Imprensa Nacional, 1846).

Lopez, Robert S., and Irving W. Raymond, trans., *Medieval Trade in the Mediterranean World: Illustrative Documents* (New York: Columbia University Press, 1955).

Loskiel, George Henry, *History of the Mission of the United Brethren among the Indians in North America* (London: Brethren's Society for the Furtherance of the Gospel, 1794).

Lowe, C. J., and M. L. Dockrill, eds., *Foreign Policies of the Great Powers* (London: Routledge, 1972).

Lucas, W. Scott, *Divided We Stand: Britain, the US and the Suez Crisis* (London: Hodder & Stoughton, 1991).

Lucier, Paul, *Scientists and Swindlers: Consulting on Coal and Oil in America* (Baltimore, MD: Johns Hopkins University Press, 2008).

Luckombe, Philip, *The Beauties of England: or, A Comprehensive View of the Antiquities of this Kingdom*, 2nd edn (London: L. Davis and C. Reymers, 1764).

Ludendorff, General, *My War Memories, 1914–1918* (London: Hutchinson & Co., 1919).

Lyon, G. F., *Journal of a Residence and Tour in the Republic of Mexico in the Year 1826* (London: John Murray, 1828).

Macdonald, George, *The Gold Coast, Past and Present: A Short Description of the Country and its People* (London: Longmans, Green, and Co., 1898).

MacMillan, Margaret, *The War that Ended Peace: How Europe Abandoned Peace for the First World War* (London: Profile Books, 2013).

Madley, Benjamin, *An American Genocide: The United States and the California Indian Catastrophe, 1846–1873* (New Haven, CT: Yale University Press, 2016).

Mahajan, Sneh, *British Foreign Policy, 1874–1914: The Role of India* (London: Routledge, 2002).

Major, John, *Prize Possession: The United States and the Panama Canal 1903–1979* (Cambridge: Cambridge University Press, 1993).

Marder, Arthur J., ed., *Fear God and Dread Nought: The Correspondence of Admiral of the Fleet Lord Fisher of Kilverstone* (London: Jonathan Cape, 1952).

——, *From the Dreadnought to Scapa Flow*, vol. 1 (Oxford: Oxford University Press, 1961).

Marlowe, John, *Anglo-Egyptian Relations 1800–1956*, 2nd edn (London: Frank Cass, 1965).

——, *Spoiling the Egyptians* (London: André Deutsch, 1974).

Marriott, James, and Mika Minio-Paluello, *The Oil Road: Journeys from the Caspian Sea to the City of London* (London: Verso, 2013).

Martinez, Anibal R., *Chronology of Venezuelan Oil* (London: George Allen and Unwin, 1969).

Marvin, Charles, *England as a Petroleum Power* (London: R. Anderson & Co., 1887).

——, *The Region of the Eternal Fire: An Account of a Journey to the Petroleum Region of the Caspian in 1883*, new edn (London: W. H. Allen & Co., 1891).

Masefield, John, ed., *Dampier's Voyages* (London: E. Grant Richards, 1906).

Massie, Robert K., *Dreadnought: Britain, Germany and the Coming of the Great War* (London: Jonathan Cape, 1992).

Mas'ūdī, *El-Mas'údí's Historical Encyclopædia, Entitled 'Meadows of Gold and Mines of Gems'*, trans. Aloys Sprenger (London: Oriental Translation Fund of Great Britain and Ireland, 1841).

Mathews, Jennifer P., *Chicle: The Chewing Gum of the Americas, from the Ancient Maya to William Wrigley* (Tucson, AZ: University of Arizona Press, 2009).

Matthew, H. C. G., *Gladstone, 1809–1898* (Oxford: Clarendon Press, 1997).

Maurice, Frederick, *Haldane 1856–1915: The Life of Viscount Haldane of Cloan* (London: Faber and Faber, 1937).

May, Gary, *Hard Oiler! The Story of Canadians' Quest for Oil at Home and Abroad* (Toronto and Oxford: Dundurn Press, 1998).

Maybee, Rolland Harper, *Railroad Competition and the Oil Trade, 1855–1873* (Ann Arbor, MI: The Extension Press, 1940).

Mayer, W., R. M. Clary, L. F. Azuela, T. S. Mota and S. Wołkowicz, eds., *History of Geoscience: Celebrating 50 Years of INHIGEO* (London: Geological Society, 2017).

McBeth, Brian S., *British Oil Policy 1919–1939* (London: Frank Cass, 1985).

——, *Dictatorship and Politics: Intrigue, Betrayal, and Survival in Venezuela, 1908–1935* (Notre Dame, IN: University of Notre Dame Press, 2008).

——, *Gunboats, Corruption and Claims: Foreign Intervention in Venezuela, 1899–1908* (Westport, CT: Greenwood Press, 2001).

——, *Juan Vicente Gómez and the Oil Companies in Venezuela, 1908–1935* (Cambridge: Cambridge University Press, 1983).

McConnell, Michael N., *A Country Between: The Upper Ohio Valley and its Peoples, 1724–1774* (Lincoln, NE: University of Nebraska Press, 1992).

McCormick, Thomas J., *China Market: America's Quest for Informal Empire, 1893–1901* (Chicago, IL: Quadrangle Books, 1967).

McDonald, David MacLaren, *United Government and Foreign Policy in Russia 1900–1914* (Cambridge, MA: Harvard University Press, 1992).

McKay, John H., *Scotland's First Oil Boom: The Scottish Shale Oil Industry, 1851 to 1914* (Edinburgh: John Donald, 2012).

McKay, John P., *Pioneers for Profit: Foreign Entrepreneurship and Russian Industrialization, 1885–1913* (Chicago, IL: University of Chicago Press, 1970).

McLaurin, John J., *Sketches in Crude-Oil*, 3rd edn (Franklin, PA: self-published, 1902).

McLoughlin, William G., *After the Trail of Tears: The Cherokees' Struggle for Sovereignty 1839–1880* (Chapel Hill, NC, and London: University of North Carolina Press, 1993).

McMeekin, Sean, *The Berlin–Baghdad Express: The Ottoman Empire and Germany's Bid for World Power, 1898–1918* (London: Allen Lane, 2010).

——, *The Russian Origins of the First World War* (Cambridge, MA: Harvard University Press, 2011).

McMurray, Jonathan S., *Distant Ties: Germany, the Ottoman Empire, and the Construction of the Baghdad Railway* (Westport, CT: Praeger, 2001).

Merk, Frederick, *Manifest Destiny and Mission in American History* (Cambridge, MA: Harvard University Press, 1963).

Middlemas, Robert Keith, *The Master Builders: Thomas Brassey, Sir John Aird, Lord Cowdray, Sir John Norton-Griffiths* (London: Hutchinson, 1963).

Middleton, P. Harvey, *Industrial Mexico* (New York: Dodd, Mead and Company, 1919).

Middleton, Richard, *Pontiac's War: Its Causes, Course, and Consequences* (Abingdon: Routledge, 2007).

Mignan, Captain R., *A Winter Journey through Russia, the Caucasian Alps, and Georgia; thence ... into Koordistaun* (London: Richard Bentley, 1839).

Mikucki, Tadeusz, *Nafta w Polsce do Połowy XIX Wieku* (Lwów: Piller-Neumann, 1938).

Miller, David, ed., *Tell Me Lies: Propaganda and Media Distortion in the Attack on Iraq* (London: Pluto Press, 2004).

Miller, Edward S., *Bankrupting the Enemy: The US Financial Siege of Japan before Pearl Harbor* (Annapolis, MD: Naval Institute Press, 2007).

Miller, Ernest C., ed., *This Was Early Oil: Contemporary Accounts of the Growing Petroleum Industry, 1848–1885* (Harrisburg, PA: Pennsylvania Historical and Museum Commission, 1968).

Miller, Geoffrey, *Straits: British Policy towards the Ottoman Empire and the Origins of the Dardanelles Campaign* (Hull: University of Hull Press, 1997).

——, *The Millstone: British Naval Policy in the Mediterranean, 1900–1914, the Commitment to France and British Intervention in the War* (Hull: University of Hull Press, 1999).

Miller, John C., *Origins of the American Revolution* (London: Faber and Faber, 1945).

Miller, William, *A New History of the United States*, rev. edn (New York: Dell Publishing, 1962).

Miner, H. Craig, *The Corporation and the Indian: Tribal Sovereignty and Industrial Civilization in Indian Territory, 1865–1907* (Norman, OK: University of Oklahoma Press, 1989).

Minorsky, V., *A History of Sharvān and Darband in the 10th–11th Centuries* (Cambridge: W. Heffer & Sons, 1958).

——, ed. and trans., *Abū-Dulaf Misʿar ibn Muhalhil's Travels in Iran (ca. A.D. 950)*(Cairo: Cairo University Press, 1955).

——, *Studies in Caucasian History* (London: Taylor's Foreign Press, 1953).

Mintz, Max M., *Seeds of Empire: The American Revolutionary Conquest of the Iroquois* (New York: New York University Press, 1999).

Mitchell, Timothy, *Carbon Democracy: Political Power in the Age of Oil* (London: Verso, 2011).

Monteiro, Joachim John, *Angola and the River Congo* (London: Macmillan & Co., 1875).

Moore, Frank, *Diary of the Revolution* (New York: Charles T. Evans, 1863).

Moore, Roy and Alma, *Thomas Jefferson's Journey to the South of France* (New York: Stewart, Tabori & Chang, 1999).

Moorey, P. R. S., *Ancient Mesopotamian Materials and Industries: The Archaeological Evidence* (Oxford: Oxford University Press, 1994).

Moreno, Federico, *Petroleum in Peru, from an Industrial Point of View* (Lima: F. Masias & Co., 1891).

Morgan, E. Delmar, and C. H. Coote, eds., *Early Voyages and Travels to Russia and Persia by Anthony Jenkinson and Other Englishmen* (London: Hakluyt Society, 1886).

Moriyama, Takeshi, *Crossing Boundaries in Tokugawa Society: Suzuki Bokushi, a Rural Elite Commoner* (Leiden: Brill, 2013).

Morón, Guillermo, *Los Orígenes Históricos de Venezuela* (Madrid: Consejo Superior de Investigaciones Científicas, Instituto 'Gonzalo Fernández de Oviedo', 1954).

Morris, A. J. A., *The Scaremongers: The Advocacy of War and Rearmament, 1896–1914* (London: Routledge & Kegan Paul, 1984).

Morris, Edmund, *Derrick and Drill* (New York: James Miller, 1865).

Morris, Richard B., *The American Revolution: A Short History* (Princeton, NJ, New York, Toronto: D. Van Nostrand, 1955).

Morton, A. L., *A People's History of England* (London: Lawrence and Wishart, 1989).

Moses, Bernard, *The Spanish Dependencies in South America* (New York and London: Harper & Brothers, 1914).

Mostashari, Firouzeh, *On the Religious Frontier: Tsarist Russia and Islam in the Caucasus* (London: I.B. Tauris, 2006).

Murphy, John McLeod, *Petroleum in Mexico* (New York: self-published, 1865).

Murray, Colonel A. M., *The 'Fortnightly' History of the War* (London: Chapman and Hall, 1916).

Nash, Alice N., and Christoph Strobel, *Daily Life of Native Americans from Post-Columbian through Nineteenth-Century America* (Westport, CT: Greenwood Press, 2006).

National Civic Federation, *Proceedings of the National Conference on Trusts and Combinations* (New York: National Civic Federation, 1908).

Needham, Joseph, *Science and Civilisation in China*, vol. 3 (Cambridge: Cambridge University Press, 1959); vol. 5 (1986).

Neiberg, Michel S., *Dance of the Furies: Europe and the Outbreak of World War* (Cambridge, MA: Harvard University Press, 2011).

Nelson, Eric, *The Royalist Revolution: Monarchy and the American Founding* (Cambridge, MA: Belknap Press of Harvard University Press, 2014).

Nester, William R., *The Great Frontier War: Britain, France, and the Imperial Struggle for North America, 1607–1755* (Westport, CT: Praeger, 2000).

Nevins, Allan, *John D. Rockefeller: The Heroic Age of American Enterprise* (New York: Charles Scribner's Sons, 1940).

Newton, Douglas, *The Darkest Days: The Truth Behind Britain's Rush to War, 1914* (London: Verso, 2014).

Ni Ni Myint, *Burma's Struggle Against British Imperialism, 1885–1895* (Rangoon: The Universities Press, 1983).

Nicolay, Helen, *Personal Traits of Abraham Lincoln* (New York: D. Appleton-Century, 1939).

Nicolson, Harold, *Sir Arthur Nicolson, Bart., First Lord Carnock: A Study in the Old Diplomacy* (London: Constable, 1930).

North, Sydney H., *Oil Fuel: Its Supply, Composition and Application* (London: Charles Griffin & Co., 1905).

Nowell, Gregory P., *Mercantile States and the World Oil Cartel, 1900–1939* (Ithaca, NY: Cornell University Press, 1994).

O'Brien, Phillips Payson, ed., *Technology and Naval Combat in the Twentieth Century and Beyond* (London: Frank Cass, 2001).

Olearius, Adam, *The Voyages & Travels of the Ambassadors from the Duke of Holstein*, trans. John Davies (London: Thomas Dring and John Starkey, 1662).

Olien, Diana Davids, and Roger M., *Oil in Texas: The Gusher Age, 1895–1945* (Austin, TX: University of Texas Press, 2002).

Ortiz, José Eugenio, Octavio Puche, Isabel Rábano and Luis. F. Mazadiego, eds., *History of Research in Mineral Resources* (Madrid: Instituto Geológico y Minero de España, 2011).

Osmaston, John, *Old Ali; or, Travels Long Ago* (London: Hatchards, Piccadilly, 1881).

Otte, T. G., *The Foreign Office Mind: The Making of British Foreign Policy, 1865–1914* (Cambridge: Cambridge University Press, 2011).

——, and Keith Neilson, eds., *Railways and International Politics: Paths of Empire, 1848–1945* (London and New York: Routledge, 2006).

Ouseley, William, trans., *The Oriental Geography of Ebn Haukal, an Arabian Traveller of the Tenth Century* (London: Wilson & Co., 1800).

Ovid, *Metamorphoses*, trans. David Raeburn (London: Penguin, 2004).

Oviedo y Valdés, Gonzalo Fernandez de, *Historia General y Natural de las Indias, Islas y Tierra-Firme del Mar Océano* (Madrid: Imprenta de la Real Academia de la Historia, 1851).

——, *Natural History of the West Indies*, trans. and ed. Sterling A. Stoudemire (Chapel Hill, NC: University of North Carolina Press, 1959).

Owen, Edgar Wesley, *Trek of the Oil Finders: A History of Exploration for Petroleum* (Tulsa, OK: American Association of Petroleum Geologists, 1975).

Partington, J. R., *A History of Greek Fire and Gunpowder* (Cambridge: W. Heffer & Sons, 1960).

Pearton, Maurice, *Oil and the Romanian State* (Oxford: Clarendon Press, 1971).

Pendelton, Wm C., *History of Tazewell County and Southwest Virginia, 1748–1920* (Richmond, VA: W. C. Hill, 1920).

Pennant, Thomas, *A Tour in Wales* (London: Henry Hughes, 1778).

Peskin, Allan, *Garfield: A Biography* (Kent, OH: Kent State University Press, 1978).

Peterson, Merrill D., *Lincoln in American Memory* (New York and Oxford: Oxford University Press, 1994).

——, ed., *Writings of Thomas Jefferson* (New York: Literary Classics of the United States, 1984).

Peto, Sir S. Morton, *The Resources and Prospects of America Ascertained During a Visit to the States in the Autumn of 1865* (London and New York: Alexander Strahan, 1866).

Philbin, Tobias R., *Admiral Von Hipper: The Inconvenient Hero* (Amsterdam: B. R. Grüner, 1982).

Philip, George, *Oil and Politics in Latin America: Nationalist Movements and State Companies* (Cambridge: Cambridge University Press, 1982).

Pickering, Joseph, *Inquiries of an Emigrant*, 4th edn (London: Effingham Wilson, 1832).

Pipes, Richard, *The Russian Revolution, 1899–1919* (London: Harvill Press, 1990).

Platt, Colin, *The English Medieval Town* (London: Book Club Associates, 1976).

Pliny, *The Natural History of Pliny*, trans. John Bostock and H. T. Riley (London: Henry G. Bohn, 1855).

Plutarch, *Plutarch's Lives*, trans. Bernadotte Perrin (London: William Heinemann, 1958).

Poley, J. P., *Eroïca: The Quest for Oil in Indonesia (1850–1898)* (Dordrecht: Kluwer Academic, 2000).

Polo, Marco, *The Book of Ser Marco Polo, the Venetian, Concerning the Kingdoms and Marvels of the East* (London: John Murray, 1903).

Porter, Glenn, *The Rise of Big Business, 1860–1910* (New York: Thomas Y. Cromwell, 1973).

Pospielovsky, Dimitry, *Russian Police Trade Unionism: Experiment or Provocation?* (London: Weidenfeld & Nicolson, 1971).

Prins, Nomi, *All the Presidents' Bankers: The Hidden Alliances that Drive American Power* (New York: Nation Books, 2014).

Pritchard, Margaret Beck, and Henry G. Taliaferro, *Degrees of Latitude: Mapping Colonial America* (Williamsburg, VA: Colonial Williamsburg Foundation, 2002).

Procopius, *Procopius*, trans. H. B. Dewing (London: William Heinemann, 1962).

Pryor, John H., and Elizabeth M. Jeffreys, *The Age of the ΔΡΟΜΩΝ: The Byzantine Navy ca 500–1204* (Leiden: Brill, 2006).

Pudney, John, *Suez: De Lesseps' Canal* (London: J. M. Dent & Sons, 1968).

Quint, Howard H., *The Forging of American Socialism: Origins of the Modern Movement* (Indianapolis, IN: Bobbs-Merrill Co., 1964).

Raleigh, Sir Walter, *The Discovery of Guiana* (London: Cassel & Co., 1887).

Rau, Jon L., *Mineral-Hydrocarbon Database and Bibliography of the Geology of East Timor* (UNESCAP, July 2002).

Rawlinson, Major-Gen. Sir Henry, *England and Russia in the East*, 2nd edn (London: John Murray, 1875).

Raymond, André, *Cairo*, trans. Willard Wood (Cambridge, MA: Harvard University Press, 2000).

Redwood, Boverton, *Petroleum*, 4th edn (London: C. Griffin & Co., 1922).

Reid, Anthony, *Southeast Asia in the Age of Commerce, 1450–1680* (New Haven, CT: Yale University Press, 1988).

——, *The Contest for North Sumatra: Atjeh, the Netherlands and Britain 1858–1898* (London: Oxford University Press, 1969).

——, ed., *Verandah of Violence: The Background to the Aceh Problem* (Singapore: Singapore University Press, 2006).

Reyerson, Kathryn L., Theofanis G. Stavrou and James D. Tracy, eds., *Pre-Modern Russia and its World* (Wiesbaden: Harrassowitz Verlag, 2006).

Rich, Claudius James, *Memoir on the Ruins of Babylon*, 3rd edn (London: Longman and John Murray, 1818).

——, *Narrative of a Residence in Koordistan* (London: James Duncan, 1836).

Richardson, James D., ed., *A Compilation of the Messages and Papers of the Presidents* (New York: Bureau of National Literature, 1897–1925).

Richter, Daniel K., *Before the Revolution: America's Ancient Pasts* (Cambridge, MA: Harvard University Press, 2011).

Ricklefs, M. C., *A History of Modern Indonesia since c.1200*, 4th edn (Basingstoke: Palgrave Macmillan, 2008).

Rintoul, William, *Spudding In: Recollections of Pioneer Days in the California Oil Fields* (San Francisco, CA: California Historical Society, 1976).

Ritter, Abraham, *History of the Moravian Church in Philadelphia* (Philadelphia, PA: Hayes & Zell, 1857).

Ritz, Dean, ed., *Defying Corporations, Defining Democracy: A Book of History and Strategy* (New York: Apex Press, 2001).

Roberts, Andrew, *Salisbury: Victorian Titan* (London: Weidenfeld & Nicolson, 1999).

Roberts, John, *The Battleship Dreadnought*, rev. edn (London: Conway Maritime Press, 2001).

Robertson, J. Ross, ed., *The Diary of Mrs. John Graves Simcoe* (Toronto: William Briggs, 1911).

Robinson, Peter, *Lincoln's Excavators: The Ruston Years, 1875–1930* (Wellington: Roundoak Publishing, 2003).

Robinson, Piers, Peter Goddard, Katy Parry and Craig Murray with Philip M. Taylor, *Pockets of Resistance: British News Media, War and Theory in the 2003 Invasion of Iraq* (Manchester: Manchester University Press, 2010).

Rockefeller, John D., *Random Reminiscences of Men and Events* (New York: Doubleday, Page & Co., 1909).

Rolt, L. T. C., *From Sea to Sea: The Canal du Midi* (London: Allen Lane, 1973).

Roosa, Ruth A., *Russian Industrialists in an Era of Revolution: The Association of Industry and Trade, 1906–1917* (London: M. E. Sharpe, 1997).

Rosenberg, Charles E., and William H. Helfand, *'Every Man His Own Doctor': Popular Medicine in Early America* (Philadelphia, PA: Library Company of Philadelphia, 1998).

Ross, Owen C. D., *Air as Fuel; or, Petroleum and Other Mineral Oils Utilized by Carburetting Air and Rendering it Inflammable*, 2nd edn (London and New York: E. & F. N. Spon, 1875).

Ross, William G., *A Muted Fury: Populists, Progressives, and Labor Unions Confront the Courts, 1890–1937* (Princeton, NJ: Princeton University Press, 1994).

Roy, William G., *Socializing Capital: The Rise of the Large Industrial Corporation in America* (Princeton, NJ: Princeton University Press, 1997).

Rüdiger, Mogens, ed., *The Culture of Energy* (Newcastle: Cambridge Scholars Publishing, 2008).

Rudin, Harry R., *Germans in the Cameroons 1884–1914: A Case Study in Modern Imperialism* (New Haven, CT: Yale University Press, 1938).

Ruggles, Richard I., *A Country So Interesting: The Hudson's Bay Company and Two Centuries of Mapping, 1670–1870* (Montreal: McGill-Queens University Press, 1991).

Rundell, Jr, Walter, *Early Texas Oil: A Photographic History, 1866–1936* (College Station, TX: Texas A&M University Press, 1977).

Ryan, David, and Patrick Kiely, eds., *America and Iraq: Policy-Making, Intervention and Regional Politics* (Abingdon: Routledge, 2009).

Sahagún, Fray Bernardino de, *Florentine Codex: General History of the Things of New Spain*, trans. Charles E. Dibble and Arthur J. O. Anderson (Santa Fe, NM: School of American Research, and Salt Lake City, UT: University of Utah, 1961).

Sahni, Kalpana, *Crucifying the Orient: Russian Orientalism and the Colonization of Caucasus and Central Asia* (Bangkok: Orchid Press, 1997).

Samuels, Warren J., and Arthur S. Miller, eds., *Corporations and Society: Power and Responsibility* (New York: Greenwood Press, 1987).

Santiago, Myrna I., *The Ecology of Oil: Environment, Labor, and the Mexican Revolution, 1900–1938* (Cambridge: Cambridge University Press, 2006).

Sarolea, Charles, *The Anglo-German Problem* (London: Thomas Nelson and Sons, 1912).

Satz, Ronald N., *American Indian Policy in the Jacksonian Era* (Norman, OK: University of Oklahoma Press, 2002).

Sayles, G. O., *The Medieval Foundations of England*, 2nd edn (London: Methuen & Co., 1950).

al-Sayyid, Afaf Lutfi, *Egypt and Cromer* (London: John Murray, 1968).

al-Sayyid Marsot, Afaf Lutfi, *A History of Egypt: From the Arab Conquest to the Present*, 2nd edn (Cambridge: Cambridge University Press, 2007).

Sazonov, Serge, *Fateful Years 1909–1916: The Reminiscences of Serge Sazonov* (New York: Frederick A. Stokes Co., 1928).

Schaefer, Robert W., *The Insurgency in Chechnya and the North Caucasus: From Gazavat to Jihad* (Santa Barbara, CA: Praeger, 2010).

Schatzker, Valerie, and Julien Hirszhaut, *The Jewish Oil Magnates of Galicia* (Montreal: McGill-Queen's University Press, 2015).

Scholes, France V., and Ralph L. Roys, *The Maya Chontal Indians of Acalan-Tixchel*, 2nd edn (Norman, OK: University of Oklahoma Press, 1968).

Schoultz, Lars, *Beneath the United States: A History of U.S. Policy Toward Latin America* (Cambridge, MA: Harvard University Press, 1998).

Scott, David, *China and the International System, 1840–1949: Power, Presence, and Perceptions in a Century of Humiliation* (Albany, NY: State University of New York Press, 2008).

Scott, Robert N., ed., *The War of the Rebellion: A Compilation of the Official Records of the Union and Confederate Armies* (Washington, DC: GPO, 1889).

Searight, Sarah, *The British in the Middle East* (London and The Hague: East-West Publications, 1979).

Seavoy, Ronald E., *The Origins of the American Business Corporation, 1784–1855: Broadening the Concept of Public Service During Industrialization* (Westport, CT: Greenwood Press, 1982).

Seely, Robert, *Russo-Chechen Conflict, 1800–2000: A Deadly Embrace* (Abingdon: Frank Cass, 2001).

Seligmann, Matthew S., Frank Nägler and Michael Epkenhans, eds., *The Naval Route to the Abyss: The Anglo-German Naval Race 1895–1914* (Farnham: Ashgate Publishing, 2015).

Selin, Helaine, ed., *Encyclopaedia of the History of Science, Technology, and Medicine in Non-Western Cultures* (Dordrecht: Springer Science+Business Media, 1997).

Seton-Watson, Hugh, *The Russian Empire, 1801–1917* (Oxford: Oxford University Press, 1988).

Severance, Frank H., *The Story of Joncaire: His Life and Times on the Niagara* (Buffalo, NY: Buffalo Historical Society, 1906).

Shaban, M. A., *Islamic History: A New Interpretation* (Cambridge: Cambridge University Press, 1976).

Shahnavaz, Shabaz, *Britain and the Opening Up of South-West Persia 1880–1914: A Study in Imperialism and Economic Dependence* (London: Routledge Curzon, 2005).

Shafiee, Katayoun, *Machineries of Oil: An Infrastructural History of BP in Iran* (Cambridge, MA: MIT Press, 2018).

Shankman, Andrew, ed., *The World of the Revolutionary American Republic: Land, Labor, and the Conflict for a Continent* (New York: Routledge, 2014).

Shaw, Tony, *Eden, Suez and the Mass Media: Propaganda and Persuasion during the Suez Crisis* (London: I.B. Tauris, 1996).

Shulman, Peter A., *Coal and Empire: The Birth of Energy Security in Industrial America* (Baltimore, MD: Johns Hopkins University Press, 2015).

Shwadran, Benjamin, *The Middle East, Oil and the Great Powers*, 3rd edn (Jerusalem: Israel Universities Press, 1973).

Siegel, Jennifer, *Endgame: Britain, Russia and the Final Struggle for Central Asia* (London and New York: I.B. Tauris, 2002).

Simmons, H. H., *The Atlantic and Great Western Railway* (New York: self-published, 1866).

Simon, William Joel, *Scientific Expeditions in the Portuguese Overseas Territories (1783–1808) and the Role of Lisbon in the Intellectual-Scientific Community of the Late Eighteenth Century* (Lisbon: Instituto de Investigação Científica Tropical, 1983).

Singer, Jonathan W., *Broken Trusts: The Texas Attorney General Versus the Oil Industry, 1889–1909* (College Station, TX: Texas A&M University Press, 2002).

Sklar, Martin J., *The Corporate Reconstruction of American Capitalism, 1890–1916: The Market, the Law and Politics* (Cambridge: Cambridge University Press, 1988).

Sluglett, Peter, *Britain in Iraq: Contriving King and Country* (London: I.B. Tauris, 2007).

Smith, Captain John, *The Generall Historie of Virginia, New-England, and the Summer Iles* (Richmond, VA: Franklin Press, 1819).

Smith, Theodore Clarke, *The Life and Letters of James Abram Garfield* (New Haven, CT: Yale University Press, 1925).

Snider, L. C., *Oil and Gas in the Mid-Continent Fields* (Oklahoma City, OK: Harlow Publishing, 1920).

Sondhaus, Lawrence, *The Naval Policy of Austria-Hungary, 1867–1918: Navalism, Industrial Development, and the Politics of Dualism* (West Lafayette, IN: Purdue University Press, 1994).

Sparks, Jared, ed., *The Writings of George Washington* (New York: Harper & Brothers, 1847).

Spellman, Paul N., *Spindletop Boom Days* (College Station, TX: Texas A&M University Press, 2001).

Staff, Gary, *German Battlecruisers of World War One: Their Design, Construction and Operations* (Barnsley: Seaforth Publishing, 2014).

——, *German Battleships 1914–18*, vol. 2 (Oxford: Osprey Publishing, 2009).

Stapleton, Thomas, ed., *Magni Rotuli Scaccarii Normanniæ sub Regibus Angliæ* (London: London Society of Antiquaries, 1844).

Stebbins, H. Lyman, *British Imperialism in Qajar Iran: Consuls, Agents and Influence in the Middle East* (London: I.B. Tauris, 2016).

Steiner, Zara S., and Keith Neilson, *Britain and the Origins of the First World War*, 2nd edn (Basingstoke: Palgrave Macmillan, 2003).

Stern, Philip J., *The Company-State: Corporate Sovereignty and the Early Modern Foundations of the British Empire in India* (Oxford: Oxford University Press, 2011).

Sternhell, Zeev, *The Birth of Fascist Ideology: From Cultural Rebellion to Political Revolution*, trans. David Maisel (Princeton, NJ: Princeton University Press, 1994).

Strabo, *The Geography of Strabo*, trans. Horace Leonard Jones (London: William Heinemann, 1930).

Stromquist, Shelton, *Reinventing 'The People': The Progressive Movement, the Class Problem, and the Origins of Modern Liberalism* (Urbana, IL: University of Illinois Press, 2006).

Sutton, John, *Wait for the Waggon: The Story of the Royal Corps of Transport and its Predecessors, 1794–1993* (Barnsley: Leo Cooper, 1998).

Swatzler, David, *A Friend Among the Senecas: The Quaker Mission to Cornplanter's People* (Mechanicsburg, PA: Stackpole Books, 2000).

Swietochowski, Tadeusz, *Russia and Azerbaijan: A Borderland in Transition* (New York and Chichester: Columbia University Press, 1995).

Symes, Michael, *An Account of an Embassy to the Kingdom of Ava, Sent by the Governor-General of India, in the Year 1795* (London: G. and W. Nicol, and J. Wright, 1800).

Taft, William Howard, *The Anti-Trust Act and the Supreme Court* (New York and London: Harper & Brothers, 1914).

Tarbell, Ida M., *All in the Day's Work: An Autobiography* (New York: Macmillan Co., 1939).

Taylor, A. J. P., *The Struggle for Mastery in Europe, 1848–1918* (Oxford: Clarendon Press, 1954).

Taylor, Alan, *The Divided Ground: Indians, Settlers and the Northern Borderland of the American Revolution* (New York: Alfred A. Knopf, 2006).

Taylor, C. Fayette, *Aircraft Propulsion: A Review of the Evolution of Aircraft Piston Engines* (Washington, DC: Smithsonian Institution Press, 1971).

Taylor, Graham D., *Imperial Standard: Imperial Oil, Exxon, and the Canadian Oil Industry from 1880* (Calgary, AL: University of Calgary Press, 2019).

Taylor, James, *Creating Capitalism: Joint-Stock Enterprise in British Politics and Culture, 1800–1870* (Woodbridge: Boydell Press for the Royal Historical Society, 2006).

Thant, Myint-U, *The Making of Modern Burma* (Cambridge: Cambridge University Press, 2001).

Thompson, A. Beeby, *Oil Pioneer: Selected Experiences and Incidents Associated with Sixty Years of World-Wide Petroleum Exploration and Oilfield Development* (London: Sidgwick and Jackson, 1961).

——, *The Oil Fields of Russia and the Russian Petroleum Industry*, 2nd edn (London: Crosby Lockwood and Son, 1908).

Thompson, J. Lee, *Politicians, the Press, & Propaganda: Lord Northcliffe and the Great War, 1914–1919* (Kent, OH: Kent State University Press, 1999).

Thomson, Sinclair, Rossana Barragán, Xavier Albó, Seemin Qayum and Mark Goodale, eds., *The Bolivia Reader: History, Culture, Politics* (Durham, NC: Duke University Press, 2018).

Thorne, Tanis C., *The World's Richest Indian: The Scandal over Jackson Barnett's Oil Fortune* (Oxford: Oxford University Press, 2003).

Thorp, Rosemary, and Geoffrey Bertram, *Peru, 1890–1977: Growth and Policy in an Open Economy* (London: Macmillan Press, 1978).

Thorsen, Niels Aage, *The Political Thought of Woodrow Wilson, 1875–1910* (Princeton, NJ: Princeton University Press, 1988).

Thwaites, Reuben Gold, ed., *The Jesuit Relations and Allied Documents* (Cleveland, OH: The Burrows Brothers Co., 1899).

Timberlake, Jr, Richard H., *The Origins of Central Banking in the United States* (Cambridge, MA: Harvard University Press, 1978).

Tinker Salas, Miguel, *Enduring Legacy: Oil, Culture, and Society in Venezuela* (Durham, NC, and London: Duke University Press, 2009).

Tirpitz, Grand Admiral von, *My Memoirs* (New York: Dodd, Mead & Co., 1919).

Tolf, Robert W., *The Russian Rockefellers: The Saga of the Nobel Family and the Russian Oil Industry* (Stanford, CA: Hoover Institution Press, 1976).

Toprani, Anand, *Oil and the Great Powers: Britain and Germany, 1914 to 1945* (Oxford: Oxford University Press, 2019).

Toutain, Jules, *The Economic Life of the Ancient World* (London: Kegan Paul, Trench, Trubner & Co., 1930).

Townsend, Camilla, *Malintzin's Choices: An Indian Woman in the Conquest of Mexico* (Albuquerque, NM: University of New Mexico Press, 2006).

Townshend, Charles, *When God Made Hell: The British Invasion of Mesopotamia and the Creation of Iraq, 1914–1921* (London: Faber and Faber, 2010).

Trinder, Barrie, *The Industrial Revolution in Shropshire*, 3rd edn (Stroud: Phillimore & Co., 2016).

Troup, Robert, *A Vindication of the Claim of Elkanah Watson, Esq. to the Merit of Projecting The Lake Canal Policy* (Geneva: James Bogert, 1821).

Tyrrell, J. B., ed., *Journals of Samuel Hearne and Philip Turnor* (Toronto: Champlain Society, 1934).

Utley, Robert M., *The Indian Frontier 1846–1890*, rev. edn (Albuquerque, NM: University of New Mexico Press, 2003).

Van Isschot, Luis, *The Social Origins of Human Rights: Protesting Political Violence in Colombia's Oil Capital, 1919– 2010* (Madison, WI: University of Wisconsin Press, 2015).

Vancouver, Captain George, *A Voyage of Discovery to the North Pacific Ocean, and Round the World* (London: G. G. and J. Robinson, Paternoster-Row; and J. Edwards, Pall-Mall, 1798).

Vassiliou, M. S., *Historical Dictionary of the Petroleum Industry* (Lanham, MD: Scarecrow Press, 2009).

Veenendaal, Jr, Augustus J., *Slow Train to Paradise: How Dutch Investment Helped Build American Railroads* (Stanford, CA: Stanford University Press, 1996).

Vickers, Adrian, *A History of Modern Indonesia*, 2nd edn (Cambridge: Cambridge University Press, 2013).

Visvanath, S. N., *A Hundred Years of Oil: A Narrative Account of the Search for Oil in India*, rev. edn (New Delhi: Vikas Publishing House, 1997).

Vitruvius, *The Ten Books on Architecture*, trans. Morris Hicky Morgan (Cambridge, MA: Harvard University Press, 1914).

Vogel, Hans Ulrich, and Günter Dux, eds., *Concepts of Nature: A Chinese-European Cross-Cultural Perspective* (Leiden: Brill, 2010).

Von Laue, Theodore H., *Sergei Witte and the Industrialization of Russia* (New York: Columbia University Press, 1963).

Wall, G. P., and J. G. Sawkins, '*Report on the Geology of Trinidad*', *Memoirs of the Geological Survey* (London: Longman, Green, Longman, and Roberts, 1860).

Wallace, Anthony F. C., *The Death and Rebirth of the Seneca* (New York: Vintage Books, 1969).

Waples, David A., *The Natural Gas Industry in Appalachia: A History from the First Discovery to the Tapping of the Marcellus Shale* (Jefferson: McFarland & Co., 2012).

Ward, James A., *Railroads and the Character of America, 1820–1887* (Knoxville, TN: University of Tennessee Press, 1986).

Ward, Matthew C., *Breaking the Backcountry: The Seven Years' War in Virginia and Pennsylvania, 1754–1765* (Pittsburgh, PA: University of Pittsburgh Press, 2003).

Warner, C. A., *Texas Oil & Gas since 1543* (Houston, TX: Gulf Publishing Co., 1939).

Wawro, Geoffrey, *Warfare and Society in Europe, 1792–1914* (London: Routledge, 2000).

Webber, William Hosgood Young, *Town Gas and Its Uses* (London: Archibald Constable & Co., 1907).

Weinberg, Steve, *Taking on the Trust: The Epic Battle of Ida Tarbell and John D. Rockefeller* (New York: W. W. Norton & Co., 2008).

Weinstein, James, *Ambiguous Legacy: The Left in American Politics* (New York: New Viewpoints, 1975).

——, *The Corporate Ideal in the Liberal State: 1900–1918* (Boston, MA: Beacon Press, 1968).

Weller, Olivier, Alexa Dufraisse and Pierre Pétrequin, eds., *Sel, eau et forêt d'hier à aujourd'hui* (Besançon: Presses universitaires de Franche-Comté, 2008).

Wells, Harwell, ed., *Research Handbook on the History of Corporate and Company Law* (Cheltenham: Edward Elgar, 2018).

Wendler, Eugen, *Friedrich List (1789–1846): A Visionary Economist with Social Responsibility*, trans. Donna Blagg (Berlin: Springer-Verlag, 2015).

Westwood, J. N., *A History of Russian Railways* (London: George Allen and Unwin, 1964).

Wheatcroft, Andrew, *The Life and Times of Shaikh Salman Bin Hamad Al-Khalifa: Ruler of Bahrain 1942–1961* (London: Kegan Paul, 1995).

Whigham, H. J., *The Persian Problem: An Examination of the Rival Positions of Russia and Great Britain in Persia* (New York: Charles Scribner's Sons, 1903).

White, B. T., *Tanks and Other Armoured Fighting Vehicles, 1900–1918*, rev. edn (London: Blandford Press, 1974).

White, Gerald T., *Formative Years in the Far West: A History of the Standard Oil Company of California and Predecessors Through 1919* (New York: Appleton-Century-Crofts, 1962).

Whiteshot, Charles A., *The Oil-Well Driller: A History of the World's Greatest Enterprise, the Oil Industry* (Mannington: Charles Austin Whiteshot, 1905).

Whitten, David O., *The Emergence of Giant Enterprise, 1860–1914: American Commercial Enterprise and Extractive Industries* (Westport, CT: Greenwood Press, 1983).

Wicker, Elmus, *The Great Debate on Banking Reform: Nelson Aldrich and the Origins of the Fed* (Columbus, OH: Ohio State University Press, 2005).

Wilkins, Mira, *The History of Foreign Investment in the United States to 1914* (Cambridge, MA: Harvard University Press, 1989).

Williams, J. J., *The Isthmus of Tehuantepec* (New York: Appleton & Co., 1852).

Williamson, Harold F., and Arnold F. Daum, *The American Petroleum Industry*, vol. 1: *The Age of Illumination, 1859–1899* (Evanston, IL: Northwestern University Press, 1959).

——, Ralph L. Andreano, Arnold R. Daum and Gilbert C. Klose, *The American Petroleum Industry*, vol. 2: *The Age of Energy, 1900–1959* (Evanston, IL: Northwestern University Press, 1963).

Williamson, J. W., *In a Persian Oil Field: A Study in Scientific and Industrial Development*, 2nd edn (London: Ernest Benn, 1930).

Wilson, Sir Arnold, *SW. Persia: A Political Officer's Diary, 1907–1914* (London: Oxford University Press, 1941).

Wilson, John F., *British Business History, 1720–1994* (Manchester and New York: Manchester University Press, 1995).

Wilson, Keith M., ed., *Decisions for War, 1914* (London: UCL Press, 1995).

——, ed., *Imperialism and Nationalism in the Middle East: The Anglo-Egyptian Experience, 1882–1982* (London: Mansell, 1983).

——, *The Policy of the Entente: Essays on the Determinants of British Foreign Policy, 1904–1914* (Cambridge: Cambridge University Press, 1985).

Wilson, Terry P., *The Underground Reservation: Osage Oil* (Lincoln, NE, and London: University of Nebraska Press, 1985).

Wilson, Woodrow, *The New Freedom* (New York, NY: Doubleday, Page & Co., 1913).

Winegard, Timothy C., *The First World Oil War* (Toronto: University of Toronto Press, 2016).

Wlasiuk, Jonathan, *Refining Nature: Standard Oil and the Limits of Efficiency* (Pittsburgh, PA: University of Pittsburgh Press, 2017).

Wonderley, Anthony, *At the Font of the Marvelous: Exploring Oral Narrative and Mythic Imagery of the Iroquois and Their Neighbors* (Syracuse, NY: Syracuse University Press, 2009).

Woodman, Dorothy, *The Making of Burma* (London: Cresset Press, 1962).

Wright, Denis, *The English Amongst the Persians: Imperial Lives in Nineteenth-Century Iran*, rev. edn (London and New York: I.B. Tauris, 2001).

Yergin, Daniel, *The Prize: The Epic Conquest for Oil, Money and Power* (New York: Simon & Schuster, 1991).

Yinke, Deng, *Ancient Chinese Inventions* (Cambridge: Cambridge University Press, 2011).

Yule, Captain Henry, *A Narrative of the Mission Sent by the Governor-General of India to the Court of Ava in 1855* (London: Smith, Elder and Co., 1858).

Zallen, Jeremy, *American Lucifers: The Dark History of Artificial Light, 1750–1865* (Chapel Hill, NC: University of North Carolina Press, 2019).

Zarlenga, Stephen A., *The Lost Science of Money* (Valatie, NY: American Monetary Institute, 2002).

Zinn, Howard, *A People's History of the United States, 1492–Present*, 3rd edn (Harlow: Pearson Education, 2003).

Zondergeld, Gjalt, *Goed en kwaad: Vijftien opstellen van fascisme tot pacifism, van Rudolf Steiner tot Colijn* (Antwerp: Garant, 2002).

ARTICLES

Akpınar, Deniz, and M. Samet Altınbilek, 'Pulk-Balıklı (Çayırlı-Erzincan) Petrol Sahasının Tarihi Süreci'/'The History of Oil Field, Pulk-Balıklı (Çayırlı-Erzincan)', *Erzincan Üniversitesi Sosyal Bilimler Enstitüsü Dergisi (ERZSOSDE)* 8, no. 1 (2015).

Alden, Rev. Timothy, 'Antiquities and Curiosities of Western Pennsylvania', *Archaeologia Americana: Transactions and Collections of the American Antiquarian Society* 1 (1820).

Alvord, Clarence W., 'Virginia and the West; an Interpretation', *Mississippi Valley Historical Review* 3, no. 1 (June 1916).

Anderson, Alexander, 'An Account of a Bituminous Lake or Plain in the Island of Trinidad', *Philosophical Transactions of the Royal Society of London* 74, pt. 1 (1789).

Ant, White, 'Practical Notes on the Yoruba Country and Its Development', *Journal of the Royal African Society* 1, no. 3 (April 1902).

Arbuckle, Robert D., 'John Nicholson and the Pennsylvania Population Company', *Western Pennsylvania Historical Magazine* 57, no. 4 (October 1974).

Ardagh, J. C., 'The Red Sea Petroleum Deposits', *Proceedings of the Royal Geographical Society* 8, no. 8 (August 1886).

Arnold-Forster, H. O., 'Notes on a Visit to Kiel and Wilhelmshaven', in Seligmann et al., eds., *The Naval Route to the Abyss* (2015).

Ascher, Abraham, 'Introduction', in Heywood and Smele, eds., *The Russian Revolution of 1905* (2005).

Aschheim, Joseph, and George S. Tavlas, 'Academic Exclusion: The Case of Alexander Del Mar', *European Journal of Political Economy* 20, no. 1 (March 2004).

Ash, Mary Ellen, 'The Friar and the Oil Spring', *Oil-Industry History* 17, no. 1 (2016).

Atwater, Caleb, 'Facts Relating to Certain Parts of the State of Ohio', *American Journal of Science and Arts* 10 (February 1826).

Ayalon, David, 'A Reply to Professor J. R. Partington', *Arabica* T. 10, Fasc. 1 (January 1963).

Ayemoti Guasu, Juan, 'The God Man', in Thomson et al., eds., *The Bolivia Reader: History, Culture, Politics* (2018).

Baecker, Thomas, 'The Arms of the Ypiranga: The German Side', *The Americas* 30, no. 1 (July 1973).

Bailey, Frank E., 'The Economics of British Foreign Policy, 1825–50', *Journal of Modern History* 12, no. 4 (December 1940).

Baker, Captain George, 'A Short Account of the Buraghmah Country', in Alexander Dalrymple, ed., *Oriental Repertory* 1, no. 2 (1791).

Bakker, H., 'Het Economisch Belang van Noord-Sumatra Tijdens de Atjehoorlog, 1873–1910', in Clemens and Lindblad, eds., *Het Belang Van De Buitengewesten* (1989).

Bakker, J. I. (Hans), 'The Aceh War and the Creation of the Netherlands East Indies State', in Ion and Errington, eds., *Great Powers and Little Wars* (1993).

Barker, Theo C., 'A German Centenary in 1986, a French in 1995 or the Real Beginnings About 1905?', in Barker, ed., *Economic and Social Effects of the Spread of Motor Vehicles* (1987).

——, 'The Spread of Motor Vehicles Before 1914', in Kindleberger and Di Tella, *Economics in the Long View* (1982), vol. 1.

Barth, B., and J. C. Whitehouse, 'The Financial History of the Anatolian and Baghdad Railways, 1889–1914', *Financial History Review* 5, no. 2 (October 1998).

Bassie-Sweet, Karen, Nicholas A. Hopkins and Robert M. Laughlin, 'History and Conquest of the Pre-Columbian Ch'ol Maya of Chiapas', in Bassie-Sweet, ed., *The Ch'ol Maya of Chiapas* (2015).

Beaton, Kendall, 'Dr. Gesner's Kerosene: The Start of American Oil Refining', *Business History Review* 29, no. 1 (March 1955).

Bennigsen Broxup, Marie, 'Introduction: Russia and the North Caucasus', in Bennigsen Broxup, ed., *The North Caucasus Barrier* (1992).

Bergman, Ted Lars Lennart, 'Medical Merchandising and Legal Procedure in Late Sixteenth-Century Spain: The Case of Petroleum as Imported Medicine', *Social History of Medicine* (2019).

Bergmann, William H., 'A "Commercial View of This Unfortunate War": Economic Roots of an American National State in the Ohio Valley, 1775–1795', *Early American Studies* 6, no. 1 (Spring 2008).

Biagini, Eugenio, 'Exporting "Western & Beneficent Institutions": Gladstone and Empire, 1880–1885', in Bebbington and Swift, eds., *Gladstone Centenary Essays* (2000).

Bielen, Adam, 'Nationalism Over Socialism in Galicia', *The Histories* 1, no. 1 (2014).

Bilder, Mary Sarah, 'The Corporate Origins of Judicial Review', *Yale Law Journal* 116, no. 3 (December 2006).

Binder, Frederick Moore, 'Gas Light', *Pennsylvania History* 22, no. 4 (October 1955).

Black, Ian, 'The "Lastposten": Eastern Kalimantan and the Dutch in the Nineteenth and Early Twentieth Centuries', *Journal of Southeast Asian Studies* 16, no. 2 (September 1985).

Blake, W. P., 'Preliminary Geological Report of the U.S. Pacific Railroad Survey, under the Command of Lieut. R. S. Williamson', *American Journal of Science and Arts* 19, no. 57 (May 1855).

Bloch, Ruth H., and Naomi R. Lamoreaux, 'Corporations and the Fourteenth Amendment', in Lamoreaux and Novak, eds., *Corporations and American Democracy* (2017).

Blumberg, Phillip I., 'Limited Liability and Corporate Groups', *Journal of Corporation Law* 11, no. 4 (Summer 1986).

Bodger, Alan, 'Russia and the End of the Ottoman Empire', in Kent, ed., *Great Powers and the End of the Ottoman Empire*, 2nd edn (1996).

Boëda, É., S. Bonilauri, J. Connan, D. Jarvie, N. Mercier, M. Tobey, H. Valladas and H. al-Sakhel, 'New Evidence for Significant Use of Bitumen in Middle Palaeolithic Technical Systems at Umm el Tlel (Syria) around 70,000 BP', *Paléorient* 34, no. 2 (2008).

Bogart, Ernest L., 'Early Canal Traffic and Railroad Competition in Ohio', *Journal of Political Economy* 21, no. 1 (January 1913).

Bolitho, Harold, 'The Echigo War, 1868', *Monumenta Nipponica* 34, no. 3 (Autumn 1979).

Boxt, Matthew A., L. Mark Raab and Rebecca B. González Lauck, 'Isla Alor: Olmec to Contact in the La Venta Hinterland', in Boxt and Dillon, eds., *Fanning the Sacred Flame* (2012).

Boyer, Deborah Deacon, 'Picturing the Other: Images of Burmans in Imperial Britain', *Victorian Periodicals Review* 35, no. 3 (Fall 2002).

Brévart, Francis B., 'Between Medicine, Magic, and Religion: Wonder Drugs in German Medico-Pharmaceutical Treatises of the Thirteenth to the Sixteenth Centuries', *Speculum* 83, no. 1 (January 2008).

Brice, William R., 'Abraham Gesner (1797–1864) – A Petroleum Pioneer', *Oil-Industry History* 3, no. 1 (2002).

——, 'Edwin L. Drake (1819–1880): His Life and Legacy', *Oil-Industry History* 10, no. 1 (2009).

——, 'Samuel M. Kier (1813–1874) – The Oft-Forgotten Oil Pioneer', *Oil-Industry History* 9, no. 1 (2008).

Bridge, F. R., 'Sir Edward Grey and Austria-Hungary', *International History Review* 38, no. 2 (2016).

Brock, Peter, 'The Fall of Circassia: A Study in Private Diplomacy', *English Historical Review* 71, no. 280 (July 1956).

Brockway, Thomas P., 'Britain and the Persian Bubble, 1888–92', *Journal of Modern History* 13, no. 1 (March 1941).

Brown, John Howard, and Mark Partridge, 'The Death of a Market: Standard Oil and the Demise of the 19th-Century Crude Oil Exchanges', *Review of Industrial Organization* 13, no. 5 (October 1998).

Brown, Jonathan C., 'Domestic Politics and Foreign Investment: British Development of Mexican Petroleum, 1889–1911', *Business History Review* 61, no. 3 (Autumn 1987).

Brown, Kaitlin M., 'Asphaltum (Bitumen) Production in Everyday Life on the California Channel Islands', *Journal of Anthropological Archaeology* 41 (2016).

——, Jacques Connan, Nicholas W. Poister, René L. Vellanoweth, John Zumberge and Michael H. Engel, 'Sourcing Archaeological Asphaltum (Bitumen) from the California Channel Islands to Submarine Seeps', *Journal of Archaeological Science* 43 (2014).

Brown, Nathan J., 'Who Abolished Corvee Labour in Egypt and Why?', *Past & Present* 144, no. 1 (August 1994).

Bryant, Lynwood, 'The Development of the Diesel Engine', *Technology and Culture* 17, no. 3 (July 1976).

——, 'The Silent Otto', *Technology and Culture* 7, no. 2 (Spring 1966).

Bucheli, Marcelo, 'Negotiating under the Monroe Doctrine: Weetman Pearson and the Origins of U.S. Control of Colombian Oil', *Business History Review* 82, no. 3 (Autumn 2008).

Büchi, Alfred, 'Modern Continental Oil Engine Practice', *Cassier's Magazine* 43, no. 3 (March 1913).

Bud-Frierman, Lisa, Andrew Godley and Judith Wale, 'Weetman Pearson in Mexico and the Emergence of a British Oil Major, 1901–1919', *Business History Review* 84, no. 2 (Summer 2010).

Burger, Pauline, Rebecca J. Stacey, Stephen A. Bowden, Marei Hacke and John Parnell, 'Identification, Geochemical Characterisation and Significance of Bitumen among the Grave Goods of the 7th Century Mound 1 Ship-Burial at Sutton Hoo (Suffolk, UK)', *PLoS ONE* 11, no. 12 (December 2016).

Burwood, Stephen, 'Debsian Socialism through a Transnational Lens', *Journal of the Gilded Age and Progressive Era* 2, no. 3 (July 2003).

Butt, John, 'Technical Change and the Growth of the British Shale-Oil Industry (1680–1870)', *Economic History Review*, new ser. 17, no. 3 (1965).

——, 'The Scottish Oil Mania of 1864–6', *Scottish Journal of Political Economy* 12, no. 2 (June 1965).

Callender, G. S., 'The Early Transportation and Banking Enterprises of the States in Relation to the Growth of Corporations', *Quarterly Journal of Economics* 17, no. 1 (November 1902).

Calvert, Peter A. R., 'The Murray Contract: An Episode in International Finance and Diplomacy', *Pacific Historical Review* 35, no. 2 (May 1966).

Campisi, Jack, and William A. Starna, 'On the Road to Canandaigua: The Treaty of 1794', *American Indian Quarterly* 19, no. 4 (Fall 1995).

Cârciumaru, Marin, Rodica-Mariana Ion, Elena-Cristina Niţu and Radu Ştefănescu, 'New Evidence of Adhesive as Hafting Material on Middle and Upper Palaeolithic Artefacts from Gura Cheii-Râşnov Cave (Romania)', *Journal of Archaeological Science* 39 (2012).

Carland, John M., 'Enterprise and Empire: Officials, Entrepreneurs, and the Search for Petroleum in Southern Nigeria, 1906–1914', *International History Review* 4, no. 2 (May 1982).

Carroll, Clint, 'Shaping New Homelands: Environmental Production, Natural Resource Management, and the Dynamics of Indigenous State Practice in the Cherokee Nation', *Ethnohistory* 61, no. 1 (Winter 2014).

Cave, Alfred A., 'Abuse of Power: Andrew Jackson and the Indian Removal Act of 1830', *The Historian* 65, no. 6 (Winter 2003).

Chapman, John W. M., 'Russia, Germany and Anglo-Japanese Intelligence Collaboration, 1898–1906', in Erickson and Erickson, eds., *Russia* (2004).

Christian, John L., 'Anglo-French Rivalry in Southeast Asia: Its Historical Geography and Diplomatic Climate', *Geographical Review* 31, no. 2 (April 1941).

Clanton, Gene, 'Populism, Progressivism, and Equality: The Kansas Paradigm', *Agricultural History* 51, no. 3 (July 1977).

Clark, K. A., S. Ikram and R. P. Evershed, 'The Significance of Petroleum Bitumen in Ancient Egyptian Mummies', *Philosophical Transactions of the Royal Society* A 374, no. 2079 (October 2016).

Clayton, T. R., 'The Duke of Newcastle, the Earl of Halifax, and the American Origins of the Seven Years' War', *Historical Journal* 24, no. 3 (1981).

Cleland, Robert Glass, 'The Early Sentiment for the Annexation of California: An Account of the Growth of American Interest in California, 1835–1846 – III', *Southwestern Historical Quarterly* 18, no. 3 (January 1915).

Clow, Archibald, and Nan L. Clow, 'Lord Dundonald', *Economic History Review* 12, nos. 1–2 (October 1942).

Coffin, Howard E., 'American Tendencies in Motor-Car Engineering – Discussion', *Proceedings of the Institution of Automobile Engineers* 6 (1 June 1912).

Colquhoun, Archibald R., 'The Strategical and Economical Effect of the Opening of the Panama Canal', *Royal United Services Institution* 52, pt. 1 (January 1908).

Conant, Charles A., 'The Economic Basis of "Imperialism"', *North American Review* 167, no. 502 (September 1898).

Condra, G. E., 'Opening of the Indian Territory', *Bulletin of the American Geographical Society* 39, no. 6 (1907).

Conlin, Jonathan, 'Debt, Diplomacy and Dreadnoughts: The National Bank of Turkey, 1909–1919', *Middle Eastern Studies* 52, no. 3 (2016).

Connan, Jacques, 'Use and Trade of Bitumen in Antiquity and Prehistory: Molecular Archaeology Reveals Secrets of Past Civilizations', *Philosophical Transactions of the Royal Society of London*, B 354 (1999).

——, and Thomas Van de Velde, 'An Overview of Bitumen Trade in the Near East from the Neolithic (c.8000 BC) to the Early Islamic Period', *Arabian Archaeology and Epigraphy* 21, no. 1 (2010).

Coogan, John W., and Peter F., 'The British Cabinet and the Anglo-French Staff Talks, 1905–1914: Who Knew What and When Did He Know It?', *Journal of British Studies* 24, no. 1 (January 1985).

Cooke, M. C., 'Naphtha', *Journal of the Society of Arts* 7, no. 352 (19 August 1859).

Coons, Anne B., 'Statistics of Petroleum and Natural Gas Production', in Day, ed., *Handbook of the Petroleum Industry* (1922), vol. 1.

Cox, Captain Hiram, 'An Account of the Petroleum Wells in the Burmha Dominions', *Asiatick Researches* 6 (1799).

Craig, Jonathan, Francesco Gerali, Fiona MacAulay and Rasoul Sorkhabi, 'The History of the European Oil and Gas Industry (1600s–2000s)', in Craig et al., eds., *History of the European Oil and Gas Industry* (2018).

Crampton, R. J., 'The Decline of the Concert of Europe in the Balkans, 1913–1914', *Slavonic and East European Review* 52, no. 128 (July 1974).

Crane, Daniel A., 'Were Standard Oil's Rebates and Drawbacks Cost Justified?', *Southern California Law Review* 85, no. 3 (March 2012).

Crawley, C. W., 'Anglo-Russian Relations, 1815–40', *Cambridge Historical Journal* 3, no. 1 (1929).

Cronin, Stephanie, 'The Politics of Debt: The Anglo-Persian Oil Company and the Bakhtiyari Khans', *Middle Eastern Studies* 40, no. 4 (July 2004).

Curtis, Thomas D., 'Land Policy: Pre-Condition for the Success of the American Revolution', *American Journal of Economics and Sociology* 31, no. 2 (April 1972).

Curzon, G., 'The Karun River and the Commercial Geography of South-West Persia', *Proceedings of the Royal Geographical Society* 12, no. 9 (September 1890).

Dahl, Eric J., 'Naval Innovation – From Coal to Oil', *Joint Force Quarterly*, no. 27 (Winter 2000).

Dann, Uriel, 'Report on the Petroliferous Districts of Mesopotamia (1905) – An Annotated Document', *Asian and African Studies*, vol. 24, no. 3 (1990).

Daum, Arnold R., 'Petroleum in Search of an Industry', *Pennsylvania History* 26, no. 1 (January 1959).

Davidson, Christopher M., 'Why Was Muammar Qadhafi Really Removed?', *Middle East Policy* 24, no. 4 (Winter 2017).

Davidson, Philip G., 'Whig Propagandists of the American Revolution', *American Historical Review* 39, no. 3 (April 1934).

Davis, Jr, Theodore H., 'Corporate Privileges for the Public Benefit: The Progressive Federal Incorporation Movement and the Modern Regulatory State', *Virginia Law Review* 77, no. 3 (April 1991).

Dawes, Anna Laurens, 'An Unknown Nation', *Harper's New Monthly Magazine* 76, no. 454 (March 1888).

Day, Mildred Leake, 'Sir Gawain and the Greek Fire: The Impact of Technology on the Heroic Imagination in "De Ortu Waluuanii"', *Arthurian Interpretations* 1, no. 1 (Fall 1986).

De La Rue, Warren, and Hugo Müller, 'Chemical Examination of Burmese Naphtha, or Rangoon Tar', *Proceedings of the Royal Society of London* 8 (1856).

Dean, Graham, 'The Scottish Oil-Shale Industry from the Viewpoint of the Modern-Day Shale-Gas Industry', in Craig et al., eds., *History of the European Oil and Gas Industry* (2018).

Deardorff, Merle H., 'The Cornplanter Grant in Warren County', *Western Pennsylvania Historical Magazine* 24, no. 1 (March 1941).

Del Curto, Davide, and Angelo Landi, 'Gas-Light in Italy between 1700s & 1800s: A History of Lighting', in Rüdiger, ed., *Culture of Energy* (2008).

Del Papa, Eugene M., 'The Royal Proclamation of 1763: Its Effect upon Virginia Land Companies', *Virginia Magazine of History and Biography* 83, no. 4 (October 1975).

Demirel, Muammer, 'Doğu Anadolu Petrol Yatakları ile İlgili Bir Belge', *Doğu Coğrafya Dergisi (Eastern Geographical Review)* 7, no. 6, (2001).

Denham, Jr, R. N., 'An Historical Development of the Contract Theory in the Dartmouth College Case', *Michigan Law Review* 7, no. 3 (January 1909).

DeNovo, John A., 'Petroleum and the United States Navy before World War I', *Mississippi Valley Historical Review* 41, no. 4 (March 1955).

Dent, Chris, '"Generally Inconvenient": The 1624 Statute of Monopolies as Political Compromise', *Melbourne University Law Review* 33, no. 2 (2009).

Destler, Chester McArthur, 'The Opposition of American Businessmen to Social Control During the "Gilded Age"', *Mississippi Valley Historical Review* 39, no. 4 (March 1953).

Dickinson, Kelvin H., 'Partners in a Corporate Cloak: The Emergence and Legitimacy of the Incorporated Partnership', *American University Law Review* 33 (1984).

Dickson, Peter R., and Philippa K. Wells, 'The Dubious Origins of the Sherman Antitrust Act: The Mouse that Roared', *Journal of Public Policy & Marketing* 20, no. 1 (Spring 2001).

DiNardo, R. L., and Austin Bay, 'The First Modern Tank: Gunther Burstyn and His Motorgeschütz', *Military Affairs* 50, no. 1 (January 1986).

Dion, Leon, 'Natural Law and Manifest Destiny in the Era of the American Revolution', *Canadian Journal of Economics and Political Science* 23, no. 2 (May 1957).

Dion, Mark, 'Sumatra through Portuguese Eyes: Excerpts from João de Barros' "Decadas da Asia"', *Indonesia* 9 (April 1970).

Dixon, Donald F., 'Petrol Distribution in the United Kingdom, 1900–1950', *Business History Review* 6, no. 1 (December 1963).

Doel, H. W. van den, 'Military Rule in the Netherlands Indies', in Cribb, ed., *Late Colonial State in Indonesia* (1994).

Domingues da Silva, Daniel B., 'The Atlantic Slave Trade from Angola: A Port-by-Port Estimate of Slaves Embarked, 1701–1867', *International Journal of African Historical Studies* 46, no. 1 (2013).

Dominian, Leon, 'Fuel in Turkey', *Transactions of the American Institute of Mining Engineers* 56 (1917).

Earle, Edward Mead, 'The Secret Anglo-German Convention of 1914 Regarding Asiatic Turkey', *Political Science Quarterly* 38, no. 1 (March 1923).

——, 'The Turkish Petroleum Company – A Study in Oleaginous Diplomacy', *Political Science Quarterly* 39, no. 2 (June 1924).

Ediger, Volkan Ş., and John V. Bowlus, 'Farewell to King Coal: Geopolitics, Energy Security, and the Transition to Oil, 1898–1917', *Historical Journal* 62, no. 2 (June 2018).

——, 'Greasing the Wheels: The Berlin-Baghdad Railway and Ottoman Oil, 1888–1907', *Middle Eastern Studies* 56, no. 2 (March 2020).

Egnal, Marc, 'The Origins of the Revolution in Virginia: A Reinterpretation', *William and Mary Quarterly*, 3rd ser. 37, no. 3 (July 1980).

Ehsani, Kaveh, 'Oil, State and Society in Iran in the Aftermath of the First World War', in Fraser, ed., *First World War and its Aftermath* (2015).

Ellis, Thomas H., 'The Ohio Company', *William and Mary College Quarterly Historical Magazine* 5, no. 2 (October 1896).

Emerson, F. V., 'A Geographic Interpretation of New York City, Part III', *Bulletin of the American Geographical Society* 41, no. 1 (1909).

Enyeart, John P., 'Revolution or Evolution: The Socialist Party, Western Workers, and Law in the Progressive Era', *Journal of the Gilded Age and Progressive Era* 2, no. 4 (October 2003).

Fages, Don Pedro, and Herbert I. Priestley, 'An Historical, Political, and Natural Description of California', *Catholic Historical Review* 5, no. 1 (April 1919).

Faraco, Marianna, Antonio Pennetta, Daniela Fico, Giacomo Eramo, Enkeleida Beqiraj, Italo Maria Muntoni and Giuseppe Egidio De Benedetto, 'Bitumen in Potsherds from Two Apulian Bronze Age Settlements, Monopoli and Torre Santa Sabina: Composition and Origin', *Organic Geochemistry* 93 (March 2016).

Ferreira, Roquinaldo, 'The Conquest of Ambriz: Colonial Expansion and Imperial Competition in Central Africa', *Mulemba* 5, no. 9 (2015).

Finkelstein, Norman, 'History's Verdict: The Cherokee Case', *Journal of Palestine Studies* 24, no. 4 (Summer 1995).

Finney, Sr, Frank F., 'The Indian Territory Illuminating Oil Company', *Chronicles of Oklahoma* 37, no. 2 (Summer 1959).

Fitzgerald, Edward Peter, 'France's Middle Eastern Ambitions, the Sykes-Picot Negotiations, and the Oil Fields of Mosul, 1915–1918', *Journal of Modern History* 66, no. 4 (December 1994).

Flaherty, Chris, 'The 1909 Turkish Machine-Gun Carrier Car', *The Armourer Militaria Magazine* (May 2013).

Flanigan, M. L., 'Some Origins of German Petroleum Policy (1900–1914)', *Southwestern Social Science Quarterly* 26, no. 2 (September 1945).

Fletcher, Max E., 'The Suez Canal and World Shipping, 1869–1914', *Journal of Economic History* 18, no. 4 (December 1958).

——, 'From Coal to Oil in British Shipping', *Journal of Transport History* 3, no. 1 (February 1975).

Foster, Ernest H., 'A Russian Petroleum Pipeline', *Cassier's Magazine* 19, no. 1 (November 1900).

Francis, Richard M., 'The British Withdrawal from the Bagdad Railway Project in April 1903', *Historical Journal* 16, no. 1 (March 1973).

Frank, Alison, 'Environmental, Economic, and Moral Dimensions of Sustainability in the Petroleum Industry in Austrian Galicia', *Modern Intellectual History* 8, no. 1 (April 2011).

——, 'The Petroleum War of 1910: Standard Oil, Austria, and the Limits of the Multinational Corporation', *American Historical Review* 114, no. 1 (February 2009).

Freyer, Tony, 'The Sherman Antitrust Act, Comparative Business Structure, and the Rule of Reason: America and Great Britain, 1880–1920', *Iowa Law Review* 74, no. 5 (July 1989).

Furber, Holden, 'The Overland Route to India in the Seventeenth and Eighteenth Centuries', *Journal of Indian History* 29, pt. 2 (August 1951).

Fursenko, A. A., 'The Oil Industry', in Rondo and Bovykin, eds., *International Banking* (1991).

Furstenberg, François, 'The Significance of the Trans-Appalachian Frontier in Atlantic History', *American Historical Review* 113, no. 3 (June 2008).

Galbraith, John S., 'British Railway Policy in Persia, 1870–1900', *Middle Eastern Studies* 25, no. 4 (October 1989).

——, 'The Trial of Arabi Pasha', *Journal of Imperial and Commonwealth History* 7, no. 3 (1979).

——, and Afaf Lutfi al-Sayyid-Marsot, 'The British Occupation of Egypt: Another View', *International Journal of Middle East Studies* 9, no. 4 (November 1978).

Galkin, Arkady I., Francesco Gerali and Irena G. Malakhova, 'Oil for Life: Russian Pioneers Chose Wisely', *AAPG Explorer* 36, no. 1 (January 2015).

Gallo, Marcus, 'Improving Independence: The Struggle over Land Surveys in Northwestern Pennsylvania in 1794', *Pennsylvania Magazine of History and Biography* 142, no. 2 (April 2018).

Gammer, Moshe, 'Russian Strategies in the Conquest of Chechnia and Daghestan, 1825–1859', in Bennigsen Broxup, ed., *The North Caucasus Barrier* (1992).

Garthwaite, Gene R., 'The Bakhtiyari Khans, the Government of Iran, and the British, 1846–1915', *International Journal of Middle East Studies* 3, no. 1 (January 1972).

Gerali, Francesco, 'A Brief Analysis of Mexican Petroleum up to the 20th Century: Environment, Economy, Politics and Technology', *Oil-Industry History* 13, no. 1 (2012).

——, 'Oil Research in Italy in the Second Half of the Nineteenth Century: The Birth of the Modern Oil Industry in Abruzzo and the Geological Contributions of Giovanni Capellini', in Ortiz et al., eds., *History of Research in Mineral Resources* (2011).

——, 'Scientific Maturation and Production Modernization; Notes on the Italian Oil Industry in the XIXth Century', *Oil-Industry History* 12, no. 1 (2011).

——, 'The Development of the Italian Oil Industry in the Emilian Apennines', *Oil-Industry History* 11, no. 1 (2010).

——, and Jenny Gregory, 'Harsh Oil: Finding Petroleum in Early 20th Century Western Australia', in Mayer et al., eds., *History of Geoscience* (2017).

——, and Jenny Gregory, 'Understanding and Finding Oil over the Centuries: The Case of the Wallachian Petroleum Company in Romania', *Earth Sciences History* 36, no. 1 (2017).

——, and Lorenzo Lipparini, 'Maiella, an Oil Massif in the Central Apennines Ridge of Italy: Exploration, Production and Innovation in the Oilfields of Abruzzo across the Nineteenth and Twentieth Centuries', in Craig et al., eds., *History of the European Oil and Gas Industry* (2018).

——, Paolo Macini and Ezio Mesini, 'Historical Study of Geosciences and Engineering in the Oilfields of the Emilia-Romagna Region in the Socio-Economic Context of Post-Unitarian Italy (1861–1914)', in Craig et al., eds., *History of the European Oil and Gas Industry* (2018).

——, and Paolo Riguzzi, 'Entender la naturaleza para crear una industria. El petróleo en la exploración de John McLeod Murphy en el Istmo de Tehuantepec, 1865', *Asclepio* 67, no. 2 (2015).

——, and Paolo Riguzzi, 'Gushers, Science and Luck: Everette Lee DeGolyer and the Mexican Oil Upsurge, 1909–19', in Mayer et al., eds., *History of Geoscience* (2017).

——, and Paolo Riguzzi, 'Los inicios de la actividad petrolera en México, 1863–1874: una nueva cronología y elementos de balance', *Boletín del Archivo Histórico de Petróleos Mexicanos* 13, no. 2 (June 2013).

Gibson, Martin, '"Oil Fuel Will Absolutely Revolutionize Naval Strategy": The Royal Navy's Adoption of Fuel Oil before the First World War', in LoCicero et al., eds., *A Military Transformed?* (2014).

Giddens, Paul H., ed., 'China's First Oil Well: Recollections of Robert D. Locke, Titusville Oil Pioneer', *Pennsylvania History* 47, no. 1 (January 1980).

Glanville, Jim, and Ryan Mays, 'The William Preston / George Washington Letters', *Smithfield Review* 18 (2014).

Goldsmid, F., 'De Morgan's "Mission Scientifique" to Persia', *Geographical Journal* 8, no. 5 (November 1896).

Gordon, Sanford D., 'Attitudes Towards Trusts Prior to the Sherman Act', *Southern Economic Journal* 30, no. 2 (October 1963).

Gorton, Gary, 'Clearinghouses and the Origin of Central Banking in the United States', *Journal of Economic History* 45, no. 2 (June 1985).

Grandy, Christopher, 'New Jersey Corporate Chartermongering, 1875–1929', *Journal of Economic History* 49, no. 3 (September 1989).

Granitz, Elizabeth, and Benjamin Klein, 'Monopolization by "Raising Rivals' Costs": The Standard Oil Case', *Journal of Law and Economics* 39, no. 1 (April 1996).

Greaves, Rose Louise, 'British Policy in Persia, 1892–1903 – I', *Bulletin of the School of Oriental and African Studies* 28, no. 1 (1965).

——, 'British Policy in Persia, 1892–1903 – II', *Bulletin of the School of Oriental and African Studies* 28, no. 2 (1965).

——, 'Sistan in British Indian Frontier Policy', *Bulletin of the School of Oriental and African Studies* 49, no. 1 (1986).

——, 'Some Aspects of the Anglo-Russian Convention and its Working in Persia, 1907–14 – II', *Bulletin of the School of School of Oriental and African Studies* 31, no. 2 (June 1968).

Griswold, A. Whitney, 'The Agrarian Democracy of Thomas Jefferson', *American Political Science Review* 40, no. 4 (August 1946).

Groen, Petra, 'Colonial Warfare and Military Ethics in the Netherlands East Indies, 1816–1941', *Journal of Genocide Research* 14, nos. 3–4 (September–November 2012).

Grossman, Richard L., and Frank T. Adams, 'Taking Care of Business: Citizenship and the Charter of Incorporation', in Ritz, ed., *Defying Corporations* (2001).

Haan, Willem de, 'Knowing What We Know Now: International Crimes in Historical Perspective', *Journal of International Criminal Justice* 13, no. 4 (September 2015).

Haldon, John, '"Greek Fire" Revisited: Recent and Current Research', in Jeffreys, ed., *Byzantine Style* (2006).

Hall, Linda B., and Don M. Coerver, 'Oil and the Mexican Revolution: The Southwestern Connection', *The Americas* 41, no. 2 (October 1984).

Hamill, Susan Pace, 'From Special Privilege to General Utility: A Continuation of Willard Hurst's Study of Corporations', *American University Law Review* 49, no. 1 (October 1999).

Harris, Christina Phelps, 'The Persian Gulf Submarine Telegraph of 1864', *Geographical Journal* 135, pt. 2 (June 1969).

Harris, Ron, 'Trading with Strangers: The Corporate Form in the Move from Municipal Governance to Overseas Trade', in Harwell Wells, ed., *Research Handbook on the History of Corporate and Company Law* (2018).

Hazlett, Thomas W., 'The Legislative History of the Sherman Act Re-examined', *Economic Inquiry* 30, no. 2 (April 1992).

Head, Judith A., 'Public Opinions and Middle Eastern Railways: The Russo-German Negotiations of 1910–11', *International History Review* 6, no. 1 (February 1984).

Heidbrink, Ingo, 'Petroleum Tanker Shipping on German Inland Waterways, 1887–1994', *Northern Mariner* 11, no. 2 (April 2001).

Helfman, Harold M., 'Twenty-Nine Hectic Days: Public Opinion and the Oil War of 1872', *Pennsylvania History* 17, no. 2 (April 1950).

Henderson, W. O., 'German Economic Penetration in the Middle East, 1870–1914', *Economic History Review* 18, nos. 1–2 (April 1948).

Henze, Paul B., 'Circassian Resistance to Russia', in Bennigsen Broxup, ed., *The North Caucasus Barrier* (1992).

Herwig, Holger H., and Christon I. Archer, 'Global Gambit: A German General Staff Assessment of Mexican Affairs, November 1913', *Mexican Studies* 1, no. 2 (Summer 1985).

Hidy, Ralph W., 'Government and the Petroleum Industry of the United States to 1911', *Journal of Economic History* 10, supp. (1950).

——, and Muriel E., 'Anglo-American Merchant Bankers and the Railroads of the Old Northwest, 1848–1860', *Business History Review* 34, no. 2 (Summer 1960).

Higonnet, Patrice Louis-René, 'The Origins of the Seven Years' War', *Journal of Modern History* 40, no. 1 (March 1968).

Hildreth, S. P., 'Observations on the Saliferous Rock Formation, in the Valley of the Ohio', *American Journal of Science and Arts* 24, no. 1 (July 1833).

Hodgson, Susan Fox, 'California Indians, Artisans of Oil', *Oil-Industry History* 4, no. 1 (2003).

Holley, Jr, I. B., 'Blacktop: How Asphalt Paving Came to the Urban United States', *Technology and Culture* 44, no. 4 (October 2003).

Holmes, James, 'Mahan, a "Place in the Sun", and Germany's Quest for Sea Power', *Comparative Strategy* 23, no. 1 (2004).

Holton, Woody, 'The Ohio Indians and the Coming of the American Revolution in Virginia', *Journal of Southern History* 60, no. 3 (August 1994).

Hopkins, A. G., 'The Victorians and Africa: A Reconsideration of the Occupation of Egypt, 1882', *Journal of African History* 27, no. 2 (1986).

Hopkins, H. L., 'The Growth of British Interest in the Route to India', *Journal of Indian History* 2 (1922–3).

Horsman, Reginald, 'American Indian Policy in the Old Northwest, 1783–1812', *William and Mary Quarterly*, 3rd ser. 18, no. 1 (January 1961).

Horwitz, Morton J., 'Santa Clara Revisited: The Development of Corporate Theory', in Samuels and Miller, eds., *Corporations and Society* (1987).

Hoskins, Halford L., 'British Policy in Africa 1873–1877: A Study in Geographical Politics', *Geographical Review* 32, no. 1 (January 1942).

——, 'The First Steam Voyage to India', *Geographical Review* 16, no. 1 (January 1926).

——, 'The Growth of British Interest in the Route to India', *Journal of Indian History* 2 (1922–3).

Hovenkamp, Herbert J., 'The Classical Corporation in American Legal Thought', *Georgetown Law Journal* 76, no. 4 (1988).

Huffaker, Shauna, 'Representations of Ahmed Urabi: Hegemony, Imperialism, and the British Press, 1881–1882', *Victorian Periodicals Review* 45, no. 4 (Winter 2012).

Hulbert, Archer Butler, and William Nathaniel Schwarze, eds., 'David Zeisberger's History of the Northern American Indians', *Ohio Archaeological and Historical Quarterly* 19, no. 1 (January 1910).

Hunt, Michael H., 'Americans in the China Market: Economic Opportunities and Economic Nationalism, 1890s–1931', *Business History Review* 51, no. 3 (Autumn 1977).

Hunter, F. Robert, 'Tourism and Empire: The Thomas Cook & Son Enterprise on the Nile, 1868–1914', *Middle Eastern Studies* 40, no. 5 (September 2004).

Huston, James L., 'The American Revolutionaries, the Political Economy of Aristocracy, and the American Concept of the Distribution of Wealth, 1765–1900', *American Historical Review* 98, no. 4 (October 1993).

Huston, Reeve, 'Land Conflict and Land Policy in the United States, 1785–1841', in Shankman, ed., *The World of the Revolutionary American Republic* (2014).

Hutson, James H., 'Benjamin Franklin and the West', *Western Historical Quarterly* 4, no. 4 (October 1973).

Ince, Onur Ulas, 'Friedrich List and the Imperial Origins of the National Economy', *New Political Economy* 21, no. 4 (2016).

Ingram, Edward, 'Great Britain's Great Game', *International History Review* 2, no. 1 (January 1980).

Ilisevich, Robert D., 'Early Land Barons in French Creek Valley', *Pennsylvania History* 48, no. 4 (October 1981).

Issawi, Charles, 'The Tabriz-Trabzon Trade, 1830–1900: Rise and Decline of a Route', *International Journal of Middle East Studies* 1, no. 1 (January 1970).

Jakle, John A., 'Salt on the Ohio Valley Frontier, 1770–1820', *Annals of the Association of American Geographers* 59, no. 4 (December 1969).

Jarausch, Konrad H., 'The Illusion of Limited War: Chancellor Bethmann Hollweg's Calculated Risk, July 1914', *Central European History* 2, no. 1 (March 1969).

Jha, Saumitra, 'Financial Asset Holdings and Political Attitudes: Evidence from Revolutionary England', *Quarterly Journal of Economics* 130, no. 3 (August 2015).

Johnson, Arthur M., 'The Early Texas Oil Industry: Pipelines and the Birth of an Integrated Oil Industry, 1901–1911', *Journal of Southern History* 32, no. 4 (November 1966).

Jones, G. Gareth, 'The Oil-Fuel Market in Britain 1900–14: A Lost Cause Revisited', *Business History* 20 (1978).

——, 'The State and Economic Development in India 1890–1947: The Case of Oil', *Modern Asian Studies* 13, no. 3 (1979).

Jonker, Joost, and Jan Luiten van Zanden, 'Searching for Oil in Roubaix', *Rothschild Archive Review of Year 2006 to 2007*.

Jung, Li, and Lien-che Tu Fang, 'An Account of the Salt Industry at Tzu-liu-ching', *Isis* 39, no. 4 (November 1948)

Kaya, Murat, 'Western Interventions and Formation of the Young Turks' Siege Mentality', *Middle East Critique* 23, no. 2 (2014).

Kedourie, Elie, 'Young Turks, Freemasons, and Jews', *Middle Eastern Studies* 7, no. 1 (January 1971).

Kellogg, Frank B., 'The Enforcement of the Sherman Anti-Trust Law', in *National Civic Federation, Proceedings* (1908).

Kelly, William J., 'Crisis Management in the Russian Oil Industry: The 1905 Revolution', *Journal of European Economic History* 10, no. 2 (Fall 1981).

——, and Tsuneo Kano, 'Crude Oil Production in the Russian Empire: 1818–1919', *Journal of European Economic History* 6, no. 2 (Fall 1977).

Kent, Donald H., and Merle H. Deardorff, 'John Adlum on the Allegheny: Memoirs for the Year 1794', *Pennsylvania Magazine of History and Biography* 84, no. 3 (July 1960).

Kent, Marian, 'Constantinople and Asiatic Turkey, 1905–1914', in Hinsley, ed., *British Foreign Policy under Sir Edward Grey* (1977).

Kessler, W. C., 'A Statistical Study of the New York General Incorporation Act of 1811', *Journal of Political Economy* 48, no. 6 (December 1940).

Klein, Benjamin, 'The "Hub-and-Spoke" Conspiracy that Created the Standard Oil Monopoly', *Southern California Law Review* 85, no. 3 (March 2012).

Klein, Daniel B., 'The Voluntary Provision of Public Goods? The Turnpike Companies of Early America', *Economic Inquiry* 28, no. 4 (October 1990).

Klein, Ira, 'British Intervention in the Persian Revolution, 1905–1909', *Historical Journal* 15, no. 4 (December 1972).

Kreike, Emmanuel, 'Genocide in the Kampongs? Dutch Nineteenth-Century Colonial Warfare in Aceh, Sumatra', *Journal of Genocide Research* 14, nos. 3–4 (September–November 2012).

Kreitner, Roy, 'Money in the 1890s: The Circulation of Politics, Economics, and Law', *UC Irvine Law Review* 1, no. 3 (September 2011).

Kudsi-Zadeh, A. Albert, 'Afghānī and Freemasonry in Egypt', *Journal of the American Oriental Society* 92, no. 1 (January–March 1972).

Kumar, Ravinder, 'The Records of the Government of India on the Berlin-Baghdad Railway Question', *Historical Journal* 5, no. 1 (1962).

Kurt, Burcu, 'Discovery of Oil and Oil-Based Environmental Pollution in Ottoman Iraq: The Incident of Mohammarah (1913–1914)', *Avrasya İncelemeleri Dergisi (Journal of Eurasian Studies)* 4, no. 2 (2015).

Lambert, Nicholas A., 'British Naval Policy, 1913–1914: Financial Limitation and Strategic Revolution', *Journal of Modern History* 67, no. 3 (September 1995).

Lambton, A. K. S., 'The Tobacco Regie: Prelude to Revolution – I', *Studia Islamica*, no. 22 (1965).

Lamoreaux, Naomi R., and William J. Novak, 'Corporations and American Democracy: An Introduction', in Lamoreaux and Novak, eds., *Corporations and American Democracy* (2017).

Layard, A. H., 'A Description of the Province of Khúzistán', *Journal of the Royal Geographical Society* 16 (1846).

Lea, Cliff, 'Derbyshire's Oil and Refining History: The James "Paraffin" Young Connection', in Craig et al., eds. *History of the European Oil and Gas Industry* (2018).

Lee, En-Han, 'China's Response to Foreign Investment in Her Mining Industry (1902–1911)', *Journal of Asian Studies* 28, no. 1 (November 1968).

Lehman, David, 'The End of the Iroquois Mystique: The Oneida Land Cession Treaties of the 1780s', *William and Mary Quarterly*, 3rd ser. 47, no. 4 (October 1990).

Leonard, Thomas C., 'American Economic Reform in the Progressive Era: Its Foundational Beliefs and Their Relation to Eugenics', *History of Political Economy* 41, no. 1 (2009).

Leslie, Christopher R., 'Revisiting the Revisionist History of Standard Oil', *Southern California Law Review* 85, no. 3 (March 2012).

Letwin, William L., 'Congress and the Sherman Antitrust Law: 1887–1890', *University of Chicago Law Review* 23, no. 2 (1955).

Lewis, G. Malcolm, 'The Recognition and Delimitation of the Northern Interior Grasslands during the Eighteenth Century', in Blouet and Lawson, eds., *Images of the Plains* (1975).

Lincoln, Benjamin, 'An Account of several Strata of Earth and Shells on the Banks of *York-River*, in *Virginia*', *Memoirs of the American Academy of Arts and Sciences* 1, pt. 2 (1783).

Lindblad, J. Thomas, 'Economic Aspects of the Dutch Expansion in Indonesia, 1870–1914', *Modern Asian Studies* 23, no. 1 (1989).

Livingstone, David, 'XVI – Explorations into the Interior of Africa', *Journal of the Royal Geographical Society of London* 25 (1855).

Lloyd, H. D., 'Story of a Great Monopoly', *Atlantic Monthly* 47, no. 281 (March 1881).

Lockhart, Laurence, 'Iranian Petroleum in Ancient and Medieval Times', *Journal of the Institute of Petroleum* 25, no. 183 (January 1939).

Loftus, William Kennett, 'On the Geology of Portions of the Turko-Persian Frontier, and of the Districts Adjoining', *Quarterly Journal of the Geological Society of London* 11, pt. 1 (1855).

Lossio, Jorge Luis, 'Del copey a las energías alternativas: Panorama histórico de las fuentes de energía en la Región Piura', *Revista Peruana de Energía*, no. 4 (December 2014).

Luter, Paul, 'Archibald Cochrane, 9th Earl of Dundonald (1748–1831): Father of the British Tar Industry', *Broseley Local History Society Journal* 28 (2006).

——, '"British Oil" – Developments in the Ironbridge Gorge during the 17th & 18th Centuries', *Broseley Local History Society Journal* 27 (2005).

Lynch, Lieut. H. B., 'The Tigris between Baghdad & Mósul', *Journal of the Royal Geographical Society* 9 (1839).

Lytle, Richard H., 'The Introduction of Diesel Power in the United States, 1897–1912', *Business History Review* 42, no. 2 (Summer 1968).

Maass, Matthias, 'Catalyst for the Roosevelt Corollary: Arbitrating the 1902–1903 Venezuela Crisis and Its Impact on the Development of the Roosevelt Corollary to the Monroe Doctrine', *Diplomacy & Statecraft* 20, no. 3 (2009).

Madureira, Nuno Luís, 'Oil in the Age of Steam', *Journal of Global History* 5, no. 1 (March 2010).

Mahan, A. T., 'Fortify the Panama Canal', *North Atlantic Review* 193, no. 664 (March 1911).

——, 'The Panama Canal and the Distribution of the Fleet', *North Atlantic Review* 200, no. 706 (September 1914).

Maier, Pauline, 'The Revolutionary Origins of the American Corporation', *William and Mary Quarterly* 50, no. 1 (January 1993).

Malley, Shawn, 'Layard Enterprise: Victorian Archaeology and Informal Imperialism in Mesopotamia', *International Journal of Middle East Studies* 40, no. 4 (November 2008).

Manross, N. S., 'Notice of the Pitch Lake of Trinidad', *American Journal of Science and Arts* 20, no. 59 (September 1855).

Martell, B., 'On the Carriage of Petroleum in Bulk on Over-Sea Voyages', *Transactions of the Institution of Naval Architects* 28 (1887).

Matveichuk, Alexander, 'Peter the Great's Plans for Russian Oil', *Oil of Russia* 3 (2010).

Maunsell, F. R., 'The Mesopotamian Petroleum Field', *Geographical Journal* 9, no. 5 (May 1897).

May, James, 'Antitrust in the Formative Era: Political and Economic Theory in Constitutional and Antitrust Analysis, 1880–1918', *Ohio State Law Journal* 50 (1989).

Mazadiego Martínez, Luis F., Octavio Puche Riart and José E. Ortiz Menéndez, 'Information about Petroleum in America Prior to the Nineteenth Century', in Ortiz et al., eds., *History of Research in Mineral Resources* (2011).

McCarthy, Tom, 'The Coming Wonder? Foresight and Early Concerns about the Automobile', *Environmental History* 6, no. 1 (January 2001).

McCormick, Richard L., 'The Discovery that Business Corrupts Politics: A Reappraisal of the Origins of Progressivism', *American Historical Review* 86, no. 2 (April 1981).

McCurdy, Charles W., 'Justice Field and the Jurisprudence of Government-Business Relations: Some Parameters of Laissez-Faire Constitutionalism, 1863–1897', *Journal of American History* 61, no. 4 (March 1975).

——, 'The Knight Sugar Decision of 1895 and the Modernization of American Corporation Law, 1869–1903', *Business History Review* 53, no. 3 (Autumn 1979).

McDonald, Grantley, 'Georgius Agricola and the Invention of Petroleum', *Bibliothèque d'Humanisme et Renaissance* 73, no. 2 (2011).

McGeoch, Lyle A., 'On the Road to War: British Foreign Policy in Transition, 1905–1906', *Review of Politics* 35, no. 2 (April 1973).

McHugh, Christine, 'Midwestern Populist Leadership and Edward Bellamy: "Looking Backward" into the Future', *American Studies* 19, no. 2 (Fall 1978).

McKay, John P., 'Baku Oil and Transcaucasian Pipelines, 1883–1891: A Study in Tsarist Economic Policy', *Slavic Review* 43, no. 4 (Winter 1984).

——, 'Entrepreneurship and the Emergence of the Russian Petroleum Industry, 1813–1883', *Research in Economic History* 8 (1983).

——, 'Restructuring the Russian Petroleum Industry in the 1890s: Government Policy and Market Forces', in Edmondson and Waldron, eds., *Economy and Society in Russia and the Soviet Union* (1992).

McLaughlin, Andrew C., 'The Court, the Corporation, and Conkling', *American Historical Review* 46, no. 1 (October 1940).

Mejcher, Helmut, 'Oil and British Policy towards Mesopotamia, 1914–1918', *Middle Eastern Studies* 8, no. 3 (October 1972).

Miller, Ernest C., 'North America's First Oil Well – Who Drilled It?', *Western Pennsylvania Historical Magazine* 42, no. 4 (December 1959).

——, 'Oily Days at Cherry Grove', *Pennsylvania History* 19, no. 1 (January 1952).

——, 'The Fountain and Hequembourg Flowing Oil Wells', *Pennsylvania History* 12, no. 3 (July 1945).

Miller, Robert Ryal, 'Arms across the Border: United States Aid to Juárez during the French Intervention in Mexico', *Transactions of the American Philosophical Society* 63, no. 6 (1973).

Mills, J. V., 'Eredia's Description of Malacca, Meridional India, and Cathay', *Journal of the Malayan Branch of the Royal Asiatic Society* 8 (1930).

Miner, H. Craig, 'The Cherokee Oil and Gas Co., 1889–1902: Indian Sovereignty and Economic Change', *Business History Review* 46, no. 1 (Spring 1972).

Mom, Gijs P. A., and David A. Kirsch, 'Technologies in Tension: Horses, Electric Trucks, and the Motorization of American Cities, 1900–1925', *Technology and Culture* 42, no. 3 (July 2001).

Montague, Gilbert Holland, 'The Later History of the Standard Oil Company', *Quarterly Journal of Economics* 17, no. 2 (February 1903).

——, 'The Rise and Supremacy of the Standard Oil Company', *Quarterly Journal of Economics* 16, no. 2 (February 1902).

Morgan, H. Wayne, '"Red Special": Eugene V. Debs and the Campaign of 1908', *Indiana Magazine of History* 54, no. 3 (September 1958).

Morris, Jane Anne, 'Sheep in Wolf's Clothing', in Ritz, ed., *Defying Corporations* (2001).

Mowat, R. C., 'From Liberalism to Imperialism: The Case of Egypt, 1875–1887', *Historical Journal* 16, no. 1 (1973).

Mt. Pleasant, Alyssa, 'Independence for Whom? Expansion and Conflict in the Northeast and Northwest', in Shankman, ed., *The World of the Revolutionary American Republic* (2014).

Munch, Francis J., 'The Anglo-Dutch-American Petroleum Industry in Mexico: The Formative Years During the Porfiriato 1900–1910', *Revista de Historia de América* 84 (July 1977).

Munro, Dana G., 'The Proposed German Petroleum Monopoly', *American Economic Review* 4, no. 2 (June 1914).

Nardella, Federica, Noemi Landi, Ilaria Degano, Marta Colombo, Marco Serradimigni, Carlo Tozzi and Erika Ribechini, 'Chemical Investigations of Bitumen from Neolithic Archaeological Excavations in Italy by GC/MS

Combined with Principal Component Analysis', *Analytical Methods* 11, no. 11 (21 March 2019).

Neilson, Keith, '"Greatly Exaggerated": The Myth of the Decline of Great Britain before 1914', *International History Review* 13, no. 4 (November 1991).

——, 'The Baghdad to Haifa Railway', in Otte and Neilson, eds., *Railways and International Politics* (2006).

Newsinger, John, 'Liberal Imperialism and the Occupation of Egypt in 1882', *Race & Class* 49, no. 2 (2008).

Noetling, Fritz, 'The Occurrence of Petroleum in Burma, and its Technical Exploitation', *Memoirs of the Geological Survey of India* 27, pt. 2 (1898).

North, Douglass G., 'Life Insurance and Investment Banking at the Time of the Armstrong Investigation of 1905–1906', *Journal of Economic History* 14, no. 3 (Summer 1954).

Notestein, Frank B., Carl W. Hubman and James W. Bowler, 'Geology of the Barco Concession, Republic of Colombia, South America', *Bulletin of the Geological Society of America* 55, no. 10 (October 1944).

Nugent, Nicholas, 'Account of the Pitch Lake of the Island of Trinidad', *Transactions of the Geological Society* 1 (1811).

Nuttall, Brandon C., 'Of Kentucky, Salt, and Oil: A History of Early Petroleum Finds Along the Cumberland River', *Oil-Industry History* 15, no. 1 (2014).

Obida, Christopher B., G. Alan Blackburn, J. Duncan Whyatt and Kirk T. Semple, 'Quantifying the Exposure of Humans and the Environment to Oil Pollution in the Niger Delta using Advanced Geostatistical Techniques', *Environment International* 111 (February 2018).

O'Brien, Patrick K., and Geoffrey Allen Pigman, 'Free Trade, British Hegemony and the International Economic Order in the Nineteenth Century', *Review of International Studies* 18, no. 2 (April 1992).

O'Brien, Phillips Payson, 'The Titan Refreshed: Imperial Overstretch and the British Navy before the First World War', *Past & Present* 172, no. 1 (August 2001).

Offer, Avner, 'The Working Classes, British Naval Plans and the Coming of the Great War', *Past & Present* 107, no. 1 (May 1985).

Öke, Mim Kemâl, 'Young Turks, Freemasons, Jews and the Question of Zionism in the Ottoman Empire (1908–1913)', *Studies in Zionism* 7, no. 2 (1986).

Oliveira, Júlia C. T., and Silva F. de M. Figueirôa, 'History of Oil Exploration in the State of São Paulo Before the Foundation of Petrobras (1872–1953)', in Figueirôa et al., eds., *History, Exploration & Exploitation of Oil and Gas* (2019).

Onley, James, 'Britain's Informal Empire in the Gulf, 1820–1971', *Journal of Social Affairs* 22, no. 87 (Fall 2005).

Onuf, Peter, 'Toward Federalism: Virginia, Congress, and the Western Lands', *William and Mary Quarterly*, 3rd ser. 34, no. 3 (July 1977).

Ormandy, William R., 'Britain and Germany in Relation to the Chemical Trade', in Gardner, ed., *The British Coal-Tar Industry* (1915).

Oron, Asaf, Ehud Galili, Gideon Hadas and Micha Klein, 'Early Maritime Activity on the Dead Sea: Bitumen Harvesting and the Possible Use of Reed Watercraft', *Journal of Maritime Archaeology* 10, no. 1 (April 2015).

Ostler, Jeffrey, '"To Extirpate the Indians": An Indigenous Consciousness of Genocide in the Ohio Valley and Lower Great Lakes, 1750s–1810', *William and Mary Quarterly* 72, no. 4 (October 2015).

Otte, T. G., 'The Foreign Office and Defence of Empire, 1856–1914', in Kennedy, ed., *Imperial Defence* (2008).

Owens, Robert M., 'Jeffersonian Benevolence on the Ground: The Indian Land Cession Treaties of William Henry Harrison', *Journal of the Early Republic* 22, no. 3 (Fall 2002).

Painter, David S., 'Oil and the American Century', *Journal of American History* 99, no. 1 (June 2012).

Palen, Marc-William, 'Protection, Federation and Union: The Global Impact of the McKinley Tariff upon the British Empire, 1890–94', *Journal of Imperial and Commonwealth History* 38, no. 3 (September 2010).

Parmenter, Jon William, 'Pontiac's War: Forging New Links in the Anglo-Iroquois Covenant Chain, 1758–1766', *Ethnohistory* 44, no. 4 (Fall 1997).

Penslar, Derek, 'Tracing the Spread of an Impious Forgery', *Forward* 104, no. 31 (12 May 2000).

Peritz, Rudolph J., 'The "Rule of Reason" in Antitrust Law: Property Logic in Restraint of Competition', *Hastings Law Journal* 40, no. 2 (January 1989).

Perkin, F. Mollwo, 'The Oil Resources of the Empire', *Journal of the Royal Society of Arts* 62, no. 3204 (17 April 1914).

Petrusic, Christopher, 'Violence as Masculinity: David Livingstone's Radical Racial Politics in the Cape Colony and the Transvaal, 1845–1852', *International History Review* 26, no. 1 (March 2004).

Pinfari, Marco, 'The Unmaking of a Patriot: Anti-Arab Prejudice in the British Attitude Towards the Urabi Revolt (1882)', *Arab Studies Quarterly* 34, no. 2 (Spring 2012).

Poggi, J., 'Account of a New Spring of Petroleum Discovered in Italy', *Philosophical Magazine* 16, no. 64 (1803).

Pollak, Oliver B., 'The Origins of the Second Anglo-Burmese War (1852–53)', *Modern Asian Studies* 12, no. 3 (1978).

Pratt, Joseph A., 'The Petroleum Industry in Transition: Antitrust and the Decline of Monopoly Control in Oil', *Journal of Economic History* 40, no. 4 (December 1980).

Prettyman, Gib, 'Gilded Age Utopias of Incorporation', *Utopian Studies* 12, no. 1 (2001).

Priest, George L., 'Rethinking the Economic Basis of the Standard Oil Refining Monopoly: Dominance Against Competing Cartels', *Southern California Law Review* 85, no. 3 (March 2012).

Purcell, Jr, Edward A., 'Ideas and Interests: Businessmen and the Interstate Commerce Act', *Journal of American History* 54, no. 3 (December 1967).

Rawlinson, Major, 'Notes on a March from Zoháb to Khúzistán and Kirmánsháh', *Journal of the Royal Geographical Society* 9 (1839).

Reader, W. J., 'Oil for the West of England, 1889–1896: A Study in Competition', *Business History Review* 35, no. 1 (Spring 1961).

Redwood, Bernard B., 'Motor Boats', *Journal of the Society of Arts* (23 March 1906).

Redwood, Boverton, 'The Future of Oil Fuel', *Scientific American*, Supplement (11 April 1914).

Reid, Anthony, 'Colonial Transformation: A Bitter Legacy', in Reid, ed., *Verandah of Violence* (2006).

Reid, Charles J., 'America's First Great Constitutional Controversy: Alexander Hamilton's Bank of the United States', *University of St. Thomas Law Journal* 14, no. 1 (2018).

Reynolds, John E., 'The Venango Trail in the French Creek Valley', *Western Pennsylvania Historical Magazine* 16, no. 1 (February 1933).

Rezneck, Samuel, 'Coal and Oil in the American Economy', *Journal of Economic History* 7, Supplement, *Economic Growth: A Symposium* (1947).

Richter, Daniel K., 'Onas, the Long Knife: Pennsylvanians and Indians, 1783–1794', in Hoxie et al., eds., *Native Americans and the Early Republic* (1999).

Riguzzi, Paolo, and Francesco Gerali, 'Los veneros del emperador. Impulso petrolero global, intereses y política del petróleo en México durante el Segundo Imperio, 1863–1867', *Historia Mexicana* 65, no. 2 (2015).

Rogers, Lore A. and Caleb W. Scribner, 'Lombard Steam Log Hauler', *American Society of Mechanical Engineers* (14 August 1982).

Roland, Alex, 'Secrecy, Technology, and War: Greek Fire and the Defense of Byzantium, 678–1204', *Technology and Culture* 33, no. 4 (October 1992).

Romero, Aldemaro, 'Death and Taxes: The Case of the Depletion of Pearl Oyster Beds in Sixteenth-Century Venezuela', *Conservation Biology* 17, no. 4 (August 2003).

Roosevelt, Theodore, 'The Trusts, the People and the Square Deal', *Outlook* 99 (18 November 1911).

Ross, John, 'Notes on Two Journeys from Baghdád to the Ruins of Al-Hadhr', *Journal of the Royal Geographical Society* 9 (1839).

Rothwell, V. H., 'Mesopotamia in British War Aims, 1914–1918', *Historical Journal* 13, no. 2 (June 1970).

Ruiz, Ramón Eduardo, 'American Imperialism and the Mexican War', in Dinnerstein and Jackson, eds., *American Vistas* (1979).

Ryan, Jr, Vincent R., 'The History of Mexican Oil and Gas Law from the Conquistadors' Conquest until 1914', *Journal of the Texas Supreme Court Historical Society* 5, no. 2 (Winter 2016).

Sabin, Paul, '"A Dive into Nature's Great Grab-bag": Nature, Gender and Capitalism in the Early Pennsylvania Oil Industry', *Pennsylvania History* 66, no. 4 (Autumn 1999).

Saikia, Arupjyoti, 'Imperialism, Geology and Petroleum: History of Oil in Colonial Assam', *Economic and Political Weekly* 46, no. 12 (19–25 March 2011).

Samuel, Marcus, 'Liquid Fuel', *Journal of the Society of Arts* 47, no. 2417 (17 March 1899).

Santiago, Myrna I., 'Class and Nature in the Oil Industry of Northern Veracruz, 1900–1938', in Boyer, ed., *Land Between Waters* (2012).

——, 'Culture Clash: Foreign Oil and Indigenous People in Northern Veracruz, Mexico, 1900–1921', *Journal of American History* 99, no. 1 (June 2012).

Santoro, Sara, 'Crafts and Trade in Minor Settlements in North and Central Italy: Reflections on an Ongoing Research Project', in Haas and Tol, eds., *Economic Integration of Roman Italy* (2017).

Saussure, M. Theodore de, 'Experiments on the Composition and Properties of the Naphtha of Amiano', *Annals of Philosophy* 10, no. 2 (August 1817).

Schölch, Alexander, 'The "Men on the Spot" and the English Occupation of Egypt in 1882', *Historical Journal* 19, no. 3 (September 1976).

Schoonover, Thomas, 'Dollars over Dominion: United States Economic Interests in Mexico, 1861–1867', *Pacific Historical Review* 45, no. 1 (February 1976).

Schwartz, Mark, and David Hollander, 'Annealing, Distilling, Reheating and Recycling: Bitumen Processing in the Ancient Near East', *Paléorient* 26, no. 2 (2000).

——, 'The Uruk Expansion as Dynamic Process: A Reconstruction of Middle to Late Uruk Exchange Patterns from Bulk Stable Isotope Analyses of Bitumen Artifacts', *Journal of Archaeological Science: Reports* 7 (June 2016).

Schwarz, P., 'Khanikin', in Houtsma et al., eds., *First Encyclopaedia of Islam* (1993), vol. 4.

Seligman, Joel, 'A Brief History of Delaware's General Corporation Law of 1899', *Delaware Journal of Corporate Law* 1, no. 2 (1976).

Seligmann, Matthew S., 'Keeping the Germans Out of the Straits: The Five Ottoman Dreadnought Thesis Reconsidered', *War in History* 23, no. 1 (2016).

Sell, George, 'Statistics of Petroleum and Allied Substances', in Dunstan, ed., *Science of Petroleum* (1938), vol. 1.

Selley, Richard C., 'U.K. Shale Gas: The Story So Far', *Marine and Petroleum Geology* 31 (2012).

Shahvar, Soli, 'Concession Hunting in the Age of Reform: British Companies and the Search for Government Guarantees; Telegraph Concessions through Ottoman Territories, 1855–58', *Middle East Studies* 38, no. 4 (October 2002).

——, 'Iron Poles, Wooden Poles: The Electric Telegraph and the Ottoman-Iranian Boundary Conflict, 1863–1865', *British Journal of Middle Eastern Studies* 34, no. 1 (April 2007).

——, 'Tribes and Telegraphs in Lower Iraq: The Muntafiq and the Baghdad-Basrah Telegraph Line of 1863–65', *Middle East Studies* 39, no. 1 (January 2003).

Shapira, Dan, 'A Karaite from Wolhynia Meets a Zoroastrian from Baku', *Iran & the Caucasus* 5 (2001).

Shepard, J., 'Closer Encounters with the Byzantine World: The Rus at the Straits of Kerch', in Reyerson et al., eds., *Pre-Modern Russia* (2006).

Shorrock, William I., 'The Origin of the French Mandate in Syria and Lebanon: The Railroad Question, 1901–1914', *International Journal of Middle East Studies* 1, no. 2 (April 1970).

Siegel, Jennifer 'The Russian Revolution of 1905 in the Eyes of Russia's Financiers', *Revolutionary Russia* 29, no. 1 (2016).

Silliman, Benjamin, 'Notice of a Fountain of Petroleum, Called the Oil Spring', *American Journal of Science and Arts* 23, no. 1 (January 1833).

Sioussat, St. George L., 'The Chevalier de la Luzerne and the Ratification of the Articles of Confederation by Maryland, 1780–1781', *Pennsylvania Magazine of History and Biography* 60, no. 3 (July 1936).

Skirius, John, 'Railroad, Oil and Other Foreign Interests in the Mexican Revolution, 1911–1914', *Journal of Latin American Studies* 35, no. 1 (2003).

Small, Melvin, 'The United States and the German "Threat" to the Hemisphere, 1905–1914', *The Americas* 28, no. 3 (January 1972).

Sorkhabi, Rasoul, 'Pre-Modern History of Bitumen, Oil and Gas in Persia (Iran)', *Oil-Industry History* 6, no. 1 (2005).

——, 'Sir Thomas Boverton Redwood (1846–1919): A Watershed in the British Oil Industry', in Craig et al., eds., *History of the European Oil and Gas Industry* (2018).

Sozański, Józef, Stanisław Kuk, Czesław Jaracz, and Piotr S. Dziadzio, 'How the Modern Oil and Gas Industry was Born: Historical Remarks', in Golonka and Picha, eds., *Carpathians and their Foreland* (2006).

Spielmann, Percy E., 'Who Discovered the Trinidad Asphalt Lake?', *Science Progress* 33, no. 129 (July 1938).

Spring, D. W., 'The Trans-Persian Railway Project and Anglo-Russian Relations, 1909–1914', *Slavonic and Eastern European Review* 54, no. 1 (January 1976).

Stagg, J. C. A., 'Between Black Rock and a Hard Place: Peter B. Porter's Plan for an American Invasion of Canada in 1812', *Journal of the Early Republic* 19, no. 3 (Fall 1999).

Steele, David, 'Britain and Egypt 1882–1914: The Containment of Islamic Nationalism', in Wilson, ed., *Imperialism and Nationalism in the Middle East* (1983).

Steffens, Lincoln, 'New Jersey: A Traitor State – Part II', *McClure's Magazine* 25, no. 1 (May 1905).

Stewart, C. E., 'Account of the Hindu Fire-Temple at Baku, in the Trans-Caucasus Province of Russia', *Journal of the Royal Asiatic Society of Great Britain and Ireland* (April 1897).

Steyn, Phia, 'Oil Exploration in Colonial Nigeria, c.1903–58', *Journal of Imperial and Commonwealth History* 37, no. 2 (June 2009).

Stone, Rufus B., 'Sinnontouan, or Seneca Land, in the Revolution', *Pennsylvania Magazine of History and Biography* 48, no. 3 (1924).

Stuart, John, 'The Panama Canal: Its Commercial and Economic Significance', *Cassier's Engineering Monthly* 44, no. 5 (November 1913).

Stucke, Maurice E., 'Does the Rule of Reason Violate the Rule of Law?', *U.C. Davis Law Review* 42, no. 5 (June 2009).

Sugden, J., 'Archibald, 9th Earl of Dundonald: An Eighteenth-Century Entrepreneur', *Scottish Economic and Social History* 8 (1988).

Sullivan, Brian R., 'Italian Warship Construction and Maritime Strategy, 1873–1915', in O'Brien, ed., *Technology and Naval Combat* (2001).

Sulman, A. Michael, 'The Short Happy Life of Petroleum in Pittsburgh: A Paradox in Industrial History', *Pennsylvania History* 33, no. 1 (January 1966).

Sumida, Jon Tetsuro, 'British Naval Operational Logistics, 1914–1918', *Journal of Military History* 57, no. 3 (July 1993).

Surlémont, Raymond, 'French Armored Cars 1902–1945', *Armored Car* 22 (March 1994).

Tallman, Ellis W., and Jon R. Moen, 'Private Sector Responses to the Panic of 1907: A Comparison of New York and Chicago', *Economic Review* (March/April 1995).

Taylor, Richard Cowling, and Thomas G. Clemson, 'Notice of a Vein of Bituminous Coal in the Vicinity of Havana, in the Island of Cuba', *London and Edinburgh Philosophical Magazine and Journal of Science*, 3rd ser. 10, no. 60 (March 1837).

Testa, Stephen M., 'The Los Angeles City Oil Field: California's First Oil Boom During the Revitalization Period (1875–1900)', *Oil-Industry History* 6, no. 1 (2005).

Thomas, James C., 'Fifty Years with the Administrative Procedure Act and Judicial Review Remains an Enigma', *Tulsa Law Journal* 32, no. 2 (Winter 1996).

Thompson, Michael J., 'The Radical Critique of Economic Inequality in Early American Political Thought', *New Political Science* 30, no. 3 (September 2008).

Thornton, Russell, 'Cherokee Population Losses During the Trail of Tears: A New Perspective and a New Estimate', *Ethnohistory* 31, no. 4 (Autumn 1984).

Tomory, Leslie, 'Building the First Gas Network, 1812–1820', *Technology and Culture* 52, no. 1 (January 2011).

Tuason, Julie A., 'The Ideology of Empire in National Geographic Magazine's Coverage of the Philippines, 1898–1908', *Geographical Review* 89, no. 1 (January 1999).

Tulucan, Alina Dana, Lucia-Elena Soveja-Iacob and Csaba Krezsek, 'History of the Oil and Gas Industry in Romania', in Craig et al., eds., *History of the European Oil and Gas Industry* (2018).

Turner, L. C. F., 'The Russian Mobilisation in 1914', in Kennedy, ed., *War Plans of the Great Powers* (1979).

——, 'The Significance of the Schlieffen Plan', in Kennedy, ibid.

Turrell, Robert Vicat, 'Conquest and Concession: The Case of the Burma Ruby Mines', *Modern Asian Studies* 22, no. 1 (1988).

Vanderlip, Frank A., 'The American "Commercial Invasion" of Europe', *Scribner's Magazine* 31 (January–March 1902).

Vansina, Jan, 'Ambaca Society and the Slave Trade c.1760–1845', *Journal of African History* 46, no. 1 (2005).

Vevier, Charles, 'The Open Door: An Idea in Action, 1906–1913', *Pacific Historical Review* 24, no. 1 (February 1955).

Vogel, Hans Ulrich, 'Bitumen in Premodern China', in Selin, ed., *Encyclopaedia of the History of Science, Technology, and Medicine in Non-Western Cultures* (1997).

——, '"That which soaks and descends becomes salty": The Concept of Nature in Traditional Chinese Salt Production', in Vogel and Dux, eds., *Concepts of Nature* (2010).

——, 'Types of Fuel Used in the Salt Works of Sichuan and Yunnan in South-Western China: A Historical Overview', in Weller et al., eds., *Sel, eau et forêt d'hier à aujourd'hui* (2008).

Von Laue, Theodore H., 'Russian Peasants in the Factory, 1892–1904', *Journal of Economic History* 21, no. 1 (March 1961).

Walker, Francis, 'Policies of Germany, England, Canada and the United States Towards Combinations', *Annals of the American Academy of Political and Social Science* 42 (July 1912).

——, 'The Oil Trust and the Government', *Political Science Quarterly* 23, no. 1 (March 1908).

Wallace, Anthony F., ed., 'Halliday Jackson's Journal to the Seneca Indians, 1798–1800', *Pennsylvania History* 19, no. 2 (April 1952).

Wärmländer, Sebastian K. T. S., Sabrina B. Sholts, Jon M. Erlandson, Thor Gjerdrum and Roger Westerholm, 'Could the Health Decline of Prehistoric California Indians be Related to Exposure to Polycyclic Aromatic Hydrocarbons (PAHs) from Natural Bitumen?', *Environmental Health Perspectives* 119, no. 9 (September 2011).

Webster, Anthony, 'Business and Empire: A Reassessment of the British Conquest of Burma in 1885', *Historical Journal* 43, no. 4 (2000).

Wendt, Carl J., and Ann Cyphers, 'How the Olmec Used Bitumen in Ancient Mesoamerica', *Journal of Anthropological Archaeology* 27, no. 2 (June 2008).

Wheeler, Joseph, and Charles H. Grosvenor, 'Our Duty in the Venezuelan Crisis', *North American Review* 161, no. 468 (November 1895).

White, Jr, John H., '"A Perfect Light is a Luxury": Pintsch Gas Car Lighting', *Technology and Culture* 18, no. 1 (January 1977).

Wik, Reynold M., 'Benjamin Holt and the Invention of the Track-Type Tractor', *Technology and Culture* 20, no. 1 (January 1979).

Wilcox, Lieutenant R., 'Memoir of a Survey of Asam and the Neighbouring Countries, Executed in 1825–6–7–8', *Asiatic Researches* 17 (1832).

Wilentz, Sean, 'America's Lost Egalitarian Tradition', *Daedalus* 131, no. 1 (January 2002).

Wilgus, H. L., 'Need of a National Incorporation Law', *Michigan Law Review* 2, no. 5 (February 1904).

Wilhelm, Samuel A., 'The Wheeler and Dusenbury Lumber Company of Forest and Warren Counties', *Pennsylvania History* 19, no. 4 (October 1952).

Williams, Beryl, 'Approach to the Second Afghan War: Central Asia during the Great Eastern Crisis, 1875–1878', *International History Review* 2, no. 1 (January 1980).

Williams, Walter L., 'United States Indian Policy and the Debate over Philippine Annexation: Implications for the Origins of American Imperialism', *Journal of American History* 66, no. 4 (March 1980).

Wilson, Keith M., 'Britain', in Wilson, ed., *Decisions for War* (1995).

——, 'Creative Accounting: The Place of Loans to Persia in the Commencement of the Negotiation of the Anglo-Russian Convention of 1907', *Middle Eastern Studies* 38, no. 2 (April 2002).

——, 'Grey and the Russian Threat to India, 1892–1915', *International History Review* 38, no. 2 (2016).

——, 'The Struggle for Persia: Sir George Clerk's Memorandum of 21 July 1914 on Anglo-Russian Relations in Persia', *Proceedings of the British Society for Middle Eastern Studies* (Oxford, 1988).

Winkler, Adam, '"Other People's Money": Corporations, Agency Costs and Campaign Finance Law', *Georgetown Law Journal* 92, no. 5 (June 2004).

Winsor, Justin, 'Virginia and the Quebec Bill', *American Historical Review* 1, no. 3 (April 1896).

Winton, Graham R., 'The British Army, Mechanisation and a New Transport System, 1900–14', *Journal of the Society for Army Historical Research* 78, no. 315 (Autumn 2000).

Wołkowicz, Stanisław, Marek Graniczny, Krystyna Wołkowicz and Halina Urban, 'History of the Oil Industry in Poland until 1939', in Mayer et al., eds., *History of Geoscience* (2017).

Woolley, C. Leonard, 'Excavations at Ur, 1927–8', *Journal of the Royal Asiatic Society of Great Britain and Ireland*, no. 3 (July 1928).

Yapp, M. A., 'British Perceptions of the Russian Threat to India', *Modern Asian Studies* 21, no. 4 (1987).

Yerbury, J. Colin, 'Protohistoric Canadian Athapaskan Populations: An Ethnohistorical Reconstruction', *Arctic Anthropology* 17, no. 2 (1980).

Yerxa, Donald A., 'The United States Navy in Caribbean Waters During World War I', *Military Affairs* 51, no. 4 (October 1987).

York, Richard, 'Why Petroleum Did Not Save the Whales', *Socius* 3 (2017).

Youngman, Anna, 'The Tendency of Modern Combination – I', *Journal of Political Economy* 15, no. 4 (April 1907), and 'The Tendency of Modern Combination – II', ibid. 15, no. 5 (May 1907).

Zelkina, Anna, 'Jihad in the Name of God: Shaykh Shamil as the Religious Leader of the Caucasus', *Central Asian Survey* 23, no. 3 (2002).

Zheng, Jiexin, and Andrew Palmer, 'Bamboo Pipelines in Ancient China (and now?)', *Journal of Pipeline Engineering* 8, no. 2 (June 2009).

Anonymous:

'Asfalto y sal gema', *Boletín de la Sociedad Mexicana de Geografía y Estadística* 6 (1858).

'Asphaltic Mastic, or Cement of Seyssel', *American Journal of Science and Arts* 34, no. 2 (July 1838).

'A Village Lighted by Natural Gas', *American Journal of Science and Arts* 17, no. 2 (January 1830).

'Barbadoes Mineral Oil, or Green Naphtha', *The Lancet* 16, no. 408 (25 June 1831).

'Coalbrookdale and Ironbridge in 1787', *Salopian Shreds and Patches, a Garland of Shropshire Specialities* 9 (31 December 1890).

'French Asphalt Roads, and American Attempts to Imitate Them', *Scientific American* (12 February 1870).

'Incorporating the Republic: The Corporation in Antebellum Political Culture', *Harvard Law Review* 102, no. 8 (June 1989).

'Proceedings of the Asiatic Society', *Journal of the Asiatic Society of Bengal* 2, no. 74 (February 1838).

'The Asphalts', *Journal of the Society of Arts* (22 November 1872).

'The Break-Up of China and Our Interest in It', *Atlantic Monthly* 84, no. 502 (August 1899).

'Trinidad Petroleum', *Journal of the Society of Arts* 16, no. 827 (25 September 1868).

THESES, WORKING PAPERS AND CONFERENCE PAPERS

Blaakman, Michael Albert, 'Speculation Nation: Land and Mania in the Revolutionary American Republic, 1776–1803', PhD thesis (Yale University, May 2016).

Brown, Warwick Michael, 'The Royal Navy's Fuel Supplies, 1898–1939: The Transition from Coal to Oil', PhD thesis (King's College London, 2003).

Caraher, Ann, 'The Construction of the Central Railway of Peru as Reflected in the 1870–1875 Company Letters', MA thesis (Loyola University, 1966).

Chen, Li-Chiao, 'British Policy on the Margins and Centre of Iran in the Context of Great Power Rivalry 1908–1914', PhD thesis (Royal Holloway, University of London, 2015).

Crane, Daniel A., 'Antitrust and Democracy: A Case Study from German Fascism', University of Michigan Public Law and Legal Theory Research Paper no. 595 (April 2018).

Dash, Michael Wynford, 'British Submarine Policy 1853–1918', PhD thesis (King's College London, 1990).

Deren, Seçil, 'German Ideas and Expectations on Expansion in the Near East (1890–1915)', PhD thesis (Middle East Technical University, Ankara, 2004).

Eichengreen, Barry, and Marc Flandreau, 'The Federal Reserve, the Bank of England, and the Rise of the Dollar as an International Currency, 1914–1939', BIS annual research conference, Lucerne (24–25 June 2010).

Fordham, Benjamin O., 'Protectionist Empire: Trade, Tariffs and United States Foreign Policy, 1890–1914', Working Paper (March 2011).

——, 'The Domestic Politics of World Power: Explaining Debates over the United States Battleship Fleet, 1890–91', Working Paper (February 2010).

Gibson, Martin William, 'British Strategy and Oil, 1914–1923', PhD thesis (University of Glasgow, 2012).

Gray, Steven, 'Black Diamonds: Coal, the Royal Navy, and British Imperial Coaling Stations, circa 1870-1914', PhD thesis (University of Warwick, March 2014).

Hamm, Geoffrey, 'British Intelligence and Turkish Arabia: Strategy, Diplomacy, and Empire, 1898–1918', PhD thesis (University of Toronto, 2012).

Haque, Jameel N., 'Conflict and Cooperation: Western Economic Interests in Ottoman Iraq 1894–1914', PhD thesis (City University of New York, 2016).

Hussain, Sarah S., 'Resistance, Pacification and Consolidation of British Rule in the Tribal and Frontier Areas of Burma, 1885–1935', PhD thesis (North Eastern Hill University, Shillong, 2010).

Kitzen, Martijn, 'The Course of Co-Option: Co-Option of Local Power-Holders as a Tool for Obtaining Control over the Population in Counterinsurgency Campaigns', PhD thesis (University of Amsterdam, 2016).

Pinelo, Adalberto José, 'The Nationalization of International Petroleum Company in Peru', PhD thesis (University of Massachusetts, Amherst, 1972).

Schenk, Joep, and Marten Boon, 'Trading Places: How Merchants Shaped the Rotterdam-Ruhr Axis in the First Global Economy, 1870–1914', Working Paper, Erasmus Centre for the History of the Rhine (2013).

Snyder, David Allan, 'Petroleum and Power: Naval Fuel Technology and the Anglo-American Struggle for Core Hegemony', PhD thesis (Texas A&M University, 2001).

Stoicescu, Maria, and Eugen Mihail Ionescu, 'Romanian Achievements in the Petroleum Industry', CBU International Conference on Innovation, Technology Transfer and Education, Prague (3–5 February 2014).

Strunk, William Theodore, 'The Reign of Shaykh Khazal ibn Jabir and the Suppression of the Principality of Arabistan: A Study in British Imperialism in Southwestern Iran 1897–1925', PhD thesis (Indiana University, 1977).

Sugden, John, 'Lord Cochrane, Naval Commander, Radical, Inventor (1775–1860): A Study of His Earlier Career, 1775–1818', PhD thesis (University of Sheffield, July 1981).

Yavuz, Enes, 'Ottoman Oil Concessions During the Hamidian Era (1876–1909)', MA thesis (İhsan Doğramacı Bilkent University, Ankara, December 2018).

GOVERNMENT AND OFFICIAL PUBLICATIONS AND DOCUMENTS

United Kingdom

Foreign Office

A Manual of Belgian Congo (London: HMSO, 1920).

British New Guinea (Papua), Handbook No. 88, Foreign Office Historical Section (London: HMSO, 1920).

Correspondence Respecting the Affairs of the Congo (London: HMSO, 1913).

Diplomatic and Consular Report: Portugal, for the year 1905, on the Trade and Commerce of Angola, no. 3704 (1906).

Diplomatic and Consular Report: Portugal, for the year 1908, on the Trade of the Province of Angola, no. 4391 (1909).

Diplomatic and Consular Reports on Trade and Finance, Roumania, 1913, no. 5326 (July 1914).

Diplomatic and Consular Reports on Trade and Finance, Russia: Trade of Batoum, 1890, no. 677 (1890).

Diplomatic and Consular Reports on Trade and Finance, Russia, for 1892, Trade of Batoum, no. 1191 (1893).

Diplomatic and Consular Reports on Trade and Finance, Russia, for 1893, Trade of Batoum, no. 1371 (1894).

Diplomatic and Consular Reports on Trade and Finance, Russia, for 1894, Trade of Batoum, no. 1562 (1895).

Diplomatic and Consular Reports on Trade and Finance, Russia, for 1895, Trade of Batoum, no. 1717 (1896).

Diplomatic and Consular Reports on Trade and Finance, Russia: Trade of Batoum, 1905, no. 3566 (1906).

Diplomatic and Consular Reports on Trade and Finance, Russia, for 1913, Trade of Batoum, no. 5296 (1914).

'Koweit and Bahrein: Oil Deposits. Water Supply', File 451/1913 pts. 1–2. British Library: IOR/L/PS/10/339, in Qatar Digital Library (http://www.qdl.qa/en/archive/81055/vdc_100000000419.0x0000ea).

'Report by Major Everett on His Recent Journey in Western Portion of Erzeroum Vilayet', 8 October 1882, 194 (FO 424/132).

'Turkey in Asia: Oil Concessions', File 3877/1912 pt. 1. British Library: IOR/L/PS/10/300, in Qatar Digital Library (http://www.qdl.qa/archive/81055/vdc_100028928516.0x00000a).

Other

Correspondence Respecting the Passage of Petroleum in Bulk through the Suez Canal, House of Commons (June 1892) C.6556.

Eele, Hancock and Portlock, 'Manufacture of Pitch, Tar, &c.', British Patent No. 330 (29 January 1694).

Report from the Select Committee on Petroleum (London: HMSO, 27 July 1894).

Report of the Royal Commission on Fuel and Engines, vol. 1 (1912), vols. 2 and 3 (1913) (NA ADM 116/1208).

Scott, J. George, *Gazetteer of Upper Burma and the Shan States* (Rangoon: Superintendent, Government Printing, 1900).

Second Report of the Royal Commission on Coal Supplies (HMSO, 1904).

Statement of the First Lord of the Admiralty Explanatory of the Navy Estimates, 1905–1906 (1905), Cd.2402.

Stewart, Col. C. E., *Report on the Petroleum Districts Situated on the Red Sea Coast* (Cairo: National Printing Office, 1888).

'The Principal Petroleum Resources of the British Empire', *Bulletin of the Imperial Institute* 2 (30 June–29 September 1904).

United States

Consular Reports

'Alcohol in Germany', *Department of Commerce and Labour, Bureau of Manufactures, Monthly Consular and Trade Reports*, July 1905, no. 298.

Consular Reports: Commerce, Manufactures, etc. 69, *no. 261, June 1902* (Washington, DC: GPO, 1902).

'Drilling for Oil in Palestine', *Daily Consular and Trade Reports*, 17th Year, no. 42 (19 February 1914).

'German Trade Conditions This Year', *Daily Consular and Trade Reports*, 16th Year, no. 204 (2 September 1913).

'Impending Petrol Famine' and 'Consumption of Petrol', in *Department of Commerce and Labor, Bureau of Manufactures, Special Consular Reports*, vol. 40, Motor Machines (Washington: GPO, 1907).

'Mexican Asphaltum', *Reports from the Consuls of the United States*, no. 28 (February 1883).

'Motor Machines, Department of Commerce and Labour, Bureau of Manufactures', *Special Consular Reports* 15, pt. 2 (Washington, DC: GPO, 1908).

'Oil Fields in Persia', *U.S. Monthly Consular Reports* (July 1905).

'Petroleum and Kerosene Oil in Foreign Countries', *U.S. Consular Reports*, no. 37, January 1884 (Washington, DC: GPO, 1884).

'Petroleum Fields of Egypt', *Reports from the Consuls of the United States*, vol. 21 (January–March 1887).

Reports from the Consuls of the United States, vol. 19, April–September 1886 (Washington, DC: GPO, 1887).

'The Russian Petroleum Trade', *Consular Reports: Commerce, Manufactures, etc.* 69, no. 261, June 1902 (Washington, DC: GPO, 1902).

Towler, John, 'Pitch Lake of Trinidad', *Reports from the Consuls of the United States*, no. 28 (February 1883).

'Trade of Mesopotamia and Irak', *Daily Consular and Trade Reports*, 16th Year, no. 256 (1 November 1913).

U.S. Geological Survey Reports

Mineral Resources of the United States, 1885 (Washington, DC: GPO, 1886).

Mineral Resources of the United States, 1893 (Washington, DC: GPO, 1894).

Mineral Resources of the United States, 1899, in *Twenty-First Annual Report of the United States Geological Survey*, pt. 6 (cont.) (Washington, DC, GPO, 1901).

Mineral Resources of the United States, 1900 (Washington, DC: GPO, 1901).

Mineral Resources of the United States, 1904 (Washington, DC: GPO, 1905).

Mineral Resources of the United States, 1911 (Washington, DC: GPO, 1912).

Mineral Resources of the United States, 1912 (Washington, DC: GPO, 1913).

Mineral Resources of the United States, 1913 (Washington, DC: GPO, 1914).

Mineral Resources of the United States, 1914 (Washington, DC: GPO, 1916).

Yerkes, R. F., H. C. Wagner and K. A. Yenne, 'Petroleum Development in the Santa Barbara Channel Region', in *Geology, Petroleum Development, and Seismicity of the Santa Barbara Channel Region, California*, Geological Survey Professional Paper 679 (Washington, DC: GPO, 1969).

U.S. Congress

Antitrust Legislation, House of Representatives, 63rd Cong., 2nd Sess., Report No. 627 (6 May 1914).

Commercial Relations of the United States with Foreign Nations, 1861, 37th Cong., 2nd Sess. (Washington, DC: GPO, 1862).

Commercial Relations of the United States with Foreign Nations, 1865, 39th Cong., 1st Sess. (Washington, DC: GPO, 1866).

Commercial Relations of the United States with Foreign Nations, 1866, 39th Cong., 2nd Sess., Ex. Doc. no. 81 (Washington, DC: GPO, 1867).

Commercial Relations of the United States with Foreign Nations, 1868, 40th Cong., 3rd Sess., Ex. Doc. no. 87 (Washington, DC: GPO, 1869).

Evacuation of Mexico by the French, House of Representatives, 39th Cong., 1st Sess., Ex. Doc. no. 93 (22 April 1866).

Hearings before the Committee on Banking and Currency, 63rd Cong., 1st Sess., Doc. no. 232, vol. 2 (Washington, DC: GPO, 1913).

Hearings Before the Industrial Commission on the Subject of Trusts and Combinations (Washington, DC: GPO, 1899).

Lands in Severalty to Indians, House of Representatives, 46th Cong., 2nd Sess., Report no. 1576 (28 May 1880).

Report of the Committee Appointed Pursuant to House Resolutions 429 and 504 to Investigate the Concentration of Control of Money and Credit, 62nd Cong., 3rd Sess., Report no. 1593 (Washington, DC: GPO, 1913).

Trusts, House of Representatives, 50th Cong., 1st Sess., Report no. 3112 (30 July 1888).

Other

Agg, John, *Proceedings and Debates of the Convention of the Commonwealth of Pennsylvania* (Harrisburg, VA: Packer, Barrett, and Parke, 1837).

Annual Reports of the Navy Department for 1913 (Washington, DC: GPO, 1914).

Annual Reports of the Navy Department for 1914 (Washington, DC: GPO, 1915).

California State Mining Bureau, *Third Annual Report of the State Mineralogist* (Sacramento, CA: State Office, 1883).

Fisher, Louis, 'Destruction of the Maine (1898)', Law Library of Congress (4 August 2009).

General Staff Report, *Organization of the Land Forces of the United States* (Washington, DC: GPO, 1912).

'Indian Affairs', *Annual Reports of the Department of the Interior, 1904* (Washington, DC: GPO, 1904).

Kinzer, Hon. Lance, 'Kansas Attorney General Opinion no. 2009–13' (9 June 2009).

Lewery, Leonard J., *Foreign Capital Investments in Russian Industries and Commerce*, Department of Commerce (Washington, DC: GPO, 1923).

Manning, Van H., *Yearbook of the Bureau of Mines, 1916* (Washington, DC: GPO, 1917).

Melville, George W., *Report of the U.S. Naval 'Liquid Fuel' Board* (Washington, DC: GPO, 1904).

Peckham, S. F., *Report on the Production, Technology, and Uses of Petroleum and its Products* (Washington, DC: GPO, 1885).

Rippy, J. Fred, 'The United States and Colombian Oil', *Information Service 5*, no. 2 (3 April 1929).

Watts, W. L., *Oil and Gas Yielding Formations of Los Angeles, Ventura, and Santa Barbara Counties*, California State Mining Bureau, Bulletin no. 11 (Sacramento: A. J. Johnston, Superintendent State Printing, 1897).

White, I. C., 'Petroleum and Natural Gas', in *West Virginia Geological Survey* 1, pt. 4 (Morgantown, VA: Post Printing House, 1899).

Wrigley, Henry E., *Special Report on the Petroleum of Pennsylvania* (Harrisburg, VA: Second Geological Survey, 1875).

Other Countries

Carne, J. E., 'Notes on the Occurrence of Coal, Petroleum and Copper in Papua', *Bulletin of the Territory of Papua*, no. 1 (Melbourne: Department of External Affairs, 1913).

Kesse, G. O., 'The Search for Petroleum (Oil) in Ghana, Ghana Geological Survey', Report no. 78/1 (17 July 1978).

Ministerio de Fomento, *Memoria Presentada á S. M. el Emperador por el Ministro de Fomento Luis Robles Pezuela de los trabajos ejecutados en su ramo el año de 1865* (México: Imprenta de J. M. Andrade y F. Escalante, 1866).

National Historic Sites Directorate, *To Confirm the Designated Place of Battle Hill National Historic Site of Canada* (Quebec: Historic Sites and Monuments Board of Canada, 2007).

Parliament of the Commonwealth of Australia, *Papua: Report for the Year Ended 30th June 1912* (20 December 1912).

Thames River Background Study Research Team, *The Thames River Watershed: A Background Study* (London: Upper Thames River Conservation Authority, 1998).

Reports and Miscellaneous

Anglo-Mexican Petroleum Products Co., *Mexican Fuel Oil* (1914).

Costa, João Carlos da, *A riqueza petrolífera d'Angola* (Lisbon: Sociedade de Geografia de Lisboa, 1908).

Hudson's Bay Company Archives, 'Instructions from James Knight to William Stewart, 27 June 1715', Archives of Manitoba, York Factory post journal, B.239/a/1, at Alberta Culture and Tourism – Alberta's Energy Resources Heritage – The Fur Trade and Alberta's Oil Sands (http://www.history.alberta.ca/energyheritage/sands/origins/the-fur-trade-and-albertas-oil-sands/ default.aspx#page-1).

Larkin, E. P., *Report of the General Superintendent to the Board of Directors of the Peruvian Petroleum Company of New York* (New York: L. H. Biglow & Co., 1866).

Silliman, B., *Professor Silliman's Report upon the Oil Property of the Philadelphia and California Petroleum Company* (Philadelphia, PA: E. C. Markley & Son, 1865).

The Derrick's Hand-Book of Petroleum (Oil City, PA: Derrick Publishing Company, 1898).

Thomas, Russell, *The Manufactured Gas Industry*, vol. 1, Historic England Research Report Series, no. 182–2020 (2020).

University of Virginia, Charlottesville, VA, 'Notes on the State of Virginia', Special Collections, A 1787.J45 (https://www.encyclopediavirginia.org/8478-0c9295cdea4dd8d/).

Index

Page numbers within parentheses indicate minor references.

Abadan, 366, 465, 540
Abbott, William, 143
Abdul Hamid II, Sultan 370
Abū-Dulaf, 7–8
Aceh *see* Dutch East Indies: Sumatra
Acosta, José de, 43
Ackroyd Stuart, Herbert, 332
Adlum, John, 98
aeroplanes
   civilian, 328, 329, 464–5
   invention of, 327–8
   military, 328–9, 452–3, 539
Afghanistan
   British-Russian rivalry over, 198,
      299, 301, 359, 367
   invasion of (2001), 236
   Second Anglo-Afghan War, 234
Agadir Crisis *see* Morocco Crisis
Agricola, Georgius, 34
Ahwaz, 358, 361
   drilling at, 465
   pipeline sabotage 540
Aitchison, Charles, 223
Akhverdov, Ivan, 251
Al-Afghani, Sayyid Jamal
      al-Din, 235
Al-Balkhi, 10

Al-Dimashqi, 9
Al-Maqrīzī, 10
Al-Mas'ūdī, 7
Al-Rammah, Al-Hasan, 9
Al-Râzî, 9
Al-Tabari, 7
Alaska, exploration in, 257, 289–90
Albania, 32
Alden, Timothy, 62
Aldrich, Nelson, 405, 406
Alexander the Great, 4, (233)
Alfaro Santa Cruz, Melchor de, 43
Algeria, oil in, 297
Alvin O. Lombard Traction Co.,
      450–1
American Asiatic Association, 278
Amherst, Jeffery, 86, 87–8
Ammianus Marcellinus, 3–4
Amoco, 404
Anderson, Alexander, 46
Andrews, Clark & Co., 123
Andrews, Samuel, 123
Anglo-American Oil Co., see
      Standard Oil Co.
Anglo-Caucasian Oil Co., 254
Anglo-Egyptian Oil Co./Oilfields
      Ltd., see Royal Dutch-Shell

Anglo-Persian Oil Co. (later Anglo-
Iranian Oil Co., now BP)
Bakhtiari Oil Co., 364, (499)
and British navy, 308, 316, 364,
366, 432, 467, 468–74,
480–504, 508–15,
529–33, 538
concession, 304–7, 495
Concessions Syndicate Ltd.,
316, 357–64
D'Arcy, William Knox, 304–8,
315–16, 357–66, 374, 375,
377, 411, 456, 461, 469, 470,
506, 514, 515
exploration: First Exploitation
Co., 307–8, 315; Concessions
Syndicate Ltd., 356–9, 361–4;
Anglo-Persian, 465
First Exploitation Co., 307–8,
315; rescue of, 315–16
formation of, 364, 456
Greenway, Charles, 457, 463–4,
468–77, 481–3, 485, 502, 506,
509, 512, 540–1; see also
Burmah Oil Co.
Marriott, Alfred, 305, 374, 375
and Mesopotamian oil
concessions, 374–7, 461–3,
468–75, 485, 486, 504–16
nationalization of, 468–9, 470–1,
480–504, 514, 529, 538
Nichols, H.E., 375
oil strike, first, at Masjid i
Suleiman, 363–4, 456, Fig. 21
origins of (from D'Arcy
concession), 304–8, 315–16,
356–67, 495
pipeline, 306–7, 363, 366, 367,
463–5, 485, 489, 491,figs.
22–3, 25

product sales/distribution, 466–7,
471, 511
production, 307, (366), (432),
(466), (471), 483, 485, (486),
(490)
refinery, 366, 367, 465, 466, 485
rescue of, by British government:
((1905) 315–16); (1913–14)
485–8, 514, 529, 538
Reynolds, George, 358, 362
and Royal Dutch-Shell rivalry,
379–80, 466–80, 482–3,
(487), 489–95, 506–14
security of, 357–9, 489–504,
529–33, 538
Strathcona of Mount Royal,
Lord, 316, 464, 465, 466
and Trinidad, 477
Wallace, Charles W., 499–500;
see also Burmah Oil Co.
Anglo-Persian War (1857),
299, 367
Anglo-Russian Convention (1907),
359–61, 365, 378, 455–8, 475,
488, 490, 495–8, 503, 517,
526–7, 528, 529–33, (535–6),
(537–8), Fig. 25
and Anglo-Persian Oil Co., 475,
488–9, 490–1, 495–7, 503,
529–33, (538), Fig. 25
and outbreak of WWI, 526–7,
528, 529–33, (535–6), (537–8)
Anglo-Saxon Petroleum Co. see
Royal Dutch-Shell
Angola
exploration in, 322, 324
pre-industrial oil, 17–18, 191
Apiaguaiki, 291–2
Arab Revolt, 516–17
Aranguren, Antonio, 341

Archbold, John D. *see* Standard
  Oil Co.
Argyropoulo, K.M., 307
Armenia, 197–8, 201, (347–56)
armoured cars/tanks *see* motor
  vehicles
Armstrong, John, 84
Arnold, Arthur, 205
Asche, Frederick D. *see* Standard
  Oil Co.
Asiatic Petroleum Co., 288–9,
  317–18, 411–12, 416, 466, 471
asphalt *see* bitumen/asphalt
Asquith, Herbert H., (533), 536
Assam *see* India
Astra Română *see* Royal
  Dutch-Shell
Austria-Hungary/Galicia
  German investment in, 412, 417
  navy *see* navy, Austrian
  production, 220–1, 287, 542
  pollution, 287
  pre-industrial oil, 37–8, 70,
    189, 221
  protectionism, 287, 414
  and Serbia, 527–8, 530, 534
  and WWI, oil during, 540, 542–3
Australian navy *see* navy, Australian
Automobile Association, 451
Azerbaijan
  in 1905 revolution, 347–56
  petroleum in: pre-19th century,
    7–8, 10, 11–12, 13–14, 194,
    195–6; 19th century, 14–15,
    196, 198–9, 200, 202–16,
    242–57, 266–7, 281–2, (304),
    306; early 20th centruy, 282,
    306, 343–56, 541, 543–4
  Russian conquest of, 195–202
  in WWI, 541, 543–4

Aztecs, use of bitumen by, 40

Babylon(ia), bitumen in, 2, 3, 4,
  (10), 14, 15
Baden-Powell, Baden F.S., 329
Bahrain, oil in, 503
Baker, George, 22
Bakhtiari, 16, (298), 300, (307),
  356–9, 360, 364–5, 366–7,
  (458), 463–5, (489), 497,
  498–500, 533, 540, Figs. 20–23
Bakhtiari Oil Co. *see* Anglo-Persian
  Oil Co.
Baku *see* Azerbaijan; Russia
Baku Oil Co., 204, 205, 208
Baku Russian Petroleum Co.,
  253, 353, 355
Balkan War
  First (1912–13), 452, 523,
    524, 525, 528
  Second (1913), 524, 525, 528
Ballin, Albert, 520
Barba, Alvaro Alonso, 44
Barbados
  and British navy, potential
    oil for, 312, 314
  early reports/use of oil, 46, (55)
  production, 290
Barber Asphalt Paving Co.,
  290, 338
Barco, Virgilio, 290
Barí tribe, 426–8
Baring, Hugo, 467, 482
Baring Brothers & Co., 105
Barings Bank, 467
Barros, João de, 20–1
Baryatinsky, Aleksandr, 201
Bassermann, Ernst, 512–13
battles
  of Fallen Timbers (1794), 98

battles – *cont'd.*
of the Great Meadows (1754), 83
Harmar campaign (1790), 97–8
of Kuruyuki (1892), 291–2
of Point Pleasant (1774), 89
Sandy Creek Expedition (1756), 84
of Sarikamish (1914–15), 541
St. Clair campaign (1791), 97–8
Sullivan, Clinton and Brodhead
campaigns (1779), 91
of Tsushima (1905), 347
Batum Oil Refining and Trading
Co., 214
Baumhauer, Edouard Henri von, 193
Bear Creek Refining Co., 219, 248
Beaulieu, Augustin de, 21
Bellamy, Edward, 170–1, 389, 392
Benjamin of Tudela, 32
Benckendorff, Count, (533)
Benson, Byron D., 152
'Benson's Folly', 151–6
Benz, Carl, 273–4
Beresford, Charles *see* navy, British
Berger, Victor, 398
Bergheim, John Simeon, 220, 293,
321–3, 324, 481–2
Bermudez Co., 425
Bernard, Charles, 222–3, 224
Bessler, Waechter & Co., 219
Bethmann Hollweg, Theobald von,
518, 534–5, 536
Beveridge, Albert J., 276
Bibi-Eybat Petroleum Co., 353
Bissell, George H., 75, 183
bitumen/asphalt
Albania, 32, 34
Angola, 17–18, 191
artistic/ritual use, 1, 40, 64, 294
Barbados, 46
Bolivia, 44
building material, 1, 2, 14, 16,
35, 37, 40, 48, 64–5, 72
Canada, 51–2, 57, 72
Central Asia, 7, 8
Colombia, 41
Cuba, 41, 48
Dead Sea, 2, 3, 7, 12
Ecuador, (41), 43, 45
France, 35, 295
fuel for steamships, 49, 50, 188
Italy, 32, 190, 295
Mesopotamia (Iraq)/Kuwait, 1–2,
3, 5, 8, 14–15, 16–17, 305,
(369), (373)
Mexico, 40, 43, 44–5, 47–8,
184–5, 294–5
Nigeria, 322
in Noah's Ark, 2
Persia/Iran, 1, 2, 3, 8, 12–13,
14–15, 16, 298, (305), 357
Peru, 45
prehistoric use, 1, 31–2, 40, 50
road surfacing / paving, 1, 35,
39–40, 48, 72, 181, (200),
290, 338, 369
Romania, 31–2
Russia, 39–40, (200)
Switzerland, 35
Trinidad, 15, 43–4, 46–7,
49–50, 187–8, 290
United Kingdom, 36
United States, 50–1, 53–4, 58,
64–5, 71–2, 182
Venezuela, 41, 290, 338–9
waterproofing, 1, 12, 17, 34, 40–1,
42–5, 46–8, 49, 51, 54, 64, 72,
185, 191, 294, 322, 369
Black, Frederick *see* navy, British
Bleichröder, Bankhaus S., 343, 412
Blériot, Louis, 329

Blood of Thyrsus, 34
Bnito (Caspian and Black Sea
Petroleum Co.) *see* Rothschild
Bodisko, K.A., 214
Boissevain, M.J., 258–9
Bolívar, Simón, 45
Bolivia
early reports of oil, 44
exploration in, 292–3
pre-industrial oil, 291
Bontius, Jacobus, 21–2
Borneo
Dutch (Kalimantan, Indonesia)
see Dutch East Indies
British (Sarawak) (Kalimantan,
Indonesia): and British navy,
potential oil for, 476, 494;
Royal Dutch-Shell in, 417,
476, 494
Boulton & Watt, 63
Bouquet, Henry, 85, 87
Bowen, Herbert, 339–40
Bowles, Thomas, 228
BP, 364, 404
Braddock, Edward, 83
Brant, Joseph, 97
Brayton, George, 327
Brazil, exploration in, 293, 295
Breckenridge Cannel Coal Co., 70
Breckenridge Coal and Oil Co.,
70, 75
Brewer, Francis B., 75, (76), 109
Brewer, Watson and Co., 74, 75,
(108)
Bright, John, 241
Britain *see* United Kingdom/Britain
British and Foreign Oil and Rubber
Trust, 477
Brodhead, Daniel, 91–2
Brough, William, 213

Buchanan, George, 497, 503, (524),
526, (527), 530–1, 535, (536)
Buchanan, William I., 342
Buckingham, Duke of, 188
Bülow, Bernhard von, 286, 517, (520)
Bunge, A., 212, 214, 216
Burma (Myanmar) *see* India
Burmah Oil Co. *see* India
Burnett, Peter, 182
Burnt House (Jenuchshadego) *see*
Seneca Indians

Cairo Syndicate, 478–9
Calhoun, William, 338, 341
California
early oil: exploration, 182–3;
reports, 50–1, 54, 58, 71–2;
strikes, 267, 268; use (pre-
industrial), 51, 54, 64, 71–2, 181
in Mexican-American War, 181–2
Native Americans, genocide
of, 182
pipelines, 267, 268
pollution, 268, 420, Fig. 24
prehistoric use of petroleum, 50
production, 267–8, 283, 284,
289, 311, (320), (416),
(441), (493)
and US navy, oil for, 441, 493
Camargo, Eugênio Ferreira de, 293
Camden, J.N., 151
Cameroon
coaling station, potential, 287
oil in, 322, 442
Canada
pre-industrial oil, 51–2, 57–8,
62–3, 65–6
production, 72–3, (179), 290
Canal du Midi, 94
Canning, Stratford, 300

Capper, John E., 328, 329
carburettor, 273, 331
Carden, Lionel, 384, 386
Cargill, David *see* Burmah Oil Co.
Carletti, Francesco, 43
Carlos III of Spain, King, 45–6
Carne, J.E., 423–4
Caribbean Petroleum Co.
    see Royal Dutch-Shell
Carruthers, Joseph, 435
Caspian Petroleum and
    Trade Co., 208
Caspian and Black Sea Petroleum
    Co. *see* Rothschild
Cassel, Ernest, 467, 468
Castro, Cipriano, 339–42, (425)
caterpillar-track vehicles,
    450–1, 452
Catherine the Great, 39, 198
Caucasus
    Russian conquest of,
        195–202, 212
    see also Armenia; Azerbaijan;
        Georgia; Persia (Iran); Russia
Caucasus & Mercury Co., 202
Central Mining and Investment
    Corp., 507
Chamberlain, Joseph, 309
Channel Tunnel (Britain-France), 445
Charlevoix, Pierre-François-Xavier
    de, 52
Chernyshov, Alexander, 199
Cherokee Indians, 55, 84, 334–7
Chester, Colby M., 379
Chevron Corp., 404
Cheyenne Indians, 333
China
    Committee on American Interests
        in (American Asiatic
        Association), 278

early reports/use of petroleum:
    earliest, 19–20; 19th century,
        27–8, 192–3
    exploration in, 257, (278)
    oil discovery, 279–80, 295
    oil imports, 138, 223, 277,
        280, 289
Chipewyan Indians, 51–2, 57
Chippewa (/Ojibwe) Indians, 57–8
Chumash Indians, 51, 54, 182
Churchill, Randolph, 223
Churchill, Winston, 383–4, 430–3,
    435, 437, 439–40, 443, 444–6,
    (448–9), 450, 453, 458–9, 478,
    (479), 480, 483–6, 489–94,
    (497), 499–500, 501–3, 509,
    513, 520–1, 529, (533), 535,
    536–7
Clark, George Rogers, 90, 93
Clark, Maurice, 122–3
Clark & Rockefeller, 122–3
Clarke, Wedworth, 185
Clemenceau, Georges, 541–2
Clerk, George, 530–2, 533
Cleveland *see* Ohio
Clinton, DeWitt, 61
coal
    coal tar: waterproofing, 36–7;
        fuel, 435–6, 441
    coal/town gas, 63, 64, 70, 117,
        270, 272–3, 435, 540
    coal-oil illuminant, 37, 38, (49),
        50, 70–1, 75, 106–7, (112),
        113, (115), 117, 119
    supplies, 20, 49, (104), 106–7,
        117, 188, 193, 206, 220,
        (231), 232, 233, 251, 266,
        267, 268, 270, 271, 275,
        278–9, 285, 287, 308, 309–10,
        313, 320–1, 324, 340, 346,

355, 385, 386, 434, 441,
444, 445, 472, 522
Cochrane
Alexander, 47
Archibald (9th Earl of
Dundonald), 36–7, (47)
Arthur, 50
Thomas (10th Earl of
Dundonald), 37, 48–50, 150
Cody, Samuel. F., 329
Colijn, Henrikus, 263
Colombia
earliest reports of oil, 41–3, (295)
oil concessions, 290, 386–7, 414
pre-industrial production, 290
Colón Development Co. see Royal
Dutch-Shell
Colonial Oil Co. see Standard
Oil Co.
Colvin, Auckland, 236, 238
Committee of Imperial Defence,
316, 376, (520), (522)
Compañía de Petróleo 'El Aguila'
S.A. see Mexican Eagle Oil Co.
Compañía del petroleo del Golfo
Mexicano, 187
Compañia Peruana de Petroleo, 183
Compañía Petrolea del Táchira, 290
Conant, Charles A., 276
Concessions Syndicate Ltd. see
Anglo-Persian Oil Co.
Constantine VII, Emperor, 6, 39
Corbett, Vincent, 342
Cornplanter, Chief (Kayéthwahkeh)
see Seneca Indians
Cornplanter Oil Co. see Oil
Region (US)
Cornplanter Township see Oil
Region (US)
corporations ('trusts')

Bureau of Corporations (US),
392–3, 402, 407
holding companies, 169, 173–4,
393, 395
New Jersey incorporation law,
163, 173, (393), 395
origins of, 160–2
rise of, in US, 139–41, 157–60,
162–75, 336, 338, 388–410
trust device, 169–75
Cortés, Hernan, 48, (383)
Cossacks
and oil, 39, 196, 199, 208, 251
in Persia, 300, 306, 364, 498
in Russian Revolution of 1905,
347–50
Costa, João Carlos da, 324
Cotte, Edouard, 305
Cowdray, Lord see Mexican
Eagle Oil Co.
Cox, Hiram, 24–5
Cox, Percy, 362–3, 465, 466, 540
Crawfurd, John, 25–7
Crédit Lyonnais, 213
Cree Indians, 52, 57
Creek Indians, 336
Creel, Bushrod, 68
Cremer, Jacob T., 260–1
Crespí, Juan, 53–4
Crewe, Lord, 463, (480), (497),
515, 518
Crimean War, 200–1, 234, 524
Crowe, Eyre, 498, 510, 515–16,
519, 533
Crusades, 8–9, 10
Cuba
pre-industrial oil, 40–1, 48
exploration in, 293
in Spanish-American War, 279,
342, 384

Cuba – *cont'd.*
US naval base in, 441
Cuban Petroleum Co., 293
Cudahy, Michael/Oil Co., 336, 392
Cuéllar, Manuel, 291, 292–3
Cuéllar y Cía, 292–3
Cueto, Manuel, 293
Cullinan, Joseph S., 269–70
Cummins, Albert B., 396
Curzon, George, 249–50, 288,
    300–2, 307, 308, (316), (317),
    (363), 368, 372–3, 459–60,
    544, 545

D'Arcy, William Knox *see* Anglo-
    Persian Oil Co.
Daimler, Gottlieb, 273–3
Dampier, William, 44–5
Daniels, Josephus, 383, 493–4
Dardanelles Strait
    oil transit through, 212, (254–5),
    438, 522–4, 527–8, 539
    see also Ottoman (Turkish)
    empire; Russia
Darwent, Walter, 188
Dawes Act / Dawes, Henry L.,
    334–5
De Dion-Bouton, 272, 274, (Fig. 16)
Dead Sea, bitumen from, 2, 3, 7, 12
Debs, Eugene, 398, 401
Delaware Indians, 53, 79, 82–9, 91
Denby, Charles, 277
Deterding, Henri *see* Royal
    Dutch Co.
Deutsche Bank, 213, 286, 317, 369,
    370, 374–9, 412, 418–19, 441,
    467, 469, 470, 472, 505,
    506–7, 511–14, (521), 532
    Gwinner, Arthur von, 286, 379,
    418, (419), 513, 521

    see also Ottoman (Turkish)
    empire: Baghdad Railway;
    Steaua Română
Deutsch-Amerikanische Petroleum-
    Gesellschaft (DAPG) *see*
    Standard Oil Co.
Díaz, Porfirio, 293, 294, 380, 381–3
Dickinson, John, 92
diesel engine/fuel *see* internal
    combustion engine
Diesel, Rudolph, 332, 447
Dilke, Charles, 237, 241
Dillon, John, 501
Dinwiddie, Robert, 80, 82–3
Dio, Cassius, 5
Dion, Marquis de, Fig. 16
Dioscorides, (12), 32
Disconto-Gesellschaft, 412
Disraeli, Benjamin, 228, 232, 233–4
distillation, of oil
    earliest, 9
    pre-industrial, 30, 35, 37–8, 40,
    (63), 66, 69–70, 75–6
    see also oil: refining
Djugashvili, Joseph *see* Stalin
Dodd, John, 192
Dodd, Samuel C.T. *see* Standard
    Oil Co.
Doheny, Edward L., 268, 294,
    380, 381–2
Dordtsche Petroleum Co., 231
Downer, Samuel, 115, (266)
Drew, George S., 184
Drake, Edwin L., 77–8, 79, 82, 85,
    106, 107, 108–9, (113), Fig. 3
Duan Chengshi, 19
Dubinin brothers, 199
Duckett, Geoffrey, 11
Dumas, Philip *see* navy, British
Dunmore, Lord, 89

Dunsterville, Lionel, 543–4
Dutch East Indies (Indonesia)
    Borneo, Dutch (Kalimantan): and
        British navy, oil for, 265, 271,
        311, 319; pollution, 261–2,
        330; 'Shell' in, 261–2, 265,
        271, 285, 319, 330
    Dutch East India Company, 21
    early exploration, 193
    Jambi/Jambi War, 259, 264–5
    Java: early exploration, 193;
        Java War (1826–30), 29;
        pre-industrial oil, 29–30;
        production, 230–1
    protectionism, 259–62, 265, 287,
        410, (414)
    Sumatra: Aceh oil, 20–1, 227–30,
        262–4; Aceh War, 228–30,
        (258–9), 262–4; early
        exploration, 227–8, 229–30;
        Perlak, 21, 228–9, 262, 264;
        pre-industrial oil, 20–2, 31,
        227, 259; production, 230,
        258–9, 264, 330

East India Company, 12, 14, 22, 25,
    161, 196, 232, 298, 300, 370
East Timor see Timor, Portuguese
    (East Timor/Timor-Leste)
Ecuador, earliest reports of oil, 41,
    43, 45
Eden, Anthony, 236
Edward VII, King, (449), (519)
Eele, Martin, 35–6
Egerton, Walter, 323
Egypt
    British invasion of (1882),
        234–42, 297, 369
    early use of petroleum in,
        2, 8–10

exploration, 297
oil for British navy, 433,
    (478–80), (482), (494)
protectionism, 414, 478–80, 482
Royal Dutch-Shell in, 417, 476,
    478–80, 482, 494
see also Suez: Canal
Eisenhower, Dwight, 236
Elgin, Lord, 322
Elibank, Lord Murray of, 507
Elizabeth I, Queen, 43, 161
Elizabeth Watts, 117
Emery, Lewis, Jr, 152, 153–4,
    168, 178, (179)
Entente Cordiale (Anglo-French
    alliance), 517, 521, 523,
    (525), 526, 537
Erédia, Manuel Godinho de, 21
Erie Canal, 74, 78, 95, 102–5, 106,
    123, 152
Erie Indians, 51
Esher, Lord Reginald, (449), 520
Esso, 404
European Petroleum Co., 253, 353
European Petroleum Union
    (EPU), 412
Evans, Lewis, 52, 57, 84
Eveleth, J.G., 75, 76
Everwijn, R., 31
Eyrinis, Eyrini d', 35
Excelsior Works, 123
Exxon/ExxonMobil, 404

Fairbank, John Henry, 117
Falimierz, Stefan, 37
Falle, Bertram, 503, 538
Farman, Elbert, 238
fascism, 408
Feigl, Arnold, 351–2
Fennema, Reinder, 227, 229

Ferdinand, Archduke Franz, 527–8,
530, 533
Fergana/Ferghana (Uzbekistan),
bitumen/oil in, 7, 8, 295
Ferris, A.C., 77
Fertig, John F. *see* Standard Oil Co.
Fidler, Peter, 57
Finlay Fleming & Co., 224
fires
oil spill, 61, 65, 69, (118), 130–5,
153–4, 336, 381, 420–1
oil storage/refinery, (118), 132–5,
283, (330), (332), 352, Fig. 5
oil tanker, 209, 217, 267
oil/gas well, 28, (60), (61), (67),
110, 114, (118), 130–1, 252,
336, 350, 352, 355, 381, Fig. 12
First Exploitation Co. *see* Anglo-
Persian Oil Co.
Fisher, John *see* navy, British
Flagler, Henry *see* Standard Oil Co.
Flannery, Fortescue, 243
Forbes, John, 84–5
Formosa (Taiwan), oil in, 192,
280, 295
forts
Detroit, 87, 93
Duquesne, 83, 85, (87), (Fig. 1)
Franklin (prev. Venango), (82),
(85), 97, 100, (Fig. 1)
Le Boeuf, 82, 83, 87, 94, Fig. 1
Machault, 82, (85), 87, (Fig. 1)
Niagara, 91, 94, (96), 99
Pitt (prev. Duquesne), (83), 87,
91, Fig. 1
Presque Isle, 82, 87, 94, Fig. 1
Stanwix, 88, 89, 94, 95, 96, 97
Venango (prev. Machault), 52,
(82), (85), 87, 97, Fig. 1
France

early petroleum in, 34–5
and Mediterranean Sea, security
of, 499, 523
protectionism, 295, 413
refining, 120, 295
shale oil, 34–5, 295, 413
Franco-Russian Alliance, 517,
525, 526, 529, 533–7
Franklin, Benjamin, 88
Franklin, John, 62
Frasch, Herman, 176
Frederick, Cesar, 10
French and Indian War (1754–63),
79–86, 88
French Creek, 52, 78, 82, 86, 91,
101, Fig. 1
Fryer, John, 12–13

Galicia *see* Austria-Hungary/Galicia
Galinée, René de Bréhant de, 51
Galey, John H., 269, 283
Garfield, James A., 121, 167–8, (235)
gas
gas engine *see* internal
combustion engine
natural gas, industrial oil era
reports/use of: California, 270;
Italy, 295; Ivory Coast, 322;
Ohio (for Detroit), 159;
Oklahoma/Indian Territory,
333; Papua, Territory of, 423;
Pennsylvania, 110, 111, 114,
130, 131, 136; Persia, 363;
Russia, 203–4, 205, 210–11;
Trinidad, 188; Britain, 295
natural gas, pre-industrial oil era
reports/use of: Burma, 26;
California, 71–2; China, 20,
27–8; Dutch East Indies, 29,
259; Japan, 25; Mesopotamia,

17; New York, 63–4; Ohio, 61, 62; Russia, 198; Texas, 269; West Virginia, 55, 60, 67, 84
town/coal gas *see* coal
Gas Light & Coke Co., 63
gasoline *see* petrol/gasoline
Gavotti, Giulio, 452
George V, King, 431, 536
Georgia, 195–202, 206, 212–13, 214–16, 344–5, 348–9, 353, 543–4
see also Russia: Batum
Gelder, Johannes Arnoldus de *see* Royal Dutch Co.
General Asphalt Co. of New Jersey, 338–9, 425
Genesis, Book of, 2
Germany
and Mesopotamia *see* Mesopotamia (Iraq)
navy *see* navy, German
oil imports, 120, 286, 331, 412, 417–19, 438, 441–2, 472, 523
oil monopoly plan, 417–19, 472, 478–9, 492–3
and oil in WWI, 540, 541–4
and Persia *see* Persia (Iran)
pre-industrial oil, 34
production, 220, 295, (469)
and Turkey *see* Ottoman (Turkish) empire: Baghdad Railway
Gerretson, Frederik, 259–60, 261–2
Gesner, Abraham, 38, 49, 70
Ghana *see* Gold Coast oil
Gil y Sáenz, Manuel, 184
Gilgamesh, 2
Gladden, Washington, 335
Gladstone

William, 228, 232–3, 236–7, 238, 240–1, 242
Henry Neville, 253–4, 265
Gold Coast (Ghana) oil, 297, 322
Golitsyn, Grigory, 344
Gómez, Juan Vicente, 341–2, 425, 426
Gordon, Arthur, 188
Granville, George, 236, (237–8), 239
Greek Fire, 5–9
Greenway, Charles *see* Anglo-Persian Oil Co.
Gresham, Walter, 389
Grey, Edward
and Anglo-Persian Oil Co., (358), (360), (377), (461–2), 468, (470), 471, 473–4, 475, (480), (483), 497, 503, 506, 515–16, 529, 530–1, 533
and Anglo-Russian Convention, 359–61, (475), 496–7, 503, 529–36, 537–8
and British navy fuel oil, 468, 471, 473–4, (483), 490, 495, 497, (503), 509, 529, 533, (535), 537–8
and Germany, 518, 520, 522, 525, 529, 535, 537
and Mexican Revolution, 383
and Ottoman Empire, (456), 458–9, 461–2, 467, 468, (470), 505–8, 509–10, 513, 515–16, 527, 529
and Turkish Petroleum Co./ Mesopotamian oil concessions, 467, 468, (470), 471, 506, 509–10, 513, 515–16
Groot (van Embden), Cornelis de, 29
Grosskopf, Paul, 371–2, 375–6
Guaraní, 291–2

Gubonin, P.I., 202, 204
Guest, Henry, 445
Guffey, James M., 283
    see also J.M. Guffey Petroleum Co.
Gulbenkian, Calouste, 371, 467,
    512–14
Gulf Oil, 284, 337, 416, 421
Guzmán, Blanco, 338
Gwinner, Arthur von see Deutsche
    Bank

Haenlein, Paul, 327
Hagelin, Karl, 355–6
Hakki Pasha, Ibrahim, 509, 513
Haldane, Richard, 518
    mission, 520–1
Hale, John, 66–7
Halifax, Earl of, 80
Hamburg-Amerika Line, 311,
    317, 377, 467, (520)
Hamilton, Alexander, 163, 277, 413
Hamilton, Horatio, 338
Hanbury, John, 80, 83
Handsome Lake (Kanyotaiyo') see
    Seneca Indians
Hankey, Maurice, 544–5
Hanna, Mark/Marcus A., 133, 401
Hanson, Thomas, 55
Hanway, Jonas, 13–14, 194
Hardinge, Arthur, 305, 306, 308
Hardinge, Charles, 353–4, (378),
    380, 456, 428, 462, 463, 518
Harlan, John, 174–5, 405
Harley, Henry, 126, 143
Harmar, Josiah, 97–8
Harmsworth, Alfred see Northcliffe,
    Lord
Harrach, Franz von, 527
Harrison, William Henry, 101
Hart, William, 63–4

Hawaii
    and Royal Dutch-Shell, 416
    US control over, 278, 281, 441
Hechler, Federico, 47
Henry, James D., 207, 417
Henry, Patrick, 89
Hernández, Manuel (Mocho), 341
Herodotus, 2, 5, (14), (16), 32
Hibbard Farm see Oil Region (US)
Hildreth, S.P., 61–2
Hillyer, C.J., 333
Hirte, Tobias, 59
Hirtzel, Arthur, 498, 510
Hit see Mesopotamia (Iraq)
Hoch, Edward W., 391–2
Hock, Julius, 273
Holland, 228, 422, 476, 495,
    522, 540
    see also Dutch East Indies
Holland, Henry, 34
Holland, John P., 327
Holland Land Company, 98, 99,
    100, 102
Hopwood, Francis see navy, British
Hornsby & Sons, 451
Hostetter, David, 147–8
House, Edward M., 399
Hoyt, Rollin, 187
Huasteca Petroleum Co., 381–2
Hubbard, Evelyn, 254
Hudson's Bay Company, 51, 57
Huerta, Victoriano, 383, 386, 388
Humboldt, Alexander von, 41
Hunt, Thomas, 72

Ibn al-Athir, 8
Ibn Hawqal, 8
Ijebu, 322
Imbert, Laurent Marie Joseph, 27–8
Imperial Oil Co. see Standard Oil Co.

Incahuasi Petroleum Syndicate, 292–3
India
   Assam: Assam Oil Co., 295;
      Assam Railways and Trading
      Co., 226, 295; oil exploration/
      production, industrial, 226,
      295; oil production/use/
      reports, pre-industrial, 28–9,
      192, 224
   Burma (Myanmar): Anglo-
      Burmese War, First (1824–6),
      25, 226; Anglo-Burmese War,
      Second (1852), 30, 191, 222;
      Anglo-Burmese War, Third
      (1885), 222–6; and British navy,
      oil for, 313, 314, 432, 438, 470,
      539; oil, pre-industrial, 22–7,
      30–1, 191–2; oil production,
      industrial, 287–8, 295, 315,
      470; and Standard Oil, 287–8,
      317, 318, 411
   Burmah Oil Co.: 222–6, 287–8,
      305, 356–7, 359, 362–6, 466,
      (468), 470, 473; and British
      navy, 313–20, 410–11, 470;
      Cargill, David, 222–4, (470–1);
      and First Exploitation Co.,
      rescue of, 315–16; Greenway,
      Charles, 317, 475–6, 477, see
      also Anglo-Persian Oil Co.;
      Wallace, Charles W., 362, see
      also Anglo-Persian Oil Co.;
      Yenangyaung-Rangoon
      pipeline, 226, 319–20
   Indian Mutiny, 234, 299, 367
   oil imports, (191), 217, 224, 226,
      242, 244, 246, 287–8; tariffs,
      287, (315), 413, 480
   protectionism, 287–8, 314,
      316–20, 410, 412, 414, 480
   routes to, strategic importance of,
      198, 201, 232–3, 249, 299,
      306–7, 315, 356, 359, 367–8,
      456–63, 482, 505, 529–30,
      531–2, 535, 537
Indian Territory (US), 289,
   333–7, (392)
Indonesia see Dutch East Indies
Inter-Allied Petroleum Conference
   (1918), 545
internal combustion engine
   in aeroplanes see aeroplanes
   alcohol fuel for, 273, 330–2, 441
   diesel, 119, 332, 431, 447–50
   invention of: petrol/gasoline,
      271–4; diesel, 332
   kerosene fuel for, 330, 332
   in marine vessels, 119, 325, 431,
      447–50; see also submarines/
      U-boats
   in motor vehicles see motor
      vehicles
   petrol/gasoline fuel for see petrol/
      gasoline
   for stationary power, 272–3, 325,
      332, 447
Iran see Persia (Iran)
Iraq see Mesopotamia (Iraq)
Iroquois (Hodenosaunee), 53, 82,
   83, 85, 88–9, 90–2, 95–6,
   99–101
Ismail, Khedive, 233, 234
Istakhri, Abu Is'haq Farsi, 8
Italian-Turkish War, 452, 528
Italy
   bitumen: Neolithic use, 32;
      19th-century use, 190
   early exploration, 190
   navy, adoption of fuel oil, 310
   pre-industrial oil, 32–4, (61)

Italy – *cont'd.*
  production, 295
Ivory Coast, oil in, 322
Izvolsky, Alexander, (525), (526)

J.M. Guffey Petroleum Co., 284, 308
J.P. Morgan, 286, 388–9, 390,
  (397–8), (406), 407–8
Jackson, Andrew, 105–6, (165)
Jagow, Gottlieb von, 512–13
Japan
  early reports/use of petroleum:
    earliest, 19, 25; 19th century,
    192
  oil imports, 138, 242, 244, 286
  production, 280, 286, 295
Jardine, Matheson & Co., 243
Jarrott, Charles, 454
Jaurès, Jean, 533–4
Java *see* Dutch East Indies
Jefferson, Thomas, 56–7, 89, 90, 93,
  94, 97, 98, 101, (103), 163, 164
Johnson, Andrew, 140
Joinville, Jean de, 8–9
Joncaire, Louis Thomas de, 52
Jones, William E. 'Grumble', 118
Juana of Castile, Queen, 43
Junghuhn, Franz Wilhelm, 193

Kalimantan *see* Dutch East Indies
Kämpfer, Engelbert, 11–2
Kansas
  Indians, 335
  Kansas Oil Producers'
    Association, 391
  refining, 283, 336, 390–2
  production, 289, 333, 335–6,
    (337), 390–2, 408, 420
Kellogg, Frank B., 394–5, 396, 402
Kelsey, Henry, 52

Kennard, Thomas, 116
Kentucky
  coal oil, 70–1, 106–7
  pre-industrial oil, 61, 65
  production, 117, (121), 176, 178
kerosene *see* internal combustion
    engine; lamp/illuminating oil
Kerosene Co. Ltd *see* Rothschild
Kerosene Oil Co., 70
Kessler, Jean Baptist August *see*
    Royal Dutch Co.
Khalafi Co., 204–5
Khanykov, A.P., 215
Kier, Samuel, 69–70, 74, 75,
    77, 107, 109
Kingsley, Charles, 188–9
Kinneir, John, 14–15, 196
Kirkuk *see* Mesopotamia (Iraq)
Kislyansky, Leonty, 12
Kitabgi, Antoine, 304–5
Kitchener, Lord, 478
Knight, James, 51–2
Kochubey, Viktor, 199
Kokorev, V.A., 202, 204
Kokovtsov, Vladimir, 354
Kühlmann, Richard von, (507), 509
Kuwait
  British interests in, 373, 377,
    463, (465), 503
  oil indications, pre-WWI,
    503, 514
  petroleum, earliest use, 1
  Sheikh Mubarak of, 373,
    463, (503)

La Follette, Robert, 397, 405
Lama, Diego de, 183
lamp/illuminating oil
  camphene, 68, 71, 116
  carbon oil, 69–70, 106, 109

kerosene/paraffin, 68–71, 106–7,
113, (115), 116, 117, 119–24,
141–4, 150, 179–80, 181, 183,
187, 189, 191, 193–4, 202–3,
206–20, (226), (229), 231,
242–9, 250–1, 254–5, 258,
(261), 265, 277–8, 280–3,
285–8, 292, 295, 296, 304,
306, 313, 315, 317, 318, 330,
332, 345–6, 355, 371, 385, 413,
419, 466, 479, 524, Fig. 13
pre-industrial, 1, 3, 9, 10, 11, 12,
14, 15, 21, 24, 26, 29–30, 32,
33–4, 35, 37–8, 46, 48, 51, 53,
61, 62, 64, 68–71, 72, 73,
74–5, 77, 181, (184), 189, 191,
196, 198, (227)
spermaceti (whale oil), 64, 68,
116, 119
Lamsdorf, Vladimir, 353
Lane, Fred, (217), 219, 242–3, 254,
288–9, 330, 379–80, 437–8,
449–50
Lane & Macandrew, 217, 219
Langley, Samuel P., 327–8
Lansdowne, Henry, 305, (306),
(307), (308), (340), (348),
(350), (352), (353), (355), 373,
375, (518)
Lansdowne Declaration, 307
see also India, routes to
Lascelles, Frank, 235, 303, (304)
Lawrence, Thomas E./of Arabia, 508
Lawson, Wilfrid, 240–1
Layard, Austen, 16–17, 300
Lazarev, M.I., 215
Le Jeune, Paul, 51
Lebrón, Jerónimo, 42
Lee, Arthur, 96
Lee, Richard Henry, (80), 89

Lee, Robert E., 118
Lee Lum, John, 290
Lenin, V.I. 343, 345, 356
Lesseps, Ferdinand de, 232–3
Lett, Lewis, 423
Lewis, Andrew, 56, 84, 89
Lianozov, G.M., 208, 215
Libby, William Herbert see Standard
Oil Co.
Libya
and Italian-Turkish War, 452,
(528)
bombing of (2011), 236
Lichnowsky, Karl, 525, 534, 535
Liebert, F.C.H., 29
Liman von Sanders, Otto/mission,
525–7, 534
Lincoln, Abraham, 140
Lincoln, Benjamin, 55
Linschoten, Jan Huygen van, 21
List, Friedrich, 277, 369, 413
Livingstone, David, 18
Loftus, William, 298, 370
Lloyd, George, 498–9
Lloyd, Henry Demarest, 159, 169
Lloyd George, David, 485,
537, (545)
London and Pacific Petroleum
Co., 268
London General Omnibus Co., 451
Loomis, George, 268
Lorimer, David, 358–9, (363), 364
Lowther, Gerard, 378–9, 380,
461–2, (467–8), (506), 507,
508, 509
Ludendorff, Erich, 542–3
Łukasiewicz, Ignacy, 38, 189
Luzerne, Chevalier de la, 90
Lynch Bros., 360, 361–2, 366,
458, 467

Lynch, H. Blosse, 15
Lynch, Henry, 360
Lyon, George, 47

M. Samuel & Co. *see* Samuel:
    M. Samuel & Co.
MacDonald, Ramsay, 491–2
Mackenzie, Alexander, 57
MacKeown & Finley, 109
Madero, Franciso, 383
Madison, James, 96, 101
Magnolia Petroleum Co., 421
Mahan, Alfred T., 276, 385, 517,
    518, 519
Maier, P.J., 29
Malacca Straits, 20, 21, 228,
    230, 330
Malet, Edward, (236), 237–8,
    239, 241–2
Mallet, Louis, 456, 463, 471,
    473–5, 509–10, 512,
    (513), 516–17
Mantashev, Alexander, 246
Marathon Oil Corp., 404
Marling, Charles, 468, (469)
Marriott, Alfred *see* Anglo-Persian
    Oil Co.
Marvin, Charles, 211–12, 213,
    224, 226
Marx, Karl, 200, (345)
Masjid i Suleiman (Masjed
    Soleyman), 16, 298, 356–9,
    361–3, 364, 461–2, 464;
    exploration, 356–7, 361–3
    oil strike, 363–4
Matos, Manuel Antonio, 340
Maunsell, F.R., 305, 371, 375
Maximilian, Ferdinand, 184, 186
Maybach, Wilhelm, 273–4
Mayo, Henry, 386

Mazut Co., see Rothschild
McGarvey, William Henry, 220,
    436, 481
McGee, James, *see* Standard
    Oil Co.
McHenry, James, 115
McKee, H.E., 334
McKinley, William, 328, 400,
    401, 414
Mediterranean Sea, security of, 499,
    500–1, 523–4, 526
Meiggs, Henry, 183–4
Mellon, William L., 178–9, (284)
Mendeleev, Dmitri, 204, 206–7
Menten, Jacobus H., 261
Merrimac Oil Co., 50
Mesopotamia (Iraq)
    and British navy, oil for, 509–15
    British occupation of, post-WWI,
        544–5
    British-Russian rivalry over, 368,
        375, 461–2
    early use of petroleum: earliest,
        1–4, 5, 10, 14, 16, 369–70;
        19th century, 15–17, 370–1;
        early 20th century, 372, 508
    Euphrates River, 2–3, 5, 10, 15,
        16, 300, 368, 370, 372, 374
    exploration, (298), 370–2,
        375–6, 377
    and Germany, 372–8, 455–62,
        467–8, 504–7, (518–19),
        532–3; *see also* Ottoman
        (Turkish) empire: Baghdad
        Railway
    Hit, 1, 2, 10, 14, 15, 298, 374,
        Fig. 26
    invasion of (2003), 236
    Kirkuk, 15, 298, 305, 370,
        371–2, 379, Fig. 2

Mosul, 1, 15, 16, 305, 370–1,
372, 373, 379, 461, 508, 515,
545, Fig. 26
Nimrud, 16, 515, Fig. 26
oil, interest in: British, (370),
(371), 373–80, 467–9, 485,
486, 504–16, 532–3; Dutch,
379–80, 417, 467–9, 476,
506–7, 509–14, 532; French,
(379), (507); German, 361,
369, 371–9, 467–73, 486,
504–15, 532–3; US, 371,
379, (402)
Tigris River, 1–2, 3, 10, 15, 16,
300, 307, 368, 371, 372, 373,
374, 375, 376, 457–8, 459,
467, 516
Tigris valley railway, 457–8, 459
and WWI, oil in, 544–5
see also Ottoman (Turkish)
empire
Mexican Eagle Oil Co. ('El Aguila'):
381–4, 386, 387, 414, 417,
486, 495, (507)
Pearson, Weetman (Lord
Cowdray), 294, 380–4, 386–7,
417, 486, 492, 495, 507
Mexican Petroleum Co., 294–5, 380
Mexico
and British navy, oil for, 383–4,
386, 387, 433, 443–4, 486,
(488), 493, (495), 539, 542
exploration, early, 184–7
Mexican-American War,
(72), 181–2
Mexican Eagle see Mexican Eagle
Oil Co.
Mexican Revolution, 382–8,
443–4, 452, 486, 488, 495
oil imports, 187, 380–1

pipelines, 381–2, (437), 542
pollution, 381, 382
pre-industrial oil, 40, 43, 44–5,
47–8, (184), 294
production, 294–5, 381–2, 387,
(411), (417), 420, 433, (436),
(437), 495, 507, (539)
Royal Dutch-Shell in, 417
and WWI, oil in, 539, 542
Miami Indians, 86
Mikhailovich, Aleksandr, 256
Mindon, King, 222
Mingo Indians see Seneca Indians
Mirzoyev, Ivan, 202, 204, 208
Mithridates VI, 5
Mobil, 404
Moeara Enim Petroleum Co.,
259–61, 265, 410
Mohammad Ali Shah, 359
Mohammerah, Sheikh of, 367,
465, 497, 533
Mole, Lancelot de, 452
Monroe, James, 96, (97), 102
Monroe Doctrine, 186, 278, 340–1,
384–7, 486, 495
Roosevelt Corollary to, 340–1,
384
Monteiro, Joachim, 191
Moore, Henry, 94
Morocco Crisis
First (1905–6), 367, 376, 518
Second/Agadir (1911), 462, 528
Morris, Robert, 99–100
Morris, Robert H., 83–4
Mossadegh, Mohammed, 236
Mosul see Mesopotamia (Iraq)
Motilon see Barí tribe
motor vehicles
armoured cars/tanks, 271–2, 327,
450–2, (454), 539, Fig. 16

motor vehicles – *cont'd.*
    electric, 272, 273, 274, 325–7,
        Fig. 15
    manufacturers: Baker Motor
        Vehicle Co., 326; Benz, 274;
        Cadillac, 258; Daimler, 274; De
        Dion-Bouton, 272, 274; Fiat,
        452; Ford, 326, 332; Gräf &
        Stift, 527; Hotchkiss, 451;
        Isotta-Frachini, 452; Olds, 326;
        Panhard & Levassor, 274, 451;
        Peugeot, 274; Thornycroft, 326
    steam-powered, 272, 273, 274,
        325–6, 327, 450, Fig. 15
    transport: civilian, 272–5, 325–7,
        329–32, 337, 450, 454–5;
        military, 327, 331, 451,
        454–5, 539
Moutoussis, Michael, 452
Mozaffar al-Din Shah, 305–6, 359
mummification, 1
Murdoch, William, 63
Murphy, John McLeod, 184–5, 186
Murray, John Hubert, 424
Muscovy Company, 11, 13, 194
Myanmar *see* India: Burma

Naser al-Din Shah, 300, 303, 304
National Bank of Turkey, 467–8,
    469, 470, 506, 510, 512, 514
National City Bank, 286, 388,
    406, 415
National Civic Federation,
    401–2, 405
navy, Australian
    fuel oil, use of, (320–1), (425),
        434–5
    ships: *Parramatta*, 434; *Yarra*, 434
navy, Austrian, fuel oil, use of,
    311, 437

navy, British (Royal Navy)
    and Anglo-Persian Oil Co., (307),
        (308), (315–16), 468–75,
        480–504, 529–30, 533, 535, 538
    Beresford, Charles, 439, 443,
        445, 446
    Black, Frederick, 437, 438–9, 484
    and Burmah Oil Co., 313–20,
        410–11, 432, 438, 470, (476),
        480, (539)
    Dumas, Philip, 472, 475
    Fisher, John, 307, 310, 311, 327,
        430–2, 433, 439, 444, 445,
        448–9, 479, 480, 517–18, 519,
        (520), 536–7
    fuel oil, adoption of, (285),
        309–11, 362, 430–46
    fuel trials: asphalt, 50, 188; diesel,
        448–50; fuel oil, 265, 266,
        270–1, 310–11
    Hopwood, Francis, 433, 478,
        481–2, 512, 522
    and Mesopotamian oil
        concessions, 468–73, 485,
        509–15
    and Mexican Eagle Oil Co., 383,
        386, (387), 433, 486, (488),
        (493), 495, (539), (542)
    oil reserve, 431–2, 434,
        439–40, 484
    oil supplies, pre-WWI: general,
        (215), 271, 311–20, 410–11,
        430–46; Borneo, 265, 271, 311,
        (312), (319), 432, 433, (434),
        (476); Burma, 313–15, 316–19,
        410, 432, 438, 470; Egypt, 433,
        (478–80), (482), (494);
        Mesopotamia (potential),
        510–15; Mexico, (383), (386),
        (387), 433, (443–4), (486),

(488), (493); Nigeria (potential), 323–4, 434, 481–2; Persia, (315–16), (366), 432, (480–504), *see also* Anglo-Persian Oil Co.; Romania, (312), 432, (436), 438, 443, (476), 522–3, 527, 537–8; Scottish shale, (313), 432, (435), 489; Trinidad, (50), (188), (290), (314), 433, (434), (476–8), (481), (482), (494); US, 312, (313), 432, 433, 436, 438, 495; Venezuela (potential), 436

oil supplies, WWI, 539–40, 541–2

oil supply crisis: pre-WWI, 431–40, 443–6, 482–4, 508–9, 522–3, 529, 537; WWI, 541–2

Pakenham, William/committee, 431, 434, 439, 440

Pretyman, Ernest, 314, 315–16

Royal Commission on Fuel and Engines, 431–50, 455, 472–3, 475–6, 478–9, 482–3, 484, 522, (536)

and Royal Dutch-Shell, 471–80, 481, 482, 489–90, 494–5, 510, 512, 513, 522, 523, 536–7, 540

and 'Shell', 265, 270–1, 311, 494–5, 523, 540

ships: *Barham*, 430; *Bedford*, 311; *Bonaventure*, 311; *Dreadnought*, 311; *Gannet*, 188; *Hannibal*, 311; *Hardy*, 448; *Malaya*, 430; *Mars*, 311; *Queen Elizabeth*, 430, (423), 439, (446); *Spiteful*, 311; *Surly*, 310; *Valiant*, 430; *Warspite*, 430

Slade, Edmond, 485, 486, 488, (490), (497), 544; commission, 485–6, (490), 503

and Standard Oil, 433

submarines, 327, (433), (455)

tankers, 320, 433, (437), 438, 439, 440, (445), 448, (479), (508), (522), (523), (536); *Isla*, 433; *Kharki*, 433; *Petroleum*, 433, 438; *Trefoil*, 448

navy, German

diesel trials, 448–50

fuel oil, adoption of, 270, 286–7, 310–11, 438, 441, 469–70

oil supplies, 311, 317, (321), (412), 418–19, 441, (442), (469), 512–13, 523, (542–3)

ships: *Derfflinger*, 448; *Derfflinger* class, 441; *Kurfürst Friedrich Wilhelm* class, 270; *Prinzregent Luitpold*, 448; *Siegfried* class, 270

tankers, 433

navy, Italian, fuel oil, adoption of, 270, (310)

navy, Romanian, fuel oil, adoption of, 311

navy, Russian

fuel oil, adoption of, 266–7, 270, 311, 320

oil supplies, 442

ships: *Potemkin*, 356; *Rostislav*, 311

navy, United States

fuel oil, adoption of, 320, 440–1

oil supplies, 312, 440–1, 493

oil trials, early, 266, 270, 311

ships: *Nevada*, 441; *Palos*, 266

Neolin, 86

Netherlands *see* Dutch East Indies; Holland

New Guinea

British *see* Papua, Territory of

New Guinea – *cont'd.*
German, 321; oil in, 442
New Trinidad Lake Asphalt Co., 339
New York & Bermudez Co., 338–42
New York state
pre-industrial petroleum, 51,
52, 58, 63–4, 66, (70)
see also Oil Region (US)
New Zealand, oil production, 296,
417, (434), (476), (477)
Nicholas II, Tsar, 345, 347, (351),
(359), 524, (525)
Nichols, H.E. *see* Anglo-Persian
Oil Co.
Nicolson, Arthur, 360–1, (380), 459,
462, 467–8, 469, 475, 505–6,
527, 530, 533, 536
Nigeria
and British navy, potential oil for,
323, 324, 411, 434, 481–2
exploration, 321–4, 411, 481–2
Oil Rivers Protectorate, 321–2
pollution, 323
protectionism, 321, 322–3,
411, 481
Nigerian Bitumen Corp., 321,
322–4, 481–2
Nimrud *see* Mesopotamia (Iraq)
nitroglycerin/torpedoes, 131–2, 336
Noah's Ark, use of bitumen for, 2
Nobel
Alfred, (207), 213, 216
Ludwig, 207
Robert, 207
Nobel Brothers Petroleum Co.,
207–10, 211, 212–15, 216–17,
219, 246, 247–8, 250, 345,
(355–6), 412, 421, 447;
tankers, 209–10, (212),
216–17, (219), 447

North West Company, 57
Northcliffe, Lord (Alfred
Harmsworth), 328, 519, 536
Novosiltsev, Ardalion, 203

Ocampo, Sebastian de, 41
O'Conor, Nicholas, 373, 376, 377
O'Day, Daniel *see* Standard Oil Co.
offshore oil production, 268–9,
Fig. 14
Ogden, David A./Ogden Land Co.,
100, 102
Ogilvy Gillanders and Co., 253
Ohio
Cleveland: 61; 'Cleveland
Massacre', 143–6, (213);
pollution, 133; refining centre,
122–4, 143–6, 148, 155, 177;
refining cartel, 143–51
Ohio Company of Virginia,
80–3, 89
Ohio Country, 79–101; petroleum
in, 52–3, 55, 56–7, 84, 91–2,
Fig. 1
pre-industrial oil, 61–2, 68–9
production (Ohio/Lima-Indiana
field), 175–6, 177, (182), 267,
270, 282, 283, 420
oil (liquid petroleum)
artistic/ritual use, 1, 34, 40, (51),
64, 184, 294, (370)
fires *see* fires
fuel, industrial, 121, 133, 175–6,
(180), (185), (193), 206,
250–1, 267–70, (281), (284–5),
290, (294), (297), 308, 355,
433, (523), 524
fuel in ships (civilian): internal
combustion, (274), 332, 447–8;
steam, (49), 121, 176, (193),

206, 250–1, 265, 266, 270–1, 284–5, 308

fuel in ships (naval): internal combustion, (362), 431, 447–50; steam, (50), (188), 265, 266–7, 270–1, 285, 286–7, (307), 309–24, 346, 430–46, 467, 468, 469–504, 512–15, 521, 522–4, 527, 529, 532–3, 538–9

fuel in steam trains, 176, (224), 250–1, 266, 268, 269, 285, 290, 308, 376, 380, 463, 466

import tariffs, (117), 207, (257), 268, 287, 288, 295, (346), 413, (415), 418, 480

lamps, 3, (10), (12), 14, 15, 21, 24, 26, 29, 32, 33, 34, 37, 38, 46, 53, 62, 64, (68), 70, 119–20, 136, 181, 189, 207; *see also* lamp/illuminating oil

lubricants: pre-industrial, 1, 19, 30, 31, 34, 35, 37, 39, 68, 70, 75, 106, 195; industrial, (75), (76), 117, 118, 121, 124, 180, 269, 281, 285, 297

market: pre-industrial, 23, 24, 26–7, 30–1; industrial, 109, (111–12), 113, 114–15, 126–7, 128–9, 132, 135–8, 141–6, 149–51, 176–80, 204–7, 212–16, 219–20, 244–8, 250–1, 268, 282–3, 306, 320, 332, 337, 339, 344, 355, 390, 413, 483–5, 489–90, 494–5, 502, 508

market-sharing agreements, 246–8, (260–1), (262), 285, 288–9, 411–12, 511, 514

medicinal use, 1, 10, 14, 21, 24, 30, 32, 33, 34, 36, 37, 39, 43, 46, (51), 52, 53, 55, 59, 63, 64, 66, (68), 71, 75, 184, 291, 370

pollution *see* pollution

price-setting, 150–1, 179

production restraint: US, 115, 129, 146, 176–7, (390); Russia, (205), (213), 246

protectionism, 259–61, 265, 286–8, 314–19, 321–4, 339, 362–4, 408–15, 456, 468–9, 470–504, 506–16, 529, 532, 538; *see also* import tariffs *above*; *individual countries*

of Modena/St Catherine, 33, 34

refining, pre-industrial, 37, 38, 69–70, 72, (75–6), 77, 198; *see also* distillation; Oil Region (US); *individual countries*

Seneca Oil, 58–9, 61–2, 64, 66, 69, 100

solvent, 31, 69, 75, 76, 117, 273, 329

of St Catherine, 33, 34

of St Quirinus, 34

supply fears, 111, 144, 147, 271, 275, 285, 308–23, 430–46, 450, 454–5, 469–515, 520–1, 522–4, 529–30, 532, 533, 537–8

waterproofing, 1, 11, 15, 17, 24, 34, 35, 36–7, 40, 41, 43–5, 46, 47, 48, 49, 51, 54, 64, 185, 191, 227, 259

weapon of war, 5–7, 8–9, 10, 19, 20, 25, 228–9, 262, (540)

see also bitumen/asphalt

Oil Creek

Association *see* Oil Region (US)

Railroad *see* railways/railroads

Oil Creek – *cont'd.*
  see also Oil Region (US)
Oil Region (US)
  Allegheny Transportation Co.,
    143, 144
  American Transfer Co., 149
  Atlantic & Great Western
    Railroad see railways/railroads
  Columbia Conduit Co., 148–9,
    151, (152), (153)
  Cornplanter Oil Co., 114
  Cornplanter Township, 75, 109
  Empire Transportation Co., 143,
    147, 148, 149
  Equitable Petroleum Co., 152
  Franklin: pre-industrial, 52, 82,
    85, 97, 99, 100, (Fig. 1);
    industrial, 111, 118, 130, 137,
    142, 143; see also forts:
    Venango
  Hibbard Farm, 75–7, (107),
    113–14
  Lake Shore Railroad see railways/
    railroads
  Meadville Railroad see railways/
    railroads
  Oil City, 113–14, 119, (120),
    125, 128, 129–30, 137,
    142, 143, 178
  Oil Creek: pre-industrial: 59–60,
    62, 74–8, 85, 97, 106, (Fig. 1);
    industrial, 108–15, (121),
    (125), 127, 129–30, (135), 144,
    (182), Fig. 6
  Oil Creek Railroad see railways/
    railroads
  Oil Creek & Allegheny River
    Railroad see railways/railroads
  Oil Creek Association,
    115, 129

Pennsylvania Railroad see
  railways/railroads
Pennsylvania Transportation Co.,
  144
Petroleum Producers' Union, 146,
  (147), 152, 157, 168
Philadelphia & Erie Railroad see
  railways/railroads
Pithole Creek, 124–7, 131, 135
pollution: pre-industrial, 61, 65,
  66–7, 69; industrial, 110–14,
  125, 129–33, 150, 153–4,
  Figs. 4–5
production, 108–15, 124, 125–6,
  128–9, 141, (144), 146, 150,
  175–6, (177), 282
refineries, 109, 112, 113, 115,
  (123), 124, 129, (143–51),
  177–8
Sunbury & Erie Railroad see
  railways/railroads
Tidewater Pipe Line Co., 152–6
Titusville: pre-industrial, 74–5,
  77–8, (79), 82, 106; industrial,
  109, 110–11, 113, 114, 115,
  118–19, (121–2), 125, 126,
  (127), 132, (135), (136), 137,
  (146), (147), (155), (177),
  178, (192)
Titusville Pipe Co., 143
transportation, 112–19, 122–7,
  142–56
Union Pipe Line Co., 147
United Pipe Line Co., 148–50,
  152, 157
Venango (Franklin), 57, 91, (99),
  (111), (118), (142), (143); see
  also forts: Venango
Waterford, pre-industrial, 78, 82,
  99, 101

Oil Rivers Protectorate *see* Nigeria
Oklahoma, 289, 333–7, (381), 408,
    416, 420, 494
Olearius, Adam, 11
OLEX, 412
Olmec, use of bitumen, 40
Olney, Richard, 340
Oneida Indians, 95–6
Organization of the Petroleum
    Exporting Countries
    (OPEC), 179
Osage Indians, 336, 494
Osmaston, John, 210–11
Ostrovskii, Mikhail, 214, 215
Ottawa Indians, 86, 87
Ottoman (Turkish) empire
    Anatolian Railway Co., 369,
        372, 376, 379, 460
    Baghdad Railway, 361, 369,
        371–8, 379, 457–63, 467, 469,
        489, 505–6, 509, 514–15,
        (521), 532, Fig. 26;
        Convention (1903), 374–5,
        506, 532
    and Britain, 198, 200–1, 367–9,
        455–63, 467–8, 504–17
    British-German rivalry over, 361,
        367, 368–9, 455–63, 504–17,
        525–7, 528, 532
    in the Caucasus, 194, 195–202
    Dardanelles (Turkish/Gallipoli)
        Strait, 198, 200, 212, 313, 361,
        438, 522–6, 539
    and France, 455–6, 458, 459,
        461, 467, 504, 507, 516, 525
    and Germany, 361, 367, 368–9,
        371, 372–9, 455–63, 504–16,
        524–5, 526–7, 534
    oil imports, (180), 371, 376, 508
    Tigris valley railway, 457–8, 459

    in WWI, 524, 539, 540,
        541, 543–5
    see also Mesopotamia (Iraq);
        India, routes to
Ovid, 1
Oviedo, Gonzalo Fernández de, 40–3

Pacific Coast Oil Co., 267, 268, 286
Paine, Thomas, 89
Pakenham, William *see* navy, British
Palashkovskii, S., 212, 214, 216, 245
Palestine, 378, 544
    oil exploration in, 461, 508
Palmerston, Lord, 200, 201, 232
Panama Canal, (278–9), (293–4),
    (320), 384–7, 476
Papen, Franz von, 542
Papua, Territory of (prev. British
    New Guinea), 422–5; oil for
    Australian navy (potential),
    425, 434–5
Paria Petroleum Co., 188
Parker, Alwyn, 506–7, (509),
    510, (513)
Paskievich, I.F., 197
Pattie, James, 64–5
Pearson, Weetman *see* Mexican
    Eagle Oil Co.
Peláez, Manuel, 382, 387
Peloponnesian War, 5
Peninsula & Oriental Steam
    Navigation Co. (P&O), 232
Pennington, Edward J., 271
Pennsylvania
    pre-industrial oil, 52–3, 55,
        58–60, 62, 64, 66, 68–70,
        74–8, 91–2
    Railroad *see* railways/railroads
    Rock Oil Co., 75–7
    see also Oil Region (US)

Perkins, George, 397–8, 407–8
Perry, Oliver, 101
Persia (Iran)
 Anglo-Russian Convention *see*
  Anglo-Russian Convention
 and British navy, oil for *see*
  Anglo-Persian Oil Co.; navy,
  British
 British-Russian rivalry over,
  298–304, 305–7, 308, 315,
  356–68, 455–8, 462–3, 439,
  482–3, 488–9, 495–9, 503,
  504, 529–33, 537–8
 Constitutional Revolution
  (1905–11), 359–60, 364–6,
  463–4, 496–7, 500
 coup (1953), 236
 exploration, (298), (302), (305),
  307, 356–9, 361–3, 465
 and Germany, 361, (376),
  456–60, 461–2, 467,
  (482–3), 518
 Karun River, 2, 300, 302, 305,
  356, 359, 360, 361, 366, 367,
  462–3, 464, 465, 497
 Karun valley railway, 457
 Khanikin, 10, 374, 375, 461,
  462, Fig. 26
 Masjid i Suleiman, 16, 298, 357,
  362, 363, 364, 464
 oil imports, 302, 304
 Persian Bank Mining Rights
  Corp., 302, 304
 Persian Gulf, strategic importance
  of *see* India, routes to
 Persian Railways Syndicate, 457
 petroleum, early use of: earliest, 1,
  2–3, 5–6, 7–8, 9, 10–14; 19th
  century, 14–16, 298, 307
 pollution, 363, 367, 465–6

 production, (298), (304), (307),
  (364), (366), 367, 432, 465–6,
  (471), (483), 485, (539), 541
 protectionism (by British), 315–16,
  (362–3), (363–4), 411, 456–7,
  468–9, 480–507, 514, 529, 538
 trans-Persian export pipeline
  (proposed), 245, 306–7,
  308, 315
 in WWI, 540–1, (543), 544
 see also Anglo-Persian Oil Co.
Peru
 pre-industrial oil, 45
 production, 183–4, 268, 289,
  290, 295, 426
 and Standard Oil, 426
Peter the Great, Tsar, 13, 39, 194
Peto, Basil, 455
Peto, Samuel Morton, 122, 126,
 131, 136
Peto & Betts, 115, 117
petrol/gasoline
 anaesthetic, 329
 Austria-Hungary/Galicia, 419
 Borneo, 262, 285, 330, 332, (450)
 engine fuel, 272, 273–5, 285, 325,
  327–32, 450–5
 heating fuel, 329
 market, (275), 285–6, 288,
  328–33, 450, 454–5
 Romania, 332, 419
 Russia/Grozny, 332, 419
 solvent, 273, 329
 Sumatra, 285–6, 330, 332, 416,
  (419), (450)
 United States, 133, 285, 286, 329,
  330, 331–2, (335–6), 337, 416,
  419, 450, (454)
 see also internal combustion engine
Philippines

production, 296
US invasion of (1898), 278–9,
280, 296
Pickering, Timothy, 96
Pico, Andrés, 72
Pierce, Henry Clay, 383
see also Waters-Pierce & Co.
pipelines
Anglo-Persian Oil Co. see Persia
(Iran) below
bamboo, 20, 28, 125, (191–2), 296
Burma (Myanmar): Yenangyaung-
Rangoon pipeline, 226, 319–20
Congo, Belgian: Matad-
Léopoldville (Kinshasa) import
pipeline, 433
Dutch East Indies (Indonesia):
Langkat-Pangkalan
Brandan pipeline, 264;
Telaga Said-Pangkalan
Brandan pipeline, 230
first industrial, see US below
leaks, 125, 153–4, (191–2), (210),
267, 268, (296)
Panama pipeline, 320
Persia (Iran): Anglo-Persian Oil
Co. pipeline, (306), 307, 363,
366, 367, 463–5, 485, 489,
491, 540–1, Figs. 22, 23, 25;
Caspian-Persian Gulf/trans-
Persian pipeline plans, 245,
306–7, 308, 315
pre-industrial, 20, (33), 63–4,
(68), 125, (191–2)
Russia: Baku-Batum pipeline,
214–16, 245–6, (250), 282,
(543), Fig. 9
US: 'Benson's Folly', 151–6;
Columbia Conduit Co., 148,
149; companies, 142–56;

Crescent Pipe Line Co., 178–9;
first industrial, 125–7; Pacific
Coast Oil Co. pipelines, 267,
268; Producers' Protective
Association pipelines, 178;
Tidewater Pipe Line Co.,
152–6; Union Oil Co., 268;
Union Pipe Line Co., 147;
United States Pipe Line Co.,
178, 179; US Navy Oklahoma-
Gulf plan, 494; see also Oil
Region (US); Standard Oil Co.
wooden, 64, 125
Pliny the Elder, 3, 5, 32, (59)
Plutarch, 4
Poincaré, Raymond, 525, 533
Poland see Austria-Hungary/Galicia
pollution see fires; Oil Region (US);
individual countries; individual
US states
Polo, Marco, 10
Pomet, Pierre, 12, 33, 34–5
Pond, Peter, 57
pood (measure), definition of, 194
Porter, James M., 166
Porter, Peter B., 102
Porter, Samuel M., 391
Portolá, Gaspar de, 53–4
Pontiac (Indian chief), 87–8
Prairie Oil & Gas Co. see Standard
Oil Co.
Preece, J.R., 357, 358, (364)
Prendergast, Harry, 223
Prentice, A.E., 183
Pretyman, Ernest see navy, British
Price's Patent Candle Co., 30, 31
Priestman engine, 273
Princip, Gavrilo, 528
Proclamation of 1763, 88
Procopius, 5–6

public relations
  National Civic Federation, 401–2
  Standard Oil, 400
Pure Oil Co., 179, 219

Quebec Act (1774), 89
Quen, Jean de, 51

Ragozin, Viktor, 205, 215
railways/railroads
  Anatolian Railway Co. see
    Ottoman (Turkish) empire
  Assam Railways and Trading Co.,
    226, 295
  Atlantic & Great Western, 74,
    115, 118, 123, 127, 142, 143
  Buffalo, New York &
    Philadelphia, 152
  Baku-Batum/Trans-Caucasus,
    212–14, 216, 245–6, 250, 282,
    348, 349, (543), Fig. 9
  Baghdad see Ottoman (Turkish)
    empire
  Baltimore & Ohio, 74, 118, 144,
    145, 148, 149, 175
  Central Railroad of New
    Jersey, 178
  Emperor Ferdinand Northern, 38
  Erie, 74, 135, 142, 143
  Grazi-Tsaritsyn, 266
  Great Western (Canada), 73, 116
  Houston, East & West Texas, 269
  Karun valley see Persia (Iran)
  Lake Shore, 74, 123, 143
  Meadville, 74
  New York & Philadelphia, 152
  New York Central, 74, 77,
    123, 148
  Oil Creek, 118–19, 125, 126,
    137, 142

  Oil Creek & Allegheny River, 144
  Pacific, 71; see also Southern
    Pacific below
  Pennsylvania, 74, 123, 142,
    143, 144, 145, 147, 148–9,
    153, (182)
  Persian Railways Syndicate, see
    Persia (Iran)
  Philadelphia & Erie (prev.
    Sunbury & Erie), (62), (74),
    (91), 93–4, (112), 115, 142
  Sunbury & Erie, 74, 112
  Southern Pacific, 267; see also
    Pacific above
  Tehuantepec, 48, (184), 294
  Tigris valley see Ottoman
    (Turkish) empire
  Trans-Caucasus see Baku-Batum
    above
  Trans-Persian, 457, 533
  Trans-Siberian, 277, 344
  Transversalbahn, 189
  Vladikavkaz, 211, 245, 253, 350
Raleigh, Walter, 43–4
Rangoon Oil Co., 191, 222–3, 410
Rathbone, William, 68
Rawlinson, Henry, 15–16, 200
'Red Flag' Act, 274
Red Jacket see Seneca Indians
Redwood, Boverton, 285, 288, 305,
  314, 315, 364, 366, 476, 477
Reerink, Jan, 193, 227
Rees, Otto van, 227, 229
refining see oil: refining
Reuter, Paul Julius von/concession,
  300, 302, 304, (305)
Revelstoke, Lord, 467
Reynolds, George see Anglo-Persian
  Oil Co.
Rich, Claudius, 14, 370

Rockefeller, John D. and William *see* Standard Oil Co.
Rockefeller & Andrews, 123
Rockefeller & Co., 124
Rohrbach, Paul, 373–4, 460
Romans, ancient, use of oil by, 5–6, (16), 27
Romania
  and British navy, oil for *see* navy, British: oil supplies
  German investment in, 190, 286, 312, 355, 376, 412, 417, 441
  navy, adoption of fuel oil *see* navy, Romanian
  pipelines, 286, 437
  Ploesti, 38, 287, 437
  pre-industrial oil, 38, 190
  prehistoric use of bitumen, 31–2
  production, 38–9, 190, 221, (286), 295, 311, (312–13), (332), (376), 413, (419), 420, 422, (432), 436, (441), (472), (476), 522, (539), 542–3
  protectionism, 227–8, 324, 326
  Royal Dutch-Shell/Astra Română in *see* Royal Dutch-Shell
  Standard Oil in *see* Standard Oil Co.
  and WWI, oil in, (539), 542–3
Româno-Americană *see* Standard Oil Co.
Romero, Matías, 185–6
Ronaldshay, Earl of, 460, 501
Root, Elihu, 385
Rothschild
  Alphonse de, 247, 317
  Bnito (Caspian and Black Sea Petroleum Co.), 212–17, 219, 242–4, 246–9, (250), (254), 285, 288–9, (295), (315), (317),

(343), (345), 351, 412, (416), (417)
  and Burma, 223, 317
  Edmond de, 347
  Kerosene Co. Ltd., 219
  N.M. Rothschild & Sons, 105
  Mazut Co., 251, 254
  and Mesopotamia, 379, 507
  and Persia, 308, 315, 411
  and Royal Dutch-Shell, 412, 417, 425, (449), 477, 494, 523
  and Suez Canal, 233, 237, 242–3, 487
  Tank Storage and Carriage Co., 219
Rouse, Henry, 109–10
Royal Commission on Fuel and Engines *see* navy, British
Royal Dutch Co.
  and Aceh War, 227–30, (258–9), 262–3
  Deterding, Henri, 258, 259, 260, 330, 412, 415, 416, 425, 436, 479, 480, 513, (536–7)
  formation of, 229
  Gelder, Johannes Arnoldus de, 229
  and Jambi War, 264–5
  Kessler, Jean Baptist August, 230, 260, (262), 265
  and 'Shell': cooperation with, 262, 285–6, 288, 317, 412; merger with, 318–19, 332, 409, 412
  Standard Oil, challenge to, 258–61, 277, 280, 285–6
  in Sumatra, 229–30, 258–65, 285–6; *see also* Aceh War
  tankers, 258, (289)
  see also Royal Dutch-Shell

Royal Dutch-Shell
  and Aceh War *see* Royal Dutch
    Co.: in Sumatra
  Anglo-Egyptian Oil Co./Oilfields
    Ltd., 478, 479, 482
  Anglo-Saxon Petroleum Co.,
    512, 513
  Astra Română, 417, 523
  in Borneo, 332, 417, 476
  and British navy *see* 'Shell'
    Transport and Trading Co.
  Caribbean Petroleum Co., 425–9
  Colón Development Co., 425–6
  Deterding, Henri *see* Royal
    Dutch Co.
  in Egypt, 417, 476, 478–80, 482
  and Jambi War *see* Royal
    Dutch Co.
  Kessler, Jean Baptist August *see*
    Royal Dutch Co.
  merger/formation of (1907),
    318–19, 332, 409, 412
  and Mesopotamian oil
    concessions, 379–80, 417,
    467–9, 476, 506–7,
    509–14, 532
  in Mexico, 417
  monopolization by, 415–17, 454,
    470–80, 489–90, 491,
    494–5, 510
  in New Zealand, 417, 476
  in Romania *see* Astra Română
  Rothschilds' investment in, 417,
    425, (477), 494, 523
  Roxana Petroleum Co., 416, 421
  in Russia, 417
  in Sarawak, 417, 476, 494
  and Standard Oil, challenge to,
    (258–61), (277), (280),
    (285–6), 409, 411–12, 415–16

  in Sumatra, (229–30), (258–65),
    (285–6), 332, 466, 476; *see also*
    Aceh War
  tankers, (258), 447, 466, 479,
    523, 536–7; diesel-powered,
    447
  in Trinidad, 417, 476–8, 481, 494
  in Venezuela, (342), 417, 425–9
  Venezuelan Oil Concessions Ltd.,
    425–6, 429
  Waley Cohen, Robert, 466, 476
  see also Royal Dutch Co.; 'Shell'
    Transport and Trading Co.
Roosevelt, Theodore, 335, 340–1,
    384, 388, 392, 400, 401,
    403–4, 407
  Corollary to the Monroe
    Doctrine, 340–1, 384
Ross, John, 15
Roxana Petroleum Co., *see* Royal
    Dutch-Shell
Royal Navy *see* navy, British
Ruffner, David and Joseph, 60
rule of capture, 111–12, 126,
    128–9, 146, 150, 390
Russia
  and Afghanistan, 198, 299,
    301, 359, 367
  Baku/Caucasus oil: earliest (when
    Persian) *see* Azerbaijan; when
    Russian, 13, 39–40, 194–6,
    198–9, 200, 202–17, (231),
    (242–5), 245–6, (246–8),
    248–57, (258), (261), (265),
    266–7, (277), 281–2, 283, (285),
    (302), (304), 306, (308), 332,
    343, 344–56, (411), 412, 417,
    (433); in WWI, 541, 543–4
  Baku-Batum pipeline *see*
    pipelines: Russia

Baku-Batum Railway *see*
    railways/railroads
Baku Petroleum Association, 214
Batum, 201, 212, 213, 214,
    215–17, 231, 242, 243, 245–6,
    249–50, 254, 255, 258, 282,
    283, 286, 344–8, 349–50, 421
British investment in, 217, 242–4,
    249–57, 261, (265), 352–5,
    (417)
and British navy, potential oil for,
    312–13, 433, (524)
Corps of Mining Engineers, (39),
    40, 198–9, 200, 202
and Dardanelles/Turkish Strait,
    198, 200, 524–7, 534, 537–8
early use of petroleum, 5–6, 7, 12,
    (13), 39–40; *see also*
    Azerbaijan: petroleum in:
    pre-19th century
exports, oil, 211–20, 221, 231,
    242–7, 248–9, (250), 254–5,
    277, 281–2, 302, 304, (306–7),
    (312–13), (315), 332, (345–6),
    350, 355, 411–12, (433), (436),
    438, (442), 524; *see also*
    foreign markets *below*
foreign markets, 212, 213,
    216–17, 221, 231, 242–9,
    254–5, 258, 261, (267–8), 277,
    281–2, 285, 302, 304, 332,
    345–6, 371, 411–12, 419,
    (433), 449, 467, (472), 508; *see*
    also exports, oil *above*
Grozny oil, 39, 196, 202, 251,
    253, 254, 282, 295, 332,
    350, 355, 421
imports, oil, 119, 202–3,
    206, 433; tariffs, 207, 257,
    (343), 413

and India, British, threat to *see*
    India, routes to; Persia (Iran):
    British-Russian rivalry over
Komi Republic/Ukhta region oil,
    39, 203
Maikop oil, (6), (39), 422, 436
monopolization, 207, 212–16,
    245, 246, 248, 251, 256
navy *see* navy, Russian
Nobel Brothers in *see* Nobel
    Brothers Petroleum Co.
Novorossiysk, 201, 215, 245,
    253, 350
and Persia, see Persia (Iran):
    British-Russian rivalry over;
    Anglo-Russian Convention
pipelines, 207–8, 253, 436. See
    also pipelines: Russia
pollution, 13, 194, 204–5, 206,
    210, 251–2, Fig. 12
production, 200, 202, 203–5,
    214–15, 219, 245, 250–3,
    281–2, 306, 343, 344,
    350–5, (382)
protectionism, 207, 249, 251,
    (257), 286, (343), 410, 413,
    414
refining, 39, 198, 202–16, 245–6,
    (250–6), 352
Revolution of 1905, 343–56,
    Figs.18, 19
Rothschilds in *see* Rothschild:
    Bnito
Royal Dutch-Shell in, 417
Sakhalin Island oil, 257, 295
Siberian oil, 12
Standard Oil's attempts to invest
    in, 286, 410, 414
and WWI: as instigator of,
    524–38; oil in, 541, 543–4

Russian Petroleum and Liquid Fuel
Co., 253–4, 353, 355
Russo-Japanese War (1904–5),
345–7, 352, (412)
Rust, Randolph, 290, 476–7

Sahagún, Bernardino de, 40
Salisbury, Lord, 235, 298, 302,
303–4, 368
Samuel
M. Samuel & Co., 242–3
Marcus see 'Shell' Transport and
Trading Co.
Samuel, 494
Sarawak see Borneo: British
(Sarawak)
Sardar Asad, Ali Quli Khan, 357, 358
Saud, Ibn, 516
Schibaieff and Co., 254, 355
Schlieffen Plan, (521), 534, 537
Scott, Thomas A., 139,
145, 182–3
Security Oil Co. see Standard
Oil Co.
Seep, Joseph see Standard Oil Co.
Selborne, Earl of, 309, 310, (311),
316, 444
Seneca Indians (Onödowá'ga:'): 52,
58, 59–60, 62, 66, 83, 86–8,
91–2, 96–100
Burnt House settlement
(Jenuchshadego), 97, 98, 99,
100, 105, 142
Cornplanter, Chief
(Kayéthwahkeh), 97, 98,
99–100, 105, 142, Fig. 2;
Township, 75, 109–10; tract,
99, 105, 113–14
Handsome Lake (Kanyotaiyo'),
100

Mingo Indians (Ohio Seneca),
83, 86, 88, 89
Red Jacket, Chief (Sagoyewatha),
96, 99
Seneca Oil, (52), 58–60, 61, 62,
64, 66, (69), 100
Seneca Oil Co., 77–8, 106, 107,
108–9, (183)
Serbia, 525, 527–8, 530,
533, 534–6
Seven Years War (1754–63), 80–6, 88
Seward, William H., 186
Seymour, Frederick, 240
shale oil, 138
see also France; United Kingdom;
United States
Shamil, Shaykh, 199, 200, 201
Shawnee Indians, 55, 79, 82–9,
90, 101
Chief Tecumseh, 101
'Shell' Transport and Trading Co.
in Borneo, 265, 271, 285, 311,
319, 330–1, 332, 346, 417,
476, 540
and British navy, 265, 271, 311,
317–19, (448), 449, 472–80,
480, 481, 482–3, (484), 489,
491, 494–5, 510, 511–12, 513,
522, 523, (536–7), 540
and Burma, 317–19, 410, 412,
475
and Standard Oil, challenge to,
265, 280, 284, 285, 288–9,
318, 329, 409, 411–12; see also
Royal Dutch-Shell; Tank
Syndicate
origins/formation of, 261; see
also Samuel: M. Samuel &
Co.; Tank Syndicate
and Russo-Japanese War, 345–6

Samuel, Marcus, 242–5, 248–9,
254, 258, 261–2, 271, 284–5,
286, 311, 317–19, 330, 346,
379–80, 412, 436, (448), 449,
454, 455, 475–6, 484, 494–5
and Sumatra, 330, 346
and Texas, 285, 308
tankers, 265, 285, (289), (346),
523; *see also* Royal Dutch-
Shell; Tank Syndicate
Waley Cohen, Robert, 466, 476
in WWI, 540, Fig. 27
see also Royal Dutch-Shell
Shen Kua, 19
Sheridan, Philip, 87
Sherman, John, 171–4
Antitrust Act, 171–4, 393,
394, 397, 405
Sicily, oil on, 32
Silliman
Benjamin Jr., 70, 75–6, 182–3
Benjamin Sr., 66, 72
Silva, Joaquim José da, 17–18
Simcoe, John Graves, 57–8
Simms, Frederick, 272
Sinclair Oil & Refining Co., 421
Slade, Edmond *see* navy, British
Slavery
Bolivia, 292
Peru, 45
Slave trade, West African-Atlantic,
17, 18, 42, 45, 321
United States, 18, (50), 66, 68,
106, 121, 140, 175
Venezuela, 41, 426
Smiley, Alfred W., 127
Smith, Henry Babington, 467,
468, 469
Smith, Herbert Knox, 407
Smith, John, 46

Smith, Hubert Llewellyn, 510–11
Smith, William A. 'Uncle Billy', 107
Snouck Hurgronje, Christiaan,
263, 265
Société Française de Pétrole Ltd., 322
Soto, Hernando De, 50–1
Sousa Coutinho, Francisco
Inocêncio de, 17
South Improvement Co., 145–6
Spanish-American War, 278–81,
384, 414
Spencer, Edmund, 199
spermaceti *see* lamp/illuminating oil
Spies Petroleum Co., 355
Spindletop *see* Texas
Spring-Rice, Cecil, 355, 359–60,
365
St Clair, Arthur, 98
Stalin, Joseph, 344–5
Standard Oil Co.:
Anglo-American Oil Co., 219,
254, 287, 317, 326, 330,
433, Fig. 13
Archbold, John D., 155, 247,
263, 285, (286)
Asche, Frederick D., 408–9, 410
in Austria-Hungary, 287
'blind tigers'/front companies,
(265), 284, 394
break-up of, 403–5, 409–10
bribery by, 150, 151–2, 159, 176,
(392), (401)
and Burma, 287–8, 317, 410
and China: as export market,
(180), (231), 242, (246–7),
(248), (261), 277–8, 279–81,
(288), 435, 385, (416); as
source of oil, potential, 257,
(279–80), (288), 296
Colonial Oil Co., 287, 317

Standard Oil Co.: – *cont'd.*
  Deutsch-Amerikanische
    Petroleum-Gesellschaft
    (DAPG), 219
  Dodd, Samuel C.T., 169
  and Dutch East Indies, (227),
    230, (248–9), 258–61, 263,
    265, 287, 409–10, 414, 416
  Fertig, John F, 263
  Flagler, Henry, 152
  formation of: Andrews, Clark &
    Co., 123; Rockefeller &
    Andrews, 123; Standard Oil
    Co, 145; trust, 169
  foreign markets, (123), (138),
    (141), 180, 187, (194), 203,
    (216–17), 217–20, (221), (227),
    (231), (242), 246–9, 254–5,
    258, 277–8, 279–81, 285–6,
    (288), (302), (304), 330, 345,
    355, 371, 380, 385, 408–9,
    415–16, 417–19, (422), 441–2,
    (495), (508), (539)
  front companies *see* 'blind tigers'/
    front companies *above*
  and German oil monopoly plan,
    417–19, 441–2
  Imperial Oil Co., 180, 290
  J.A. Bostwick & Co., 150–1
  legal action against, 173, 283–4,
    390–4, 396–7, 403–5, (409–10)
  Libby, William Herbert, 180, 216,
    220, 258, 409–10
  McGee, James, 278
  and Mesopotamia/Ottoman
    empire, 371, 379, 507–8
  and Mexico, 187, 380, 383
  monopolization by, 124, 143–59,
    169–74, 175–80, 181, 194,
    203, (207), (216), (242–3),

    246–9, (259–61), (277),
    (285–6), 287–8, 389–405,
    408–10, (410–17), 417–19,
    441–2, (454), (472), 477, 489,
    (502), (510), Fig. 8
  O'Day, Daniel, 155, 390, 392
  and Persia, 244, 249, 287
  pipelines: American Transfer Co.,
    149; Buckeye Pipe Line Co.,
    175–6, 177–8; Connecting Pipe
    Line, 178; Cygnet Pipe Line,
    178; Eureka Pipe Line, 178;
    Indiana Pipe Line, 178;
    National Transit Co., 155–6,
    175, 176, 178, 216; Mid-
    Continent system, 337, 390–2;
    Southern Pipe Line, 178; United
    Pipe Line Co., 148–50,
    152, 157
  pollution, 133, 153–4
  Prairie Oil & Gas Co., 337,
    390–1, 421
  public relations, 400
  Rockefeller, John D., 122–4, 128,
    133, 143–6, 149, 159, 170,
    172, 173, (174), (175), (207),
    (213), (220), (247), 278, (285),
    317, 346, (362), (371), 388–9,
    390, 400, 401, 403, (404), 406,
    407, (414), 415, (416), (418),
    (419), (492), 507–8
  Rockefeller, William, 124, 388
  in Romania/Româno-Americană,
    287, 433
  and Russia; attempt to invest in,
    286, 410, 414; 1905
    revolution, 355
  and Russo-Japanese War, (346)
  Security Oil Co., 284
  Seep, Joseph/Agency, 150–1, 179

and Spanish-American War,
277–81, (414)
tankers, (218–19), 219, (231),
258, 280
Teagle, Walter, 418–19
Vacuum Oil Co., 287
Warden, W.G., 175, 220
Waters-Pierce & Co. (affiliate),
187, 270, 284, 380–1, 383
Stauss, Emil von, 512
Steaua Română, 286, 376
Stevens, Patrick, 348–53
Steward, C.E., 267
Stewart, Charles, 203–4, 210
Stollmeyer, Conrad, 49–50, 188
Stone Calf, Cheyenne chief, 333
Stoop, Adriaan, 227, 230–1, 257
Strabo, 3, 4, 32
Strathcona of Mount Royal, Lord
see Anglo-Persian Oil Co.
Suart, Alfred, 217, 218, 249, 251,
253, 265, 270
submarines/U-boats, 327, 332, (433),
445, 447, 455, 500–1, 539
Suez
Canal: British government stake
in, 233–4, (456), 487, 492;
strategic importance of, 228,
231–4, 237, 241, 346, (363),
438, 460; tanker route, 217,
231, 242–4, 302, 332, 438,
467, 474, 501, 523; see also
tankers, oil
Crisis (1956), 236
Sullivan, John, 91
Sumatra see Dutch East Indies
Sumerian use of bitumen, 1–2, 16
Sutton Hoo, 7
Sykes, Frederick H., 452–3
Sykes, Mark, 376, 503

Symes, Michael, 22–3
Syria 7, 120, 375, 376, 456, 460, 461
exploration, (461)
oil imports, 120, 246
petroleum, earliest use of, 1, 3;
pre-WWI, 461, 512, 516
Syrian Exploration Co., 461

Tachibana Nankei, 25
Tacitus, 3
Taft, William, 393, 397, 400,
402, (404)
Tagieff and Co., 215, 254
Taiwan see Formosa (Taiwan)
Tank Storage and Carriage Co. see
Rothschild
Tank Syndicate, 242–5, 248–9,
258, 261
challenge to Standard Oil, 242–5,
248–9, 258
tankers: Bullmouth, Conch,
Cowrie, Clam, Murex,
Spondilus, Turbo Trocas,
Volute, 243–4
see also 'Shell'
tankers, oil
accidents, 209–10, 217, 218
Atlantic, 117, 209, 218–19,
254–5, 270, 284, 320, (382),
523, 541–2
Baltic, 212, 518
Black Sea, 206, 210, (212–13),
216–17, (218), 242–3, 248–9,
253–5, 437, 438, 522–3, 523,
524, (539)
British navy, see navy, British
Caspian, (206), 209, (216), (353)
first: sail, 117, 218, 219; steam,
(206), 209, 216–17; diesel,
447, 448

tankers, oil – *cont'd.*
 German navy *see* navy, German
 lack of: pre-WWI, 285, 320, 382,
  433, 437–8, 485, 508, (536–7);
  in WWI, 541–2
 Mexican Petroleum Co. (Edward
  L. Doheny), 382
 Nobel Bros., 209–10,
  (212), 447
 Pacific, 267, 268, 280, 289
 ships: *Andromeda*, 218; *Atlantic*,
  117; *Baku Standard*, 255, 270;
  Beacon Light, 254; *Bakuin*,
  217; *Broadmayne*, 254;
  Buddha, 209; *Charles*, 209;
  Chigwall, 218; *Circassian
  Prince*, 254; *Crusader*, 218;
  Dyelo. 447; *Elbruz*, 254;
  Fanny, 218; *Fergusons*, 217;
  George Loomis, 268;
  Glückauf, 218, 219; *Great
  Western*, 117; *Haliotis*, 270;
  Jan Mayn, 218; *Lindernoes*,
  218; *Lux*, 249; *Moses*, 210;
  Nederland, 209;
  Nordenskjöld, 209, 210;
  Nordkyn, 218; *Petriana*, 217;
  Petrolea, 217; *Phosphor*, 254,
  Ramsey, 117; *Rocklight*, 255;
  Stat, 218; *Sviet*, 217;
  Switzerland, 209; *Trefoil*, 448;
  Vaderland, 209; *Vandal*, 447;
  Ville de Calais, 217; *Vulcanus*,
  447; *W.L. Hardison*, 267;
  Zoroaster, 209, Fig. 7; *see also*
  Tank Syndicate
 trade, early 219
 see also Dardanelles Strait; Lane
  & Macandrew; 'Shell'; Suez:
  Canal; 'Shell'

tanks/armoured cars *see* motor
 vehicles
Tano Syndicate Ltd., 322
Tarbell, Ida, 178, 389–90
Tassart, L.C., 379
Taube, Johann, 34
Teagle, Walter *see* Standard Oil Co.
teamsters/horse-drawn oil transit,
  112–13, 115, 124–7, 204, 206,
  208, 219, (292), 326
 ox-drawn, 22–3, (24), 193
Tecumseh, Chief *see* Shawnee
 Indians
Téenek, 294, 382
Tewfik, Khedive, 234–5, 236,
  238, 239
Texaco, origin of, 284
Texas
 and British navy, oil for, (312),
  313, (320), 432, (433), 438; *see
  also* navy, British: oil supplies,
  pre-WWI: US
 in Mexican-American War, 182
 pipelines, 269, 284
 pollution, 269, 283
 pre-industrial oil, 51, 269
 production, 269–70, 283–4, 308,
  311, 337, 382, 420
 Spindletop, 283, 284, (294), 308
Texas Co. (Texaco), 284, 337,
  416, 421
Thibaw, King, 222–5
Thomas Cook & Son, 241
Thomas, Garnet, 423
Thompson, Arthur Beeby, 251–3,
  255, 256–7
Thucydides, 5
Tidewater Pipe Line Co., 151–6
Tilford, W.H. *see* Standard
 Oil Co.

Timor, Portuguese (East Timor/
  Timor-Leste)
  oil, 296–7, 320–1, 434–5; and
    Australian navy, potential oil
    for, 434–5
Timor Petroleum Concessions
  Ltd., 435
Tirpitz, Alfred von, 441, 448,
  450, 517
Titusville see Oil Region (US)
Tlazolteotl, 40
toluol, 540
Tongva Indians, 54, 182
torpedoes, oil well/nitroglycerin,
  131–2, 336
Totonac Indians, 187
Townsend, James M., 75–7
Transcaspian Trading Co., 202
treaties
  Berlin (1828), 202
  Big Tree (1797), 99
  Bucharest (1812), 195
  Canandaigua (1794), 99
  Easton (1758), 85–6, 88
  Fort Stanwix (1768), 88–9, 94
  Fort Stanwix (Schuyler) (1784), 96
  Greenville (1795), 98
  Gulistan (1813), 195, 196, 197
  Hünkâr İskelesi (1833), 198
  Lancaster (1744), 81
  Logstown (1752), 81–2
  Paris (1763), 86
  Paris (1783), 92
  Paris (1856), 201
  San Stefano (1878), 214
  Turkmenchai (1828), 197
  Walking Purchase (1737), 82
Trinidad
  asphalt production, 49–50,
    290, 338–9

and British navy, oil for see navy,
  British: oil supplies, pre-WWI:
  Trinidad
exploration, 50, 187–8, 290
pre-industrial oil, 43–4, 46–7,
  49–50
protectionism, 314–15, 338–9,
  414, 476–8, 481
Royal Dutch-Shell in, 417, 476–8,
  481, 494
Trinidad Asphalt Co. of New
  Jersey, 290, 338
Trinidad Oilfields Ltd., 476
Trinidad Petroleum Co., 50, 187
Triple Alliance, 458, 526, 529, 535
Triple Entente, 462, 518, 526, 529,
  535
Tripp, Charles N., 72–3
Tsitsianov, Paul, 195
Turkey
  earliest use of petroleum in, 1, 3,
    6; see also Dead Sea, bitumen
    from; Mesopotamia (Iraq);
    Syria
  see also Ottoman (Turkish)
    empire
Tower of Babel, 2
Turkish Petroleum Co., 467–75,
  (482), 486, 504–16, 532–3
Tweedle, Herbert, 214, 215

Ukraine see Austria-Hungary/
  Galicia
Ulster Volunteer Force, 451
Union Oil Co. (later Unocal), 267,
  268, 320, 420
United Kingdom/Britain
  Bank of England, 115, (254)
  Board of Trade, 80, 87, (115),
    510, 512, 514

United Kingdom/Britain – *cont'd.*
and Burma *see* India: Burma (Myanmar)
and Egypt *see* Egypt; Suez: Canal
early petroleum, 35–7
and Nigeria *see* Nigeria
oil imports: civilian, 30–1, 117, 120, 138, 209, 217, 219, (248), 254–5, 282–3, 285–6, 308, 320, 329, 330–2, (413), 433, (436), 444, 450, 454–5, (472), 522–3, 523, 524; naval *see* navy, British
oil strikes, 435
and Persia (Iran), see Persia (Iran)
and Mesopotamia (Iraq), see Mesopotamia (Iraq)
refining, 117, 120, (209), (454), (455)
shale oil: for British navy, (313), 432, 435, 489, 539–40; Dorset, 37, 435, 476; Scotland, 37, 117, 220, 295, 313, 432, 435, 489, 539–40; Wales, 37
and WWI, oil dependency in, 539–40, 541–2, 544–5
United States
American Revolution, 55, 90–2, (128)
Appalachian oilfields, 176, 177, 282, 283, 420; see also New York; Oil Region (US); Pennsylvania; West Virginia; Kentucky; Ohio
Billingsley Bill, 176
and British navy, oil for *see* navy, British: oil supplies, pre-WWI: US
Bureau of Corporations, 392–3, (402), (407)
Civil War, (108), 116, 117–18, 119, 121–2, 122, 123, 128, 139–40, (186), (397–8)
and German navy, oil for, 418–19, 441–2, 523
Interstate Commerce Act, 171
and Persia (Iran), see Standard Oil Co.; Persia (Iran)
pre-industrial reports/use of petroleum *see individual US states*
Mid-Continent oilfields, 289, 332–7, 390, 420–1
navy *see* navy, United States
Ohio-Indiana/Lima oilfields, 175–6, 177–8, (182), 267, 270, 282, 283
oil exports, 115–16, 117, 119–22, 137–8, 141–2, (155), 178, 180, 202, 218–20, 231, 242–3, 258, 277–8, 280–1, 281–2, 284, 285, 329, 355, 380–1, 408–9, 412, 417–19, 441–2, 450; *see also* Standard Oil Co.: foreign markets
oil imports, (267–8), 288, 289, 329, (416); tariffs, 268, (277), 413
oil production, (179–80), 281, (343), (393), 419–20; *see also* Appalachian oilfields *above*; Mid-Continent oilfields *above*; Ohio-Indiana/Lima oilfields *above*; Oil Region (US); *individual US states*
'Open Door' policy, 221, 305, (327), (328)
pipelines *see* pipelines: US
Producers' Protective Association, 176–8

refining, 69–70, 72, (76), 77, 109,
112, 113, 115, 122–4, 126, 129,
133, 141–56, 175–80, 181, 218,
219, 248, 266, 270, 280, 281–3,
284, 289, 336, 337, 382, 390–4,
411, 413, 416, 421, 477, (493);
see also Ohio: Cleveland; Oil
Region (US)
shale oil, (113)
see also Indian Territory (US);
Standard Oil Co.; individual
US states
United States-Venezuela Co., 339–40
Unocal, origin of, 267
Urabi, Ahmed, 235–42
Uzbekistan see Fergana/Ferghana
(Uzbekistan)

Vacuum Oil Co. see Standard Oil Co.
Van Alkemade, I.A. Van Rijn, 265
Van Daalen, Gotfried, 263
Van Heutsz, Joannes B., 263–5
Van Syckel, Samuel, 125–6,
127, (142)
Vancouver, George, 58
Vanderbilt, William H., 155, 159
Vanderlip, Frank, 406, 415
Venango (Franklin) see Oil
Region (US)
Venezuela
asphalt production, 290, 338–41
earliest reports of oil, 41–2, 290
exploration, (342), 425–9
pre-industrial oil production, 290
Royal Dutch-Shell in, 417, 425–9
Venezuelan Crisis (1902), 340
Venezuelan Oil Concessions Ltd. see
Royal Dutch-Shell
Vigas, Andrés J., 341
Virginia see West/western Virginia

Vitruvius, 3, 32
Viviani, René, 533
Vorontsov, Mikhail, 40
Voskoboinikov, Nikolai, 40

Waley Cohen, Robert see 'Shell'
Wallace, Charles W. see Burmah Oil
Co.; Anglo-Persian Oil Co.
War of 1812, 101–2
War of the Pacific (1879–84), 184
Ward, Thomas de, 507
Warden, W.G. see Standard Oil Co.
Warner-Quinlan Asphalt Co., 338–41
Washington, George, 55–6, (80), 83,
84, 88, 89, 91, 92, 93–4, 96,
97, (103)
Waters-Pierce & Co., 187, 270, 284,
380–1, 383; see also Pierce,
Henry Clay; Standard Oil Co.
Watson, Elkanah, 95, 98–9
Watson, Jonathan, (74), (75), 108
Wayne, 'Mad' Anthony, 98
wells, drilled
for oil, earliest: Russia, 40, 200,
202; US, (85), 107
pre-industrial oil, from brine/
water: China, 20, 27; US, 60,
65, 66–7, 74, (107)
Wells, H.G., 329
West India Petroleum Co., 290, 312
West Indies Petroleum Co., 187–8
West/western Virginia,
Burning Spring, 55–6, 84, 117–18
pre-industrial petroleum, 52,
55–6, 60, 66–7, 84, 107
production, (117–18), (175),
178, 282
whale oil see lamp/illuminating oil:
spermaceti
White, Edward, 404

White, Herbert Thirkell, 225
Wilcox, R., 28–9
Wilhelm II, Kaiser, 372,
    418, (514)
William III of the Netherlands,
    229
Williams, James M., 73
Wilson, Arnold, 361–3
Wilson, James, 98
Wilson, William, 100
Wilson, Woodrow, 279, 383, 397,
    398–9, 400, (404), 414–15,
    (541–2)
Witsen, Nicholas, 39
Witte, Sergei, 245, 251, 256, 257,
    306, 343, 345
Wolf, Simon, 235
Wolff, Henry Drummond, 301, 302,
    303, 304
Wolseley, Garnet, 241
Woodson, Archelaus, 396–7
World War I
    oil in, 539–45

outbreak of: causes, 517–38; oil
    as factor in, 522–4, 527,
    529–30, 530–3, 537–8
Wright, Wilbur and Orville, 328–9
Wyandot Indians, 86

Xenophon, 5

Yāqūt, 8
Yaraghi, Muhammad al-, 197
Yermolov, Aleksei, 196–7
Yoruba, 322
Young, D.M., 500
Young, James, 38
Yule, Henry, 30–1

Zakynthos (Zante), oil on, 32,
    34, 190
Zeh, Jan, 38
Zeisberger, David, 52–3
Zijlker, Aeilko Jans, 227, 229
Zoroastrian fire temple, Baku, 13,
    203–4, 205–6, 210